S0-CBI-343

Methods in Enzymology

Volume 266
COMPUTER METHODS FOR MACROMOLECULAR SEQUENCE ANALYSIS

METHODS IN ENZYMOLOGY

EDITORS-IN-CHIEF

John N. Abelson   Melvin I. Simon

DIVISION OF BIOLOGY
CALIFORNIA INSTITUTE OF TECHNOLOGY
PASADENA, CALIFORNIA

FOUNDING EDITORS

Sidney P. Colowick and Nathan O. Kaplan

Methods in Enzymology

Volume 266

Computer Methods for Macromolecular Sequence Analysis

EDITED BY

Russell F. Doolittle

CENTER FOR MOLECULAR GENETICS
UNIVERSITY OF CALIFORNIA, SAN DIEGO
LA JOLLA, CALIFORNIA

QP 601
C 71
V. 266
1996

ACADEMIC PRESS

San Diego   New York   Boston   London   Sydney   Tokyo   Toronto

This book is printed on acid-free paper. ∞

Copyright © 1996 by ACADEMIC PRESS, INC.

All Rights Reserved.
No part of this publication may be reproduced or transmitted in any form or by any means, electronic or mechanical, including photocopy, recording, or any information storage and retrieval system, without permission in writing from the publisher.

Academic Press, Inc.
A Division of Harcourt Brace & Company
525 B Street, Suite 1900, San Diego, California 92101-4495

United Kingdom Edition published by
Academic Press Limited
24-28 Oval Road, London NW1 7DX

International Standard Serial Number: 0076-6879

International Standard Book Number: 0-12-182167-6

PRINTED IN THE UNITED STATES OF AMERICA
96 97 98 99 00 01 MM 9 8 7 6 5 4 3 2 1

Table of Contents

Section I. Databases and Resources

Section II. Searching through Databases

v

Section III. Multiple Alignment and Phylogenetic Trees

Contributors to Volume 266

Article numbers are in parentheses following the names of contributors.
Affiliations listed are current.

STEPHEN F. ALTSCHUL (27), *National Center for Biotechnology Information, National Library of Medicine, National Institutes of Health, Bethesda, Maryland 20894*

PATRICK ARGOS (8), *European Molecular Biology Laboratory, 69117 Heidelberg, Germany*

MARCELLA ATTIMONELLI (17), *Dipartimento de Biochimica e Biologia Molecolare, Università di Bari, 70125 Bari, Italy*

WINONA C. BARKER (3, 4), *National Biomedical Research Foundation, Washington, District of Columbia 20007*

GEOFFREY J. BARTON (29), *Laboratory of Molecular Biophysics, University of Oxford, Oxford OX1 3QU, United Kingdom*

PEER BORK (11), *European Molecular Biology Laboratory, D-69012 Heidelberg, Germany; and Max-Delbrück-Center for Molecular Medicine, Department of Bioinformatics, D-13122 Berlin-Buch, Germany*

JAMES U. BOWIE (35), *Department of Chemistry and Biochemistry and DOE Laboratory of Structural Biology and Molecular Medicine, University of California, Los Angeles, Los Angeles, California 90095*

STEVEN E. BRENNER (37), *Medical Research Council Centre Laboratories of Molecular Biology, Cambridge CB2 2QH, United Kingdom*

GRAHAM N. CAMERON (1), *European Molecular Biology Laboratory Outstation—the European Bioinformatics Institute, Hinxton, Cambridge CB10 1RQ, United Kingdom*

CYRUS CHOTHIA (37), *Medical Research Council Centre Laboratories of Molecular Biology and Cambridge Centre for Protein Engineering, Cambridge CB2 2QH, United Kingdom*

JEAN-MICHEL CLAVERIE (14), *Laboratory of Structural and Genetic Information, E.P. 91—Centre National de la Recherche Scientifique, 13402 Marseille, France*

MARC DELARUE (40), *Immunologie Structurale Institut Pasteur, 75015 Paris, France*

RUSSELL F. DOOLITTLE (21), *Center for Molecular Genetics, University of California, San Diego, La Jolla, California 92093*

DAVID EISENBERG (35), *Department of Chemistry and Biochemistry and DOE Laboratory of Structural Biology and Molecular Medicine, University of California, Los Angeles, Los Angeles, California 90024*

JONATHAN A. EPSTEIN (10), *National Center for Biotechnology Information, National Library of Medicine, National Institutes of Health, Bethesda, Maryland 20894*

THURE ETZOLD (8), *European Molecular Biology Laboratory, 69117 Heidelberg, Germany*

SCOTT FEDERHEN (33), *National Center for Biotechnology Information, National Library of Science, National Institutes of Health, Bethesda, Maryland 20894*

JOSEPH FELSENSTEIN (24), *Department of Genetics, University of Washington, Seattle, Washington 98195*

DA-FEI FENG (21), *Center for Molecular Genetics, University of California, San Diego, La Jolla, California 92093*

JEAN GARNIER (32), *Unité de Bioinformatique Biotechnologies, INRA, 78352 Jouy-en-Josas, Paris, France*

DAVID G. GEORGE (3, 4), *National Biomedical Research Foundation, Washington, District of Columbia 20007*

JEAN-FRANÇOIS GIBRAT (32), *Unité de Bioinformatique Biotechnologies, INRA, 78352 Jouy-en-Josas, Paris, France*

TOBY J. GIBSON (11, 22), *European Molecular Biology Laboratory, 69012 Heidelberg, Germany*

WARREN GISH (27), *Department of Genetics, Washington University School of Medicine, St. Louis, Missouri 63108*

MICHAEL GRIBSKOV (13), *San Diego Supercomputer Center, La Jolla, California 92093*

XUN GU (26), *Human Genetics Center, Sph, University of Texas, Houston, Texas 77225*

DANIEL GUSFIELD (28), *Computer Science Department, University of California, Davis, Davis, California 95616*

ROBERT A. L. HARPER (1), *European Molecular Biology Laboratory Outstation—the European Bioinformatics Institute, Hinxton, Cambridge CB10 1RQ, United Kingdom*

JOTUN HEIN (23), *Department of Ecology and Genetics, Institute of Biological Sciences, Aarhus University, DK-8000 Aarhus, Denmark*

JORJA G. HENIKOFF (6), *Fred Hutchinson Cancer Research Center, Seattle, Washington 98104*

STEVEN HENIKOFF (6), *Howard Hughes Medical Institute, Fred Hutchinson Cancer Research Center, Seattle, Washington 98104*

DESMOND G. HIGGINS (22), *European Molecular Biology Laboratory Outstation—the European Bioinformatics Institute, Hinxton, Cambridge CB10 1RQ, United Kingdom*

LIISA HOLM (39), *European Molecular Biology Laboratory Outstation—the European Bioinformatics Institute, Hinxton, Cambridge CB10 1RQ, United Kingdom*

TIMOTHY J. P. HUBBARD (37), *Medical Research Council Centre Laboratories of Molecular Biology and Cambridge Centre for Protein Engineering, Cambridge CB2 2QH, United Kingdom*

LOIS T. HUNT (3), *National Biomedical Research Foundation, Washington, District of Columbia 20007*

MARK S. JOHNSON (34), *Molecular Modelling and Biocomputing Group, Turku Center for Biotechnology, University of Turku, FIN-20521 Turku, Finland*

JONATHAN A. KANS (10), *National Center for Biotechnology Information, National Library of Medicine, National Institutes of Health, Bethesda, Maryland 20894*

ANTHONY R. KERLAVAGE (2), *The Institute for Genomic Research, Gaithersburg, Maryland 20850*

EUGENE V. KOONIN (18), *National Center for Biotechnology Information, National Library of Medicine, National Institutes of Health, Bethesda, Maryland 20894*

ERIC S. LANDER (19), *Whitehead Institute for Biomedical Research and Department of Biology, Massachusetts Institute of Technology, Cambridge, Massachusetts 02142*

WEN-HSIUNG LI (26), *Human Genetics Center, Sph, Health Science Center, University of Texas, Houston, Texas 77225*

CRAIG D. LIVINGSTONE (29), *Genomics Support Group, SmithKline Beecham Pharmaceuticals, New Frontiers Science Park, Harlow, Essex CM19 5AW, United Kingdom*

ANDREI LUPAS (30), *Abteilung Molukulare Strukturbiologie, Max-Planck-Institut für Biochemie, D-82152 Martinsried, Germany*

THOMAS L. MADDEN (9), *National Center for Biotechnology Information, National Library of Medicine, National Institutes of Health, Bethesda, Maryland 20894*

ALEX C. W. MAY (34), *Department of Crystallography, Birkbeck College, University of London, London WC1E 7HX, United Kingdom*

RICHARD J. MURAL (16), *Biology Division, Oak Ridge National Laboratory, Oak Ridge, Tennessee 37831*

ALEXEY G. MURZIN (37), *Medical Research Council Centre Laboratories of Molecular Biology and Cambridge Centre for Protein Engineering, Cambridge CB2 2QH, United Kingdom*

HITOMI OHKAWA (10), *National Center for Biotechnology Information, National Library of Medicine, National Institutes of Health, Bethesda, Maryland 20894*

CHRISTINE A. ORENGO (36), *Department of Biochemistry and Molecular Biology, University College, London WC1E 6BT, England*

JOHN P. OVERINGTON (34), *Computational Chemistry, Pfizer Central Research, Sandwich, Kent CT13 9NJ, United Kingdom*

LASZLO PATTHY (12), *Institute of Enzymology, Biological Research Center, Hungarian Academy of Sciences, Budapest H-1113, Hungary*

WILLIAM R. PEARSON (15), *Department of Biochemistry, University of Virginia, Charlottesville, Virginia 22908*

GRAZIANO PESOLE (17), *Dipartimento di Biochimica e Biologia Molecolare, Università di Bari, 70125 Bari, Italy*

FRIEDHELM PFEIFFER (4), *Martinsried Institute for Protein Sequences, Max Planck Institute for Biochemistry, Martinsried 82152, Germany*

OLIVIER POCH (40), *UPR 9002 du Centre National de la Recherche Scientifique, I.B.M.C. du Centre National de la Recherche Scientifique, 67084 Strasbourg, France*

BARRY ROBSON (32), *Dirac Foundation, Bioinformatics Laboratory, Royal Veterinary College, University of London, London NW1 0TU, United Kingdom*

MICHAEL A. RODIONOV (34), *Molecular Modelling and Biocomputing Group, Turku Centre for Biotechnology, University of Turku, FIN-20521 Turku, Finland; and Institute of Bioorganic Chemistry, Belarus Academy of Sciences, Minsk-141, Republic of Belarus 220141*

BURKHARD ROST (31), *Protein Design Group, European Molecular Biology Laboratory, 69012 Heidelberg, Germany*

KENNETH E. RUDD (18), *National Center for Biotechnology Information, National Library of Medicine, National Institutes of Health, Bethesda, Maryland 20894*

CECILIA SACCONE (17), *Dipartimento di Biochimica e Biologia Moleculare, Università di Bari and Centro di Studio sui Mitocondri e Metabolismo Energetico, CNR, 70125 Bari, Italy*

NARUYA SAITOU (25), *Laboratory of Evolutionary Genetics, National Institute of Genetics, Mishima-shi, Shizuoka-ken, 411, Japan*

CHRIS SANDER (39), *European Molecular Biology Laboratory Outstation—the European Bioinformatics Institute, Hinxton, Cambridge CB10 1RQ, United Kingdom*

GREGORY D. SCHULER (10), *National Center for Biotechnology Information, National Library of Medicine, National Institutes of Health, Bethesda, Maryland 20894*

BENNY SHOMER (1), *European Molecular Biology Laboratory Outstation—the European Bioinformatics Institute, Hinxton, Cambridge CB10 1RQ, United Kingdom*

RODGER STADEN (7), *Medical Research Council Centre Laboratories of Molecular Biology, Cambridge CB2 2QH, United Kingdom*

P. STELLING (28), *Computer Science Department, University of California, Davis, Davis, California 95616*

JENS STØVLBÆK (23), *Department of Ecology and Genetics, Institute of Biological Sciences, Aarhus University, DK-8000 Aarhus, Denmark*

MARK BASIL SWINDELLS (38), *Department of Molecular Design, Institute for Drug Discovery Research, Yamanouchi Pharmaceutical Company, Ltd., Tsukuba 305, Japan*

ROMAN L. TATUSOV (9, 18), *National Center of Biotechnology Information, National Library of Medicine, National Institutes of Health, Bethesda, Maryland 20894*

WILLIAM R. TAYLOR (20, 36), *Division of Mathematical Biology, National Institute for Medical Research, London NW7 1AA, United Kingdom*

JULIE D. THOMPSON (22), *European Molecular Biology Laboratory, 69012 Heidelberg, Germany*

EDWARD C. UBERBACHER (16), *Computer Sciences and Mathematics Division, Oak Ridge National Laboratory, Oak Ridge, Tennessee 37831*

ANATOLY ULYANOV (8), *European Molecular Biology Laboratory, 69117 Heidelberg, Germany*

STELLA VERETNIK (13), *San Diego Supercomputer Center, La Jolla, California 92093*

OWEN WHITE (2), *The Institute for Genomic Research, Gaithersburg, Maryland 20850*

MATTHIAS WILMANNS (35), *European Molecular Biology Laboratory, 69001 Heidelberg, Germany*

JOHN C. WOOTTON (33), *National Center for Biotechnology Information, National Library of Medicine, National Institutes of Health, Bethesda, Maryland 20894*

CATHY H. WU (5), *Departments of Epidemiology and Biomathematics, University of Texas Health Center at Tyler, Tyler, Texas 75710*

YING XU (16), *Computer Sciences and Mathematics Division, Oak Ridge National Laboratory, Oak Ridge, Tennessee 37831*

TAU-MU YI (19), *Whitehead Institute for Biomedical Research and Department of Biology, Massachusetts Institute of Technology, Cambridge, Massachusetts 02142*

JINGHUI ZHANG (9), *National Center for Biotechnology Information, National Library of Medicine, National Institutes of Health, Bethesda, Maryland 20892*

KAM ZHANG (35), *Division of Basic Sciences, Fred Hutchinson Cancer Center, Seattle, Washington 98104*

Preface

Volume 183 of *Methods in Enzymology* dealing with the computer analysis of protein and nucleic acid sequences has proved very popular with molecular biologists and biochemists. Computers and computer programs evolve rapidly, however, and can become outmoded very quickly. As a result, there was pressure to issue an updated volume that covers much the same general subject areas.

Like the earlier volume, this one is divided into several sections, the first of which deals with databases and some aspects related to their holdings. Also, there have been some relocations of major databases. GenBank is now centered at the National Center for Biotechnology Information (NCBI) at the National Library of Medicine in Bethesda, Maryland, and the EMBL Database has relocated to the European Bioinformatics Institute (EBI) at a site just outside Cambridge, England. More than ever, of course, geographic location is becoming moot, thanks to the World Wide Web (WWW) and extended hyperlink access.

There is some new vocabulary in this volume that did not appear in Volume 183. The use of neural nets, for example, is discussed in several places, including chapters dealing with the classification of sequences, on the one hand, and with predicting secondary structure, on the other. The kinds of databases are also changing. For instance, it has been found that the fragmentary data known as Expressed Sequence Tags (EST) are extremely useful.

Searching newly determined sequences remains the first order of business. More often than not, a simple search of a new sequence provides both functional and structural information. New pattern searching programs have greatly extended the power of this approach so that very distant relatives of well-characterized families can be identified.

The multiple alignment of protein sequences continues to have a prominent role in protein characterization. Whether the sequences are of the "same" protein from different organisms or are paralogs that have resulted from gene duplications, the alignment problems are the same. Interestingly, the most popular algorithms have not changed much, but the amino acid substitution tables that support them have. This is chiefly the result of there being so much comparative data in the current databases that empirical measures of relationships can be obtained by simply tallying the occurrences of the amino acids in blocks of obviously aligned sequences. As discussed in Chapter [6] by Henikoff and Henikoff, these BLOSUM tables have been remarkably effective.

Among their many uses, multiple alignments are used to construct profiles for more sensitive searching than is possible by single-searching. They are also used in the consensus mode for better predictions of secondary structure and for three-dimensional searches. And, of course, they are used in the construction of phylogenetic trees.

Recent advances have led to some changes in emphasis in some of the sections. Most of the chapters focus on protein sequences, even though the vast majority of those are determined by DNA sequencing. Accordingly, a section on RNA folding that appeared in the earlier volume has been dropped, and instead a number of chapters that relate to the secondary structure and three-dimensional aspects of proteins have been added.

Indeed, three-dimensional searching is following the course of sequence searching a decade ago. As a new protein structure is characterized, the first matter of general interest is to determine whether the fold resembles that of any that were reported previously. The remarkable thing is that not only are most new structures falling into well-defined families, but often there is no hint in advance on the basis of either structure or function. The problems associated with structure searching are similar to those experienced by sequence searchers in the past: a burgeoning data bank (PDB is the Protein Data Bank), choices of search programs, and, finally, the problem of judgment on how significant a resemblance may be. Many of these problems are addressed in Section V of this volume.

As with Volume 183, authors were encouraged to make their programs or databases available to readers. Many chapters make reference to a WWW home page or an Internet email address from which additional information can be extracted.

Finally, I thank all the authors who wrote such interesting and informative chapters under a very strict and compressed timetable. Academic Press, and especially our editor, Shirley Light, outdid themselves in getting the manuscripts through the publication process in record time. As in the case of the previous volume dealing with this topic, I must also acknowledge that the task could not have been accomplished without the help of my assistant, Karen Anderson. Her relentless but always gentle prodding of authors to produce manuscripts and her remarkable organizational skills that kept the courier traffic flowing in the right direction were indispensable.

RUSSELL F. DOOLITTLE

METHODS IN ENZYMOLOGY

Section I

Databases and Resources

[1] Information Services of the European Bioinformatics Institute

By Benny Shomer, Robert A. L. Harper,
and Graham N. Cameron

Introduction

The European Bioinformatics Institute (EBI) was established in September 1994 as a new outstation of the European Molecular Biology Laboratories (EMBL). The new outstation is located at Hinxton Hall, Cambridgeshire, United Kingdom. Its main tasks are management of databases for molecular biology, bioinformatics services, and research and development in these fields.[1]

The move of the bioinformatics services from the EMBL headquarters in Heidelberg, Germany, to the EBI had various implications, including considerable expansion in the computer power and the number of staff. The computers are used for management of the principal databases, and for providing network servers. The outstation provides excellent communications channels to the scientific and research community throughout Europe, and a specialized user support group ensures that all the services are properly maintained and functional.

Various new services (which will be reviewed in this chapter) have been established, and this has been due to the fact that there has been an increase in both computational power and manpower at the EBI. The inspiration for these new services has come from the various research and development (R&D) teams now operating at the EBI, who do research on managing sequence databases and studying the interrelationships between various kinds of data. The main thrust of this work is to provide novel ways to access the data and to provide interfaces that are intuitive and easy to use for the EBI user community.

This chapter is divided into two sections. The first section is devoted to describing the various current and future databases and resources that are being developed in-house, and the second section describes the various interfaces and network connections that EBI provides for the scientific community globally. A glossary is provided at the end of this chapter that gives a brief description of common terms.

---

[1] D. B. Emmert, P. J. Stoehr, G. Stoesser, and G. N. Cameron, *Nucleic Acids Res.* **22,** 3445 (1994).

Copyright © 1996 by Academic Press, Inc.
All rights of reproduction in any form reserved.

EBI Databases and Resources

EMBL Nucleotide Sequence Database

The EMBL Nucleotide Sequence Database is a comprehensive database of DNA and RNA sequences either collected from the scientific literature and patent applications or submitted directly from researchers and sequencing groups.[2] The database is produced in a collaboration between the EMBL, GenBank (Washington DC, USA), and the DNA Data Bank of Japan (DDBJ, Mishima, Japan). Each entry that is created at any of these databases is automatically exchanged between the other two databases. This allows almost complete synchronization between the databases.

Currently, there is a 75% annual growth rate of the nucleotide sequence database. The total number of entries and bases for different taxonomic divisions can be seen in Table I. With further technological advancements, the rate of growth of the databases will increase even more.

The nucleotide database is maintained in the relational database management system (RDBMS) ORACLE, running on a DEC Alpha VMS cluster. Each entry in the database is assigned an accession number, which is a permanent unique identifier. The entry is represented externally as an ASCII "flat file." The flat file (see Fig. 1) is composed of lines beginning with a two-character tag and followed by an associated text. The header information ("annotation") is followed by the sequence itself. The sequence entry ends with the unique identifier "//." Table II summarizes the meaning of the two-character line tags.

The EBI maintains a very high level of quality assurance of the sequence data in the EMBL database. Each new entry is carefully reviewed by a team of annotators, and, when necessary, direct communication with the submitting author is initiated to clarify ambiguities. Rapid data turnaround is essential; we guarantee to process well-formed submissions within 1 week, although in practice entries are created within 2–3 days after receipt.

Development of the next generation of the sequence database is one of the R&D group activities. This group concentrates on various means of ensuring database integrity and developing state-of-the-art implementations of the data. The latest release (Release 45, December 1995) contains 622,566 entries, comprising 427,620,278 nucleotides.

SWISS-PROT Protein Sequence Database

The SWISS-PROT Protein Sequence Database is a database of protein sequences.[3] This database is produced and maintained in a collaboration

---

[2] C. M. Rice, R. Fuchs, D. G. Higgins, P. J. Stoehr, and G. N. Cameron, *Nucleic Acids Res.* **21**, 2967 (1993).

[3] A. Bairoch and B. Boeckmann, *Nucleic Acid Res.* **22**, 3578 (1994).

TABLE I

NUMBERS OF ENTRIES AND BASES IN EMBL
NUCLEOTIDE SEQUENCE DATABASE[a] ACCORDING
TO TAXONOMIC DIVISION

Division[b]	Entries	Nucleotides
Bacteriophage	1066	1,493,417
EST	123,526	39,332,522
Fungi	8420	19,940,449
Invertebrates	13,831	27,610,495
Organelles	8195	9,364,254
Other mammals	6272	6,976,315
Other vertebrates	7041	8,144,622
Plants	11,105	14,145,431
Primates	35,290	36,665,648
Prokaryotes	21,427	37,074,154
Rodents	23,626	26,850,022
STS	7232	2,288,477
Synthetic	8597	4,295,284
Unclassified	6082	3,577,630
Viruses	21,496	24,801,066
Subtotal	303,206	262,559,786
Other patents	6686	2,507,063
Total	309,892	265,066,849

[a] Data are total numbers of entries and bases
in the EMBL nucleotide database at the time
of freezing the database for building Release
42.
[b] EST, Expressed sequence tags; STS,
sequence tagged sites.

between Dr. Amos Bairoch from the University of Geneva and the EBI. The data in SWISS-PROT arise from several sources; they are derived from translations of sequences from the EMBL Nucleotide Sequence Database, adapted from the Protein Identification Resource (PIR) collection, extracted from the literature, and directly submitted by researchers. The database contains high-quality annotation, is nonredundant, and is cross-referenced to several other databases, notably the EMBL nucleotide Sequence Database, PROSITE pattern database, and Protein Data Bank (PDB). The latest release (Release 32, November 1995) contained 49,340 sequence entries comprising 17,385,503 amino acids abstracted from 43,056 references.

As in the nucleotide sequence database, SWISS-PROT entries are represented externally as an ASCII flat file. The main difference between both flat files is in the feature table, which in SWISS-PROT describes the

```
ID   CLOSGDHG   standard; DNA; PRO; 1636 BP.
XX
AC   Z11747; S35943;
XX
DT   28-FEB-1992 (Rel. 31, Created)
DT   30-JUN-1993 (Rel. 36, Last updated, Version 6)
XX
DE   C.symbiosum gdh gene encoding glutamate dehydrogenase.
XX
KW   gdh gene; glutamate dehydrogenase.
XX
OS   Clostridium symbiosum
OC   Prokaryota; Bacteria; Firmicutes; Endospore-forming rods and cocci;
OC   Bacillaceae; Clostridium.
XX
RN   [1]
RP   1-1636
RX   MEDLINE; 92267007.
RA   Teller J.K., Smith R.J., McPherson M.J., Engel P.C., Guest J.R.;
RT   "The glutamate dehydrogenase gene of Clostridium symbiosum.
RT   Cloning by polymerase chain reaction, sequence analysis and
RT   over-expression in Escherichia coli.";
RL   Eur. J. Biochem. 206:151-159(1992).
XX
RN   [2]
RP   1-1636
RA   Teller J.K.;
RT   ;
RL   Submitted (26-FEB-1992) to the EMBL/GenBank/DDBJ databases.
RL   Teller J.K., University of Sheffield, Molecular Biology and
RL   Biotechnology, Western Bank, Sheffield, United Kingdom, S10 2UH
XX
DR   SWISS-PROT; P24295; DHE2_CLOSY.
XX
FH   Key             Location/Qualifiers
FH
FT   source          1..1636
FT                   /organism="Clostridium symbiosum"
FT                   /clone="pGS516"
FT   RBS             189..194
FT                   /citation=[1]
FT   CDS             204..1556
FT                   /gene="gdh"
FT                   /EC_number="1.4.1.2"
FT                   /product="Glutamate Dehydrogenase"
FT                   /evidence=EXPERIMENTAL
FT                   /citation=[1]
FT                   /note="pid:g49280"
XX
SQ   Sequence 1636 BP; 474 A; 329 C; 416 G; 417 T; 0 other;
     aacgtcgatc gtgcacgttt gcgctgtaac aattataatg ctaattcaat ttgcttatat        60
     aagtgaaatg cgttataata aaaccagaac agaaaatttc acaaaaacat agatcgtgag       120
<.....>
     aagaccggca gctattattt aataacaatt gcataagcgg ttgtctgaat gattggggct      1620
     gctgcattaa gtatat                                                       1636
//
```

TABLE II
TWO-LETTER CODES HEADING EACH LINE OF THE FLAT FILE AND THEIR MEANING[a]

Code	Meaning
ID	An identifier line, containing the accession code, type, and length of molecule
AC	Accession number(s)
XX	A blank separator
DT	Creation and update dates
DE	Description of the sequence
KW	Keywords
OG	Organelle
OS	Organism species
OC	Organism classification
RN	Reference number
RP	Reference page
RX	Cross-reference
RA	Reference authors
RT	Reference title
RL	Reference location (publication source)
RC	Reference comments
DR	Databases cross-reference line
FH	Features header
FT	Feature table and qualifiers lines
SQ	Sequence
//	Terminator

[a] A full descriptive reference can be found in the EMBL user manual, available on request.

characteristics of the protein sequence. The database is currently maintained using a mix of MS-DOS and UNIX systems, but current research and development work may result in the migration of the database to a relational database management system.

A relatively recent development is translation of the EMBL nucleotide sequence database, which will act as an unannotated supplement to SWISS-PROT. This new subgroup will consist of several sections containing entries derived from patent data, synthetic sequences, immunoglobulins, and T-cell receptors (IMGT database). These are in addition to a section containing translations of all naturally occurring sequences.

FIG. 1. Typical EMBL database entry, represented in the flat file format. Note that a large part of the sequence has been omitted (designated by <.....>).

External Databases Repository

The EBI provides a range of external databases on a caveat emptor basis. The databases are maintained by scientists throughout the world, who take responsibility and credit for the accuracy and currency. For a comprehensive list of the various databases, see Table III.[4-44]

[4] S. Pascarella and P. Argos, *Protein Eng.* **5,** 121 (1992).

[5] J. Jurka and T. Smith, *Proc. Natl. Acad. Sci. U.S.A.* **85,** 4775 (1988).

[6] T. Specht, *et al.*, *Nucleic Acids Res.* **19,** 2189 (1991).

[7] P. Rodriguez-Tome, EMBL–EBI (1995).

[8] J. C. Wallace and S. Henikoff, *CABIOS* **8,** 249 (1992).

[9] M. Cherry, Massachusetts General Hospital, Boston (1992).

[10] F. Larsen, *et al.*, *Genomics* **13,** 1095 (1992).

[11] K. Wada, *et al.*, *Nucleic Acids Res.* **20,** 2111 (1992).

[12] M. Olson, L. Hood, C. Cantor, and D. Botstein, *Science* **254,** 1434 (1989).

[13] M. Kroger, *et al.*, *Nucleic Acids Res.* **20,** 2119 (1992).

[14] A. Bairoch, *Nucleic Acids Res.* **21,** 3155 (1993).

[15] P. Bucher and E. N. Trifonov, *Nucleic Acids Res.* **14,** 10009 (1986).

[16] The FlyBase Consortium, *Nucleic Acids Res.* **22,** 3456 (1994).

[17] E. G. D. Tuddenham, *Nucleic Acids Res.* **22,** 3511 (1994).

[18] F. Giannelli, P. M. Green, S. S. Sommer, D. P. Lillicrap, M. Ludwig, R. Schwaab, P. H. Reitsma, M. Goossens, A. Yoshioka, and G. G. Brownlee, *Nucleic Acids Res.* **22,** 3534 (1994).

[19] J. G. Bodmer, S. G. Marsh, E. D. Albert, W. F. Bodmer, B. Dupont, H. A. Erlich, B. Mach, W. R. Mayr, P. Parham, and T. Sasazuki, *Tissue Antigens* **44,** 1 (1994).

[20] M. P. Lefranc, V. Giudicelli, C. Busin, A. Malik, I. Mougenot, P. Dénais, and D. Chaume, *Ann. NY Acad. Sci.* **764,** 47 (1995).

[21] E. A. Kabat, *et al.*, Technological Inst., Northwestern University, Evanston, Illinois (1992).

[22] G. Keen, G. Redgrave, J. Lawton, M. Sinkosky, S. Mishra, J. Fickett, and G. Burks, *Math. Comput. Modelling* **16,** 93 (1992).

[23] R. Dölz, M. D. Mossé, A. Bairoch, P. P. Slonimski, and P. Linder, *Nucleic Acids Res.* **24,** 66 (1994).

[24] M. Nelson and M. McClelland, *Nucleic Acids Res.* **19,** 2045 (1991).

[25] M. Hollstein, *Nucleic Acids Res.* **22,** 3551 (1994).

[26] S. K. Hanks and A. M. Quinn, *Methods Enzymol.* **200,** 38 (1991).

[27] T. K. Attwood, M. E. Beck, A. J. Bleasby, and D. J. Parry-Smith, *Nucleic Acids Res.* **22,** 3590 (1994).

[28] E. Sonnhammer and D. Kahn, *Protein Sci.* **3,** 482 (1994).

[29] A. Bairoch, *Nucleic Acids Res.* **20,** 2013 (1992).

[30] B. L. Maidak, *et al.*, *Nucleic Acids Res.* **22,** 3485 (1994).

[31] A. Bairoch, University of Geneva, Geneva (1991).

[32] R. Eberhard, *Genetic Analysis: Techniques and Applications (GATA)* **10,** 49 (1993).

[33] R. J. Roberts and D. Macelis, *Nucleic Acids Res.* **20,** 2167 (1992).

[34] J. Jurka, *et al.*, *J. Mol. Evol.* **35,** 286 (1992).

[35] H. Lehrach, *Genome Analysis* **1,** 39 (1990).

[36] J. M. Neefs, Y. Van de Peer, P. De Rijk, S. Chapelle, and R. De Wachter, *Nucleic Acids Res.* **21,** 3025 (1993).

[37] S. Pongor, Z. Hátsági, K. Degtyarenko, P. Fábián, V. Skerl, H. Hegyo, J. Myrvai, and V. Bevilacqua, *Nucleic Acids Res.* **22,** 3610 (1994).

[38] S. Gupta and R. Reddy, *Nucleic Acids Res.* **19,** 2073 (1991).

[39] C. Zwieb and N. Larsen, *Nucleic Acids Res.* **20,** 2207 (1992).

Software Repository

The EBI also maintains a repository of software for molecular biology applications. The programs are provided by scientists throughout the user community and are also provided on a caveat emptor basis. That is, the EBI takes neither responsibility nor credit for their quality. Most programs are in a compressed format, using worldwide accepted formats of compression utilities (e.g., zip, gnuzip, compress, stuffit, and compact-pro). Most UNIX programs are archived as tar files, and Macintosh programs are encoded in BinHex 4.0 format.

The software repository is arranged according to the platform for which the program is intended. The whole repository is hierarchically arranged under the subdirectory "software," with subdirectories according to the platform (DOS, Mac, Unix, VAX, VMS). The programs in the software repository are included in the software BioCatalog that is now maintained at the EBI.

BioCatalog

The BioCatalog[7] is an ongoing project, started in 1993 by Généthon and the CEPH-Fondation-Jean-Dausset with the support of the RESIG project (Networks of Computer Servers for Genomes) and a grant from the GREG (Groupement pour la Recherche et l'Etude des Genomes). The main aims of the project are collecting and maintaining a software directory of general interest in molecular biology and genetics, and distributing it on the Internet.

The catalog is categorized according to common topics (termed domains), as follows: DNA, proteins, alignments, genetics, mapping, molecular evolution, molecular graphics, database, servers, and miscellaneous. Each of the domains contains further subdivisions. Each entry in the catalog contains (where available) information about the program, its description, bibliographic references, programming languages, and hardware and software requirements. The original site from which the program can be downloaded is cited, and in the HTML (Hypertext Markup Language) version it is also linked for a direct ftp session. The author details and means of contact are included.

The BioCatalog is now maintained, distributed, and further developed at the EBI on a collaborative basis. It is available as a full text version for

[40] D. Ghosh, *Nucleic Acids Res.* **20,** 2091 (1992).
[41] S. Steinberg, A. Misch, and M. Sprinzl, *Nucleic Acids Res.* **21,** 3011 (1993).
[42] E. Wingender, *J. Biotechnol.* **35,** 273 (1994).
[43] C. Brown, *Nucleic Acids Res.* **21,** 3119 (1993).
[44] S. Liebl and E. Sonnhammer, MIPS, Germany, and Sanger Centre, UK (1994).

TABLE III
EXTERNAL DATABASES PROVIDED BY EBI[a]

Database	Comment	Ref.
3d_ali	Database merging related protein structures and sequences	4
alu	Alu sequence database	5
berlin	RNA databank of 5 S rRNA and 5 S rRNA gene sequences	6
bio_catal	Catalog of molecular biology programs	7
blocks	Sequence blocks database	8
codonusage	Tables of codon frequencies, calculated for different organisms	9
epgisle	Human CpG-island database	10
cutg	Tables of codon frequencies in a tabulated format	11
dbEST	EST (expressed sequence tags) database	12
dbSTS	STS (sequence tagged sites) database	12
ecdc	*Escherichia coli* database collection	13
enzyme	Enzymes database	14
epd	Eukaryotic promoter database	15
flybase	*Drosophila melanogaster* set of databases	16
haema	Mutations in factor VIII gene associated with hemophilia A	17
haemb	Mutations/deletions associated with hemophilia B	18
HLA	Alignments of HLA (human leukocyte antigen) class I and II nucleotide and protein sequences	19
IMGT	Immunogenetics database	20
kabat	Database of sequences of proteins of immunological interest	21
limb	Listing of molecular biology databases	22
lista	Nucleotide sequences encoding proteins from yeast *Saccharomyces*	23
methyl	List of effects of site-specific methylation on methylases and restriction enzymes	24
p53	Database of p53 somatic mutations in human tumors and cell lines	25
pkcdd	Protein kinase catalytic domain database	26
prints	Database of protein motif fingerprints	27
prodom	Homologous domains database of nonfragment protein sequences	28
prosite	Database of known, specific sites in proteins	29
rdp	Database and programs of the Ribosomal Database Project	30
reflist	Reference lists with relevance to molecular biology	31
relibrary	Different restriction enzyme files for sequence analysis programs	32
rebase	Restriction enzymes database, including commercial sources	33
RepBase	Repetitive elements from different eukaryotic species	34
RLDB	Reference Library DataBase of various sequence libraries	35
rRNA	Databases of small and large ribosomal subunit rRNA sequences	36
sbase	Collection of annotated protein domain sequences	37
smallrna	Compilation of small RNA sequences	38
srp	Signal recognition particle database from eukaryotes and Archaea	39
tfd	Transcription factor database	40
trna	tRNA database	41
transfac	Eukaryotic cis-acting regulatory DNA elements and trans-acting factors	42
transterm	Translational termination signal database	43
yeast	Complete DNA sequences of yeast chromosomes	44

[a] Through the ftp server, the WWW, and gopher servers and on the CD-ROM releases.

ftp. It is also indexed by the WAIS (wide area indexing system) and SRS (sequence retrieval system) indexing systems and thus searchable, when accessed through the EBI World Wide Web (WWW) server.

Immunogenetics Database: IMGT

The IMGT database is an integrated database of immunological inter-est,[20] under development through collaboration coordinated by the Laboratoire d'Immunogénétique Moléculaire (LIGM). The IMGT database will contain nucleotide and protein sequences of immunoglobulins (Ig) and T-cell receptors (TCR), detailed expert annotation of these sequences, mapping data, and the results of comparative sequence analysis. Further collaboration with ICRF (Imperial Cancer Research Foundation) London (J. Bodmer) will allow integration of human leukocyte antigen (HLA) proteins and genes, and that with IFG (Institute for Genetics) Cologne (W. Mueller) will permit integration of murine alignments in the IMGT database. The LIGM-DB is part of the IMGT database developed by the LIGM (Montpellier, France), IFG (Cologne, Germany), ICRF (London, UK), and EMBL outstation EBI (Cambridge, UK).

The objectives for the IMGT database are to contain information about immunoglobulins and T-cell receptors from all species, specifically, to contain all sequences and alignments, allele information, sequence tagged sites (STS) and polymorphism, genomic maps, molecular modeling information, and information about the relations with diseases and hybridomas. Software will be developed for facilitating the annotation process, for classification of sequences, and for molecular modeling. The aims include developing a user-friendly graphical interface, stabilizing keywords used in immunogenetics, and incorporating results of sequence alignments and translation of sequences to amino acid sequences. The database will provide a detailed morphological and functional analysis of immunoglobulins and T-cell receptors. The data are already indexed by the SRS system. It can be obtained from the EBI ftp server in the databases section. It can also be obtained and searched through via the EBI WWW server. The database team can be contacted at the following address: IMGT@ebi.ac.uk.

Interfaces between EBI and User Community

Submission Systems

Submission of Sequence Data. There are three main ways to submit sequence data to the EBI sequence databases. The first two refer to the nucleotide sequence and SWISS-PROT databases, while the third one (WWW submissions) refers only to nucleotide sequences.

MANUAL EDITING OF ELECTRONIC SUBMISSION FORM. A text (ASCII) submission form can be filled using any text editor. The editing task can be complex and error prone, especially for inexperienced users. Furthermore, because no data validation can be carried out in real-time, the user receives no feedback on possible errors or omissions.

The submission form can be obtained by various methods: (1) by an E-mail request from

datalib@ebi.ac.uk

(2) by ftp from ftp.ebi.ac.uk in the directory

/pub/doc/emblsub.form

or (3) from the EBI gopher server gopher.ebi.ac.uk (port 70) from the menu selection

EMBL Nucleotide Sequence database/
Nucleotide Sequence Submissions/Updates/

When using ftp, the file type must be set to ASCII before downloading.

Once the text version of the submission has been prepared, it can be sent by E-mail to datasubs@ebi.ac.uk, or it can be sent on a diskette via regular mail to the EBI postal address at The EMBL Outstation—The European Bioinformatics Institute, Hinxton Hall, Hinxton, Cambridge CB10 1RQ, United Kingdom.

AUTHORIN PROGRAM. Authorin is an interactive program to help the user to prepare a submission. The program exists for Macintosh and IBM-compatible machines. Authorin works interactively with the submitter, to prepare the submission while validating data as they are entered. At the end of the submission process, the program produces a text file in a special format that can be interpreted by software at the EBI. The output from Authorin can be sent on a diskette or by E-mail the same way as the submission form is sent.

Currently Authorin is a good way to create automatically processed direct submissions, but new tools aimed at overcoming some of its disadvantages are under development. In particular we aim to obviate the need to actually install the program on your own machine, to deal with new data items that are not handled by Authorin, and to create tools to run on modern hardware that is at present incompatible with Authorin.

The Authorin program can be downloaded from the EBI ftp server:

ftp.ebi.ac.uk

The version for DOS operated machines is under

/pub/software/dos/authorin.exe

The version for Macintosh computers (*not* PowerPC) is under

/pub/software/mac/authorin.hqx

WORLD WIDE WEB BASED SEQUENCE SUBMISSION SYSTEM. A complete data submission system, based on a WWW server, has been developed at the EBI. The system provides a user with the ability to submit sequence data in a direct and easy way. The only requirement on the user's side is to install a WWW browser that can handle forms. The system has a few major advantages. First, in contrast to a stand-alone program, EBI constantly maintains and updates the program. This means that the user is always working with the latest version of the program. Second, if the WWW client is already installed, the user doesn't have to waste time, effort, and disk space on installation of a program on the computer. Third, the program uses the EBI database resources (like the list of previous submitters, or journals) to enable more user-friendly interface by avoiding the lengthy business of entering information already available. Finally, the user may freeze a submission session for a very long time.

The system breaks the complicated task of sequence submission into a set of interactive forms which check the user's input and present the following forms according to the input. The system is compatible with the various WWW browsers currently available, on all platforms. An effort was made to reduce the need for typing to a minimum, for example, by providing mechanisms to load automatically the personal details (where available) of the submitter according to an accession number of a previously submitted sequence. If more than one sequence is to be submitted, the system enables reuse of most data items that had been already typed in. Each submission cycle can present a practically unlimited number of features and qualifiers sets. At the end of the submission process the system mails to the submitter the data entered, formatted into the EMBL flat file format, which can be reviewed again by the submitter.

The submission system has a "crash recovery" mechanism. If the submitter's computer (or the WWW browser) has crashed during the submission process, the system can resume the submission at the stage where it was abandoned, based on a unique identifier provided with each submission.

The WWW submission system can be accessed from the EBI home page, or it can be directly accessed at the following URL (uniform resource locator):

http://www.ebi.ac.uk/subs/emblsubs.html

Submission of Software to the Software Repository

To submit software that has been written or developed for molecular biology, the author should send an E-mail message to the address allocated for this purpose:

software@ebi.ac.uk

The message should contain information about the program, what it does, what platform is it intended to run on, and what are the hardware requirements. It should note whether the source code is included and whether it is a demo/shareware/freeware; any known problems and full details of the submitting author should also be included.

The EBI software team will then contact the author to finalize the means of providing the program. In most cases, the program is either UUencoded or converted to BinHex 4.0 and is sent by E-mail. If the program is very large, EBI will provide the author with a temporary user login and password to enable upload to the EBI ftp server.

The authors should also provide detailed information about the program to be included in the BioCatalog. Information can be submitted using the WWW BioCatalog submission form (accessible through the EBI WWW server), or authors can send the information to biocat@ebi.ac.uk.

Although staff at the EBI will carry out simple checks on the program such as for obvious viruses or compilation failures, we have neither the resources nor the expertise to do detailed quality control. Thus submitting authors must understand that they are assumed to have tested the software appropriately and that they may be contacted by users encountering problems with the software.

Providing Information and Retrieval Systems

CD-ROM Distribution of Databases. The EBI databases on CD-ROM provide a snapshot of all the databases at a specified time. Quarterly releases of the sequence databases are distributed in CD-ROM format. The disks contain the EMBL database, the SWISS-PROT database, their index files, and search utilities for Macintosh and IBM-compatible computers. The disks also contain more than 20 related databases prepared by collaborators.

Usage of the search programs requires the presence of at least one CD-ROM drive, but it is preferred that the system be equipped with two CD-ROM drives. If only one drive is present, the system's hard disk must have (currently) at least 150 Mb free space. As the EMBL database currently has an annual growth rate of about 70%, the index files of the next releases are likely to occupy much more disk space. Users can order single CD-ROM releases or subscribe indefinitely or for several releases.

To order the EBI CD-ROM set, send an E-mail request to datalib@ebi.ac.uk or use the special form that appears in various WWW pages (EMBL, SWISS-PROT, documentation and software) that lets the user subscribe on-line.

ftp Server. The ftp server of the EBI can be accessed by opening an ftp session:

$$\text{ftp} \quad \text{ftp.ebi.ac.uk}$$

Login as "anonymous" (lowercase) and type your E-mail address as a password.

The session starts by default in the /pub directory. The /pub directory is organized as follows:

README	(file)
/contrib	(directory)
/databases	(directory)
/doc	(directory)
/help	(directory)
ls-lR.Z	(file)
/software	(directory)

The file README contains a general description of the ftp server. The file ls-1R.Z contains (UNIX) compressed information of all the directories and files of the ftp system. The directory "databases" contains the updates of the EMBL and SWISS-PROT databases, and all the external databases that are provided on CD-ROM. The directory "doc" contains documentation and forms. The directory "help" contains various information files about the directories and databases on the ftp server. The directory "software" contains various demo, shareware, and freeware programs for DOS, Macintosh, UNIX, VAX, and VMS platforms in the following directories accordingly: "dos, mac, unix, vax, and vms." There is also a "tools" subdirectory that contains tools which help the user to communicate with the EBI. All the ftp directories and files are also accessible through the EBI gopher and WWW servers.

Gopher Server. Although the most facile access to EBI services is via the WWW server, a gopher server provides a last resort for users limited to text based access. The gopher server provides access to the nucleotide and SWISS-PROT databases (documentation and data), the ftp server for databases and software, the BioCatalog software directory (excluding its search utility), EMBnet gopher servers, and searches in gopherspace using VERONICA. There is a simple text based program for the WWW called lynx, and we recommend that if you are limited to text based systems then use lynx to connect to EBI's WWW server. To connect with EBI's gopher use the following address:

$$\text{gopher.ebi.ac.uk}$$

World Wide Web Server. The World Wide Web (WWW) server is currently the main interface of the EBI with the scientific community. The advantages of the WWW as a system which provides the combination of text, graphics, and the ability of collecting data from the user by using forms enables EBI to use it as an optimal mechanism for providing and collecting information. The EBI WWW home page can be logically divided into several major topics as follows.

MAIN DATABASES

EMBL Nucleotide Sequence Database Area. The home page introduces the user to the EMBL Nucleotide Sequence Database. It provides the user with the updated database release information, information for submitters, information about the various methods of data submission, contact addresses, and the feature table definition. There is a link to a form providing an easy means of updating the database with minor corrections. The corrections are provided in a noninteractive manner, as free text. There is also a link to the new WWW based sequence submission system described above. Users who wish to subscribe to the database may do so on-line, using a WWW based subscription system, linked to the EMBL page.

SWISS-PROT Protein Sequence Database area. The home page of the SWISS-PROT Protein Sequence Database provides users with access to documentation, including release notes and the user manual for the database. There is a link to the new "protein machine." This is a form based on a script which translates a nucleic acid sequence to the protein product attempting to deal with all the complexities and exceptions such as unusual translation tables. Users can also subscribe on-line if they wish to receive the database on CD-ROM.

The SWISS-PROT home page provides links to a wide range of retrieval services, related databases, and search services: retrieval by accession number or entry name, SRS (sequence retrieval system) access, links to dbEST and dbSTS (see Table III), and FASTA, BLITZ, BLAST, and PROSITE searches. A huge advantage of the WWW interface is that a rich range of services can be offered without making the user interface overcomplex.

SEQUENCE-RELATED OPERATIONS

Sequence query and retrieval. The most simple and direct retrieval system is operated by providing the server with an accession number (e.g., X58929) or an entry name (e.g., SCARGC). Although this method is limited to cases where the user knows the identity of the entry (e.g., when an accession number is cited), it is the fastest method of obtaining a sequence from the database. Users may retrieve sequences directly from the EMBL, SWISS-PROT, PROSITE, and PDB databases.

If the sequence is found in the database, it is returned to the user formatted as a linked HTML document. Where applicable, the MEDLINE cross-reference is linked to the MEDLINE entry containing the reference abstract and publication details. When the entry has a database cross-reference, it is linked to the appropriate database entry as well. For instance, the nucleotide sequence with accession number J00231 has a cross-reference line:

DR   SWISS-PROT: P01860; GC3_HUMAN.

The SWISS-PROT accession number P01860 appears as a hypertext entry, linked to the actual SWISS-PROT file of P01860, and then it is a simple matter to click on this hypertext link to call up the SWISS-PROT entry.

Sequence Retrieval System. The sequence retrieval system (SRS) is a robust indexing system, developed by Thure Etzold and Gerald Schäfer in collaboration with Reinhard Dölz from the Biozentrum in Basel.[45,46] The SRS enables a fast and efficient search for keywords and definitions through various databases. Currently, there are 33 database systems indexed by the SRS on the EBI server (see Table IV). An interface to search mechanisms of the SRS indexes is provided as a WWW form.

The SRS allows flexible selection of which databases to search, which fields in the database should be searched, the target keywords to be sought (including trailing wildcards), and the fields to be presented in displaying the search "hits." Complex searches can be built up using the usual Boolean operators, rendering the entire system powerful, flexible, and easy to use. Indeed, SRS is the most popular access method supported by the EBI.

Expressed sequence tags and sequence tagged sites. The two specialist sequence libraries dbEST (database of expressed sequence tags) and dbSTS (sequence tagged sites, Ref. 12), developed by the National Center for Biotechnology Information (NCBI), are mirrored by EBI. dbEST is a database of sequence and mapping data on expressed sequence tags, which are partial, "single pass" cDNA sequences, whereas dbSTS contains sequence and mapping data on short genomic landmark sequences or sequence tagged sites. Both databases are completely searchable by using the SRS described above.

SEQUENCE SIMILARITY SEARCHES

Nucleic acid homology searches. The WWW server enables an easy submission of homology searches of nucleotide and amino acid sequences in the EMBL and SWISS-PROT databases, by using the FASTA program. FASTA performs searches of the database for sequence homology against a provided target, using the FASTA algorithm.[47] The WWW form enables

[45] T. Etzold and P. Argos, *Comput. Appl. Biosci.* **9,** 49 (1993).
[46] T. Etzold and P. Argos, *Appl. Biosci.* **9,** 59 (1993).
[47] W. R. Pearson and D. J. Lipman, *Proc. Natl. Acad. Sci. U.S.A.* **85,** 2444 (1988).

TABLE IV
DATABASES SEARCHABLE THROUGH THE SEQUENCE
RETRIEVAL SYSTEM

Name	Entries[a]	Library group
EMBL	422,829	Sequence
EMNEW	18,579	Sequence
SWISSPROT	43,470	Sequence
SWISSNEW	3804	Sequence
PIR	71,995	Sequence
NRL3D	4153	Sequence
NRSUB	248	Sequence
PDB	3588	Protein structure
HSSP	3248	Protein structure
DSSP	3143	Protein structure
FSSP	557	Protein structure
ALI	84	Protein structure
SWISSDOM	28,224	Sequence related
PRODOM	23,105	Sequence related
FLYGENE	7126	Sequence related
ECDC	3894	Sequence related
ENZYME	3556	Sequence related
REBASE	2486	Sequence related
EPD	1252	Sequence related
PIRALN	1183	Sequence related
PROSITE	1029	Sequence related
CPGISLE	965	Sequence related
IMGT	885	Sequence related
PROSITEDOC	786	Sequence related
BLOCKS	770	Sequence related
MEDLINE	179,262	Literature
SEQANALREF	2579	Literature
LIMB	120	Others
TFSITE	4042	Transcriptional factors
TFFACTOR	1412	Transcriptional factors
DBEST	241,909	Tagged sites
DBESTNEW	7025	Tagged sites
DBSTS	12,890	Tagged sites
DBSTSNEW	11	Tagged sites

[a] Data are numbers of entries as of July 1995.

an easy way for selecting the target library for searches and selecting the level of sensitivity (ktup), the number of matched sequences to be listed, and the number of aligned sequences to be listed. After typing or copying the sequence in the appropriate window, one initiates the search by the system, and the results are sent back to the user by E-mail.

Protein sequence homology searches: BLITZ database searches. The WWW server enables submission of sequences for a BLITZ search. BLITZ uses the MPsearch program of Shane Sturrock and John Collins.[48] MPsearch allows sensitive and extremely fast comparisons of protein sequences against the SWISS-PROT protein sequence database using the Smith and Waterman best local similarity algorithm.[49] It runs on the MasPar family of massively parallel machines. A typical search time for a query sequence of 400 amino acids is approximately 40 sec, which covers a search of the entire SWISS-PROT database. Additional time is required to reconstruct the alignments depending on the number of alignments requested. MPsearch is the fastest implementation of the Smith and Waterman algorithm currently available on any machine.

PROSITE database searches. The PROSITE database search is a WWW interface to Mail-PROSITE based on the ppsearch software derived from the MacPattern program developed by R. Fuchs.[50] It allows a rapid comparison of a new protein sequence against all patterns stored in the PROSITE pattern database.[51] The WWW form is very simple to use. The user needs to provide only a title for the search and the amino acid sequence in question. Thus, it saves the use of an E-mail submission and retrieval of the search results. Because the database being searched is relatively small, the results are returned in real time directly to the WWW client.

BLAST searches. There are two pointers for a form based interface with the two BLAST search servers. The BLAST program searches in SBASE 3.1, a collection of annotated protein domains. One server is located in Trieste, Italy, and the other at the NCBI (Bethesda, MD). The main difference between both servers is that the NCBI server provides a very straightforward search form with predetermined search parameters, whereas the one in Trieste calls for a thorough knowledge of the program parameters but enables more freedom of operation. The NCBI server will return the results of the search directly on-line, and the server at the International Centre for Genetic Engineering and Biotechnology in Trieste returns the results by E-mail. The interface provides a convenient manner of setting the various variables needed for the analysis, including the type of matrix to be used, the genetic code (for nucleic acid sequences), and the format of the output to be provided.

[48] S. S. Sturrock and J. F. Collins, MPsrch version 1.3. Biocomputing Research Unit. University of Edinburgh, UK (1993).
[49] T. F. Smith and M. S. Waterman, *J. Mol. Biol.* **147,** 195 (1981).
[50] R. Fuchs, *Comput. Appl. Biosci.* **10,** 171 (1994).
[51] A. Bairoch, *Nucleic Acids Res.* **21,** 3097 (1993).

DOCUMENTATION AND VARIOUS SERVICES. The documentation area of the WWW server provides some documentation of general interest, like documentation of the EBI services and a reference list for authors.

BioCatalog. The BioCatalog is a database of computer programs for molecular biology and genetics. This project was initiated by Généthon and the CEPH-Fondation-Jean-Dausset. The EBI now supports the maintenance, development, and distribution of the BioCatalog as part of the ongoing research and development scheme.

The BioCatalog is divided logically into various areas of interest, called domains. The domains available are DNA, proteins, alignments, genetics, mapping, molecular evolution, molecular graphics, database, servers, and miscellaneous.

The BioCatalog exists on the EBI server as two versions: a text based version, available for downloading through the ftp server (under /pub/databases/bio-catal) and through the gopher server, and a WAIS indexed version. The indexed version can be searched by using a specialized query form on the WWW server. The query form supports several search possibilities: a full text search, according to a BioCatalog known accession number, by name, by description, or by author name or by bibliographic information. The user may define the logical operator to be used (either AND or OR), how many successful search results to display, and whether to display them as full records or only as short informative headers. An SRS indexed version also exists, and it is searchable through the WWW SRS searches interface.

A very important aspect of the BioCatalog is that the users actively update the database by announcing new programs or updating existing ones. There is a special WWW form for announcements on new programs. Not only does the form enable an easy way of providing the information, but it also enables the database maintainers to direct the submitting authors to provide the most appropriate information to describe the program.

EBI netnews filtering system. One of the major problems of modern scientists is keeping up to date with news in related fields of interest and maintaining communications with colleagues. The Usenet network news system helps to overcome this problem. However, the volume of information that flows through the news groups constantly increases, and it is now a problem to filter the relevant messages.

The idea behind the EBI Netnews filtering system is to allow users to provide a search profile that identifies the topics they are interested in. A special program will scan the Usenet groups and will mark out the articles with relevance to the user according to the search profile provided. The profile itself may contain Boolean operators to provide a more stringent search. The user can set a certain threshold to increase the filtering power of the program. A higher threshold provides less articles, with a higher index of relevance to the search profile.

The search program runs on a regular basis at predetermined intervals. It indexes all the Usenet articles and sends results by E-mail. Each search hit contains the first few lines from the message (the number of lines can be determined by the user).

The WWW based form that enables a user to submit a search profile requires the user to provide a password, enabling discretion and concealing of the user's fields of interest. An end-user can submit as many profiles as desired to the system, but it is good practice to test run each profile before submitting it. Test runs can give an estimate of how efficient the keywords in the profile are before the profile is submitted. Each subscription is given an ID number. All the ID numbers can be listed, and canceled at any time.

NETWORK NAVIGATION RELATED OPERATIONS. Several documents and services are provided to aid the users in finding network resources related to their fields of interest. The "Bio-wURLd" is a home page that contains a list of links submitted by biologists. This is an interesting service, because users have the possibility to add new sites of interest to the list. In essence Bio-wURLd is actively maintained by the user community. Another method for the discovery of network resources is to look at "clickable maps." The EBI WWW server has clickable maps for the whole of Europe and for the United Kingdom in particular.

In a similar manner there is also "Career Connection," which allows users to advertise job opportunities. Again, this service is end-user driven since all the jobs being listed have been contributed through the EBI WWW server.

EBI-CUSI search. There are many search engines that allow users to explore the WWW. The EBI-CUSI interface is a compilation of some of the best search engines available, and they are all collected under one page to allow ease of access. Users can find resources by searches. There is a special multiform page that will help users to submit search requests to many search servers. There are a few search groups that can be accessed: searches through selected indexes of WWW pages, searches through indexes generated by special search robots, other non-WWW based Internet search engines (e.g., VERONICA, WAIS), various methods of searching for software, finding people and places on the network, dictionaries available on the Internet, and other documents of general interest.

FINDING OUT MORE ABOUT EBI SERVICES. A verbal description of WWW services, such as given here, does not do justice to their ease of use. By exploring the EBI home page you will find that all this information is very easy to access and understand. Even details on where the institute is located geographically can be found, and information about staff members is also available on-line. There is no better way than simply to try it.

Electronic Mail Operated Servers

A complete list of E-mail addresses can be found in Table V.

Electronic Mail Server. The automatic file server provides users who are limited only to E-mail communication with a convenient way of obtaining sequences and software through E-mail messages. Indeed, experienced users often find this the most convenient way to access some services. The user sends the server a message that contains a command, or set of commands, in a precise syntax. In response, the server will send the user the requested information.

The most basic operation is to send the server a message that contains the word "HELP" either in the subject line or in the body of the message. In response, the server will send the user a help file that leads the user through all the steps required to use the service. The user can ask for a sequence, either by accession number or entry name. In response, the sequence will be sent to the user's E-mail address in EMBL flat file format. Computer programs and other binary files will be sent in a UUencoded format (see Glossary), which calls for extra steps on the part of the user

TABLE V

ADDRESSES FOR COMMUNICATING WITH EBI

Mail address:	
EMBL Outstation, The European Bioinformatics Institute, Hinxton Hall, Hinxton, Cambridge CB10 1RQ, UK	
Telephone	+44 1223 494400
Fax	+44 1223 494468
E-mail addresses for human readable messages	
(Any) Sequence data submission	DataSubs@EBI.ac.uk
(Any) Sequence updates	Update@EBI.ac.uk
Inquiries and user support	DataLib@EBI.ac.uk
Software support	software@EBI.ac.uk
Networking and server problems	nethelp@EBI.ac.uk
Network addresses of computer automated servers	
Network file server	Netserv@EBI.ac.uk
ftp server	ftp.EBI.ac.uk (anonymous ftp)
Gopher server	gopher.EBI.ac.uk
WWW server	http://WWW.EBI.ac.uk
Mail FASTA server	FASTA@EBI.ac.uk
BLITZ searches mail server	BLITZ@EBI.ac.uk

but provides a good solution for users who do not have any access other than E-mail.

To get started with the network mail server, the user needs only to send an E-mail message that contains the word "HELP" to the following address:

netserv@ebi.ac.uk

Mail FASTA Nucleotide Homology Search Server. The mail Fasta nucleotide homology search server is a fast and convenient interface to the FASTA program. It enables users to submit homology searches for a given sequence against the whole database, using the FASTA algorithm.[47] The search sequence must be presented following a precise syntax. The results of the search are sent back to the user by E-mail. To get started with the mail FASTA server, the user should send an E-mail message containing the word "HELP" to the address

FASTA@ebi.ac.uk

Mail BLITZ Amino Acid Homology Search Server. The BLITZ mail server enables an easy access to the MPsearch program in a manner very similar to that of the mail FASTA server, as described above. The MPsearch program also uses the Smith and Waterman algorithm, but for searching through protein sequences. Here, too, the results of the search are sent back to the user by E-mail. To get started with the mail BLITZ server, users have to send an E-mail message containing the word "HELP" to the address

BLITZ@ebi.ac.uk

NetNews Mail Server. The NetNews filtering service described above can also be accessed through a mail server. Users can submit a search profile through E-mail, by sending a message that contains the word "HELP" (without the quotes) to the address

netnews@ebi.ac.uk

As a result, a help file will be sent to the user with step-by-step instructions on how to make the most of the NetNews server.

Support from EBI

There are three main groups of individuals at the EBI who can provide solutions to various technical problems.

User Support Group. The user support group provides answers to problems associated with the various on-line servers (WWW, gopher, ftp), helps to incorporate newly established or updated databases onto the EBI servers, and also answers questions of a general nature. The user support group serves as the front end for EBI's relations with the scientific community and can be contacted at the following address:

datalib@ebi.ac.uk

Software Group. The software group maintains the molecular biology software repository. Any communication regarding uploading of new software, requests for help, and technical problems with software should be addressed to this group, using the address

software@ebi.ac.uk

Network Operations Group. The network operations group provides help and technical support with issues regarding networking problems, mail, and search servers. Such problems should be addressed to

nethelp@ebi.ac.uk

Summary

The scope of the EBI is focused on providing better services to the scientific community. Technological advancements in the hardware area provide EBI with means of producing data much faster than before, and with greater accuracy since there is now a better technical ability to produce more exhaustive searches through larger indices. Hand in hand with the technological developments, research and development work is continuing on better indexing systems and more efficient ways of establishing and maintaining the future databases. The existing links of communication between EBI and the user community are exploited to study the needs of the scientific community, to provide better services, and to enhance the quality of databases by interpreting user feedback and updates. A very important goal is to enhance the awareness of the scientific (and, maybe even more, the nonscientific) public of the importance of the modern field of bioinformatics and to introduce special meetings and courses, in which more specific subjects will be studied in depth.

Another aspect of this goal is to help in constructing special bioinformatics programs in university faculties. In such programs, in contrast to the existing layout, students will pursue studies in a combined environment that provides basic training in biology and in computation. Currently, one of the main problems in the field is that scientists are either biologists, who are self-educated in the field of computers and programming, or computer scientists without sufficient knowledge of biology. It is hoped that a combined program will provide a high level of education in both fields of interest at the appropriate ratios.

Building an efficient and friendly interface between the EBI and the user community is the basis for any future development. This aim is achieved by using the most modern server systems while continuously researching newer and better systems and interfaces. This task can never be complete without involvement of the user community by providing feedback to any of EBI's services. A better bioinformatics community is a necessity for any future development of the biological research aiming at a better society.

Glossary

Anonymous ftp   Many computers on the network provide various data items or files for the public, without the need for a special account (i.e., a user name and a password). There is a common way of providing such data by using the ftp protocol (see below). Following creation of the ftp connection to the required computer, the user types "anonymous" (in lowercase) instead of the user name, and provides his or her E-mail address instead of the password. This enables a restricted, but very useful, access to some areas of the computer file system

ASCII: American Standard Code for Information Exchange   The ASCII is a standard that assigns a numeric value to each character, enabling different computer systems to exchange data. Although the standard includes a set of control characters and graphic characters, it is commonly used in the computation jargon as a synonym for plain text documents (as opposed to binary formats)

BinHex 4.0   BinHex 4.0 is a special file format that enables transfer of Macintosh files across networks. The need for the BinHex format arises because Macintosh files are divided into two parts (a fork). This enables association of various data items with the file (such as the icon and the file attributes) but creates a problem of transferring the file as a whole piece. The BinHex 4.0 format encodes both parts of the fork into a single ASCII file. The encoded file can be transferred over networks and included in E-mail messages. It is then decoded back into a Macintosh file by a compatible utility. Many utilities have BinHex converters (e.g., BinHex itself, stuffit, compact-pro, fetch, and some WWW browsers)

browser   A common name for a WWW client, a browser program is capable of interacting with a WWW server. It can present hypertext documents, graphics, and forms. The browser is capable of interpreting a special standard language, called HTML (see below), and presenting the hypertext accordingly. It is also capable of sending the server requests for information according to the user's selections, and of providing the server information typed by the user into a form. These properties turn the browser into a sophisticated, generic tool of interaction and exchange of information across networks

E-mail: Electronic Mail   E-mail provides a very fast means of communication between computer users. Each user on the network is identified by a unique address. The address combines two parts, separated by the address character ("@"). An electronic message can be typed directly into the computer (this is the most common usage), or it can be composed of attached files. Until relatively recently, the only file type that could be attached was a plain text file. There is a relatively new standard, called MIME, that enables sending images, sounds, and other nontext information in E-mail messages. After being sent, the message travels between the various network nodes (connection points) according to the domains (network areas) specified in the E-mail address. When the message reaches the addressed computer, it is stored in a special mail file (called a spool) and is ready to be read

EMBnet: European Molecular Biology Network   The EMBnet is a network of nationally mandated nodes. Each node maintains copies of the EMBL database distribution (and of other databases) and provides search utilities and local technical assistance with bioinformatics associated issues. A local EMBnet node is the first and probably the most convenient place to call for assistance before looking elsewhere

freeware   Freeware is a computer program free for use and distribution. The author(s) of the program does not charge any payment for using the program. Although being free of charge, freeware is normally copyrighted and is distributed under various legal terms. It is a common demand that freeware will be distributed as a package which includes all the documentation and legal notifications

ftp: File Transfer Protocol   The ftp enables a very fast means of transferring data between different computers and operating systems across the network. The protocol enables transfer of binary data and of text files. When transferring text files between different platforms, the ftp program performs the required translation of the characters that control the end of line and paragraph, which vary between the different operating systems

gopher   The gopher data transfer system resembles the WWW but does not make use of hypertext or graphcis. The data are organized in a tree structure and menus. Gopher was the first easy-to-operate data-providing system on the network. Although currently the WWW is much more popular, many users still find good use for gopher, especially in domains that are limited to using text-based terminals (e.g., vt100)

HTML: Hypertext Markup Language   HTML is a collection of styles that define the various components of a WWW document. It is based on the SGML standard. The HTML code is expressed as special tags, inserted into the text. These are interpreted by the browser, which forms the final presentation accordingly

http   The http is a scheme of identifying a file on a WWW server. Most WWW addresses take that general form, for example, http://www.ebi.ac.uk/. This address means the user is asking the WWW server at ebi.ac.uk to provide the user with its index document (also called the home page). Other schemes can be gopher, ftp, file, WAIS, telnet, and news

lynx   The lynx WWW browser (client) is text based. Although not capable of presenting the graphics associated with a HTML document, lynx will run on a text terminal (e.g., vt100) and provide users who are limited to this environment with the ability of browsing WWW documents and finding information. The HTML language includes a special tag that provides lynx with textual information about the graphic image that is normally presented. The lynx browser will show this text instead of the graphic image

shareware   A computer program that is not free for use, but is free for distribution and initial evaluation, is called shareware. The author of a shareware program allows the user to install the program on the computer and evaluate its usefulness for a given period. If, following that time, the user wants to use the program further, a registration and a fee are requested. Many people confuse shareware with freeware (see above). Shareware programs are not free

URL: Uniform Resource Locator   The URL is the WWW standard of specifying the location of a file or resource to a WWW server and has the following syntax: scheme://

host.domain [:port]/path/filename. The scheme may be http, gopher, ftp, file, WAIS, telnet, or news. The port number is optional and is normally omitted

UUencoding   The UUencoding method transforms binary data (e.g., executable programs, graphics, and sound files) into plain text. This transformation enables sending the files through normal E-mail (i.e., not MIME type). On the receiving side, the process is reversed by a UUdecoder. Most mainframes are installed with both an encoder and a decoder, and these programs also exist for personal computers

VERONICA   The utility VERONICA searches for indexed text within gopher documents and is one of gopher's great advantages. VERONICA is very efficient and robust and in many cases provides just the right answer in a search. In a typical VERONICA search, the user provides a keyword (or a set of keywords and logical operators where the service is provided) and launches a search. The result is a collection of pointers ready to be selected, which are associated with the search keyword(s). VERONICA can be accessed through the EBI gopher server

WAIS: Wide Area Indexing System   The WAIS indexes text documents from keywords. These keywords are then searchable. WAIS indexing is widely used in gopher VERONICA searches as well as in many other text search utilities

webmaster   The jargon expression webmaster indicates the person (or group) who maintains the WWW documents and server on a specific site. Most sites maintain an E-mail address unique for WWW associated queries and problem reports, which takes the following form: webmaster@machine.domain (e.g., webmaster@ebi.ac.uk), so users can send mail to the webmaster if they know the domain part of the address. In most cases, the address can be guessed even if it is not specified explicitly

wildcard   Special characters called wildcards can be used in text searches to replace one or more characters in a word. The most commonly used wildcards include "?" to replace a single character in the word and "*" to replace more than one character in a word

WWW: World Wide Web   The WWW system is capable of providing and presenting hypertext, graphics, and sound linked documents over networks. It operates on a server–client basis, using a special language called HTML (see above). The documents are requested from the server by the client, using a special format, the URL (see above). EBI currently uses the WWW as the main tool for providing fast and easy access to the information it provides and for collecting information from the user community

[2] TDB: New Databases for Biological Discovery

By OWEN WHITE and ANTHONY R. KERLAVAGE

Introduction

The successful scaleup of automated, high-throughput DNA sequencing technologies has made a dramatic change in the science of discovery in

Copyright © 1996 by Academic Press, Inc.
All rights of reproduction in any form reserved.

biology. Whereas it typically required years of work to identify a single gene or protein having a particular function of interest or associated with a particular phenotype, now a small (<2 Mbp) genome can be sequenced in 1 year or less. Two entire bacterial genomes have been completed,[1,2] and the genomes of *Saccharomyces cerevisiae* and *Caenorhabditis elegans* will be completed within a few years. An initial assessment of human gene diversity and expression patterns has also been made based on large-scale cDNA sequencing.[3] In addition, major sequencing efforts will soon be underway for human chromosomes. Instead of asking questions about individual genes, we can now ask questions about genome organization and whole genome evolution. We can also look closely at patterns of expression, correlate them with functions in the cell, and speculate on the minimal set of functions required for a living organism.

The sheer volume of data being generated has opened new avenues for research; however, this extraordinary amount of data also presents new problems to be overcome. We are challenged with new ways to deal with data accuracy, sequence redundancy, inconsistent nomenclature, and functional classification. New analysis tools will be needed to help answer ques-

[1] R. D. Fleischmann, M. D. Adams, O. White, R. A. Clayton, E. F. Kirkness, A. R. Kerlavage, C. J. Bult, J.-F. Tomb, B. A. Dougherty, J. M. Merrick, K. McKenney, G. Sutton, W. FitzHugh, C. Fields, J. D. Gocayne, J. Scott, B. Shirley, L.-I. Liu, A. Glodek, J. M. Kelley, J. F. Weidman, C. A. Phillips, T. Spriggs, E. Hedbloom, M. D. Cotton, T. R. Utterback, M. C. Hanna, D. T. Nguyen, D. M. Saudek, R. C. Brandon, L. D. Fine, J. L. Fritchman, J. L. Fuhrmann, N. S. M. Geoghagen, C. L. Gnehm, L. A. McDonald, K. V. Small, C. M. Fraser, H. O. Smith, and J. C. Venter, *Science* **269**, 496 (1995).

[2] C. M. Fraser, J. D. Gocayne, O. White, M. D. Adams, R. A. Clayton, R. D. Fleischmann, C. J. Bult, A. R. Kerlavage, G. Sutton, J. M. Kelley, J. L. Fritchman, J. F. Weidman, K. V. Small, M. Sandusky, J. Fuhrmann, D. Nguyen, T. R. Utterback, D. M. Saudek, C. A. Phillips, J. M. Merrick, J.-F. Tomb, B. A. Dougherty, K. F. Bott, P.-C. Hu, T. S. Lucier, S. N. Peterson, H. O. Smith, C. A. Hutchison III, and J. C. Venter, *Science* **270**, 377 (1995).

[3] M. D. Adams, A. R. Kerlavage, R. D. Fleischmann, R. A. Fuldner, C. J. Bult, N. H. Lee, E. F. Kirkness, K. G. Weinstock, J. D. Gocayne, O. White, G. Sutton, J. A. Blake, R. C. Brandon, C. Man-Wai, R. A. Clayton, R. T. Cline, M. D. Cotton, J. Earle-Hughes, L. D. Fine, L. M. FitzGerald, W. M. FitzHugh, J. L. Fritchman, N. S. M. Geoghagen, A. Glodek, C. L. Gnehm, M. C. Hanna, E. Hedbloom, P. S. Hinkle, Jr., J. M. Kelley, J. C. Kelley, L.-I. Liu, S. M. Marmaros, J. M. Merrick, R. F. Moreno-Palanques, L. A. McDonald, D. T. Nguyen, S. M. Pelligrino, C. A. Phillips, S. E. Ryder, J. L. Scott, D. M. Saudek, R. Shirley, K. V. Small, T. A. Spriggs, T. R. Utterback, J. F. Weidman, Y. Li, D. P. Bednarik, L. Cao, M. A. Cepeda, T. A. Coleman, E. J. Collins, D. Dimke, P. Feng, A. Ferrie, C. Fischer, G. A. Hastings, W. W. He, J. S. Hu, J. M. Greene, J. Gruber, P. Hudson, A. Kim, D. L. Kozak, C. Kunsch, J. Hungjun, H. Li, P. S. Meissner, H. Olsen, L. Raymond, Y. F. Wei, J. Wing, C. Xu, G. L. Yu, S. M. Ruben, P. J. Dillon, M. R. Fannon, C. A. Rosen, W. A. Haseltine, C. Fields, C. M. Fraser, and J. C. Venter, *Nature (London)* **377** (Suppl.), 3–174 (1995).

tions we could not previously ask about development, disease, and evolution. These challenges will have to be solved by the entire biological community, not just by large sequencing or informatics laboratories. In our judgment, the best way to facilitate progress in this area is to devise new tools and data representations that allow the community to carry out this work. These include databases that allow complex queries against the data, as complements to databanks that now archive deposited sequences. Toward that end, we have developed at The Institute for Genomic Research (TIGR, Rockville, MD) the TIGR Database (TDB) as a collection of tools and databases designed to facilitate discovery in biology. Currently, TDB contains information about human genes and transcripts, bacterial genes and complete genomes, and links between sequences and sample collection and source materials.

Expressed Sequence Tags

The expressed sequence tag (EST) approach[4] has proved to be the most effective means to identify transcribed sequences. Of the 3×10^9 base pairs in the human genome, only a few percent code for the approximately 65,000 genes[5] that are transcribed into messenger RNA (mRNA) molecules and are ultimately translated into proteins. In the laboratory, complementary DNA (cDNA) can be synthesized from mRNAs isolated from tissue samples, and the resulting collection of cDNAs (or "library") reflects the fraction of the human genome that encodes genes. ESTs are generated by single-pass sequencing of either end of randomly selected clones from a cDNA library (Fig. 1). ESTs represent only a portion of each transcript but are long enough (~300–500 bp) to determine if the sequence is similar to that of a known or previously undefined gene. EST sequencing projects have increased the number of known human genes at a rate much greater than current genomic sequencing strategies.

The first large-scale random cDNA sequencing project began at the National Institutes of Health (Bethesda, MD) using a single automated sequencer and resulted in the published description of a total of 380 human sequences.[4] Since that time, many laboratories around the world have produced more than 300,000 ESTs from 44 different organisms and deposited them in dbEST,[6] a special database archive at the National Center for Biotechnology Information (NCBI, Bethesda, MD). High throughput

[4] M. D. Adams, J. M. Kelley, J. D. Gocayne, M. Dubnick, M. H. Polymeropoulos, H. Xiao, C. R. Merril, A. Wu, B. Olde, R. Moreno, A. R. Kerlavage, W. R. McCombie, and J. C. Venter, *Science* **252,** 1651 (1991).
[5] C. Fields, M. D. Adams, O. White, and J. C. Venter, *Nat. Genet.* **7,** 345 (1994).
[6] M. S. Boguski, T. M. J. Lowe, and C. M. Tolstoshev, *Nat. Genet.* **4,** 332 (1993).

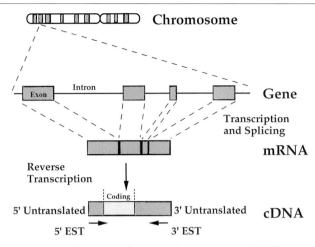

FIG. 1. Derivation of expressed sequence tags (ESTs).

generation of ESTs has defined segments of approximately half of the genes in the human.[3] These ESTs represent a great deal of complex gene information that is available to the biological community. However, as tens of thousands of new ESTs are added to the public databases, it becomes increasingly more difficult to use the information because of the high degree of redundancy, increasingly inconsistent nomenclature, and occasional problems with data quality. Provided that the data can be effectively utilized, ESTs will lead to the kinds of discoveries that were imagined at the inception of the Human Genome Initiative.

A major use of EST sequence databases is for search comparison to assign putative functions to new cDNA or genomic sequences that are generated in the laboratory. Retroactive analysis of EST databases has also been used to identify genes associated with disease states such as colon cancer.[7,8] peroxisome biogenesis disorder,[9] and Alzheimer's dis-

[7] N. Papadopoulos, N. C. Nicolaides, Y.-F. Wei, S. R. Ruben, K. C. Carter, C. A. Rosen, W. A. Haseltine, R. D. Fleischmann, C. M. Fraser, M. D. Adams, J. C. Venter, S. R. Hamilton, G. M. Peterson, P. Watson, H. T. Lynch, P. Peltomäke, J.-P. Mecklin, A. de la Chapelle, K. W. Kinzler, and B. Vogelstein, *Science* **263**, 1625 (1994).

[8] N. C. Nicolaides, N. Papadopoulos, S. R. Ruben, K. C. Carter, C. A. Rosen, W. A. Haseltine, R. D. Fleischmann, C. M. Fraser, M. D. Adams, J. C. Venter, M. Dunlop, S. R. Hamilton, G. M. Peterson, A. de la Chapelle, B. Vogelstein, and K. W. Kinzler, *Nature (London)* **371**, 75 (1994).

[9] G. Dodt, N. Braverman, C. Wong, A. Moser, H. W. Moser, P. Watkins, D. Valle, and S. J. Gould, *Nat. Genet.* **9**, 115 (1995).

ease.[10] EST data can also aid in areas of biological inquiry beyond gene identification. For example, organismal development requires an orchestration of multiple levels of gene expression that is spatially and temporally regulated during the entire life span of the organism. mRNA levels vary in a tissue- and time-specific manner, and they vary within the same tissue isolated at different stages of development. Elucidation of gene expression typically involves Northern or Western hybridization of probe sequences against electrophoretically separated materials, or *in situ* hybridization for localization of probe sequences against sectioned tissue samples. In either case, the experiments are labor intensive and confined to individual gene sequences that have been isolated and cloned. However, cDNA libraries are representative of the *in vivo* levels of large-number gene transcripts in the source tissues at the time of isolation,[11] and large-scale EST analysis of cDNA clones is a rapid and efficient analytical measure of the gene expression associated with formation of a complex organism. We have linked over 350,000 ESTs to their source tissues and collapsed the overall sequence redundancy inherent in these data by building assemblies of ESTs. The resulting "gene anatomy" information is available through two of TIGR's databases, the Expressed Gene Anatomy Database (EGAD) and the TIGR Human cDNA Database (TIGR HCD).

Expressed Gene Anatomy Database

Deriving expression information from EST sequences requires that ESTs must be consistently assigned to known database sequences. We have constructed the Expressed Gene Anatomy Database (EGAD) by curatorial extraction of GenBank[12] data to facilitate robust putative identification of ESTs. The development of the EGAD data set of human transcript (HT) sequences was crucial to the construction of assemblies of ESTs and their subsequent annotation.

Sequences derived from at least three methodologies are stored in the public sequence archives such as GenBank. Genomic sequences, derived by direct determination of the DNA of an organism, are typically the longest and can contain coding regions, introns (intervening sequences that

---

[10] E. Levy-Lahad, E. Wasco, P. Poorkaj, D. M. Romano, J. Oshima, W. H. Pettingell, C.-E. Yu, P. D. Jondro, S. D. Schmidt, K. Wang, A. C. Crowley, Y.-H. Fu, S. Y. Guenette, D. Galas, E. Nemens, E. M. Wijsman, T. D. Bird, G. S. Schellenberg, and R. E. Tanzi, *Science* **269**, 973 (1995).

[11] N. H. Lee, K. G. Weinstock, E. F. Kirkness, J. A. Earle-Hughes, R. A. Fuldner, S. Marmaros, A. Glodek, J. D. Gocayne, M. D. Adams, A. R. Kerlavage, C. M. Fraser, and J. C. Venter, *Proc. Nat. Acad. Sci. U.S.A.* **92**, 8303 (1995).

[12] D. Benson, D. J. Lipman, and J. Ostell, *J. Nucleic Acids Res.* **21**, 2963 (1993).

are removed during transcript maturation), promoters and other regulatory regions, repetitive elements, and intergenic regions. cDNA sequences are roughly 1–10 kb in length and represent the sequence derived from transcribed mRNA molecules. ESTs are also derived from cDNAs but, for the sake of efficiency, usually come from a single sequencing gel experiment that results in approximately 300 to 500 bp of sequence. GenBank may contain a considerable amount of information for each sequence. Some data, such as the coordinates for protein coding regions or the genus and species name of the organism, are located in designated fields of each GenBank entry. Other types of data (e.g., the chromosomal map location of a gene or laboratory strain of virus) are nonuniformly placed in the entry and are only useful when viewing an individual GenBank accession; they are not readily accessed for automated uses of the data. Thus, from the point of view of how the sequences are described electronically, there are attributes of genes that must be meaningfully organized. We were motivated to construct EGAD in order to collect certain features of GenBank information for the purpose of robust computational analyses. Some of the data types curated in EGAD that are crucial in the representation of gene information are listed below.

Accessions

It is desirable to track sequence records that are associated with a single gene for simplified management, searching, and retrieval. However, sequences belonging to the same gene are not consistently linked together in GenBank. Determining all relevant entries is difficult because sequences associated with the same gene can be found in separate entries that are partial or full-length cDNAs, alternative splice forms, exon fragments, genomic sequences, and large genomic sequences (i.e., cosmids or larger segments). In EGAD, previously unlinked GenBank entries from the same transcript are linked, and pointers to relevant accessions are saved. This consolidation of GenBank records has reduced sequence redundancy so that EGAD contains a unique set of HT sequences. To date, 31,202 GenBank entries from human sequences have been linked to create 4417 different EGAD transcripts.

Common Names

Common names embody what is known about the functional, physical, phenotypic, or physiological aspects of a gene product [e.g., alcohol dehydrogenase, HSP70 (heat-shock protein 70), wingless, and GABA (γ-aminobutyric acid) receptor, respectively] but are not effective ways to retrieve gene sequences consistently. Gene nomenclature is essentially a historical record of the method used to identify a gene rather than the result of a consistent vocabulary. As an example, a gene contained in a large segment of DNA is often annotated with the name of that segment rather than the name of the gene. Common names are dynamic entities because what is known about genes, and thus the way genes are named, is rapidly changing. Names in EGAD were curated for semantic consistency, and naming conventions were adopted to make queries by common name more effective.

Expression Data

Expression information was assigned to the HT sequences by searching them against ESTs. Exact matches between an EST and an HT sequence is considered to be experimental evidence that the GenBank sequence is expressed in the tissue source of the cDNA library. In addition, when the information has been recorded in GenBank, the tissues from which GenBank transcript sequences were derived were also stored in EGAD. Links between 345,000 EST sequences, representing data from 37 different tissues, and 4417 HT sequences that point to over 31,000 GenBank sequence accessions have been established in EGAD.

Cellular Roles

Biological role classifications were created and linked to genes in EGAD. A list of role categories represented in EGAD is shown in Table I. This list was designed to represent the broadest range of structural and biochemical functions possible. More than one role may be assigned to an individual gene in cases where the activity of a gene varies in different tissues or developmental stages. Roles that are unique to alternatively spliced genes are also represented. We view role assignment to be an ongoing curation process and realize that new assignments will need to be made as new biological information is obtained.

Sequences

Genes are encoded in genomic sequences that are transcribed into mRNAs with coding, intronic, and 5' and 3' untranslated regions. During maturation, the pre-mRNA molecule is polyadenylated, the introns are removed, and the mRNA is transcribed into proteins in the cell cytoplasm. Multiple splicing pathways of exons and introns during the maturation of transcription units can result in multiple alternate splice forms of an individual gene, where each splice form may encode distinct proteins, presumably with an altered function. The HT sequences in EGAD represent mRNA transcripts as they would appear in the cell cytoplasm (i.e., mature mRNAs).

TABLE I
BIOLOGICAL ROLES IN EXPRESSED GENE
ANATOMY DATABASE

Cell division
  General
  Apoptosis
  Cell cycle
  Chromosome structure
  DNA synthesis/replication
Cell signaling/cell communication
  Cell adhesion
  Channels/transport proteins
  Effectors/modulators
  Hormone/growth factors
  Intracellular transducers
  Metabolism
  Protein modification
  Receptors
Cell structure/motility
  General
  Cytoskeletal
  Extracellular matrix
  Microtubule-associated proteins/motors
Cell/organism defense
  General
  Homeostasis
    General
    DNA repair
    Carrier proteins/membrane transport
    Stress response
  Immunology
Gene/protein expression
  RNA synthesis
    RNA polymerases
    RNA processing
    Transcription factors
  Protein synthesis
    Posttranslational modification/targeting
    Protein turnover
    Ribosomal proteins
    tRNA synthesis/metabolism
    Translation factors
Metabolism
  General
  Amino acid
  Cofactors
  Energy/TCA cycle
  Lipid
  Nucleotide
  Protein modification
  Sugar/glycolysis
  Transport
Unclassified

The known splice forms of a gene are represented individually and linked together with their gene. The HT sequences were created from GenBank accessions by addressing redundancy at multiple stages. Genomic and/or cDNA sequences belonging to the same gene splice form were stored as the assembly of aligned, overlapping sequences.

Candidate sequences for an HT were first searched against all human accessions in GenBank using BLAST.[13] GenBank sequences with greater than 98% similarity to the query sequence were used to generate a set of aligned sequences determined to belong to the same gene. The consensus sequence derived from the aligned sequences along with pointers to each element were stored in EGAD. As part of an automated loading process, sequences of greater than 98% identity for over 100 nucleotides were compared, and the longest cDNA was stored. Coding sequences shorter than 30 nucleotides were not loaded. Separate sequences were stored for each known alternative splice form of a gene. The HT data set is updated with each new release of GenBank. Approximately 4400 human sequences that encode mRNAs have been annotated in EGAD with links to over 31,000 related sequences.

Tentative Human Consensus Sequences

The large number of human EST sequences determined in the past several years represent a significant amount of redundant information. We have combined over 160,000 human ESTs sequenced at TIGR and Human Genome Sciences (HGS) with 185,000 human ESTs from dbEST in TIGR's Human cDNA Database. To reduce the redundancy and thus make the data more useful, we have assembled these sequences into tentative human consensus sequences (THCs).[3] We developed an assembly algorithm (TIGR Assembler[14]) to accomplish this task for such a large number of sequences. The EGAD HT set was included in the assembly process. The consolidation of EST and HT sequences significantly improves the quality of both EST and HT data by (i) extending the length of the transcript information, (ii) improving accuracy by increasing sequence depth, (iii) providing better identification and annotation of ESTs, (iv) linking expression information with transcripts, and (v) identifying new alternative splice forms and potentially polymorphic sequences. In the most dramatic example of reduction of redundancy, elongation factor 1 alpha (EF1α), which is abundantly

[13] S. F. Altschul, W. Gish, W. Miller, E. W. Myers, and D. J. Lipman, *J. Mol. Biol.* **215,** 403 (1990).

[14] G. G. Sutton, O. White, M. D. Adams, and A. R. Kerlavage, *Genome Sci. Technol.* **1,** 9 (1995).

transcribed and found in most tissues, has over 200 separate ESTs in the public archives in addition to 2400 sequences identified at TIGR and HGS. These ESTs were assembled into a 1804-bp consensus sequence, representing a tremendous decrease in redundant data. The total EST data set (345,000 ESTs) was assembled into 40,000 THC sequences with 80,000 singleton ESTs remaining.

Tentative human consensus sequences are linked to the EGAD gene attributes assigned to the HT in the THC. Once the sequence level links have been established, biological role, common names, and expression information are linked across EGAD HT sequences and GenBank. However, many EST and THC sequences do not have identical matches to sequences that are contained in the public archives. Roles have been assigned to the ESTs or THC sequences that are similar (but not exact matches) to either an HT sequence or a nonhuman sequence. These are typically labeled as isologs of those sequences. ESTs or THC sequences with no known database match are assigned unique identifiers for database tracking, and are linked only to expression data.

TIGR Human cDNA Database

We have developed the TIGR Human cDNA Database (HCD) as a tool to allow researchers to search the most complete set of EST data in unique ways. The HCD contains THC and singleton EST sequence data, as outlined above. An initial assessment of these data has been published.[3] New sequences from the dbEST are constantly being added to the HCD, assembled, and linked to expression, role, and putative identification data. The HCD may be searched by unique identifier (EST or THC number), by putative identification, or by sequence similarity to a user-supplied nucleotide or peptide sequence. A unique algorithm is used to produce alignments that span potential frameshifts in either the user-supplied or HCD sequence. THC reports (Fig. 2) contain alignments of the ESTs that were used to create each consensus sequence and show the tissues in which the sequence is expressed. Clones from TIGR ESTs that are found in the THC are available from the American Type Culture Collection (ATCC, Rockville, MD). The information contained in these reports allows easy identification of 3' untranslated regions for cDNA mapping experiments.

All of the data in HCD have been subjected to rigorous quality control measures, include trimming of sequences containing vector and ambiguous nucleotides, length checks, and tests for the presence of bacterial contaminants.[15] ESTs shorter than 100 nucleotides are discarded, and 3' ends are

---

[15] O. White, T. Dunning, G. Sutton, M. Adams, J. C. Venter, and C. Fields, *Nucleic Acids Res.* **21,** 3829 (1993).

```
>THC50790
CTCGTsrrTGGAGsyrGCTGAGTCGCGCGCTCTGCTCCACCCGACGGGGCTGTGTGTGCTGGGCCTGGCTCGCGGCGAAC
CGAGATGGCAGAGCAGTCGGACGAGGCCGTGAAGTACTACACCCTAGAGGAGATTCAGAAGCACAACCACAGCAAGAGCA
CCTGGCTGATCCTGCACCACAAGGTGTACGATTTGACCAAATTTyTGGAAGAGCATCCTGGTGGGGAaGAAGTTTTAAGG
GAACAAGCTGGAGGTGACGCTACTGAGAACTTTGAGGATGTCGGGCACTyTACAGATGCCAGGGAAATGTCCAAAACATT
CATCATTGGGGAGCTCCATCCAGATGACAGACCAAAGTTAAACAAGCCTCCGGAAAcTyTTATCACTACTATTGATTCTA
GTTCCAGTTGGTGGACCAACTgGGTGATCCCTGCCATCTCTGCAGTGGCCGTCGCCTTGATGTATCGCCTATACATGGCA
GAGGACTGAACACCTCCTCAGAAGTCAGCGCAGGAAGAGCCTGCTTTGGACACGGGAGAAAGAAGCCATTGCTAACTAC
TTCAACTGACAGAAACCTTCACTTGAAAACAATGATTTTAATATATCTCTTTCTTTTTCTTCCGACATTAGAAACAAAAC
AAAAAGAACTGTCCTTTCTGCGyyCAAATTTTTCGAGTGTGCCTTTTTATTTCATCTACTTTATTTTGAkGTTTCCTTAA
TGtGTAATTTACTtATTATArGCAkGATCTTTTAAAAATATATTkGGCTTTTAAAGTAwAAAAAAAAG
```

Putative ID: cytochrome b-5

```
1================================THC50790================================788
----------1------------>    -------------11------------->
----------------2------------------>    -------------15------------->
----------------3---------------->    ----------13---------->
----------4------------->    --------------------12-------------------->
------------------5------------------>
----------6------------>    -------------------14------------------->
-----------7----------->
---------------8---------------->
--------------------9-------------------->
-------------10------------>
```

#	EST#	GB#	ATCC#	left	right	library
1	A EST49086			1	259	Liver
2	B EST151809			6	401	Lung
3	A EST68730			8	350	Lung
4	B EST151070			9	294	Lung
5	B EST133756			21	388	Embryo
6	A EST44888			30	290	Skin
7	B EST166494			42	293	White blood cells
8	B EST139460			61	390	Skin
9	F	T57569		168	602	Ovary
10	A EST51564			216	495	Gall bladder
11	A EST18446			293	594	Liver
12	E HT1344			343	779	
13	B EST130286			379	608	Embryo
14	B EST148349			386	788	Kidney
15	A EST76367	T29319	105932	440	716	Brain

Sequence source codes: A = TIGR, B = HGS, F = WashU/Merck, E = EGAD

Fig. 2. Human cDNA Database THC report. The report provides the sequence of the THC, its putative identification, a diagrammatic representation of the component ESTs and THCs along with their coordinates on the consensus sequence, and a list of the component sequences and their source tissues. GenBank and ATCC accession numbers are provided when available. The source laboratory for each EST is also provided. In the WWW representation of this report, each EST number is linked to a report for that EST in HCD, each HT number is linked to an HT report in EGAD, GenBank accession numbers are linked to the Genome Sequence Data Base (GSDB) record at the National Center for Genome Resources, and each ATCC accession number is linked to the ATCC Repository information.

trimmed until the sequence contains less than 3% unknown nucleotides (Ns). Sequences are evaluated for vector by searching against a set of vector sequences. Any traces of vector, polyadenylation, or sequence artifacts [e.g., poly(CT)] are trimmed accordingly. Sequences containing human mitochondrial genes and ribosomal RNAs are discarded.

Sequences, Sources, Taxa

The Institute for Genomic Research has developed Sequences, Sources, Taxa (SST), a relational database which links molecular sequence data with source, collection, and taxonomy data. Sources are the voucher specimens, clones, or registered cell lines from which molecular sequences (DNA, RNA, or protein) were obtained. Collection site information is currently available for some sources and will be added for new sequences whenever available. Alternate taxonomic classifications and synonyms can be maintained within SST. Links have been developed between SST and EGAD that provide access to molecular sequences and annotation such as expression and role information, based on a taxonomic search. SST should prove to be a valuable tool in systematics and biodiversity studies.

Databases for Bacteria

The Institute for Genomic Research has created a database of nonredundant bacterial proteins (NRBP) specifically for the purpose of aiding in the annotation of bacterial genomes. The same principles described above for the development of the HT set were applied to all available bacterial protein and nucleotide sequences, including the large segments of *Escherichia coli* in GenBank. The database currently contains 21,445 protein sequences from 1099 taxa and represents each occurrence of a gene in a distinct species. The ability to search against this database and view hits to homologs from several species has been invaluable in annotating the *Haemophilus influenzae* and *Mycoplasma genitalium* genomes.[1,2]

We have also developed interactive databases for the *H. influenzae* and *M. genitalium* genomes. These are available at our World Wide Web (WWW) site and allow searching by unique gene identifier, putative gene product name, cellular role, or user-supplied query sequence. We have also implemented the ability to view and retrieve any user-specified segment of the complete genome, which cannot be done with the GenBank sequence.

TIGR Database System

All of the underlying data in EGAD, HCD, and SST are maintained at TIGR in relational databases using the Sybase RDBMS (relational database

management system). Entities in each database are linked by stable identifiers. This structure allows the most flexible query capability to access the
data. Complex, ad hoc queries may be constructed across all three databases.

A World Wide Web access interface to the three databases has been
developed and is available at the URL (uniform resource locator): http://
www.tigr.org/tdb/tdb.html. Extensive hypertext links have been incorpo-

SEQUENCE DATA

```
EGAD sequence HT1041
sequence name: tyrosine hydroxylase, alt. splice 1
sequence type: cdna
coding sequence length: 528
transcript sequence length: 1921

HT1041 expression data
```

ACCESSION DATA
```
HT:1041 is derived from accession(s):
M17589 (cDNA)
```

ALTERNATIVE SPLICE INFORMATION
```
alternative splice forms for this gene:
HT:3957
HT:3958
HT:3959
```

SEQUENCE

```
nucleotide:

cactgagccatgcccacccccgacgccaccacgccacaggccaagggcttccgcagggcc
gtgtctgagctggacgccaagcaggcagaggccatcatggtaagagggcagggcgccccg
.
.
.
gccccaatcaccgtcacaataaaagaaactgtggtctctaaaaaaaaaaaaaaaaaaaaaa


protein:

MPTPDATTPQAKGFRRAVSELDAKQAEAIMVRGQGAPGPSLTGSPWPGTAAPAASYTPTP
.
.
.
RIQRPFSVKFDPYTLAIDVLDSPQAVRRSLEGVQDELDTLAHALSAIG
```

Fig. 3. Expressed Gene Anatomy Database HT report. The report provides the name of
the transcript, its length, the length of the coding region, and nucleotide and translated protein
sequences (shown truncated for brevity). Links are available to expression data in HCD, the
alternative splice forms from the same gene, and the GenBank accessions that were used to
derive the HT (linked directly to the GSDB WWW page containing those accessions).

rated to allow browsing of the complete set of data, even if they are contained in different databases.

As an example, a user may search HCD with a new peptide sequence that he/she has determined. A THC having a high degree of similarity may be discovered in the HCD data set, indicating that the two sequences may possibly be members of the same gene family. The user may obtain a report on that THC and discover that it is expressed in only a single tissue (different than the one from which the user's sequence was obtained), indicating that it may be a tissue-specific isolog of the query sequence.

As another example, a user may query EGAD by a particular role category and obtain an HT report (Fig. 3) for a sequence of interest within that role category. The user may then link directly to known alternative splice forms for that gene and to a THC report in HCD to obtain its expression pattern.

Finally, for a systematics study a user may query SST for several species and obtain a listing, for example, of cytochrome c sequences obtained from those species. The user may then find where each of the samples was collected, where the sample is currently stored, and who the relevant contacts are. In addition, for each sequence the user may link to an EGAD report, GenBank reports, and Genome Data Base (GDB)[16] reports for mapping information.

We have developed the TIGR databases as a collection of unique databases having a robust representation of sequences and associated information. The databases have consistent semantics and nomenclature, have minimized redundancy compared with sequence archives, and have persistent unique identifiers for the data, allowing extensive links to be established among the data. These databases should play a key role in the analysis and interpretation of data from DNA sequencing projects in many species.

[16] A. J. Cuticchia, K. H. Fasman, D. T. Kingsbury, R. J. Robbins, and P. L. Pearson, *Nucleic Acids Res.* **21,** 3003 (1993).

[3] PIR-International Protein Sequence Database

By DAVID G. GEORGE, LOIS T. HUNT, and WINONA C. BARKER

Introduction

Green *et al.*[1] have stated that "Understanding the functions and structures of the array of proteins expressed in living organisms is a fundamental goal of molecular biology." From its origin, the Protein Sequence Database has been an essential tool for the analysis, identification, and understanding of both nucleic acid and protein sequence data.[2-6] As such, it has a fundamental role in the evaluation and storage of information generated by the Human Genome Initiative (HGI),[7] and, along with other bioinformatics resources, it must respond to the increasing demands stimulated by the HGI. Robert Robbins[8] stated, "If the information side is not handled well, the HGI could spend billions of dollars and the researchers might still find it easier to obtain data by repeating the experiments."

As the demands on and uses of the databases change, so must the design strategy. Macromolecular sequence databases of the current generation were designed primarily to support research using computerized sequence database searching methods. This methodology has been highly effective for discovering relationships among proteins and has resulted in major contributions to research in the study of carcinogenic processes, acquired immunodeficiency syndrome (AIDS), and many other viral, genetic, and immunological diseases. The research scientists at the National Biomedical Research Foundation (NBRF) pioneered in the development of this approach and have contributed many scientific discoveries in this area. Because of the high sensitivity of this methodology, errors in the sequence data can be tolerated. Because of the relatively small size of the current database collections, redundancy is only recently beginning to be recognized to interfere significantly. As a result, although the databases strive for

[1] P. Green, D. Lipman, L. Hillier, R. Waterson, D. States, and J.-M. Claverie, *Science* **259,** 1711 (1993).
[2] R. F. Doolittle, this series, Vol. 183, p. 99.
[3] R. F. Doolittle, "Of URFs and ORFs: A Primer on How to Analyze Derived Amino Acid Sequences." University Science Books, Mill Valley, California, 1987.
[4] W. R. Pearson, *Genomics* **11,** 635 (1991).
[5] D. J. Lipman and W. R. Pearson, *Science* **227,** 1435 (1985).
[6] R. F. Doolittle, *Protein Sci.* **1,** 1563 (1992).
[7] T. D. Yager, D. A. Nickerson, and L. E. Hood, *Trends Biochem. Sci.* **16,** 545 (1991).
[8] K. A. Frenkel, *Commun. ACM* **34,** 41 (1991).

Copyright © 1996 by Academic Press, Inc.
All rights of reproduction in any form reserved.

accuracy in recording reported sequence data and their associated citations to published and unpublished work, consolidating overlapping information has not been considered to be a high priority. Nor has high priority been given to collecting other information (annotation) associated with the sequence; hence, these data are incomplete, inconsistent, ambiguous, and even incorrectly recorded.

The critical aspect of sequence database searching is the analysis and interpretation of the results. Because of the relatively small sizes of the current databases, the role of the annotation in this process has not been fully appreciated. When evaluating the results of database searches, researchers generally examine the annotation corresponding to each matching sequence on an individual basis; often it is required to refer to the original publication before determining whether a potential match is meaningful. There continues to be a need for more efficient algorithms and more powerful computer systems to allow sequence database searches to be accomplished in reasonable time.[9] Concurrently there is a need for the development of more intelligent database searching protocols that involve some level of preinterpretation of the results, if the interpretation phase is to scale accordingly. Such methods will rely directly on the annotations and can be effective only if consistently annotated databases, organized within a uniform architecture that effectively represents the properties and interrelationships among the sequence data, are made available.

The emergence of an information age in biology implies that information in the databases will be accessible through complex queries that can be generated by computational processes. Sequence searching is only one example of these classes of queries; most others involve interpretation of the associated information. Consider the query "return the calcium-binding domains of all calcium-regulated protein kinases," for example. Satisfying this query requires that all calcium-regulated protein kinases have been identified and labeled as such and that the calcium-binding domains contained within these sequences have been identified and have been represented in a formalism that allows them to be located within the molecules unambiguously. As a source of information, the macromolecular sequence databases rely on data submitted to the database centers directly from research investigators or on data extracted from the published literature. In general, these data are produced and processed independently and reflect the current state of biological knowledge at the time they were entered into the database. At the time of data deposition, the properties and characteristics of the sequences are generally not known. It is only on comparison with other related forms and further experimentation that these properties come to be understood. Unless a directed effort is made to pool the data

---

[9] E. S. Lander, R. Langridge, and D. M. Saccocio, *Commun. ACM* **34,** 32 (1991).

(combining overlapping reports), cross-compare them, and represent them using uniform conventions and standardized terminologies, they will remain a redundant collection of incomplete, inconsistent, independent data elements. Such a collection cannot satisfy the query given above. Probabilistic query techniques, such as those employed in document-based information retrieval systems,[10] cannot compensate effectively for incompletely, inconsistently, and/or incorrectly recorded information. Developing databases capable of resolving complex queries will require a renewed effort to improve significantly the quality, completeness, uniformity, and correctness of the information represented in the databases.

The PIR-International Protein Sequence Database was initiated in the early 1960s, with the pioneering work of the late Margaret O. Dayhoff,[11–17] to provide a data set for the support of research on the interrelationships and evolution of proteins. Because the database has from its origins been operated by scientists as a tool to support research, there has been an ongoing commitment to completeness and biological correctness of the information in the database. Since 1984, the growth in sequences represented in the database has been exponential (the current doubling time is about 2.4 years), yet the financial resources in equivalent dollars available to the Protein Information Resource (PIR) have decreased. A critical step in coping with this increasing workload in the face of diminishing resources was taken in 1987 with the establishment of PIR-International.[18–20] PIR-International is an association of database centers dedicated to the development and maintenance of a single, comprehensive, complete, high-quality

[10] W. B. Frakes and R. Baeza-Yates, "Information Retrieval: Data Structures & Algorithms." Prentice Hall, Englewood Cliffs, New Jersey, 1992.

[11] M. O. Dayhoff, R. V. Eck, M. A. Chang, and M. R. Sochard, "Atlas of Protein Sequence and Structure." National Biomedical Research Foundation, Silver Spring, Maryland, 1965.

[12] R. V. Eck and M. O. Dayhoff, "Atlas of Protein Sequence and Structure." National Biomedical Research Foundation, Silver Spring, Maryland, 1966.

[13] M. O. Dayhoff and R. V. Eck, "Atlas of Protein Sequence and Structure." National Biomedical Research Foundation, Silver Spring, Maryland, 1967–1968.

[14] M. O. Dayhoff, "Atlas of Protein Sequence and Structure," Vol. 4. National Biomedical Research Foundation, Silver Spring, Maryland, 1969.

[15] M. O. Dayhoff, "Atlas of Protein Sequence and Structure," Vol. 5. National Biomedical Research Foundation, Washington, D.C., 1972.

[16] M. O. Dayhoff, "Atlas of Protein Sequence and Structure," Vol. 5, Suppl. 3. National Biomedical Research Foundation, Washington, D.C., 1979.

[17] M. O. Dayhoff, W. C. Barker, and L. T. Hunt, this series, Vol. 91, p. 524.

[18] W. C. Barker, D. G. George, H.-W. Mewes, and A. Tsugita, *Nucleic Acids Res.* **20,** 2023 (1992).

[19] W. C. Barker, D. G. George, H.-W. Mewes, F. Pfeiffer, and A. Tsugita, *Nucleic Acids Res.* **21,** 3089 (1993).

[20] D. G. George, W. C. Barker, H.-W. Mewes, and A. Tsugita, *Nucleic Acids Res.* **22,** 3569 (1994).

protein sequence database. These centers include the PIR at the NBRF, the Martinsried Institute for Protein Sequences (MIPS) at the Max Planck Institute for Biochemistry in Martinsried, Germany, and the Japan International Protein Information Database (JIPID) at the Science University of Tokyo.

PIR-International emerged from the concept that biological data are neither generated nor used within any single nation; hence, scientists worldwide have a legitimate stake in the development of biomolecular databases. The formation of this association has allowed the costs of maintaining the resource to be shared on an international level. Moreover, the distributed approach provides for cross-fertilization of ideas and concepts generated by scientists throughout the world and thus leads to a better quality product. The work reported here is the result of the full cooperation of our partners in PIR-International who have shared in these developments. The PIR-International centers collaborate on all aspects of the database project including database design and documentation; data input, processing, analysis, and maintenance; software and database system development; and data distribution.

Over the past 5 years PIR-International has established an infrastructure for keeping pace with the influx of sequence data. Given limited resources, this was accomplished at the expense of completeness in the information associated with the sequences (the annotation). The essential challenge facing the Protein Sequence Database project is to develop a strategy that can continue to handle the high volume of data input while renewing efforts to maintain the high level of accuracy, uniformity, completeness, and organization required to satisfy effectively the evolving needs of the user community. In this chapter we describe the scientific basis of our data processing strategy and demonstrate how it will be able to allow the database project to meet effectively the increasing demands for bioinformatics at realistic cost. The current status of the implementation of this strategy is discussed.

Protein Sequence Database

The PIR-International Protein Sequence Database uses homology and comparative analysis to provide consistency and allow inferences to be made. Comparative analysis is one of the foremost tools in biology and underlies many computerized methods of sequence analysis. Those analyses that do not directly compare sequences are largely empirical; they utilize patterns, motifs, or other modes of analysis that have been derived from comparative studies. Most of the properties attributed to proteins in the published literature have been assigned by homology. For example, among the bacterial ferredoxins, an unusually well-studied group, the three-dimensional structures of only five representatives have been analyzed, and for only a handful of others have direct chemical methods been employed to locate the iron–sulfur binding sites; yet, on the basis of homology, these sites are believed to be well established for all of the 60–70 sequences now known. Thus, homology is used (in the absence of direct evidence) as the basis for identification of proteins, standardization of protein names, and consistent and comprehensive assignment of properties and characteristics.

The Protein Sequence Database contains information concerning all naturally occurring, wild-type proteins whose primary structure (the sequence) is known. In addition to sequence data, the database contains information (called annotation) concerning (1) the name and classification of the protein and the organism in which it naturally occurs; (2) references to the primary literature, including information concerning the sequence determination; (3) the function and general characteristics of the protein, including gene expression, posttranslational processing, and activation; and (4) sites and regions of biological interest within the sequence. Entries in the database are cross-referenced to other related databases, including GenBank,[21] the European Molecular Biology Laboratories (EMBL) Nucleotide Sequence Database,[22] the DNA Data Bank of Japan (DDBJ),[23] the human Genome Database (GDB),[24] and MEDLINE.[25] Work is currently underway to cross-reference the *Drosophila* genome database (FlyBase),[26] the Brookhaven Protein Data Bank (PDB),[27] and the Complex Carbohydrate Structure Database (CCSD) of the international CarbBank project.[28]

An example of a sequence entry is shown in Fig. 1. An entry can be divided into several record sections. The first record section in the entry contains a single record, the ENTRY record. This record gives a unique identification code for the entry and describes the type of sequence. This section is followed by the Header Record Section, the Reference Record

[21] D. A. Benson, M. Boguski, D. J. Lipman, and J. Ostell, *Nucleic Acids Res.* **22,** 3441 (1994).

[22] D. B. Emmert, P. J. Stoehr, G. Stoesser, and G. N. Cameron, *Nucleic Acids Res.* **22,** 3445 (1994).

[23] T. Tateno, Y. Ugawa, Y. Yamazaki, H. Hayashida, N. Saitou, and T. Gojobori, *CODATA Bull.* **23**(4), 74 (1991).

[24] K. H. Fasman, A. J. Cuticchia, and D. T. Kingsbury, *Nucleic Acids Res.* **22,** 3462 (1994).

[25] J. B. Courteau, *NCBI News* **1**(1), 3 and 7. National Center for Biotechnology Information, NLM, NIH, Bethesda, Maryland, 1991.

[26] J. Merriam, M. Ashburner, D. L. Hartl, and F. Kafatos, *Science* **254,** 221 (1991).

[27] E. E. Abola, F. C. Bernstein, S. H. Bryant, T. F. Koetzle, and J. Weng, *in* "Crystallographic Databases—Information Content, Software Systems, Scientific Applications" (F. H. Allen, G. Bergerhoff, and R. Sievers, eds.), p. 107. Data Commission of the International Union of Crystallography, Cambridge, 1987.

[28] S. Doubet, *CODATA Bull.* **23**(4), 56 (1991).

{

---- Entry Record Section ----

{ ENTRY   YRHU1   #type complete }

---- Header Record Section ----

{ TITLE   monophenol monooxygenase (EC 1.14.18.1) precursor
- human }
{ ALTERNATE_NAMES cresolase; monophenol oxidase; phenolase; tyrosinase }
{ ORGANISM   #formal_name Homo sapiens #common_name man }
...

---- Reference Record Section ----

{ REFERENCE A38444 #authors Giebel L.B.; Strunk, K.M.; Spritz, R.A.
   #journal   Genomics ##volume 9 ##pages 435-445 ##year 1991
     ##title   Organization and nucleotide sequences of the human tyrosinase
                 gene and a truncated tyrosinase-related segment
   #cross_references MUID:91236163
   #accession   A38444 ##molecule_type DNA ##residues 1-529
              ##label GIE ##cross_references GB:M60296 }
...
{ REFERENCE A60149 #authors Wittbjer, A.; Odh, G.; Rosengren, A.M.;
               Rosengren, E.; Rorsman, H.
   #journal   Acta Dermatol. Venereol. ##volume 70 ##pages 291-294
              ##year 1990 ##title Isolation of soluble tyrosinase from
              human melanoma cells
   #accession A60149 ##molecule_type protein ##residues 19-23,'X',25-28
             ##label WIT ...}
...

---- General Property Record Section ----

{ GENETICS   #gene GDB:TYR #map_position 11q21
              #introns 273/3; 346/1; 395/2; 456/1 ##status experimental }
{ CLASSIFICATION #superfamily monophenol monooxygenase }
{ KEYWORDS albinism; copper; glycoprotein; melanin biosynthesis;
              monooxygenase; oxidoreductase; transmembrane protein }

---- Feature Record Section ----

{ DOMAIN   1-18   #description signal sequence ... #status experimental ...}
{ PRODUCT   19-529   #description monophenol monooxygenase
                        #status experimental #label MAT }
{ DOMAIN   474-500 #description transmembrane ... #status predicted ...}
{ BINDING_SITE 86;111;161;230;337;371 #residues ASN #bond_class
                   covalent #ligand carbohydrate #status predicted }

---- Sequence Record Section ----

{ SUMMARY #length 529 #molecular_weight 60393 #checksum 3879 }
{ SEQUENCE   5   10   15   20   25
            1 M L L A V L Y C L L W S F Q T S A G H F P R A C V ... }

}

FIG. 1. PIR-International Protein Sequence Database entry. Ellipses (...) indicate information that has been omitted for display purposes. The entry is shown in the Sequence Database Definition Language (SDDL) representation.[29]

Section, the General Property Record Section, the Feature Record Section, and the Sequence Record Section.

The Header Record Section gives general descriptive information including protein name, biological source, and tracking information, such as creation and modification dates. The Reference Record Section contains citations to the primary literature (full bibliographic citation information is retained including MEDLINE UID) and descriptive information pertaining to each report (including the sequence as reported). The General Property Record Section gives information concerning function, genetics, and any other characteristics not explicitly linked to sites or regions within the sequence. The Features Record Section gives information explicitly linked to sites or regions within the sequence. In addition to the sequence itself, the Sequence Record Section gives summary properties of the sequence.

Entries are composed of records that are delimited by braces, { }. Within each record data elements are preceded by identifiers specifying the type of information. Pound signs, #, precede record subidentifiers to distinguish them from data elements. Records are structured hierarchically; the number of pound signs preceding the subidentifiers indicates the level of nesting. For example, in Fig. 1 the GENETICS record contains the subrecords gene, map_position, and introns, while the introns subrecord contains the subrecord status. The hierarchical structure of the records allows information to be grouped and associated to express interdependencies among the data elements. For example, the information in the status subrecord refers explicitly to the introns subrecord and indicates that the positions of introns relative to the encoded protein sequence have been determined experimentally.

Data Processing Strategy

The information in the Protein Sequence Database originates from a wide variety of experimental methods as depicted in Fig. 2.[29] Experimental methods of sequence determination are error-prone. Most protein sequences are not determined directly but are inferred from the sequences of the corresponding nucleic acid coding regions, introducing additional uncertainty. As the mechanisms of gene expression are not always clearly understood, such inferences may result in serious errors in the sequence data. Annotation information (e.g., features) is obtained by a wide variety of experimental techniques, each with its own intrinsic limitations.

[29] D. G. George, B. C. Orcutt, H.-W. Mewes, and A. Tsugita, *Protein Sequences Data Anal.* **5,** 357 (1993).

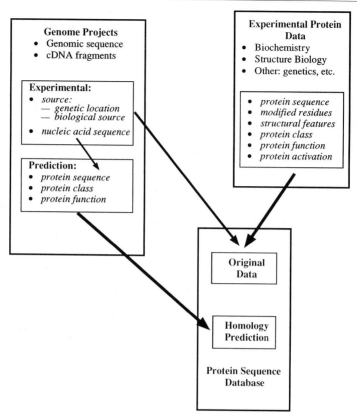

Fig. 2. Sources of experimental information in the PIR-International Protein Sequence Database.

The only information available from genomic sequencing is the sequence itself, the genetic location of the sequence, and its biological source. Sequencing of mRNA (or derivatives such as cDNA) yields additional information concerning the coding region but does not always provide enough information to clearly establish its boundaries or location within the messenger. In many cases the coding region is established only by homology with known protein sequences or by other predictive methods. Additional experiments are required to establish accurately the complete primary structure of the protein. Information concerning posttranslational modification of the protein sequence and the structural and functional properties is not available from nucleic acid sequencing studies. It is common practice to infer this information by homology with related forms whose properties have been determined by experimental methods. These inferences are stored as information in the Protein Sequence Database; hence, it is critical that the information that has been established experimentally be distinguished from that which has been inferred. The status subrecords of the feature records and of the introns subrecord of the GENETICS record (see Fig. 1) specify whether the associated information has been predicted or established by direct experimentation.

All information concerning the same protein molecule is combined into a single entry. When the information is available, the precursor form (that corresponding to the sequence translated from the mRNA) of the molecule is represented in the database. This requires that the originally reported sequence data be transformed into the precursor form and be combined (or merged) into a single canonical form. The REFERENCE records contain citations to the sequence reports. Within the REFERENCE record individually reported sequences are listed within each accession subrecord. The accession number is associated with the reported sequence. Within the accession subrecord, the type of sequence originally determined is given in the molecule_type subrecord. The residues subrecord contains the residues specification, which is an instruction for transforming the canonical sequence given in the entry into the originally reported sequence. For example, the second reported sequence depicted in the entry shown in Fig. 1 (accession A60149) can be reconstructed by extracting the residues at positions 19–23, appending an X (designating an undetermined amino acid), and appending the residues at positions 25–28. The paper cited reports the amino-terminal fragment of the mature protein; hence, the signal sequence is absent. The residue at position 24 could not be determined. The residues specification allows the reported sequence to be represented nonredundantly without loss of information.

Incongruities among closely related sequences often reveal errors such as peptide transpositions, reading-frame shifts, or the use of inappropriate genetic codes. Hence, the procedure of comparing and merging of overlapping sequence data is a vital strategy for increasing the reliability of the data. When sufficient information is available to ascertain the correct sequence, the reported data are corrected accordingly. In these cases, the original information is preserved by noting the nature of the discrepancy and the rationale for the correction. In most cases, the discrepancies cannot be resolved. They may originate from experimental error in the sequence determination; often they are due to differences in the biological sources of the data, but this cannot be ascertained because the source has not been specified in sufficient detail (i.e., the sequences may be from different strains or varieties or from different genes or alleles, but the reports do not specify this information). In these cases, the residue specifications indicate the positions at which various reports disagree and thus provide a direct measure of certainty for the user. Positions corroborated by several independent reports are established with a high degree of certainty, whereas those exhibiting discrepancies are questionable. As a result of the merging process, the sequence entries become canonical representations of biological entities rather than individual experimental sequence reports.

The sequences in the database are organized further into families and superfamilies of evolutionarily related proteins. As closely related sequences are expected to exhibit significant sequence similarities, particularly in well-conserved regions, deviations observed in specific sequences from the similarity patterns exhibited by the family or superfamily provide an additional check on sequence reliability.

Figure 3 depicts the hierarchical association of the information in the database. A source document is defined as any tangible medium that conveys sequence or sequence-related data (annotation), including electronic

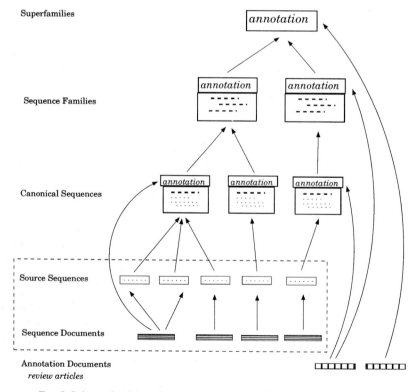

Fig. 3. Information hierarchy among sequence and sequence-related data.

data depositions residing at one or more data collection centers, traditional published works (journals, monographs, etc.), or other nonelectronic, unpublished bodies of work. We distinguish two types of source documents: those reporting sequence data and those reporting annotation information with no original sequence data. It is common for a sequence document to report more than one sequence; hence, the documents must be registered independently of each reported (source) sequence. The data are organized at three levels: (1) source sequence data, extracted from independent source documents, concerning the same sequence are compared and merged to generate a canonical sequence constructed from the various reported forms; (2) the canonical sequences are clustered into groups of evolutionarily closely related sequence families; (3) families of closely related sequences are further clustered into superfamilies.

Verification and Inference of Information via Superfamily Classification

The most time-intensive task in the preparation of entries in the Protein Sequence Database is the processing, verification, and standardization of the annotation, that is, the information associated with the protein sequences. Well over 80% of the biological information concerning protein sequences reported in the published literature has been inferred by homology. Superfamily and family classification provides an effective architecture for the intercomparison, correlation, and analysis of information associated with the sequences within a homology class. Provided that there is justification for the direct inference (or spreading) of information among members of a homology class, new sequence entries can directly inherit the annotation information associated with existing homologous sequences. Moreover, as new experimental information becomes available it can be applied to the entire class of homologous sequences, rather than to sequences on an entry-by-entry basis.

By processing entries in classes rather than individually, critical processing-time factors are related to the number of superfamilies (and families) and the amount of information known about these superfamilies rather than to the number of individual source sequences. It was originally proposed by Margaret Dayhoff[30] that an upper limit existed to the number of protein superfamilies. More recent attempts to refine this estimate[1,31,32] agree reasonably well with Dayhoff's original estimate and vary from less

[30] M. O. Dayhoff, *Fed. Proc.* **35,** 2132 (1976).
[31] G. H. Gonnet, M. A. Cohen, and S. A. Benner, *Science* **256,** 1443 (1992).
[32] C. Chothia, *Nature* (*London*) **357,** 543 (1992).

than a thousand to several thousand. Such estimates are largely dependent on the criteria for superfamily membership. We now adhere to much more conservative criteria (see [4] in this volume); hence, although we have not made a careful projection, we expect the ultimate limit on superfamilies to be on the order of several tens of thousands. The numbers of families and superfamilies are expected to remain small relative to the number of sequences and to exhibit a slow, stable growth as opposed to the exponential growth expected of the sequence data.

A second important consideration is superfamily coverage, that is, what fraction of the sequences are expected to fall within families or superfamilies with more than one member and therefore can be processed more efficiently according to this strategy. We estimate that 80–90% of source sequences in the current database will cluster into families or superfamilies containing at least two members. This assessment reflects the current tendency toward directed sequencing (nucleic acids corresponding to genes of which homologs are already known are preferentially being sequenced) and may be biased accordingly. The results from the yeast chromosome sequencing projects indicate that a considerable fraction of the proteins deduced from genomic sequencing may not fall into homologous groups; rather, these will correspond to species-specific proteins (e.g., regulatory proteins) that are nonhomologous by design (Hans-Werner Mewes, director of the informatics coordination center for the European yeast genomic sequencing effort, personal communication, 1993). Nevertheless, nearly all available annotation information is associated with sequences that fall within homology classes. Although undirected nucleic acid sequencing methods may yield sequences with no identifiable homologs, experimental efforts generally are not directed toward understanding the properties of the products of such potential coding regions until additional homologs have been discovered. Hence, in practice, there is little information available concerning nonhomologous sequences, and little effort is required by the database staff for their annotation.

This situation may change with the European proposals to launch a directed functional analysis program aimed at elucidating this information. However, experimental determination and verification of the characteristics and behaviors of proteins and protein sequences remain time-intensive tasks, and the rate at which annotation information is elucidated will not approach the rate of sequence determination in the foreseeable future. In conclusion, it can be reasonably expected that the development of a methodology based on processing homologous classes of sequences will make the annotation problem tractable, thus allowing the database to be effectively maintained by a staff that is stable and limited in size.

Figure 4 illustrates the change in data processing strategy that has begun

Strategy Prior to 1987
processing by individual source sequence

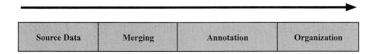

Developing Strategy
processing by protein class

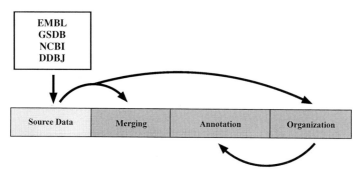

FIG. 4. Evolution of data processing strategy. EMBL, European Molecular Biology Laboratories; GSDB, Genome Sequence Data Base; NCBI, National Center for Biotechnology Information; DDBJ, DNA Data Bank of Japan.

as a result of this rationalization. Prior to about 1987 source data were entered into the database, information from overlapping reports was merged, and the entries were annotated and finally classified into families and superfamilies. Under the new processing strategy, candidates for merging are selected simultaneously with the selection of candidates for addition to families and superfamilies. Annotation proceeds after the entries have been merged or classified.

Validation of Methodology

Assigning sites of biological interest (features) by homology requires the construction of a multiple sequence alignment. It has been argued by many that the construction of multiple sequence alignments is an inherently subjective process. Because mathematically rigorous multiple sequence alignment algorithms cannot guarantee biologically realistic alignments, it is common practice to adjust algorithmically generated alignments by hand. Nevertheless, from these alignments (in the published literature) we infer important biological information that is represented in the Protein Sequence Database. The multiple sequence alignments represented in the database are best understood as hypotheses: a statement of the relationships among specific residues within the related sequences. As such, they are independent of the methods used to establish them. They are critical pieces of information because they provide a detailed, explicit record of the information that was used in making a feature assignment. The existence of conserved biological features within the alignment provides biological evidence for the correctness of the alignment.

We have observed that, among sequences and subsequences that are greater than about 50 residues in length and less than 50–65% different, the major features of the alignment are reproduced by a wide variety of algorithms. This is the realm of the closely related. Because few gaps are required in such alignments, relatively few decisions are required for their placement. Nevertheless, there are localized regions, particularly at the ends of the alignments (the boundaries of sequence domains), where the results of various methods disagree. Fortunately, these localized differences among the generated multiple sequence alignments are generally of little biological significance. This realm of closely related sequence domains occurs within Doolittle's sequence distance region prior to the boundary with the twilight zone.[3] Significantly, within this realm, alignments derived by comparison of three-dimensional structures also agree well with those derived solely by sequence comparison methods.[33,34]

We take a threshold approach to multiple sequence alignment and protein classification. Closely related sequences are grouped into protein families and are aligned by standard methodologies. Such families (including families containing single members) are further clustered into superfamilies. It is our premise that fully objective and fully automatic multiple sequence alignment methods that reflect the biological understanding of the sequences can be developed at the family level. Further classification at the superfamily level is often a straightforward task. We are developing procedures that allow superfamily classification to proceed efficiently for the straightforward cases. When corrected for complications introduced by multiple domain proteins, these classifications agree well with those found using automated clustering techniques derived from information theory (J. Pardowitz, Max Planck Institute for Experimental Medicine,

---

[33] C. Sander and R. Schneider, *Proteins: Struct. Funct. Genet.* **9,** 56 (1991).
[34] L. Holm, C. Ouzounis, C. Sander, G. Tuparev, and G. Vriend, *Protein Sci.* **1,** 1691 (1992).

Göttingen, Germany, personal communication, 1992) and neural network approaches.[35,36]

The threshold approach is novel and is based on sound biological reasoning. Moreover, it provides a practical, objective basis for assigning and verifying protein sequence features. There is good evidence that sequence structures are conserved within protein families.[33] Assigning features across family classes requires an additional level of inference that cannot be justified in the absence of additional biological information. We employ alignments as a basis for self-consistently assigning and verifying features among families of domains for which there is experimental evidence for the existence of the site or region of interest for one or more of the member homologs. The consistency of conserved sites and regions (and other annotations) among families within superfamilies is examined as a verification measure. Proteins within the same superfamily are not expected to exhibit widely varying properties. Verification efforts can be made much more efficient by focusing staff time on the resolution of discrepancies detected within superfamilies and otherwise allowing the data to pass automatically.

Component Model of Database

We have adopted a component model of the database. Much of the ancillary information concerning the proteins is stored in separate database components. These components serve as repositories for the authoritative version of the information that may be shared by a number of sequence entries (e.g., citations or formal names of organisms). The component model was adopted to reduce redundancy and, perhaps more importantly, to promote a separation of concerns; partitioning allows database activities to be focused on specific aspects of the data at hand.

The database components fall into six general categories: (1) source data components; (2) canonical sequence component; (3) classification component; (4) nomenclature standardization components; (5) sequence and feature verification components; and (6) database interoperability components. The relationships among the various components are depicted in Fig. 5. The Protein Sequence Component contains the canonical representations of the protein sequences; it corresponds to the current PIR1 and PIR2

[35] C. Wu, G. Whitson, J. McLarty, A. Ermongkonchai, and T.-C. Chang, *Protein Sci.* **1**, 667 (1992).
[36] C. Wu, M. Berry, Y.-S. Fung, and J. McLarty, "Proceedings of the First International Conference on Intelligent Systems for Molecular Biology" (ISMB-93), p. 429. AAAI Press, Menlo Park, California, 1993.

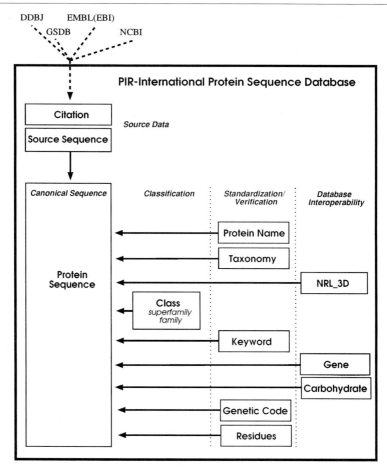

Fig. 5. Component model of the PIR-International Protein Sequence Database.

database sections. Because the distinctions that originally justified dividing the database into these sections no longer apply, we are combining and reformulating them as the Protein Sequence Component. Entries in this component are linked to the auxiliary components, and some of the information in the sequence entries is extracted or derived from that present in the auxiliary components.

The first three categories of database components reflect the data processing strategy. The Citation Component contains citations to all source documents of interest to the project. Some of these citations may never appear explicitly in the Protein Sequence Component, and often a single source document may report several distinct source sequences. The Source Sequence Component serves as a repository for all available source sequence data. Entries in this component correspond to individual sequence reports. Sequences come into the Protein Sequence Database from a variety of overlapping sources including each of the four nucleic acid database centers: the National Center for Biotechnology Information (NCBI),[21] the European Bioinformatics Institute (EBI) of the European Molecular Biology Laboratories (EMBL),[22] the Genome Sequence Data Base (GSDB) of the National Center for Genome Resources, and the DNA Data Bank of Japan (DDBJ).[23] Each of these centers employs specifically different data processing strategies and conventions. Hence, PIR-International is required to maintain a source sequence collection as a stable interface to these other data repositories. At this level outright duplication (multiple representation of the same source information by different database centers) is detected and eliminated.

The Class Component contains the family and superfamily classification scheme. Family Classes are represented by alignments; Superfamily Classes are represented as classes of families (or individual sequence elements for single-member families). The Class entries are structured hierarchically and may include annotation that is common among family and superfamily members.

The standardization components consist of repositories for maintaining semantic consistency and standardization within the database. The Protein Name Component contains standardized protein names. The Taxonomy Component contains standardized species names organized in a taxonomic hierarchy. The Keyword Component contains standard keywords that are associated with Protein Sequence Entries.

The Genetic Code Component contains alternate (special) genetic codes used in organism-specific protein translation. The Residues Component contains standardized modified amino acid abbreviations and other associated information such as *Chemical Abstracts* Registry Numbers. These Components are used for verification of sequence and feature data presented in the Protein Sequence Component and to ensure that the data are represented in a standardized form.

Effective linking of information across databases requires that conceptual entities common to the linked databases be identified and represented in a formalism that can be cross-mapped; in general, this information is specific to specific pairs of databases. The most straightforward approach is to employ a common nomenclature to represent common data elements. The database interoperability components contain sets of information common to the Protein Sequence Database and other nonsequence and nonbibliographic databases. The NRL_3D Component cross-links the Protein Sequence Database to the Brookhaven National Laboratory Protein Data Bank (PDB). The Gene Component contains information linking PIR-International Genetics Records to the genetic mapping databases, such as GDB and FlyBase. The Carbohydrate Component contains information cross-linking PIR-International Feature records to the Complex Carbohydrate Structure Database.

Current Status

The transition to a more efficient data processing strategy and toward the component architecture of the database has been underway for several years. The Source Sequence and Citation Components comprise the ARCHIVE/INQ system for importing and characterizing reported sequences and their citations as previously described.[18-20] As indicated above the PIR1 and PIR2 database sets are in the process of being combined into a single logical set (although the data may remain physically separated to facilitate ease of data access). Entries from the PIR3 collection are being rapidly processed into the Protein Sequence Component, and this data store will be emptied.

The elements of the Class Component are described in [4], this volume. Currently, this information is represented as placement numbers, in the alignment database,[19] in the family classification developed at MIPS, and in the domain and superfamily classifications developed at PIR. A uniform architecture is under development to consolidate these data. The Taxonomy, Keyword, Genetic Code, Residues, and NRL_3D data components have been maintained for a number of years and are made publicly available.[18-20] In collaboration with PDB, we expect to expand the role of the NRL_3D component to more directly cross-link the data and to serve as a vehicle for cross-verification of information in the two data collections. The Gene data set is currently employed by the database staff and is being developed in collaboration with GDB and with FlyBase. The Carbohydrate Component is under development in collaboration with CarbBank.

Availability of Database

The PIR-International Protein Sequence Database is public domain and is made unconditionally available. Hence, it is widely redistributed, is integrated into other public data sets including the SWISS-PROT Protein Sequence Database maintained by the EBI[22] and those assembled by the NCBI, and is incorporated into commercial products such as the Genetics Computer Group (GCG) package.

The database is made available directly from the PIR-International Centers and from a variety of sites on the World Wide Web. For specific information contact the PIR Technical Services Coordinator, National Biomedical Research Foundation, 3900 Reservoir Road NW, Washington, DC 20007; telephone +1 202 687-2121; FAX +1 202 687-1662; electronic mail PIRMAIL@nbrf,georgetown.edu. In Europe, contact MIPS: Martinsried Institut für Proteinsequenzen, Max-Planck-Institut für Biochemie, D-82152 Martinsried bei München, Germany; telephone +49 89 8578 2657; FAX +49 89 8578 2655; electronic mail MIPS@ehpmic.mips.biochem.mpg.de. In Asia or Australia, contact JIPID: Japan International Protein Information Database, Science University of Tokyo, 2669 Yamazaki, Noda 278, Japan; telephone +81 471 239778; FAX +81 471 221544; electronic mail TSUGITA@JPNSUT31.bitnet or EX5292@JPNSUT30.bitnet.

Acknowledgments

We acknowledge the contributions of all of the scientists and staff of the PIR-International centers whose work we report here. This publication was supported in part by Grant P41 LM05798 from the National Library of Medicine. Its contents are solely the responsibility of the authors and do not necessarily represent the official views of the National Library of Medicine.

[4] Superfamily Classification in PIR-International Protein Sequence Database

By WINONA C. BARKER, FRIEDHELM PFEIFFER, and DAVID G. GEORGE

Brief History of Superfamily Concept

In the mid 1970s, Dayhoff proposed that all naturally occurring proteins would cluster into families and superfamilies whose members have diverged from common ancestral forms.[1,2] A similar proposal was made by Emil Zuckerkandl.[3] Estimates of the number of protein superfamilies were in the low thousands. Using a variety of criteria for superfamily membership,

[1] M. O. Dayhoff, P. J. McLaughlin, W. C. Barker, and L. T. Hunt, *Naturwissenchaften* **62,** 154 (1975).
[2] M. O. Dayhoff, *Fed. Proc.* **33,** 2314 (1976).
[3] E. Zuckerkandl, *J. Mol. Evol.* **7,** 1 (1975).

Copyright © 1996 by Academic Press, Inc.
All rights of reproduction in any form reserved.

more recent estimates of the same order of magnitude have been reported.[4-6]

Although superfamily relationships sometimes are so ancient as to preclude recognition solely on the basis of sequence similarity, our group has
employed sequence similarity as the main criterion for partitioning the
Protein Sequence Database into independent, nonoverlapping groups. In
1976, the nearly 500 completely sequenced proteins then known were each
assigned to one of 116 superfamilies.[7] At that time, there were no examples
in the database of complete precursor sequences, of polyproteins, or of
products of alternative splicing of mRNA. Most of the known sequences
were of mature forms of peptides or proteins. There were only a few
examples of multidomain proteins.

Within a few years of the introduction of the superfamily concept, it
became evident that many protein sequences contain homology domains,
namely, regions of local similarity contained in otherwise unrelated proteins. Such domains are often responsible for similar properties (such as
calcium binding, DNA binding, or catalytic activity) shared by diverse
proteins. Evidence from X-ray crystallography and chemical studies revealed that these domains often correspond with compact regions of the
structure or with easily cleaved fragments. More surprising was the discovery that the genes for many proteins contain noncoding regions (introns)
that divide the protein coding region into exons that sometimes approximately correspond with the domains as defined by structure or protein
sequence. "Exon shuffling" among genes is now recognized as one mechanism in the evolution of "new" proteins.

Although the term protein superfamily is widely used, its meaning is
not well defined when applied to multidomain proteins. In the literature,
terms such as "the immunoglobulin superfamily"[8,9] have come to mean the
collection of all proteins that contain the named domain. This usage of the
term, however, does not allow the database to be unambiguously partitioned
into superfamilies. Multidomain proteins such as the platelet-derived
growth factor receptor can be placed in the protein kinase superfamily, or
in the immunoglobulin superfamily, or in a different superfamily of se-

[4] P. Green, D. Lipman, L. Hillier, R. Waterson, D. States, and J.-M. Claverie, *Science* **259,**
1711 (1993).
[5] G. H. Gonnet, M. A. Cohen, and S. A. Benner, *Science* **256,** 1443 (1992).
[6] C. Chothia, *Nature* (*London*) **357,** 543 (1992).
[7] M. O. Dayhoff, W. C. Barker, and L. T. Hunt, *in* "Atlas of Protein Sequence and Structure"
(M. O. Dayhoff, ed.), Vol. 5, Suppl. 2, p. 9. National Biomedical Research Foundation,
Washington, D.C., 1976.
[8] T. Hunkapiller and L. Hood, *Nature* (*London*) **323,** 15 (1986).
[9] A. F. Williams and A. N. Barclay, *Annu. Rev. Immunol.* **6,** 381 (1988).

quences that contain both of these types of domains. We have developed a formal model of the protein superfamily concept that encompasses the most common usages and integrates both homology at the domain level and homology at the level of complete proteins.[10,11] This model preserves the ability to fully partition the Protein Sequence Database, permits the organization of the database in a structured way, and introduces a more precise and unambiguous definition for the term protein superfamily.

Superfamily Concept Revised and Generalized

The concepts of superfamily and family have been generalized to encompass any scheme for classifying proteins (or regions within proteins) that partitions the proteins (or protein regions) into hierarchically nested sets that are closed under transitivity.[10,11] A superfamily is a union over families. Families are sets within the superfamily hierarchy for which the members meet a threshold level of relatedness.

In the Protein Information Resource (PIR)-International Protein Sequence Database we classify sequence homology domains and apply the terms superfamily and family to these units of information. A homology domain is a sequence region found in diverse proteins that is likely to be derived from a common evolutionary ancestor. Homology domains differ from patterns or motifs (that may be contained within them) in that they are demonstrably similar along their entire extents as observed by multiple sequence comparison and alignment. Generally, they are greater than 50 residues in length. The most common homology domains noted in the June 1995 release of the Protein Sequence Database are listed in Table I.

A homology domain may encompass all of a protein sequence or may represent a subsequence within it. Because the insertion (or deletion) of exogenous loops of sequence within individual domains is a commonly observed evolutionary event, the subsequence composing a homology domain may not be contiguous. Protein domains may be complex, that is, composed of more than one distinct domain. Complex domains may be formed by coalescense of two or more originally independently evolving domains; after concatenation, the domains evolve as a unit. Domains not composed from other identifiable domains are called simple domains. Examples of some complex domains are listed in Table II. Protein sequences may contain a single simple domain or may be mosaics composed of a

---

[10] D. G. George, 1993, unpublished.
[11] W. C. Barker, F. Pfeiffer, and D. G. George, *in* "Methods in Protein Structure Analysis" (M. Z. Atassi and E. Appella, eds.), p. 473. Plenum, New York, 1995.

TABLE I
MOST COMMONLY OCCURRING HOMOLOGY DOMAINS

Homology domain[a]	Number of entries[b]	Taxonomic distribution[c] (%)				
		Animals[d]	Plants[e]	Fungi	Protists	Prokaryotes[f]
Immunoglobulin	2922	100	0	0	0	0
Protein kinase	623	85	5	7	3	0
Homeobox	531	96	2	2	0	0
Calmodulin repeat	428	86	6	3	5	0
Trypsin	339	98	0	0	0	2
Kazal proteinase inhibitor	191	100	0	0	0	0
Myosin head	176	88	3	3	6	0
Cytochrome c	166	32	23	9	3	33
SH3 (src homology 3)	158	89	0	8	3	8
Ribonucleoprotein repeat	149	72	15	12	1	0
H^+-transporting ATP synthase α chain	146	16	41	10	1	32
Translation elongation factor Tu	145	16	19	12	5	48
EGF (epidermal growth factor)	134	99	0	1	0	0
MalK protein	131	34	2	5	4	55
SH2 (src homology 2)	129	100	0	0	0	0

[a] In the database, the names of homology domains end with the word "homology," which has been omitted in these tables for brevity.
[b] Number of entries containing domain found in PIR-International Protein Sequence Database, Release 45.0 (June 1995), with sections PIR1 and PIR2 totaling 66,573 entries.
[c] Percentage of entries found (column 2).
[d] Includes animal viruses.
[e] Includes plant viruses.
[f] Includes bacteriophages.

number of simple and/or complex domains. Some proteins that contain five or more homology domains are listed in Table III.

The domain that represents the entire protein is called the homeomorphic domain. Because the precursor form (the form initially translated from the mRNA) of the protein is generally represented in the database, homeomorphic domains include portions of the sequence removed during maturation. Two proteins belong to the same homeomorphic superfamily when they show homology over the length of their entire sequences; hence, two members of the same homeomorphic superfamily contain the same homology domains in the same order.

Within a homology domain superfamily, more closely related domains

TABLE II
COMPLEX HOMOLOGY DOMAINS

Complex domain	Constituent domains	Proteins[a]
Cytochrome b	Cytochrome b_6 Plastoquinol–plastocyanin reductase 17K protein	FcbH bifunctional protein [B32382] Cytochrome b [CBHU]
Carbamoyl-phosphate synthase (ammonia)	Carbamoyl-phosphate synthase (glutamine-hydrolyzing) large chain Carbamoyl-phosphate synthase (glutamine-hydrolyzing) small chain TrpG	Carbamoyl-phosphate synthase (ammonia) I [SYRTCA] Pyrimidine synthesis protein CAD [A23443]
HisI bifunctional enzyme	HisI protein Histidinol dehydrogenase	HisI–hisD trifunctional enzyme [SHNC] HisI bifunctional enzyme [YNECHI]
Leukocyte common antigen cytosolic domain	Protein-tyrosine-phosphatase (2 copies)	Leukocyte antigen-related protein [TDHULK] Leukocyte common antigen [A46546]
Osteonectin homology	Agrin inhibitor-like repeat Kazal proteinase inhibitor Calmodulin repeat	Matrix glycoprotein SC1 [GERTX1] Osteonectin [GEHUN]

[a] Entry code or accession number is given in square brackets.

are grouped into families. For practical reasons we place domains into the same protein family if they show at least 50% sequence identity (see [3], this volume). When the data are fragmentary, they may be classified provided that one or more homologs exist that are related closely enough to reasonably assume that the missing data conform with those of the homologs.

Partitioning of the families and superfamilies is achieved by treating homology domains containing overlapping regions independently; in other words, complex domains and the simple domains from which they are composed are separately classified and treated as independent entities. This redundancy allows the entire database to be partitioned into homeomorphic families while simultaneously allowing the entire (nonredundant) collection of simple domains to be independently classified. A system is under development for maintaining the relationships among simple and complex domains separate from the superfamily classification scheme.

TABLE III
PROTEINS WITH FIVE OR MORE TYPES OF HOMOLOGY DOMAINS

PIR entry	Superfamily name	Simple homology domains
XYRTFA, fatty-acid synthase (EC 2.3.1.85)—rat	Rat fatty-acid syn-thase	3-Oxoacyl-[acyl-carrier-protein] synthase I Acyl carrier protein Long-chain alcohol dehydrogenase Oleoyl-[acyl-carrier-protein] hydrolase Short-chain alcohol dehydrogenase [Acyl-carrier-protein] S-malonyltransferase
BVBYA1, ARO1 protein—yeast (*Saccharomyces cerevisiae*) BVASA1 aroM protein—*Emericella nidulans*	Aro1 protein	3-Dehydroquinate dehydratase 3-Dehydroquinate synthase 3-Phosphoshikimate 1-carboxyvinyltransferase Shikimate dehydrogenase Shikimate kinase
KFHU12, coagulation factor XIIa (EC 3.4.21.38) precursor—human)	Coagulation factor XII	EGF (epidermal growth factor) Fibronectin type I repeat Fibronectin type II repeat Kringle Trypsin
QZFF, rudimentary protein—fruit fly (*Drosophila melanogaster*) A23443, pyrimidine synthesis protein CAD—golden hamster QZBYU2, pyrimidine synthesis protein URA2—yeast (*Saccharomyces cerevisiae*) QZDOP3, pyrimidine synthesis protein PYR1-3—slime mold (*Dictyostelium discoideum*) (fragments)	Rudimentary enzyme	Aspartate/ornithine carbamoyltransferase *Bacillus* dihydroorotase Carbamoyl-phosphate synthase (ammonia) Carbamoyl-phosphate synthase (glutamine-hydrolyzing) large chain Carbamoyl-phosphate synthase (glutamine-hydrolyzing) small chain trpG

Defining Homeomorphic Families and Superfamilies

For purposes of partitioning the databases, the defining relationship is said to be sequence homology, that is, the logical inference of common ancestry. However, homology is not a quantifiable characteristic: there either was common ancestry or there was not. There are no degrees of homology; thus, no thresholds can be established reliably to detect homology or to partition sequences within homology groups. In practice, we use similarity as an indicator of homology and establish threshold levels of similarity on which to base the partitioning.

Methods of detecting and quantifying sequence similarity range from fairly straightforward to rather sophisticated and computationally expensive. The latter methods, however, are needed only for the difficult cases—those where the sequence similarity is not well conserved. Although these very distant relationships are scientifically of great interest, for the purposes of database operations (classification, standardization of annotation within classes) such clusterings are less useful than clusterings of closely related sequences. There is much empirical evidence suggesting that a threshold can be established whereby it can be inferred that closely related proteins share common biological properties. Because closely related groups of proteins, which generally are homologs of the same protein in various species or products of recent gene duplications, can be expected to share many structural and functional characteristics, new sequences can directly inherit annotation information associated with existing closely related sequences. This provides a mechanism for comprehensive and consistent annotation within large areas of the database.

Formerly, our approach to classification was to group sequences into superfamilies, which were then subdivided into families, subfamilies, entries, and subentries according to percentage of sequence identity.[7] This has the disadvantage that it may involve making difficult (and time-consuming) decisions prior to making simple and very useful decisions that are amenable to automation.

We now initially classify proteins into homeomorphic protein families.[11] To facilitate the development of semiautomated procedures, we have adopted a working definition of a homeomorphic protein family as a set of sequences that can be aligned end to end without major discrepancy by standard multiple sequence alignment methods. In practice, such sequences generally will have an overall sequence identity of at least 50% and the same domain architecture. Classification into protein families is one of the first steps in annotation.

Protein families for which the members have similar overall architecture (i.e., the same domains in the same order) are then clustered into homeomorphic superfamilies. Sequences representing different families within a superfamily are alignable end to end but with less certainty and with considerably more latitude (e.g., we permit terminal regions and interdomain regions that are sufficiently dissimilar that the alignment is essentially arbitrary, some variability in the number of repeats of a domain and in the size of regions of restricted composition, and the absence of a domain because of alternative splicing or use of an alternative initiator).

Clustering Sequences into Protein Families

The FASTA program permits rapid comparison of a query sequence to a sequence database.[12] The Martinsried Institute for Protein Sequences (MIPS), the European branch of PIR-International, has generated the FASTA database.[13] This database contains results of FASTA searches of each entry in the database against the entire PIR-International Protein Sequence Database (plus the PATCHX database[14] of sequences available elsewhere but not yet processed by PIR-International). The FASTA database is updated dynamically as new entries are added to the database.

The FASTA results can be used either to select candidates for an existing protein family from among as yet unclassified entries or to search for the protein family to which a single unclassified entry belongs.[11] Candidate sequences are aligned automatically, and the aligned pairs are screened for congruence of length and threshold level of similarity. Those that meet rather stringent requirements are routinely classified; others are examined and classified by scientific staff. Multiple sequence alignments are computed for all protein families containing more than two members, using the Feng and Doolittle algorithm[15] as implemented in the Genetics Computer Group PILEUP program.

As of release 45.0 of the Protein Sequence Database (June 1995), 70% of all database entries have been classified into protein families. About 7% of all sequences are not classifiable, generally because the sequences are too short or are fragmentary. About 18% of all classified entries are unique representatives for a protein family. Routinely, when a protein family contains members that have been assigned to a superfamily, all members of the family are automatically classified into that superfamily.

Placement Group Classification

Placement is defined as the ordering of the sequences (homeomorphic domains) in the database. We strive to place homeomorphic superfamilies into broader categories: electron-transfer proteins, enzymes, enzyme inhibitors, etc. Within these categories, we often cluster superfamilies that contain one or more related domains, have some functional relationship (e.g., participation in the same physiological process), or possess some other biological similarity.

[12] W. R. Pearson and D. J. Lipman, *Proc. Natl. Acad. Sci. U.S.A.* **85,** 2444 (1988).
[13] S. Liebl, *et al.*, 1992, unpublished.
[14] W. C. Barker, D. G. George, H. W. Mewes, F. Pfeiffer, and A. Tsugita, *Nucleic Acids Res.* **21,** 3089 (1993).
[15] D. F. Feng and R. F. Doolittle, *J. Mol. Evol.* **25,** 351 (1987).

A series of five numbers is used to specify the placement of each sequence entry. The first number identifies the placement (homeomorphic superfamily) group, the second identifies the family within the superfamily group, and the subsequent numbers identify subfamily, entry, and subentry.[7] The subfamily and entry numbers were originally employed as an indication of threshold levels of sequence similarity (more than 80% and 95% identical, respectively). Currently, they are used as a convenience in ordering large superfamilies, without strict adherence to the numerical thresholds. The number 0.0 is assigned at the family level and below to indicate a null assignment when a strict ordering within the superfamily or family has not been established. Our goal is to classify all classifiable database entries at the family and superfamily levels. However, classification will be tentative for entries containing fragmentary data or for peptides derived from much larger precursors. In each database release, unclassified sequences follow the classified sequences; they are sorted first by taxonomic class and then ordered alphabetically by entry title.

The placement numbers are not stable identifiers for the superfamilies. When a new superfamily or subgroup within a superfamily is required, we may use a fractional number. From time to time, the database is renumbered and all numbers are converted to consecutive whole numbers. Therefore, it is recommended that the superfamily name (see below) be used when referring to a superfamily.

Defining Homology Domains

The essential concept of homology domains predated the development of the superfamily concept; it arose from the observation that duplicated regions within homologous proteins showed more similarity to the corresponding regions of their homologs than to the duplicated region within the same protein sequence. This evidence of gene duplication was first described for ferredoxins.[16] In the Protein Sequence Database, the term "homology regions" was first applied in the study of the evolution of the C (constant) regions of the classic immunoglobulins.[17] In the mid 1970s the National Biomedical Research Foundation (NBRF) introduced a specialized program, RELATE, for detecting sequence duplication. All modern database searching programs are designed to detect "local" sequence similarities and are well suited for detecting unusually similar regions or domains

[16] R. V. Eck and M. O. Dayhoff, *Science* **152**, 363 (1966).
[17] W. C. Barker, P. J. McLaughlin, and M. O. Dayhoff, *in* "Atlas of Protein Sequence and Structure" (M. O. Dayhoff, ed.), Vol. 5, p. 31 and D-378. National Biomedical Research Foundation, Washington, D.C., 1972.

in otherwise unrelated sequences. Homology domains are now commonly observed by researchers examining the results of database searches.

Elucidating the relationships among domains is a more difficult subset of the task of discovering distant protein relationships. There is a large literature on methods for examining such relationships, and these methods continue to be studied and improved (see, e.g., Pearson[18]). A full analysis of the significance of proposed domain relationships can require the use of a number of the most sophisticated tools available for computer analysis of sequences. Confirmation of proposed structural and functional relationships may require nuclear magnetic resonance (NMR) or x-ray crystallographic studies, site-directed mutagenesis, and analysis of the properties of chemically modified or engineered sequences. Therefore, the full characterization of any type of protein domain is beyond the scope of the activities of a sequence database. All of the homology domains annotated in the Protein Sequence Database are understood to have been deduced on the basis of sequence similarity, and only sometimes confirmed by other evidence. When we "define" a homology domain, we are not claiming to have characterized it; we are simply recording the criteria that we use to determine that such a domain does exist in the protein.

From the beginning, we have defined a homology domain by constructing a multiple sequence alignment of the proposed homologous segments. When refining and evaluating such an alignment, we also consider other information, such as the identity and location of known functional residues. Because homology domains of a given type can be very distantly related (less than 20% identical in some cases), the details of such alignments, outside of regions that are well conserved, are highly tentative. We periodically review our domain alignments in the light of new examples that have been discovered and of structural and functional relationships that have been elucidated by experimental research.

When a pattern of conserved residues or sequence characteristics is clear, the major decision to be made is the assignment of the boundaries of the domain. Often the similarity is weak at one or both ends, and the length of intradomain regions in proteins where the domain is repeated is variable. Thus, the boundary assignments may be rather arbitrary and differ from author to author. We take the view that domain boundaries are approximate and to some extent arbitrary. For users of the database, the presence of a domain is more useful information than are estimates of its "real" boundaries. Rather than trying to portray the maximum extent of a domain, which may involve aligning regions of little similarity, we often choose boundaries in relation to conserved features that occur close to the

---

[18] W. R. Pearson, *Protein Sci.* **4,** 1145 (1995).

ends of the domain. This strategy has several advantages: the choice is more often obvious, and little time need be spent making boundary decisions; homologous domains can be reasonably aligned using widely available multiple alignment software; and computer-assisted methods for detecting additional examples of the known homology domains can be used effectively.[11] For the maintenance of a database of rapidly increasing size, these practical considerations are very important.

We attempt to annotate self-consistently all occurrences of a homology domain in the feature records using a semiautomated procedure. A computer program[19] examines the PIR-International Protein Sequence Database and extracts each annotated homology domain. With the exception of signal sequences, transit peptides, short sequences, and short interdomain regions (<50 residues), all sequences and subsequences not represented in this data set are also extracted. These segments and the identified homology domains are cross-compared and the intercomparison scores are stored in a FASTA database. These data are examined to ensure consistency in domain assignments and to detect previously unidentified members of existing domain families. Segments that show high overall sequence similarity with a known homology domain are automatically aligned with the most similar member of the domain family. If the alignment shows high conservation at both ends, does not contain large gaps, and shows a uniform distribution of sequence similarity, then the segment is automatically classified as an additional member of the domain family. These criteria aid in preserving the homology domain boundaries as originally established. A multiple alignment of the entire family is then computed and the domain boundaries are refined accordingly. Additional segments overlooked by this procedure may be included in the family after examination of the FASTA scores by database staff.

Naming of Superfamilies and Domains

Names are assigned to each homeomorphic superfamily and to each homology domain. These names are given in the Protein Sequence Database on the Superfamily record. If the sequence has been classified into a homeomorphic superfamily, its name appears first on the record. The names of homology domains are distinguished from homeomorphic superfamily names by including the word "homology" at the end of the name.

Usually the superfamily or domain is initially named for the first or most well-known member protein characterized. We often must assign a name when function is poorly characterized and it is not known how wide-

---

[19] F. Pfeiffer, 1993, unpublished.

spread is the occurrence of the protein or domain. We strive to make the names recognizable, either because they are descriptive (hevein chitin-binding domain homology) or because they are commonly used (kringle homology). A domain that is repeated in the protein for which it is named is labeled a repeat homology. Sometimes the name of a superfamily or of a homology domain is later changed to a more suitable name or to one that has become widely used.

Conclusions

Since the origin of the Protein Sequence Database, the classification of protein sequences into superfamilies has provided a biologically meaningful organization of the data. As discussed in [3] in this volume, this classification provides a systematic scheme for verification of the information in the database and for inferring additional information by homology in a controlled way. Information generated from large-scale sequencing projects is incomplete and not well understood. The major task of computational biology is to assign biological meaning to these data. Homology is the major operating principle employed in these analyses. The superfamily classification provides a useful architecture for self-consistent and objective examination of sequence data by homology.

Availability of Data

The classifications as presented in this chapter are available from the nodes of PIR-International in various ways. The Superfamily record and homology domains annotated as features are integral parts of the Protein Sequence Database entries. Retrieval programs available from PIR-International, ATLAS and PSQ,[20] allow selection of sequence entries by super-family name or placement number. On-line access is possible both at NBRF and at MIPS. These sites also operate full-function network file servers that handle database queries, sequence searches, and sequence submissions, in addition to file server requests.

For further information, please contact the PIR Technical Services Coordinator, National Biomedical Research Foundation, 3900 Reservoir Road NW, Washington, DC 20007; telephone +1 202 687-2121; FAX +1 202 687-1662; electronic mail PIRMAIL@nbrf.georgetown.edu. In Europe, contact MIPS: Martinsried Institute for Protein Sequences, Max Planck Institute for Biochemistry, D-82152 Martinsried near Munich, Germany; telephone

---

[20] D. G. George, W. C. Barker, H. W. Mewes, F. Pfeiffer, and A. Tsugita, *Nucleic Acids Res.* **22**, 3569 (1994).

+49 89 8578 2657; FAX +49 89 8578 2655; electronic mail MIPS@ehpmic. mips.biochem.mpg.de. In Asia or Australia, please contact JIPID: Japan International Protein Information Database, Science University of Tokyo, 2669 Yamazaki, Noda 278, Japan; telephone +81 471 239778; FAX +81 471 221544; electronic mail TSUGITA@JPNSUT31.BITNET.

The Protein Sequence Database is also accessible on the Internet over the World Wide Web (WWW) (web address for MIPS: http://www.mips.bio-chem.mpg.de/). The results from classification into protein families are also available via the WWW at MIPS. As of June 1995, about 6000 multiple alignments including 40,000 database entries were available. The multiple alignments are enriched by addition of the consensus sequence and of all features annotated in the database entries. In addition, 230 multiple alignments of homology domains, representing about 10,000 individual homology domains, can be inspected. Additional WWW services will be implemented in the near future; contact the MIPS WWW site or any of the PIR-International nodes for further information.

Acknowledgments

We gratefully acknowledge the contributions of all of the scientists and staff of the PIR-International centers whose work we report here. This work was supported in part by Grant P41 LM05798 from the National Library of Medicine. Its contents are solely the responsibility of the authors and do not necessarily represent the official views of the National Library of Medicine. MIPS is supported by the Max-Planck-Gesellschaft, the Forschungszentrum f. Umwelt und Gesundheit (GSF), and the European Economic Community BRIDGE Programme Grants BIOT-CT-0167 and BIOT-CT-0172.

[5] Gene Classification Artificial Neural System

By Cathy H. Wu

Introduction

As technology improves and molecular sequencing data accumulate exponentially, continued progress in the Human Genome Project will depend increasingly on the development of advanced computational tools for rapid annotation and easy organization of genomic sequences. Currently, a database search for sequence similarities represents the most direct computational approach to decipher the codes connecting molecular sequences with protein structure and function.[1] A sequence classification method can

---

[1] R. F. Doolittle, this series, Vol. 183, p. 99.

Copyright © 1996 by Academic Press, Inc.
All rights of reproduction in any form reserved.

be used as an alternative approach to the database search/organization problem with several advantages: (1) speed, because the search time grows with the number of sequence classes (families), instead of the number of sequence entries; (2) sensitivity, because the search is based on information of a homologous family, instead of any sequence alone; and (3) automated family assignment.[2]

There are many different types of classification learning systems, broadly categorized into parametric and nonparametric models.[3] The nonparametric classifiers do not require *a priori* recognition of certain specific patterns and make few assumptions in characterizing the samples. Examples include the nearest-neighbor families of statistical methods and neural network algorithms, such as back-propagation[4] and counterpropagation.[5] As a technique for computational analysis, neural network technology has been applied to many studies involving sequence data analysis,[6] including protein structure prediction,[7] identification of protein-coding sequences,[8,9] and prediction of promoter sequences.[10]

Several sequence classification methods have been devised, including our back-propagation neural network method,[11] a multivariate statistical technique,[12] a binary similarity comparison followed by an unsupervised learning procedure,[13] and Kohonen's self-organized feature map.[14] All of these classification methods are very fast, and thus applicable to the large sequence databases. The major difference between our and other approaches is that the back-propagation neural network is based on supervised learning, whereas the others are unsupervised. The supervised learning can be performed using training sets compiled from any existing second

[2] C. H. Wu, *Comput. Chem.* **17,** 219 (1993).

[3] S. M. Weiss and C. A. Kulikowski, "Computer Systems That Learn: Classification and Prediction Methods from Statistics, Neural Nets, Machine Learning, and Expert Systems." Morgan Kaufmann, San Mateo, California, 1991.

[4] D. E. Rumelhart and J. L. McClelland (eds.), "Parallel Distributed Processing: Explorations in the Microstructure of Cognition. Volume 1: Foundations." MIT Press, Cambridge, Massachusetts, 1986.

[5] R. Hecht-Nielsen, *Appl. Opt.* **26,** 4979 (1987).

[6] J. D. Hirst and M. J. E. Sternberg, *Biochemistry* **31,** 7211 (1992).

[7] N. Qian and T. J. Sejnowski, *J. Mol. Biol.* **202,** 865 (1988).

[8] E. C. Uberbacher and R. J. Mural, *Proc. Natl. Acad. Sci. U.S.A.* **88,** 11261 (1991).

[9] R. Farber, A. Lapedes, and K. Sirotkin, *J. Mol. Biol.* **226,** 471 (1992).

[10] M. C. O'Neill, *Nucleic Acids Res.* **20,** 3471 (1992).

[11] C. H. Wu, *in* "The Protein Folding Problem and Tertiary Structure Prediction" (K. Merz and S. LeGrand, eds.), p. 279. Birkhauser, Boston, 1994.

[12] M. van Heel, *J. Mol. Biol.* **220,** 877 (1991).

[13] N. Harris, L. Hunter, and D. States, *in* "Proceedings of 10th National Conference on Artificial Intelligence," p. 837. AAAI Press, Menlo Park, California, 1992.

[14] E. A. Ferran, B. Pflugfelder, and P. Ferrara, *Protein Sci.* **3,** 507 (1994).

generation database (i.e., database organized according to family relationship) and used to classify new sequences into the database according to the predefined organization scheme of the database. The unsupervised system, on the other hand, defines its own family clusters and can be used to generate new second generation databases.

This chapter describes the gene classification artificial neural system (GenCANS), the neural network system we have developed for genetic sequence classification, and discusses future system enhancement and applications. GenCANS is a combined version of ProCANS (protein classification artificial neural system)[15] and NACANS (nucleic acid classification artificial neural system).[11] ProCANS was designed for the automatic classification of protein sequences according to the PIR (Protein Identification Resource) superfamilies[15] and was extended into a full-scale system for classification of more than 3300 protein superfamilies.[16] In parallel, NACANS was developed for nucleic acid sequence classification by employing similar design principles. A network was implemented for classification of ribosomal RNAs (rRNAs) according to RDP (Ribosomal Database Project) phylogenetic classes.[17]

Gene Classification Artificial Neural System Algorithm

The neural network system was designed to classify new (unknown) sequences into predefined (known) classes. It involves two steps, sequence encoding and neural network classification, to map molecular sequences (input) into gene families (output) (Fig. 1).

Sequence Encoding Schema

The sequence encoding schema, used in the preprocessor, converts molecular sequences (character strings) into input vectors (numbers) of the neural network classifier (Fig. 1). An ideal encoding scheme should satisfy the basic coding assumption so that similar sequences are represented by close vectors. There are two different approaches for sequence encoding. One can either use the sequence data directly, as in most neural network applications of molecular sequence analysis, or use the sequence data indirectly, as in Uberbacher and Mural.[8] Where sequence data were encoded directly, most studies[7,9] used an indicator vector to represent each molecular residue in the sequence string; that is, they use a vector of 20 input units

[15] C. H. Wu, G. Whitson, J. McLarty, A. Ermongkonchai, and T. Chang, *Protein Sci.* **1**, 667 (1992).

[16] C. H. Wu, M. Berry, S. Shivakumar, and J. McLarty, *Machine Learning* **21**, 177 (1995).

[17] C. H. Wu and S. Shivakumar, *Nucleic Acids Res.* **22**, 4291 (1994).

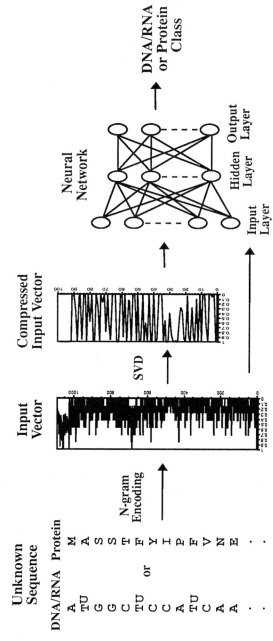

FIG 1. The gene classification artificial neural system (GenCANS) for molecular sequence classification. The sequence strings are first converted by an n-gram sequence encoding method into input vectors of real numbers. Long n-gram input vectors can be compressed by a SVD (singular value decomposition) method to reduce vector size (dimension). The neural network then maps the sequence vectors into predefined classes according to sequence information embedded in the neural interconnections after network training. The neural networks used are three-layered, feed-forward networks that employ the back-propagation or counterpropagation learning algorithm.

(among which 19 have a value of zero, and 1 has a value of one) to represent an amino acid, and a vector of 4 units (3 are zeroes and 1 is one) for a nucleotide. This representation, however, is not suitable for sequence classification where long and varied-length sequences are to be compared.

n-Gram Method. We have been using an *n*-gram hashing method[2,15] that extracts and counts the occurrences of patterns (terms) of *n* consecutive residues (i.e., a sliding window of size *n*) from a sequence string. Unlike the FASTA method,[18] which also uses *n*-grams (*k*-tuples), our search method uses the counts, not positions, of the *n*-gram terms along the sequence. Therefore, our method is length-invariant, provides certain insertion/deletion invariance, and does not require the laborious sequence alignments of many other database search methods. The counts of the *n*-gram terms from each encoding method are scaled to fall between 0 and 1 and used as input vectors for the neural network, with each unit of the vector representing an *n*-gram term. The size of the input vector for each *n*-gram extraction is m^n, where *m* is the size of the alphabet. The original sequence string can be represented by different alphabet sets in the encoding. The alphabet sets used for protein sequences include the 20-letter amino acids and the 6-letter exchange groups derived from the PAM (accepted point mutation) matrix. The alphabet for nucleic acid sequences is the 4-letter AT(U)GC.

The major drawback of the *n*-gram method is that the size of the input vector tends to be large. This indicates that the size of the weight matrix (i.e., the number of neural interconnections) would also be large because the weight matrix size equals *w*, where *w* is input size times hidden size plus hidden size times output size. This prohibits the use of larger *n*-gram sizes; for example, the trigrams of amino acids would require 20^3 or 8000 input units. Furthermore, accepted statistical techniques and current trends in neural networks favor minimal architecture (with fewer neurons and interconnections) to avoid data overfitting and provide better generalization capability.[19]

Singular Value Decomposition Method. The singular value decomposition (SVD) method[16] is used to reduce the size (i.e., the number of dimensions) of the *n*-gram vectors and to extract semantics from the *n*-gram patterns. The method was adopted from the Latent Semantic Indexing analysis used in the field of information retrieval and information filtering. The approach is to take advantage of implicit high-order structure in the association of terms with documents in order to improve the detection of

[18] W. R. Pearson and D. J. Lipman, *Proc. Natl. Acad. Sci. U.S.A.* **85,** 2444 (1988).

[19] Y. Le Cun, J. Denker, and S. Solla, *in* "Advances in Neural Information Processing Systems 2," p. 598. Morgan Kaufmann, San Mateo, California, 1990.

relevant documents which may or may not contain actual query terms. In SVD, the n-gram term matrix (i.e., term-by-sequence matrix) is decomposed into a set of k orthogonal factors from which the original matrix can be approximated by linear combination. The reduced model can be shown by

$$\mathbf{X} \cong \mathbf{Y} = \mathbf{TSP}' \tag{1}$$

where $\mathbf{X}$ is the original term matirx, $\mathbf{Y}$ is the approximation of $\mathbf{X}$ with rank k, $\mathbf{T}$ and $\mathbf{P}$ are the matrices of left and right singular (s) vectors corresponding to the k-largest s values, and $\mathbf{S}$ is the diagonal matrix of the k-largest s values. Note that if $\mathbf{X}$ is used to represent the original term matrix for training sequences, then $\mathbf{P}$ becomes the reduced matrix for the training sequences.

The representation of unknown sequences is computed by "folding" them into the k-dimensional factor space of the training sequences. The folding technique, which amounts to placing sequences at the centroid of their corresponding term points, can be expressed by

$$\mathbf{P}_u = \mathbf{X}_u'\mathbf{TS}^{-1} \tag{2}$$

where $\mathbf{P}_u$ and $\mathbf{X}_u$ are the reduced and original term matrices of unknown sequences, $\mathbf{T}$ is the matrix of left s vectors computed from Eq. (1) during the training phase, and $\mathbf{S}^{-1}$ is the inverse of $\mathbf{S}$, which reflects scaling by reciprocals of corresponding s values. It has been shown that, as with the n-gram sequence encoding method, the SVD method also satisfies the basic coding assumption.[16]

Neural Network Paradigm

Back-Propagation Networks. The back-propagation (BP) neural networks in GenCANS are three-layered, feed-forward networks[15] (Fig. 1). A feed-forward calculation (change of state function) is used to determine the output of each neuron as in

$$\mathbf{O}_i = f(\sum_j \mathbf{O}_j\mathbf{W}_{ij}) = f(\text{net}_i) = 1/[1 + e^{(-\text{net}_i+\theta)}] \tag{3}$$

where $\mathbf{O}_j$ represents the output from neuron j in the preceding layer, $\mathbf{W}_{ij}$ represents the connection weight between neurons i and j, net_i is the net input to each neuron, θ is a bias term, and f is a nonlinear activation (squashing) function. The back-propagation learning applies the generalized delta (δ) rule to recursively calculate the error signals and adjust the weights [Eqs. (4)–(6)]. The error signal at the output layer is given by

$$\delta_i = f'(\text{net}_i)(\mathbf{T}_i - \mathbf{O}_i) \tag{4}$$

where $\mathbf{T}_i$ is the target value, and f' (net$_i$) is the first derivative of the activation function. The error signal at the hidden layer is given by

$$\delta_j = f'(\text{net}_j) \, \Sigma_i \, \mathbf{W}_{ij} \delta_i \tag{5}$$

The error signals are then used to modify the weights by

$$\Delta \mathbf{W}_{ij(t+1)} = \eta \mathbf{O}_j \delta_i + \alpha \Delta \mathbf{W}_{ij(t)} \tag{6}$$

where η is the learning rate and α is the momentum term.

In the BP network, the size of the input layer (i.e., number of input units) is dictated by the sequence encoding schema chosen. The size is m^n with n-gram encoding; the size is the reduced dimension (k), if the n-gram vector is compressed by SVD. The output layer size is determined by the number of classes represented in the network, with each output unit representing one class. The hidden size is determined heuristically, usually being a number between input and output sizes. In GenCANS, the networks are trained using weight matrices initialized with random weights ranging from -0.3 to 0.3. Other network parameters include a learning factor of 0.3, a momentum term of 0.2, and a constant bias term of -1.0.

Pattern Selection Strategy. One major problem frequently encountered when using BP neural networks is the long training time. In BP, all patterns in the training set are usually presented equally to the neural network (i.e., uniform presentation). To speed up the BP training, a pedagogical pattern selection strategy[17] is used to favor the selection of patterns producing high error values to the disadvantage of the patterns already mastered by the network. In GenCANS, a modified EDR (error-dependent repetition) method is used, in which training proceeds as described[20] for the first 400 iterations, followed by another 100 iterations using all training patterns with uniform presentation.

Counterpropagation Networks. A modified counterpropagation (CP) algorithm[21] with supervised LVQ (learning vector quantizer) and dynamic node allocation is another learning paradigm used. The forward-only CP network has three layers (an input layer, a Kohonen layer, and a Grossberg outstar conditioning layer). As in BP networks, the sizes of the input and output (Grossberg) layers are the size of the input vector and the number of output classes, respectively. The size of the hidden (Kohonen) layer is configured dynamically during the course of training. In our modified algorithm, n (number of classes) Kohonen nodes are allocated initially, with their weights initialized to be the average input vector of each class. During the training phase, nodes are added dynamically when any given

[20] C. Cachin, *Neural Networks* **7**, 175 (1994).
[21] C. H. Wu, H. L. Chen, and S. C. Chen, *Applied Intelligence* in press.

training pattern fails to be assigned to the Kohonen units that represents its class. The weight vector of the newly added node is then assigned with the input vector of the untrainable pattern.

The network training involves two steps, to select a winner and to update the weight vector of the winner. The winner, selected on the basis of a spherical arc distance measure [Eq. (7)], is the Kohonen unit whose weight vector is closest to the input vector (i.e., with the highest weighted sum or smallest angle). The spherical arc distance between the normalized, unit-length input vector ($\mathbf{X}$) and weight vector ($\mathbf{W}_j$) is computed by

$$S_j = \Sigma \, x_i w_{ij} = \mathbf{X} \cdot \mathbf{W}_j = \cos \theta_j \tag{7}$$

where x_i is the activation level of input unit i, w_{ij} is the weight from input unit i to Kohonen unit j, S_j is the weighted sum for Kohonen unit j, and θ_j is the angle between X and W_j. The sum equals the dot product of the input and weight vectors.

A supervised learning that selects two winners with punishment mechanisms is used for weight update. If the first winner is correct, it is updated with positive weight [Eq. (8)]; otherwise, the winner is punished with negative weight [Eq. (9)] and the second winner is chosen. If the second winner is correct, it is updated with positive weight; if incorrect, a new unit is allocated. The weight adjustments for correct and incorrect winners are given by the Kohonen learning law:

$$w_{ic}(t + 1) = w_{ic}(t) + \alpha(x_i - w_{ic}) \tag{8}$$

$$w_{ic}(t + 1) = w_{ic}(t) - \alpha(x_i - w_{ic}) \tag{9}$$

where α is the learning rate ($0 < \alpha \leq 1.0$), and $w(t)$ and $w(t + 1)$ are the weights from the previous iteration and the current iteration, respectively. The Kohonen learning rate for weight adjustment is chosen as 0.2.

Implementation of Gene Classification Artificial Neural System

GenCANS Versions

Presently, we have implemented two versions of GenCANS, Gen-CANS_PIR for full-scale classification of protein sequences into PIR super-families and GenCANS_RDP for classification of rRNAs according to RDP phylogenetic classes.

GenCANS_PIR. The system was trained with the PIR database[22] (Release 44.0, March 31, 1995) using a modular network architecture[2] that

[22] D. G. George, W. C. Barker, H.-W. Mewes, F. Pfeiffer, and A. Tsugita, *Nucleic Acids Res.* **22**, 3569 (1994).

involves multiple independent neural networks to partition different protein functional groups. The database has three sections, PIR1 for annotated and classified entries, PIR2 for annotated but unclassified or tentatively classified sequences, and PIR3 for unverified entries. The classification in PIR is based on the superfamily concept. A superfamily is a group of proteins that share sequence similarity due to common ancestry, and sequences within a superfamily have a less than 10^{-6} probability of similarity by chance. The training set for the GenCANS_PIR consisted of all PIR1 entries in 3462 superfamilies, partitioned into 15 modules (Table I). During the training phase, each network module was trained separately using the sequences of known superfamilies; during the prediction phase, unknown sequences were classified on all modules with classification results combined. All PIR2 entries, which had tentative superfamily assignment and were at least 30 amino acids long, were used in the prediction set for system evaluation.

GenCANS_RDP. The system was trained with the RDP database[23] (Release 5.0, May 17, 1995). There are three collections of rRNA sequences, the SSU_Prok (prokaryotic small subunit), SSU_Euk (eukaryotic small subunit), and LSU (large subunit). A total of 220 SSU (177 Prok + 43 Euk) classes, containing 3285 (2,849 Prok + 436 Euk) sequence entries, were derived directly from the SSU_Prok.phylo and SSU_Euk.phylo phylogenetic listing files in the RDP database. Similarly, 15 LSU classes containing 72 sequences were derived from the LSU.phylo file. In the SSU classification scheme, no one class at the nonleaf level had more than 50 sequence entries (i.e., any class that had more than 50 entries at one given level was further subdivided into classes at its sublevel, unless it reached the leaf level). The classes may be at different levels in the tree.[17] The finest level of classification would be to use one neural network class to represent every single leaf node on the phylogenetic tree. This would yield a total of 287 classes for SSU sequences. The LSU sequences, on the other hand, were classified to the leaf level. Table II gives a partial list of the phylogenetic classes of SSU and LSU sequences. To test the system performance, a 3-fold cross-validation method[21] was used to partition the data set. The data were randomly divided into three approximately equal sized sets, and in each trial one set was used for prediction and the union of the others for training.

Program Structure

As shown in the GenCANS structure chart (Fig. 2), the system software has three major components: a preprocessor to create from input sequence

---

[23] B. L. Maidak, N. Larsen, M. J. McCaughey, R. Overbeek, G. J. Olsen, K. Fogel, J. Blandy, and C. R. Woese, *Nucleic Acids Res.* **22**, 3485 (1994).

TABLE I

NEURAL NETWORK MODULES OF GenCANS_PIR FOR PARTITIONING SUPERFAMILIES

Network module	Protein functional groups	Superfamilies		Number of entries[a]	
		Begin–end	Total	PIR1	PIR2
EO	Electron transfer proteins, oxidoreductases	1–209	209	1227	1880
TR	Transferases	210–478	269	926	828
HY	Hydrolases	479–739	261	972	1161
LI	Lyases, isomerases, ligases	740–934	195	763	446
PG	Protease inhibitors, growth factors, hormones, toxins	935–1185	251	1154	1493
IH	Immunoglobulin-related, heme carrier, chromosomal, ribosomal proteins	1186–1378	193	1794	2022
FL	Fibrous, contractile system, lipid-associated proteins, miscellaneous	1379–1632	254	1151	1810
PM	Plant, membrane, organelle proteins	1633–1838	206	533	862
BP	Bacterial proteins	1839–2059	221	432	599
AD	Bacteriophage, plasmid, yeast, animal DNA viral proteins	2060–2309	250	686	247
LH	Large DNA viral proteins, herpesvirus	2310–2481	172	463	70
LA	Large adenovirus, vaccinia proteins	2482–2640	159	314	87
AR	Animal RNA viral proteins	2641–2864	224	1378	557
PP	Plant viral, phage proteins	2865–3069	205	493	146
PH	Phage T, hypothetical proteins	3070–3462	393	286	494
	Total	1–3462	3462	12,572	12,702

[a] The sequence entries in PIR1 (Release 44.0) were used for training, and those in PIR2 for prediction.

files the training and prediction patterns, a neural network program to classify input patterns, and a postprocessor to summarize classification results. The preprocessor has subprograms for n-gram encoding (ngram), SVD computation (svd), and combining different sequence encoding algorithms (comb). The neural network program has subprograms for standard

TABLE II
PHYLOGENETIC CLASSES IN GenCANS_RDP[a]

Class number	Number of sequences	Class ID number[b]	Class name[b]
Small subunit (SSU)			
1	8	1.1.1	Archaea.Euryarchaeota.Methanococcales
2	19	1.1.2	Archaea.Euryarchaeota.Methanobacteriales
.	.	.	.
.	.	.	.
.	.	.	.
10	39	1.2	Archaea.Crenarchaeota
11	6	2.1	Bacteria.Thermophilic_Oxygen_Reducers
.	.	.	.
.	.	.	.
135	95	2.15.1.12.2	Bacteria.Gram_Positive_Phylum.High_G+C_Subdivision.Mycobacteria
.	.	.	.
.	.	.	.
.	.	.	.
177	10	2.15.5.16	Bacteria.Gram_Positive_Phylum.Bacillus-Lactobacillus-Streptococcus_Subdivision.Alicyclobacillus_Group
178	17	3.1.1.1	Eukaryotes.Eumycota.Ascomycotina.Plectomycetes
179	2	3.1.1.2	Eukaryotes.Eumycota.Ascomycotina.Loculoascomycetes
.	.	.	.
.	.	.	.
220	10	3.21	Eukaryotes.Microsporidia
Large subunit (LSU)			
1	6	1.1	Archaea.Euryarchaeota
2	4	1.1.1	Archaea.Euryarchaeota.Extreme_Halophiles
.	.	.	.
.	.	.	.
5	2	2.1	Bacteria.Flexibacter-Cytophaga-Bacteroides_Phylum
.	.	.	.
.	.	.	.
12	2	2.4.2	Bacteria.Gram_Positive_Phylum.Clostridial_Subdivision
13	11	3.1	Eukaryotes.Fungi_and_Protists
14	7	3.2	Eukaryotes.Plants
15	10	3.3	Eukaryotes.Animals

[a] Partial listing.

[b] The class ID number and name are directly taken from SSU_Prok.phylo, SSU_Euk.phylo, and LSU.phylo, the phylogenetic listing files in the RDP database, Release 5.0.

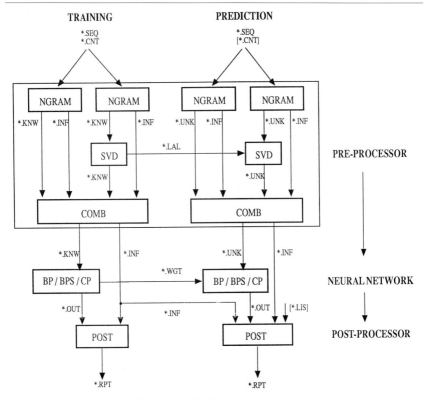

Fig. 2. GenCANS structure chart.

back-propagation (bp), back-propagation with pattern selection strategy (bps), and counterpropagation (cp). The postprocessor ranks all classification scores and provides identifications for top-ranking classes. All programs were coded in C language and implemented on the Cray Y-MP8/864 super-computer of the University of Texas System. The programs have also been ported to several other UNIX computer platforms, including a DEC alpha workstation and a Silicon Graphics workstation.

GenCANS (Version 3.0, 1996)[24] is a software package that provides various functions needed to use the classification systems in an integrated environment. Through either the command line or a menu interface, the user can work on either protein or DNA/RNA sequences, either training or prediction mode, either single-module or multiple-module neural networks,

[24] H. L. Chen and C. H. Wu, "GenCANS User's Guide." University of Texas Health Center at Tyler, 1996 (available on WWW server).

and either a single encoding algorithm or several combined algorithms. There are also many utility functions for tracking system performance, maintaining a system log, and converting file formats.

Program Availability

GenCANS is available on the Internet via the World Wide Web (WWW) (http://diana.uthct.edu) or by anonymous FTP from diana.uthct.edu. A distribution version of the GenCANS_RDP can be found in the compressed "tar" file (gencans_rdp.tar.Z). The version consists of all source codes needed for network prediction, the trained network weight files, sample input and output files, and README. The source programs were written in ANSI C and should be able to compile and run on any UNIX machines with a C/C++ compiler. During installation, a total of about 200 megabytes disk space is needed, after which some intermediate files would be removed (by the makefile), and the total disk space needed to store all files is less than 60 megabytes. The present distribution version can be used for speedy on-line rRNA sequence classification with the weight files obtained from off-line training. A distribution version for GenCANS_PIR will be prepared and made available from the same site. Future releases may contain programs for network training and prediction to provide both on-line training and classification capabilities.

Furthermore, our WWW server allows users to perform direct PIR or RDP sequence classifications by entering a sequence using a WWW client program, such as Mosaic or Netscape. The query will be configured to automatically generate and submit an appropriate query to be searched on our WWW server. The search results are then returned to the client as a Hypertext Markup Language (HTML) document (Table III).

Evaluation of Gene Classification Artificial Neural System

Evaluation Mechanism

The system performance is measured by its speed (CPU time) and predictive accuracy. The predictive accuracy is expressed with three terms: the total number of correct patterns (true positives), the total number of incorrect patterns (false positives), and the total number of unidentified patterns (false negatives). A sequence entry is considered to be accurately classified if one of its top-ranking classes matches the target class with a classification score above the threshold (i.e., the cutoff value). The classification score ranges from 1.0 for perfect match to 0.0 for no match. In this study, results for correct patterns are reported using two measures, the first

TABLE III
DIRECT GenCANS_RDP SEARCH ON WORLD WIDE WEB SERVER (http://diana.uthct.edu)

(A) Input to GenCANS_RDP (entered via WWW client)
Sequence Name/identifier:
rdp/Mc.jannasc Methanococcus jannaschii str.JAL-1.
Sequence[a]:
AUUCCGGUUGAUCCUGNNGGAGGCCACUGCUAUCGGGGUCCGACUAAGCCAUGCGAG
UCAAGGGGCUCCCUUCGGGGAGCACCGGCGCACGGCUCAGUAACACGUGGCUAACCUA
..........UUGCACACACCGCCCGUCACGCCACCCGAGUUGAGCCCAAGUGAGGCCCUGUC
CGCAAGGGCAGGGUCGAACUUGGUAACAAGGAACCUGGAUCACCUCC

(B) Classification result from GenCANS_RDP (returned as HTML document)
Sequence length: 1437

Class_ID	Class_Score	Class_Name
1.1.1.0.0	0.98	Archaea.Euryarchaeota.Methanococcales
1.1.4.0.0	0.48	Archaea.Euryarchaeota.Thermococcales
1.1.3.5.0	0.47	Archaea.Euryarchaeota.Methanomicrobacteria_and_Relatives.Archaeoglobales
1.1.5.0.0	0.45	Archaea.Euryarchaeota.Methanopyrales
1.1.2.0.0	0.45	Archaea.Euryarchaeota.Methanobacteriales

[a] Only a partial input sequence is shown.

fit (the class with the highest score matches the target) and the first n-fits (one of the classes with the n highest scores matches the target).

Performance of GenCANS_PIR

The result of the GenCANS_PIR version is shown in Table IV. The training of 12,572 sequences on the 15 network modules took a total of 2.2 Cray CPU (central processing unit) hours, which averaged to 0.63 sec per sequence. (Without the pedagogical pattern selection strategy, the training would be about five times longer.) Among all training patterns, 99.68% were trained into appropriate classes after 500 iterations. The majority of the remaining untrainable patterns belonged to single-membered or double-membered superfamilies, as one would expect. The prediction of an unknown sequence on all 15 networks took an average of 0.62 CPU second on a DEC alpha workstation. (The equivalent BLAST[25] search took 18 sec, or approximately 30 times longer, on the same workstation).

At a threshold of 0.1, 80.55% patterns were correctly classified as first fit. If we consider the top ten fits (among a total of 3462 possible classes)

[25] S. F. Altschul, W. Gish, W. Miller, E. W. Myers, and D. J. Lipman, *J. Mol. Biol.* **215**, 403 (1990).

TABLE IV
TRAINING AND PREDICTION RESULTS OF GenCANS

System	Encoding method[a]	Network algorithm[b]	Network configuration[c]	CPU time[d]		Predict accuracy	
				Train	Predict	First fit	n-fits[e]
PIR	a23e4_100	BPS	100 × 100 × (159–393)	0.63	0.62	80.55	87.72
RDP	n8_100	BPS	100 × 100 × 220	1.50	0.28	96.07	97.64
(SSU)	n8_100	CP	100 × (339–393) × 220	0.75	0.28	94.91	98.53
		Avg(BPS, CP)		—	0.29	96.26	98.74
RDP	n8_40	BPS	40 × 30 × 15	0.48	0.34	92.86	98.57
(LSU)	n8_40	CP	40 × (17–18) × 15	0.40	0.32	94.29	98.57
		Avg(BPS, CP)		—	0.35	92.86	100.00

[a] The a23e4_100 method used the a23e4 n-gram method, followed by the SVD method to generate an input vector of 100 dimensions. The a23e4 n-gram method concatenated three separate n-gram vectors, namely, a2 (bigrams of amino acids), a3 (trigrams of amino acids), and e4 (tetragrams of exchange groups), and formed a vector of 9696 units (i.e., $9696 = 20^2 + 20^3 + 6^4$). The n8_100/40 method used the n8 (octagrams of nucleic acids) n-gram method, followed by the SVD compression to reduce the input vector from 65,536 (i.e., 4^8) to 100/40 dimensions.

[b] BPS, Back-propagation with pattern selection strategy; CP, counterpropagation; Avg(BPS, CP), classification results obtained from averaging the BPS and CP classification scores.

[c] The PIR system has 15 neural network modules, each with a configuration of $100 \times 100 \times n$, where n is the number of protein superfamilies in the module, ranging from 159 to 393 (Table I). The hidden size of the CP networks used in the RDP system was configured dynamically and varied for different data sets in ranges as indicated.

[d] The CPU time indicated is the average time per sequence and includes the total time for preprocessing, neural network learning/mapping, and postprocessing. The training time unit is Cray CPU second (on a Cray Y-MP), and the prediction time unit is DEC CPU second (on a DEC 3000 workstation).

[e] n-fits is the top ten fits (out of 3462 superfamilies) for the PIR system, five fits (out of 220 phylogenetic classes) for the RDP SSU system, and two fits (out of 15 classes) for the RDP LSU system.

as correct classification, then the predictive accuracy of the full-scale system was 87.72%. To make the "megaclassification" helpful, one can use a much higher threshold to reduce the number of false positives. At a threshold of 0.9, although only 51.54% patterns were correctly classified, the incorrectly classified entries were reduced to 0.23%. The remaining entries were not classified at this threshold. Again, most of the entries that failed to be classified (correctly predicted) by the neural nets were those in single-membered or double-membered superfamilies, which account for one-half and one-sixth of all PIR superfamilies, respectively. When only the 50 largest superfamilies were used, more than 98% of the sequences were correctly classified as the first fit at the threshold of 0.1; close to 90% of the entries were classified with a classification score of more than 0.9, with no false positives.[16] It should also be noted that not all entries "misclassified" by the neural network are truly misclassified. Indeed, a few erroneous superfamily placements have been identified (W. C. Barker, National Biomedical Research Foundation-PIR, personal communication, 1995) by the neural network using the high threshold of 0.9.

Performance of GenCANS_RDP

The training of the entire RDP system (for 3285 SSU and 72 LSU sequences on both BPS and CP networks) took about 1.4 Cray CPU hours. The training time for individual networks is shown in Table IV. The prediction of an unknown sequence took approximately 0.3 DEC CPU second (Table IV), about an order of magnitude faster than other search methods, including BLAST and Similarity Rank.[17] The predictive accuracy measured using 3-fold cross-validation indicates that the BP and CP networks had similar accuracy, and their combination yielded best results, close to 99% for SSU and 100% for LSU prediction (Table IV). The time required to obtain the combined results [i.e., Avg(BPS, CP)] is about the same as that for individual BPS or CP networks, because the actual network prediction took very little time. As observed before,[17] for query sequences whose correct class appeared as second or third fit, their first- and second-fit classes were also closely located on the phylogenetic tree.

Discussion

This chapter describes two neural network sequence classification systems, GenCANS_PIR for PIR superfamily placement of protein sequences and GenCANS_RDP for RDP phylogenetic classification of rRNA sequences. The major applications of the classification neural networks are rapid sequence annotation and automated family assignment.

Currently, the classification speed of GenCANS is about an order of magnitude faster than other methods. The rate gap will continue to widen with the accelerated growth of molecular sequence databases. Unlike most other sequence comparison or database search methods in which search time is directly proportional to the number of sequence entries in the database, neural network classification should lead to search times that are expected to remain low even if there is a 100- to 1000-fold increase of sequence entries. This is because the total network classification time is dominated by the preprocessing time (i.e., more than 95% of the total time), and the n-gram and SVD computation time required for preprocessing query sequences is determined by the size of the n-gram vectors but is independent of the number of training sequences.

The full-scale PIR classification system can be used as a filter program for other database search methods to minimize the time required to find relatively close relationships. Similarly, classification of unknown rRNA sequences into predefined classes on a phylogenetic tree, without sequence alignment, is a rapid means of sequence annotation. Ribosomal RNA sequences are now the method of choice for surveying the biosphere. The rRNA classification system can be used to screen large collections of short rRNA sequences, such as those produced by environmental characterizations, in order to identify those worthy of more complete sequence determination.

The neural classification system can also be used to automate family assignment. The tool is generally applicable to any molecular databases that are developed according to family relationships because the neural networks employ supervised learning algorithms. An automated classification tool is especially important for the organization of databases according to family relationships and for handling the influx of new data in a timely manner. Among all entries in the PIR database, only a small fraction are classified and placed in superfamilies. The neural network system is currently being used by the PIR database for superfamily identification of unclassified sequences. Likewise, the size of the RDP database continues to grow, having doubled over the past 5 years, and the phylogenetic tree placement gradually lags. The neural network tool can be used by the RDP database to automate family assignment and help build the phylogenetic tree.

There are two major directions for future research: improving the classification accuracy and extending the current designs for developing a gene identification system. The current *n*-gram sequence encoding method, although effective in preserving sequence similarity, encodes global information only. The capability to encode local information conserved in the motif regions, however, is essential, since information content from the entire sequence string is not equal. We have developed a new database search algorithm, termed MOTIFIND (Motif Identification Neural Design),[26] for rapid and sensitive protein family identification. The method employs an *n*-gram term weighting algorithm for extracting motif patterns and integrated neural networks for combining global and local information, and has shown a significant improvement in predictive accuracy. The design of the neural system can be easily extended to classify other nucleic acid sequences. Preliminary studies have been conducted to classify DNA sequences (containing both protein-encoding regions and intervening sequences) directly into protein superfamilies with satisfactory results. It is therefore possible to develop a gene identification system that can classify indiscriminately sequenced DNA fragments.

[26] C. H. Wu, S. Zhao, H. L. Chen, C. J. Lo, and J. W. McLarty, CABIOS, in press.

Acknowledgments

This research is supported in part by Grant R29 LM055234 from the National Library of Medicine, and by the University Research and Development Grant Program of Cray Research, Inc.

[6] Blocks Database and Its Applications

By Jorja G. Henikoff and Steven Henikoff

Introduction

The detection of homology between a newly determined sequence and a sequence in a databank is often the most important clue to the function of a gene, and so the homology search has become a standard tool for molecular biologists. This tool increases in power as data banks expand with representatives from a large percentage of all protein families,[1] so that most newly discovered sequences now have recognizable homologs in current data banks.[2] Such successes have fueled large-scale sequencing projects, including those involving single-pass sequencing of cDNAs[3] and those of model organisms with high gene densities.[4] As a result of these activities, the sequence data banks have become more complex, and this complexity can complicate the interpretation of homology search results.

Here we discuss a homology searching system that is based on a database of protein family representations rather than sequences. The representations consist of blocks of aligned segments derived from the most highly conserved regions of proteins. This concentration of information from conserved alignments can improve the detection of relationships in difficult situations. Furthermore, because the number of protein families is increasing slowly relative to the number of sequences, this system does not become substantially more complex with time; rather, the families become more informative.

Since its introduction in 1991,[5] the system has been improved in several

---

[1] P. Green, *Curr. Opin. Struct. Biol.* **4,** 404 (1994).

[2] E. V. Koonin, P. Bork, and C. Sander, *EMBO J.* **13,** 493 (1994).

[3] M. D. Adams, J. M. Kelley, J. D. Gocayne, M. Dubnick, M. H. Polymeropoulos, H. Xiao, C. R. Merril, A. Wu, B. Olde, R. F. Moreno, A. R. Kerlavage, W. R. McCombie, and J. C. Venter, *Science* **252,** 1651 (1991).

[4] S. G. Oliver, Q. J. M. van der Aart, M. L. Agostoni-Carbone, M. Aigle, L. Alberghina, D. Alexandraki, G. Antoine, R. Anwar, and J. P. G. Ballesta, *Nature* (*London*) **357,** 38 (1992).

[5] S. Henikoff and J. G. Henikoff, *Nucleic Acids Res.* **19,** 6565 (1991).

Copyright © 1996 by Academic Press, Inc.
All rights of reproduction in any form reserved.

ways. The searching program was modified to incorporate a strategy for the detection of multiple blocks representing a group[6] and to utilize improved sequence weights to reduce redundancy within a protein family.[7] Furthermore, the system has been applied to large DNA sequences, such as those of whole chromosomes,[8] and has been made accessible via electronic mail (E-mail)[9] and the World Wide Web (WWW).[10] The WWW server also provides sequence logos[11] for intuitive graphical display of blocks. In addition to its use in homology searching, the database itself has been utilized in the construction of amino acid substitution matrices[12] and in the evaluation of sequence weighting methods.[7]

Motifs to Blocks

Most protein families can be characterized by sets of locally similar sequence segments, typically referred to as motifs. Within a single sequence, a motif consists of a contiguous stretch of amino acids shared by a group of proteins and conserved for function. This definition includes motifs found in families of related proteins as well as in collections of unrelated proteins that share a common structural feature. An example of the second kind of motif is the helix–turn–helix DNA-binding motif that is thought to have converged to form a similar 20-residue structure with weak sequence constraints.[13] The terms signature and pattern are often used as synonyms for motif. Motifs represent local as opposed to global features of a protein, and a protein sequence can contain multiple motifs.

Representations of a motif are based on a local multiple alignment of the group of proteins that share it in the region containing the motif. We call a local multiple alignment made without insertions or deletions (gaps) in any of the sequences a protein block. Its width is that of the conserved region, and its depth is the number of sequences in the group that share the motif. The Blocks Database contains blocks representing motifs found in known families of related proteins. Currently, the families represented in the Blocks Database are those documented in the Prosite Database,[14]

[6] S. Henikoff and J. G. Henikoff, *Genomics* **19,** 97 (1994).

[7] S. Henikoff and J. G. Henikoff, *J. Mol. Biol.* **243,** 574 (1994).

[8] S. Henikoff and J. G. Henikoff, *Proc. 27th Hawaii Int. Symp. Systems Sci.,* 265 (1994).

[9] S. Henikoff, J. G. Henikoff, S. Agus, and J. C. Wallace, *in* "Automated DNA Sequencing and Analysis Techniques" (M. D. Adams, C. Fields, and J. C. Venter, ed.), p. 313. Academic Press, San Diego, 1993.

[10] S. Henikoff, J. G. Henikoff, W. J. Alford, and S. Pietrokovski, *Gene* **163,** GC17 (1995).

[11] T. D. Schneider and R. M. Stephens, *Nucleic Acids Res.* **18,** 6097 (1990).

[12] S. Henikoff and J. G. Henikoff, *Proc. Natl. Acad. Sci. U.S.A.* **89,** 10915 (1992).

[13] I. B. Dodd and J. B. Egan, *Nucleic Acids Res.* **18,** 5019 (1990).

[14] A. Bairoch, *Nucleic Acids Res.* **20,** 2013 (1992).

but the concept of cataloging blocks for other families and for unrelated groups of proteins that share a motif is completely general.

The Blocks Database provides actual multiple alignments, as do the PRINTS[15] and ProDom[16] databases. These differ from some other databases in general use, such as those that provide consensus sequences,[17] profiles,[14,18] patterns,[14] and fingerprints.[19] An advantage of providing actual alignments is that they contain all of the information needed to derive any of these other representations. The Blocks Database is constructed automatically once the protein groups are identified, whereas some other searchable databases of motif representations have been constructed using a combination of manual and automatic techniques.[15,20] The ProDom database[16] is similar to the Blocks Database in that it is automated and provides multiple alignment representations; however, the groups represented in ProDom are themselves derived automatically by clustering the database into groups on the basis of sequence similarity,[17] whereas the Blocks Database uses protein families identified manually on the basis of multiple criteria.

Constructing the Blocks Database

The two-step PROTOMAT system constructs the Blocks Database using a fully automated procedure.[5] The only input necessary for PROTOMAT to produce blocks is a group of protein sequences that presumably have one or more motifs in common. Currently, the Blocks Database is based on the Prosite catalog of protein families[14] and its corresponding SWISS-PROT Database of protein sequences.[21] Each entry in Prosite includes a list of known protein sequences in SWISS-PROT for the family. For each Prosite entry, PROTOMAT makes a set of one or more blocks from this list of sequences. Prosite also includes a pattern for each family, but this pattern is not used in any way by PROTOMAT; one of the blocks constructed by PROTOMAT may contain all or part of the Prosite pattern, but this is not guaranteed since the PROTOMAT blocks are found by an

[15] T. K. Attwood and M. E. Beck, *Protein Eng.* **7,** 841 (1994).

[16] E. L. L. Sonnhammer and D. Kahn, *Protein Sci.* **3,** 482 (1994).

[17] R. F. Smith and T. F. Smith, *Proc. Natl. Acad. Sci. U.S.A.* **87,** 118 (1990).

[18] M. Gribskov, A. D. McLachlan, and D. Eisenberg, *Proc. Natl. Acad. Sci. U.S.A.* **84,** 4355 (1987).

[19] J. T. L. Wang, T. G. Marr, D. Shasha, B. A. Shapiro, and G.-W. Chirn, *Nucleic Acids Res.* **22,** 2767 (1994).

[20] S. Pongor, V. Skerl, M. Cserzo, Z. Hatsagi, G. Simon, and V. Bevilacqua, *Nucleic Acids Res.* **21,** 3111 (1993).

[21] A. Bairoch and B. Boeckmann, *Nucleic Acids Res.* **20,** 2019 (1992).

entirely different procedure than that used to derive Prosite patterns. The high quality of Prosite and its detailed documentation make it an outstanding source of related groups on which to base the Blocks Database.

Starting with a list of sequences from related proteins, PROTOMAT first generates a large number of candidate blocks using a motif-finding algorithm. The MOTIF algorithm of Smith *et al.*[22] is used for this step, but other motif finders such as the Gibbs sampler[23] could be used, and may be used in the future, to make the Blocks Database. MOTIF scans the sequences exhaustively, looking for spaced triplets of amino acids out to a maximum distance in at least a subset of the sequences. Examples of spaced triplets are Ala-Ala-Ala and Val-x-x-x-Ala-x-Cys, where x represents any amino acid. A spaced triplet found in enough sequences anchors a local multiple alignment against which sequences lacking the triplet are aligned to maximize a block score. All parameters for MOTIF are determined empirically using the characteristics of the sequences. To maximize the sensitivity of MOTIF, we allow the subset of sequences that must contain a spaced triplet to be so small that a few triplets are found even when the sequences are shuffled by randomly permuting each amino acid.

Candidate blocks are passed on to the MOTOMAT block assembly program. At this stage the candidates may overlap one another in all or some of the sequences. MOTOMAT refines them by merging those that overlap consistently in all of the sequences and extending them in both directions until similarity falls off, out to a maximum width. After refinement, MOTOMAT uses graph theory techniques to assemble a best set of blocks that occur in the same order within a sequence, without overlapping, in a significant number of sequences. Sequences that do not conform are dropped from the set of blocks.

After a block is made, weights are computed for each sequence segment in it. These weights are intended to compensate for overrepresentation of some of the sequences. Low weights are given to redundant sequences, and high weights to diverged sequences. There are many methods for computing sequence weights,[24] but the Blocks Database currently uses position-based sequence weights, which are easy to compute and have been shown to perform well in searching applications[7] (see Fig. 6). These are simple weights derived from the number of different residues and their frequency in each position of the block.

[22] H. O. Smith, T. M. Annau, and S. Chandrasegaran, *Proc. Natl. Acad. Sci. U.S.A.* **87,** 826 (1990).

[23] C. E. Lawrence, S. F. Altschul, M. S. Boguski, J. S. Liu, A. F. Neuwald, and J. C. Wootton, *Science* **262,** 208 (1993).

[24] M. Vingron and P. R. Sibbald, *Proc. Natl. Acad. Sci. U.S.A.* **90,** 8777 (1993).

No gaps are inserted in the block alignments by PROTOMAT, and some of the sequences may be imperfectly aligned, especially near the edges of the blocks. PROTOMAT was designed to make blocks for database searching applications, and occasional misalignments can be tolerated so long as the block is correctly aligned for the large majority of sequences. In such cases a contribution from misaligned segments will be diluted out.

Blocks 8.0 was derived from Prosite 12 and SWISS-PROT 29 and contains 2884 blocks representing 770 different protein families, an average of 3.75 blocks per family. These blocks contain from 2 to 507 sequences, and they range in width from 4 to a maximum of 55 amino acids. Figure 1 contains a sample entry from Blocks 8.0.

Users can make blocks from their own group of protein sequences using PROTOMAT implemented on the Block Maker server. Block Maker-generated blocks are appropriate for searching sequence data banks for new family members[25] and for designing PCR (polymerase chain reaction) primer.[26]

Searching the Blocks Database

The Blocks Database is used primarily to aid in the detection and interpretation of protein sequence homology. A protein or DNA sequence is compared to the protein blocks to see if the sequence belongs to any family represented in the Blocks Database. The rationale behind searching a database of blocks is that information from multiply aligned sequences is present in a concentrated form, reducing background and increasing sensitivity to distant relationships. If a hit is found, the Prosite documentation for the family may provide insights as to the function of the sequence.

A block can be searched either as a query compared with a target database of sequences or as an entry in a target database that is compared with a query sentence. In either case, the block must be converted to a representation that can be scored against a sequence, such as a pattern, consensus sequence, or profile. In this representation, the block is aligned with a sequence at every possible position and its score computed.

The Blocksearcher system uses the BLIMPS (blocks improved searcher) program followed by the BLOCKSORT analysis program to perform a search of a query sequence against the Blocks Database,[6] although other methods have also been used.[27] Each block in the database is converted to a position-specific scoring matrix (PSSM), which is similar to a profile.[18]

[25] S. Henikoff, *New Biol.* **4,** 382 (1992).
[26] M. D'Esposito, G. Pilia, and D. Schlessinger, *Hum. Mol. Genet.* **3,** 735 (1994).
[27] R. Fuchs, *CABIOS* **9,** 587 (1993).

```
ID    GLUTAREDOXIN; BLOCK
AC    BL00195C; distance from previous block=(4,14)
DE    Glutaredoxin proteins.
BL    VIG motif; width=31; seqs=11; 99.5%=581; strength=2254
GLRX_ECOLI (   45) KEDLQQKAGKPVETVPQIFVDQQHIGGYTDF    74

  THIO_BPT4 (   51) LLTKLGRDTQIGLTMPQVFAPDGSHIGGFDQ  100

  YRUB_CLOPA (   37) KEREEMRSLSKQSGVPVINIDGNIIVGFNKA   94

  GLRX_BOVIN (   55) EIQDYLQQLTGARTVPRVFIGQECIGGCTDL   27
  GLRX_HUMAN (   55) EIQDYLQQLTGARTVPRVFIGKDCIGGCSDL   25
    GLRX_PIG (   55) EIQDYLQQLTGARTVPRVFIGKECIGGCTDL   26
  GLRX_RABIT (   55) EIQDYLQQLTGARTVPRVFLGKDCIGGCSDL   28

  GLRX_VACCC (   56) ELRDYFEQITGGRTVPRIFFGKTSIGGYSDL   31
   GLRX_VARV (   56) KLHDYFEQITGGRTVPRIFFGKTSIGGYSDL   34

  GLRX_YEAST (   60) EIQDALEEISGQKTVPNVYINGKHIGGNSDL   39
  YCD5_YEAST (   61) DIQAALYEINGQRTVPNIYINGKHIGGNDDL   50
```

FIG. 1. Sample entry from Blocks 8.0 for the glutaredoxin family of proteins as documented in Prosite 12.0. The protein sequences used to make the block were extracted from SWISS-PROT 29. The ID, AC, and DE lines are adapted from Prosite; the ID and DE lines contain a short and longer description of the family, respectively. The AC line contains the Blocks Database accession number, in this case BL00195C, which is adapted from the Prosite AC, PS00195. The "C" indicates that this is the third block for the family in the Blocks Database; the preceding blocks will end in "A" and "B." The comment "distance from previous block = (4,14)" means that among the 11 sequences included in BL00195C, the minimum distance from the end of BL00195B to the beginning of BL00195C was 4 residues and the maximum was 14. The BL line contains the name of the initial motif from the MOTIF program around which this block was constructed, the width of the block (31), the number of sequences in the block (11), the 99.5th percentile score of presumed true-negative sequences when the block was searched against SWISS-PROT 29 (581), and the median calibrated score of known true-positive sequences, called strength, from the same search (2254). During a search of a sequence against the Blocks Database, the raw score is divided by the 99.5% score so that diverse blocks can be compared, and the strength score is reported as an aid to evaluating a possible hit. Following the BL line are the aligned segments of the 11 sequences included in the block. The SWISS-PROT sequence name is followed by the location of the first residue in parentheses, the segment, and the position-based sequence weight. The sequence weights are normalized so that the most distant sequence receives a weight of 100. The segments are clustered into groups separated by blank lines. Segments appear in the same cluster if any two have at last 80% identical residues.

A PSSM has as many columns as there are positions in the block, and 20 rows, one for each amino acid. It also contains additional rows for other characters that may be encountered in a protein or translated DNA sequence: B (D or N), Z (E or Q), X (unknown residue), - (gap), and * (stop codon). Each PSSM entry consists of numeric scores that are based on the ratio of the observed frequency of an amino acid in a block column to its expected overall frequency in SWISS-PROT (odds ratio). The observed

```
ID   GLUTAREDOXIN; MATRIX
AC   BL00195C; distance from previous block=(4,14)
DE   Glutaredoxin proteins.
MA   VIG motif; width=31; seqs=11; 99.5%=581; strength=2254
```

A	B	C	D	E	F	G	H	I	K	L	M	N	P	Q	R	S	T	V	W	X	Y	Z	*	-
0	6	0	11	33	0	0	0	0	41	13	0	0	0	0	0	0	0	0	0	0	0	20	0	0
0	0	0	0	33	0	0	0	44	0	22	0	0	0	0	0	0	0	0	0	0	0	20	0	0
0	7	0	12	0	0	0	13	0	0	0	0	0	0	40	20	0	14	0	0	0	0	16	0	0
7	25	0	46	17	0	0	0	0	20	9	0	0	0	0	0	0	0	0	0	0	0	10	0	0
11	0	0	0	14	0	0	0	0	0	10	0	0	0	17	0	0	0	0	0	0	48	15	0	0
0	0	0	0	0	15	13	0	0	0	19	36	0	0	17	0	0	0	0	0	0	0	7	0	0
0	0	0	15	0	0	0	0	12	0	0	0	0	24	34	0	0	0	0	0	0	14	19	0	0
10	10	0	19	14	0	0	0	0	0	0	0	0	43	0	13	0	0	0	0	0	0	25	0	0
0	0	0	0	0	0	13	0	36	0	28	0	0	0	0	0	22	0	0	0	0	0	0	0	0
0	5	0	0	0	0	0	0	13	0	0	12	0	25	0	19	30	0	0	0	0	10	0	0	0
0	0	0	0	0	42	0	21	19	0	0	17	0	0	0	0	0	0	0	0	0	0	0	0	0
15	0	0	0	0	0	25	0	0	0	0	0	0	48	0	0	0	12	0	0	0	19	0	0	0
0	0	0	0	14	0	0	0	0	8	13	0	0	0	49	16	0	0	0	0	0	0	9	0	0
0	0	0	0	0	0	15	0	0	0	0	0	0	0	0	0	84	0	0	0	0	0	0	0	0
0	0	0	0	0	0	0	0	0	0	0	39	0	0	0	0	0	60	0	0	0	0	0	0	0
0	0	0	0	0	0	0	0	0	0	0	0	0	99	0	0	0	0	0	0	0	0	0	0	0
0	8	0	0	0	0	0	0	0	0	0	18	0	39	29	0	0	13	0	0	0	15	0	0	0
0	0	0	0	0	0	0	0	57	0	0	0	0	0	0	0	0	42	0	0	0	0	0	0	0
0	7	0	0	0	63	0	0	0	0	0	16	0	0	0	0	0	0	0	0	20	0	0	0	0
14	0	0	0	0	18	0	51	0	3	0	0	0	0	0	0	0	12	0	0	0	0	0	0	0
0	28	0	33	0	0	25	0	0	0	0	21	20	0	0	0	0	0	0	0	0	0	0	0	0
0	11	0	20	0	0	27	0	0	26	0	0	0	26	0	0	0	0	0	0	0	10	0	0	0
0	15	0	10	8	0	14	0	0	15	0	0	21	0	18	0	0	11	0	0	0	0	12	0	0
0	0	34	0	0	0	0	42	10	0	0	0	0	0	0	0	13	0	0	0	0	0	0	0	0
0	0	0	0	0	0	0	36	63	0	0	0	0	0	0	0	0	0	0	0	0	0	0	0	0
0	0	0	0	0	0	59	0	23	0	0	0	0	0	0	0	0	18	0	0	0	0	0	0	0
0	0	0	0	0	0	99	0	0	0	0	0	0	0	0	0	0	0	0	0	0	0	0	0	0
0	5	37	0	0	15	9	0	0	0	0	12	0	0	0	0	0	0	0	0	27	0	0	0	0
0	14	0	9	0	25	0	0	0	0	0	21	0	0	0	22	22	0	0	0	0	0	0	0	0
0	45	0	83	0	0	0	0	16	0	0	0	0	0	0	0	0	0	0	0	0	0	0	0	0
14	0	0	0	0	22	0	0	0	0	33	0	0	0	29	0	0	0	0	0	0	11	0	0	0

FIG. 2. Position-specific scoring matrix (PSSM) computed for the block in Fig. 1 using position-based weights and odds ratios. The PSSM is rotated for display, with the 31 columns of the block appearing as rows and one column for each possible amino acid, plus values for B, Z, X, -, and *. The conserved P in the 16th column and G in the 27th receive the maximum score of 99. During a search of a sequence against the Blocks Database, the block is aligned with the query sequence and the alignment scored by adding values from a PSSM such as this.

frequencies are weighted for sequence redundancy using the sequence weights in the block, and pseudocounts (described below) are added to compensate for nonobserved amino acids. Blocksearcher uses a variation of the data-dependent pseudocount method.[28] PSSM entries are normalized so that each entry is an integer between 0 and 99 (Fig. 2).

During a search, the query sequence is aligned with each block at all possible positions, and an alignment score is computed by adding the scores for each position. Each block is scored individually, including multiple blocks for the same family, and the top scores are saved. If the sequence is DNA, BLIMPS translates it in all six possible frames and scores all six translated sequences against the block. BLIMPS scores all possible

[28] R. L. Tatusov, S. F. Altschul, and E. V. Koonin, *Proc. Natl. Acad. Sci. U.S.A.* **91,** 12091 (1994).

alignments and saves all high-scoring ones, so that repeated motifs can be detected.

Before scores resulting from a search of a query sequence against the Blocks Database can be used to rank the blocks relative to one another, the blocks must be calibrated to provide standardized scores that are comparable between blocks. This is because the blocks in the database vary both in width and in the number of sequences they contain. For example, the raw score for a wide block is likely to be larger than that for a narrow block just because the wide block has more columns to add up. Each block in the Blocks Database has two standard scores for comparison (Fig. 1). They are obtained by searching each block against the version of SWISS-PROT from which the sequences making the block were extracted and then analyzing the raw scores. Using Prosite's list of known members of the block's family as the set of true-positive sequences for the block and assuming all other sequences are true negatives, we compare the score distributions of these two sets of sequences. Perfect separation of the two distributions, in which the highest-scoring true-negative sequence scores lower than the lowest scoring true-positive sequence, is not always obtained, sometimes because of the presence of uncatalogued true positives. To allow for errors such as this, the 99.5th percentile of the true-negative sequence scores is chosen as a "lower" calibration score. Scores above this are likely to be interesting, and those below are likely to be true negatives. Blocksearcher divides the raw score by this lower calibration score, multiplies the result by 1000, and ranks the search results by this calibrated score. Calibrated scores above 1000 can therefore be considered higher than 99.5% of scores for known true-negative sequences.

The median calibrated score for the known true-positive sequences is reported as the strength of the block. Strength is a measure of how highly true-positive sequences score against the block for comparison with the query sequence score. A low value for strength indicates that the block is weak and may fail to exclude chance alignments. Alternatively, a block that is too strong may be so specific that it will exclude distant relatives.

The BLOCKSORT analysis program provides additional aids for evaluating the results of a search against the Blocks Database.[6] BLOCKSORT processes the BLIMPS search results and groups together all blocks from the same family. The calibrated BLIMPS score along with strength are useful for evaluating the local similarity between a query sequence and a single block, but it is important to interpret them in terms of a reasonable model of chance. To assess both the effectiveness of the calibrated score and its significance, we took 7082 sequences from SWISS-PROT that were not represented in the Blocks Database, shuffled each one by random permutation of each amino acid, and searched each shuffled sequence against the Blocks Database. The resulting distribution of calibrated scores for first-ranking hits showed that all blocks were about equally likely to score at or above a given level, demonstrating that the calibration step is effective. It also provided BLOCKSORT with a percentile value to assign to the calibrated score for a real search.

If a query sequence obtains a good score for more than one block from the same family in the Blocks Database, and if these blocks are in the same order and similar distances apart as they are in the sequences in the blocks, then this provides strong confirmatory evidence that the query may belong to the family.[25] A multiple block hit contains information about global similarity between the query and members of the family. It is very difficult to construct a realistic theoretical model for scoring multiple block hits because different protein groups include different numbers of members, because groups are represented in the Blocks Database by different numbers of blocks with diverse properties, and because not necessarily all of the blocks for a group may score high in a search. Therefore, BLOCKSORT computes an "expectant value" on the basis of an intuitive model that has been verified empirically to quantify the degree of global similarity seen in a multiple block hit.[6]

A nomogram showing the frequency of first-ranking hits with various expectant values observed in the 7082 searches of shuffled sequences is provided with the BLOCKSORT output to assist evaluation (Fig. 3). The nomogram presents a level of confidence given a combination of a local score (or percentile) and global expectant value. Although the nomogram is based on shuffled sequences, searches carried out with the same 7082 sequences, but not shuffled, provided a very similar distribution of scores, once the uncatalogued true-positive sequences were pruned from the results list. This confirms that confidence levels based on search results for randomized sequences are realistic and can be applied to the evaluation of search results for real sequences.

An example illustrates how the nomogram can be used to judge whether a hit reflects significant similarity (Fig. 4). The protein encoded by the *Saccharomyces cerevisiae INO2* gene, which is involved in the regulation of phospholipid metabolism, has been shown by extensive genetic and biochemical evidence to contain a helix–loop–helix DNA-binding motif.[29] Nevertheless, the level of similarity to other proteins that contain this motif is so low that BLAST[30] and Smith–Waterman[31] searches with the INO2

[29] D. M. Nikoloff and S. A. Henry, *J. Biol. Chem.* **269**, 7402 (1994).

[30] S. F. Altschul, W. Gish, W. Miller, E. W. Myers, and D. J. Lipman, *J. Mol. Biol.* **215**, 403 (1990).

[31] T. F. Smith and M. S. Waterman, *J. Mol. Biol.* **147**, 195 (1981).

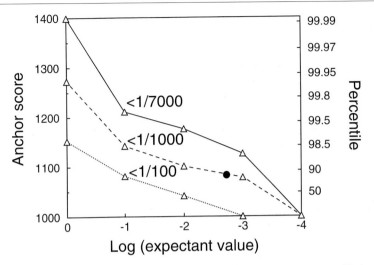

FIG. 3. Nomogram returned with search results that is used to evaluate how likely a first hit could have occurred by chance. The anchor block is the single highest-ranking block in the set of blocks from a single family, and it is provided with a percentile score. A nomogram of this type is general for any sequence query and any version of the Blocks Database, but it is specific for the methods used to construct PSSMs representing blocks in the database (Fig. 2). This nomogram is for PSSMs that are based on position-based weights and pseudocounts derived from BLOSUM 62 (J. G. Henikoff and S. Henikoff, 1995, unpublished results), which is implemented in Block Searcher; it differs from the nomogram reported previously for PSSMs based on 80% clustering weights and odds ratios.[6] The solid circle shows the intersection of the 80th percentile and $p = 0.002$ found for hit 2 using INO2_YEAST as query (see Fig. 4).

protein sequence fail to detect any of the 103 members of this family catalogued in Prosite 8.0. Consistent with such a low level of similarity, Blocksearcher ranks the helix–loop–helix block A (BL00038A, the "anchor" block) in second place at only the 80th percentile (Fig. 4), high enough to be reported but not high enough to be considered interesting. However, the 23rd-ranking segment in the search aligns with BL00038B at a compatible distance (10 amino acids) downstream of BL00038A, providing an expectant value (approximating a p value) of $p < 0.0021$ for BL00038B in support of BL00038A. The combined probability of obtaining one block at the 80th percentile and other blocks in support at $p < 0.0021$ ($=10^{-2.7}$) is seen from the nomogram (filled symbol in Fig. 3) to occur on the average as the first hit in only 1 search in 1000, consistent with the experimental evidence that this region is truly a helix–loop–helix DNA-binding domain.

This example also illustrates a potential complication of interpreting

```
Query=INO2_YEAST ,
 Size=304 Amino Acids
Database=mats.dat, Blocks Searched=2884

1.-----------------------------------------------------------------------
Block        Rank Frame Score Strength   Location (aa) Description
BL00482A      1    0    1084  1419        149-     161 Dihydroorotase proteins.

1084=86.22th percentile of anchor block scores for shuffled queries
P not calculated for single block BL00482A
                        |--- 107 amino acids---|
   BL00482 AAA:::.B:::::::::::::::::::::::::::::::::::::::::::::::::::........CC
INO2_YEAST AAA

BL00482A      <->A     (11,1486):148
PYR1_DICDI 1437        DVHVHLREPGATH
                       |  | | |    |
INO2_YEAST 149         esHLHiRSPKkqH


2.-----------------------------------------------------------------------
Block        Rank Frame Score Strength   Location (aa) Description
BL00038A      2    0    1076  1277        237-     260 Myc-type, 'helix-loop-he
BL00038A    389    0     922  1277        248-     271 Myc-type, 'helix-loop-he
BL00038B     23    0     979  1264        271-     291 Myc-type, 'helix-loop-he

1076=80.80th percentile of anchor block scores for shuffled queries
P<0.0021 for BL00038B in support of BL00038A
                   |--- 38 amino acids---|
   BL00038 AAAAAAAAAAAAAAAAA:::.·..........................BBBBBBBBBBBBBBB
INO2_YEAST AAAAAAAAAAAAAAAAA:::::::BBBBBBBBBBBBBBB
INO2_YEAST <         AAAAAAAAAAAAAAAAA

BL00038A      <->A     (0,667):236
CBF1_YEAST 223        RKDSHKEVERRRRENINTAINVLS
                      | |  || |  |     |     |
INO2_YEAST 237        RKwKHVqMEKIRRiNtKEAFERLi

BL00038B    A<->B     (4,46):10
AST4_DROME 143        HKKISKVDTLRIAVEYIRSLQ
                    ·  | |  |    |    | |
INO2_YEAST 271        GKRIPKhILLTcVMNdIKSIR
```

Fig. 4. Actual search results returned by E-mail (or WWW server) when the sequence of INO2_YEAST was submitted, showing the top two (of seven) hits reported. For each anchor block, BLOCKSORT analyzes the highest-scoring set of supporting blocks that are in the correct order and are separated by distances along the query sequence comparable to those along the sequences comprising the blocks. If such supporting blocks exist, then the probability that they support the anchor block is reported. Maps of the database blocks and query sequence are shown, where "AAA" represents the first block roughly in proportion to its width, colons represent the minimum distance between blocks in the database, periods represent the maximum distance between blocks in the database, and "⟨ ⟩" indicate the sequence has been truncated to fit the page. The query map is aligned on the highest scoring block. Multiple block hits that are consistent with the highest scoring block are separated by colons. Block hits that are not consistent are mapped below. The alignment of the query sequence with the sequence closest to it in the Blocks Database is depicted below the map. The distance between detected blocks is listed as "(min, max):" for the database entry followed by the distance in the query. The distance to the first detected block is also shown. Uppercase type in the query indicates at least one occurrence of the residue in that column of the block. No supporting blocks were detected for the first hit, and so a probability was not calculated. The second hit consists of an anchor block (BL00038A) and one supporting block (BL00038B) representing the helix–loop–helix family of DNA-binding proteins. Note that this hit falls near the 1/1000 border of the nomogram (middle line in Fig. 3); however, because it is the second-ranking hit, confidence that it is a true positive must be less than if it were a first-ranking hit. Each numbered result consists of one or more blocks from a Prosite group detected by the query sequence. This search was returned in 2 min after receipt; longer protein sequences and DNA sequences, which must be translated in all six frames, will require proportionally longer times.

hits in which the anchor block is not the first-ranking block. To allow for the possibility that a sequence might belong to more than one family, BLOCKSORT does not explicitly penalize hits that do not rank first, even though the nomogram is based on only first-ranking blocks. Therefore, confidence that a hit is real must be reduced whenever the anchor block is outranked by a presumed false-positive hit, such as in the example. Here, the helix–loop–helix block A ranks a close second (the first-ranking block is found at the 86th percentile), so our confidence in the result remains high. In general, confidence in a hit decreases with decreasing rank, to an extent that depends on the available evidence.

For each multiple block hit representing a family in the Blocks Database, BLOCKSORT prints a map of the blocks from the database and compares it with a similar map of the blocks in the query sequence. For each block included in the hit, the alignment of the query with the sequence from the block in the Blocks Database with which it shares the most identical residues is shown to further assist evaluation (Fig. 4). If a family is documented in Prosite as having a maximum of r repeated motifs within a sequence, then BLOCKSORT will map alignments from the search up to r repeats found in a single sequence.

The Blocksearcher E-mail server was inaugurated in 1992 and currently averages about 700 searches per month. Over 5000 different users have contacted this server since its inception. In 1994 Blocksearcher was also made available via the World Wide Web (WWW), and subsequently a sequence logo[11] option was added.

Other Applications of the Blocks Database

Amino Acid Substitution Matrices

Methods for analysis of sequence similarity typically use scores from an amino acid substitution matrix. A substitution matrix is a 20 by 20 symmetric array of numerical values that provide scores that are applied to each position in an alignment. These scores can be derived from presumed true-positive alignments, an approach pioneered by Margaret Dayhoff.[32] In such cases, scores are based on the odds that one amino acid may be substituted by another. Large positive values in the matrix indicate frequent substitution, and large negative values indicate that the substitution is rare. Although other scoring schemes have been proposed, Dayhoff's

---

[32] M. O. Dayhoff and R. V. Eck, "Atlas of Protein Sequence and Structure," p. 33. National Biomedical Research Foundation, Silver Spring, Maryland, 1968.

mutation data matrices[33] were for many years considered the standard for pairwise alignment and searching programs. In the Dayhoff model, substitution rates are derived from global alignments of pairs of proteins sequences that are at least 85% identical and then extrapolated to more distant sequences using a Markov model. As pointed out by Altschul,[34] the basis of a substitution matrix is its target frequencies, which are the observed frequencies of substitution for each pair of amino acids. The aligned segments in the Blocks Database provide ample data for estimating the target frequencies in regions of local similarity, which we used to construct a series of amino acid substitution matrices from Blocks 5.0.[12]

Rather than estimate substitution rates from closely related sequences and extrapolate as in the Dayhoff approach, target frequencies were obtained directly from alignments of segments in blocks by counting pairs of aligned amino acids at each position within a block. The total counts for all of the 210 possible amino acid pairs obtained for every position of every block in the Blocks Database were converted to target frequencies, and odds ratios were calculated by dividing them by the corresponding expected frequencies for each pair.

To make a series of substitution matrices using this approach, sequences were clustered within each block, and each cluster was weighted as a single sequence in counting aligned amino acid pairs. A clustering percentage was specified in which sequence segments that are identical for at least that percentage of amino acids were grouped together. For example, if the percentage was set at 35% and sequence segment A was identical to sequence segment B in at least 35% of the aligned positions, then A and B were clustered. If C was identical to either A or B in at least 35% of aligned positions, it was also clustered with them, and the contributions of A, B, and C were averaged in counting aligned amino acid pairs. Pairs were counted only between clusters, not within clusters. If all of the segments in a block clumped into one cluster, then the block provided no counts. We refer to the series of matrices constructed in this way as the BLOSUM series (for blocks substitution matrix). BLOSUM35, therefore, is based on observed substitutions between segments that are less than 35% identical, BLOSUM62 on segments that are less than 62% identical, BLOSUM100 on segments that are less than 100% identical, and so forth. BLOSUM100 is based on over 11 million observed substitutions from 2106 blocks, BLOSUM62 on 1.3 million aligned amino acid pairs from 1572 blocks, and BLOSUM35 on 0.1 million pairs from 439 blocks. The pairs counted for

[33] M. Dayhoff, "Atlas of Protein Sequence and Structure," p. 345. National Biomedical Research Foundation, Washington, D.C., 1978.
[34] S. F. Altschul, *J. Mol. Biol.* **219,** 555 (1991).

	C	S	T	P	A	G	N	D	E	Q	H	R	K	M	I	L	V	F	Y	W
C	9																			
S	-1	4																		
T	-1	1	5																	
P	-3	-1	-1	7																
A	0	1	0	-1	4															
G	-3	0	-2	-2	0	6														
N	-3	1	0	-2	-2	0	6													
D	-3	0	-1	-1	-2	-1	1	6												
E	-4	0	-1	-1	-1	-2	0	2	5											
Q	-3	0	-1	-1	-1	-2	0	0	2	5										
H	-3	-1	-2	-2	-2	-2	1	-1	0	0	8									
R	-3	-1	-1	-2	-1	-2	0	-2	0	1	0	5								
K	-3	0	-1	-1	-1	-2	0	-1	1	1	-1	2	5							
M	-1	-1	-1	-2	-1	-3	-2	-3	-2	0	-2	-1	-1	5						
I	-1	-2	-1	-3	-1	-4	-3	-3	-3	-3	-3	-3	-3	1	4					
L	-1	-2	-1	-3	-1	-4	-3	-4	-3	-2	-3	-2	-2	2	2	4				
V	-1	-2	0	-2	0	-3	-3	-3	-2	-2	-3	-3	-2	1	3	1	4			
F	-2	-2	-2	-4	-2	-3	-3	-3	-3	-3	-1	-3	-3	0	0	0	-1	6		
Y	-2	-2	-2	-3	-2	-3	-2	-3	-2	-1	2	-2	-2	-1	-1	-1	-1	3	7	
W	-2	-3	-2	-4	-3	-2	-4	-4	-3	-2	-2	-3	-3	-1	-3	-2	-3	1	2	11
	C	S	T	P	A	G	N	D	E	Q	H	R	K	M	I	L	V	F	Y	W

FIG. 5. The BLOSUM62 amino acid substitution matrix in half-bit units. The cluster percentage of 62 provides a good general purpose matrix for local alignment applications, with relative entropy 0.7 bit and an expected value of -0.5 bit. Matrices in the Blosum series are available in various formats and scales by anonymous ftp.

BLOSUM35 are from segments less similar than those from BLOSUM100, so that BLOSUM35 has correspondingly lower relative entropy[34] (0.21 bit) than BLOSUM100 (1.45 bits).

Extensive testing using the BLAST searching algorithm, which depends on ungapped local alignments,[30] has shown that the Blosum series outperforms other matrices for this task.[12,35] BLOSUM62 (Fig. 5) with relative entropy of 0.7 bit performs the best for database searching, and it has been adopted as the default for BLAST and other similar applications. Information theory predicts that other matrices in the series might be better suited for other applications.[34] For example, BLOSUM45 was found to perform especially well in profile searches.[36] It is likely that gap penalties employed by many programs have an important effect. There is currently no accepted theory of gap penalties, so that determination of the best matrix to use for an application that employs gap penalties should be guided by empirical testing.[37]

[35] S. Henikoff and J. G. Henikoff, *Proteins: Struct. Funct. Genet.* **17,** 49 (1993).
[36] R. Luthy, I. Xenarios, and P. Bucher, *Protein Sci.* **3,** 139 (1994).
[37] W. R. Pearson, *Protein Sci.* **4,** 1145 (1995).

Dirichlet Mixtures

Position-specific scoring matrix representations of multiple alignments are based on the observed frequencies of amino acids in each position of the alignment. It is typical that most amino acids do not appear at all in a position, and so it must be decided how to deal with these nonoccurrences. One approach is to assume that, since a residue is not observed in a position, it should never occur there; this is unrealistic if the alignment contains only a few sequences. Another is to use a substitution matrix to arrive at a weighted average score for the unobserved amino acid[18]; this will reduce specificity of the PSSM if the alignment is well represented.

A statistical approach to the problem of nonoccurrences adds imaginary pseudocounts to the observed counts of each amino acid at a position, based on some belief about amino acids expected to occur there. Brown et al.[38] have utilized Dirichlet mixtures to calculate pseudocounts for a hidden Markov model application. They noted that each position in a multiple alignment can be thought of as a sample from a multinomial distribution and that the Dirichlet is the natural conjugate family of prior distributions. Once the many parameters are estimated, the Dirichlet priors provide elegant pseudocounts that take into consideration amino acid inter-relationships. Estimation of the parameters is laborious, however, and requires many sample aligned positions. Initially, Brown et al. estimated parameters from the HSSP database of multiple alignments on the basis of structure,[39] but this group has also found alignments from the Blocks Database to be useful for this purpose (K. Sjolander, personal communication, 1994).

Position-Specific Scoring Matrix Evaluations

The Blocks Database together with the lists of known true-positive sequences for the families in it (from Prosite) has been used to evaluate alternative ways of computing PSSMs. The general approach is to compute a PSSM from each block in the database using two different methods, then search the competing PSSMs against SWISS-PROT, and finally compare the two score distributions of known true-positive and true-negative sequences. This procedure is similar to the calibration searches described above; however, when testing searching methods, it is more effective to search a sequence database that contains some true-positive sequences which were

[38] M. P. Brown, R. Hughey, A. Krogh, I. S. Mian, K. Sjolander, and D. Haussler, in "Proceedings of the First International Conference on Intelligent Systems for Molecular Biology," p. 47. AAAI Press, Menlo Park, California, 1993.

[39] C. Sander and R. Schneider, Proteins: Struct. Funct. Genet. 9, 56 (1991).

not used to make the blocks. Because Prosite and SWISS-PROT are maintained in tandem, we accomplish this by searching blocks made from an older version of SWISS-PROT against a newer version of SWISS-PROT that contains more sequences, and using the corresponding newer version of Prosite to provide the lists of true-positive sequences.

One study using this general approach compared several different sequence weighting schemes.[7] Three methods were found to perform best among those tested: position-based weights,[7] Voronoi weights,[40] and branch proportional weights.[41] An update of this study is shown in Fig. 6.[42-45] Currently, we are comparing different methods of computing pseudocounts, extending the analysis of Tatusov *et al.*[28] to large numbers of protein families.

Access

The Blocks Database, the Blosum series of matrices, and associated software are available by anonymous ftp from the NCBI repository:

ftp ncbi.nlm.nih.gov
cd repository blocks

However, most users will find E-mail and WWW servers to be preferable to installation of the software. E-mail server addresses are blocks@howard.fhcrc.org for searching the Blocks Database or retrieving a set of blocks and corresponding Prosite files, and blockmaker@howard.fhcrc.org for making blocks from user-defined protein sequences. Send the word "help" in the subject line of a blank message or as the only word in the body of a message to obtain help files for either server. Both servers are able to interpret FASTA, GenBank, EMBL (European Molecular Biology Laboratories), PIR (Protein Identification Resource), GCG (Genetics Computer Group), and Genepro sequence formats. In addition, Blocksearcher will decide whether the submitted sequence is protein or DNA and, in the case of DNA, will search it in all six reading frames, putting together multiple hits from the same strand.[6] Blockmaker accepts up to 250 protein sequences in the same format concatenated into a single message. DNA sequences or messages with fewer than 3 sequences should not be sent to Blockmaker.

The WWW is an especially efficient method for utilizing tools described

[40] P. R. Sibbald and P. Argos, *J. Mol. Biol.* **216,** 813 (1990).
[41] J. D. Thompson, D. G. Higgins, and T. J. Gibson, *CABIOS* **10,** 19 (1994).
[42] M. Gerstein, E. Sonnhammer, and C. Chothia, *J. Mol. Biol.* **236,** 1067 (1994).
[43] S. R. Eddy, G. Mitchison, and R. Durbin, *J. Comput. Biol.* **2,** 9 (1995).
[44] M. Vingron and P. Argos, *CABIOS* **5,** 115 (1989).
[45] S. F. Altschul, R. J. Carroll, and D. J. Lipman, *J. Mol. Biol.* **207,** 647 (1989).

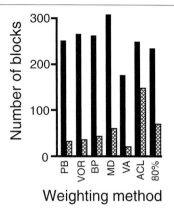

Fig. 6. Evaluation of sequence weighting methods. A set of 1673 blocks from Blocks 5.0 (based on Prosite 8.0 keyed to SWISS-PROT 22) were converted to PSSMs using various sequence weighting methods and SWISS-PROT amino acid frequencies to compute odds ratio scores. Each block was then used to search SWISS-PROT 29 (which corresponds to Prosite 12.0), from which lists of true-positive sequences were obtained. All 1673 blocks had been updated with new sequences in Prosite 12.0, so that these new sequences were necessarily absent from the block queries. Sequence weighting methods tested: PB, position-based[7]; VOR, modified Voronoi[24,40]; BP, branch proportional[41,42]; MD, maximum discrimination[43]; VA, Vingron–Argos[44]; ACL, Altschul, Carroll, and Lipman[45]; 80%, 80% clustering.[12] All test PSSMs are compared against a PSSM made with equal sequence weights. Solid bars represent the number of test PSSMs for which performance was better than performance of the corresponding equal-weighted PSSMs. Cross-hatched bars represent the number of equal-weighted PSSMs for which performance was better than for test PSSMs. Here, performance of a PSSM was measured as the "equivalence number,"[37] which is the point at which the number of true-positive sequences equals the number of true-negative sequences, so that a lower equivalence number reflects better separation of true positives from true negatives in the vicinity of the twilight zone. These results are very similar to our results reported previously using different search and evaluation criteria,[7] except that maximum discrimination weights[43] were not available at that time.

in this chapter. The Uniform Resource Locator (URL) for the Blocks home page is http://www.blocks.fhcrc.org. The WWW server allows protein and DNA queries to be searched against the current Blocks Database by placing a sequence in the text box provided on the Blocksearcher page. Results are returned with hypertext links to Prosite and SWISS-PROT, and from there to other databases, including EMBL/GenBank and MEDLINE. Because of the hypertext capability of the WWW, results can be evaluated rapidly with minimum effort on the part of the user. The WWW server also allows retrieval and browsing, returning the full set of blocks and Prosite documentation. An added feature of the WWW server is the provision of a sequence logo[11] for each block in the Blocks Database, further aiding the evaluation of search results. Logos are also provided with the WWW Blockmaker server.[10]

Summary

Protein blocks consist of multiply aligned sequence segments without gaps that represent the most highly conserved regions of protein families. A database of blocks has been constructed by successive application of the fully automated PROTOMAT system to lists of protein family members obtained from Prosite documentation. Currently, Blocks 8.0 based on protein families documented in Prosite 12 consists of 2884 blocks representing 770 families. Searches of the Blocks Database are carried out using protein or DNA sequence queries, and results are returned with measures of significance for both single and multiple block hits. The database has also proved useful for derivation of amino acid substitution matrices (the Blosum series) and other sets of parameters. WWW and E-mail servers provide access to the database and associated functions, including a block maker for sequences provided by the user.

[7] Indexing and Using Sequence Databases

By RODGER STADEN

Introduction

Given reliable Internet connections and the existence of good remote servers it can be argued that there is little need to provide local access to the data held in sequence libraries. However, there are also strong arguments in favor of laboratories having their own copies of the files, and this chapter describes methods we have devised for this purpose. There are three components: (a) the files involved and their organization; (b) the programs for creating index files; and (c) the programs for using the indexes to provide rapid searching and access to the data files.

Our index creation methods can handle EMBL (European Molecular Biology Laboratories), SWISS-PROT, and GenBank libraries in their distributed formats, and in the split format where sequences and annotations are divided into separate files. This latter type includes both the ASCII and binary forms of GCG (Genetics Computer Group)[1] sequence library

---

[1] J. Devereux, P. Haeberli, and O. Smithies, *Nucleic Acids Res.* **12,** 387 (1984).

METHODS IN ENZYMOLOGY, VOL. 266

Copyright © 1996 by Academic Press, Inc.
All rights of reproduction in any form reserved.

format. With the proviso explained in the Discussion, our routines can also index PIR (Protein Identification Resource) and NRL3D in CODATA format. Our searching and browsing routines can read all the above-mentioned formats, and, for reasons that will become apparent, we provide similar access to the PROSITE motif library documentation and data files.

There are two approaches to extracting particular entries from a large file. Either the file is scanned from the beginning until the entry name is matched, or an index is used in which the location (or offset) of the start of each entry is recorded. In the latter case, especially if the index is ordered in a particular way, such as alphabetically, the entry name and its associated offset can be found quickly, and then only the necessary parts of the data file read. The same is true for the provision of text searches of the annotation data. Either the whole of the text is scanned each time a search is requested, which even with fast matching routines can still take a long time and will put a heavy load on a machine, or for each release of the library, indexes are created that contain all possible search keys and lists of the entries that contain them. For both tasks we have adopted the latter approach, which produces extremely rapid searches and entry extraction.

The applications that we concentrate on here are searching the libraries on author names, species, and keywords; and we discuss entry extraction based on accession numbers and entry names. Note that keywords means all the textual information in the data files, not just the entries in special keyword records. Our methods for searching the sequences in the libraries for patterns of motifs have changed little since they were last described in this series.[2]

An important advantage of our methods over many others is that they do not require any changes to the distributed forms of the sequence libraries. We simply create indexes to the data as distributed, saving a great deal of computer time and disk space. In addition, being able to create indexes to the data files means that it is straightforward to provide comprehensive access to between-release updates of the sequence libraries.

Data Files, Indexes, and Their Organization

The data in the sequence libraries is divided into entries. Each entry has an entry name, at least one accession number, a set of annotations including lists of author names, sequence description, taxonomic information, and finally the sequence. Related entries are collected together into division files. At present EMBL and GenBank each have 15 division files

[2] R. Staden, this series, Vol. 183, p. 193.

and SWISS-PROT only 1. As mentioned above, we use precalculated indexes for entry extraction and text searches.

The indexing method used is the same as that employed on the EMBL CD-ROM. All the indexes contain binary data. I describe the relevant components below, taking the EMBL library in distributed format as an example. Figure 1 lists the files involved. The division files are numbered 1 to 15. For example, the file mam.dat, which contains mammalian data, is given the number 4. The correspondence between these division numbers and their associated division files is stored in the file Division_lookup. The division lookup file gives independence between the index files (see below) and the data files that they refer to, and hence allows us to name and locate the division files however we please.

The file Entryname_index contains a record for each entry in the library. Each record contains the entry name, annotation offset, sequence offset, and division code of the entries. The records are sorted alphabetically on the entry names. In conjunction with the division lookup file, this file enables programs to locate entries based on entry name, whence either the sequence or its annotation can be extracted.

The file Brief_index contains a useful summary for each entry. It consists of a single record for each entry in the library and contains the entry name, primary accession number, sequence length, and an 80-character description of the entry. Like the entry name index file, it is sorted alphabetically on the entry name so identical record numbers in each file refer to the same entries.

The rest of the index files are arranged in pairs of target and hit files, where the target is a possible search string such as an author name, and the hits identify the entries that contain the target. All target files have the same structure consisting of the following: first hit record, number of hits, and target string. The files are sorted alphabetically on the target strings. The hit files consist entirely of records containing the record number of entries in the entry names index. To find the names of the entries that contain the name of a particular author we do the following. Locate the name of the author in the target file and get the corresponding first hit record (fhr) and number of hits (noh). Then, in the hit file, the records from fhr to fhr + noh − 1 contain the record numbers of the required entries in the entry name index file (and also the file Brief_index). The types of data for which target and hit files exist are accession numbers, authors, text, and species. The text index contains data from all textual components of the annotation sections of the entries, not just the keyword records. The species index is, by default, called the taxon index. The accession numbers are stored in this way because an entry can contain several

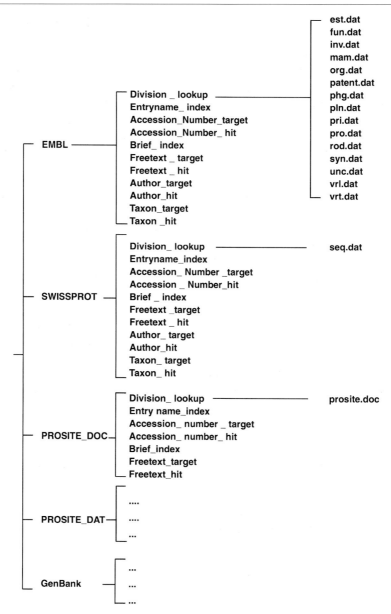

Fig. 1. Organization of the files used by the analytical software to locate and process the sequence libraries.

accession numbers and any accession number can be divided between several entries.

As is explained later we create identical indexes to those described above for all the different library formats. This means that we leave the division files unchanged, and hence that we have to use format-specific routines for reading the annotation and sequence data, but we can employ a single set of routines for doing the more complex job of using the index files. Clearly there is sufficient information in the files described above to allow application programs to perform efficient searches of the data, and now we turn to the task of organizing the files so that the application programs can find and use them.

The programs that use the libraries need to know which libraries are available, what their format is, where their data files are, where their division lookup file is, and where their indexes are. To achieve this the libraries are defined at several levels using a treelike structure and a number of additional files. Figure 1 shows a schematic of the relationship of the files. Each of the three levels in the tree corresponds to a particular type of file. Environment variables are used extensively.

The top level file contains a list of the available libraries. Using the symbols A–E, it defines the library format type (EMBL/SWISS-PROT, GenBank, PIR, PROSITE documentation, and PROSITE data) so that the programs know how to read the data files; it names a file that contains the next level of information about each library; and it provides messages to appear on the screen of the user. An example of this type of file is shown in Fig. 2. The libraries have types A, B, C, D, and E. The names such as EMBLFILES are actually environment variables linked to real files that

```
A EMBLFILES EMBL nucleotide library ! comment

C GENBFILES GenBank nucleotide library

A SWISSFILES SWISSPROT protein library

B PIRFILES PIR protein library

B NRL3DFILES NRL3D protein library

D PROSITEDATFILES Prosite data

E PROSITEDOCFILES Prosite documentation
```

FIG. 2. Contents of the file containing data about the available libraries and their formats.

contain the next level of information. The prompts are EMBL nucleotide library, SWISSPROT protein library, etc. Anything to the right of a "!" is a comment. (Note that we also include the PROSITE library here because it has EMBL CD-ROM style indexes which can be handled identically to those for the sequence libraries.) The environment variable SEQUENCE-LIBRARIES is associated with this file. Below we follow the branch of the tree that deals with the EMBL library.

The next level of files defines the indexes and division lookup file for a particular library. For the EMBL library the file has the environment variable EMBLFILES. This file uses environment variable EMBLDIV-PATH for the directory that contains the division lookup file, EMBLIND-PATH for the directory containing the index files. An example is shown in Fig. 3. The software that searches the library uses the types, A, B, C, ..., K to know which files contain which type of information. For example, if it needs the entry name index it looks for a line starting with the letter B and then reads the file EMBLINDPATH/entrynam.idx.

As has been explained above, the next level of files associates the

```
A EMBLDIVPATH/embl_div.lkp

B EMBLINDPATH/entrynam.idx

C EMBLINDPATH/acnum.trg

D EMBLINDPATH/acnum.hit

E EMBLINDPATH/brief.idx

F EMBLINDPATH/freetext.trg

G EMBLINDPATH/freetext.hit

H EMBLINDPATH/author.trg

I EMBLINDPATH/author.hit

J EMBLINDPATH/taxon.trg

K EMBLINDPATH/taxon.hit
```

FIG. 3. Contents of the file containing data about the index files available for the EMBL library.

division numbers in the entryname index with actual file names stored on disk. For the EMBL library it is called EMBLDIVPATH/embl_div.lkp; for the distributed form of the library, it contains the information shown in Fig. 4.

Note that the EMBL CD-ROM contains a division lookup file for the data on the CD-ROM, but this is not the one used by our software. We rewrite it so the directory structure and file names can be chosen locally and also to deal with the case of split files. Although it goes against our philosophy of leaving the sequence libraries in their distributed formats,

```
 1 EMBLPATH/est.dat

 2 EMBLPATH/fun.dat

 3 EMBLPATH/inv.dat

 4 EMBLPATH/mam.dat

 5 EMBLPATH/org.dat

 6 EMBLPATH/patent.dat

 7 EMBLPATH/phg.dat

 8 EMBLPATH/pln.dat

 9 EMBLPATH/pri.dat

10 EMBLPATH/pro.dat

11 EMBLPATH/rod.dat

12 EMBLPATH/syn.dat

13 EMBLPATH/unc.dat

14 EMBLPATH/vrl.dat

15 EMBLPATH/vrt.dat
```

FIG. 4. Contents of the division lookup file which is used to associate the division numbers in the index files with actual files stored on disk.

our application programs can also handle division files that have been split into separate annotation and sequence components (e.g., as for the GCG package). This variation is managed at the level of the division lookup files by writing them in the form shown below, and by the parsing software being able to recognize and handle both formats. The example shows the division lookup file record for mammalian data when it is split into a sequence file mam.seq and an annotation file mam.ref.

4 EMBLPATH/mam.seq EMBLPATH/mam.ref

Creating Index Files

Because all the sequence libraries use different formats, there are two types of problems involved with creating the index files described above: the first is to devise a method that works for any particular library format, and the second is to make it work for them all. Obviously the indexes are created by reading the division files (although see below about the GCG entry name index) and parsing the appropriate records. To reduce the problem of dealing with different formats all the tasks are divided into several steps that create intermediate files. Then only the first program for a given task needs to be specific to any particular library format.

The details of how the indexes are created are given in Staden and Dear.[3] In summary, a set of programs written in the C language are executed under the control of a family of scripts written in the C shell. Several sorting steps are employed.

For the case of the split libraries in GCG format, which can contain packed binary data, we have to use a different strategy to calculate the entry name index. The binary segments could contain any bit pattern so we cannot parse the files. Instead we have to assume that GCG style indexes, created as part of the reformatting process, are available; from these we extract the necessary fields, concatenate and sort them, and write them in our standard index format. To calculate all the other indexes our scripts parse the annotation files.

Application of Indexes

Programmers will recognize the potential of the indexes described above for providing fast and efficient searches of the data. Because of the arrangement of the records in the indexes, binary searches give rapid access to target strings, and these point directly to the required entries.

---

[3] R. Staden and S. Dear, *DNA Sequence* **3**, 99 (1992).

By using the file organization described above our application programs can find out which libraries are available, and which types of interaction can be provided. If a full set of indexes are available our application programs provide searches based on author names, text, and species. In its current implementation each individual search can use up to five search strings. The results of each search can be combined, using the logical operators AND, OR, and NOT, with the results of previous searches. The current set of hits can be displayed on the screen. The information shown is taken from the brief directory index, and hence shows entry name, primary accession number, sequence length, and an 80-character description of each entry. The associated entry extraction routines will return a sequence or allow users to browse through the annotation sections. Note that this access to the sequence libraries is available from within all of our analytical programs.

Because, for the PROSITE data and documentation files, EMBL creates indexes of identical format to those described above, our application programs can treat them like a sequence library. Hence they can perform text searches and annotation browsing on the very useful PROSITE files.

The algorithms used by our programs that search the sequence libraries for patterns of motifs[2] are necessarily made slightly more complicated by having to deal with division files that contain both annotation and sequence. The strategy is to open all the division files and then use the entry name index to get the offsets to each of the sequence segments. These programs can also work with lists of entry names, which makes this method of access particularly efficient because only the listed entries are accessed.

Discussion

I have described the creation and use of sequence library indexes compatible with those distributed on the EMBL CD-ROM. The strategy outlined avoids the onerous and time-consuming task of reformatting the libraries. Obviously for those who only use the standard releases of the EMBL and SWISS-PROT libraries, the application programs described can use the files on the CD-ROM directly or they can simply be copied to hard disk for greater speed. For those who want to use between-release updates (which do not include the necessary indexes), or who want to use other sequence libraries, or indeed libraries of their own construction, the indexing programs described provide the means to create indexes, and the application programs a way of using them. The use of precomputed indexes of text, authors, and species gives very much faster searches than can be achieved by scanning through the data each time a search is requested.

Note that the parsing routines used by the index creation programs assume that records in the division files use fixed column positions for items of data. This is true of EMBL, SWISS-PROT, and GenBank, but not of CODATA. Consequently if the column positions used in the CODATA versions of PIR and NRL3D change from one release to the next, the index creation routines will not work in the present form. The package does not currently include routines to calculate taxon indexes or to create indexes for PROSITE.

The routines described can be obtained as part of our sequence analysis package.[4] They run on DEC alpha machines under Digital UNIX (formerly known as OSF), and on SUN machines using either SunOS 4.x or Solaris 2.x, and probably on other UNIX systems. For further information see http://www.mrc-lmb.cam.ac.uk/pubseq/.

[4] R. Staden, *Methods Mol. Biol.* **25,** 9 (1994).

[8] SRS: Information Retrieval System for Molecular Biology Data Banks

By THURE ETZOLD, ANATOLY ULYANOV, and PATRICK ARGOS

Introduction

The explosion of efforts in the field of molecular biology and biotechnology has made both the acquisition of information from data banks and deposit of new data increasingly critical. The types of data are diverse and can range from the protocol of an experiment to the sequence of an entire chromosome. The specificity and precision of data such as the exact coordinates of a protein atom in three-dimensional space (Protein Data Bank, PDB[1]) or the free text description of a genetic disease (MIM[2]) are also highly variant. These contrasting data types have been compiled into data banks by individuals or organizations and are often available to the scientific community for retrieval as well as updating and addition of information. The organization of the data banks also varies greatly, ranging from simple text files to relational or object-oriented systems. Few of these information

[1] E. E. Abola, F. C. Bernstein, and T. F. Koetzle, *in* "Computational Molecular Biology: Sources and Methods for Sequence Analysis" (A. M. Lesk, ed.), p. 69. Oxford Univ. Press, Oxford, 1988.

[2] P. L. Pearson, C. Francomano, P. Foster, C. Bocchini, P. Li, and V. A. McKusick, *Nucleic Acids Res.* **22,** 3470 (1994).

Copyright © 1996 by Academic Press, Inc.
All rights of reproduction in any form reserved.

sources are completely isolated, as each assimilates or references information in other data banks.

It is not surprising that in this complex system of coexisting data banks built and maintained with many different philosophies, the *lingua franca* is still the text file in a flat file format where entries follow in sequential order. The contents of the file are read directly by humans or by an innumerable number of parsing routines that make the data available for further analysis or for incorporation into more advanced database management systems. We present here a retrieval system called SRS (Sequence Retrieval System[3,4]) that acts on data banks in a flat file or text format. It provides a homogeneous interface to about 80 biological databanks for accessing and querying their contents and for navigating among them.

Indexing and Parsing

A fast query system usually relies on indices created before query time. SRS has its own indexing system and treats entire data banks as sequences of entries, each with different data fields. The contents of the data fields are parsed, and selected words are extracted and inserted into an index. There is usually a separate index created for each data field.

In data banks such as the EMBL (European Molecular Biology Laboratories) collection of nucleotide sequences,[5] this separation is important even if some information in different data fields appears redundant at first glance. For instance, the keywords, the definition, the reference title, and feature table data fields in a given sequence entry may contain the word "kinase." The least consistent use of "kinase" is in the reference title field, where it must be viewed within its context and might only indirectly describe the sequence in the entry as, for example, in "... is phosphorylated by casein kinase II ...." The intent of the definition field is to render a short description of the entry in free text format. However, no discernible convention need be applied; for instance, the same protein or enzyme can be described by more than one name or even the same name spelled differently in different entries. In the case of entries containing a complete genome, the definition field may not even mention the protein names associated with the genes therein. The keywords data field relies on a controlled dictionary which classifies, rather than describes, the entry. Entries of related sequences should be attributed with the same, or at least largely overlapping,

[3] T. Etzold and P. Argos, *Comput. Appl. Biosci.* **9**, 49 (1993).

[4] T. Etzold and P. Argos, *Comput. Appl. Biosci.* **9**, 59 (1993).

[5] C. M. Rice, R. Fuchs, D. G. Higgins, P. J. Stoehr, and G. N. Cameron, *Nucleic Acids Res.* **21**, 2967 (1993).

list of keywords. The feature table, as described later, defines mostly subsequences with particular characteristics (exons or introns, repeats, domains, etc.).

As the information requests gain specificity, more discrimination is required in the choice of the index and the search word applied in a query. A search for appropriate entries in biological data banks is often a quest for the right search word(s) which can be obtained by testing different queries with successive refinement. A good starting point for such a procedure is request to the AllText index of SRS, which is not related to a real data field but rather represents all indices of all data fields containing textual information such as those previously mentioned.

SRS allows the indexing of any word in an entry. An index search usually requires only a fraction of a second but takes longer if the query contains regular expressions or wildcards at the beginning of the search word. Numbers such as the date, sequence length, resolution, or molecular weight can also be indexed in special structures that allow the retrieval of numeric ranges; for instance, all sequences with a length between 200 and 400 or all protein structures with a resolution greater than 2 Å.

It is evident that SRS requires much information to parse and extract data fields and indexable items from data banks that often have their own unique format and syntax. Two languages have been designed to describe to the system a data bank whose information is usually deposited in a single file. The ODD (Object Design and Definition) language is used to detail the physical structure (data fields, files), whereas the other, an extended Backus–Naur notation, describes the syntax of the data fields. This results in a highly flexible environment where data bank representations can be easily added or changed.

In a system with many data banks, many indices (one per data bank per data field) have to be created and updated. The link indices (see below) add substantially to the complexity of the maintenance since links manifest interdependences between individual data banks. A program has been written that analyzes the state of the overall system and writes a command script in case indices need to be rebuilt. The index building is completely automated, and the program runs automatically at frequent intervals to check if a new version of a data bank exists. The storage size for the indices is relatively small, totaling about 10 to 20% of the size of the actual indexed data banks.

Sequence Retrieval System Query Language

SRS has its own query language to represent index searches, the combining of sets with Boolean operators, and links between data banks.

Most interfaces to SRS (see below) shield the user completely or at least to some extent from the query language expressions, often through a query form.

The query language is set oriented, where a set is a list of entries from one or more data banks. The operators require sets as operands and produce sets of entries as the result. A set can be an entire data bank as denoted by the data bank name or a portion of it resulting from a query or a retrieval command. A retrieval command must specify the data bank name, the name of the index to be searched, and the query string or a numeric range. For example, the query

[embl-organism : human]

retrieves all entries in EMBL with the word "human" in the organism data field. The query

[embl-seqlength#400 : 500]

retrieves all sequences from the EMBL nucleotide databank with length between 400 and 500 base pairs. The two queries can be combined with the logical operator AND, denoted "&," resulting in a list of overlapping entries obtained from the separate commands:

[embl-organism : human]&[embl-seqlength#400 : 500]

SRS provides the three logical operators AND, OR, and BUTNOT (combination of binary AND and unary NOT) as well as two further link operators described below. Many retrieval criteria can be specified and combined by logical operators in a single expression. Only indexed information can be used for specifying constraints, and only entire data banks can be searched.

Retrieval of Subentries

Retrieval systems usually consider the entire entry as the only retrievable unit, a useful simplification that, however, completely ignores its internal structure. This approach may be acceptable for data banks such as Prosite,[6] where each entry describes one amino acid sequence pattern, but not for nucleotide sequence data banks such as GenBank[7] or EMBL where an entry may contain a single exon but also a complete genome or chromosome with several hundred genes. Genes, or rather their parts such as the promoter, coding sequence, introns, exons, and the like, are characterized

---

[6] A. Bairoch and P. Bucher, *Nucleic Acids Res.* **22**, 3583 (1994).

[7] D. A. Benson, M. S. Boguski, D. J. Lipman, and J. Ostell, *Nucleic Acids Res.* **22**, 3441 (1994).

in the feature table data field, which increasingly forms the principal portion of the annotation in EMBL and GenBank. Feature information can also be found in the protein sequence data banks SWISS-PROT[8] and PIR (Protein Identification Resource),[9] which document subentries such as transmembrane segments, calcium binding sites, and disulfide bridges. The feature table introduces a new structural element into the entry, namely, a list of subentries. A subentry can be viewed independent of its parent entry. For instance, the coding sequences (CDS features) from a DNA sequence entry can be extracted from the parent entry by using the sequence location information and subsequently translated and aligned with others as complete entries describing cDNA sequences.

SRS allows the indexing of the feature table such that the whole entry or the subentry (sequence feature) can be retrieved after an appropriate query. Each word in the feature description can be indexed and thereby associated with the feature itself and not just the parent entry. Figure 1 shows four subentries resulting from the query in SWISS-PROT for the transmembrane regions of the acetylcholine receptor sequence entry ACHA_BOVIN. The resulting entries contain the ID line of the parent entry, the relevant part of the feature table, and the subsequence extracted from the parent sequence as defined by the terminal positions in the feature annotation.

It is also possible to redefine the begin and end positions so that a fragment with a "safety" region around them will be provided. This is useful when extracting, for instance, transmembrane segments which are mostly putative with much uncertainty as to the termini. Another example is the retrieval of intron/exon boundaries.

Linking Data Banks

Most data banks provide cross-references to others. One of the best examples is SWISS-PROT, which maintains links to more than 20 other data banks. A natural usage of these cross-references, as is popular with many browsing systems in the World Wide Web (WWW), results in displayed entries with cross-references marked as hypertext links where the user need only click on the highlighted text to view the associated information or entry. Figure 2 shows a SWISS-PROT entry as displayed by the SRSWWW server (see below). A hypertext link can access any WWW server in the world where the related data are contained, resulting in a

[8] A. Bairoch and B. Boeckmann, *Nucleic Acids Res.* **22,** 3578 (1994).
[9] D. G. George, W. C. Barker, H.-W. Mewes, F. Pfeiffer, and A. Tsugita, *Nucleic Acids Res.* **22,** 3569 (1994).

```
ID    ACHA_BOVIN        STANDARD;        PRT;    457 AA.
DE    ACETYLCHOLINE RECEPTOR PROTEIN, ALPHA CHAIN PRECURSOR.
FT    TRANSMEM      231     255
>P1;ACHA_BOVIN
     Length: 25
PLYFIVNVII PCLLFSFLTG LVFYL*


ID    ACHA_BOVIN        STANDARD;        PRT;    457 AA.
DE    ACETYLCHOLINE RECEPTOR PROTEIN, ALPHA CHAIN PRECURSOR.
FT    TRANSMEM      263     281
>P1;ACHA_BOVIN
     Length: 19
MTLSISVLLS LTVFLLVIV*


ID    ACHA_BOVIN        STANDARD;        PRT;    457 AA.
DE    ACETYLCHOLINE RECEPTOR PROTEIN, ALPHA CHAIN PRECURSOR.
FT    TRANSMEM      297     316
>P1;ACHA_BOVIN
     Length: 20
YMLFTMVFVI ASIIITVIVI*


ID    ACHA_BOVIN        STANDARD;        PRT;    457 AA.
DE    ACETYLCHOLINE RECEPTOR PROTEIN, ALPHA CHAIN PRECURSOR.
FT    TRANSMEM      429     447
>P1;ACHA_BOVIN
     Length: 19
ILLAVFMLVC IIGTLAVFA*
```

FIG. 1. Transmembrane regions of the SWISS-PROT entry ACHA_BOVIN extracted by SRS. Each of the four subentries includes the ID and definition (DE) lines of the parent entry and the relevant part of the feature table (FT) together with the subsequence it describes (see also the parent entry in Fig. 2).

truly distributed system with different and yet intertwined services. Indeed, this has been so successful that most people think of navigation between data banks in terms of hypertext links. Though they are easy to comprehend and very powerful in making accessible related information, they are not without limitations.

First, hypertext links are unidirectional unless the linked entries provide a cross-reference back to the referencing entry. For instance, EMBL and SWISS-PROT are mutually referenced, but only SWISS-PROT points to PDB (Protein Data Bank, containing atom coordinates and associated information for known protein three-dimensional structures), which has no cross-references of any kind. There are many cases where the referencing is not mutual, or is undesirable if, for instance, private or temporary data banks are involved.

Second, it is always possible, but may be very inconvenient, to collect all entries of one type referenced by a single entry. A PROSITE entry,

```
ID   ACHA_BOVIN      STANDARD;       PRT;      457 AA.
AC   P02709;
DT   21-JUL-1986 (REL. 01, CREATED)
DT   21-JUL-1986 (REL. 01, LAST SEQUENCE UPDATE)
DT   01-FEB-1994 (REL. 28, LAST ANNOTATION UPDATE)
DE   ACETYLCHOLINE RECEPTOR PROTEIN, ALPHA CHAIN PRECURSOR.
OS   BOS TAURUS (BOVINE).
OC   EUKARYOTA; METAZOA; CHORDATA; VERTEBRATA; TETRAPODA; MAMMALIA;
OC   EUTHERIA; ARTIODACTYLA.
RN   [1]
RP   SEQUENCE FROM N.A.
RM   84039794
RA   NODA M., FURUTANI Y., TAKAHASHI H., TOYOSATO M., TANABE T.,
RA   SHIMIZU S., KIKYOTANI S., KAYANO T., HIROSE T., INAYAMA S., NUMA S.;
RL   NATURE 305:818-823(1983).
CC   -!- FUNCTION: AFTER BINDING ACETYLCHOLINE, THE ACHR RESPONDS BY AN
CC       EXTENSIVE CHANGE IN CONFORMATION THAT AFFECTS ALL SUBUNITS AND
CC       LEADS TO OPENING OF AN ION-CONDUCTING CHANNEL ACROSS THE PLASMA
CC       MEMBRANE.
CC   -!- SUBUNIT: PENTAMER OF TWO ALPHA CHAINS, AND ONE EACH OF THE BETA,
CC       DELTA, AND GAMMA CHAINS.
CC   -!- SUBCELLULAR LOCATION: INTEGRAL MEMBRANE PROTEIN.
CC   -!- SIMILARITY: BELONGS TO THE LIGAND-GATED IONIC CHANNELS FAMILY.
DR   EMBL; X02509; BTACHRA1.
DR   PIR; A03169; ACBOA1.
DR   PROSITE; PS00236; NEUROTR_ION_CHANNEL.
KW   RECEPTOR; POSTSYNAPTIC MEMBRANE; IONIC CHANNEL; GLYCOPROTEIN; SIGNAL;
KW   TRANSMEMBRANE.
FT   SIGNAL         1      20
FT   CHAIN         21     457      ACETYLCHOLINE RECEPTOR PROTEIN, ALPHA.
FT   DOMAIN        21     230      EXTRACELLULAR.
FT   TRANSMEM     231     255
FT   TRANSMEM     263     281
FT   TRANSMEM     297     316
FT   DOMAIN       317     428      CYTOPLASMIC.
FT   TRANSMEM     429     447
FT   DISULFID     148     162
FT   DISULFID     212     213      ASSOCIATED WITH RECEPTOR ACTIVATION.
FT   CARBOHYD     161     161      PROBABLE.
SQ   SEQUENCE    457 AA;   51947 MW;   1201373 CN;
```

FIG. 2. SWISS-PROT entry ACHA_BOVIN as displayed by SRSWWW. The hypertext links, underlined and in boldface type, refer to related entries in other data banks. Note that each item in the feature table contains a hypertext link which leads to the respective subsequence.

which provides a subsequence pattern or motif shared by a protein family, also lists all members of that family, of which there can be hundreds. To obtain all related members of such a family, the user needs to click as many times as there are members and save each accessed entry as a single file. A growing number of data banks such as PROSITE, ProDom,[10] and HSSP,[11] provide references to sequence groups related by family membership or homology, and often the inconvenience of hypertext links prohibits their full exploitation.

[10] E. L. L. Sonnhammer and D. Kahn, *Protein Sci.* **3,** 482 (1994).
[11] C. Sander and R. Schneider, *Nucleic Acids Res.* **22,** 3597 (1994).

Third, linking entire sets of single entries or even entire data banks is impractical, as each single entry in the set must be accessed, and then the hypertext link must be located and the linked entries, if any, accessed and saved. In effect, hypertext links are too cumbersome to ask straightforward questions such as which member of a set of lysozymes has a known tertiary structure or which entries of the current set belong to the family of amino-transferases.

Finally, often a desired link between two data banks does not exist; for instance, it is impossible in the absence of cross-reference information to navigate directly from an entry in SWISS-PROT to an associated entry in EPD (Eukaryotic Promoter Database[12]) which describes the promoter of the gene coding for the protein whose amino acid sequence is given in the SWISS-PROT entry. However, the link can be found by first accessing a related entry in EMBL which provides links to EPD. In general, there are many cases where data banks are not directly but indirectly linked. Exploitation of this requires of the user intimate and extensive knowledge regarding the intertwining of all data banks, and in the process of repeated linking, it is likely the user will get lost in the Web.

The SRS linking algorithm allows the user to overcome facilely these barriers. The cross-references are processed before query time such that the available data banks are scanned for cross-references and the results are inserted into special link indices. For each link between two data banks an index is built. In case of mutual referencing, it is sufficient to read the references from only one source, alleviating the necessity to scan all data banks when building a network with many data banks. Once indexed, a link can be used bidirectionally. All the links must be defined in the ODD language (see above) and are therefore known to the system. This link information is used to build a graph structure in memory, where the data banks are nodes and the links the edges. In this system a link is defined as the shortest or optimal path that can be found between any desired pair of nodes or data banks assuming bidirectional edges. Figure 3 shows the situation for the data banks installed at EMBL. To add a new data bank to the network, it is sufficient to add a single link to any of its nodes. Though the usefulness of the link operation depends on the quality of the cross-references in the respective data banks, it is also valuable in locating wrong or missing cross-reference information.

The SRS query language has, apart from the logical operators previously described, two further and unique operators: link left and link right. Figure 4 illustrates their functionality where the two data banks A and B are linked through explicit cross-references in the entries of A used to create

[12] P. Bucher and E. N. Trifonov, *Nucleic Acids Res.* **14,** 10009 (1986).

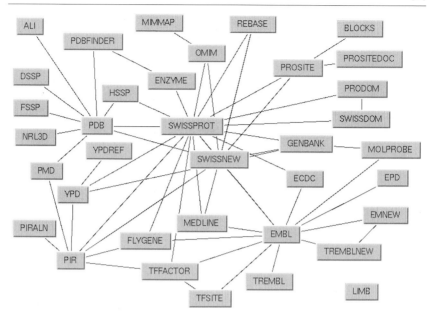

Fig. 3. Network of data banks installed at the EMBL in Heidelberg, Germany. This diagram can also be called from the SRSWWW server (http://www.embl-heidelberg.de/srs/srsc?-np), and each box can be clicked to obtain further information about the respective data bank. Note that LIMB [Listing of Molecular Biology databases, G. Keen, G. Redgrave, J. Lawton, M. Cinkosky, S. Mishra, J. Fickett, and C. Burks, *Math. Comput. Modelling* **16,** 93 (1992)] is the only databank without a link since it provides information about all the others.

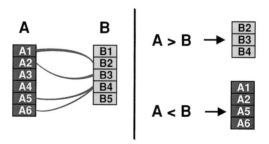

Fig. 4. Function of the SRS link operators exemplified on the two data banks A and B with respective numbered entries. See text for further explanations.

a link index. Operation (1) A > B returns those entries in B that are cross-referenced by entries in A. Operation (2) A < B returns entries in A that cross-reference entries in B.

Since all link indices can be used in both directions, it is of no consequence whether entries in A reference B or entries in B reference those in A. It is therefore valid to state that entries in A are linked to entries in B, either by references in A or B. The results of the above expressions can now be redefined as the evaluation of (1) returns entries in B linked to A and (2) returns entries in A linked to B. The link operator can be viewed as an arrow pointing to the set [link right (1) and link left (2)] from which the result will be taken. Reversing the order of operands has the same effect; hence, A > B and B < A yield the same result, which is entries in B linked to A. The operands of the link operators can be a data bank name or a set returned by a retrieval command or any other SRS query language expression (see above).

Links can also be specified between data banks that are not direct neighbors in the link diagram as shown in Fig. 3. Consider the example pdb > embl which retrieves, for all proteins with solved tertiary structure, the list of DNA sequences that encode their protein sequences. Algorithmic examination of the data bank network reveals that the shortest path between PDB and EMBL is to link PDB first to SWISS-PROT and then to EMBL, which can then be automatically carried out. It may occur that two alternative paths exist. The data bank administrator should give priority to one path by assigning a lower cost value to the link object as specified in the ODD language.

It is also possible to give a path explicitly by specifying a succession of links. The expression pdb > hssp > swissprot forces the system not to take the shortest route from PDB to SWISS-PROT but a deviation via HSSP. The result is a set of all SWISS-PROT protein sequences with solved tertiary structure and, in addition, all protein sequences in SWISS-PROT that are homologous to them. The expression takes advantage of the fact that HSSP entries list, for every entry in PDB, the SWISS-PROT entries that are similar above a given percentage residue identity. The link via HSSP can be understood as an amplification of a set of entries to all their related counterparts as defined by the data bank chosen. In lieu of HSSP, other data banks such as Prosite and Prodom, which collect and multiply align related protein sequences of subsequences, could have been chosen.

It has been discussed previously that subentries can be retrieved by searching in the feature table index. The result of such a query can be converted to a set of parent entries by a link to a predefined entity "parent." The result of the conversion can be used to link to other data banks. The following finds all PDB entries with proteins that have calcium binding sites as annotated in SWISS-PROT:

[swissprot-features:ca_bind] > parent > pdb

The link in the reverse direction gives all calcium binding sites with solved tertiary structure:

pdb > swissprot > [swissprot-features:ca_bind]

Note that the "parent" keyword is not needed here since the link from PDB to SWISS-PROT already results in a list of SWISS-PROT entries needed to link to the appropriate SWISS-PROT subentries or features.

Interfaces to Sequence Retrieval System

Sequence Retrieval System is written in the programming language ANSI C and runs on almost all UNIX platforms, as well as VMS, IBM-compatible, and Macintosh computers. Several interfaces have been written for SRS including graphical, character terminal-based, and World Wide Web (WWW) interfaces. Discussed subsequently are the availability of SRS, the application programmers interface (API), the command line interface, and the WWW server.

Availability

The SRS program and its interfaces are provided together with the SRS distribution file, which can be obtained via anonymous ftp from the server felix.embl-heidelberg.de in the directory pub/software/unix/srs. The current release of SRS is 4.08, with new releases made available every few months. A news group, bionet.software.srs, provides a forum for discussion and informs the community of bug fixes, new features, and releases. SRS is also evolving into a collaborative effort, most of the interfaces have been developed by other groups, and many of the data bank format descriptions have been written by individuals maintaining SRS services at their sites.

A large part of the SRS functionality will be accessible through the program "lookup" of the GCG program package for sequence analysis.[13] It will be distributed starting with release 8.1 of GCG.

SRS Programming Interface

The core of SRS consists of a library of functions to perform queries and extract and parse entries from flat file data banks in their original

[13] J. Devereux, P. Haeberli, and O. Smithies, *Nucleic Acids Res.* **12**, 387 (1984).

format. Entries can be processed in sequential order or by random access. Full entries or individual data fields can be extracted from about 80 different flat file data banks. A number of functions inform the user of the status and structure of the indexed data banks and the data bank network. Emphasis is on the extraction and output of sequences and subsequences in various formats. The function library is documented and is particularly suited for extending interpreted languages such as Perl, Tcl, and Python. SRS relies on a number of powerful tools such as the parser and the indexer, which can be directly used through the API.

Command Line Interface

The command line interface, the program getz, is a UNIX style command which gives access to all the retrieval power of SRS. Although graphical interfaces are more intuitive to use, getz provides a much more direct mode of operation and can be called from other programs or, through piping, can be combined with other UNIX commands. Also of interest is an enhanced version, hgetz, that accesses local as well as remote SRS servers through the HASSLE protocol.[14]

World Wide Web Server

The most popular access to SRS is through its World Wide Web server. The server combines the power of SRS links with the ease of WWW hypertext navigation. In contrast to most other WWW services, it maintains state, that is, previous user actions and selections are remembered so that all queries in the session can be reinspected or combined by SRS query language expressions.

Figure 5 illustrates the structure of the server. The top page allows the selection of the data banks to be queried and assigns a user ID on first entry. The next page presents a query form and options for the display of single entries or the entry list. A successful query leads to the entry list page, which gives access to the single entries and options to link the entire result set or single entries to other data banks, which must be selected on a separate page. Cross-references in single entries are, whenever possible, converted to hypertext links, which lead to entries served by the local or a remote SRSWWW installation or other servers such as the one from GDB (Genome Data Base).[15] The query manager page lists all successful queries conducted during the session and lets the user combine them by SRS query language expressions. Auxiliary pages, one for each data bank,

[14] R. Doelz, *Comput. Appl. Biosci.* **10**, 31 (1994).
[15] K. H. Fasman, A. J. Cuticchia, and D. T. Kingsbury, *Nucleic Acids Res.* **22**, 3462 (1994).

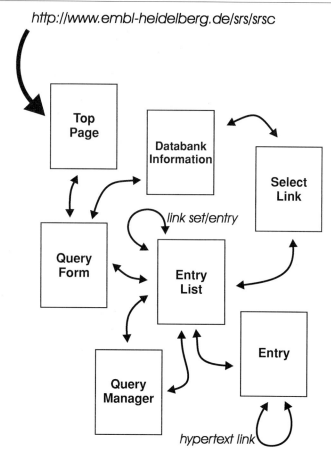

http://www.embl-heidelberg.de/srs/srsc

FIG. 5. Structure of the WWW server. Only the main pages are illustrated; arrows indicate entry paths. Top Page is the entry point of SRSWWW and can be obtained with the specified URL (Uniform Resource Locator). See text for further details.

inform the user about the indexed data banks, often with links to the distributor, and also allow the user to explore the data bank network. All pages without exception are built during run time so that changes in the system configuration, such as adding a new data bank, are immediately reflected by the interface.

The SRSWWW server is currently installed at many sites throughout the world. The number of data banks installed on a single node varies but can reach as many as 50. A special program polls all these servers once a day and retrieves the name, the number of entries, the release number, and the indexing date for each data bank installed at the site. The information is then compiled in a report which lists all data banks served by all nodes. This list includes 13 sites with a total of 80 data banks. Figure 6 shows the entry for the SWISS-PROT nucleotide data bank, which also includes links to the data bank information page and relates status information for all nodes serving that data bank. This information is useful both to the user who can select the closest server to avoid slow internet connections or the one with the most recent release of a given data bank and to the node administrators who can verify the currency of their system.

Discussion

Sequence Retrieval System is an integrated system that provides a homogeneous interface to all flat file data banks retained in their original format. It is a retrieval system that allows access to, but not the depositing of, data. Several elements are combined into a system that extends the power of normal retrieval systems and which rivals that of real databases, such as a relational system, without compromising speed. These elements include languages for data bank and syntax definition, a programmable parser, an indexing system, support for subentries, a novel system for exploiting links between data banks, and a query language. The database linking is a unique feature that considerably extends the capability of hypertext links.

SWISSPROT

ABC, Hungary			40292	21-Feb-1995	?!
BEN, Brussels			43470	07-Jun-1995	?!
BiO, Oslo	31.0		43470	28-Jun-1995	?!
Biozentrum, Basel			43470	22-Mar-1995	?!
CAOS/CAMM, Nijmegen	31.0		43470	31-May-1995	?!
CSC, Finland	31.0		43470	08-May-1995	?!
EBI, Hinxton, UK			43470	18-Jun-1995	?!
EMBL, Heidelberg			43470	22-Mar-1995	?!
IUBio Archive Indiana	31.0		43470	15-Apr-1995	?!
INSERM, France			43470	02-Jun-1995	?!
Sanger, Hinxton, UK	31.0		43470	05-Jun-1995	?!
Skirball Inst., NY			43470	13-Jun-1995	?!
Weizmann, Israel			43470	01-Jun-1995	?!

FIG. 6. Status of the data bank SWISS-PROT as installed on 13 SRSWWW servers. The entry has been taken from the status page (http://www.embl-heidelberg.de/srs/status.html) describing all data banks served by all 13 SRSWWW nodes. From left to right are listed the following: a hypertext link to the server; if specified, the release number of the installed version of SWISS-PROT; the number of entries; the indexing date; and the hypertext link that leads to a page with further information about SWISS-PROT as installed at the respective site.

SRS nonetheless has its limitations. To benefit from the linking capabilities, all data banks in the network must be installed at one site. This is in contrast to World Wide Web hypertext links, which very naturally allow navigation from one site to another. All information needs to be indexed before it can be accessed such that addition of new entries requires reindexing of the entire data bank. Work is in progress to overcome these limitations. It is also intended to enhance SRS from a server of textual information to an object server so that the retrieved data can be conveniently submitted to analysis programs such as sequence analysis tools.

Acknowledgment

The authors are grateful for financial support from the European Union (Grant Gene-CT-93-0043) under the Biomed I program.

Section II

Searching through Databases

[9] Applications of Network BLAST Server

By Thomas L. Madden, Roman L. Tatusov, and Jinghui Zhang

Introduction

The sequence databases have seen phenomenal growth in the last several years, with the number of base pairs in GenBank doubling every 21 months.[1] The composition of GenBank has also been changing, with over half of the sequences in the database now coming from high-throughput sequencing.[2] These trends present new challenges to sequence analysis tools. Database homology searches now report so many high-scoring hits that it is tedious to sort through them by hand. In addition, many of the hits are to sequences from high-throughput projects, and these records typically carry only minimal annotation. As almost nothing is known about the biological function of high-throughput sequences, researchers will want to use database searches to identify homologs between these sequences and already characterized genes.

To help meet these challenges, a sequence analysis strategy needs to combine a search tool such as BLAST[3] (Basic Local Alignment Search Tool) with processing of the query and results. Preprocessing of the query can mask repetitive elements, such as ALU sequences[4] or low-complexity regions,[5,6] thereby reducing the number of trivial hits. Postprocessing can further reduce the output by restricting the hits to a specific organism class, for example, or by applying a second pass with a tool having a higher sensitivity than BLAST. Some of this strategy is already implemented and available with programs such as the sequence filter seg.[5,6] However, recent changes in the databases have led to the development of an environment where different sequence analysis tools can more easily work together. In this chapter, we discuss a new BLAST service that will allow this. First we give a short description of BLAST.

[1] D. Benson, private communication (1995).
[2] M. S. Boguski, *Trends Biochem. Sci.* **20,** 295 (1995).
[3] S. F. Altschul, W. Gish, W. Miller, E. W. Myers, and D. J. Lipman, *J. Mol. Biol.* **215,** 403 (1990).
[4] J.-M. Claverie and W. Makalowski, *Nature* (*London*) **371,** 752 (1994).
[5] J. C. Wootton and S. Federhen, *Comput. Chem.* **17,** 149 (1993).
[6] J. C. Wootton and S. Federhen, this volume [33].

TABLE I
BLAST FAMILY OF PROGRAMS

Program	Description
blastn	Nucleotide database is compared with both strands of a nucleotide query. Program is optimized for speed, not for finding weak similarities encoded in nucleic acids
blastp	Protein database is compared with a protein query. Default scoring matrix is BLOSUM62
blastx	Nucleotide query is translated in six frames and compared to the protein database. Default scoring matrix is BLOSUM62
tblastn	Protein query is compared to a nucleotide database, translated in six frames. Default scoring matrix is BLOSUM62
tblastx	Nucleotide query, translated in six frames, is compared to a nucleotide database, translated in six frames. Because of time required, databases are restricted to dbEST, dbSTS, and alu

Description of BLAST

BLAST performs comparisons by concentrating on local alignments. This approach has been very successful, as frequently only localized regions, such as active sites, are preserved between sequences. Local alignments are also amenable to rigorous statistical analysis,[7,8] which allows a user to decide whether a hit is statistically significant. The BLAST algorithm takes advantage of these rigorous results to minimize the time spent on alignments unlikely to exceed the threshold score at which chance similarities occur. This makes BLAST much faster than many other heuristic algorithms. The recent addition of "sum" statistics,[9] which sums the highest scores if a sequence has multiple hits, has increased the sensitivity of BLAST.

Five different programs are available: blastn, blastp, blastx, tblastn, and tblastx. Essentially, all combinations of DNA or protein query sequences with searches against DNA or protein databases are possible. Table I describes these programs.

The central unit of BLAST output is the high-scoring segment pair (HSP). An HSP consists of (equal) lengths of two sequences that have a locally maximal alignment, according to the scoring system being used. With command-line parameters the user may control the significance of reported HSPs and database sequences. A typical query, for blastp, would be

blastp spdb aa.query S=70 S2=70 -filter seg

[7] S. Karlin and S. F. Altschul, *Proc. Natl. Acad. Sci. U.S.A.* **87**, 2264 (1990).
[8] S. F. Altschul and W. Gish, this volume [27].
[9] S. Karlin and S. F. Altschul, *Proc. Natl. Acad. Sci. U.S.A.* **90**, 5873 (1993).

```
BLASTP 1.4.8MP [19-Dec-94] [Build 09:57:45 May  1 1995]

Reference: Altschul, Stephen F., Warren Gish, Webb Miller, Eugene W. Myers,
and David J. Lipman (1990).  Basic local alignment search tool.  J. Mol. Biol.
215:403-10.

Query= RAE2_HUMAN
       (656 letters)

Database: SWISS-PROT Release 31.0
          43,470 sequences; 15,335,248 total letters.
Searching.................................................done
```

		High Score	Smallest Sum Probability P(N)	N
Sequences producing High-scoring Segment Pairs:				
sp|P26374|RAE2_HUMAN	RAB PROTEINS GERANYLGERANYLTRANSFERA...	3130	0.0	1
sp|P24386|RAE1_HUMAN	RAB PROTEINS GERANYLGERANYLTRANSFERA...	1679	1.3e-284	2
sp|P37727|RAE1_RAT	RAB PROTEINS GERANYLGERANYLTRANSFERA...	1690	7.7e-278	2
sp|P39958|GDI1_YEAST	SECRETORY PATHWAY GDP DISSOCIATION I...	165	1.7e-43	3
sp|P21856|GDIS_BOVIN	GDP DISSOCIATION INHIBITOR FOR RAB3A...	158	3.4e-41	3
sp|P32864|RAEP_YEAST	RAB PROTEINS GERANYLGERANYLTRANSFERA...	164	1.5e-32	3
sp|P24588|AK79_HUMAN	A-KINASE ANCHOR PROTEIN 79 (AKAP 79)...	76	0.057	1
sp|P37747|YEFE_ECOLI	HYPOTHETICAL 43.0 KD PROTEIN IN RFC-...	71	0.25	1

FIG. 1. The one-line descriptions from a blastp run for the human choroideremia protein are shown. The output consists of a FASTA identifier and a definition of the database sequence. The probability [listed under P(N)] expresses the likelihood of the match occurring by chance, with lower probabilities, or *p* values, indicating greater significance.

where the database is spdb, which contains SWISS-PROT and PDB (Protein Data Bank) sequences, and aa.query is the filename of the query sequence. The parameter S=70 limits the reported database sequences to those with at least one HSP that has a score of at least 70. The parameter S2=70 limits the reporting of individual HSPs to those with a score of at least 70 (a database sequence may have more than one HSP). Finally, -filter seg invokes the seg program, which masks regions of low compositional complexity (for amino acid queries) that might otherwise result in many irrelevant matches.[5,6]

The BLAST report consists of three parts: a histogram, a list of one-line descriptions of the hits that includes (truncated) definitions, and a set of alignments. The example in Fig. 1 shows a portion of the blastp output for a search of the human choroideremia protein (RAE_2), the product of a gene implicated in hereditary blindness.[10] The first hit in the list is the query sequence (which is already in the database). Alignments for the fourth and eighth related hits are shown in Fig. 2. These two hits include the yeast homolog of the choroideremia protein and an uncharacterized protein YefE from *Escherichia coli*. Note that the two alignments, even

[10] M. C. Seabra, M. S. Brown, and J. L. Goldstein, *Science* **259**, 377 (1993).

```
>sp|P39958|GDI1_YEAST SECRETORY PATHWAY GDP DISSOCIATION INHIBITOR.
          Length = 451

  Score = 132 (60.7 bits), Expect = 1.7e-43, Sum P(3) = 1.7e-43
  Identities = 24/59 (40%), Positives = 39/59 (66%)

Query:    3 DNLPTEFDVVIIGTGLPESILAAACSRSGQRVLHIDSRSYYGGNWASFSFSGLLSWLKE 61
            + + T++DV+++GTG+ E IL+   S  G++VLHID + +YGG  AS + S L    K+
Sbjct:    4 ETIDTDYDVIVLGTGITECILSGLLSVDGKKVLHIDKQDHYGGEAASVTLSQLYEKFKQ 62


>sp|P37747|YEFE_ECOLI HYPOTHETICAL 43.0 KD PROTEIN IN RFC-RFBX INTERGENIC
          REGION.
          Length = 367

  Score = 71 (32.6 bits), Expect = 0.28, P = 0.25
  Identities = 13/38 (34%), Positives = 25/38 (65%)

Query:    9 FDVVIIGTGLPESILAAACSRSGQRVLHIDSRSYYGGN 46
            +D +I+G+GL ++ A   +  ++VL I+ R++ GGN
Sbjct:    2 YDYIIVGSGLFGAVCANELKKLNKKVLVIEKRNHIGGN 39
```

FIG. 2. Two local alignments from the blastp run discussed in Fig. 1. Here the complete definition line of the database sequences is shown, as well as some statistics about the match. The alignments show identity between residues by printing the one-letter symbol between the sequences. Similarity between two different residues, according to the BLOSUM 62 matrix, is shown by a "+" between the sequences. The dinucleotide-binding motif (see text) is in boldface type and underlined in both sequences.

though one is highly statistically significant and the other has only a marginal significance, include the same portion of the choroideremia protein (underlined in Fig. 2). Subsequent analysis using the CAP[11] and MoST[11] programs (described below) showed that this segment contains a dinucleotide-binding motif, with implications for the functions of the choroideremia protein.[12]

To obtain more information about using BLAST, send an E-mail message to blast-help@ncbi.nlm.nih.gov. The BLAST World Wide Web (WWW) server may be accessed through the home page of the National Center for Biotechnology Information (NCBI) at http://www.ncbi.nlm.nih.gov.

BLAST Network Service

BLAST is an example of a client–server application. The client is a program on the user's machine and the server is a program on machines at the NCBI that provides access to the search programs and databases. A service designated "experimental" has been offered since 1992 to allow users to search databases at the NCBI. This service transfers data between the client and the server as the BLAST report. Unfortunately this limits

[11] R. L. Tatusov, S. F. Altschul, and E. V. Koonin, *Proc. Natl. Acad. Sci. U.S.A.* **91,** 12091 (1994).
[12] E. V. Koonin, *Nature Genetics,* in press (1996).

the flexibility of BLAST clients. It is difficult to write robust applications that parse text, as minor or even inadvertent changes to the BLAST report (that may not be considered important by providers of the BLAST service) can force a major rewrite of parsers. A new "network" service, now being introduced by the NCBI, transfers structured data between the client and server. This structured data is based on a specification called Abstract Syntax Notation 1 (ASN.1) and can be reliably and easily parsed. A software developer can now concentrate on the client user interface or additional processing programs rather than the parsing of the BLAST report. BLAST client software may be developed by building the portable NCBI toolbox into the application. An Application Programming Interface (API) in the toolbox facilitates access to BLAST. The NCBI toolbox, and the API, may be obtained from the ncbi.nlm.nih.gov anonymous FTP site in the /toolbox/ ncbi_tools directory.

The network BLAST service allows clients more flexibility in the presentation of BLAST results. Here we present four BLAST client projects that take advantage of the network service to handle BLAST results in innovative new ways. Readers interested in obtaining these clients should follow the instructions at the end of each project description. The NCBI will continue to support a client that prepares the traditional BLAST report. This client is part of the NCBI toolbox and runs under many different operating systems, including such popular ones as Apple Macintosh, Microsoft Windows, as well as a number of flavors of UNIX.

BLAST_ORG: Filter by Organism

Genome researchers interested in assembling the sequence map of a particular organism need to filter out sequences that are not from their organism. With the traditional BLAST interface it is necessary to use the unreliable approach of checking either the locus name or the definition line from the text output. A more reliable alternative is to look up the organism field in GenBank records. Doing this manually is quite inconvenient for a small data set; it is impossible for a large volume of data, such as the 6000 GenBank/EMBL (European Molecular Biology Laboratories)/ SWISS-PROT records of *Escherichia coli*.

blast_org was originally written to parse the traditional BLAST report, fetch the sequence record from *Entrez*[13] and then filter out records that did not match the organism specified by the user. Minor changes to the BLAST report, such as the addition of the gi, required that blast_org be modified. blast_org now uses structured output coming from the network

---

[13] G. D. Schuler, J. A. Epstein, H. Ohkawa, and J. A. Kans, this volume [10].

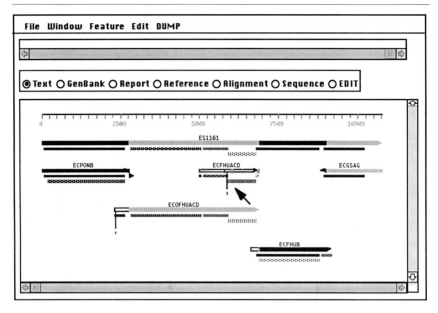

FIG. 3. Graphical view of the one-to-many alignment between the contig sequence ES1161 (from EcoSeq6) and five other related *E. coli* entries in GenBank. The alignment is computed by the program sim2 as a postprocess of the BLAST search after filtering by organism to the *E. coli* K12 strain. The vertical line attached to a sequence indicates an insertion, which can be observed in ECOFHUACD and ECFHUACD. A gap is presented as a line connecting two boxes (⬜–⬜), which can be observed in ECFHUACD. A mismatch is indicated by a red line (gray in black–white printouts) in sequence (⬜—⬜). The coding region features are marked underneath each sequence (▬). The arrow points to a frameshift in a coding region of sequence ECFHUACD.

BLAST service. From this output it is simple to obtain the sequence identifier; the sequence record can then be obtained directly. blast_org also now incorporates the sim2[14] algorithm to compute gapped alignments of all the hits. The structured output from BLAST supplies the orientation and approximate range as input to sim2, which makes the sim2 computation more efficient than an alignment using the entire subject sequence. blast_org was used in the (genome curating) EcoSeq[15] project to find only hits to *E. coli*. The related entries are aligned to an *E. coli* contig sequence to give a complete view of all the sequence data. Figure 3 presents a graphical view of the result. Under an *E. coli* contig sequence, sequence variations such as gaps, insertions, and mismatches can be observed.

[14] K.-M. Chao, J. Zhang, J. Ostell, and W. Miller, *Comput. Appl. Biosci.* **11,** 147 (1994).
[15] K. E. Rudd, *ASM News* **59,** 335 (1993).

blast_org can be obtained by contacting J. Zhang at zjing@ncbi.nlm. nih.gov.

ALUBLAST: Filter for Repetitive Regions

A homology search of very long genomic sequences (over 100 kb) can be a formidable task if one uses the traditional BLAST interface. The high background noise can obscure truly meaningful hits. Repetitive sequences, such as ALU sequences in primates, can have hundreds of hits.[4] Low complexity regions can also have thousands of random matches.[5,6] The new network API, and the NCBI toolbox, facilitates writing software tools that filter out the background noise before sending the query to the server. Only regions that are of interest to a researcher are sent to the server and searched against the database, making more efficient use of the BLAST service. An example of this approach is the alublast program developed for internal use at the NCBI. alublast scans through the sequence to filter out repetitive regions, such as ALU and MER sequences; low complexity regions are identified by the dust[16] algorithm, and the corresponding residues are masked; finally, the filtered data are sent to the BLAST server using the BLAST API.

alublast has been tested with a 235,701-bp human genomic sequence.[17] The filtering steps significantly reduce the number of extraneous hits and improve the quality of the resulting list. When the repetitive sequences are filtered, without the dust step, in the region from 1505 to 4334 bp, there are 3000 BLAST hits. Most of these are aligned to low complexity regions. Using the traditional BLAST interface, one must raise the cutoff score to 600 to get rid of this background noise. The drawback is obvious: sensitivity on the whole sequence is sacrificed for the selectivity of one particular region. In contrast, when the low complexity regions are filtered, the cutoff score can be as low as 50 and only seven hits are found in this region. The alublast search reveals five potential coding regions in this entry; most of these are identified by homology to cDNAs or expressed sequence tags (ESTs). One is over 70,000 bp long, starting from bp 163,070 and extending to the end of sequences on the minus strand. It includes 30 exons, and the conceptually translated protein shows significant homology to the clathrin heavy chain proteins from various organisms, such as yeast, slim mold, worm, fruit fly, rat, and human. However, as shown in Fig. 4, the sequence difference is significant enough to regard it as a new class of clathrin protein. Without filtering the background noise, it would be almost impossible to

---

[16] R. L. Tatusov and D. J. Lipman, in preparation.
[17] B. Roe, personal communication (1995).

```
BruceRoe_1   523   QALVELFESFKSYKGLFYFLGSIVNFSQDPDVHLKYIQAACKT
CLH_HUMAN    137   .S.I........FE...................F.........
   gi 29983  137   .S.I........FE...................F.........
   gi 434761 696   .S.I........FE...................F.........
    CLH_RAT  696   .S.I........FE...................F.........
   gi 203302 696   .S.I........FE...................F.........
CLH_DROME    697   K..ID...G....D......S..........E..F........
   gi 7722   697   K..ID...G....D......S..........E..F........
CLH_CAEEL    698   DK.I.M..NH...E................E..F.....TR.
   gi 458481 698   DK.I.M..NH...E................E..F.....TR.
CLH_DICDI    697   E.IIAM....RL.E..YLY.TQV.VT.TS.E..F...E..A.I
   gi 167688 697   E.IIAM....RL.E..YLY.TQV.VT.TS.E..F...E..A.I
CLH_YEAST    702   ST.IK...DYNATE..Y.Y.A.L..LTE.K..VY...E..A.M
   gi 3536   702   ST.IK...DYNATE..Y.Y.A.L..LTE.K..VY...E..A.M
```

FIG. 4. Alignment of the conceptually translated protein sequence BruceRoe_1 and 13 other clathrin heavy chain protein sequences in GenBank. BruceRoe_1 is derived from an alublast search and comprises 30 exons. For each sequence in the alignment, both the sequence name (either the NCBI gi number or the SWISS-PROT locus name) and the position are labeled. The aligned sequences are from different organisms: human (CLH_HUMAN, gi 29983, gi 434761), rat (CLH_RAT, gi 203302), fruit fly (CLH_DROME, gi 7722), worm (CLH_CAEEL, gi 458481), slime mold (CLH_DICDI, gi 167688), and yeast (CLH_YEAST, gi 3536). Mismatched residues are labeled, whereas the sequence identities are represented with dots.

identify all the exons scattered in such a wide range. The identified exons agree perfectly with the cDNA sequencing data from this research project.

alublast can be obtained by contacting J. Zhang at zjing@ncbi.nlm. nih.gov.

CAP: Consistent Alignment Parser

Especially difficult to interpret in BLAST results are statistically insignificant but potentially biologically important similarities. One approach in such a case is to use a motif search tool to detect subtle sequence similarities. MoST[11] is such a tool that takes an alignment block as input and performs iterative scans of the sequence database to find blocks of conserved segments. CAP[11] (Consistent Alignment Parser) performs a BLAST run and produces alignment blocks as input for MoST. These blocks are optimized for a motif search and group together similar HSPs from different database sequences. The CAP algorithm performs an almost exhaustive search of all possible blocks that could be constructed from BLAST output. The only heuristic used is that segments with high scores are more probable to form a block. One may specify that blocks include no less than a certain percentage of the sequences found by BLAST, the default being 50%; one may also specify that only alignments that have a minimum BLAST score be included.

Figure 5 presents CAP output from the blastp run of the human choroideremia protein against SWISS-PROT that was discussed in Figs. 1 and 2.

```
Query= RAE2_HUMAN (656 letters)
Database: SWISS-PROT Release 31.0 43,470 sequences; 15,335,248 total letters.
Total number of segments: 17
Limit height  7 of 9 seq
Limit width   7
Limit overlap 0
Criterion     Height * Width

height: 8/9 (89%),  width: 38,    *266   Score: 1698

                              .....:....|....:....|....:....|....:...
RAE2_HUMAN                  9 FDVVIGTGLPESILAAACSRSGQRVLHIDSRSYYGGN 46
>sp|P26374|RAE2_HUMAN       9 FDVVIGTGLPESILAAACSRSGQRVLHIDSRSYYGGN 46   S = 3130,  P = 0
>sp|P24386|RAE1_HUMAN       9 FDVVIGTGLPESIIAAACSRSGRRVLHVDSRSYYGGN 46   S = 410,   P = 0
>sp|P37727|RAE1_RAT R       9 FDVIVGTGLPESIIAAACSRSGQRVLHVDSRSYYGGN 46   S = 350,   P = 0
>sp|P39958|GDI1_YEAST      10 YDVIVLGTGITECILSGLLSVDGKKVLHIDKQDHYGGE 47  S = 132,   P = 1.69557e-43
>sp|P32864|RAEP_YEAST      47 VDVLIAGTGMVESVLAAALAWQGSNVLHIDKNDYYGDT 84  S = 131,   P = 1.5e-32
>sp|P21856|GDIS_BOVIN       5 YDVIVLGTGLTECILSGIMSVNGKKVLHMDRNPYYGGE 42  S = 118,   P = 3.39997e-41
>sp|P37747|YEFE_ECOLI       2 YDYIIVGSGLFGAVCANELKKLNKKVLVIEKRNHIGGN 39  S = 71,    P = 0.25

                           6 0.8 .DV..GTGL.E.I.A...S..G..VLH.D...YYGG.
                           4 0.6 F..III....P.S.L.AAC.RS..R...I.SRS....N
                           4 0.6 ...V...............................
                           3 0.4 Y...............L....KK.....K.........
                           3 0.4 ...................Q...............
```

Fig. 5. CAP output for the example discussed in Figs. 1 and 2. Here the block was calculated with a parameter set that has been found useful: blocks include at least 70% of the sequences found by BLAST, and included sequences have a minimum BLAST score of 70.

The left-hand column lists the accessions of the sequences; the middle column presents the alignment ("block"); the right-hand column shows scores and p values found by blastp. To assist in analysis each block is accompanied by consensus information, where the most conservative letters are shown along with their frequencies.

CAP and MoST can be obtained by anonymous FTP to ncbi.nlm.nih.gov (cd to pub/most).

BLANCE: BLAST Report Summary

BLAST can produce an overwhelming amount of data that can hide hits of interest. Voluminous output might be due to sequence runs with unusual composition or to the multiple domain structure of a query sequence, both of which lead to a large number of hits. To address these problems, BLANCE was developed to present a graphical summary of BLAST results.

The BLAST matches are organized on the basis of mutual similarity. The output is divided into groups containing similar sequences by adding them one at a time to a group of sequences, beginning with the first one in the output (usually the most similar). Sequences are added to a group if they contain HSPs that overlap by a certain percentage, which is an adjustable parameter. A graphical (or pseudographical in text mode) presentation helps one understand the structure of the BLAST hits by a single glance at the output. This allows the user to see quickly where different database sequences share similarities with the query, and possibly with each other. Figure 6 presents BLANCE output of the blastp run discussed in Figs. 1 and 2.

BLANCE can be obtained by anonymous FTP to ncbi.nlm.nih.gov (cd to pub/tatusov/blance).

RAE2_HUMAN

8 sequences
4 groups (70% clustering)

```
                                                              length: 656
 ┌────────────────────────────────────────────────┐          ..3 ● RAE 2 HUMAN
 ┌──┐········┌────┐·······┌────┐···················            ..3 ● GDI1 YEAST
      ·····┌────────┐·····························             ..1 ● AK79 HUMAN
 ·┌──┐······································                    ..1 ● YEFE ECOLI
```

FIG. 6. BLANCE output for the example discussed in Figs. 1 and 2. The bars schematically mark regions where hits occur. Each line represents a group of similar hits, with the number of sequences in the group listed to the right of the bars (e.g., the first line represents three similar hits) and the most similar sequence identifier, for each group, listed on the right (e.g., RAE2_HUMAN). Note the overlap of the leftmost bar of the GDI1_YEAST group and the only bar of YEFE_ECOLI, which contains the dinucleotide-binding motif underlined in Fig. 2.

Summary

   The sequence databases continue to grow at an extraordinary rate. Contributions come from both small laboratories and large-scale projects, such as the Merck EST project.[2] This growth has placed new demands on computational sequence comparison tools such as BLAST. Even now it is no longer practical to evaluate some BLAST reports manually; it is necessary to filter the output by, for example, organism, source, or degree of annotation. The new network BLAST service makes such tools possible. It is also possible to present BLAST output in different formats, such as BLANCE. Perhaps most important of all, it becomes simple to call BLAST from another application, making it one step within an integrated system. This makes the automated preparation of sequence evaluations that include BLAST runs possible. In the near future we expect to see a number of applications that use the network BLAST interface to help molecular biologists search against a database that is growing not only in size but in biological richness.

Acknowledgments

   We thank Dr. Mark Boguski and Dr. Dennis Benson for critical reading of the manuscript and helpful discussions. We thank Dr. Eugene Koonin for suggesting examples and reviewing the manuscript. We thank Dr. David Lipman for suggesting the idea of the BLANCE algorithm.

[10] *Entrez:* Molecular Biology Database and Retrieval System

By Gregory D. Schuler, Jonathan A. Epstein, Hitomi Ohkawa, and Jonathan A. Kans

Introduction

   The process of discovery in science often involves making serendipitous connections between seemingly unrelated observations. For researchers in the biomedical arena, this process is made difficult, above all, by the explosive rate at which these observations are being made and reported in the scientific literature. By any of several possible measures—numbers of literature records in MEDLINE, DNA and protein sequences in the public databases, solved three-dimensional molecular structures, and mapped hu-

Copyright © 1996 by Academic Press, Inc.
All rights of reproduction in any form reserved.

man genes—the rate of growth has been and continues to be exponential.[1] A number of developments of the past decade, notably CD-ROM technology, the Internet, and the World Wide Web (WWW), have made it significantly easier and less expensive to distribute large numbers of scientific documents by electronic means. The ensuing explosion in the volume of available biomedical information has produced an expansive and varied information landscape that must be traversed in the quest for new discoveries. While it has become relatively easy to collect and distribute large amounts of information, making effective use of that information remains a challenge. In effect, we require some sort of road map to keep from getting lost in the information landscape, some means of establishing a path from one observation to another.

In the field of molecular biology, sequence comparison, both at the DNA and the protein level, has become an increasingly important means of establishing connections among diverse biological systems, often opening up entirely new fields of inquiry.[2] One of the first examples of this was the discovery that the sequences of the viral oncogene v-*sis* and human platelet-derived growth factor (PDGF) were virtually identical, an observation which shed new light on a human disease process.[3,4] Moreover, connections between sequences and the corresponding three-dimensional protein structures often provide additional clues to biological function, since comparison of three-dimensional structures often reveals relationships that cannot be discerned at the sequence level. Many of the cases of structural similarity reported to date have been recognized by visual inspection, but computational methods are now becoming more widely available.[5] For both sequences and structures, links to the relevant scientific literature are critical to providing an intellectual framework for understanding any connections that may be detected by computational means. Furthermore, decades of research in automatic text analysis provide a statistical means for establishing connections between related papers, providing yet another avenue to discovery.[6]

Entrez is a molecular biology database and retrieval system developed by the National Center for Biotechnology Information (NCBI)[7] that pre-

[1] M. S. Boguski, *Curr. Opin. Genet. Dev.* **4,** 383 (1994).

[2] S. F. Altschul, M. S. Boguski, W. Gish, and J. C. Wootton, *Nat. Genet.* **6,** 119 (1994).

[3] R. F. Doolittle, M. W. Hunkapiller, L. E. Hood, S. C. Davare, K. C. Robbins, and S. A. Aaronson, *Science* **221,** 275 (1983).

[4] M. D. Waterfield, G. T. Scrace, N. Whittle, P. Stroobant, A. Johnsson, A. Wasteson, B. Westermark, C. H. Heldin, J. S. Huang, and T. F. Duel, *Nature* (*London*) **204,** 35 (1983).

[5] L. Holm and C. Sander, *Proteins: Struct. Funct. Genet.* **19,** 165 (1994).

[6] G. Salton, *Science* **253,** 974 (1991).

[7] D. Benson, M. Boguski, D. J. Lipman, and J. Ostell, *Genomics* **6,** 702 (1990).

sents an integrated view of biomedical data and their interrelationships (Fig. 1). Database documents containing biological sequences, three-dimensional structures, or abstracts from the scientific literature can be retrieved using simple Boolean queries. However, documents identified in this manner are not end points in themselves. Instead, they serve as entry points for further exploration (hence the name *Entrez*, French for enter). Hypertext links may then be used to navigate through the information space using a simple point-and-click interface. Some of these links are simple cross-references, for example, between a sequence and the abstract of the paper in which the sequence was reported, or between a protein sequence and the DNA sequence of the gene which encodes it. Others, however, are based on computed similarities among the documents (described in more detail below), thereby providing the nearest "neighbors" of any given document. The mode of navigation encouraged by *Entrez*—directed query followed by neighborhood browsing—is analogous to the way one might use a library: consultation of the card catalog sends the user to the stacks in search of a specific book, only to find many other interesting books in its neighborhood on the shelf.

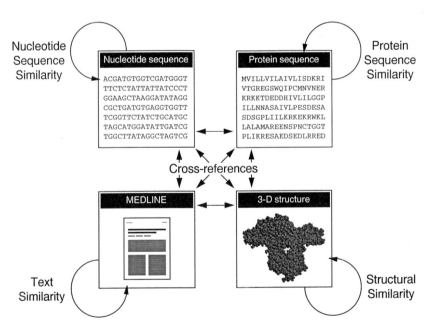

Fig. 1. Different classes of documents available for browsing in *Entrez* are linked by both intrinsic cross-reference information and computed relationships.

Building the *Entrez* Information Resource

Biological Sequence Data

DNA and protein sequence data in *Entrez* are drawn from all publicly available sources. A primary source is GenBank, a comprehensive repository for annotated nucleotide and protein sequence data that is built and distributed by the NCBI.[8] Data collection is international in scope and is managed by the NCBI together with collaborators at the European Bioinformatics Institute (EBI)[9] and the DNA Data Bank of Japan (DDBJ),[10] with whom information is exchanged on a daily basis. Two special classes of data, expressed sequence tags (ESTs) and sequence tagged sites (STSs), are handled through the specialized databases dbEST[11] and dbSTS,[12] respectively. In addition, the *Entrez* database has been augmented with nucleotide and protein sequences from SWISS-PROT,[13] PIR-International (Protein Identification Resource),[14] Brookhaven Protein Data Bank (PDB),[15] the Protein Research Foundation (PRF) database, and the Genomic Sequence Database (GSDB). Data from these sources enter the integrated sequence database of the NCBI through a process that involves (1) assignment of unique sequence identifiers, (2) conversion to a uniform data representation, (3) extensive content validation, (4) matching literature citations against MEDLINE, and (5) matching organism names against the GenBank Taxonomy.

Accurate taxonomic classification is essential to the effective retrieval of sequences from specific organisms or from larger taxonomic groups.[16] The GenBank Taxonomy was developed for this purpose by merging and

---

[8] D. Benson, M. Boguski, D. J. Lipman, and J. Ostell, *Nucleic Acids Res.* **24,** 1 (1996).

[9] D. B. Emmert, P. J. Stoehr, G. Stoesser, and G. N. Cameron, *Nucleic Acids Res.* **22,** 3445 (1994).

[10] H. Kitakami, Y. Yamazaki, K. Ikeo, Y. Ugawa, T. Shin-i, N. Saito, T. Gojobori, and Y. Tateno, *in* "Advances in Molecular Bioinformatics" (S. Schulze-Kremer, ed.), p. 123. IOS Press, Amsterdam, 1994.

[11] M. S. Boguski, T. M. Lowe, and C. M. Tolstoshev, *Nat. Genet.* **4,** 332 (1993).

[12] NCBI Creates New Database, New GenBank Division for STS Data, *NCBI News* **February** (1994).

[13] A. Bairoch and B. Boeckman, *Nucleic Acids Res.* **22,** 3578 (1994).

[14] D. G. George, W. C. Barker, H.-W. Mewes, F. Pfeiffer, and A. Tsugita, *Nucleic Acids Res.* **22,** 3569 (1994).

[15] E. E. Abola, F. C. Bernstein, S. H. Bryant, T. F. Koetzle, and J. C. Weng, *in* "Crystallographic Databases: Information Content, Software Systems, Scientific Applications" (F. H. Allen, G. Bergerhoff, and R. Sievers, eds.), p. 107. International Union of Crystallography, Bonn, Chester, Cambridge, 1987.

[16] D. M. Hillis, *Syst. Biol.* **43,** 461 (1994).

unifying taxonomic data from a variety of sources.[17] It encompasses all of the taxonomic designations used in the source sequence databases, so it is a simple matter to assign sequences to nodes in the phylogenetic tree. Furthermore, specialized taxonomies have been supplied by the International Committee on Taxonomy of Viruses (ICTV), the U.S. Department of Agriculture (USDA), FlyBase, the National Oceanic and Atmospheric Administration/National Oceanographic Data Center (NOAA/NODC), the American Type Culture Collection (ATCC), and the Ribosome Data Project (RDP). The GenBank Taxonomy is curated by a staff of taxonomists at the NCBI with significant contributions and advice from a panel of taxonomic specialists.

The finding that a gene of interest has a homolog in a model organism can immediately open many possibilities for further experimentation. A typical way to achieve this phylum hopping is to make use of a fast search program, such as FASTA[18] or BLAST,[19] to compare a query sequence against an entire sequence database. Patterns of sequence conservation revealed by these methods can provide important clues to the structure, function, and evolution of the gene under study. Today, no researcher would seriously consider publishing a new sequence without having first performed a database search. Indeed, this type of search is so common that it becomes worthwhile to perform all possible searches in advance and store the results for rapid retrieval.

The sequence neighbors provided by *Entrez* are generated by comparing all database sequences among themselves using the BLAST algorithm and saving a ranked list of the best 100 matching sequences. Because BLAST seeks alignments that may be inexact and possibly confined to short, localized sequence segments, an important question is whether a given alignment reported by BLAST represents a true relationship or could be due simply to chance.[20] Statistical methods developed by Karlin and Altschul[21,22] allow the statistical significance of any BLAST hit to be evaluated using a simple calculation involving only its score and certain properties of the search context, such as the size of the database and the substitution matrix in use. Because the operation of BLAST is fundamentally based on the Karlin–Altschul model, it is possible to specify the sensitivity of the search in terms

[17] Improving GenBank's Taxonomy, *NCBI News* **February** (1994).
[18] W. R. Pearson and D. J. Lipman, *Proc. Natl. Acad. Sci. U.S.A.* **85**, 2444 (1988).
[19] S. F. Altschul, W. Gish, W. Miller, E. W. Meyers, and D. J. Lipman, *J. Mol. Biol.* **215**, 403 (1990).
[20] S. F. Altschul and W. Gish, this volume [27].
[21] S. Karlin and S. F. Altschul, *Proc. Natl. Acad. Sci. U.S.A.* **87**, 2264 (1990).
[22] S. Karlin and S. F. Altschul, *Proc. Natl. Acad. Sci. U.S.A.* **90**, 5873 (1993).

of a statistical expectation E, which may be thought of as the number of hits likely to occur by chance.

Entrez protein neighbors are generated at a significance level of $E = 0.2$, which keeps false positives to a low rate while still allowing some relatively distant relationships to be detected. Although the "ideal" substitution matrix for amino acid sequence comparison depends on the degree to which the proteins have diverged (an unknown in this case), the PAM120 matrix has been found useful for general-purpose searching.[23] It has been recognized that dispersed regions of biased amino acid composition are a significant source of false positives in database searches. These features, also called low-complexity regions (LCRs), are surprisingly common in proteins, with more than half of the proteins in the database containing at least one such region.[24] Although LCRs may well have some interesting functional and structural importance, they present problems for sequence comparisons, and a common practice is to mask these regions by replacing their amino acids with the ambiguity character "X," thereby preventing them from contributing to the alignment score. Two programs, SEG[25] and XNU,[26] have been developed for detecting and masking LCRs, both of which are employed in the generation of protein neighbors for *Entrez*.

For nucleotide sequence comparisons, the scoring system is a simple match/mismatch scheme, with a score of 5 being awarded for a match and -4 for a mismatch. A greater degree of statistical significance ($E = 0.000001$) is demanded for DNA than in the case of proteins, with the assumption that the most useful type of links will be among genomic sequences that overlap to form a contig or between mRNA and genomic sequences corresponding to the same gene. Finding subtle patterns at the DNA level is difficult because the smaller alphabet size (4 bases as opposed to 20 amino acids) results in more background "noise." One common source of false positives is the presence of repetitive DNA, such as Alu repeats. To circumvent this problem, an additional constraint is imposed that requires the region of similarity to extend essentially the full length of the overlap between the two sequences being compared. Isolated regions of similarity found centrally often correspond to repetitive elements. More sophisticated measures for dealing with repetitive elements are currently under development.

[23] S. F. Altschul, *J. Mol. Biol.* **219,** 555 (1991).
[24] J. C. Wootton and S. Federhen, this volume [33].
[25] J. C. Wootton and S. Federhen, *Comput. Chem.* **17,** 149 (1993).
[26] J.-M. Claverie and D. J. States, *Comput. Chem.* **17,** 191 (1993).

Molecular Structure Data

The amount of available three-dimensional data for biological macro-molecules has increased rapidly as a result of advances in X-ray crystallography and nuclear magnetic resonance (NMR) spectroscopy. These data describe biological macromolecules in atomic detail, often, as in the case of protein–ligand complexes, providing a clear picture of the structural basis for biological function. The primary reason for including these data in *Entrez* is to allow biologists to determine easily whether a three-dimensional structure is available for a protein of interest, and to examine it readily. This information can be quite valuable even when the structure of a particular protein is unknown, since the structures of homologs identified by sequence neighboring may be available. In this case, conservation of structure[27,28] may allow one to infer readily the locations of active-site residues and other functional properties. A secondary reason for including structural data in *Entrez* is to allow easy identification of structural neighbors, proteins with statistically significant similarities of three-dimensional structure. Many structural neighbors are homologs that cannot be detected by sequence comparison, because of the more rapid evolutionary divergence of sequence than structure. By including structural neighbors the scope of similarity relationships detectable in *Entrez* is expanded.

Three-dimensional structure data appear in *Entrez* as a distinct class of documents that may be presented in a variety of views. A simple text summary provides a convenient view for browsing molecule names, structural features, and comments provided by the original authors. However, structure records are most usefully examined in a three-dimensional graphic display. The NetEntrez application (see below) supports direct three-dimensional visualization and allows simultaneous highlighting of selected residues in the structure and sequence displays. In addition, the data may be rendered in the Kinemage Format used by the Mage[29] viewer and in PDB Format which is used by the RasMol[30] application (both programs are publicly available for a variety of platforms). Structure documents are cross-referenced to the corresponding MEDLINE citations, so that one may examine abstracts of the original structure reports. They are also linked to the PDB-derived protein and nucleotide sequences, which have been generated from the atomic coordinates.

Structural data in *Entrez* are derived from the latest available quarterly release of PDB.[15] These data are changed in form, however, and undergo

---

[27] C. Chothia and A. M. Lesk, *EMBO J.* **5,** 823 (1986).
[28] A. C. W. May and T. L. Blundell, *Curr. Opin. Struct. Biol.* **5,** 355 (1994).
[29] D. C. Richardson and J. S. Richardson, *Protein Sci.* **1,** 3 (1992).
[30] M. A. Saqi and R. Sayle, *Comput. Appl. Biosci.* **10,** 545 (1994).

a number of validation checks in preparation of MMDB, the Molecular Modeling Database that is used by *Entrez*.[31] The main purpose of the validation checks is to prepare a description of covalent structure that is strictly consistent with the reported three-dimensional structure, as is required for homology modeling software. The results of these checks are for the most part invisible to the *Entrez* user, but one should be aware that the content of the PDB Format files produced by *Entrez* may sometimes differ from the source PDB file with respect to noncoordinate data, when validation has necessitated changes to the biopolymer sequence or bonded connectivity of nonpolymer groups.

The *Entrez* structural database also differs from PDB in that it provides precomputed "views" of each model at different levels of details and certain machine-derived annotation. The three views available are a single-coordinate-per-residue model, a single-coordinate-per-atom model, and an ensemble/correlated-disorder model. The first simple view is intended for rapid transmission of backbone-only coordinates over the Internet. The second provides full atomic detail but omits description of alternative, statistically defined atom locations. The last corresponds to the complete coordinate data from PDB, and it should be selected only when the available computer and network connection are capable of supporting the many megabytes of data that sometimes result. Currently the only added annotation is an automatic definition of linear secondary structure elements, as is used in calculation of structural neighbors.[32,33] This supplements PDB-defined secondary structure descriptions, which are available for some but not all structures. The complete MMDB dataset is available by FTP from ncbi.nim.nih.gov in the /pub/mmdb directory.

The strategy used to identify structurally similar proteins is based on the observation that related proteins often show significant conservation of internal "core" substructures even though the length and configuration of surface loops may vary.[27,28] The process begins by automatically identifying the β strands and α helices that comprise the core of the protein. The spatial orientation and chain connectivity of these regions are then compared to produce a similarity score.[32,33] The neighbors that are included in *Entrez* are those with similarity scores that stand out clearly from the background distribution of scores expected for randomly chosen globular proteins. It should be noted, however, that structural similarity may be reported for some proteins with no obvious evolutionary relationship or

[31] H. Ohkawa, J. Ostell, and S. H. Bryant, *in* "Proceedings of The Third International Conference on Intelligent Systems for Molecular Biology." Cambridge, 1995.

[32] T. Madej, J.-F. Gibrat, and S. H. Bryant, *Protein: Struct. Funct. Genet.* **23**, 356 (1995).

[33] J.-F. Gibrat, T. Madej, and S. H. Bryant, in preparation.

functional similarity, including the so-called superfolds.[5,34] These neighbors may represent either very ancient divergence or possibly convergent evolution, but these possibilities cannot be distinguished by structural comparison alone.

MEDLINE Bibliographic Data

Links to the scientific literature can be critical to understanding potential relationships detected at the sequence or structural level, and thus *Entrez* makes available the abstracts for articles in the area of molecular genetics. The expressiveness of the natural language present in the abstract provides far greater insight for the human user than any controlled database field. Furthermore, as discussed below, it is possible to compute a measure of similarity for text documents consisting of plain text, thereby providing ranked neighbor lists. Consequently, neighborhood browsing within the literature provides yet another route to discovery.

MEDLINE is an international bibliographic database maintained by the National Library of Medicine (NLM), containing more than 8 million records drawn from roughly 4000 journal titles. Each record contains the title, authors, and journal citation for the article, and most also have the full text of the abstract. Additionally, each record bears a set of MeSH (Medical Subject Heading) identifiers that have been assigned by expert indexers. The MeSH system of controlled subject headings, which is published each year by the NLM, is hierarchical in nature, so, for example, the heading Genetics contains Immunogenetics, which in turn contains Antibody Diversity.

Since plain text constitutes the bulk of the data in the computers, there has been a long history of research in the area of text-based document retrieval.[6] A conventional approach would involve assigning keywords or index terms to each document and making use of them, combined with the Boolean operators AND, OR, and NOT to query the database. For MEDLINE, the assigned MeSH terms may serve as index terms. A full-text retrieval system, in addition to assigned keywords, uses all of the words contained in a document as retrieval terms (excluding certain stop terms, such as "if," "but," "many," and so forth). However, rather than treating all terms as equally valuable, significant retrieval power can be gained by weighting terms based on some measure of their importance in determining content. In a statistical text retrieval system, frequencies of term occurrences in both the database as a whole and in individual documents are used to calculate term weights automatically.

[34] C. A. Orengo, D. T. Jones, and J. M. Thornton, *Nature (London)* **372,** 631 (1994).

Through the use of statistical text methods, it is possible to develop a similarity metric for pairs of MEDLINE records that can be used to generate ranked neighbor lists for browsing in *Entrez*. It has been found useful to weight the individual terms on the basis of some measure of their value in determining subject matter. Most methods for calculating text similarity take this as the product of two values: a global weight and a local weight.[35-38] The global weight is meant to measure the overall usefulness of a term and depends on the fraction of the "documents" in the database that contain it. Less frequently occurring terms, dystrophin for instance, would have high global weights, whereas those that occur very frequently, such as "important" or "result," would be discounted. The local weight is scaled by the frequency of the term in the particular document. A more frequent term in the document is more likely to reflect the content of that document.

Entrez neighboring makes use of local weights, but the global weighting method is unique.[39,40] A global weight is calculated on the basis of the fraction of the documents that contain the term as well as an estimate of the importance of the term in relating documents, called the strength of the term. Statistics are kept on the closely related pairs of documents that have been found in previous processing. From these statistics, estimates are made of the probability that when a term occurs in one document it will also occur in a closely related document. This probability is called the strength of the term. We actually calculate two global weights for each term, a positive inclusive and a negative exclusive global weight. The similarity of two documents is judged based on the sum of the inclusive weights for all terms they have in common plus the sum of all exclusive weights for those terms that occur in only one of the documents. The larger the sum of weights, the more similar is the pair of documents.

Because of the large size and wide scope of MEDLINE, it has become necessary (particularly with the CD-ROM distribution) to work with various subsets of the bibliographic records. A Biosequence subset has been defined that consists of MEDLINE records which are cited in sequence database records. The Biosequence associated subset additionally includes records that are close neighbors of those in the Biosequence subset. These two groups, currently totaling roughly 200,000 abstracts, are supplied in

[35] G. Salton, "Automatic Text Processing," Addison-Wesley, Reading, Massachusetts, 1989.
[36] S. K. M. Wong and V. V. Raghaven, *in* "Research and Development in Information Retrieval" (C. J. van Rijsbergen, ed.), p. 167. Cambridge Univ. Press, Cambridge, 1984.
[37] G. Buckley and A. F. Lewit, *in* "Proceedings of the Eighth International ACM Conference on Research and Development in Information Retrieval," p. 97. ACM Press, New York 1985.
[38] H. Turtle and B. Croft, *ACM Trans. Information Systems* **9,** 187 (1991).
[39] W. J. Wilbur, *Proc. Am. Soc. Inf. Sci.* **29,** 216 (1992).
[40] W. J. Wilbur and Y. Yang, *Comput. Biol. Med.* in press.

the CD-ROM distribution. However, on the Internet it is possible to make available a much larger set, which currently consists of approximately 1.2 million records. It is roughly equivalent to those records having a MeSH term that falls below the heading Genetics in the MeSH hierarchy, but augmented as necessary to ensure that all records cited by sequences are included. The size of this segment may be expanded in the future as resources allow.

Accessibility of *Entrez* Information

For maximal impact on the research community, the data must be made accessible in a convenient and timely manner. Rapid advances in computer technology continually provide new means of achieving this goal. As discussed below, three mechanisms for the delivery of *Entrez* data are currently available (Table I).

Entrez on CD-ROM

The original *Entrez* package, CdEntrez, is based on CD-ROM technology and contains the complete DNA and protein sequence databases along with a small subset of MEDLINE. It is available on a subscription basis from the U.S. Government Printing Office, updated six times annually. The supplied software runs on many computer platforms, including the Apple Macintosh and Microsoft Windows systems, as well as a variety of workstations. Since its introduction in October, 1992, the CdEntrez distribution has grown from a single disk to a set of five. This increase was largely fueled by the explosive growth in the EST segment of GenBank. Although it is still possible to use the system with fewer than five CD-ROM drives,

TABLE I

THREE IMPLEMENTATIONS OF *ENTREZ*

Implementation	Requirements	Benefits
CdEntrez	*Entrez* CD-ROM subscription	Stand-alone system
	1 to 5 CD-ROM drives	Can support local area network users
NetEntrez	Internet connection	No subscription charges
	NetEntrez client software	Expanded MEDLINE subset
		Three-dimensional structures
		Efficient use of network bandwidth
WebEntrez	Internet connection	No subscription charges
	WWW browser software	Expanded MEDLINE subset
		Hyperlinks to external databases
		Usable on character-only terminals

it is at the expense of convenience, since many disk swaps will be required during the course of a single session. To get around the limitations associated with current CD-ROM technology and to provide more convenient and timely access to the data, two Internet-based solutions have been implemented.

Client–Server Implementation of Entrez

The network-enabled *Entrez* application NetEntrez was introduced in 1993. It has the same look and feel of the original CdEntrez (Fig. 2a) but obtains its data over the Internet instead of from CD-ROM. NetEntrez is an example of what is called a client–server application. The client portion is the program running on the local machine, while the server is a process running on one or more large computers at the NCBI, providing access to the databases. The NetEntrez client software (again, available for many platforms) may be obtained by FTP from ncbi.nlm.nih.gov (look below the entrez/network directory for a README file containing downloading and installation instructions). The only requirement for using it is a direct connection to the Internet (an E-mail only connection is not sufficient). Many academic institutions provide hard-wired (TCP/IP) Internet connections, and a number of commercial carriers offer dial-up (PPP or SLIP) Internet access. The NetEntrez protocol is also customized to provide rapid performance and waste minimum bandwidth by transmitting data across the Internet only on an as-needed basis. Depending on the speed of the connection, NetEntrez may rival or exceed CdEntrez in responsiveness.

NetEntrez has several distinct advantages over CdEntrez. Since there are no subscription charges, access to the data is effectively free (although connection charges to an Internet access provider may be required, depending on the site). Furthermore, without the limitations of CD-ROM capacity, there is the potential for virtually limitless growth. The data can also be updated on a more frequent basis. Currently, the sequences are updated daily and MEDLINE weekly. Finally, centralized storage makes it very easy to add new data sources or enlarge existing ones. For example, with the CD-ROM distribution, it is not practical to supply large subsets of MEDLINE or complete three-dimensional structure data, but it is possible to make these data available on the Internet.

Entrez on World Wide Web

The World Wide Web (WWW) is an Internet-based information retrieval system that was originally developed at CERN in Geneva,[41] but it has

---

[41] T. J. Berners-Lee, R. Calliau, J. F. Groff, and B. Pollermann, *Electronic Networking: Research, Applications, and Policy* **2,** 52 (1992).

grown to be a ubiquitous global medium for the distribution and hypertext browsing of electronic documents ("pages") which may contain text, graphics, and multimedia.[42] The hypertext nature of the WWW provides a natural fit for *Entrez*, which emphasizes navigation based on connections between related documents. A series of WWW pages providing the *Entrez* functionality, called WebEntrez, was introduced in 1994 and has increased dramatically in popularity. The WWW also operates under a client–server model, but the client in this case is a generalized WWW browser program (e.g., Mosaic or Netscape Navigator) that makes use of the hypertext transfer protocol (HTTP). Therefore, the system adopts the look and feel of the particular browser in use (Fig. 2b). A request for information on the WWW makes use of a uniform resource locator (URL), which is effectively a network address for the data of interest. The URL for the NCBI home page on the WWW is http://www.ncbi.nlm.nih.gov/.

Many of the advantages described above for NetEntrez apply to WebEntrez as well, since it too makes use of data stored on a centralized server, which can provide large and frequently updated versions of the databases. Furthermore, it is possible within WebEntrez to include hypertext links to external data sources. For example, certain *Drosophila* sequence records contain links to FlyBase.[43] Conversely, any WWW site around the world can include hyperlinks to documents in *Entrez* (instructions on how to do this may be found on the NCBI WWW server).

Entrez Usage Statistics

The history of *Entrez* usage reflects its general popularity as well as the relative popularity of the underlying technologies that the three different versions employ (Fig. 3). During 1994 the CD-ROM subscriptions plateaued, partly because of the additional CD-ROMs needed and partly due to the increased popularity of the Internet versions. WebEntrez usage, in particular, has continued to grow during 1995 due to the success of the World Wide Web. Comparing usage of the three types of *Entrez* is complex because there is no way to know how many users are actually using the *Entrez* CD-ROM and its derivative hard-disk copies. Furthermore, there is no exact analogy between NetEntrez sessions and WebEntrez URLs. However, a more detailed study suggests that an average Network *Entrez* session corresponds to roughly eight WebEntrez URLs.

Strategies for Using *Entrez*

Although the usage details differ between WebEntrez and the other two implementations, the basic concepts remain the same. The user begins

---

[42] B. R. Schatz and J. B. Hardin, *Science* **265**, 895 (1994).
[43] The FlyBase Consortium, *Nucleic Acids Res.* **22**, 3456 (1994).

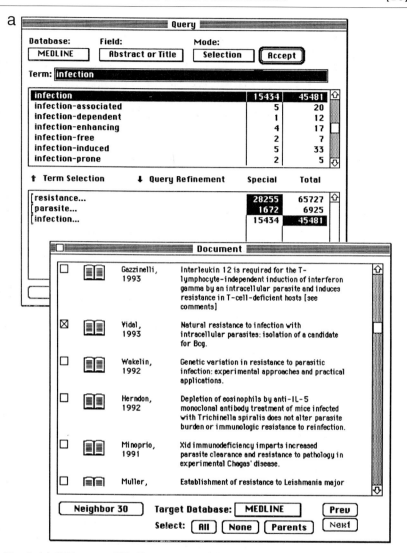

FIG. 2. (a) CdEntrez and NetEntrez are specialized applications developed by the NCBI for browsing biomedical information. (b) WebEntrez makes use of generalized WWW browser software (Netscape Navigator in this case).

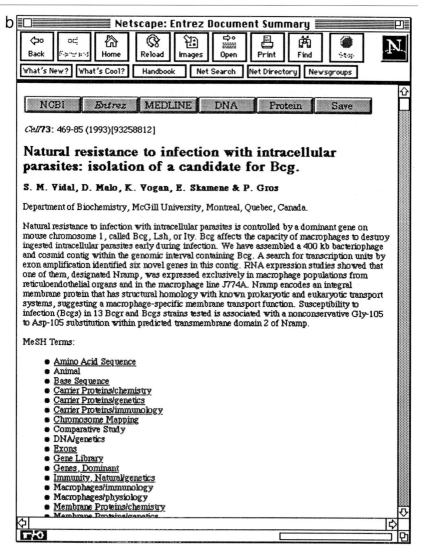

FIG. 2. (*continued*)

an *Entrez* session by choosing one of the available databases (MEDLINE, protein, nucleotide, or structure) and composing a Boolean query designed to select a small set of documents. The terms comprising the query may be drawn from either plain text or any of several more specialized fields, such as gene symbol or E/C number (Table II). One field, organism, has special properties as a consequence of its hierarchical nature. For instance,

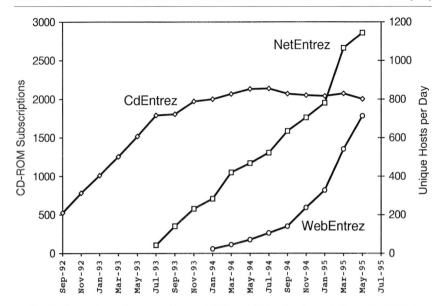

Fig. 3. Comparison of usage patterns of different *Entrez* implementations. For CdEntrez, the number of subscriptions is shown. Usage of NetEntrez and WebEntrez is presented in terms of unique host machines per day.

the use of the organism term Eumycota (or the common name fungi) would retrieve all fungal sequences, regardless of whether the term actually appears in the record. The number of documents containing each of the terms selected is displayed, along with the number satisfying the query as a whole. The query can be refined by adding or merging terms, and the number of documents satisfying the current query is continually updated. Once a satisfactory query is selected, a list of summaries (brief descriptions of the document contents) may be viewed. Successive rounds of neighboring (to related documents of the same database type) or linking (to associated records in other databases) may be performed using this list. At any step it is possible to view complete MEDLINE, sequence, or structure documents, each in a variety of formats. The records may also be printed and saved to disk files for future reference or for use by other software.

Example: Entrez Session

It is perhaps easiest to illustrate the use of *Entrez* through an example (Fig. 4), based on mechanisms of resistance to parasites. To retrieve relevant articles from the MEDLINE subset of *Entrez*, perform an initial Boolean query with the terms resistance, parasite, and infection. When using NetEn-

TABLE II
CATEGORIES OF INDEX TERMS

		Document types[a]			
Category	Description	ML	NT	AA	ST
Text	Plain text drawn from all text-containing fields	√	√	√	√
Author	Author's last name and initials	√	√	√	√
Journal title	Journal title abbreviation	√	√	√	√
Date	Year of publication	√	√	√	√
MeSH	Medical Subject Heading[b]	√			
Organism	Taxonomic and common organism name[c]		√	√	
Gene symbol	Gene symbol and gene name	√	√	√	
Protein name	Protein name and description	√	√	√	√
EC Number	Enzyme Commission number	√	√	√	√
Substance	Chemical substance name	√	√	√	√
Keyword	Sequence database keywords		√	√	
Feature key	GenBank feature key symbol (e.g., CDS)		√		
Properties	Miscellaneous properties (e.g., partial)		√	√	

[a] ML, MEDLINE record; NT, nucleotide sequence; AA, protein sequence; ST, molecular structure.
[b] In NetEntrez, a browsing facility for the MeSH tree is included.
[c] A hierarchical view of the taxonomic tree is available for browsing.

trez, the initial query may be satisfied by a large number of documents. Then the query may be refined (reducing the number of articles to a more manageable level) by requiring some of the words to be in the title of articles. The goal is to find one article of particular interest and then use *Entrez* neighboring to find others, rather than trying to formulate a Boolean query that would give complete coverage to a topic. The results of the initial query for this example are shown in Fig. 2a.

One of the retrieved documents is entitled, "Natural resistance to infection with intracellular parasites: isolation of a candidate for Bcg."[44] This article discusses a cloned sequence in mouse, NRAMP (for natural resistance associated macrophage protein), a candidate for Bcg, which is an autosomal dominant gene that affects the ability of macrophages to destroy intracellular parasites early in infection. Given an interesting article, a typical request to a librarian is, "Find more papers like this." *Entrez* neighbors incorporate this functionality, and they can be retrieved with a click of the computer's mouse. The nearest neighbors of the NRAMP article are indeed other papers about Bcg or NRAMP. (Bcg is also known as Ity

[44] S. M. Vidal, D. Malo, K. Vogan, E. Skamene, and P. Gros, *Cell (Cambridge, Mass.)* **73,** 469 (1993).

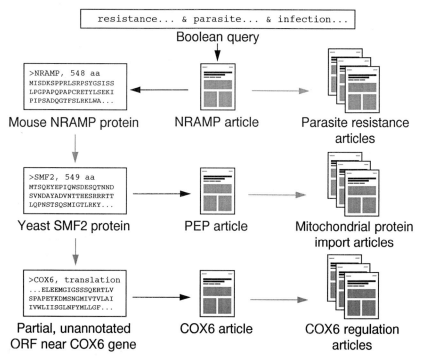

FIG. 4. Exploration of a topic with *Entrez* following the example presented in the text. The process begins by selecting a few terms to serve as the query. A truncation mode (indicated by an ellipsis in the user interface) allows matching on related words that share the same stem, for example, "infection" and "infections." Solid arrows indicate links (cross-references to a different database), and shaded arrows indicate neighbors (computed similarities within a single database). The connection between SMF2 protein and the translated *COX6* sequence was identified using TBLASTN, which may be executed directly from NetEntrez.

or Lsh and influences resistance to infection by *Leishmania* and other important parasites, so this is a popular area of study.)

Another mouse click in *Entrez* links from the NRAMP article to the associated protein sequence record (in this case the conceptual translation from a GenBank record's coding region feature). This record contains the sequence of an integral membrane protein 548 amino acids in length. Neighboring on the NRAMP protein (the equivalent of running a BLAST search) results in a number of protein sequence records. Among these are homologs from mouse, human, nematode, fruit fly (malvolio, involved in taste behavior), and yeast (SMF1 and SMF2).

Given a protein BLAST hit, only a limited amount of biological context can be gleaned from the sequence record itself. A large amount of pertinent information is in the abstract of the associated journal article. Because *Entrez* links are reciprocal, one can move from MEDLINE to sequence and sequence to MEDLINE with equal ease. The article in which SMF1 and SMF2 were published is entitled, "Two related genes encoding extremely hydrophobic proteins suppress a lethal mutation in the yeast mitochondrial processing enhancing protein."[45] A glance at the abstract of this article reveals that the SMF1 and SMF2 cloned genes complement a defect in the yeast PEP gene (for processing enhancing protein). PEP is involved in removal of signal peptides from precursor proteins being imported into mitochondria, and an assay for its activity is already available. This immediately raises the possibility that the assay could be used to detect other NRAMP genes, or to screen for antiparasitic drugs that might influence the effectiveness of the natural resistance mechanism. Interestingly, the MEDLINE neighbors of this article all discuss mitochondrial protein processing, but do not include the original NRAMP article or its domain of literature. So these two fields might have remained unaware of their intimate underlying biological relationship if it were not for the connection made by sequence similarity.

An epilog to this example is that when SMF2 is run through TBLASTN (a variant of BLAST that translates the nucleotide database in all six reading frames and searches the results with a protein sequence), it detects an unannotated similarity adjacent to the *COX6* gene in yeast. The article in which this was published discussed *COX6* and two uncharacterized open reading frames.[46] *COX6*, a nuclear gene, encodes one of the subunits of cytochrome oxidase, which is translated in the cytoplasm and then transported through the mitochondrial membrane. The presence of a potential regulator of the transport step next to its potential target suggests that complementation experiments on this region may need to be reinterpreted.

Using Entrez Functionality in Other Software

Commercial and academic users may wish to develop their own software which can access the power of *Entrez*, including access to all of the databases, neighbors, and links provided by *Entrez*. Such software may be developed by building the portable NCBI Software Development Toolkit[47] into the developer's application software. Access to either the CdEntrez or NetEntrez databases may be achieved via the use of a single application

[45] A. H. West, D. J. Clark, J. Martin, W. Neupert, F. U. Hartl, and A. L. Horwich, *J. Biol. Chem.* **267,** 24625 (1992).

[46] R. M. Wright, B. Rosenzweig, and R. O. Poyton, *Nucleic Acids Res.* **17,** 1103 (1989).

[47] J. M. Ostell, *in* "Nucleic Acid and Protein Analysis: A Practical Approach" (M. Bishop and C. Rawlings, eds.), in press. IRL Press, Oxford, 1996.

programming interface (API). The complete NCBI Software Development Toolkit, including source code for CdEntrez and NetEntrez, may be freely obtained from the ncbi.nlm.nih.gov anonymous FTP site in the /toolbox/ ncbi_tools directory.

Conclusions and Future Directions

Entrez is a biomedical information resource that has been designed to facilitate the discovery process by providing connections among biological sequences, molecular structures, and abstracts. Since it must be anticipated that the amount of data will continue to grow at phenomenal rates, the Internet would seem to be the most practical medium for future dissemination of this information. However, besides these quantitative changes, several trends promise to alter qualitatively the nature of the nucleotide sequence database.

One trend whose impact has already been felt is the rise in popularity of EST projects, which generate large volumes of single-pass cDNA sequence. The rationale for this survey sequence approach is to generate a large amount of gene sequence rapidly and inexpensively that, despite its incompleteness and high error rate, can nevertheless be useful for identifying novel genes, making molecular probes, and developing gene-based mapping reagents.[48] Merck & Co., in collaboration with Washington University Genome Center, launched a major initiative to generate some 400,000 new EST sequences over a period of 18 months. At 6 months into the project, fully one-third of all entries in the current release of GenBank had resulted directly from the Merck initiative, with about 1500 new sequences being added each day. Besides sheer bulk, one problem that this presents for users of *Entrez* is the high degree of redundancy that can result from abundantly expressed genes. This causes the nucleotide sequence neighbor lists to accumulate many ESTs, which often do not provide any biological insight for the user other than the fact that a sequence may be expressed. We are currently working toward partitioning these sequences into unique clusters that can be used to eliminate redundancy and generate more useful views of the data in *Entrez*.

As a result of the Human Genome Initiative,[49] the trend toward very large-scale genomic sequencing will be increasingly important in the coming years. Much of the progress to date has been in the development of high-resolution genetic and physical maps of the human genome, together with the genomes of experimentally important model organisms. However,

---

[48] O. White and A. R. Kerlavage, this volume [2].
[49] F. S. Collins and D. Galas, *Science* **262,** 43 (1993).

with this initial groundwork approaching completion, attention is shifting toward high-throughput sequencing of genomic regions that may be megabases in size, perhaps encompassing whole chromosomes in some organisms.

Clearly, it is necessary to be prepared for the expected volume of data, but changes to the *Entrez* user interface may also be needed to make effective use of it. For example, imagine finding a sequence of interest and asking for its sequence neighbors only to be presented with a complete chromosome sequence. Would such a neighbor be useful? Certainly it could be if it were presented in the right way. Researchers studying a disease gene mapped by linkage analysis to a specific region of the chromosome for example, would want to see all sequences that map to that region (i.e., neighbors would need to include locations instead of being simply pairs of documents). On the other hand, those primarily interested in studying the function and regulation of single genes would prefer to view the data in smaller, gene-sized units.

What is needed is the ability to examine the data at varying levels of detail. The data model and software tools used by the NCBI to represent and manipulate sequences were designed from the beginning to accommodate exactly these needs.[50] Sequences can be stored in segments of manageable size but include the instructions needed by the software to reassemble the complete sequence on-demand. Furthermore, the data model can with equal ease be used to represent simple maps that contain no actual sequence at all (a map can be thought of as a low-resolution sequence) or composites of sequences and maps. Consequently, it will be possible in future versions of *Entrez* to incorporate some of the high-quality genetic and physical map information that has been generated by the scientific community. A new graphical viewer in *Entrez* will allow the user to view the genomic landscape from different vantage points and make connections to the sequences, structures, and abstracts relevant to specific chromosomal regions. Through the unification of these diverse information sources, *Entrez* provides an information infrastructure to support biomedical research well into the twenty-first century.

Acknowledgments

Renata McCarthy provided data on Entrez CD-ROM subscriptions, and Rose Marie Woodsmall provided numbers for MEDLINE record count and journal titles indexed. For

---

[50] J. M. Ostell, *IEEE Engineering in Medicine and Biology Magazine* **14,** 730 (1995).

critical review of the manuscript and many helpful suggestions we thank Steve Bryant, John Wilbur, Dennis Benson, Mark Boguski, and Jim Ostell.

The production of *Entrez* requires the efforts of many talented individuals. There is insufficient space to list them all here, but virtually the entire staff of the NCBI is involved in one way or another. We thank them all for a job well done.

[11] Applying Motif and Profile Searches

By Peer Bork and Toby J. Gibson

Introduction

The demonstration of homology, meaning descent from a common ancestor, is an essential tool in gaining understanding of gene function, whether one wants to obtain an overview of the functions for all the genes described during a genomic sequencing project or to focus on a particular protein. With the expansion of sequence databases, similarity searches have a steadily increasing chance of providing a clue toward functional characterization. The likelihood of identifying homologs is currently higher than 80% for bacteria, 70% for yeast, and about 60% for animal sequence queries.[1,2]

On the basis of experience with large-scale sequence analysis,[3–5] we estimate that at present about 10–20% of identifiable similarities cannot be retrieved automatically by standard database search programs such as BLASTP[6,7] and FASTA[8] alone. The proportion of missed similarities is even higher when considering modular proteins that are composed of several (often small) functionally and structurally independent domains. The significance of "twilight zone" matches (i.e., tempting pairwise similarities below widely used thresholds of approximately 25% identity, depending

[1] P. Bork, C. Ouzounis, and C. Sander, *Curr. Opin. Struct. Biol.* **4**, 393 (1994).

[2] E. V. Koonin, R. L. Tatusov, and K. E. Rudd, this volume [18].

[3] P. Bork, C. Ouzounis, C. Sander, M. Scharf, R. Schneider, and E. Sonnhammer, *Protein Sci.* **1**, 1677 (1992).

[4] E. V. Koonin, P. Bork, and C. Sander, *EMBO J.* **13**, 493 (1994).

[5] P. Bork, C. Ouzounis, G. Casari, R. Schneider, C. Sander, M. Dolan, W. Gilbert, and P. M. Gillevet, *Mol. Microbiol.* **16**, 955 (1995).

[6] S. F. Altschul, W. Gish, W. Miller, E. W. Myers, and D. J. Lipman, *J. Mol. Biol.* **215**, 403 (1990).

[7] S. F. Altschul, M. S. Boguski, W. Gish, and J. C. Wootton, *Nat. Genet.* **6**, 119 (1994).

[8] W. R. Pearson and D. Lipman, *Proc. Natl. Acad. Sci. U.S.A.* **85**, 2444 (1988).

Copyright © 1996 by Academic Press, Inc.
All rights of reproduction in any form reserved.

on length) has to be assessed using information provided by inspection of multiple alignments of protein families, as well as by deploying motif and profile alignment strategies based on these alignments. Searches with these tools, in turn, often lead to the identification of even more divergent homologs. The purpose of this chapter is to outline the main strategies currently in use, making clear both their powers and pitfalls, and to demonstrate their usage with two well-known domains as examples.

Terminology: Motifs, Patterns, and Profiles

The meaning of the terms motif, block, pattern, and profile need to be clarified. The search for motifs, namely, small conserved regions within larger entities, implies that only some of the information contained in a protein or domain is used. Sometimes, motifs are applied to certain functional features (e.g., glycosylation sites, SH3-binding sites) that develop independently from the surrounding context; for this (minority of) motifs the concept of homology may be irrelevant. As insertions and deletions (gaps) within a motif are not easy to handle from the mathematical point of view, a more technical term, alignment block, has been introduced that refers to conserved parts of multiple alignments containing no insertions or deletions. Patterns can be used in a more broader sense, as they can describe small motifs or larger regions containing several motifs and can also contain gaps. (Some authors use the term pattern to mean compositionally biased segments or low complexity regions, e.g., runs of aspartate residues; this usage is not meant here.) Profile is usually used to mean a full representation of features in the aligned sequences and normally implies position-dependent weights/penalties for all 20 amino acids (as well as for insertion and deletion).

Thus, there need be no contradiction of the terms motif and profile, as profiles can also be restricted to smaller regions. Nevertheless, the terms motif and profile mirror two different ideologies in the field of using family information for improving the sensitivity of homology searches: (1) restriction to key conserved features to reduce the higher "noise" level of the more variable regions, in contrast to (2) inclusion of all possible information to maximize the overall signal of the entity (protein/domain). Both approaches are valid, as documented by many successful applications of each. It is worth noting that motifs are usually harnessed to fast word search algorithms, which can be used despite limited resources, whereas profiles often use exhaustive but slow dynamic programming algorithms. Therefore it is best to use the method most appropriate given the resources and the nature of the protein family under study. It can also be advisable to do the earlier searches using motifs first, and to follow up with the slower profiles at the end.[9]

Numerous methods exist for both motif and profile searches, and the different methods grade into each other. No exact formula can be given for the choice of the method, as each protein family has a different conservation pattern. Different functional and structural constraints lead to a certain distribution of conserved and variable regions within a multiple alignment that may better suit one of the approaches.

Note that we have specifically excluded here methods that can be combined under the term "threading," that is, that try to derive potentials (which can be translated into profiles) from known three-dimensional structures, in order to recognize the compatibility of a given sequence to one of the known three-dimensional folds. Such methods have been reviewed extensively (e.g., Ref. 10 and references therein). Although they should have great potential in finding distant homologies, as yet only a few predictions based on such approaches have been published. Such successes as there have been could, in our view, have been achieved with conventional motif and profile searches using solely sequence information.

Procedures

In the hope of predicting a function for a protein under study, fast homology search programs are almost universally used. The current standard seems to be the BLAST series of programs, accessible via several World Wide Web (WWW) servers, although both FASTA and BLITZ are also frequently used. These programs undertake a database search for a query sequence and are usually the sole search undertaken. However, the results of such searches make a logical starting point for motif and profile searches (Fig. 1). These can be divided into two steps: (1) derivation of a pattern and (2) searching for the pattern (motif/profile). For the first step, programs such as CAP (consistent alignment parser; see Ref. 11 and references therein) have been developed that are able to parse outputs of "one against all" initial database search programs and that automatically create multiple alignments of conserved regions shared by the query sequence and some of the database proteins. Other methods have been developed that can be used to find conserved regions in a set of unaligned sequences (e.g., Gibbs sampler[12]; blockfinder: S. Henikoff, personal communication,

[9] P. Bork, J. Gellerich, H. Groth, R. Hooft, and F. Martin, *Protein Sci.* **4,** 268 (1995).

[10] F. Eisenhaber, B. Persson, and P. Argos, *Crit. Rev. Biochem. Mol. Biol.* **30,** 1 (1995).

[11] R. L. Tatusov, S. F. Altschul, and E. V. Koonin, *Proc. Natl. Acad. Sci. U.S.A.* **91,** 12091 (1994).

[12] C. E. Lawrence, S. F. Altschul, M. S. Boguski, J. S. Liu, A. F. Neuwald, and J. C. Wootton, *Science* **262,** 208 (1993).

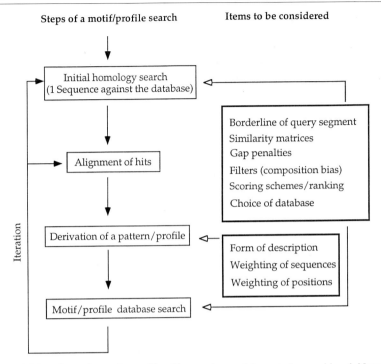

Steps of a motif/profile search   Items to be considered

FIG. 1. Flowchart of steps in motif/profile searches and items to be considered. Note that some items may not apply to all the methods (e.g., gap parameters are not needed in MoST). Also, some of the programs combine several steps (e.g., in SOM, a starting alignment is not necessary because the detection of the conserved features in a set of sequences is part of the pattern recognition algorithm itself, Ref. 33a).

see Fig. 2; pattern extraction[13]; SOM: Ref. 33a, see Fig. 2). Even if such motifs are defined, it remains non-trivial to develop a proper description of these regions in order to be as sensitive as possible in the second step, the pattern search itself.

Available Programs

It is impossible to give a comprehensive overview of the numerous methods that exist for pattern (motif/profile) searches, especially given the explosion of WWW activities. The Internet provides access to a broad variety of programs/servers from simple string searches to sophisticated profile descriptions (for a small collection of recommended servers, see

---

[13] R. F. Smith and T. F. Smith, *Proc. Natl. Acad. Sci. U.S.A.* **87,** 118 (1990).

WWW: searchable motif and pattern databases

PROSITE	[Geneva's Expasy]	http://expasy.hcuge.ch/sprot/prosite.html
Motif search, ICR	[Kyoto]	http://www.genome.ad.jp/SIT/MOTIF.html
Scan of profiles, ISREC	[Lausanne]	http://ulrec3.unil.ch/software/PFSCAN_form.html
BLOCKS, Hutchinson	[Seattle]	http://www.blocks.fhcrc.org/
PRINTS , UC	[London]	http://www.biochem.ucl.ac.uk/~attwood/PRINTS/PRINTS.html
PIMA, BCM	[Houston]	http://dot.imgen.bcm.tmc.edu:9331/seq-search/protein-search.html
PRODOM	[Toulouse]	http://protein.toulouse.inra.fr/prodom.html

WWW: Motif and profile searches

Regular expressions,Univ	[Washington]	http://ibc.wustl.edu/fpat/
PROFILE, Weizmann	[Tel Aviv]	http://sgbcd.weizmann.ac.il/Bic/ExecAppl.html
PATSCAN motif search	[Argonne]	http://www.mcs.anl.gov/home/papka/ROSS/patscan.html
PatternFind, ISREC	[Lausanne]	http://ulrec3.unil.ch/software/PATFND_mailform.html
Pmotif (protein >DNA)	[Minneapolis]	http://alces.med.umn.edu/pmotif.html
HMM descriptions	[St. Louis]	http://genome.wustl.edu/eddy/hmm.html
Discover (email server!)	[New York]	http://hertz.njit.edu/~jason/help.html

FTP: addresses for some motif and profile search programs

Barton's flexible patterns	[Oxford]	ftp://geoff.biop.ox.ac.uk
Propat (property pattern)	[Berlin]	ftp://ftp.mdc-berlin.de/pub/makpat
SOM (neural networks)	[Berlin]	ftp://ftp.mdc-berlin.de/pub/neural
SearchWise	[Oxford]	http://www.ocms.ox.ac.uk/~birney/wise/topwise.html
PROFILE, EBI	[Cambridge]	ftp://ftp.ebi.ac.uk/pub/software/unix
TPROFILESEARCH, EBI	[Cambridge]	ftp://ftp.ebi.ac.uk/pub/vax/egcg
MoST (motif search tool)	[Bethesda]	ftp://ncbi.nlm.nih.gov/pub/koonin/most
CAP (blast output parser)	[Bethesda]	ftp://ncbi.nlm.nih.gov/pub/koonin/cap

FIG. 2. Some programs that are accessible on the Internet via WWW pages or ftp addresses. Note that this list is not comprehensive but that the authors have tested the programs referred to in this table. A more detailed list and respective pointers can be obtained via the WWW from http://www.embl-heidelberg.de/~bork/pattern.html.

Fig. 2, or go directly to the WWW at http://www.embl-heidelberg.de/~bork/ pattern.html). In this section we will briefly summarize proven methods that have been extensively and successfully used in the identification of distant homology.

Most of the methods currently used have their origins in approaches and ideas developed in the 1970s and 1980s. The thorough review by Taylor[14] presents the earlier history of motif and profile searches. The template pattern matching method of Taylor and colleagues is based on combinations of profiles of amino acid properties and secondary structure propensities implemented in a flexible controlled environment.[15] Although the template methods require a fairly high level of user understanding and are not being widely used, they did achieve a notable success. On the basis of very weak similarities to the subdomain fold of aspartic proteases that were picked up by the template method, Pearl and Taylor[16] built a model structure for the human immunodeficiency virus (HIV) protease. The model was refuted

[14] W. R. Taylor, *Protein Eng.* **2,** 77 (1988).
[15] W. R. Taylor, *Prog. Biophys. Mol. Biol.* **54,** 159 (1989).
[16] L. H. Pearl and W. R. Taylor, *Nature (London)* **328,** 351 (1987).

by the first solved HIV protease structure, but the second solved structure found the first one to have been incorrectly built, demonstrating that the prediction was essentially correct.[17]

Searches with Regular Expressions

The simplest method to search for a motif makes use of regular expressions that can be combined with logical operators to identify a repetitive or especially conserved motif in other proteins. This method is usually very fast, is implemented in various software packages, and can be recommended as a first scan in order to estimate the number of occurrences, as well as the background noise, for a motif of interest. As the search does not provide any significance estimates, conclusions drawn from matches have to be evaluated very carefully.

Motif Databases

Signatures, based on regular expressions, for numerous protein families and functional sites are stored in the PROSITE database[18]; the database annotation provides an excellent description of the respective families/ motifs. As a simple description by regular expressions has obvious limitations, more flexible descriptors have been developed,[19] and thus the PRO-SITE database will become even mor useful in the near future. Another approach to increase the utility of the collected motifs was taken by Henikoff and Henikoff,[20] who reconstructed alignments from the PROSITE entries, providing a database of core alignment "blocks" which is searched with tools that are more sophisticated than regular expressions. The Blocks server can be accessed by E-mail or WWW (Fig. 2).

Consensus Patterns

Patthy has developed and applied a method[21,22] that assigns a pattern to an alignment using the concept of a consensus sequence. The pattern is a string of amino acids that are conserved according to a user-defined threshold, are separated by "unimportant" positions, and are given position-dependent gap penalties. Although this method does not use the full

[17] A. Wlodawer, M. Miller, M. Jaskólski, B. K. Sathayanarayana, E. Baldwin, I. T. Weber, L. M. Selk, L. Clawson, J. Schneider, and S. B. H. Kent, *Science* **245**, 616 (1989).
[18] A. Bairoch and P. Bucher, *Nucleic Acids Res.* **22**, 3583 (1994).
[19] P. Bucher, K. Karplus, M. Mooeri, and K. Hofmann, *Comput. Chem.* **20**, in press (1996).
[20] S. Henikoff and J. G. Henikoff, *Nucleic Acids Res.* **19**, 6565 (1991).
[21] L. Patthy, *J. Mol. Biol.* **198**, 567 (1987).
[22] L. Patthy, this volume [12].

information of the multiple alignment (learning set), the success of the method in finding distant relationships[22] shows that filtering important positions can be used to increase the signal-to-noise ratio (although the signal itself becomes weaker).

Property Patterns

The program PROPAT, developed by Bork and Grunwald[23] and later improved by Rohde and Bork,[24] has been applied many times in the detection of distant homologies. This method is able to generalize a pattern, even from a rather small learning set, by automatically deriving distinct combinations of physicochemical properties for each position; a vector of such properties is assigned to each amino acid (for details, see Ref. 24). It can be used for a single motif, combinations of motifs, or for whole domains, and is already a step toward profile searching, since a vector of weights (in this case penalties) is assigned to each position of the alignment (including gaps). PROPAT can search six-frame translations of DNA databases.

Flexible Patterns

The flexible patterns of Barton and Sternberg[25] combine features of motifs and profiles. The patterns can be set up in various ways but are essentially permutations of conserved blocks, separated by gaps of specified ranges, and are compared to sequences using a dynamic programming approach. The Barton approach has been applied, for example, in a survey of the DHR domain distribution.[26]

Classical Profile Method

Profile analysis as implemented by Gribskov et al.[27,28] is used to perform exhaustive alignment by dynamic programming of a family-based scoring matrix against test sequences. The profile is comprised of two components for each position in the alignment: (1) scores for the 20 amino acids and (2) variable gap opening and extension penalties. The amino acid substitution scores are created by summing Dayhoff exchange matrix values according

[23] P. Bork and C. Grunwald, Eur. J. Biochem. **191,** 347 (1990).
[24] K. Rohde and P. Bork, Comput. Appl. Biosci. **9,** 183 (1993).
[25] G. J. Barton and M. J. E. Sternberg, J. Mol. Biol. **212,** 389 (1990).
[26] C. P. Ponting and C. Phillips, Trends Biochem. Sci. **20,** 102 (1995).
[27] M. Gribskov, A. D. McLachlan, and D. Eisenberg, Proc. Natl. Acad. Sci. U.S.A. **84,** 4355 (1987).
[28] M. Gribskov and S. Veretnik, this volume [13].

to the observed amino acids in each column of the alignment. Gap penalties are reduced at positions with gaps, according to the length of the longest insertion spanning that point in the alignment. The programs PRO-FILEMAKE, PROFILESEARCH, and PROFILEGAP are widely available through the GCG (Genetics Computer Group) sequence analysis package,[29] making them the most frequently used programs in the field of motif and profile searches. However, the GCG PROFILESEARCH (including version 8.0) does not handle current database sizes and fails to warn the user that the search is incomplete. The TPROFILESEARCH version [available from P. Rice, EBI (European Bioinformatics Institute), Hinxton, UK] corrects this problem.

Improved Profile Methods

A number of modifications have been suggested for improving the creation of profiles that increase the sensitivity of the method. Several of these improvements have been incorporated into programs such as PRO-FILEWEIGHT[30] and the method of Lüthy *et al.*[31] For example, an alignment often consists of many closely related sequences together with a few rather divergent ones. The closely related sequences in the multiple alignment (learning set) offer little additional information, yet bias the profile residue scores. Sequence weighting schemes which upweight divergent sequences while downweighting closely related groupings have been found to improve profile sensitivity.[30,31]

Noise is also reduced in database searches by gap excision[30] as long insertions are sites of breakdown in homology within the family and typically lack meaningful conservation. Release 2 of PROFILEWEIGHT will also bring in new gap penalty reductions based on average gap length, rather than the single longest sequence, to better match observed gap properties in alignments.

Both TPROFILESEARCH (P. Rice, EBI, Hinxton, see Fig. 2) and the PairWise/SearchWise package (E. Birney, J. Thompson, and T. Gibson, see Fig. 2) are able to perform protein profile alignments to six-frame translations of DNA sequences. The latter programs use an extension to dynamic programming to compare the profile simultaneously to the three translation frames of a DNA strand, allowing frame jumping.[32]

---

[29] J. Devereux, P. Haeberli, and O. Smithies, *Nucleic Acids Res.* **12,** 387 (1984).
[30] J. D. Thompson, D. G. Higgins, and T. J. Gibson, *Comput. Appl. Biosci.* **10,** 19 (1994).
[31] R. Lüthy, I. Xenarios, and P. Bucher, *Protein Sci.* **3,** 139 (1994).
[32] T. J. Gibson, E. Birney, M. Hyvönen, A. Musacchio, and M. Saraste, *Trends Biochem. Sci.* **19,** 349 (1994).

Automated Iterative Motif Search

The motif search method MoST (for Motif Search Tool[2,11]) follows the BLAST strategy (in which gaps are not treated) in order to be able to handle the resulting alignment blocks in a proper mathematical sense. These blocks are converted to a position-dependent weight matrix following a log-odds, Bayesian-based approach and incorporating prior residue probabilities calculated from a mixture of Dirichlet distributions. MoST combines an extremely fast block search with good sensitivity.[11] In addition, automatic iterations have been incorporated, that is, database sequences scoring above a user-defined threshold are incorporated in the block alignment and the weighting for the next iteration is adapted to the new alignment. To allow for the different behaviors of protein families, manual intervention is possible at several levels. The drawback of excluding gaps will be circumvented in an improved version that handles the statistics of several blocks (E. V. Koonin, personal communication). Thus, this method is highly recommended, particularly if functionally conserved residues are surrounded by semiconserved but structurally important positions, as can be observed in distantly related enzymes.

Other Methods

In addition to improvements in techniques related to the ones mentioned above, other approaches that might prove valuable in identifying distant homologies include the application of neural networks[33,33a] and methods that try to tackle the automatic iteration of the search procedure.[34] The use of hidden Markov models (HMMs) offers the prospect of a more formal mathematical treatment for profiles. For example, using a profile description that incorporates HMMs,[19] Bucher and colleagues have been able to identify several divergent intracellular domains. A number of groups are currently working on the application of HMMs to profile searches (see Fig. 2 for a WWW pointer to one of the HMM alignment descriptions).

Parameters and Pitfalls

Multiple factors influence the choice for a certain program, such as computer resources, sensitivity, local availability, and user-friendliness. Ideally, one should know the powers and pitfalls of each of them and also how to handle the optional parameters. To help the user set up and run

---

[33] D. Frischman and P. Argos, *J. Mol. Biol.* **228**, 951 (1993).
[33a] J. Hanke, G. Beckmann, P. Bork, and J. G. Reich, *Protein Sci.* **5**, in press.
[34] T. M. Yi and E. S. Lander, *Protein Sci.* **3**, 1315 (1993).

pattern searches effectively, we have provided a checklist of points, which we routinely apply (Table I).

The GCG profile programs illustrate why it is important for the user to review the parameter set up. The default normalization for database entry length, which downweights the scores against long sequences, can be highly deleterious for domains in large proteins. The default normalization for amino acid bias also lowers sensitivity as it downweights characteristic residues in database entries.

Another important point to consider is the report of only the optimal hit; that is, it is easy to overlook internal repeats. Most of the programs discussed here have ways to handle such internal repeats. With the GCG programs, the user must remember to look for repeats themselves, for example, by iteratively searching the regions in a protein on either side of a match against a domain profile. Dot-plot self-comparison can provide a second independent check.

Domain Borders

As in database searches with a single sequence, choice of the proper length of the query is of great importance. Three-dimensional structures increasingly reveal how to refine the domain borders, adjustment of which often leads to increased sensitivity. One particular problem is the phasing of successive repeats; in the literature there are numerous examples in which artificial domains are published spanning the C terminus of one repeat and the N terminus of the next. These "repeats" will thus never be detected when occurring as a single domain in other proteins. Contrary to popular perception, intron boundaries are extremely unreliable guides to intracellular domain boundaries, because traceable exon shuffling appears to involve extracellular domains exclusively. Intron boundaries have been more useful in demarcating extracellular modules, but they must show cross-consistency between repeats (or proteins), as the intron positions have often rearranged subsequently, particularly in *Caenorhabditis elegans* (see review by Patthy[35] for discussion of exon shuffling and module boundaries).

Gap Treatment and Sequence Weights

Gapped regions of alignments usually contain little useful information. They are typically deleted in motif searches and specified as allowed insertion ranges. In profiles, they are given reduced gap penalties and may also be deleted. Gaps vary in their tolerance of long insertions, and this ought to be reflected in the cost assigned to each gap. As yet there is no satisfactory

---

[35] L. Patthy, *Curr. Opin. Struct. Biol.* **4**, 383 (1994).

TABLE I
CHECKLIST FOR MOTIF AND PROFILE SEARCHES

Point to be considered	Reason
Is multiple alignment correct?	Correct alignment is prerequisite for any search method: misalignments and frameshifts seriously degrade the signal
Is given set of sequences representative?	Some programs lack effective downweighting schemes for close relatives in learning set: sometimes one should omit very redundant sequences and select a representative set of divergent sequences
Are borderlines correctly defined or does enlargement/reduction of the studied segment make more sense?	Arbitrarily truncated segments will have a weaker signal, whereas artificially enlarged ones add noise
Is appropriate method being used for the sequence family?	For highly gapped alignments, block searches are not advisable; profiles spanning full proteins are sometimes less sensitive than restriction to a few short conserved motifs, when there is no other strong conservation
Is chosen amino acid substitution matrix appropriate?	Depending on the family, another matrix might lead to clearer/different results. For example, "soft" matrices tend to be inappropriate for short motifs
Are gaps considered appropriate?	Large insertions might occur between more conserved regions; small extracellular domains with cores mainly of disulfide bridges have more freedom for insertions/deletions than, for example, enzymes
For profile searches, have the gap penalties been optimized on trial runs?	Appropriate penalties vary with divergence of query set. Set too strong: gaps cannot be crossed. Set too weak: query profile spreads out over false positives, giving higher scores
Are apparently essential positions (e.g., required for catalysis) set to be required in pattern/profile?	Weight/penalty for such positions is often not high enough in available programs and additional information should be manually included, e.g., by stronger weight/penalties
Have all databases been searched?	Many programs are unable to search in DNA databases or six-frame translations which usually harbor additional hits; some network servers might offer out-of-date databases
Is output interpreted correctly?	Be familiar with parameters and scoring systems, to be sure of the resulting scores (e.g., normalized Z scores in PROFILESEARCH are misleading when searching for small domains in larger proteins as they upweight small sequences)
Is knowledge about putative target sequences applied appropriately?	Searching with core metabolic enzyme suggests downweighting hits with extracellular proteins, as biological context is different, but be careful as all kinds of exceptions exist and proteins with unrelated functions can be homologous

TABLE I (*continued*)

Point to be considered	Reason
Have databases been searched with putative novel members before inclusion into alignment for next iteration?	If putative novel member belongs to a family that is well-characterized and distinct or which has different conserved regions, it is likely a false positive
Has reciprocity of detections been checked?	If profile of family A identifies family B as similar, does profile of family B find family A? Caution is needed as in some cases profiles of two artificially aligned families might identify both families before noise

way of calculating these costs. The user should conduct several trial searches varying the gap parameters in order to optimize alignment of the query profile with the database proteins (especially to eliminate dramatic spreading of the profile against false positives) and the detection signal-to-noise ratio.

Sequence weighting should be used in preparing profiles. See Higgins *et al.*[36] for a discussion of the resulting benefits.

Appropriate Use of Amino Acid Similarity Matrices

New residue substitution matrices, particularly the BLOSUM[37] and Gonnet series,[38] have been found to improve the signal-to-noise ratio in profiles.[30,31] BLOSUM 45, a moderate to high divergence matrix, works well and is a good starting point.[31] However, several matrices should always be tested. As in all sequence searches, the length of the query affects the tolerance of the noise introduced by high divergence matrices. Small domains may need more stringent matrices such as BLOSUM 62 (or Gonnet Pam120/Pam160), whereas profiles for larger domains are often most sensitive at higher divergence, in which case the Gonnet series in the range Pam250–Pam350 is useful. The older Dayhoff Pam matrices and the GCG default normalized matrix perform considerably less well. Gap penalties need to be recalibrated whenever a different matrix is used.

Errors and Expressed Sequence Tags

Errors in sequence databases, particularly shifts in the translation frame, can be caused by sequencing mistakes and wrongly predicted introns and

[36] D. G. Higgins, J. D. Thompson, and T. J. Gibson, this volume [22].
[37] S. Henikoff and J. G. Henikoff, *Proc. Natl. Acad. Sci. U.S.A.* **89,** 10915 (1992).
[38] S. A. Benner, M. A. Cohen, and G. H. Gonnet, *Protein Eng.* **11,** 1323 (1994).

exons. Frameshift errors are known to be widespread, and it is likely that some 5% of entries in SWISS-PROT, the most carefully annotated sequence database, have frameshifted segments. They are so common that we now expect to find frameshifts in each new protein family being investigated, as was true for both test examples given below.

If there are frameshifts in the learning set, they can lead to the introduction of gaps at erroneous positions and affect the apparent residue conservation. Both of these effects will degrade the query pattern. If there are frameshifts (due to errors or introns) in database entries, at the protein level only part of a pattern will match, while at the DNA level the frames jump. SearchWise can track the frame jump and reveal these in the search outputs, whereas its interactive partner PairWise allows suspicious sequences to be investigated in more detail.

Another increasing problem concerns fragmented and truncated entries. Truncated open reading frames (ORFs) are often adjacent to a targeted gene sequence. Similarly, the ends of cosmid entries from genomic sequencing projects have unreliable annotation. Often, cDNAs are reported as full length but are truncated at their N termini (and occasionally C termini). Even more complicated to deal with are the EST entries (expressed sequence tags), single gel reads (usually ~300 bases) of random cDNAs. Not only is there a high frameshift error rate ($\geqslant$10% of entries, sometimes multiple), but the hits are, by definition, fragments of the query and so search scores will be less than for full-length sequences. To deal with such a situation, parameters might be specially tuned as, for example, implemented in SearchWise (and applied by Aasland et al.[39]), in which case the EST databases should be searched separately.

Significance Assessments

Having discussed major sources for improvement of sensitivity, the assessment of significances for tempting hits remains problematic. Most of the programs provide some significance estimates, but all these calculations have to make certain assumptions that lead in practice to limitations. The calculation of Z scores (normal deviates) is estimated from the quality of the input sequences compared to the total distribution of scores. However, the score distribution does not follow a Gaussian distribution, as assumed in many Z-score calculations; Dirichlet mixtures are currently used to describe phenomena such as tails in the distributions. Another scoring system is based on p values that give the probability of a match by chance. The probability may not take into account sequence and residue biases, and is

---

[39] R. Aasland, T. J. Gibson, and A. F. Stewart, *Trends Biochem. Sci.* **20**, 56 (1995).

dependent on database size, so that a given value (e.g., 10^{-6}) will become less significant in time as the databases expand.

The better the statistical description, the more reliable will be the resulting scores. (Thus, here is another hidden parameter that has an influence on the results but differs from family to family.) Deficiencies in the current schemes are widely acknowledged, and efforts are being made to improve statistical assessments. Claverie[40] has suggested statistics aimed at better discrimination for the threshold between true and false hits for ungapped profiles of fixed length, assuming that random matches behave according to the extreme value distribution. The distribution of the family members is considered in MoST when estimating the ratio of the expected number of sequence segments with a given score to the observed number (parameter r in MoST), yielding a rather accurate estimate for a given alignment block.

In any event, the significance values of the different methods only try to eliminate false-positive hits. In this they are largely successful for globular sequences but are less reliable for reduced complexity sequences. As judged from the numerous very similar three-dimensional structures without obvious sequence similarity, many homologs do not meet the significance criteria of the particular method and are likely to escape attention when applying motif and profile searches.

Assessing Statistically Insignificant Hits

The way to assess weak candidate hits is to consider what are the constraints which must apply between sequences as a consequence of homology between them. These constraints are predominantly structural in nature, and yet they can often be applied even in the absence of a solved structure. Table II provides a checklist designed to help in weeding out the false hits: again we routinely use these checks. This logic was used, for example, in detecting highly divergent PH domains.[41] Conserved hydrophobic residues were assumed to be in the core, whereas runs of hydrophilic residues tolerating gaps were assumed to be in exposed loops. From the periodicity of the conserved residues it was possible to infer the number of α helices and β strands in the domain. Borderline hits from profile searches were then only accepted when the sequence was fully compatible with all the predicted secondary structures, the hydrophobic core residues, the predominantly hydrophilic surface residues, and the absence (or rarity) of Pro and Gly in helices and strands. Later, solved PH domain structures

[40] J.-M. Claverie, *Comput. Chem.* **18,** 287 (1994).
[41] A. Musacchio, T. J. Gibson, P. Rice, J. Thompson, and M. Saraste, *Trends Biochem. Sci.* **18,** 343 (1993).

TABLE II

CHECKLIST FOR EVALUATING PLAUSIBILITY OF WEAK HITS

Examinations on hit	Reason
Is amino acid distribution consistent with globular (cytosolic, extracellular), integral membrane, coiled-coil, fibrous, or random-coil structure?	These are mutually exclusive structural classes that should not overlap within a domain (although they can be juxtaposed in multidomain proteins). High scoring random coil is not a good indicator of homology
Is there structural information known for query or hit?	Knowledge of three-dimensional structure greatly facilitates evaluation, as constraints from hydrophic core, catalysis, etc., can be included
Is there a partial overlap of hit with established domain class in reciprocal search?	Partial overlap immediately rules out potential similarity. By definition, globular domains do not overlap (although they can be inserted into loops in other domains)
Is full domain potentially present?	Globular structure is stabilized by interactions in hydrophobic core. The presence of only half a globular domain is thus very rarely observed
Is there match to all conserved alignment blocks?	Conserved blocks usually indicate secondary structural elements
Is there match to all highly conserved hydrophobic residues?	Conserved hydrophobic residues are essential to given hydrophobic core. Very few exceptions are tolerated
Do most positions that are aligned to unconserved positions in query have hydrophilic residues?	Surface residues are usually hydrophilic and are unconserved unless binding other molecules. Multiple mismatched hydrophobic residues are contrary indicators. (Surprisingly frequently, transmembrane regions are erroneously aligned to cytosolic proteins)
Has Pro been aligned to position in block where it was not seen before?	Pro is favored in the N-terminal 3 residues of α helices. Any deeper and it breaks H bonds. It is allowed on edge β strands. It breaks H bonds on internal strands. Exceptions are rare and cannot be arbitrarily invoked for weak hits
Has Gly been aligned to position in block where it was not seen before?	The lack of a side chain reduces helix and strand stability. Gly aligned to small hydrophobic residues (Ala, Val, Cys) may indicate a plausible tight packing arrangement; otherwise, only occasional exceptions may be tolerated
Is segment rich in Gly, Pro, Asn, Ser aligned to block poor in these residues?	Such alignment indicates that a loop region is erroneously aligned to a secondary structure element

TABLE II (*continued*)

Examinations on hit	Reason
Are matches to blocks consistent with block secondary structure?	Secondary structures of matched blocks should be identical. In addition to above rules, amino acid preferences may be indicative: e.g., aligning a sequence composed of β-preferring residues like Ile, Val, Thr, Ser onto an α helix would be highly implausible (unless these were already favored in aligned sequences)
Have new insertions/deletions appeared in conserved regions?	Alignment blocks are usually conserved due to structural or functional constraints; therefore, large or frequent insertions and deletions are unlikely
For Cys-rich sequences, do Cys patterns match?	Number and spacing of Cys residues distinguish between classes of extracellular disulfide-rich modules, as well as (often with His) intracellular zinc fingers, e.g., GAL4
Are functions of hits compatible?	On one hand do not overinterpret results to fit a tempting functional context; on the other hand, some functional aspects should be considered (e.g., query proteins are extracellular, hit is a metabolic enzyme)
Does additional functional or biochemical information provide some clues for homology?	Already identified catalytic residues, disulfide bridges, mutation data, etc., add constraints that can be helpful in excluding false positives

vindicated the structural assignments, thereby lending credence to the PH domain detections.[32]

Secondary structure predictions can help in clarifying whether two or more sequence families are structurally related. It is now recognized that secondary structure predictions based on multiple alignments are often quite accurate. The server of Rost and Sander[42] usually gives good results, in particular when the reliability scores are taken into account. For example, pattern searches with a phosphate-binding motif present in certain β/α-barrel enzymes, such as glycolate oxidase, weakly picked up a surprising number of different enzyme families with unknown structures and, mostly, poorly investigated mechanisms. These families had no apparent sequence similarity to each other. Secondary structure predictions,[42] based on alignments of the individual families, gave alternating β/α predictions, consistent with the β/α-barrel structure. This information, together with the length of the proteins and the C-terminal location of the phosphate-binding motif, was critical to assigning homology between a large set of enzymes acting

[42] B. Rost and C. Sander, *J. Mol. Biol.* **232**, 584 (1993).

on heterocyclic compounds in diverse pathways, including histidine biosynthesis, purine metabolism, and thiamin biosynthesis.[9]

To illustrate the strategies explained above, we present two examples in which we are able to identify new members of well-established domain families using several different pattern search programs. In the process, we identified two frameshifted entries, and one with a wrongly predicted spliced exonic sequence, which undoubtedly hindered earlier detection of the domains.

Retrieving GAL4 Domains

Our first example is the retrieval of fungal zinc binuclear cluster domains, a specific DNA-binding domain found in numerous fungal transcription factors. The three-dimensional structure has been determined for the corresponding domains in PPR1 and in GAL4 (see Marmorstein and Harrison[43] and references therein). GAL4 activates galactose catabolism in yeast and is the best characterized member of the family. GAL4-like DNA-binding domains are extremely widespread in fungi and are involved in numerous transcription regulation pathways, and yet not a single one of these domains has been found in any other eukaryote, nor in bacteria. With more than 50% of the genomic sequence of yeast available in public databases, an up-to-date collection would yield a first realistic estimate of their total number. Another interesting problem is the mode of spreading of this domain: Is it simple gene duplication and subsequent modification, or, as suggested by the considerable number, is a sort of domain shuffling the reason for their frequency? In the latter case, the location of the domain should vary within the proteins. So far, it has been mostly found at the N terminus of transcription regulators, including GAL4 itself.

Given the defined GAL4 domain borderlines (verified by the three-dimensional structure) and assuming a nonbiased amino acid distribution within the domain, a first quick homology search can be performed by BLASTP using only the domain itself as the query, in order to reduce the noise caused by matches of other parts of the GAL4 protein. A scan of the well-annotated SWISS-PROT database (release 31) records 9 hits with significant matches below a probability of matching by chance of $p = 10^{-7}$. (Although this is a strict value for assessing BLAST outputs, the first false positive, a viral coat protein, had a p value of 9.4×10^{-6}.)

These sequences will normally be aligned in the next step, an option either provided within the motif/search/programs or performed with a multiple alignment program. The GAL4 alignment (Fig. 3) shows six invariant cysteines, essential for structural zinc binding, and strong conservation of

[43] R. Marmorstein and S. C. Harrison, *Genes Dev.* **8**, 2504 (1994).

several other residues, in particular positively charged residues between the second and third cysteine, that are important in DNA binding. Only one gap was opened in the starting set. We compared here three methods (profile/SearchWise, PROPAT, MoST); all three collect the same number of sequences in the protein databases, with no false positives in SWISS-PROT (a few difficult cases are discussed below). All methods performed much better than iterative BLASTP database searches. Thus, another application of motif and profile search programs becomes apparent with the presence of large families in databases, namely, the fast collection of all members. In SWISS-PROT release 31 we found 42 GAL4-like domains, whereas only 37 of them will be found using BLASTP (with the threshold $p = 10^{-7}$). Even if the remaining 5 can be pulled out of the twilight zone by various output evaluation procedures, preparation and evaluation of many BLASTP runs require a considerable effort compared, for example, to a single run (after the initial BLASTP search with GAL4) of MoST.

When extending the search to other data resources, including nucleotide sequence databases, 60 sequences were retrieved (Fig. 3). Fifty entries contain annotation hinting at the presence of GAL4-like domains: of the remainder, only two had been published at the time of writing, but the domain presumably escaped attention in several other cases. During the course of the searches, two erroneous sequences were detected, both of which were already entered in SWISS-PROT: (1) Yhl6_Yeast lacks the N-terminal GAL4 domain because it contains a frameshift (detected by SearchWise) and so was not placed in the predicted ORF YHR056c by automatic translation procedures (Fig. 4a), and (2) Alcr_Emeni has a missing exon (detected by several programs) due to incorrect interpretation of the splicing pattern (Fig. 4b).

Thus, the example of GAL4 underlines some of the points discussed. The pattern searches led to the identification of previously undetected GAL4 family members and also revealed two common kinds of errors (one in the sequence itself and one in the interpretation of the raw DNA sequence). From the 44 GAL4 domains we found in yeast, spread over all the chromosomes, we can extrapolate to the whole genome and expect more than 80 GAL4 domains in total. The majority of the proteins seem to be colinear, sharing another (more weakly) conserved domain (P. Bork, unpublished results), so that the frequency in yeast seems to be the result of extensive gene duplication rather than domain shuffling.

Jak Kinases Contain SH2 Domains

We also conducted pattern searches with SH2, a well-known domain in signaling proteins that recognizes and binds phosphotyrosine-containing

```
AFLR_ASPFL   23  RKLRDSCTSCASSKVRCTKEKP- 1-CARCIERGL--ACQYMVSKRMGRN  P41765
ALCR_EMENI    6  RRQNHSCDPCRKGKRRCDAPEN-11-CSNCKRWNK--DCTFNWLSSQRSK  P21228   a
AMDR_ASPOR   13  GNGSAACIHCHRRKVRCDARIV- 3-CSNCRSAGK-ADCRIHEKKKRLAV  Q06157
AMDR_EMENI   13  GNGSAACVHCHRRKVRCDARLV- 3-CSNCRSAGK-TDCQIHEKKKKLAV  P41044
ARG2_YEAST   15  AKTFTGCWTCRGRKVKCDLRHP- 1-CQRCEKSNL--PCGGYDIKLRWSK  P05085
CAT8_YEAST   64  YRIAQACDRCRSKKTRCDGKRP- 1-CSQCAAVGF--ECRISDKLLRKAY  P39113
CB32_YEAST    8  LKSKHPCSVCTRRKVKCDRMIP- 0-CGNCRKRGQDSECMKSTKLITASS  P40969
CYP1_YEAST   58  NRIPLSCTICRKRKVKCDKLRP- 1-CQQCTKTGVAHLCHYMEQTWAEEA  P12351
CZF1_CANAL  312  KRSRMGCLTCRQRKKRCCETRP- 1-CTECTRLRL--NCTWPKPGTEHKN  P28875
DA81_YEAST  144  GNLMGSCNQCRLKKTKCNYFPD- 3-CLECETSRT--KCTFSIAPNYLKR  P21657
GAL4_YEAST    5  SSIEQACDICRLKKLKCSKEKP- 1-CAKCLKNNW--ECRYSPKTKRSPL  P04386
LAC9_KLULA   89  EVMHQACDACRKKKWKCSKTVP- 1-CTNCLKYNL--DCVYSPQVVRTPL  P08657
LEUR_YEAST   31  RKRKFACVECRQQKSKCDAHER- 4-CTKCAKKNV--PCILKRDFRRTYK  P08638
LY14_YEAST  153  KYSRNGCSECKRRRMKCDETKP- 1-CWQCARLNR--QCVYVLNPKNKKR  P40971
MA3R_YEAST    2  TLVKYACDYCRVRRVKCDGKKP- 0-CSRCIEHNF--DCTYQQPLKKRGS  P38157
MA6R_YEAST    2  GIAKQSCDCCRVRRVKCDRNKP- 0-CNRCIQRNL--NCTYLQPLKKRGP  P10508
NIRA_EMENI   36  RCVSTACIACRRRKSKCDGNLP- 1-CAACSSVYH-TTCVYDPNSDHRRK  P28348
NIT4_NEUCR   47  RCVSTACIACRRRKSKCDGALP- 1-CAACASVYG-TECIYDPNSDHRRK  P28349
PDR1_YEAST   40  SKVSKACDNCRKRKIKCNGKFP- 0-CASCEIYSC--ECTFSTRQGGARI  P12383
PDR3_YEAST    9  SKVSTACVNCRKRKIKCTGKYP- 0-CTNCIYDC--TCVFLKKHLPQKV  P33200
PPR1_YEAST   28  SKSRTACKRCRLKKIKCDQEFP- 1-CKRCAKLEV--PCVSLDPATGKDV  P07272
PUT3_YEAST   28  QRSSVACLSCRKRHIKCPGGNP- 0-CQKCVTSNA--ICEYLEPSKKIVV  P25502
QA1F_NEUCR   70  QRVSRACDQCRAAREKCDGIQP- 1-CFPCVSQGR--SCTYQASPKKRGV  P11638
QUTA_EMENI   43  QRVSRACDSCRSKKDKCDGAQP- 1-CSTCASLSR--PCTYRANPKKRGL  P10563
SUC1_CANAL    7  APYTRPCDSCSFRKVKCDMKTP- 0-CSRCVLNNL--KCTNNRIRKKCGP  P33181
THI1_SCHPO   33  RRVFRACKHCRQKKIKCNGGQP- 1-CISCKTLNI--ECVYAQKSQNKTL  P36598
UGA3_YEAST   11  KYSKHGCITCKIRKKRCSEDKP- 1-CRDCRRLSF--PCIYISESVDKQS  P26370
UME6_YEAST  765  TRSRTGCWICRLRKKKCTEERP- 1-CFNCERLKL--DCHYDAFKPDFVS  P39001   d
YAF1_YEAST   60  NRILFVCQACWKSKTKCDREKP- 1-CGRCVKHGL--KCVYDVSKQPAPR  P39720
YB00_YEAST  101  SRVTKACDYCRKRRIKCTEIEP- 4-CRNCIKYNK--DCTFHFHEELKRR  P38114
YB89_YEAST   34  KNTNVACVNCRRLHVSCEAKRP- 0-CLRCISKGLTALCVDAPRKKSKYL  P38140
YB90_YEAST   24  GRTFTGCWACRFKKRRCDENRP- 1-CSLCAKHGD--NCSYDIRLMWLEE  P38141
YBG6_YEAST   51  HRPVTSCTHCRQHKIKCDASQN- 4-CSRCEKIGL--HCEINPQFRPKKG  P34228
YBO3_YEAST   50  KKASHACDQCRRKRIKCRFDKH- 3-CQGCLEVGE--KCQFIRVPLKRGP  P38073
YCO1_YEAST   11  SKAFKTCLFCKRSHVVCDKQRP- 0-CSRCVKRDIAHLCREDDIAVPNEM  P19541
YCZ6_YEAST    9  PRLRLVCLQCKKIKRKCDKLRP- 1-CSRCQQNSL--QCEYEERTDLSAN  P25611
YE14_YEAST   12  SRVTKACDRCHRKKIKCNSKKP- 0-CFGCIGSQS--KCTYRNQFREPIE  P39961
YHL6_YEAST    9  VRKPPACTQCRKRKIGCDR!KP- 1-CGNCVKYNK-PDCFYPDGPGKMVA  P38781   b
YHX8_YEAST   16  TTELYSCARCRKLKKKCGKQIP- 1-CANCDKNGA--HCSYPGRAPRRTK  P38699
YIN0_YEAST   15  RRVTRACDECRKKKVKCDGQQP- 0-CIHCTVYSY--ECTYKKPTKRTQN  P40467
YJ16_YEAST   41  GRAHRACIACRKRKVRCSGNIP- 0-CRLCQTNSY--ECKYDRPPRNSSV  P39529
YJX9_YEAST   14  KSIQTACEFCHTKHIQCDVGRP- 1-CQNCLKRNIGKFCRDKKRKSRKRI  P42950
YK44_YEAST   13  HRITVVCTNCKKRKSKCDRTKP- 0-CGTCVRLGDVDSCVYLTDSSGQPE  P36023
YKD8_YEAST   41  TKASRACDQCRKKKIKCDYKDE- 3-CSNCQRNGD--RCSFDRVPLKRGP  P32862
YKW2_YEAST   18  RKPAKSCHFCRVRKLKCDRVRP- 1-CGSCSSRNR-KQCEYKENTSAMED  P35995
MALX_YEAST    7  TCAKQACDCCRIRRVKCDGKRP- 0-CSSCLQNSL--DCTYLQPSRKRGP  L12223_1
PRIB/LINED   14  VRGARACTTCRAAKMKCVGAED- 4-CQRCKRANV--QCIFEKHRRQRK  D14489_1
PAH2/PICAN    ?  RKVGAACVICHRRKIKCDIGTA- 3-CSKCKELKVESQCVLHKRRRKTDG  U22930_1  d
SIP4/YEAST   40  VRKAHACDRCRLKKIKCDGLKP- 1-CSNCAKIDF--PCKTSDKLSRRGL  U17643_1
UAY/ASPNI    61  FRNVSACNRCRQRKNRCDQRLP- 1-CQACEKAGV--RCVGYDPITKREI  X84015_1  d
YX1/CANPE   194  KGNPNPCDHCRRRQIKCITVPN- 3-CVQCETKGI--KCTHSESPSNPAL  X02903_3  c
YFX1/YEAST    2  ARNRQACDCCCIRRVKCDRKKP- 0-CKCCLQHNL--QCTYLRPLKKRGP  D50617
YLX1/YEAST   38  NKSKTGCDNCKRRRVKCDEGKP- 1-CKKCTNMKL--DCVYSPIQPRRRK  U19027_9  d
YLX2/YEAST    9  VKPSFVCLRCRKRIKCDKLWP- 1-CSKCKASSS--ICSYEVEPGRINK  Z47973_17 d
YLX3/YEAST   35  KGRSRSCLLCRRRKQRCDHKLP- 1-CTACLKAGI--KCVQPSKYSSSTS  U17243_1  d
YMX1/YEAST   81  LRVQKACELCKKRKVKCDGNNP- 0-CLNCSKHQK--ECRYDFKATNRKR  Z49211_7
YMX2/YEAST   25  RKVIKSCAFCRRKLKCSQARP- 1-CQQCVIKRL-PQCVYTEEFNYPLS  U17244_9
YMX3/YEAST   70  KRNSFACVCCHSLKQKCEPSDV- 7-CRRCLKHKK--LCKFDLSKRTRKR  Z46373_6  d
YOX1/YEAST  130  KRVSKACDHCRKRKIRCDEVDQ- 4-CSNCIKFQL--PCTFKHRDEILKK  X83121_8  d
YPX1/YEAST    2  SIVRSQCDCCRVRRVKCDRNRP- 0-CDRCRQRNL--RCTYLQPLRKRGP  U25841_11
```

FIG. 3. GAL4 domain alignment. From left to right: first column, names (SWISS-PROT codes are given when available); second column, position of the first amino acid; third column, sequence; fourth column, database accession number (an underscore indicates the number of the ORF in larger cosmids). Numbers within the alignments indicate omitted amino acids

peptides. Solved structures of SH2/peptide complexes[44] have revealed the SH2 core and functional residues, of which the most conserved is a phosphate-binding arginine. We began with a search of SWISS-PROT (release 31) using BLASTP and the archetypal chicken Src SH2 domain. The BLASTP search detected numerous SH2 domains before the highest false positive ($p = \sim 10^0$), but many were well below our very conservative threshold of $p = 10^{-7}$. Many of the top hits were also tyrosine kinases almost identical to Src, which would add little or no value to the patterns. Therefore, a representative set of 27 SH2 domains above the threshold of 10^{-7} was used to initiate the different motif and profile searches. Again, all three methods performed well, with the profiles being slightly ahead as they can cope best with many (sometimes large) insertions (see alignment of SH2 domains in Higgins *et al.*[36]). Although the number of iterations and the computer time used are different for each program, all programs could detect known divergent members of the family such as the only described yeast SH2 domain (Spt6), its *C. elegans* homolog Emb5, and the STAT family of transcription factors.

In addition to the expected hits, both PROPAT and SearchWise indicated the presence of divergent SH2 domains in the Jak group of tyrosine kinases, which were not assigned SH2 domains in the annotation. SH2 in Jak kinases would be of clear biological significance since the great majority of cytoplasmic tyrosine kinases possess the domains. Both programs picked up the Jak entries more strongly in the later iterations, as increasingly divergent SH2 domains were added to the alignments. The Jak entries scored better than the STATs. The core structural and surface functional positions, including the critical conserved phosphate-binding positive charge, were all satisfied by the Jaks. Reciprocal searches (as recommended in Table I) with SearchWise and a profile of the putative Jak kinase SH2 domains listed 55 SH2-containing entries before the first false positive in SWISS-PROT. Therefore, the Jak kinases fulfill the criteria in Table II for divergent homologs.

[44] G. Waksman, S. E. Shoelson, N. Pant, D. Cowburn, and J. Kuriyan, *Cell* (*Cambridge, Mass.*) **72,** 779 (1993).

---

in a loop region. Cys residues essential for the binuclear cluster are highlighted in boldface type. Marker letters: a, sequence differs from the protein in SWISS-PROT release 31 due to splicing revision (see Fig. 4a); b, symbol ! indicates frameshift, and sequence differs in SWISS-PROT release 31 (see Fig. 4; both errors have been corrected in subsequent SWISS-PROT releases, on the basis of this report); c, this protein (and the frameshifted YHL6) had not been reported to contain a GAL4 domain; d, GAL4 domains are not annotated in the database entries and thus some of these GAL4 domains were probably not detected earlier.

a

```
Score 9750
Aligned Ranges:
10-40 (profile)
5445-5541 (sequence)
Showing backward strand


                       **  *****  **   !  *   *  *     *  *
Gal4 Profile:      10 ACDRCRKRKVKCDG+++KRPpCSRCAKRG.LECTY 40
DNA Translation:      ACTQCRKRKIGCDR^^^^KPICGNCVKYNKPDCFY
Embl:SCH8025:  5541' gtactaaaaagtgaGCCaacatgatgataacgttt
                      cgcaggagatggag   actggagtaaaacagta
                      tccacggagcgccg   agatgtccgtcggcttt 5445'
```

b

```
Score 9190
Aligned Ranges:
1-36 (profile)
1080-1277 (sequence)


                       *       **  ***  *   **
Gal4 Profile:      1 KRVSTACDRCRKRKVKCDGKRP...........................
DNA Translation:    RRQNHSCDPCRKGKRRCDAPVGCRYRLPSVCTDSR*DVTQENRNEANENG
Embl:ANALCR:   1080 cccacatgctcagacctggcggtctcccagtagactggacgaaaggagag
                    ggaaaggacggagagggacctgggagtcgtgcagggatcaaagaacaaag
                    acgttcctctcgcgacttcgatcatgcccgctctcataagatacgctacc


                       ** *        **
Gal4 Profile:     23 ..pCSRCAKRGLECTY 36
DNA Translation:    WVSCSNCKRWNKDCTF
Embl:ANALCR:   1230 tgtttatactaagtat
                    gtcgcagaggaaagct
                    gtgtatcgtgcgttcc 1277
```

Fig. 4. (a) Highest scoring local alignment produced by PairWise for the GAL4 profile against all three reverse frames of the yeast genomic cosmid embl:SCH8025. A single base insertion has caused a frameshift in the middle of a GAL4 domain, which would otherwise belong to the N terminus of the adjacent ORF YHR056c (predicted range −5426 to −2828). Translated DNA codons are shown in lowercase, and nucleotides spanned by the frameshift site are in uppercase and capped by the ^ symbol. Cys residues essential for the binuclear cluster are highlighted in boldface type. A WWW server that allows fast detection of frameshifts even in large cosmids (J. Boyle, N. P. Brown, and P. Bork, unpublished) will be available under http://www.embl-heidelberg.de/~boyle/errors/. (b) Highest scoring local alignment produced by PairWise for the GAL4 profile against the three forward frames of the *Aspergillus nidulans AlcR* gene in embl:ANALCR. An in-frame insertion including a stop codon separates two halves of a GAL4 domain, with only the latter part being in the predicted *AlcR* translation. Good matches to fungal splice donor, acceptor, and branch-point consensus sequences are shown in boldface italic type and reveal an overlooked exon containing the N-terminal portion of the GAL4 domain. Incorrect splicing inferences are not uncommon. In-frame introns (especially when lacking stop codons) are usually overlooked in fungi such as yeast due to the perception that introns are rare.

Surprisingly, two separate mouse Jak3 entries had different scores. One was well detected, but the other was below a number of false hits. Comparison of this sequence,[45] using the frameshifting algorithm of PairWise, to human, rat and the other mouse Jak3 revealed that it possessed seven frameshifted regions, of which two fell in the SH2 domain, accounting for the reduced profile score.

The exercise here reveals that the pattern methods can detect SH2 domains in Jak tyrosine kinases, which makes good sense, implying either inter- or intramolecular phosphotyrosine peptide recognition by these kinases. A review of the literature reveals that the Jak SH2 domains had been proposed earlier[46] yet were apparently rejected within the field, presumably to the detriment of Jak kinase research. The alignment of Jak and other SH2 domains is presented elsewhere in this volume in the Clustal alignment chapter,[36] where it is used as a divergent alignment test case.

Conclusion

There are numerous examples in which predictions based on motif and profile searches were useful as guides in further research, while themselves being verified by various experimental approaches. On the debit side, there are also a considerable number of "black sheep" that have caused resources to be squandered in exploration of wrong hypotheses. We wish to stress the two checklists provided here in Tables I and II, one for setting up and running the searches and the other for evaluating the borderline top hits. If these are followed, most of the suggestive, yet false, hits should be identified and discarded. Ultimately, solved three-dimensional structures are the arbiters of truth for homology predictions. Currently, the identification of new protein domains is followed almost immediately by the determination of the three-dimensional structure by NMR or X-ray. For more than 50% of the known intra- or extracellular modules (mostly defined by motif and profile searches), a three-dimensional structure is already available for at least one member of the family.[47] In addition to the identification of numerous domains[47] (e.g., the discovery of a nuclear domain superfamily present in cyclin, transcription factor IIB (TFIIB), and retinoblastoma proteins[48]), experimentally verified predictions with functional implications from our own work include the prediction of a type X polymerase in yeast,[3]

[45] S. G. Rane and E. P. Reddy, *Oncogene* **9**, 2415 (1994).

[46] A. G. Harpur, A.-C. Andres, A. Ziemiecki, R. R. Aston, and A. F. Wilks, *Oncogene* **7**, 1347 (1992).

[47] P. Bork and A. Bairoch, *Trends Biochem. Sci.* **20**, poster, March issue (1995).

[48] T. J. Gibson, J. D. Thompson, A. Blocker, and T. Kouzarides, *Nucleic Acids Res.* **22**, 946 (1994).

ATPase activities of several prokaryotic cell cycle proteins,[49] and homodimerization of the Norrie's disease protein via a specific disulfide bridge.[50]

Two additional points should be noted. (1) Soon essentially all sequences will have homologs in public databases. (2) Motif and profile search methods are being actively developed at a number of institutions and can be expected to be significantly improved. As a result these methods will continue to be very valuable tools.

Acknowledgments

We are indebted to Philipp Bucher, Eugene Koonin, and Joe Lewis for helpful comments on the manuscript and to the many colleagues who have also been involved in our cited work.

[49] P. Bork, A. Valencia, and C. Sander, *Proc. Natl. Acad. Sci. U.S.A.* **89,** 7290 (1992).
[50] T. Meitinger, A. Meindl, P. Bork, B. Rost, C. Sander, M. Haasemann, and J. Murken, *Nat. Genet.* **5,** 376 (1993).

[12] Consensus Approaches in Detection of Distant Homologies

By Laszlo Patthy

Introduction

Recognition of homologies may provide important hints about the structure and function of proteins; therefore, there is a growing interest in methods of sequence comparison. The FASTA and FASTP programs use a rapid sequence comparison algorithm which, by identifying proteins with high similarity scores, is useful for the detection of related sequences.[1,2] When sequence similarity is low, however, it is difficult to decide whether this similarity is due to common ancestry or whether it merely reflects chance similarity of unrelated proteins. In such cases, statistical tests are used to decide whether two sequences are more similar than would be expected by chance. In this approach, the similarity score of the actual comparison is compared with the distribution of the scores determined for pairs of a large number of random permutations of the two sequences, and the standard deviation of the comparison above the mean of the randomized comparisons is calculated. The use of high cutoff values increases the confi-

[1] D. J. Lipman and W. R. Pearson, *Science* **227,** 1435 (1985).
[2] W. R. Pearson and D. J. Lipman, *Proc. Natl. Acad. Sci. U.S.A.* **85,** 2444 (1988).

Copyright © 1996 by Academic Press, Inc.
All rights of reproduction in any form reserved.

dence that similarity reflects genuine homology, but more distant relationships with lower scores may still be buried in a background of irrelevant chance similarities. A clear separation of low scoring related and unrelated sequences cannot be achieved purely on statistical grounds because "an intrinsic problem of the application of statistics to biological data is the lack of correspondence between statistical and biological significance."[3]

Experience with FASTA and FASTP clearly shows that even their more sensitive versions are unable to distinguish the lowest scoring related sequences from the highest scoring unrelated sequences.[4,5] In the zone of marginal protein sequence similarity, the search for homologies thus maneuvers between the Scylla and Charybdis of false positives and false negatives: unrelated proteins with fortuitous high scores are sometimes claimed to be related, and distantly related sequences with low scores may be missed. In the opinion of Pearson, "Very few genes have been successfully identified on the basis of marginal similarity scores, but there are many examples of mistaken identity."[5] Our experience has shown that many true homologies were missed by usual sequence comparison procedures.[6–16]

To solve these problems one must first understand what distinguishes related and unrelated sequences that otherwise have identical scores with a query sequence. Clearly, it is not the sum of the scores, but the pattern of the scores. In the case of a true homolog the scores come from positions that are characteristically conserved in the family of the query sequence, whereas in the case of false positives the scores come from positions that show little conservation during the evolution of that family. To eliminate false positives one has to use a search protocol that concentrates on scores coming from positions conserved in the family of the query sequence but suppresses scores originating from variable positions. The consensus sequence procedure I developed utilizes this principle to detect distant homologies. This procedure proved to be useful for the detection of distant homologies that were missed by conventional search procedures.[6–16]

---

[3] S. Karlin and V. Brendel, *Science* **257,** 39 (1992).
[4] W. R. Pearson, *Genomics* **11,** 635 (1991).
[5] W. R. Pearson, *Curr. Opin. Struct. Biol.* **1,** 321 (1991).
[6] L. Bányai, A. Váradi, and L. Patthy, *FEBS Lett.* **163,** 37 (1983).
[7] L. Patthy, *J. Mol. Biol.* **198,** 567 (1987).
[8] L. Patthy, *J. Mol. Biol.* **202,** 689 (1988).
[9] L. Patthy, *Biochem. J.* **253,** 309 (1988).
[10] L. Patthy, *Mol. Immunol.* **26,** 1151 (1989).
[11] L. Patthy, *Cell (Cambridge, Mass.)* **61,** 13 (1990).
[12] L. Patthy, *FEBS Lett.* **289,** 99 (1991).
[13] L. Patthy, *J. Biol. Chem.* **266,** 6035 (1991).
[14] L. Patthy, *FEBS Lett.* **298,** 182 (1992).

The reliability of the consensus sequence procedure is supported by the fact that subsequent studies on the function, three-dimensional structure, and evolutionary history of these proteins, and/or exon–intron organization of their genes, have confirmed our conclusions. This procedure was instrumental in the recognition of distant homologies of modules of mosaic proteins and thus contributed significantly to the formulation of the principles of modular protein evolution.[17,18]

Rationale of Consensus Sequence Procedure

Distantly related members of a protein family may have vastly divergent functions but still have the same overall three-dimensional structure. In multiple alignments of such distantly related sequences, regions essential for the structural integrity of the protein fold (e.g., regions that form regular secondary structures, cysteines forming disulfide bonds) are more likely to be conserved, whereas regions that correspond to external loops connecting structural motifs may be more tolerant to variation in sequence and to gap events. Consequently, each protein fold may be characterized by a unique pattern of accepted mutations, and this pattern may be formulated in some way to distinguish conserved positions, variable positions, and gap positions.

Construction of Consensus Sequences Characteristic of Protein Fold and Their Use in Database Searches

In the consensus sequence procedure, multiple alignment of the sequences of a protein family and construction of consensus sequences are carried out by a progressive iterative alignment procedure that forces the alignment of conserved residues.[7] During the course of this procedure consensus sequences are determined that incorporate characteristic features of the related sequences: conserved residues are indicated by the 1-letter code of amino acids, variable positions are marked with x, and positions that tolerate gap events are identified by g.

Scoring of real sequence–consensus sequence alignments follows the rule used in scoring real sequence alignments, except that x positions should be scored with reduced weight and g positions must receive reduced gap penalties. Such a scoring approach ensures that alignment of conserved

---

[15] N. Behrendt, M. Ploug, L. Patthy, G. Houen, F. Blasi, and K. Dano, *J. Biol. Chem.* **266,** 7842 (1991).

[16] L. Patthy and K. Nikolics, *Trends Neurosci.* **16,** 76 (1993).

[17] L. Patthy, *Cell (Cambridge, Mass.)* **41,** 657 (1985).

[18] L. Patthy, *Curr. Opin. Struct. Biol.* **1,** 351 (1991).

residues are forced and gaps are clustered in regions known to be preferred sites of insertions–deletions. The method of Dayhoff *et al.*[19] was modified by adding the following rules to the log odds scoring matrix: x or g of the consensus sequence versus an amino acid of the real sequence is scored 0.0. A 1-residue gap of the real sequence versus a conserved residue or x of the consensus sequence, and a 1-residue gap of the consensus sequence versus an amino acid of the real sequence, are assigned a value of -0.5. A 1-residue gap of the real sequence versus g of the consensus sequence is given a score of 0.0. Note that the presence or absence of a residue in g position is scored 0.0.

The procedure is carried out in the following steps.

The sequences are compared pairwise, and the most closely related sequences are grouped on the basis of similarity scores. The rationale for this step is that the most closely related sequences can be aligned with the least ambiguity. Gaps (if any) introduced at this stage are correctly located in the sense that they are most likely to identify regions tolerant to gaps, such as surface loops.

In each group of closely related sequences alignments are surveyed to identify conserved positions, variable positions, and the preferred location of gaps. From these data consensus sequences are deduced for each group.

By aligning the consensus sequences of the various groups, a unified consensus sequence characteristic of the majority of sequences is determined.[7] The sequences are aligned with the unified consensus sequence and the resulting multiple alignment is used to deduce consensus sequences.

Throughout this process, consensus sequences are determined from alignments as follows: x is written in positions where an amino acid is present in more than a fraction k of the sequences of the multiple alignment; g is written in positions where an amino acid is present in k or less than k fraction of the sequences; x positions are surveyed, and where the most frequent amino acid (or amino acid type) is found in f or more than f fraction of the sequences, x is replaced by the 1-letter code sign of the most frequent amino acid (or the most frequent representative of the dominant amino acid type). Similar amino acids are grouped on the basis of the mutation probability matrix of Dayhoff *et al.*[19] The most commonly used grouping was V, I, L, M; Y, F, W; D, E, N, Q; H; R, K; S, T; A; P; G; C. In the case of more distantly related families, broader definitions of amino acid similarities may be useful.

Depending on the choice of the parameters (values of f, k, identity or similarity of conserved amino acids, definition of similar amino acids), a variety of consensus sequences are determined. If stringent criteria (e.g.,

---

[19] M. O. Dayhoff, W. C. Barker, and L. T. Hunt, this series, Vol. 91, p. 524.

$f = 1.0$, identity of amino acids) are used, then many conserved residues will be hidden in the x positions in the resulting consensus sequence. The distinction between variable and conserved positions will also be obscured if the criteria are too relaxed (by using a too broad definition of amino acid similarity, low f values, etc.). The consensus sequences most characteristic of the given protein fold may be selected from different consensus sequences determined by varying the above parameters. Whether a consensus sequence properly separates conserved (high scoring) and variable (low scoring) positions may be checked by the analysis described in the following section.

A practical test of the quality of a consensus sequence is its utility in database searches: when used to search the database, known members of the family should be clearly separated from even the highest scoring unrelated sequences. Such consensus sequences are also expected to retrieve some previously unidentified members of the family.[7-16] If a database search identifies additional members of a family, their inclusion necessitates the construction of a new multiple alignment and new consensus sequences. If the new members are distantly related, they may have significant effect on the consensus sequence; therefore, the database search with the new consensus sequence should be repeated. This search may yield some even more distantly related members of the family which may again necessitate a new round of the database search, etc. Such iterative database searches permit a homology walking whereby more and more distantly related members may be identified. Table I lists distantly related members of some module families that have been identified by such iterative searches.

Consensus Sequences and Distribution of Similarity Scores

The typical evolutionary "behavior" of the distribution of similarity scores in a highly divergent protein family may be illustrated with the sequences of complement B-type modules (also known as short consensus repeats). These domains of about 55–60 residues are found in numerous mosaic proteins of the complement system, constituents of the extracellular matrix, receptors, etc.[18] More than 150 members of this module family are known, permitting the analysis of the distribution of scores over a wide sequence similarity range. The consensus sequence determined with $f = 0.5$, $k = 0.5$ was used to analyze the distribution of scores in conserved (c) and variable ($v = x + g$) positions of the family. The scores along the entire length of the aligned sequences (t), in consensus (c) positions, and in variable (v) positions were expressed as percent maximal possible scores in t, c, v positions, respectively to give t^*, c^*, and v^*. (Maximal possible scores were determined by self-alignment.) The distribution of scores in

TABLE I
DISTANTLY RELATED MEMBERS OF MODULE FAMILIES DETECTED BY
ITERATIVE DATABASE SEARCHES

Module family[a]	Protein	h values[b]
TSP	Squid light organ peroxidase[c]	1.05
UPAR	Intra-acrosomal protein SP-10[d]	1.05
LMNA	NG2 chondroitin sulfate proteoglycan[e]	0.95, 0.97
B	Product K07E12.1[f]	1.02
vWA	Product F09G8.8[g]	0.92, 0.95
SAP	Product T07C4.4[g]	1.02

[a] TSP, Thrombospondin type I module; UPAR, urokinase receptor/Ly6 module; LMNA, laminin A module; B, complement B-type module; vWA, von Willebrand factor type A module; SAP, saposin module.

[b] The h values were calculated as described in the text. An h value close to 1.00 indicates that the consensus sequence characteristic of a protein family is also valid for the new sequence.

[c] S. I. Tomarev, R. D. Zinovieva, V. M. Weis, A. B. Chepelinsky, J. Piatigorsky, and M. J. McFall-Ngai, *Gene* **132,** 210 (1993). The TSP module is in the C-terminal part (residues 838–891) of squid light organ peroxidase Lo4.

[d] R. M. Wright, E. John, K. Klotz, C. J. Flickinger, and J. C. Herr, *Biol. Reprod.* **42,** 693 (1990). The UPAR module is in the C-terminal part (residues 188–265) of the intra-acrosomal protein SP-10.

[e] A. Nishiyama, K. J. Dahlin, J. T. Prince, S. R. Johnstone, and W. B. Stallcup, *J. Cell Biol.* **114,** 359 (1991). Two segments (residues 44–202 and 218–390) of this proteoglycan belong to the LMNA module family.

[f] R. Wilson, R. Ainscough, K. Anderson, C. Baynes, M. Berks, J. Bonfield, J. Burton, M. Connell, T. Copsey, J. Cooper, A. Coulson, M. Craxton, S. Dear, Z. Du, R. Durbin, A. Favello, L. Fulton, A. Gardner, P. Green, T. Hawkins, L. Hillier, M. Jier, L. Johnston, M. Jones, J. Kershaw, J. Kirsten, N. Laister, P. Latreille, J. Lightning, C. Lloyd, A. McMurray, B. Mortimore, M. O'Callaghan, J. Parsons, C. Percy, L. Rifken, A. Roopra, D. Saunders, R. Shownkeen, N. Smaldon, A. Smith, E. Sonnhammer, R. Staden, J. Sulston, J. Thierry-Mieg, K. Thomas, M. Vaudin, K. Vaughan, R. Waterston, A. Watson, L. Weinstock, J. Wilkinson-Sproat, and P. Wohldman, *Nature (London)* **368,** 32 (1994). The N-terminal part (residues 64–158) of the large K07E12 protein of *Caenorhabditis elegans* belongs to the B-module family. The protein also contains a number of immunoglobulin-like, epidermal growth factor-like, and fibronectin type III modules.

[g] J. Sulston, Z. Du, K. Thomas, R. Wilson, L. Hillier, R. Staden, N. Halloran, P. Green, J. Thierry-Mieg, L. Qiu, S. Dear, A. Coulson, M. Craxton, R. Durbin, M. Berks, M. Metzstein, T. Hawkins, R. Ainscough, and R. Waterston, *Nature (London)* **356,** 37 (1992). Two segments (residues 46–238 and 254–444) of the F09G8.8 protein of *C. elegans* belong to the vWA module family. Residues 71–143 of the T07C4.4 protein of *C. elegans* belong to the SAP module family.

175 pairwise comparisons was analyzed by plotting c^* and v^* as a function of t^* (Fig. 1).

It is clear from Fig. 1 that as we move from perfect identity ($t^* = 100$) to lower values of t^*, scores are preferentially lost from v positions, the scores in c positions are much less affected. It is important to note that in the case of distantly related sequences ($t^* < 30$) similarity in the variable positions is indistinguishable from that expected by chance; the scores reflecting homology are concentrated in the conserved positions. Accordingly, when distantly related proteins are compared the conserved positions of the consensus sequence are the really sensitive indicators of homology, whereas similarity in variable positions is expected to be negligible. By disregarding similarity in variable positions the signal-to-noise ratio may be significantly improved, permitting the more sensitive detection of distantly related members of a protein family. When searching the database with a consensus sequence, variable positions should therefore be masked to suppress scores that could lead to false positives.

Obviously, if we determine the pattern of scores for a protein family

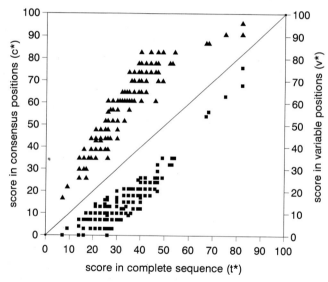

Fig. 1. Typical pattern of scores in a divergent protein family. The distribution of scores in pairwise comparisons of complement B-type modules was analyzed with the consensus sequence of this module family. Percent maximal scores in consensus positions (c^*) and in variable positions (v^*) were plotted as a function of the percent maximal scores of the complete sequences (t^*). Note that in the range of distant homologies ($t^* < 30$) scores are concentrated in consensus positions, whereas scores in the variable positions correspond to the background noise level. The c^* values are marked by triangles) and v^* values by rectangles.

over a wide sequence similarity range, this pattern may be used to predict the probable distribution of scores of any new member of the family. Because the distribution of sequence similarity scores is unique for a protein fold, such analyses may be used to evaluate the biological significance of low similarity scores.

Evaluation of Biological Significance of Marginal Sequence Similarities

If some sequence comparison procedure detects a marginal similarity of proteins A_x and Y, the significance of this finding may be evaluated by determining whether the pattern of similarity scores agrees or conflicts with that expected on the basis of homology. The analysis is performed in the following steps.

(1) The pattern of scores characteristic of the protein family(ies) is determined. If at least one of the proteins (A_x) has known homologs $(A_1, \ldots, A_n)$, multiple alignments of the homologous sequences $A_x, \ldots, A1, \ldots, A_n$ are constructed and consensus sequences are determined. Sequences $A_x, \ldots, A_1, \ldots, A_n$ are compared pairwise, and parameters c^* and v^* are plotted versus t^* to determine the distribution of scores. In the case of distant homologs $(t^* < 30)$, scores are concentrated in c positions, and c^* must be higher than the corresponding t^* (see Fig. 1). Excess scores of consensus positions $(E_A = c^* - t^*)$ are calculated. If Y also has known relatives $(Y_1, \ldots, Y_n)$, the same steps may also be performed with members of the Y family.

(2) The distribution of scores in comparisons of the new candidate is determined. The sequences of proteins $A_x, \ldots, A_1, \ldots A_n$ are compared with the sequence of the new candidate Y, and the distribution of scores is analyzed with the consensus sequence used for the analysis of protein family A; scores c^* and v^* are plotted versus t^* to determine the distribution of scores. Excess scores of consensus positions $(E_Y = c^* - t^*)$ are calculated. If Y also has known relatives $(Y_1, \ldots, Y_n)$ all members of the Y family may be compared with all members of the A family and data analyzed as described above.

(3) The distribution of scores observed in (1) and (2) are compared. This comparison is quite straightforward if the two data sets overlap (i.e., if data at similar t^* values are available from both groups of comparisons). In this case excess score determined for a given t^* in comparisons of Y with $A_x, \ldots, A_1, \ldots, A_n$ (E_Y) is divided by excess score determined in comparisons of $A_x, \ldots, A_1, \ldots, A_n$ (E_A) at the same t^* value to calculate the parameter $h = E_Y/E_A$. If Y versus $A_x, \ldots, A_1, \ldots, A_n$ comparisons give lower t^* values than any of the $A_x, \ldots, A_1, \ldots, A_n$ comparisons (i.e.,

Y is suspected to be a more distantly related member of the A family), we must extrapolate to these lower t^* values from score-distribution plots (Fig. 1). In this case excess score determined at a given t^* in comparisons of Y with $A_x, \ldots, A_1, \ldots, A_n$ (E_Y) is divided by the excess scores calculated for A proteins at that t^* value (E_A) to calculate the parameter $h = E_Y/E_A$.

(4) The scores must be evaluated. If the new candidate Y belongs to the same protein family as $A_x, \ldots, A_1, \ldots, A_n$, its scores must follow the rules characteristic of the protein fold of protein family A; therefore, E_Y should be similar to that determined for E_A. The parameter $h = E_Y/E_A$ should thus be close to 1.00, indicating that the same consensus sequence is valid for the new member of the family.

If Y is unrelated to A_x, that is, their protein folds are dissimilar, then the consensus sequence of protein family A is meaningless for Y versus $A_x, \ldots, A_1, \ldots, A_n$ comparisons. In this case scores could come with equal or lower probability from c than t positions: excess score E_Y and thus $h = E_Y/E_A$ could be close to or below 0.00.

The utility of such analyses may be illustrated by the following examples.

Similarity of Segment of Tissue Plasminogen Activator to Finger Domains of Fibronectin

We have previously found that a short (38 residue) segment of tissue plasminogen activator (t-PA) shows 20–27% identity with finger units of fibronectin.[6] Because of the low degree of sequence similarity of just a very short segment, gaps in the alignments, etc., this similarity was necessarily missed by conventional sequence comparison procedures. Despite this low sequence similarity, we concluded that it reflects common ancestry, primarily because most of the residues conserved in all finger domains of fibronectin were also conserved in t-PA. In other words, the "consensus sequence" of fibronectin fingers was also valid for this short segment of t-PA.[6] We have therefore proposed that finger domains of fibronectin and this short segment of t-PA have a common evolutionary origin and similar three-dimensional structures, and we suggested that exon shuffling was the mechanism whereby these otherwise unrelated proteins acquired homologous finger domains.[6,17]

Analysis of the distribution of scores in comparisons of the finger domains of fibronectin with one another was used to establish the evolutionary behavior of this protein fold using the consensus sequence determined with $f = 0.8$, $k = 0.5$ (Fig. 2). Comparison of the 38-residue segment of t-PA with the twelve finger domains of fibronectin revealed that the distribution of scores fits the pattern characteristic of fibronectin fingers, with the two

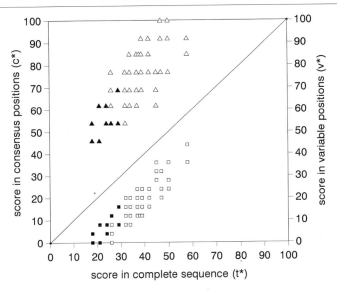

Fig. 2. Pattern of scores in the fibronectin type I repeat (finger module) family. The distribution of scores was analyzed with the consensus sequence determined for fibronectin fingers. Data collected from pairwise comparisons were analyzed by plotting percent maximal scores in consensus positions (c^*) and in variable positions (v^*) as a function of percent of maximal scores of the complete sequences (t^*). The c^* values are marked by triangles and v^* values by rectangles; open symbols denote comparisons of fibronectin-fingers and filled symbols comparisons of tissue plasminogen activator (t-PA) with fibronectin fingers. Note that the distribution of scores in the two types of comparisons fit the same pattern, indicating that t-PA has a true finger module.

sets of data partially overlapping in the $t^* = 26$–29 range (Fig. 2). In the case of the fibronectin–t-PA comparisons the values calculated for $t^* = 26$ were $c^* = 58$, $v^* = 7$, and $E_Y = 32$; comparisons of fibronectin fingers give a similar value, $E_A = 33$, for $t^* = 26$. The parameter $h = E_Y/E_A$ at $t^* = 26$ is thus 0.97, indicating that the pattern of similarity scores agrees with that expected on the basis of homology of fibronectin and t-PA. In other words, the consensus sequence characteristic of the finger domains of fibronectin is also valid for the N-terminal segment of t-PA, thus t-PA has a true finger domain.

Subsequent studies on t-PA and fibronectin have confirmed our conclusion that the 38-residue segment of t-PA is homologous with fibronectin fingers. Studies on the genes of fibronectin and t-PA have shown that the finger domains are encoded by homologous exons, supporting our hypothesis that exon shuffling is responsible for the exchange of these domains.[20,21] Moreover, nuclear magnetic resonance (NMR) studies on the structures of finger domains of tissue plasminogen activator and fibronectin have revealed that their protein fold is strikingly similar, confirming our claim that they are homologous.[22,23]

Similarity of Domain of Cytokine Receptors to Fibronectin Type III Domains

We have shown that a 90-residue segment of the extracellular portion of growth hormone receptor and other members of the cytokine receptor family show 14–23% sequence identity with type III modules of fibronectin, tenascin, twitchin, etc.[11] Despite the low sequence similarity of a relatively short segment, we suggested that it reflects common ancestry, primarily because the residues conserved in known members of the fibronectin type III module family were also conserved in cytokine receptors. We have therefore proposed that the type III domains of fibronectin, tenascin, etc., and these segments of cytokine receptors are of common evolutionary origin. We have predicted that these domains have similar three-dimensional structures, and the exon–intron organization of their genes might also be similar.[11]

The distribution of scores in comparisons of type III domains of fibronectin, tenascin, twitchin, neural adhesion molecule L1, leukocyte common antigen-related protein LAR, etc., was analyzed with the consensus sequence of type III domains determined at $f = 0.7$, $k = 0.5$ (Fig. 3). The parameters calculated for $t^* = 20$ were $c^* = 52$, $v^* = 12$, and $E_A = 32$. Analysis of the distribution of scores in comparisons of type III domains with cytokine receptors with the same consensus sequence revealed that the pattern of scores is superimposable with that observed in comparisons of known type III domains (Fig. 3). The parameters calculated for $t^* = 20$ were $c^* = 51$, $v^* = 13$ and $E_Y = 31$, the parameter $h = E_Y/E_A$ is thus 0.97. In other words, the consensus sequence characteristic of the type III domains of fibronectin is valid for a segment of the extracellular part of the cytokine receptors.

Our conclusion that a part of cytokine receptors is homologous with type III domains has been confirmed by subsequent studies on these proteins. Thus, X-ray crystallography and NMR studies on growth hormone

[20] T. Ny, F. Elgh, and B. Lund, *Proc. Natl. Acad. Sci. U.S.A.* **81,** 5355 (1984).

[21] R. S. Patel, E. Odermatt, J. E. Schwarzbauer, and R. O. Hynes, *EMBO J.* **6,** 2565 (1987).

[22] M. Baron, D. Norman, A. Willis, and I. D. Campbell, *Nature (London)* **345,** 642 (1990).

[23] A. K. Downing, P. C. Driscoll, T. S. Harvey, T. J. Dudgeon, B. O. Smith, M. Baron, and I. D. Campbell, *J. Mol. Biol.* **225,** 821 (1992).

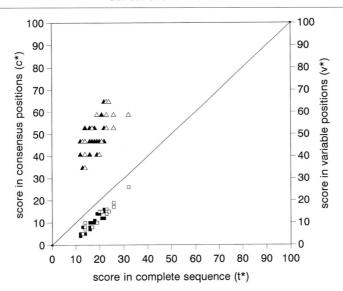

Fig. 3. Pattern of scores in the fibronectin type III module family. The distribution of scores was analyzed with the consensus sequence of fibronectin type III modules. Data collected from pairwise comparisons were analyzed by plotting percent maximal scores in consensus positions (c^*) and in variable positions (v^*) as a function of percent of maximal scores of the complete sequences (t^*). The c^* values are marked by triangles and v^* values by rectangles. Open symbols denote comparisons of known members of the fibronectin type III module family and filled symbols comparisons of known type III modules with segments of cytokine receptors. Note that the distribution of scores in the two types of comparisons fit the same pattern, indicating that cytokine receptors have typical fibronectin type III domains.

receptor[24] and type III domains of fibronectin[25,26] and tenascin[27] have shown a remarkable similarity of the three-dimensional structures of the proposed type III domains. Moreover, the exon–intron organization of cytokine receptor genes was found to agree with that predicted on the basis of homology with fibronectin type III domains.[28,29]

[24] A. M. de Vos, A. A. Ultsch, and A. A. Kossiakoff, *Science* **255,** 306 (1992).

[25] M. Baron, A. L. Main, P. C. Driscoll, H. J. Mardon, J. Boyd, and I. D. Campbell, *Biochemistry* **31,** 2068 (1992).

[26] A. L. Main, T. S. Harvey, M. Baron, J. Boyd, and I. D. Campbell, *Cell* (*Cambridge, Mass.*) **71,** 671 (1992).

[27] D. J. Leahy, W. A. Hendrickson, I. Aukhil, and H. P. Erickson, *Science* **258,** 987 (1992).

[28] H. Shibouya, M. Yoneyama, Y. Nakamura, H. Harada, M. Hatakeyama, S. Minamoto, T. Kono, D. Takeshi, R. White, and T. Taniguchi, *Nucleic Acids Res.* **18,** 3697 (1990).

[29] C. M. Pleiman, S. D. Gimpel, L. S. Park, H. Harada, T. Taniguchi, and S. F. Ziegler, *Mol. Cell. Biol.* **11,** 3052 (1991).

Sequence Similarity of Surfactant Protein B to Kringle Domains?

This type of analysis is also able to establish if a marginal sequence similarity is due to chance and not to homology. This point may be illustrated by the sequence similarity of surfactant protein B (SP-B) and the kringle domain of factor XII. SP-B was found to exhibit 26% sequence identity with the kringle of coagulation factor XII, and one of the intrachain disulfide bonds of surfactant protein B was also suggested to be analogous with a disulfide bond of kringles.[30] On the basis of these similarities it has been suggested that kringles and surfactant protein B might be related.[30] To decide whether the pattern of scores is consistent with this assumption, the distributions of scores were analyzed with consensus sequences determined for both the saposin module family (to which SP-B belongs[13]) and the kringle module family. Comparison of SP-B with kringles, or the kringle of factor XII with members of the saposin family, gave distributions of scores that were inconsistent with the assumption that the two families are related. Analysis of the SP-B/factor XII alignment scores with the kringle consensus sequence gave $h = 0.03$; analysis of alignment scores with the consensus sequence of saposin modules gave $h = 0.08$. These data indicate that neither the kringle-consensus sequence nor the saposin consensus sequence is valid for the SP-B/factor XII similarity; their sequence similarity therefore does not reflect homology.

It should be noted that the only member of the kringle family which showed moderate similarity with SP-B was the kringle of factor XII. Also, the only member of the saposin family that showed some similarity with the kringle of factor XII was SP-B. The fact that this similarity could be shown to be due to chance cautions that if similarity is restricted to just one member of a populous family it is unlikely to reflect true homology. It is easy to see why: if scores are concentrated primarily in the conserved positions (as expected for distant relatives) similarity should not be the privilege of just one member of the family. This conclusion contradicts the advice given by others that "if a distant relationship is suspected between two families of sequences, it is best to compare all possible pairs between the two sets ... evolutionary memory and statistics may indicate a match in only one or a few of the pairs ... ."[31]

Comparison of Consensus Sequence Procedure with Other Consensus Approaches

A common feature of the various consensus approaches is that they all exploit the evolutionary information present in multiple alignments of

[30] J. Johansson, T. Curstedt, and H. Jörnvall, *Biochemistry* **30,** 6917 (1991).
[31] P. Argos, M. Vingron, and G. Vogt, *Protein Eng.* **4,** 375 (1991).

homologous sequences. There are, however, significant differences in the way the information is extracted and utilized. The method of Altschul and Lipman[32] is unique in that it searches a database for multiple alignments, whereas most other consensus approaches (including the one discussed in this chapter) search for pairwise similarities to a consensus sequence (profile, template, pattern) derived from a predefined multiple alignment.[5,7,33–36]

The multiple alignment problem has been approached from two main directions: by standard dynamic programming techniques and by consensus sequence methods.[37] The methods using dynamic programming guarantee a mathematically optimal alignment but require excessive amounts of computer time. It may be argued that much of this time is wasted since the biologically optimal multiple alignment (i.e., one which faithfully reflects any similarities in three-dimensional structure) is not necessarily found through mathematical optimalization.

Heuristics may significantly reduce computation time by using various shortcuts. The heuristic consensus sequence-based methods are much faster than the dynamic programming methods and can align hundreds of sequences within a reasonable time. Common to these methods is the principle that a consensus sequence is defined against which other sequences are aligned, thus reducing the problem to a series of repeated cycles of pairwise alignments followed by consensus definition, etc. Most consensus procedures also use various biological rationales to find the biologically optimal multiple alignments. Most importantly, since alignment of the most closely related sequences is the most accurate, sequences are aligned progressively according to the order of branching in a hypothetical phylogenetic tree.[37,38] This approach ensures that greater weight is attached to gaps that result from the comparison of closely related sequences.

Once a biologically valid multiple alignment is constructed, the evolutionary information characterizing the given family must be condensed into a consensus sequence, a pattern, template, or profile.[7,34–36] The different pattern-matching methods differ significantly in their decision as to what features of the multiple alignment are emphasized and to what degree certain features of the related sequences are exaggerated. At one extreme we find procedures[34,35] that retain all the global information present in the multiple alignment, the other extreme is represented by procedures that concentrate only on the most highly conserved local similarities (blocks,

[32] S. F. Altschul and D. J. Lipman, *Proc. Natl. Acad. Sci. U.S.A.* **87,** 5509 (1990).
[33] W. R. Taylor, *Protein Eng.* **2,** 77 (1988).
[34] W. R. Taylor and D. T. Jones, *Curr. Opin. Struct. Biol.* **1,** 327 (1991).
[35] G. J. Barton, this series, Vol. 183, p. 403.
[36] M. Gribskov, R. Lüthy, and E. Eisenberg, this series, Vol. 183, p. 146.
[37] S. C. Chan, A. K. C. Wong, and D. K. Y. Chiu, *Bull. Math. Biol.* **54,** 563 (1992).
[38] D.-F. Feng and R. F. Doolittle, this series, Vol. 183, 375 (1990).

motifs, signatures). The procedure discussed in this chapter occupies an intermediate position: it condenses global similarities of protein folds into consensus sequences but, unlike the procedures of Barton[35] and Gribskov et al.,[36] reduces the information to a character string in which low scoring positions are suppressed. In our experience this step is advantageous not only because it reduces computational time but also because it sharpens key features of the family. Since nonconserved sites are not represented in the consensus sequence, "bad hits" with nonhomologous sequences are eliminated, thereby increasing the discriminating power of the consensus sequence.

[13] Identification of Sequence Patterns with Profile Analysis

By Michael Gribskov and Stella Veretnik

Introduction

In the mid 1980s, it was rare for more than one or two sequences belonging to a homologous family to be known. Today, the situation is dramatically different, and we are quickly approaching the time when it will be rare that a newly determined sequence does not fall into a known family. This amazing growth of molecular sequence data has brought us to the paradoxical point where a database search with a new sequence now may reveal so many significantly related sequences that it becomes difficult to decipher what they have in common. The increasingly overwhelming nature of sequence data has led to a number of efforts to organize information into libraries of motifs (e.g., PROSITE,[1] BLOCKS,[2] and ProDom[3]), and more recently into more sophisticated mathematical models of protein families based on hidden Markov models.[4,5] At the same time, improvements in our ability to determine realistic three-dimensional models of proteins based on the structures of homologous proteins has lead to an increased interest in the conserved regions of families of sequences because the highly conserved regions correspond to structurally conserved regions.

[1] A. Bairoch and P. Bucher, *Nucleic Acids Res.* **22**, 3583 (1994).
[2] S. Henikoff and J. G. Henikoff, *Proc. Natl. Acad. Sci. U.S.A.* **89**, 10915 (1992).
[3] E. L. Sonnhammer and D. Kahn, *Protein Sci.* **3**, 482 (1994).
[4] A. Krogh, M. Brown, I. S. Mian, K. Sjolander, and D. Haussler, *J. Mol. Biol.* **235**, 1501 (1994).
[5] P. Baldi, Y. Chauvin, T. Hunkapiller, and M. A. McClure, *Proc. Natl. Acad. Sci. U.S.A.* **91**, 1059 (1994).

Copyright © 1996 by Academic Press, Inc.
All rights of reproduction in any form reserved.

This link between conserved regions in sequence families and the core of three-dimensional structures makes methods, such as profile analysis, that allow the information present in sequence alignments of protein families to be used in homology modeling increasingly important.

The idea of a profile is straightforward and easy to understand. The profile is a weight matrix that, for each position in a group of aligned sequences, assigns a score for each of the 20 possible amino acid residues. At its simplest, the profile can be thought of as merely a convenient data structure capable of encoding the character of conserved residues seen in a group of related sequences. At a more sophisticated level, however, each profile can be seen as a mathematical model for a group of protein sequences. This model is more complex than a simple weight-matrix model in that it contains position specific information on insertions and deletions in the sequence family, and is quite closely related to the one employed in hidden Markov models for sequences. The profile approach has turned out to be quite flexible, with applications ranging from describing DNA sequence motifs, to the characterization of protein families, and to the mapping of sequences onto three-dimensional structures.[6–8]

Methods

Profile Analysis

The profile (Fig. 1) is a two-dimensional weight matrix in which the rows correspond to aligned positions in a group of sequences, and the columns correspond to each of the 20 possible amino acid residues (or four DNA bases). The profile uses a similarity-based scoring system where positive values indicate that the residue represented by the column in which the value occurs is similar to the corresponding residues in the aligned sequences, and negative values indicate dissimilarity. Profiles differ from generic weight matrices in having two additional columns that specify position specific weights for gap penalties. The two additional columns represent weights on the gap opening penalty and the gap extension penalty.

Profiles can be matched with sequences using an extension of standard dynamic programming sequence alignment techniques, usually the algorithm of Smith and Waterman.[9] A linear gap penalty (often called an affine gap penalty) consisting of a length independent (gap opening) and length

[6] J. U. Bowie, R. Luthy, and D. Eisenberg. *Science* **253,** 164 (1991).
[7] M. Wilmanns and D. Eisenberg, *Proc. Natl. Acad. Sci. U.S.A.* **90,** 1379 (1993).
[8] K. Y. Zhang and D. Eisenberg, *Protein Sci.* **3,** 687 (1994).
[9] T. F. Smith and M. S. Waterman, *J. Mol. Biol.* **147,** 195 (1981).

Cons	A	C	D	E	F	G	H	I	K	L	M	N	P	Q	R	S	T	V	W	Y	Gap	Len
G	9	-38	4	-2	-44	52	-22	-28	-15	-36	-28	3	-6	-14	-23	8	-2	-18	-53	-45	100	100
P	10	19	0	-1	-35	3	-8	-6	-10	-16	-10	0	30	-3	-9	8	7	0	-46	-26	100	100
H	-15	-60	39	37	-56	-16	53	-37	-4	-44	-32	26	-13	42	-3	-8	-16	-36	-69	-32	100	100
I	-21	-54	-53	-44	-8	-50	-53	91	-50	26	19	-43	-42	-43	-45	-41	-15	46	-94	-28	100	100
V	-15	-55	-56	-46	-19	-49	-53	60	-53	28	20	-47	-43	-43	-48	-44	-15	72	-105	-36	100	100
V	-16	-55	-63	-51	-30	-54	-65	69	-62	16	14	-53	-51	-54	-56	-49	-15	78	-125	-49	100	100
A	66	-68	-16	-18	-75	37	-61	-56	-52	-71	-57	-17	-14	-42	-63	6	-3	-29	-100	-80	100	100
T	-50	-106	-84	-88	-130	-75	-109	-97	-98	-120	-101	-75	-74	-99	-105	-50	**117**	-89	-46	-126	100	100
P	-16	-81	-61	-55	-110	-53	-62	-83	-72	-83	-85	-56	**115**	-46	-61	-26	-40	-62	-117	-104	100	100
G	-99	-15	-104	-110	-151	**102**	-5	-9	-125	-141	-10	-107	-111	-121	-8	-101	-110	-123	-38	-26	100	100
! 11																						
R	-98	-144	-7	-10	-141	-105	-68	-98	0	-111	-60	-56	-91	-59	**129**	-75	-74	-110	-84	-69	100	100
L	-24	-70	-62	-48	5	-56	-47	34	-51	69	39	-47	-39	-30	-46	-45	-24	32	-94	-21	100	100
L	-25	-65	-55	-44	8	-51	-42	30	-46	70	41	-42	-35	-26	-41	-41	-23	20	-86	-17	100	100
D	-42	-94	105	33	-94	-34	-19	-63	-33	-78	-68	14	-48	0	-52	-32	-41	-66	-106	-74	100	100
L	-17	-38	-28	-25	35	-32	-9	16	-26	47	32	-21	-20	-14	-22	-23	-15	10	-30	33	100	100
L	-23	-61	-51	-41	1	-48	-41	33	-44	69	38	-40	-33	-26	-40	-38	-21	22	-93	-26	100	100
Q	-8	-52	27	35	-53	-11	16	-25	17	-32	-12	12	-7	37	14	-4	-7	-26	-46	-40	100	100
K	-26	-62	-8	-9	-65	-32	1	-29	59	-41	2	8	-18	7	51	-8	-8	-38	-34	-62	100	100
G	8	-26	13	5	-38	21	-4	-18	-3	-28	-17	16	0	-3	-9	17	9	-14	-41	-32	100	100
T	4	-15	6	2	-25	1	0	-4	6	-10	-1	6	2	0	3	7	11	-3	-23	-19	100	100
! 21																						
V	-3	-37	33	21	-40	-7	-11	20	-15	-7	-1	8	-17	0	-25	-9	-3	36	-74	-36	22	22
T	11	-29	10	6	-39	16	-13	-7	-7	-24	-12	7	0	-4	-18	12	25	-3	-57	-35	22	22
K	-21	-61	-10	-13	-50	-32	-14	-18	60	-21	11	8	-21	0	32	-6	3	-26	-48	-53	22	22
G	9	-25	4	0	-29	26	-16	2	-13	-9	-4	1	0	-8	-19	5	4	8	-56	-29	22	22
L	-10	-37	-28	-24	16	-27	-23	27	-25	47	33	-22	-18	-17	-23	-20	-8	25	-52	3	22	22
K	-8	-36	10	3	-40	-9	7	-15	32	-21	0	19	-6	7	26	1	0	-18	-26	-33	100	100
L	-24	-63	-52	-42	9	-49	-39	23	-44	71	41	-41	-33	-25	-40	-39	-24	17	-79	-13	100	100
K	-5	-38	18	13	-44	-4	7	-17	25	-25	-4	17	-4	12	20	2	0	-19	-32	-36	100	100
K	-10	-40	5	3	-42	-13	16	-17	34	-23	1	10	-4	13	34	-1	-1	-20	-19	-36	100	100
V	-6	-9	-38	-32	-14	-32	-42	52	-39	27	23	-32	-29	-33	-36	-25	-4	58	-89	-26	100	100
! 31																						
K	-20	-58	4	0	-62	-23	1	-27	56	-38	1	15	-16	8	33	-4	-4	-34	-40	-56	100	100
L	-17	-38	-34	-30	40	-35	-20	21	-29	46	36	-25	-24	-21	-27	-25	-15	15	-27	40	100	100
L	-27	-70	-58	-46	11	-54	-42	27	-48	72	41	-44	-36	-27	-43	-43	-26	17	-80	-12	100	100
V	-44	-89	-95	-83	-61	-85	-93	38	-94	-11	-13	-86	-82	-84	-86	-84	-45	103	-150	-77	100	100
L	-48	-105	-92	-71	-9	-82	-65	8	-74	89	40	-69	-57	-41	-67	-70	-50	0	-126	-46	100	100
D	-76	-59	**124**	-13	-60	-74	-61	-98	-72	-42	-34	-19	-84	-43	-21	-69	-78	-101	-72	-39	100	100
E	-83	-57	-14	**122**	-60	-79	-62	-106	-77	-116	-29	-48	-89	-34	-12	-79	-86	-108	-69	-120	100	100
A	**99**	-111	-76	-75	-128	-53	-107	-102	-99	-119	-101	-74	-63	-94	-106	-45	-45	-80	-141	-125	100	100
D	-76	-59	**124**	-13	-60	-74	-61	-98	-72	-42	-34	-19	-84	-43	-21	-69	-78	-101	-72	-39	100	100
R	-40	-76	-23	-24	-79	-49	-3	-38	56	-55	-4	-1	-26	2	82	-13	-18	-51	-28	-85	100	100
! 41																						
M	-30	-79	-63	-50	2	-59	-47	24	-47	72	78	-49	-39	-28	-45	-47	-29	15	-103	-28	100	100
L	-47	-100	-90	-70	-5	-80	-63	8	-73	88	34	-67	-56	-40	-66	-69	-50	0	-110	-37	100	100
D	-14	-59	71	46	-66	-8	-2	-37	-10	-52	-41	30	-22	12	-28	-2	-12	-39	-73	-49	100	100
L	-12	-46	-31	-26	5	-31	-26	26	-22	52	51	-25	-20	-16	-23	-22	-10	23	-71	-12	100	100
G	1	-73	-8	-17	-68	79	-55	-61	-45	-68	-64	-9	-30	-39	-62	0	-26	-40	-97	-81	100	100
F	-63	-76	-81	-77	120	-81	-54	-19	-77	8	-1	-65	-68	-67	-70	-65	-62	-37	-50	26	100	100
G	-1	-20	10	9	-24	0	9	-8	10	-8	0	7	1	12	11	0	0	-7	-17	-18	100	100
Q	-5	-45	47	44	-51	-1	5	-25	-2	-36	-26	24	-10	19	-14	-1	-7	-26	-57	-37	100	100
E	-9	-62	47	50	-62	-10	16	-33	-3	-41	-29	18	-13	42	-10	-8	-14	-33	-68	-43	100	100
I	-15	-50	-43	-35	2	-41	-40	64	-38	41	32	-34	-31	-30	-35	-32	-11	44	-82	-17	100	100
! 51																						
D	-1	-39	29	27	-39	1	12	-17	5	-24	-13	15	-2	19	0	1	-1	-17	-43	-27	100	100
Q	-9	-47	18	18	-48	-11	22	-22	24	-29	-7	17	-6	31	22	-1	-4	-24	-35	-36	100	100
I	-20	-53	-55	-45	-19	-50	-58	94	-52	21	17	-44	-44	-46	-47	-42	-14	49	-109	-41	100	100
L	-13	-47	-38	-31	11	-37	-32	42	-32	46	37	-30	-26	-24	-30	-27	-10	40	-66	-5	100	100
K	-3	-28	4	1	-37	-4	1	-12	29	-19	1	8	-1	5	25	3	4	-13	-23	-33	100	100
L	-4	-18	-4	-4	1	-9	3	10	-6	23	17	-3	-4	-1	-4	-6	-2	9	-28	1	100	100
L	-24	-70	-62	-48	3	-55	-47	34	-51	69	40	-48	-39	-30	-46	-45	-23	33	-98	-23	100	100
*	16	2	34	18	15	21	7	31	17	44	12	4	9	8	21	8	14	24	0	3		

dependent (gap extension) term is widely considered to be the most useful for sequence alignments, and the two gap weights included in the profile allow these terms to be separately modified. A formal description of the alignment of profiles and sequences has been presented[10] and will not be repeated here.

The Profile Analysis package of programs[11] provides a suite of tools for creating profiles and matching them with sequences. These programs have been described in detail in earlier papers and we will describe them only briefly here: PROFILEMAKE[10,12] is used to create profiles from groups of aligned sequences. PROFILEGAP[10,13] is used to align a profile and one or more sequences. A useful new function is the ability to produce a multiple alignment of a group of sequences to a profile, in addition to the pairwise alignments available previously. PROFILESEARCH[10,14] uses a profile as a query in a database search, normalizes the results for systematic dependence on length, and converts the scores to Z scores (standardized scores). The previously independent PROFILENORMAL program has been incorporated into PROFILESEARCH. A supercomputer implementation of PROFILESEARCH called PROFILE-SS is available[15] for the CRAY C90. PROFILESCAN[16] compares a single sequence to a library of statistically characterized profiles. The current library consists of 650 protein sequence motifs.

[10] M. Gribskov, R. Lüthy, and D. Eisenberg, this series, Vol. 183, p. 146.

[11] The Profile Analysis package is available from the authors, although the programs are in the midst of conversion from FORTRAN to C programming languages. Please contact Michael Gribskov at gribskov@sdsc.edu for details on the status of implementation. Current program source code and libraries of profiles are available by FTP from ftp.sdsc.edu/pub/sdsc/biology. The Profile Analysis package is also distributed by the Genetics Computer Group, Madison, Wisconsin, as part of their sequence analysis package.

[12] M. Gribskov, A. D. McLachlan, and D. Eisenberg, *Proc. Natl. Acad. Sci. U.S.A.* **84,** 4355 (1987).

[13] M. Gribskov and D. Eisenberg, *in* "Techniques in Protein Chemistry" (T. E. Hugli, ed.), p. 108. Academic Press, San Diego, 1989.

[14] M. Gribskov, *Methods Mol. Biol.* **25,** 247 (1994).

[15] PROFILE-SS is available from the Pittsburgh Supercomputing Center. Contact Alex Ropelewski, ropelews@psc.edu for details. Using World Wide Web, access http://pscinfo.psc.edu/general/software/packages//profiless/profiless.html.

[16] M. Gribskov, M. Homyak, J. Edenfield, and D. Eisenberg, *CABIOS* **4,** 61 (1988).

FIG. 1. Evolutionary profile calculated for the sequences shown in Fig. 4. Each row corresponds to a column of the aligned sequence. The consensus sequence shown at the left represents the highest scoring column in each row and can be used as a cross-reference to Fig. 4. The most conserved regions of the sequence have the consensus sequence and the corresponding column shown in boldface type.

Average Profiles

We refer to the original method for calculating profiles as the average method. Briefly, a single fixed scoring table (e.g., the PAM 250 table,[17]) is used as the basis of the profile. This table can be thought of as specifying 20 model residue frequency distributions, one for each possible ancestral residue. The model distributions are combined into a mixture distribution with the components weighted by the relative frequencies in the observed distribution:

$$\text{Profile}_{ij} = \sum_{k=1}^{20} f_{ik} M_{jk} \tag{1}$$

where f_{ik} is the relative frequency of residue k at position i in the aligned sequences, and M_{jk} is the comparison score for residues j and k in the basis scoring table. Note that the relative frequencies f_{ik} may be adjusted in various ways to account for sampling error or bias (see Gribskov *et al.*,[10] and below).

Evolutionary Profiles

We have developed a finite mixture method based on the Dayhoff model of protein evolution[17] for the calculation of profiles. We refer to this method as the evolutionary profile method to distinguish it from the earlier average profile method. The evolutionary profile method models each position in the observed group of sequences as arising from one or more ancestral residues, each possibly at a different evolutionary distance. In adopting this approach we are attempting to explicitly include biologically relevant prior information about the rate and kind of change occurring at each position in a homologous family.

The evolutionary profile method requires two steps at each position in the aligned sequences: first, the Dayhoff evolutionary model is used to generate a series of model distributions for each of the 20 possible ancestral residues at various evolutionary distances (typically 1, 2, 4, 8, 16, 32, 64, 128, 256, 512, 1024, and 2048 PAM distances). The evolutionary distance that minimizes the cross entropy, *H*, of the model and observed distributions is chosen for each possible ancestral residue. A different evolutionary distance is fit for each possible ancestral residue, giving 20 model distributions in all:

[17] M. O. Dayhoff, R. M. Schwartz, and B. C. Orcutt, *in* "Atlas of Protein Sequence and Structure" (M. O. Dayhoff, ed.), Vol. 5, Suppl. 3, p. 345. National Biomedical Research Foundation, Washington, D.C., 1978.

$$H = -\sum_{a=1}^{20} f_a \ln p_a \qquad (2)$$

where f_a are the observed residue frequencies and p_a are the predicted frequencies in the model distribution at a specific evolutionary distance.

In the second step, the 20 model distributions determined above are mixed based on the probability that each one could give rise to the observed frequency distribution. This is done by determining a weight (mixture coefficient) for each of the 20 model distributions such that the weight corresponds to the extent to which the model predicts the observed distribution.

The probability of the model distribution, M_a, for a given ancestral amino acid residue, a, giving rise to the observed residue frequency distribution F is given by Eq. (3):

$$P(M_a|F) = \frac{P(M_a) \times P(F|M_a)}{\sum_{a=1}^{20} P(M_a) \times P(F|M_a)} \qquad (3)$$

where $P(M_a)$ is the prior distribution for the amino acid residues, normally the amino acid residue frequencies in the database, and

$$P(F|M_a) = \prod_{a=1}^{20} p_a^{f_a} \qquad (4)$$

Mixture coefficients, W, are given by $W_a = P(M_a|F) - P(M_{\mathrm{random}}|F)$ where M_{random} is the residue frequency distribution of random sequences, i.e., the amino acid residue frequencies in the database. Typically only a few residues will have weights appreciably greater than zero (see Fig. 2). Because M_a at very long PAM distances is identical to M_{random}, the weights are guaranteed to be positive. Finally the profile is calculated as the log-odds ratio for the weighted sum of the mixture components:

$$\mathrm{Profile}_{ij} = \ln \left[\sum_{a=0}^{20} (W_{ai} p_{aij}) \Big/ p_{\mathrm{random}\, j} \right] \qquad (5)$$

where W_{ai} is the weight on the ancestral residue, a, at position i, and p_{aij} is the frequency of residue j in the ancestral residue frequency distribution a at position i.

Sequence Weighting

Families of sequences are nearly always highly biased. This is primarily due to bias in the selection of organisms to sequence, typically focusing on mammals, yeast, *Escherichia coli,* and *Drosophila.* In any case, it is common

G 128 (0.53)	S 256 (0.12)	A 256 (0.11)	K 512 (0.08)				
P 256 (0.34)	A 256 (0.31)	S 256 (0.13)	C 512 (0.12)	V 256 (0.08)			
H 128 (0.21)	Q 128 (0.19)	D 128 (0.18)	E 128 (0.17)	N 128 (0.12)	K 512 (0.07)		
I 32 (0.36)	V 128 (0.32)	L 256 (0.27)					
V 64 (0.51)	L 256 (0.32)	I 64 (0.16)					
V 64 (0.59)	I 64 (0.23)	L 256 (0.16)					
A 32 (0.51)	G 128 (0.36)	S 128 (0.07)					
T 1 (0.86)	A 128 (0.06)	S 128 (0.06)					
P 16 (0.81)	A 128 (0.10)						
G 1 (0.94)							
R 1 (0.79)	K 128 (0.18)						
L 128 (0.69)	V 128 (0.18)	I 128 (0.08)					
L 128 (0.72)	V 256 (0.15)	I 128 (0.08)					
D 16 (0.59)	E 128 (0.19)	N 128 (0.07)					
L 256 (0.59)	F 256 (0.17)	Y 256 (0.07)	V 512 (0.07)				
L 128 (0.69)	V 256 (0.18)	I 128 (0.10)					
K 256 (0.27)	E 128 (0.21)	Q 128 (0.16)	D 256 (0.13)	N 256 (0.06)	H 256 (0.05)		
K 128 (0.63)	R 128 (0.17)	Q 256 (0.05)					
G 256 (0.28)	S 128 (0.18)	K 512 (0.10)	A 256 (0.10)	N 128 (0.09)	T 256 (0.07)	D 256 (0.06)	
T 256 (0.33)	K 512 (0.31)	S 256 (0.09)	A 512 (0.08)				
V 128 (0.34)	D 128 (0.24)	E 256 (0.17)	I 256 (0.07)	A 512 (0.06)			
A 256 (0.23)	T 128 (0.22)	G 256 (0.22)	S 256 (0.09)	D 256 (0.08)	E 256 (0.05)		
K 128 (0.71)	L 512 (0.11)	T 128 (0.06)					
G 256 (0.40)	A 256 (0.25)	V 256 (0.20)					
L 256 (0.57)	V 256 (0.22)	I 256 (0.10)	F 256 (0.07)				
K 256 (0.54)	N 128 (0.13)	R 256 (0.08)	D 512 (0.05)				
L 128 (0.72)	V 256 (0.14)	I 256 (0.06)					
K 256 (0.39)	D 256 (0.11)	E 256 (0.10)	N 128 (0.07)	R 256 (0.07)	S 256 (0.06)	G 512 (0.06)	Q 256 (0.06)
K 256 (0.55)	R 256 (0.18)	H 256 (0.07)	N 256 (0.06)	Q 256 (0.06)			
V 128 (0.55)	L 256 (0.20)	I 128 (0.18)					
K 128 (0.59)	R 256 (0.11)	E 256 (0.05)	Q 256 (0.05)				
L 256 (0.52)	F 256 (0.22)	Y 256 (0.10)	V 256 (0.08)	I 256 (0.06)			
L 128 (0.76)	V 256 (0.11)	F 256 (0.05)	I 128 (0.05)				
V 16 (0.78)	L 256 (0.11)	I 64 (0.10)					
L 64 (0.88)	V 256 (0.06)						
D 1 (0.80)	E 128 (0.10)						
E 1 (0.83)	D 128 (0.09)						
A 4 (0.88)							
D 1 (0.80)	E 128 (0.10)						
K 128 (0.55)	R 64 (0.37)	M 32 (0.08)					
L 128 (0.76)	V 256 (0.10)						
L 64 (0.87)	V 256 (0.06)						
D 64 (0.37)	E 128 (0.23)	N 128 (0.11)	G 512 (0.07)	S 128 (0.06)			
L 256 (0.69)	V 256 (0.15)	M 128 (0.08)	I 256 (0.07)				
G 64 (0.78)	A 256 (0.09)	S 256 (0.05)					
F 16 (0.68)	L 256 (0.20)	Y 256 (0.09)					
K 512 (0.46)	Q 256 (0.12)	E 512 (0.11)	R 512 (0.08)	D 512 (0.07)			
D 128 (0.25)	E 128 (0.25)	G 512 (0.11)	N 128 (0.08)	K 512 (0.07)	Q 256 (0.07)	A 512 (0.05)	
E 128 (0.33)	D 128 (0.22)	Q 128 (0.19)	K 512 (0.08)	N 256 (0.07)	H 256 (0.05)		
L 256 (0.48)	V 128 (0.27)	I 64 (0.22)					
E 256 (0.29)	D 256 (0.23)	K 512 (0.15)	N 256 (0.12)	Q 256 (0.11)	H 512 (0.06)		
K 256 (0.34)	Q 128 (0.13)	E 256 (0.12)	D 256 (0.11)	R 256 (0.09)	H 256 (0.09)	N 128 (0.07)	
I 32 (0.39)	V 128 (0.38)	L 256 (0.21)					
L 256 (0.56)	V 128 (0.21)	I 128 (0.14)	F 256 (0.05)				
K 256 (0.47)	R 256 (0.09)	S 256 (0.09)	T 256 (0.09)	G 512 (0.08)	A 512 (0.05)		
L 512 (0.62)	V 512 (0.14)	I 512 (0.07)					
L 128 (0.69)	V 128 (0.18)	I 128 (0.08)					

FIG. 2. Mixture density components for the evolutionary profile of the DEAD region of helicases, corresponding to the profile shown in Fig. 1 and the alignment shown in Fig. 4. Each line shows, in rank order, the components of the mixture model at that position. Components are given as $A\ D\ (W)$, where A is the ancestral residue, D is the fit PAM distance, and W is the weight of the component in the mixture distribution. Note that the component of the mixture with the highest weight does not necessarily correspond to the highest scoring column in the profile.

to find in a group of sequences that several are nearly identical, and several others are as little as 20% identical when aligned. It is clear in this case that each of the nearly identical sequences contributes much less information than each of the 20% identical sequences. Weighting procedures seek to correct for this sampling bias, hopefully correcting the observed residue counts so that they correspond to a random sample. In this work, sequences have been weighted using the approach of Felsenstein.[18] However, rather than basing the weights on a completely resolved phylogenetic tree, the weights are based on an approximate tree where the distances are the percentage of differing residues in pairwise alignments. Briefly, a single-linkage multifurcating tree is constructed in which branches are joined at nodes representing discrete distance thresholds. For instance, all sequences with less than 10% differing residues are joined at a single node. This node is joined with all sequences with less than 20% differing residues at the next node, and so on. Because these trees are approximate, they are robust to small errors in alignments and insensitive to the fine points of tree topology.

Evaluating Matching with Receiver Operating Characteristic Analysis

The receiver operating characteristic (ROC) is a widely used technique for evaluating the performance of clinical tests and treatments (for a review, see Zweig and Campbell[19]). ROC analysis has several advantages over other techniques. The ROC is a function of both the sensitivity of an assay (what fraction of the true positives are detected) and the specificity of the assay (how well are the true positives separated from the true negatives). One of its major advantages is threshold independence; the entire distribution of scores is examined rather than just scores over an arbitrary significance threshold.

The analysis involves the construction of an ROC plot (Fig. 3). The plot is constructed by examining each observation, in this case, each sequence in the results of a database search, and plotting the fraction of true positives (homologous family members) and true negatives (unrelated sequences) with equal or higher scores on the ordinate and abscissa, respectively. The area under the curve of the ROC plot measures the probability of correct classification and is a simple statistic that can be used to compare searches using different query sequences or conditions (higher values indicate better performance in detecting the homologous family). We have introduced the ROC_{50} for the evaluation of sequence database searches.[20,21] The ROC_{50} is the area under an ROC curve where the list of results is truncated after observing 50 negative sequences, that is, the number of true negatives is exactly 50.

[18] J. Felsenstein, *Am. J. Hum. Genet.* **25**, 471 (1973).

[19] M. H. Zweig and G. Campbell, *Clin. Chem.* **39**, 561 (1993).

[20] M. Gribskov and N. L. Robinson, *Comput. Chem.* **20**, 25 (1996).

[21] M. Gribskov, *in* "Protein Folds—A Distance-Based Approach" (H. Bohr and S. Brunak, eds.), pp. 71–79, CRC Press, Boca Raton, Florida, 1996.

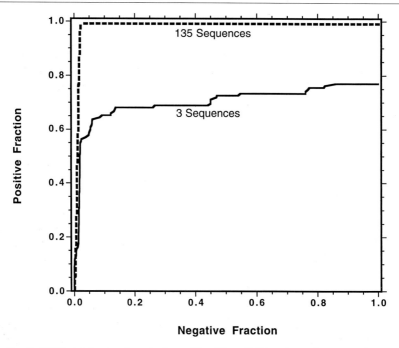

FIG. 3. ROC plot for two ferredoxin profiles. The solid line shows the curve of a profile calculated by the average method using only three sequences ($ROC_{50} = 0.71$), an example of a relatively poor discrimination. The dashed line shows the ROC plot for an evolutionary profile based on all 134 ferredoxin sequences in the database ($ROC_{50} = 0.99$), an example of nearly perfect discrimination.

Results

4Fe-4S Ferredoxins

The 4Fe-4S ferredoxins are small proteins involved in electron transfer (for review, see Beinert[22]). Ferredoxin-like molecules also function in photosynthesis and are found in a variety of enzymes involved in oxidation–reduction reactions (e.g., succinate, fumarate, and glycerol-3-phosphate dehydrogenases, dimethyl sulfoxide reductase, formate hydrogenase, and sulfite reductases). These ferredoxins bind a 4Fe-4S cluster at a highly conserved 12-residue sequence. The core of the conserved region has the consensus pattern C-X-X-C-X-X-C-X-X-X-C-[PEG], and insertions or deletions are not usually required to align this region. In SWISS-PROT

[22] H. Beinert, *FASEB J.* **4**, 2483 (1990).

release 31.0 there are 134 4Fe-4S ferredoxins (including the false-negative members of the family not detected by the PROSITE signature). Most of the members of this family have two copies of the characteristic 12-residue repeat and bear two 4Fe-4S centers. For the work described here we generated profiles for the 19 residues beginning 2 residues before the first conserved cysteine and ending 6 residues after the last conserved cysteine.

Table I shows a comparison of the efficacy of profiles made by the average and evolutionary methods in detecting the 4Fe-4S family. Note that because the sequences have been limited to the most highly conserved region, we are not taking full advantage of the ability of the profile to distinguish conserved and unconserved regions. Similarly, because matching to this region does not require gaps, we obtain no advantage from the position-specific gap penalties that can be encoded in a profile, an important feature when matching to distantly related sequence families (see Gribskov et al.[10]). The differences between the average and evolutionary profile methods therefore correspond only to their respective abilities to generalize the sequence pattern typical of the family from the observed set of sequences. The evolutionary profile consistently performs better than the average profile. When all sequences are included, the evolutionary profile has only about one-third the average classification error (1-ROC_{50}) as the average profile, a small but important difference.

TABLE I

PERFORMANCE OF AVERAGE AND EVOLUTIONARY PROFILES ON 4Fe-4S
FERREDOXIN FAMILY[a]

Profile method	Number of sequences included in profile			
	3	6	12	134
Average	82.2 (5.6)	93.0 (2.0)	93.6 (1.5)	95.2
Evolutionary	84.2 (8.0)	95.2 (1.0)	95.6 (0.6)	98.3

[a] Profiles were generated from aligned sequences selected at random from the 4Fe-4S ferredoxin family. Subsets of sizes 3, 6, and 12 as well as the entire family of sequences (134 members) were used to produce profiles. Searches of the SWISS-PROT database (release 31.0, Feb. 1995) using the program PROFILESEARCH were then performed with each of the profiles. The ability of the profile to identify sequences in the family was evaluated using ROC_{50} method. Values in the table are the ROC_{50} times 100 and represent mean and the standard deviation (in parentheses) of 10 replicates of subsets of the indicated size.

ATP-Dependent RNA Helicases

The members of the ATP-dependent RNA helicase family are involved in ATP-dependent unwinding of nucleic acids. This family is also known as the DEAD helicase family due to the presence of a highly conserved sequence [LIVM]-X-X-D-E-A-D-[RKEN] at what is thought to be part of the ATP binding site. The proteins comprise a number of conserved blocks distributed across the length of the sequences. As an example, we have focused on the conserved block containing the DEAD signature (Fig. 4). As can be seen, this conserved block requires that some insertion/deletion be allowed to get the proper alignment and therefore represents a more

```
            1                                                     57
an3_xenla   GCHLLVATPGRLVDMMERGK....IGLDFCKYLVLDEADRMLDMGFEPQIRRIVEQD
p54_human   TVHVVIATPGRILDLIKKGV....AKVDHVQMIVLDEADKLLSQDFVQIMEDIILTL
p68_human   GVEICIATPGRLIDFLECGK....TNLRRTTYLVLDEADRMLDMGFEPQIRKIVDQI
db73_drome  KADIVVTTPGRLVDHLHATK...GFCLKSLKFLVIDEADRIMDAVFQNWLYHLDSHV
dbp1_yeast  GCDLLVATPGRLNDLLERGK....VSLANIKYLVLDEADRMLDMGFEPQIRHIVEEC
dbp2_schpo  GVEICIATPGRLLDMLDSNK....TNLRRVTYLVLDEADRMLDMGFEPQIRKIVDQI
dbp2_yeast  GSEIVIATPGRLIDMLEIGK....TNLKRVTYLVLDEADRMLDMGFEPQIRKIVDQI
dbpa_ecoli  APHIIVATPGRLLDHLQKGT....VSLDALNTLVMDEADRMLDMGFSDAIDDVIRFA
DEAD_ecoli  GPQIVVGTPGRLLDHLKRGT....LDLSKLSGLVLDEADEMLRMGFIEDVETIMAQI
DEAD_klepn  GPQIVVGTPGRLLDHLKRGT....LDLSKLSGLVLDEADEMLRMGFIEDVETIMAQI
ded1_yeast  GCDLLVATPGRLNDLLERGK....ISLANVKYLVLDEADRMLDMGFEPQIRHIVEDC
dhh1_yeast  TVHILVGTPGRVLDLASRKV....ADLSDCSLFIMDEADKMLSRDFKTIIEQILSFL
drs1_yeast  RPDIVIATPGRFIDHIRNSA...SFNVDSVEILVMDEADRMLEEGFQDELNEIMGLL
glh1_caeel  GATIIVGTVGRIKHFCEEGT....IKLDKCRFFVLDEADRMIDAMGFGTDIETIVNY
if41_human  APHIIVGTPGRVFDMLNRRY....LSPKYIKMFVLDEADEMLSRGFKDQIYDIFQKL
if42_mouse  APHIVVGTPGRVFDMLNRRY....LSPKWIKMFVLDEADEMLSRGFKDQIYERVQKL
if4a_caeel  GIHVVVGTPGRVGDMINRNA....LDTSRIKMFVLDEADEMLSRGFKDQIYEVFRSM
if4a_drome  GCHVVVGTPGRVYDMINRKL....RTQYIKLFVLDEADEMLSRGFKDQIQDVFKML
if4a_orysa  GVHVVVGTPGRVFDMLRRQS....LRPDYIKMFVLDEADEMLSRGFKDQIYDIFQLL
if4a_rabit  APHIIVGTPGRVFDMLNRRY....LSPKYIKMFVLDEADEMLSRGFKDQIYDIFQKL
if4a_yeast  DAQIVVGTPGRVFDNIQRRR....FRTDKIKMFILDEADEMLSSGFKEQIYQIFTLL
if4n_human  GQHVVAGTPGRVFDMIRRRS....LRTRAIKMLVLDEADEMLNKGFKEQIYDVYRYL
me31_drome  KVQLIIATPGRILDLDMKKV....ADMSHCRILVLDEADKLLSLDFQGMLDHVILKL
ms16_yeast  RPNIVIATPGRLIDVLEKYS...NKFFRFVDYKVLDEADRLLEIGFRDDLETISGIL
p110_mouse  GCHLLVATPGRLVDMMERGK....IGLDFCKYLVLDEADRMLDMGFEPQIRRIVEQD
pr05_yeast  GTEIVVATPGRFIDILTLND.GKLLSTKRITFVVMDEADRLFDLGFEPQITQIMKTV
pr28_yeast  GCDILVATPGRLIDSLENHL....LVMKQVETLVLDEADKMYDLGFEDQVTNILTKV
rhlb_ecoli  GVDILIGTTGRLIDYAKQNH....INLGAIQVVVLDEADRMYDLGFIKDIRWLFRRM
rhle_ecoli  GVDVLVATPGRLLDLEHQNA....VSLDQVEILVLDEADRMLDMGFIHDIRRVLTKL
rm62_drome  GCEIVIATPGRLIDFLSAGS....TNLKRCTYLVLDEADRMLDMGFEPQIRKIVSQI
spb4_yeast  RPQILIGTPGRVLDFLQMPA....VKTSACSMVVMDEADRLLDMSFIKDTEKILRLL
srmb_ecoli  NQDIVVATTGRLLQYIKEEN....FDCRAVETLILDEADRMLDMGFAQDIEHIAGET
vasa_drome  GCHVVIATPGRLLDFVDRTF....ITFEDTRFVVLDEADRMLDMGFSEDMRRIMTHV
ybz2_yeast  SGQIVIATPGRFLELLEKDN.TLIKRFSKVNTLILDEADRLLQDGHFDEFEKIIKHL
yhm5_yeast  KPHIIIATPGRLMDHLENTK...GFSLRKLKFLVMDEADRLLDMEFGPVLDRILKII
yhw9_yeast  KPHFIIATPGRLAHHIMSSGDDTVGGLMRAKYLVLDEADILLTSTFADHLATCISAL
yk04_yeast  GCNFIIGTPGRVLDHLQNTKVIKEQLSQSLRYIVLDEGDKLMELGFDETISEIIKIV
yn21_caeel  RPHIIVATPGRLVDHLENTK...GFNLKALKFLIMDEADRILNMDFEVELDKILKVI

Consensus   GPHIVVATPGRLLDLLQKGTVTKGLKLKKVKLLVLDEADRMLDLGFGQDEDQILKLL
```

FIG. 4. Alignment of helicases in the region of the conserved DEAD sequence. The most conserved regions are shown in boldface type. The bottom row, labeled consensus, represents the highest scoring column in the evolutionary profile shown in Fig. 1.

challenging case than that of the ferredoxins. The character of the conservation in these sequences is more variable than in the case of the ferredoxins making it a more interesting subject for profile analysis.

Table II shows the ROC_{50} for the helicase family as a function of the size of the subset of sequences used to generate the profile. It is clear that both the average and evolutionary profile methods are able to extract a large amount of useful information from a fairly small set of sequences. Highly discriminatory profiles can be generated from as few as two to six sequences in the case of the evolutionary method. The evolutionary profile method is distinctly better than the average method for the helicase example; it is not until subsets of at least twelve sequences are used that the average profiles equal the performance of the evolutionary profiles calculated from only two sequences.

Discussion

Comparison to Single Sequences

One can interpret the ROC_{50} statistic as the probability that a randomly selected positive sequence will score higher than a randomly selected negative sequence. As the negative sequences are limited to only the highest scoring ones, the top 50 for the ROC_{50}, one could say that it is the probability that a truly homologous sequence will score higher than the most likely false positives. Table II shows that for database searches using single sequences as

TABLE II

PERFORMANCE OF AVERAGE AND EVOLUTIONARY PROFILES ON ATP-DEPENDENT HELICASE FAMILY[a]

Method	Single sequence PAM 250	Number of sequences included in profile				
		2	3	6	12	38
Average	78.0 (16.4)	86.6 (7.0)	91.2 (4.4)	95.6 (1.8)	97.4 (0.9)	97.7
Evolutionary	—	97.2 (1.4)	98.2 (0.9)	99.2 (0.9)	99.3 (0.09)	99.3

[a] Profiles were generated from aligned sequences selected at random from the helicase family sequences shown in Fig. 4. Subsets of size 2, 3, 6, and 12 as well as the entire family of sequences (38 members) were used to produce profiles. Searches of the SWISS-PROT database (release 31.0) using the program PROFILESEARCH were then performed with each of the profiles. The ability of the profile to identify sequences in the family was evaluated using ROC_{50} method. Values in the table are the ROC_{50} times 100 and represent mean and the standard deviation (in parentheses) of 10 replicates of the indicated numbers of sequences. For comparison, the average ROC_{50} for the 38 individual sequences in Smith-Waterman database searches using the PAM-250 scoring matrix is shown.

queries there is an average chance of about 20% that a homologous sequence will score below unrelated sequences, leading to a high chance of missing a homolog or misclassifying a sequence as related to a false positive. The high variability of matching to single sequences is also seen in the large standard deviation for the single sequence value. Adding information from as little as one or two additional sequences greatly improves the discriminatory power and, as importantly, greatly reduces the variability. These are important concerns in the development of motif descriptions suitable for the automatic annotation of sequences or homology-based structural models.

Average Profiles versus Evolutionary Profiles

The average profile method seeks to extract information from a single set of prior information embodied in the scoring table used in the averaging process. When the scoring table is based on the observed mutational exchanges between amino acid residues, as is the PAM 250 table typically used, it represents a superposition of all of the chemical similarities between the residues. A heuristic way to view the average method is that it seeks to discover, from all the superposed chemical similarities, the one property that is common at an aligned position. The average profile achieves this because only residues that are chemically similar to one another will end up with high scores after the averaging process (see Gribskov[14] for examples).

Average profiles have been shown to be excellent discriminators for classifying protein families, usually achieving perfect or nearly perfect classification at unambiguous significance levels, for example, Z scores of 7.5 and above. This classification ability is usually accompanied by a lower level of false positives than is found with regular expression methods (results not shown), and by a greater ability to detect distantly related sequences that may lack residues that, up to then, were absolutely conserved. These properties have led to the high level of interest in the further development of profile and profilelike models (e.g., Bairoch and Bucher[1]). The average profile method, however, clearly does not adequately emphasize positions that are highly conserved. Consider, for example, a residue that is absolutely conserved in every sequence in a family of 100 sequences. Such a position is required, often participating in critical structures or functions such as the active site of an enzyme. However, the average profile represents such a position with a row of values identical to the corresponding row for the conserved residue in the scoring table on which the profile is based [Eq. (1)]. This inability to properly model highly conserved positions gave us the impetus to develop the evolutionary profile method.

The idea of the evolutionary profile method is to make a much more detailed and biologically relevant model of protein sequence families. There are two basic observations that guided the development of the evolutionary profile approach. First, it is well known that the amount of conservation among protein sequences varies widely from position to position. Thus it can be said that the positions in a sequence evolve at different rates. Second, the type of conservation varies widely from position to position, that is, there are different allowed residues at each position in a sequence, a constraint that arises primarily from the three-dimensional structure.

The evolutionary profile method selects the set of matching residues at each position by fitting the observed distribution of residues to distributions predicted for all possible ancestral residues and PAM distances according to the Dayhoff evolutionary model. This generates a model of a sequence family in which each position can be interpreted in a biologically sensible and intelligible way as a small set of preferred residues and evolutionary rates. A comparison of the alignment shown in Fig. 4 with the mixture components shown in Fig. 2 shows that the model closely corresponds to biological intuition. The highly conserved positions are modeled as mixtures of only one or two components at short evolutionary distances, while less conserved positions are modeled as mixtures of several components generally at longer evolutionary distances. It is noteworthy that evolutionary profiles can be easily scaled to longer or shorter evolutionary distances by simply multiplying or dividing the PAM distances fit during the modeling process. For instance, by simply multiplying all the fit PAM distances in Fig. 2 by a constant, and then recalculating the log-odds matrix, we can generate a model of the family at a greater evolutionary distance. We have not yet investigated this feature in detail, but it has the potential for allowing one to extend a model based on relatively closely related sequences to very distant members of a family.

Evolutionary profiles perform better than average profiles in generating discriminators for sequence classification (Tables I and II). Clearly, using this approach, models with very good discriminatory power can be generated from very small numbers of sequences, a sharp contrast to the fairly large numbers of sequences required to train hidden Markov models. This ability to generalize from a small set of observed sequences is due to the incorporation of a strong biological model of sequence conservation. In the near future we will examine the possibility of incorporating other biologically relevant prior information, such as known patterns of chemical similarity between the amino acid residues and predicted secondary structure, within the same mixture model framework used for the evolutionary profile.

Comparison to Hidden Markov Models

The underlying model represented by a profile bears a close similarity to hidden Markov models (HMMs) introduced for describing protein families.[4,5] Each row in the profile can be regarded as a match state, and the values in the row as the emission probabilities for each of the 20 possible amino acid residues. The position specific gap weights represent transition probabilities for moving to an insert or delete state from a match state. The main difference between the profile model and the most common HMM is that the profile model requires that the transition from a match state to an insert state and the transition from a match state to a delete state have the same probability. Because an insertion in one sequence can be viewed as a deletion in another (hence, the common term "indel"), the requirement of the profile model that the insert and delete transitions be equal seems reasonable (note, however, that the original profile model[12] did not have this requirement, making it more similar to an HMM).

Profile Libraries

We are actively engaged in extending the available profile libraries. We currently have a library of 650 protein motifs based on release 10 of PROSITE. These profiles were generated by locating the signature sequence for each of the PROSITE families in the annotated true-positive sequences, extending these sequences by 20 residues on both sides of the signature, multiply aligning the sequences, and producing average profiles. Each of these profiles has been validated by database searches and is available for use with PROFILESCAN. These profiles will be updated in early 1996 using the evolutionary profile method.

Acknowledgments

This work was supported by the National Science Foundation through cooperative agreement ASC-8902825 with the San Diego Supercomputer Center, and by NIH Grant P41 RR08605. Any opinions, findings, and conclusions or recommendations expressed in this publication are those of the author and do not necessarily reflect the views or policies of the National Science Foundation, the National Institutes of Health, or other supporters of the San Diego Supercomputer Center.

[14] Effective Large-Scale Sequence Similarity Searches

By JEAN-MICHEL CLAVERIE

Introduction

Computational biologists and genomics researchers, as well as molecular biologists involved with cloning and sequencing genes, are all confronted

Copyright © 1996 by Academic Press, Inc.
All rights of reproduction in any form reserved.

with a worsening situation when interacting with sequence databases. The main difficulties arise from (i) the decreasing quality (both in term of errors and redundancy) of the data banks and (ii) their burgeoning sizes.

The onset of multiple large-scale sequencing projects worldwide has dramatically increased the pace at which new sequences are submitted to primary repository data banks like GenBank[1] and EMBL.[2] This has forced the database administration to abandon any serious human validation of the submitted data and to rely mostly on automated checking procedures. Judging from the poor quality of data bank entries,[3-10] this scheme apparently fails to detect any but the most obvious errors. Combined with the fact that automated sequencing has become accessible to all types of laboratories, this has resulted in increasing pollution of the current databases with erroneous entries. In the absence of scientific curation, the sequence redundancy of the data bank is also rapidly increasing.

The absolute size of the databases has made them impossible to work with or search in a reasonable amount of time, except for a few leading centers in the world. Those centers usually provide a "search server" which does little to alleviate the *de facto* monopoly on large-scale database search and analysis. While useful, the E-mail or World Wide Web (WWW) service offered to the public consists only of a selected subset of the fastest search algorithms (to be used within a narrow range of options). They also limit the submission of queries from each client to a small, "reasonable" number. Scientific assessments, involving very large-scale comparisons (such as entire database versus themselves), exotic algorithms, or unusual parameter setting, are not possible in this context. In addition, the lack of confidentiality of this mode of operation can be worrisome to some laboratories, and is definitely not acceptable to the private biotechnology industry.

In this chapter, we review two concepts, sequence masking and distributed processing, that we think are key to the local (and secure) implementation of effective and flexible large-scale sequence comparison.

[1] D. A. Benson, M. Boguski, D. J. Lipman, and J. Ostell, *Nucleic Acids Res.* **22,** 3441 (1994).
[2] C. M. Rice and G. N. Cameron, *Methods Mol. Biol.* **24,** 355 (1994).
[3] J.-M. Claverie, *Genomics* **12,** 838 (1992).
[4] R. Lopez, T. Kristensen, and H. Prydz, *Nature (London)* **355,** 211 (1992).
[5] J. Posfai and R. J. Roberts, *Proc. Natl. Acad. Sci. U.S.A.* **89,** 4698 (1992).
[6] E. D. Lamperti, J. M. Kittelberger, T. F. Smith, and L. Villa-Komaroff, *Nucleic Acids Res.* **20,** 2741 (1992).
[7] T. Kristensen, R. S. Lopez, and H. Prydz, *DNA Sequence* **2,** 343 (1992).
[8] J.-M. Claverie, *J. Mol. Biol.* **234,** 1140 (1993).
[9] C. Savakis and R. Doelz, *Science* **259,** 1677 (1993).
[10] J.-M. Claverie and W. Makalowski, *Nature (London)* **371,** 752 (1994).

Concept of Sequence Masking

Making Matches: The Necessity of Filtering

A number of important scientific contexts involve the comparison of a large number of query sequences against an entire sequence database. The recognition of exons in human genomic sequence by database similarity search,[11,12] in particular using the rapidly growing collection of partial cDNA sequences[13,14] (the so-called expressed sequence tags or ESTs), is clearly one of them. It will serve as an example throughout this chapter.

At the present rate of several thousand ESTs a day, a nearly complete sampling of all transcripts should be available shortly (with the possible exception of tightly developmentally regulated or tissue-specific genes). It should thus become possible to locate all exons, including those (invisible to all other methods) within 5' and 3' untranslated regions (UTRs).

However, any attempt to compare a human genomic sequence with EST data quickly reveals that this promising method, in its simplest form, does not provide meaningful results, for both artifactual and biological reasons. As an example, we analyzed a 67-kb sequence from the p22.3 region of chromosome X, one extremity of which has been shown to contain two small coding exons (225 and 142 nucleotides) and the 3' UTR (4 kb) of the gene responsible for Kallmann's syndrome.[15,16] Using the program ORFDB (J.-M. Claverie, unpublished, 1991), a collection of all possible open reading frames (ORFs) longer than 90 nucleotides was constituted, representing a total of 1343 distinct putative peptide-encoding sequences originating from either strand in approximately the same number.

This collection of putative exons, of which we expect at most 10 to be real, was then compared to the current public EST data bank, dbEST[17] (containing approximately 150,000 ESTs of human origin), using the popular local similarity search program BLAST.[18] We used both BLASTN (ORF nucleotide sequence versus EST nucleotide sequence) and TBLASTN[19] (putative ORF peptide translation versus EST nucleotide sequence). The

---

[11] J.-M. Claverie, *Genomics* **23**, 575 (1994).

[12] J.-M. Claverie, *in* "Methods in Gene Mapping" (J. Boultwood, ed.), in press. Humana Press, Totowa, New Jersey, 1996.

[13] M. D. Adams, *et al., Science* **252**, 1651 (1991).

[14] J. M. Sikela and C. Aufray, *Nat. Genet.* **3**, 189 (1993).

[15] R. Legouis, *et al., Cell* (*Cambridge, Mass.*) **67**, 423 (1991).

[16] B. Franco, *et al., Nature* (*London*) **353**, 529 (1991).

[17] M. S. Boguski, T. M. Lowe, and C. M. Tolstoshev, *Nat. Genet.* **4**, 332 (1993).

[18] S. F. Altschul, W. Gish, W. Miller, E. W. Myers, and D. J. Lipman, *J. Mol. Biol.* **215**, 403 (1990).

[19] W. Gish and D. J. States, *Nat. Genet.* **3**, 266 (1993).

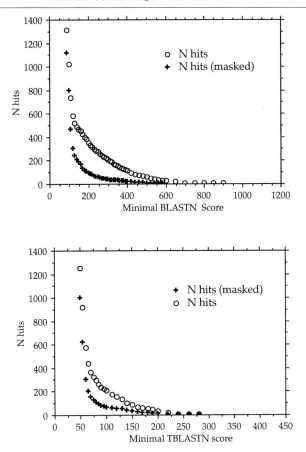

FIG. 1. Number of ORFs matching EST sequences.

number of putative ORFs reported to match at least one EST is plotted in Fig. 1 (top, BLASTN; bottom, TBLASTN, N hits curves). The default scoring system was used ($M = 5, N = -4$ for BLASTN, and the BLOSUM62 matrix for TBLASTN).

As expected, the fraction of ORFs matching at least one EST rapidly increases as the required minimal score decreases. For BLASTN scores at or below 100 or TBLASTN scores up to 50, almost all ORFs have a match. This behavior is in agreement with the Karlin–Altschul statistics,[20] which predict a very high probability value (~1) for random matches associated to those low scores. However, as we impose larger minimal scores and thus

[20] S. Karlin and S. F. Altschul, *Proc. Natl. Acad. Sci. U.S.A.* **87,** 2264 (1990).

more stringent local similarity, the number of matching ORFs remains much higher than expected. For instance (Fig. 1), at a minimal BLASTN score of 200, corresponding to an average probability of less than 10^{-8} (where no false positive is expected), more than 354 "candidate" exons are "identified" by their match to a human EST (Fig. 1, top). The situation is very similar with TBLASTN (Fig. 1 bottom), where a minimal score of 100 ($p \leq 2 \times 10^{-7}$) still falsely identifies more than 210 putative exons. Such an extremely high fraction of false-positive identification makes the direct EST lookup method of exon identification totally impractical until we better understand the nature of the problem.

The Karlin–Altschul model apparently fails because it assumes a randomness in sequence data that is not actually valid. Besides a small fraction of well-behaved regions, actual sequences, both genomic or ESTs, are constituted of many different repeats, from simple mononucleotide segments to "regular looking" repeats such as the ubiquitous Alu, accounting for up to 13% in the Kallmann's syndrome gene contig discussed here. Repetitive DNA represents over 50% of the human genome. It has been classified in various categories such as retroposons, satellites, and medium reiteration frequency repeats (MERs). These repeats, present both in the target EST data bank and most of our putative ORF queries, dramatically increase the chance of a fortuitous match. Eventually, this noise can obscure the few alignments with biological significance.

As a general solution to this problem, and as a prerequisite to all large-scale sequence comparison, we have introduced the concept of sequence masking.[11,21,22] It simply consists of delineating the various type of repeats and other *a priori* troublesome segments by *ad hoc* programs, and then replacing the corresponding positions with a special character neutral to the specific scoring scheme (traditionally "X" for proteins and "N" for DNA sequences). Figure 1 shows the spectacular statistical effect that masking for the most frequent repeats (Alu and simple sequences) can have on the distribution of the number of hits for a given minimal score.

Simple versus Other Repeats in Human Genome and Data Banks

The contiguous repeat of a small motif made of 1, 2, 3, or 4 nucleotides is a frequent occurrence in the human genome. These are referred to as "simple" repeat sequences because, following Kolmogorov's notion of information, the recipe to generate them is simple. Taking into account

[21] J.-M. Claverie, *in* "Automated DNA Sequencing and Analysis Techniques" (M. Adams, C. Fields, and J. C. Venter, eds.), Chap. 36, p. 267. Academic Press, San Diego, 1994.
[22] J.-M. Claverie, *in* "Advances in Computational Biology" (H. Villar, ed.), Vol. 2, p. 161. Jai Press, London, 1996.

strand symmetry (A=T, C=G) and circular permutation (ATC=CAT= TCA), simple repeat sequences can be constituted of the following distinct motifs: [A, C], [AC, AG, AT, CG], [AAC, AAG, AAT, ACC, ACG, ACT, AGC, AGG, ATC, CCG], as well as 33 four-nucleotide elements.[23,24] Sequences containing numerous repeats of higher order motifs are (fortunately) rare. By reference to Shannon's notion of information, simple repeat sequences are also referred to as "low entropy."

Simple repeat segments are very abundant in EST data banks, as they are frequently found in the 3' UTR of genes. Figure 2 A shows various examples of actual ESTs exhibiting one-, two-, and three-nucleotide simple repeats. These ESTs (along with hundreds of others) have an increased random probability of matching any unrelated genomic region having a similar compositional bias. Translated into putative peptide sequences they will also induce matches with unrelated protein sequences (e.g., when using TBLASTN, or BLASTX). It is worth noticing that segments of simple sequences are even more frequent in the data banks of sequence-tagged sites (STSs), namely, small uniquely polymerase chain reaction (PCR)-amplifiable sequences associated with a genetic or physical map location. Many of these STSs were designed to target the repeats, in order to make use of the length polymorphism often characterizing those sequences.[25] Furthermore, the frequency of length polymorphism seems to be increased in the vicinity of Alu elements,[26] causing STS databases to be enriched in the latter also. As a consequence, the electronic mapping of genomic or EST sequences by direct similarity search against an STS data bank is even more plagued by false alignments than are attempts to identify exons.

Figure 3 shows examples of Alu repeats in EST sequences. Alu repeats may constitute a large part or even all of the EST sequence. In addition to being a well-known contaminant of cDNA libraries, Alu sequences are also frequent in 3' UTRs. Alu sequences have been (and continue to be[10]) incorporated in pseudoprotein sequences and have mistakenly found their way into protein sequence data banks.[3,22] As a consequence, Alu-containing ESTs can be erroneously identified as *bona fide* protein-encoding transcripts. The comparison of Alu-containing genomic regions with ESTs and protein data banks is thus error prone.

In addition to the most frequent Alu repeats, vertebrate genomes (and EST data) contain many more "regular entropy" (i.e., normal looking) repeats (such as Line-1, MER, etc.), which constitute additional sources of

[23] J. S. Beckmann and J. L. Weber, *Genomics* **12,** 627 (1992).
[24] G. Bell and D. Torney, *Comput. Chem.* **17,** 185 (1993).
[25] D. Tautz and M. Renz, *Nucleic Acids Res.* **17,** 6463 (1989).
[26] G. Zuliani and H. H. Hobbs, *Am. J. Hum. Genet.* **46,** 963 (1990).

A

>T65117 EST yc74b11.s1 H.sapiens   <T>
TTTTTTTTTTTTTTTTTTTTTTTTTTTTTCATTGAGAAAAATCTTTATTGATTGACTACACCAT
CCACAAGAAAAAGAAACTTAATTTAAGAAG

>T41390 EST ph4b6_19/1TV H.sapiens   <C>
GAATTCCCCCCCCCCCCCCCCCCCCCCCCCCCCCATCTCAAAAAAAAAAAAAACCTAAGATGATA
ACATTGACAGCTAGTGGGTGGAGATGCTGCTAAACATCCTTATAATGCACAGGACAGATC
CTTTCTCCTCCTCTTCCCCGATCTCACAAACAAAGAATTACTGGGACTAAATGTCAATTG
TGCTAAGGCGAGAAGCTCTTTTTCAGAGAGAGACC

>F03797 HSC2AC042 EST H.sapiens   <CA>
CTACACCCCAGAAAAANAATTGAGGAAGGNTTCTNCCCGCTACAAGTCAGACCAAAGGCA
ATCATCATTTNCCTTGGATACACACACACACACACACACACACACACACACACCAGTC
ACCCTGCTCTTCGGNTGAGNTTAAAAAATTCTACAGGCATAAACTGTTTAAGTAT

>T27311 EST hbc2596 H.sapiens   <CT>
GGGTGCAGGAATTCGGCACGAGTCTCTCTCTCTCTCTCTCTCTCTCTCTCTCTCTCTC
TCTCTCTCTCTCTCTCTCTCTCTCTCTCTCCAGAAATCTAAG

>T66755 EST ya49g02.s2 H.sapiens   <TA>
TTTTTTTTTAGAGTTCTTTCGTATATATTTTAAATATACATGTGCAATACATTAAAATAT
ATAGAGATGTGTGAATATATATTTACATATATATATATATATATATATATATATATATAT
ATATATATATATNCACAGATGGGTTACAT

>F02337 EST H.sapiens   <TGT>
AAGAATATAAGTCATTTAATACTGTTAATTTTATAGCACAAAATAAAACAAGCTATGATC
CCCAAAAATAATTTTAAAAGCTTACACAGNAAATATTATTGCCTGAAGTTTATGATCTTT
AAGTTACAGGNCAAAAGNGTTTTATGTTGTTGTTGTTGTTGTTGTTGTTGTTGTTTTAAA
ACACACAGTGA

>T32809 EST EST54688 H.sapiens   <TTA>
GGAAGACACTGGTCAAAGGTTTATTATTATTATTATTATTATTATTATTATTTTAAATTT
TATTTCTTTTTAAAATGTGAGTTCCAATAAAATTTAAAAATTAGATTCCAACCNGTAGAT
TAAAATGAATTAAAAAATACAAAATCATATACAACGNTCTCTTCACAGGGATGTTACATG
CCATGAGAACAGGTTTCTCTGTCATGTGATATGAAACAGGGAAGCAGATGCCTNTAAGTC
AGGATGGGGACAGG

>U21467 HSU21467 EST H.sapiens   <CAC>
GAATTCCGAAGAGGCGCTCCTCCCCCACCACCTTCAAGAGCTCCCACAGCTGCACCTCCA
CCACCTCCACCACCACCACCACCACCACCACTACCACCACCACCACTCACAATCATG
GCAAGCCAACGCACTT

>T30449 EST EST16856 H.sapiens   <CAG>
TTTCTCACCTGTNCAGTCTCATCTAACCTTCCAATGTCTGATGTTCCTGCCAAATTCCTG
CCTGATTCTGGGTCCGTCCTGACCTCCAAAGGTCAGCTTGGTGCTTGAGGTCTCCCTGCT
CTTGGTGGCAGTGGTAGCAGCAACAGCAGCAGCAGCAGCAGCAGCAGCAGAGACCTC
TCCACTTTCCCTTAGCCCCTCTGCTGGGTAGAGAGGCACTTTCAGGGACTTCCCTCCAGC
TGCCTCTTCATCTGGGAATGAGCTAAGCAAGGCTGAGNCTNCTCCTGTTGCTTTGAAATA
ATGATGATATTAAAGG

FIG. 2. Examples of EST sequences containing simple repeats (A) and EST sequences after masking for simple repeats (B).

B

>T65117 EST yc74b11.s1 H.sapiens
nnnnnnnnnnnnnnnnnnnnnnnnnnnnnnCATTGAGAAAAATCTTTATTGATTGACTACACCAT
CCACAAGAAAAAGAAACTTAATTTAAGAAG

>T41390 EST ph4b6_19/1TV H.sapiens
GAATTnnnnnnnnnnnnnnnnnnnnnnnnnnnnATCTCAAAAAAAAAAAAAACCTAAGATGATA
ACATTGACAGCTAGTGGGTGGAGATGCTGCTAAACATCCTTATAATGCACAGGACAGATC
CTTTCTCCTCCTCTTCCCCGATCTCACAAACAAAGAATTACTGGGACTAAATGTCAATTG
TGCTAAGGCGAGAAGCTCTTTTTCAGAGAGAGACC

>F03797 HSC2AC042 EST H.sapiens
CTACACCCCAGAAAAANAATTGAGGAAGGNTTCTNCCCGCTACAAGTCAGACCAAAGGCA
ATCATCATTTNCCTTGGnnnnnnnnnnnnnnnnnnnnnnnnnnnnnnnnnnnnnnnnCAGTC
ACCCTGCTCTTCGGNTGAGNTTAAAAAATTCTACAGGCATAAACTGTTTAAGTAT

>T27311 EST hbc2596 H.sapiens
GGGTGCAGGAATTCGGCACGAGnnnnnnnnnnnnnnnnnnnnnnnnnnnnnnnnnnnnnnnn
nnnnnnnnnnnnnnnnnnnnnnnnnnnnnnnnnnnnCAGAAATCTAAG

>T66755 EST ya49g02.s2 H.sapiens
TTTTTTTTTAGAGTTCTTTCGTATATATTTTAAATATACATGTGCAATACATTAAAATAT
ATAGAGATGTGTGAATATATATTnnnnnnnnnnnnnnnnnnnnnnnnnnnnnnnnnnnnnnn
nnnnnnnnnnnnnNCACAGATGGGTTACAT

>F02337 EST H.sapiens
AAGAATATAAGTCATTTAATACTGTTAATTTTATAGCACAAAATAAAACAAGCTATGATC
CCCAAAATAATTTTAAAAGCTTACACAGNAAATATTATTGCCTGAAGTTTATGATCTTT
AAGTTACAGGNCAAAAGNGTTTTAnnnnnnnnnnnnnnnnnnnnnnnnnnnnnnnnnnnnAAA
ACACACAGTGA

>T32809 EST EST54688 H.sapiens
GGAAGACACTGGTCAAAGGTnnnnnnnnnnnnnnnnnnnnnnnnnnnnnnnnnnnnnnnAAATTT
TATTTCTTTTTAAAATGTGAGTTCCAATAAAATTTAAAAATTAGATTCCAACCNGTAGAT
TAAAATGAATTAAAAAATACAAAATCATATACAACGNTCTCTTCACAGGGATGTTACATG
CCATGAGAACAGGTTTCTCTGTCATGTGATATGAAACAGGGAAGCAGATGCCTNTAAGTC
AGGATGGGGACAGG

>U21467 HSU21467 EST H.sapiens
GAATTCCGAAGAGGCGCTCCTCCCCCACCACCTTCAAGAGCTCCCACAGCTGCACCTnnn
nnnnnnnnnnnnnnnnnnnnnnnnnnnnnnnnnnnnnnnnnnnnnnnnnnnnnnnnnnnnnnCTCACAATCATG
GCAAGCCAACGCACTT

>T30449 EST EST16856 H.sapiens
TTTCTCACCTGTNCAGTCTCATCTAACCTTCCAATGTCTGATGTTCCTGCCAAATTCCTG
CCTGATTCTGGGTCCGTCCTGACCTCCAAAGGTCAGCTTGGTGCTTGAGGTCTCCCTGCT
CTTGGTGGCAGTGGTnnnnnnnnnnnnnnnnnnnnnnnnnnnnnnnnnnnnnnnAGACCTC
TCCACTTTCCCTTAGCCCCTCTGCTGGGTAGAGAGGCACTTTCAGGGACTTCCCTCCAGC
TGCCTCTTCATCTGGGAATGAGCTAAGCAAGGCTGAGNCTNCTCCTGTTGCTTTGAAATA
ATGATGATATTAAAGG

Fig. 2. (*continued*)

>F03973 HSC2FE112 EST H.sapiens
GTATTTTAAGTAGAGATGGGGTTTCGCCATGTTGGCCATGCTGGTCTCAAACTCCTGACC
TCAGGTGATCTACCCACCTTGGCCTTCCAAAGTGCTGGAATTACTTGTGTAAGCCACCGT
ACCTGGCCTATTGGAACCCCCTTNTTGAAACACCTGTGTGAGCACCCTGNTTACCATCTA
CCAGTGACCACACCCACATAGGGTTCTATTTCAGTGAGACTACAGAGCCAACAAAGTGCA
CTAGTTAGCTGTGATGGAAAA

>F03973 HSC2FE112 EST H.sapiens   <Alu Masked>
nnnnnnnnnnnnnnnnnnnnnnnnnnnnnnnnnnnnnnnnnnnnnnnnnnnnnnnnnnnnn
nnnnnnnnnnnnnnnnnnnnnnnnnnnnnnnnnnnnnnnnnnnnnnnnnnnnnnnnnnnnn
nnnnnnnnnTATTGGAACCCCCTTNTTGAAACACCTGTGTGAGCACCCTGNTTACCATCTA
CCAGTGACCACACCCACATAGGGTTCTATTTCAGTGAGACTACAGAGCCAACAAAGTGCA
CTAGTTAGCTGTGATGGAAAA

>F03332 HSC1TG072 EST H.sapiens
AACATTGTAACATTAACATTTATTTAGAATCTCTTTTGAAGCTGTCCTTGACCTGTTTCT
CAAGACTGGAATTCTGCTGGGTGCCGTGGTGCATGCTTGTAATCCCAGCACTTTGGGAGG
CTGACACAGGAGGATGACTTGAGGCCATGAGTTCAAGACTAGCTTGCCGNACAACATAGC
AAGATTCTGTCTCTGATGTTGCTTGCCTTGTGTTTCTAGGCAGAGGACCTGCCCATGAGC
TTCTGCAACCAGTTCACTCTC

>F03332 HSC1TG072 EST H.sapiens <Alu Masked>
AACATTGTAACATTAACATTTATTTAGAATCTCTTTTGAAGCTGTCCTTGACCTGTnnnn
nnnnnnnnnnnnnnnnnnnnnnnnnnnnnnnnnnnnnnnnnnnnnnnnnnnnnnnnnnnnn
nnnnnnnnnnnnnnnnnnnnnnnnnnnnnnnnnnnnnnnnnnnnnnnnnnnnnnnnnnnGC
AAGATTCTGTCTCTGATGTTGCTTGCCTTGTGTTTCTAGGCAGAGGACCTGCCCATGAGC
TTCTGCAACCAGTTCACTCTC

>T07954 EST EST05845 H.sapiens
CTCTTTTTTTTTTTTTTTNAGATGGAGTTTCGNTCTNGTTGCCCAGGCTGGAGTGCAATG
GCACGANCTTGGCTCACTGCAACCTCCATCTCCTGGGTTCAAGAGACTCTCCTGCCTCAG
NCTCCTGAGCAGCTGGGATTACAAGCATGCACCTGGCTAATTTTGTATTTTNAGTAGAGA
CAGAGTTTCTCCATGTTGGTCAGGCTGGTCTCAAACTCCCGACCTCAGGTGATCCCCNGA
CTCGGCCTTCCAAAGTGCTGG

>T07954 EST EST05845 H.sapiens <Alu Masked>
nnnnnnnnnnnnnnnnnnnnnnnnnnnnnnnnnnnnnnnnnnnnnnnnnnnnnnnnnnnnn
nnnnnnnnnnnnnnnnnnnnnnnnnnnnnnnnnnnnnnnnnnnnnnnnnnnnnnnnnnnnn
nnnnnnnnnnnnnnnnnnnnnnnnnnnnnnnnnnnnnnnnnnnnnnnnnnnnnnnnnnnnn
nnnnnnnnnnnnnnnnnnnnnnnnnnnnnnnnnnnnnnnnnnnnnnnnnnnnnnnnnnnnn
nnnnnnnnnnnnnnnnnnnnnn

FIG. 3. Examples of Alu-containing EST sequences before and after masking.

spurious ORF identification. Until a sizable fraction of the human genome is sequenced, the definition and classification of these repeats will remain in a state of flux.

Delineating Simple Repeats in DNA Sequences

Protein sequences consist of 20 letters, representing the 20 different classic amino acids. DNA sequences, on the other hand, consist of a 4-letter text. Because of this fundamental difference, the range of local information content is much greater for protein sequences than for DNA. This makes it much easier to delineate simple repeat sequence segments in protein than in DNA. Two algorithms, one based on the periodicity of sequence similarities (XNU[27]), the other (SEG[28]) on an explicit computation of the local linguistic complexity, have been successfully applied to the location of potentially troublesome segments for the alignment of proteins sequences.

Although the XNU and SEG algorithms are incapable of correctly delineating low entropy segments in nucleotide sequences, we have found that this can be readily achieved using a BLASTN search against a small database (Simple.db) containing 26 different simple sequence prototypes. All sequences in Simple.db are 25 nucleotides long. They are made up of the two 1-letter motifs, the four 2-letter motifs, and the 10 3-letter motifs cited above, plus the 10 4-letter motifs most frequently found in the human genome[23]: AAAC, AAAG, AAAT, AAGG, AATC, AATG, ACAT, ACAG, AGAT, ATCC. Delineating and masking the simple repeat segments in a sequence (e.g., "est") is performed by a search with BLASTN, the output of which is parsed by the XBLAST program[21,27]:

blastn Simple.db est S=110 B=30 -overlap | xblast + est "n"

Details regarding the syntax of XBLAST have been given elsewhere.[21,22] Here, the minimal score $S = 110$ requires that 24 out of every 25 nucleotides of a simple segment must match identically. Note that the -overlap option (or -span in newer BLAST versions) is essential. XBLAST then parses the output and generates a new query in which the matching segments are replaced by the neutral symbol n. Figure 2B shows the accuracy with which the simple repeat are masked from the original EST sequences (Fig. 2A) using this simple procedure.

An alternative strategy is to translate the DNA sequence into a putative peptide sequence, and use the XNU or SEG programs to delineate and mask the resulting low entropy segments. This approach makes good sense when working with ORFs if the ultimate large-scale similarity search must involve protein sequences (e.g., BLASTP, BLASTX, or TBLASTN).

Delineating Alu (and Other Nonsimple) Repeats in DNA Sequences

Sequences with Alu repeats are polymorphic "high entropy" sequences. We have previously demonstrated the effectiveness of a filtering scheme that uses the BLAST search/XBLAST combination on a reference Alu data bank.[21,22,27] This has become standard procedure in most large-scale human sequencing projects, including EST production.

---

[27] J.-M. Claverie and D. J. State, *Comput. Chem.* **17**, 191 (1993).
[28] J. C. Wootton and S. Federhen, *Comput. Chem.* **17**, 149 (1993).

We have assembled and tested a new subset of 327 Alu family members for improved sensitivity. This subset was built as follows. The 4887 Alu elements contained in repbase[29] were progressively clustered using pairwise BLASTN comparisons at decreasing score thresholds, and a greedy algorithm was used to retain the longest prototype sequence for each cluster. In this process, 22 atypical Alu sequences (e.g., encompassing flanking coding regions or pseudogene segments) have been discovered and eliminated. The final set contained 319 sequences, to which we added the 8 consensus sequences of the 8 main Alu types recognized.[29] The sequences in this reference set have sizes ranging from 180 to 397 nucleotides and do not contain any ambiguous symbol. To be used in the context of protein sequence comparison, we also build the PAlu.db database from the corresponding 6-frame conceptual translations.

The masking of Alu-like segments in DNA or protein sequences is performed using a combination of BLAST searches/XBLAST filtering with the proper parameters. For instance, for an EST:

blastn Alu.db est S=178 B=30 | xblast + est "n"

and for a peptide sequence (such a translated ORF):

blastp PAlu.db orf S=65 S2=65 -filter "XNU" | xblast + orf "x"

For slightly better results, the default matrix BLOSUM62 can be replaced by Blosum62.msk in which the strong negative penalties for matching a stop ("*") are set to zero. Alternatively, one can use the -altscore option in the BLAST command line.

Using BLASTN, the percentages of known Alu sequences detected at various score thresholds are 99.3% at $S = 150$, 97.3% at $S = 178$, and 94.3% at $S = 207$. Using BLASTP, BLASTX, or TBLASTN and the Blosum62.msk scoring matrix, those numbers are 98.7% at $S = 65$, 96.7% at $S = 70$, and 93.7% at $S = 75$.

Net Result of Masking Junk Segments

The masking of simple repeats and Alu elements prior to a database search has dramatic effects on the distribution of N Hits versus Score, as shown in Fig. 1, for both nucleotide/nucleotide (BLASTN, top) and protein/nucleotide (TBLASTN, bottom) comparisons. Both of the distributions for masked matches now exhibit a much sharper transition ($S \geq 200$, Fig. 1 top; $S \geq 90$, Fig. 1 bottom) between the significant and nonsignificant range of scores. Overall, the volume of the output is also greatly reduced by the

[29] J. Jurka, J. Walichiewicz, and A. Milosavljevic, *J. Mol. Evol.* **35,** 286 (1992).

masking procedure. For $S \geq 300$ (BLASTN) and $S \geq 200$ (TBLASTN), 70% of the EST/ORF matches correspond to the actual coding exons and 3' UTR of the known Kallmann's syndrome gene. The other matches might define exons of an unrelated gene but are not yet confirmed. The masking for simple repeats and Alu elements has dramatically improved the situation, so that exon detection by EST lookup becomes a viable method. However, there is still an excess of matches in the ranges 200–300 (BLASTN, Fig. 1 top) and 90–200 (TBLASTN, Fig. 1 bottom). We suspect that many of these are induced by less frequent, less well-characterized repeats such as those of the various MER families.

Exon detection by EST lookup is only one of the many contexts of large-scale sequence comparison for which "junk" masking is a prerequisite. "Electronic mapping," where genomic or EST sequences are linked to genetic or physical locations by comparison to an STS data bank, requires the same meticulous masking steps in order to produce meaningful results. The same is also true of the exhaustive pairwise comparisons performed within an EST data bank to estimate the number of independent clusters and eventually assemble genes from overlapping pieces. Without the proper masking of the ubiquitous simple repeats and Alu segments, unrelated ESTs are more likely to be improperly merged in the same homology cluster.

The study of individual protein sequences can also benefit from masking. Without precaution, database "mining" (i.e., the search of all the entries related to a given query) can result in a huge redundant output from which biologically significant information is impossible to extract. This, for example, is the case when studying multidomain protein families (such as vertebrate cell-surface receptors). If one of the domains is highly represented in the data bank (such as immunoglobulin or fibronectin domains), subtle local alignments in other part of the molecule (and eventually key for functional assignment) will be lost in a 1000 page output.

In each context, the general concept of sequence masking can be adapted to a specific repertoire of junk sequences (Alu, vector, immunoglobulin domain, etc.), to the optimal alignment program (BLASTN, BLASTP, etc.), and to any value of the stringency parameters.

Parallel Processing for Large-Scale Sequence Comparison

Necessity of Parallel Processing

With the size of searchable data banks reaching several hundred megabytes and still doubling every other year, large-scale sequence comparisons (many queries against GenBank, the EST data bank against itself, the EST data bank against GenBank, etc.) are becoming very lengthy when run on the type of workstation traditionally available to biological laboratories (100 Mips or approximately 60 SPECint92). Even though it is a fast algorithm, a BLASTN query against the whole EST data bank takes as long as 5 sec, while a BLASTX query of the protein data bank SWISS-PROT[30] may take about 30 sec. In the same context, average run times for FASTA[31] will be of the order of several minutes, and much longer for a full dynamic programming alignment.[32] While those processing times may be acceptable for the analysis of a single or a few sequences, they are incompatible with large-scale sequence comparisons. At this rate, comparing the content of the whole EST data bank to SWISS-PROT (e.g., for annotation purpose) will take 3 months using the fastest BLASTX. Many large-scale experiments of the same magnitude are desirable for computational biology laboratories. They include clustering and assembling large collections of ESTs, self-comparison of data banks in the context of evolutionary studies,[33] monitoring for the purpose of decreasing data bank redundancy, etc. Finally, the shotgun sequencing, interpretation, and assembly of a whole microbial genome might also involve a similar amount of comparison.

Even though the required supercomputing power (e.g., 2000–3000 Mips) is available in a few service centers throughout the world, they are usually saturated and/or do not allow their clients to run large-scale studies.

Specialized hardware is commercially available to accelerate some of these algorithms (such a full dynamic programming or exact matches), but they tend to be expensive and cannot implement the wide variety of data bank scanning algorithms (such as block[34] and profile searches[35]) or serve other computational purposes (such as numerical computations). History has also shown that specialized hardware will rapidly lose its advantage before the rapid progress of general purpose computers.

General purpose multiprocessor machines are the current state of the art at the level of supercomputing required by large-scale sequence comparisons. They are typically made of 4 to 32 processors sharing 256 to 1024 Mb of RAM. This generation of "coarse-grain" parallel machines follows a not very successful era of experimental machines made of several hundred, sometimes thousands of exotic processors. The extensive adaptation and recoding of existing algorithms required to take advantage of "fine-grain"

[30] A. Bairoch and B. Boeckmann, *Nucleic Acids Res.* **22**, 3578 (1994).
[31] W. R. Pearson, this series, Vol. 183, p. 63.
[32] T. F. Smith and M. S. Waterman, *J. Mol. Biol.* **147**, 195 (1981).
[33] P. Green, D. J. Lipman, L. Hillier, R. Waterston, D. States, and J.-M. Claverie, *Science* **259**, 1711 (1993).
[34] S. Henikoff and J. G. Henikoff, *Genomics* **19**, 97 (1994).
[35] M. Gribskov, R. Lüthy, and D. Eisenberg, this series, Vol. 183, p. 146.

parallelism were major impediments and reasons behind the commercial failure of this concept.

A key feature of modern coarse-grain parallel machines is that they are constituted of the same UNIX-compatible processors found in the popular workstation, and are thus capable of running any existing software on each of their individual processors. In a multiuser environment, these machines are often simply used as a collection of separate processors performing unrelated tasks for different users. In an intermediary mode of operation, different users may run the same program and access a single memory-resident copy of a common database. This would be the case when different users are each running a different BLASTN query against GenBank. In the fully parallel mode of operation, a single user is allowed to recruit all the processors and, for instance, run n different BLASTN queries against the same target database. Such multiprocessing is represented in Fig. 4. The net gain in speed will be directly proportional to the number of processors, provided not too many collisions occur while accessing the shared memory. Finally, in another mode of parallel operation the query program could actually clone itself in as many processors as are available (as does BLAST[18,19]), and each run the same query against a fraction of the central database. This last mode of operation achieves the highest possible level of gain in performance. However, it requires the explicit encoding of parallel features into the program (usually using some message passing libraries like MPI or PVM) to achieve the necessary coordination.

Affordable Solution: Distributed Parallel Processing

In the range of performance needed for large-scale sequence comparisons, multiprocessor machines are still very expensive, and not truly affordable to many computational biology laboratories. Although workstations can be purchased as financially attractive all-inclusive packages, multiprocessor machines are proportionally much more costly to buy, maintain, and upgrade. This context makes the notion of distributed parallel processing very attractive. It consists of simply recruiting an ensemble of workstations linked by a fast network (e.g., ethernet) to constitute a "virtual" multiprocessor machine, as shown in Fig. 4 (bottom).

This type of "shared-nothing" architecture has been previously presented[36] as the optimal design for relational database systems. We also found it very effective and natural for the parallelization of large-scale database searches. In the shared-nothing design, each memory and disk is owned by a processor that acts as a server for that data. In the context of

[36] D. Dewitt and J. Gray, *Commun. ACM* **35,** 85 (1992).

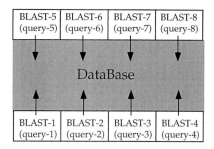

Multi- Processing

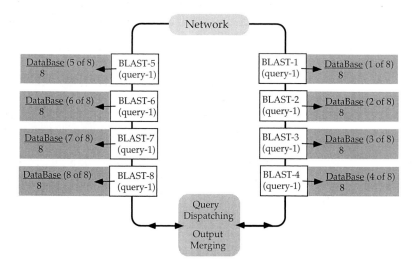

Distributed Parallel Processing

FIG. 4. Multiprocessing versus distributed processing of similarity searches.

database similarity search, each processor is set to run the same alignment program on a fraction of the total database. Because each of the processors are in fact those of standard UNIX workstations, any popular alignment method can be run (BLAST, FASTA, profile or full dynamic, etc.), an advantage over specialized hardware. While other parallel architecture may have to move large quantities of data through the interconnection network, the design shown in Fig. 4 need only move queries and answers through the network. Intensive memory and disk access are performed locally in a processor, and only reduced traffic (query dispatching and output retrieving) is sent across the network. In our case, the same query is simultaneously dispatched to all processors, and the partial output sent back for merging to a master machine (which can be any one of the pool). Given that the time required to complete a partial search on each machine is of the order of seconds, any standard local area network is fast enough to implement this distributed architecture. An added benefit of this architecture is to ease the constraint of storing a very large database into the memory of a single processor. Instead, the storage is easily achieved by distributing it among several workstations each with only medium-sized memory (typically 64–128 Mb).

Using this distributed architecture, the most efficient parallelization is reached without any modification of the code of the similarity search program. Because the various clones of the program run without any interprocess communication, there is no need to implement PVM or MPI library calls. A simple shell script can be used to dispatch the current query from a master machine to each participating processor, to locally start the search against the partial database, and to write back the output on the master machine. A virtual parallel machine of this type, constituted of up to 12 networked Silicon Graphics Indy workstations (each with 128 Mb of memory), is used in our laboratory. As expected, the observed performances for large-scale BLAST or FASTA searches are exactly proportional to the number of processors recruited in the pool. For one-tenth of the cost of a comparable multiprocessor, the power of a 1200 Mips (and 300 Mflops) machines with an apparent central memory of 1.6 Gb can thus be harnessed by a single laboratory to perform any large-scale sequence analysis experiment of its choice.

Acknowledgments

I thank Dr. Richard Goold for introducing me to electronic mapping and for useful comments, and Dr. Ingrid Akerblom for reading the manuscript.

[15] Effective Protein Sequence Comparison

By WILLIAM R. PEARSON

Introduction

Over the past few years, our ability to extract information from protein and DNA databases has improved dramatically: databases are larger and more comprehensive, computers are faster, and comparison algorithms are

Copyright © 1996 by Academic Press, Inc.
All rights of reproduction in any form reserved.

more effective, in large part because of the incorporation of statistics for local similarity scores in both heuristic and rigorous sequence comparison programs. This chapter examines programs and search strategies for identifying distantly related protein sequences.

Searching for Homologies

BLASTP, BLITZ, FASTA, SSEARCH, and other programs for searching protein sequence databases are most effective when used to identify sequences that are homologous (i.e., that share a common ancestor). Homology is the most informative inference that one can draw from a database search, both because it is the most reliable (homology can be inferred from statistically significant similarity scores with high confidence) and because it is the most informative (homologous sequences always share a common three-dimensional fold and secondary structure). If a newly characterized sequence is homologous to an entry in the Protein Data Bank (PDB) of three-dimensional structures, it is straightforward to build a model of the new sequence by molecular replacement. Surprisingly, this implication is not universally recognized; sometimes strong similarities are observed with sequences in the SWISS-PROT or GenPept database, but NRL_3D (a database of sequences with known three-dimensional structures) is not examined. Homologous sequences may also share a binding or catalytic activity, but they may not. Even when two homologous sequences have different functions, they can be relied on to share a common three-dimensional fold.

We infer protein homology by calculating sequence similarity. Similarity is a quantity (two sequences share 30 or 15% identity) while homology is an inference (two proteins are either homologous or they are not). In general, statistically significant similarity scores can be used to infer homology with high levels of confidence; the major exception is low complexity repeated domains in proteins.[1] The converse is not true; absence of significant similarity does not guarantee nonhomology. Many diverse protein families contain members with low pairwise sequence similarity, but they can be identified because homology is transitive: if A is related to B and B is related to C, A must be related to (homologous to) C, even if they do not share significant sequence similarity.

Conversely, if two sequences are not homologous, no conclusions can be drawn. In general, similar, but nonhomologous, sequences do not share common structures or other features. With the exception of hydrophobic transmembrane domains and α-helical coiled-coil structures, similarity without homology provides little reliable information.

---

[1] S. F. Altschul, M. S. Boguski, W. Gish, and J. C. Wootton, *Nat. Genet.* **6**, 119 (1994).

Improvements in Sequence Comparison Methods

Today, database similarity searches are more likely to detect homologous sequences because databases are larger and because sequence comparison methods are more sensitive and more selective. In addition, current programs use modern scoring matrices and gap penalties that perform significantly better than earlier parameters. Protein and DNA sequence databases have expanded 3- to 10-fold over the period 1990–1995, from 14,000 to 43,000 sequences in the SWISS-PROT databases and from 33,000 to 435,000 sequences in GenBank.[2] Thus, one is more likely to find the yeast homolog of a human sequence because a larger fraction of the yeast genome is available in the sequence databases. Fortunately, computers have kept pace with, and often exceeded, this explosion in biological sequence data. The searches reported in Volume 183 of this series took about 1 min on a high-performance UNIX workstation of the time; today, those same searches take about 30 sec with more sensitive algorithms on a 3-fold larger database.

Sensitivity and Selectivity

Most of this chapter focuses on the evaluation of distant alignments that have ambiguous similarity scores. This is the major problem facing an investigator with a new protein sequence who wishes to assign a function, or perhaps class of functions, based on homology. Often such an inference is obvious. For example, the mouse class-mu glutathione transferase GTB1_MOUSE sequence can be aligned with the *Drosophila* GT2_DROME with a BLOSUM50 similarity score of 164, which is expected to be found by chance only once in 100,000 searches of a database the size and composition of SWISS-PROT,[3] and the two sequences share 25% identity over 202 amino acids. In contrast, the relationship between GTB1_MOUSE and human elongation factor EF1γ is unclear; the similarity score for the alignment is high, but not statistically significant, and the percent identity is high (26%) but the alignment shorter (162 amino acids). The relationship between GTB1_MOUSE and EF1γ can be established only by examining other members of the family that share significant similarity to both sequences.

In any database search, there is always a library sequence with the best score, regardless of whether that sequence shares common ancestry with the query sequence. In sequence comparison, there is a trade-off between

---

[2] The number of sequences in GenBank rose by 50% from Feb. 1995 to June 1995 because of large-scale expressed sequence tag (EST) sequencing projects.

[3] A. Bairoch and B. Boechmann, *Nucleic Acids Res.* **19**(Suppl.), 2247 (1991).

sensitivity (the ability to identify distantly related sequences) and selectivity (the avoidance of false positives, i.e., unrelated sequences with high similarity scores). The perfect sequence comparison method would be both sensitive and selective; it would rank all the members of a protein family that share a common ancestor above all the sequences that are similar but nonhomologous.

The selectivity of an algorithm or scoring matrix is reflected in the statistical estimates for a search. These estimates indicate the probability that a similarity score would be obtained by chance by a "random" query sequence with the same length and amino acid composition. Thus, it is the distribution of similarity scores from unrelated sequences that determines the expectation values for high-scoring sequences. If the statistical estimates are accurate, the highest scoring unrelated sequences should have expectation values around 1.0 (e.g., 0.5–2.0). A more selective algorithm rarely calculates high scores for unrelated sequences, so that lower "raw" similarity scores would still have high statistical significance. This is often seen with BLASTP,[4] where low similarity scores (e.g., 55–70), are statistically significant. With the Smith–Waterman algorithm,[5] alignments typically require scores over 120 for significance. Expectation values allow one to compare the performance of different algorithms with different scoring matrices, and they are used extensively in this chapter.

Statistics of Local Similarity Scores

The major technical improvement in protein sequence comparison has been the incorporation of statistical estimates in widely used similarity searching programs. Such estimates were first widely available with the BLAST programs[4] and have been incorporated into the FASTA and SSEARCH (Smith–Waterman) programs. Accurate statistical estimates greatly decrease the likelihood that an investigator will call sequences homologous when they are not. Although statistics can prevent erroneous inferences of homology, they cannot be used to demonstrate nonhomology. Homologous proteins in diverse families often lack significant similarity.

The statistics of local similarity scores for alignments without gaps have been described by Karlin and Altschul.[6] Local similarity scores are described by the extreme value distribution. Using the parameters λ and K, which can be derived from the scoring matrix and the amino acid composi-

---

[4] S. F. Altschul, W. Gish, W. Miller, E. W. Myers, and D. J. Lipman, *J. Mol. Biol.* **215,** 403 (1990).

[5] T. F. Smith and M. S. Waterman, *J. Mol. Biol.* **147,** 195 (1981).

[6] S. Karlin and S. F. Altschul, *Proc. Natl. Acad. Sci. U.S.A.* **87,** 2264 (1990).

tion of the query sequence, the probability that a normalized similarity score[1,6]:

$$S' = \lambda S - \ln Kmn \qquad (1)$$

where m is the length of the query sequence and n is the length of the library sequence can be calculated as

$$P(S' \geq x) = 1 - \exp(-e^{-x}) \qquad (2)$$

Because a database search usually involves thousands of pairwise comparisons, the expectation of finding a score $S' \geq x$ for a search of D sequences is $E(S' \geq x) = PD$.

The relationship between the raw similarity score S and the normalized score S' in Eq. (1) shows that scores for alignments without gaps between random sequences increase as $\ln Kmn$. Because K and m are fixed for each search, S is expected to increase with $\ln n$. This is seen empirically for alignments that contain gaps.[7,8] For local similarity scores, the variance of the score should be independent of library sequence length. Figure 1 shows the increase in similarity scores for unrelated sequences with library sequence length. As a result of this increase, lower scoring related sequences (marked L in Fig. 1A) are "buried" under a background of long, high-scoring, unrelated sequences. If the raw similarity scores S are rescaled to remove this length effect (Fig. 1B), the low-scoring related sequences are ranked higher than the unrelated sequences. Thus, normalization of similarity scores by fitting a line to the relationship of similarity score to $\ln n$ makes it possible to detect more distant relationships.[9]

For alignments with gaps, low gap penalties can cause the similarity scores to lose their local character,[8] so that extreme-value distribution-based probability estimates are not valid. Searches with randomly shuffled sequences can be used to test whether the statistical estimates for a search are reasonable. Because there should be no sequences in the database with significant similarity to a random sequence, the expectation values for the highest scoring library sequences in a search with a random query sequence should be between 0.2 and 2.0. The FASTA package includes the RANDSEQ program to produce a randomly shuffled sequence from a genuine one.

Searching Protein Sequence Databases: An Example

Most similarity searches are done to identify unknown proteins. However, to demonstrate the strengths and limitations of similarity searching,

[7] J. F. Collins, A. F. W. Coulson, and A. Lyall, *Comput. Appl. Biosci.* **4,** 67 (1988).
[8] R. Mott, *Bull. Math. Biol.* **54,** 59 (1992).
[9] W. R. Pearson, *Protein Sci.* **4,** 1145 (1995).

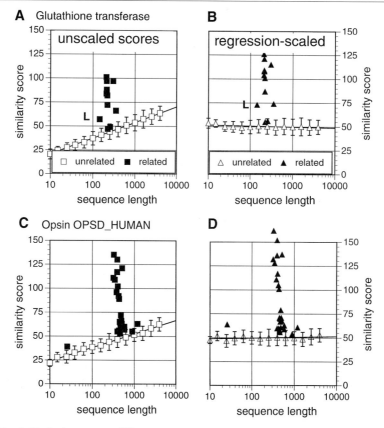

Fig. 1. Similarity scores and library sequence length. The distribution of Smith–Waterman similarity scores is plotted as a function of log(n), where n is the length of the library sequence. Filled symbols indicate individual related sequences (only the most distantly related sequences are shown); open symbols show the average and standard error of similarity scores for unrelated sequences.

we will use two protein families as examples, the glutathione transferases and the G-protein-coupled receptors. Both families were chosen because they are large and very diverse. No single member of either family shares a statistically significant similarity with every other member of the family in the databases. The glutathione transferases are soluble proteins that are found in animals and plants as glutathione conjugating enzymes, but they also occur in animals as crystallins, in plants as small heat-shock proteins, and in bacteria as dehalogenases.[10] Remarkably, elongation factor EF1γ

[10] R. M. Armstrong, *Adv. Enzymol.* **69,** 1 (1994).

```
alpha0 65% fasta
 using matrix file BLOSUM50
 fasta 2.0x1 [May, 1995] searches a sequence data bank
Please cite:
W.R. Pearson & D.J. Lipman PNAS (1988) 85:2444-2448

test sequence file name: gtb1_mouse.aa

Choose sequence library:
     P: NBRF Protein database (complete)
     D: NRL_3d structure database
     G: GENPEPT Translated Protein Database (rel 88.0)
     S: Swiss-Prot Release 31

Enter library filename (e.g. prot.lib), letter (e.g. P)
or a % followed by a list of letters (e.g. %PN): s
ktup? (1 to 2) [2] ____
use optimized scores? [yes]: ____
>GT8.7 transl. of pa875.con, 19 to 675: 217 aa
vs Swiss-Prot Release 31 library
searching /seqlib/lib/swiss.seq 5 library
```

Fig. 2. Sample FASTA search.

contains a well-conserved glutathione transferase domain.[11] The glutathione transferase fold appears to be very ancient, and the family provides a good example of a diverse family of globular proteins.

Figure 2 shows an example of a search of the SWISS-PROT protein sequence database, using a murine glutathione transferase as the query sequence. After the program is started, the query sequence file name (gtb1_mouse.aa in Fig. 2) is requested and the sequence library selected (s for SWISS-PROT). To search all of the available databases in one run, one would specify %PDGS.

Next, the *ktup* parameter (2 in this case by default) and optimization (on by default) are selected. The *ktup* parameter determines the speed and sensitivity of the search: *ktup* = 2 is about 4 times faster than *ktup* = 1 but not as sensitive.

In this example, the user is prompted for the file names and *ktup* parameter, but each of these entries can be specified on the command line. For example, the command "fasta -q gtb1_mouse.aa s 1" would do the

[11] E. V. Koonin, A. R. Mushegian, R. L. Tatusov, S. F. Altshul, S. H. Bryant, P. Bork, and A. Valencia, *Protein Sci.* **3,** 2045 (1994).

same search with $ktup = 1$ but without prompting for any information after the program starts.

The histogram of scores (Fig. 3) shows how well the distribution of normalized similarity scores ("$==$") fits to the expected extreme value distribution ("*"). In general, the fit will be extremely close, as seen in Fig. 3. However, if the query sequence contains a low-complexity repeated domain or if the gap penalties are set too low, there will be an excess of high-scoring unrelated sequences with expectation values below 0.1. If there appears to be a substantial excess of high-scoring sequences (z scores from 80 to 110 have 3–5 times as many scores as expected), the search should be done again after the low-complexity portion of the query has been removed or the gap penalties have been raised.

After the search is finished and the distribution of similarity scores has been displayed, the program summarizes the statistics for the search (Fig. 4) and asks for a file name for the results (gtb1.ok2). If a file name is specified, the histogram, the list of top-scoring library sequences, and the sequence alignments are written to the file as well as to the display terminal. If the results are written to a file, the sequence alignments are not shown at the terminal. The program then asks for the number of sequences to be displayed initially.

Next, FASTA lists the highest scoring sequences in the library, their similarity scores, and the expectation that the score would be obtained by chance in a search of a database this size (Fig. 5). The expectation value is the expected number of times that a sequence would obtain a *Z-opt* score as high or higher by chance. Thus, the similarity score for a FASTA alignment of GTB1_MOUSE with GT1_MUSDO would be expected to occur only 0.027 times in a search of SWISS-PROT. In this search, there are 10 members of the glutathione transferase family that have high similarity scores that are not statistically significant ($E > 0.05$), for example, GTTR_RAT ($E = 1.5$) and GT31_MAIZE ($E = 9.9$). (For the purpose of illustration, many high-scoring related sequences have been omitted from Fig. 5, and the high-scoring unrelated sequences are highlighted in boldface type.)

Finally, FASTA displays alignments for the high-scoring sequences (Fig. 6). In contrast to earlier versions, the Smith–Waterman algorithm is used to produce this alignment, and there are no limits on the number of gaps that can be inserted. This can be seen in Fig. 6, where the optimized score, which limits the gaps to a band of 16 residues (for $ktup = 2$; $ktup = 1$ uses a 32-residue wide band) around the best initial region, is 161; whereas the Smith–Waterman score, which allows more gaps, is 164. The optimal alignment in Fig. 6 requires a band of at least 24 residues.

The glutathione transferase search in Figs. 2–6 shows the common result

```
one = represents 67 library sequences
for inset = represents 2 library sequences
   z-opt E()
< 20    135     0 :===
  22      0     0 :
  24      5     0 :=
  26      5     1 :*
  28     16    10 :*
  30     66    60 :*
  32    229   230 :===*
  34    675   625 :=========*=
  36   1540  1284 :====================*===
  38   2386  2121 :==============================*====
  40   3213  2959 :==========================================*===
  42   3823  3617 :==================================================*====
  44   3980  3990 :=======================================================*
  46   3892  4064 :=======================================================*
  48   3432  3891 :=================================================        *
  50   3163  3551 :=============================================        *
  52   3004  3122 :=========================================== *
  54   2574  2666 :===================================*
  56   2058  2227 :==============================   *
  58   1733  1829 :========================= *
  60   1572  1481 :=======================*=
  62   1193  1187 :==================*
  64    983   944 :==============*
  66    811   746 :===========*=
  68    656   587 :========*=
  70    506   460 :======*=
  72    383   360 :=====*
  74    331   280 :====*
  76    227   218 :===*
  78    199   170 :==*
  80    133   132 :=*
  82    106   101 :=*
  84     83    80 :=*
  86     71    62 :*=
  88     46    48 :*
  90     47    37 :*
  92     34    29 :*          :==============*===
  94     22    22 :*          :==========*
  96     22    17 :*          :========*==
  98      9    13 :*          :===== *
 100      6    10 :*          :=== *
 102      6     8 :*          :===*
 104      4     6 :*          :==*
 106      7     5 :*          :==*=
 108      4     4 :*          :=*
 110      3     3 :*          :=*
 112      4     2 :*          :*=
 114      2     2 :*          :*
 116      3     1 :*          :*=
 118      0     1 :*          :*
>120     62     1 :*          :*=============================
```

FIG. 3. Glutathione transferase search: distribution of scores.

```
15335248 residues in 43470 sequences
statistics extrapolated from 20000 to 43267 sequences
Kolmogorov-Smirnov statistic: 0.0242 (N= 29) at 42
results sorted and z-values calculated from opt score
18981 scores better than 43 saved, ktup: 2, variable pamfact
BLOSUM50 matrix, gap penalties: -12,-2
joining threshold: 36, optimization threshold: 24, width: 16
 scan time:  0:00:29
Enter filename for results : gtb1.ok2
```

FIG. 4. Glutathione transferase search: summary.

when the query sequence is related to a large family of proteins in the database. Mammalian glutathione transferases can be grouped into four classes—alpha, mu, pi, and theta; members of the same class share 80–95% amino acid sequence identity, whereas interclass alignments show 20–30% sequence identity.[12] Members of three of the four glutathione transferase classes share significant similarity with the class-mu GTB1_MOUSE; the highest scoring class-theta sequences have expectation values [E(), E values] near 1.0 with FASTA. Although many members of the family have E values much less than 0.01, more distantly related members have E values from 1 to 10 and higher.

In this search, the highest scoring unrelated sequence (spectrin β-chain, SPCB_HUMAN) has a similarity score with an expectation value of 1.8, suggesting that the statistical estimates are accurate for this search. Other methods for evaluating statistical significance are discussed later in this chapter.

When related library sequences have similarity scores that are much lower than 0.02, while the highest scoring unrelated sequences have scores with expectation values of at least 0.5, one can be confident that a homologous sequence has been found. Alternatively, if a search does not reveal any sequences with expectation values below 0.1, then the search has failed to identify sequences with significant similarity, and additional searches with more sensitive algorithms (e.g., FASTA, *ktup* = 1 or SSEARCH) should be performed.

Methods for Searching Protein Sequence Databases

Investigators seeking to characterize a newly determined protein by searching a sequence database are faced with an intimidating variety of algorithms (local or global), comparison programs (BLAST, BLITZ, FASTA, SSEARCH), databases (SWISS-PROT, PIR, GenBank), and

[12] B. Mannervik and U. H. Danielson, *Crit. Rev. Biochem.* **23**, 2831 (1988).

The best scores are:		initn	init1	opt	Z-opt	E(43267)
GTB1_MOUSE	Glutathione S-transferase GT8.7	1490	1490	1490	1798.4	0
GTB1_RAT	GST YB1 (class-mu)	1406	1406	1406	1697.6	0
GTB3_MOUSE	GST GT9.3	1242	1242	1299	1569.2	0
GTB2_MOUSE	GST 5	1240	1240	1259	1521.2	0
GTB2_RAT	GST YB2	1222	1222	1241	1499.6	0
GTM1_HUMAN	GST HB4	1235	1235	1235	1492.4	0
GT2_CHICK	GST 2	954	954	954	1155.0	0
GT26_FASHE	GST 26 kd	661	567	703	853.8	0
GT27_SCHMA	GST 26 kd	598	512	621	755.4	0
GTP_MOUSE	GST P (class-pi)	298	153	361	443.6	10^{-17}
GTP_HUMAN	GST P	292	155	356	437.6	10^{-18}
GTA2_MOUSE	GST YA (class-alpha)	106	57	229	284.8	10^{-9}
GTA1_MOUSE	GST GT41A	82	57	218	271.6	10^{-8}
SC11_OMMSL	S-crystallin SL11	183	87	217	270.9	10^{-8}
GTC_MOUSE	GST YC	130	91	215	268.0	10^{-8}
GT28_SCHHA	GST 28 kd	164	127	203	253.9	10^{-7}
GTH2_HUMAN	GST A2-2 2.5	42	42	198	247.6	10^{-7}
GTH1_HUMAN	GST A1-1 2.5	36	36	198	247.6	10^{-7}
GT28_SCHBO	GST 28 kd 2.	157	120	197	246.7	10^{-7}
GTC_RAT	GST YC	42	42	191	239.2	10^{-6}
GT5A_MOUSE	GST GST 5.7	113	113	183	229.5	10^{-6}
GTA8_RAT	GST	118	118	179	224.7	10^{-5}
GT28_SCHJA	GST	123	96	169	213.3	10^{-5}
GT2_DROME	GST	60	60	161	202.4	10^{-4}
GTAC_CHICK	GST, CL-3	56	56	144	182.6	0.001
SC1_OCTDO	S-crystallin 1 (OL1).	78	78	132	168.6	0.0061
GT1_MUSDO	GST	88	52	122	156.8	0.027
SC1_OCTVU	S-crystallin 1.	58	58	121	155.4	0.033
SC2_OCTVU	S-crystallin 1.	78	78	121	155.4	0.033
ARP_TOBAC	auxin-regulated protein	57	57	117	150.4	0.062
SC3_OCTDO	S-crystallin 3 (OL3).	76	47	111	143.4	0.15
GT1_DROME	GST 1-1 (class-theta)	50	50	100	130.3	0.81
GTTR_RAT	GST YRS-YRS (class-theta)	89	55	97	125.7	1.5
GT_PROMI	GST GST-6.0	37	37	95	124.5	1.7
SPCB_HUMAN	β-spectrin	92	44	108	124.0	1.8
ARP2_TOBAC	auxin-induced protein	47	47	93	121.5	2.5
YHC9_YEAST	hypothetical 77.8 kd prot.	33	33	96	117.5	4.2
GT_ECOLI	GST	32	32	88	116.2	5
VMP_SOCMV	movement protein	38	38	90	115.8	5.3
PPA5_RAT	tartrate-resistant acid phos.	38	38	90	115.3	5.6
ARP3_TOBAC	auxin-induced protein	34	34	87	114.3	6.4
YJJV_ECOLI	hypothetical 23.7 kd prot.	68	68	86	113.5	7.1
GUX1_PHACH	exoglucanase I	52	52	91	113.3	7.2
ARP4_TOBAC	auxin-induced protein	59	59	85	112.0	8.6
PPA5_MOUSE	tartrate-resistant acid phos.	29	29	87	111.7	8.9
GT31_MAIZE	GST III	40	40	84	110.8	9.9

FIG. 5. Glutathione transferase search: highest scoring sequences.

other scoring parameters (PAM250, BLOSUM62, with or without gaps). This section compares briefly several of the widely used search programs, scoring parameters, and gap penalties.

Protein versus DNA Sequence Comparison

For effective sequence identification, one should always search protein sequence databases, not DNA sequence databases. Protein sequence com-

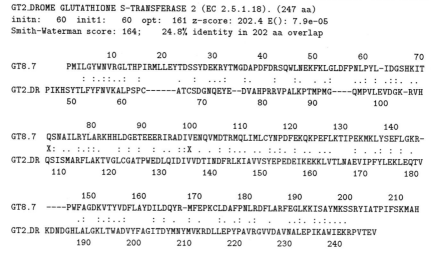

```
GT2_DROME GLUTATHIONE S-TRANSFERASE 2 (EC 2.5.1.18). (247 aa)
initn:   60  init1:   60  opt:  161 z-score: 202.4 E(): 7.9e-05
Smith-Waterman score: 164;    24.8% identity in 202 aa overlap

                10        20        30        40        50        60        70
GT8.7     PMILGYWNVRGLTHPIRMLLEYTDSSYDEKRYTMGDAPDFDRSQWLNEKFKLGLDFPNLPYL-IDGSHKIT
          : :.::..:  :         .  :   ...: :.   :   :.  . .:    ..: : .::.. ..
GT2_DR PIKHSYTLFYFNVKALPSPC------ATCSDGNQEYE--DVAHPRRVPALKPTMPMG----QMPVLEVDGK-RVH
          50        60              70        80        90        100

                80        90        100       110       120       130       140
GT8.7     QSNAILRYLARKHHLDGETEEERIRADIVENQVMDTRMQLIMLCYNPDFEKQKPEFLKTIPEKMKLYSEFLGKR-
          X: .. :.::.   : : :   : .. ::X . .: :.... .. :.:. : ... .    : ..: : .
GT2_DR QSISMARFLAKTVGLCGATPWEDLQIDIVVDTINDFRLKIAVVSYEPEDEIKEKKLVTLNAEVIPFYLEKLEQTV
          110       120       130       140       150       160       170       180

                150       160       170       180       190       200       210
GT8.7     ----PWFAGDKVTYVDFLAYDILDQYR-MFEPKCLDAFPNLRDFLARFEGLKKISAYMKSSRYIATPIFSKMAH
          .:  :.:..:    : : .  : .   :. .: .:   ... .:.....
GT2_DR KDNDGHLALGKLTWADVYFAGITDYMNYMVKRDLLEPYPAVRGVVDAVNALEPIKAWIEKRPVTEV
          190       200       210       220       230       240
```

Fig. 6. Alignment of GTB1_MOUSE and GT2_DROME. Identical amino acids are indicated by a colon (":"); a dot (".") shows conservative replacements. The X's denote the boundaries of the best initial region. In alignments, the pairs (*ktup* = 2) of identical residues that bound the initial region may no longer be aligned after the introduction of gaps. When this occurs, the "X" is replaced by "v" and "^".

parisons routinely identify sequences that shared a common ancestor more than 1 billion years ago (e.g., the glutathione transferases shared by animals and plants). In contrast, it is often difficult to detect homology in noncoding DNA sequences that diverged 200 million years ago. Even for protein coding DNA sequences, it is rare to detect significant DNA sequence similarity for sequences that diverged more than 600 million years ago, whereas significant similarities are detected between protein sequences that diverged more than 2.5 billion years ago.[13]

Table I compares the statistical significance of protein and DNA sequence similarity scores. When DNA sequences are used, the alignment of mouse class-mu glutathione transferase GTB1_MOUSE and human class-alpha GSTA1 (HUMLGTH1) does not obtain a statistically significant score; in contrast, the protein sequence similarity score high enough to be expected only once in 10,000 searches of the GenPept database. DNA sequences are far less informative than protein sequences both because DNA sequences can change without changing the encoded protein sequence, and because nucleotide substitutions that cause amino acid changes

[13] R. F. Doolittle, D. F. Feng, M. S. Johnson, and M. A. McClure, *Cold Spring Harbor Symp. Quant. Biol.* **51,** 447 (1986).

TABLE I
DNA versus Protein Sequence Comparison[a]

Sequence	Source	Score	E(DNA)	E(prot)
MUSGLUTA	Mouse glutathione *S*-transferase class mu	5625	0	0
MUSGSTA	Mouse glutathione transferase GT9.3 mu	3953	0	0
HUMGSTAA	*Homo sapiens* glutathione transferase	1257	0	0
MAMGLUTRA	*Mesocricetus auratus* mu class GST	399	10^{-11}	0
RATGSTYD	Rat glutathione *S*-transferase Yb subunit	399	10^{-11}	0
HSGSTM4	*H. sapiens* GSTM4 gene for GST	390	10^{-10}	0
RATGSTY	*Rattus norvegicus* GST	372	10^{-9}	0
HSGSTM1B	*H. sapiens* GSTM1b gene for GST	358	10^{-9}	0
HSGSTMU3	Human GSTmu3 gene for a GST	322	10^{-7}	
HSGST145	Human GST-1 gene for GST	308	10^{-6}	
BTGST	Bovine GST mRNA for GST	249	0.0002	10^{-16}
HSGSTPI1	Human mRNA for anionic GST	237	0.0008	10^{-17}
MUSGTF	*Mus musculus* GST mu	196	0.06	
CRUGSTP	Chinese hamster GST	196	0.06	10^{-16}
CRUGSTPIE	*Cricetulus griseus* GST pi	196	0.06	10^{-16}
HAMGSTPIE	*Mesocricetus auratus* GSP pi	191	0.1	10^{-16}
RRGTS8	*Rattus rattus* mRNA for GST	182	0.2	
HUMKAL2	**Human glandular kallikrein gene**	**170**	**0.6**	
HUMTROPI01	**Human troponin I, slow-twitch isoform**	**170**	**0.8**	
RNGSTYC2F	*R. norvegicus* GST Yc2	170	0.8	10^{-7}
MMGLUT	*M. musculus* mRNA for GST	168	1.1	10^{-7}
MUSTHYGP	**Mouse Thy-1.2 glycoprotein**	**163**	**1.3**	
HUMLGTH1	Human liver glutathione *S*-transferase	157	3.4	10^{-5}
ATCON430S1	***Rattus norvegicus* connexin**	**155**	**3.6**	
HUMA1AR2	**Human a-1-antitrypsin-related protein**	**154**	**3.6**	
HUMVLDLR	**Human VLDL protein receptor**	**152**	**4.5**	
RABGSTB	*Oryctolagus cuniculus* glutathione *S*-transferase	153	5.1	10^{-9}
HUMHSF1	**Human heat-shock factor 1 (TCF5)**	**151**	**5.5**	
RATRIIA	**Rat type I reg. subunit of cAMP**	**151**	**5.9**	
RNGSTYC1F	*R. norvegicus* GST Yc1	148	8.5	10^{-6}
RATGSTYC	Rat liver glutathione *S*-transferase Yc	148	8.6	10^{-6}
MUSCX43GA	**Mouse Cx43 gene, exon 1**	**147**	**11**	
HUMTAN1	**Human TAN-1 mRNA**	**142**	**12**	
OCDHPR	**Rabbit mRNA for dihydropyridine (DHP)**	**142**	**12**	
AO1444	**Human DNA for 4.6 kb retinoblastoma**	**142**	**12**	
HUMGSTB	Human glutathione *S*-transferase	144	14	
HUMGSTH	Human glutathione *S*-transferase	144	14	10^{-6}
HUMGST2	Human glutathione *S*-transferase 2	144	14	10^{-6}
S49975	Human glutathione transferase A1-1	144	14	10^{-6}

[a] Expectation values are given for searches against DNA [score, E(DNA)] and protein databases [E(prot)]. A mouse glutathione transferase (GST) cDNA sequence (MUSGLUTA) was used to search the primate, rodent, mammalian, and vertebrate divisions of the GenBank DNA sequence database for the DNA sequence comparisons (77,772 comparisons). Protein expectations [E(prot)] were calculated from a search the translated cDNA sequence against the GenPept sequence database, which includes all of the translated GenBank (126,212 comparisons). Unrelated sequences are set in boldface type; E(prot) values for unrelated sequences are much greater than 100.

lack the biochemical information that is retained in an amino acid substitution matrix, like PAM250 or BLOSUM62.

Differences in the performance of sequence comparison algorithms, scoring matrices, or gap penalties are insignificant compared to the loss of information in DNA sequence comparison. The most important lesson from this chapter is that if the biological sequence of interest encodes a protein, protein sequence comparison should always be done.

Search Programs: Smith–Waterman, FASTA, and BLASTP

Comparisons: Local and Global. Virtually all sequence similarity searching today is done with algorithms that calculate a local similarity score (i.e., a score that identifies the most similar regions shared by two proteins or DNA sequences without requiring the similarity to extend to the ends of the sequences). Methods that calculate local sequence similarity are the best because they can detect homologous protein domains that are embedded in different sequence environments and because they can be used with partial sequences. Thus, the alignment

```
VVVVVASCDEFGYYYYY
 : . : . : : :
QQQQQATCEEFGLLLLL
```

that begins with A and ends with G has an optimal local similarity score because an extension in either direction would reduce the score. A "global" alignment score for these sequences would require that the alignment begin with the first V–Q and end at the last Y–L.[14] Clearly, such an alignment is inappropriate in this case, where only the internal domain is conserved. Local methods perform better than methods that calculate a global similarity score, even with full-length protein sequences.[9]

Three algorithms are widely used for calculating local similarity scores for database searches: BLASTP,[4] FASTA,[15] and Smith–Waterman[5] (SSEARCH, BLITZ). The Smith–Waterman algorithm is guaranteed to calculate the optimal local similarity score given a pair of sequences, a scoring matrix (e.g., PAM250), and gap penalties. BLASTP and FASTA are heuristic algorithms that do not examine every possible alignment of a query sequence with a library sequence. BLASTP and FASTA are 10–50 times faster than Smith–Waterman; yet, for many protein families, they

---

[14] W. R. Pearson and W. Miller, this series, Vol. 210, p. 575.
[15] W. R. Pearson and D. J. Lipman, *Proc. Natl. Acad. Sci. U.S.A.* **85,** 2444 (1988).

TABLE II
ALGORITHM SEARCH SENSITIVITY[a]

		FASTA		Smith–Waterman
Sequence	BLASTP	ktup = 2	ktup = 1	
OPSD_HUMAN	379	362	387	394
GTB1_MOUSE	66	60	61	63

[a] Numbers of family members with expectation values below 2.0 [BLASTP $P(N)$ < 0.86] in a search of SWISS-PROT (release 31) using a human opsin (OPSD_HUMAN) or mouse glutathione transferase (GTB1_MOUSE) query sequence are given.

are just as effective in identifying members of the family (Table II). Table III summarizes the time required to search the SWISS-PROT protein database using BLASTP, FASTA, and Smith–Waterman with an average length query sequence on several different computers.

The Smith–Waterman Algorithm. A rigorous algorithm for calculating local similarity scores was first described by Smith and Waterman.[5] Current versions of the algorithm use a more efficient implementation[16] that requires $O(MN)$ comparison operations, where M is length of the query sequence and N is the length of the library. The similarity between two sequences is calculated using a scoring matrix and gap penalties. While the optimal similarity score is unique, there may be several alternative alignments that produce the same score.

The distribution of Smith–Waterman similarity scores follows the extreme value distribution[8]; one can estimate the statistical significance of a score from parameters derived from calculating the scores for a large number of unrelated sequences,[7,8] or from multiple local alignments from a single pair of sequences,[17] or by calculating similarity scores for shuffled sequences (see below). Similarity estimates based on the relationship of similarity score to sequence length for unrelated sequences have been incorporated into the SSEARCH program, which is an implementation of the Smith–Waterman algorithm distributed as part of the FASTA package.

For full-length query sequences, the Smith–Waterman algorithm with modern scoring matrices, optimized gap penalties, and normalization to remove the effect of library sequence length performs significantly better than FASTA and BLASTP.[9] With partial length sequences, FASTA (with

[16] O. Gotoh, *J. Mol. Biol.* **162**, 705 (1982).
[17] M. S. Waterman, *Bull. Math. Biol.* **56**, 743 (1994).

TABLE III
LIBRARY SEARCH TIMES[a]

Computer	BLASTP	FASTA		Smith–Waterman
		ktup = 2	ktup = 1	
DEC Alpha 2100	0.5	0.5	1.7	10.1
Sun Sparc10	1.6	2.3	6.4	55.7
Macintosh 8100/80		3.2	9.5	65.8
Intel Pentium/120		2.1	5.3	37.5

[a] Times in minutes are given for a search of a 347-amino acid query sequence (OPSD_HUMAN) with SWISS-PROT (release 31, with 15,335,248 residues in 43,470 sequences) using various algorithms.

appropriate matrices, gap penalties, and score normalization) performs as well as Smith–Waterman.

A rapid parallel version of the Smith–Waterman algorithm is available from the BLITZ E-mail server (blitz@ebi.ac.uk). As of this writing, BLITZ does not provide the length normalization available in SSEARCH and does not provide gap penalties of the form −12 for the first residue and −2 for each additional residue; a constant penalty for each residue in a gap (e.g., −7) is required. Without length-dependent normalization to reduce the scores of long library sequences, BLITZ is not expected to perform significantly better than BLASTP or FASTA.

FASTA. The FASTA program[15,18] was the first widely used method for rapid searching of protein and DNA sequence databases. FASTA is available on both UNIX workstations and microcomputers, and it can be used to search a wide variety of sequence database formats.[19] With version 2.0 (March, 1995), FASTA was modified to improve significantly both its sensitivity and selectivity. Sensitivity was improved by calculating an "optimized" score for a majority of the sequences in library search; selectivity was improved by correcting similarity scores for the expected increase due to library sequence length.

In addition to the scoring matrix and gap penalties used by the Smith–Waterman algorithm, FASTA uses the *ktup* parameter to vary the speed and sensitivity of a search. When FASTA searches a protein database with the standard *ktup* = 2, only alignments between the query and library sequences where there are a pair of identical residues are examined. Thus, the region

[18] D. J. Lipman and W. R. Pearson, *Science* **227,** 1435 (1985).
[19] W. R. Pearson, this series, Vol. 183, p. 63.

```
QSNAILRYLARKHHLDGETEEERIRADIV
::  ..  :.::.     :  :  :     :  ..  :::
QSISMARFLAKTVGLCGATPWEDLQIDIV
```

would be considered with *ktup* = 2, because of the QS and IV identical matches at the ends of the region, but

```
QSNAILRYLARKHHLDGETEEERIRADIVENQVMDTRMQLIMLCYNPDFEKQKPEFLKTIPEKMKLYSEFL
::  ..  :.::.     :  :  :     :  ..  :::  . .  :....  ..  :.::.  :  ...     :  ..:  :  :
QSISMARFLAKTVGLCGATPWEDLQIDIVVDTINDFRLKIAVVSYEPEDEIKEKKLVTLNAEVIPFYLEKL
```

is considered only if *ktup* = 1 because there is only one identity at the right end of the longer alignment. The latter alignment contains the former, but extends it more than twice as far and produces an initial similarity score that is twice as high. This increased sensitivity comes at a price: searches with FASTA *ktup* = 1 are typically 3–4 times slower than searches with *ktup* = 2 (Table III).

The FASTA program is available as part of the FASTA package, via anonymous ftp from ftp.virginia.edu in the directory pub/fasta. Versions are available for UNIX, Macintosh, and DOS/Windows. The FASTA package includes other programs for searching sequence databases (SSEARCH, TFASTA), programs for evaluating statistical significance by shuffling (PRSS, PRDF), and programs for finding repeated domains in protein sequences (LALIGN, LFASTA). The major programs in the FASTA package are listed in Appendix A, and FASTA/SSEARCH program options are noted in Appendix B.

BLASTP. The BLAST package of programs,[4] including BLASTP, BLASTN, BLASTX, and TBLASTN, provided the first rapid sequence comparison programs to incorporate estimates of statistical significance based on an analytical theory for the statistics of similarity scores without gaps (high-scoring segment pair or HSP scores).[6,20] Like FASTA, BLASTP uses a word-based method for identifying rapidly conserved regions in proteins; unlike FASTA, BLASTP considers words with amino acid substitutions as well as identities. In addition, BLASTP examines the local similarity around every word match; FASTA does not. Current versions of BLASTP use a sophisticated scheme for calculating the statistics for sequences that contain several high-scoring regions.[21]

Tests comparing BLASTP using the default BLOSUM62 matrix and the Smith–Waterman algorithm with the conventional PAM250 matrix and uncorrected similarity scores show that BLASTP can perform as well as the much slower Smith–Waterman algorithm.[9] BLASTP performs significantly

[20] S. Karlin, P. Bucher, V. Brendel, and S. F. Altschul, *Annu. Rev. Biophys. Biophys. Chem.* **20**, 175 (1991).
[21] S. Karlin and S. F. Altschul, *Proc. Natl. Acad. Sci. U.S.A.* **90**, 5873 (1993).

better than versions of FASTA that do not calculate optimized scores for most library sequences (versions before 2.0). However, if modern scoring matrices and optimal gap penalties are used with length-corrected similarity scores, Smith–Waterman and FASTA perform significantly better than BLASTP.

Although current versions of BLASTP are very effective at identifying distantly related sequences, BLASTP does not produce biologically mean-ingful alignments because it does not allow gaps. Distant sequence relation-ships (<30% identity) typically extend over entire protein sequences or long protein domains and require gaps to include the entire homologous region. Because of its restriction on gaps, BLASTP may break up a long homologous domain into several HSPs without gaps, which, when com-bined, have significant similarity. Thus, although BLASTP is effective at identifying distant relationships, a Smith–Waterman alignment should al-ways be used when BLASTP matches are analyzed and displayed.

Search Parameters: Scoring Matrices and Gap Penalties

In addition to comparison algorithms, investigators can choose from different scoring matrices and gap penalties. Although the PAM250 matrix developed by Dayhoff and colleagues[22] has been widely used in the past, modern scoring matrices allow more effective search performance.[9]

Modern scoring matrices have been derived using two different ap-proaches. The first method, which was used to develop the PAM250 ma-trix,[22] estimates the transition frequencies for a small amount of sequence change (typically 1%) and then extrapolates the transition frequencies by successive multiplication to a matrix that models the distribution of amino acid substitutions after 120% (PAM120), 200% (PAM200), or 250% (PAM250) amino acid substitution. Although it seems surprising to consider alignments where two sequences have changed by 250%, in fact, such se-quences are expected to remain about 15% identical and are thus in the twilight zone that contains distantly related and high-scoring unrelated sequences.[13] A modern set of extrapolated matrices has been described by Jones et al.[23]

Henikoff and Henikoff have used an alternative approach to calculate the BLOSUM series of matrices from aligned blocks of conserved resi-dues.[24] Here, transition frequencies are not extrapolated; they are observed

[22] M. Dayhoff, R. M. Schwartz, and B. C. Orcutt, *in* "Atlas of Protein Sequence and Structure" (M. Dayhoff, ed.), Vol. 5, Suppl. 3, p. 345. National Biomedical Research Foundation, Silver Spring, Maryland, 1978.

[23] D. T. Jones, W. R. Taylor, and J. M. Thornton, *Comput. Appl. Biosci.* **8,** 275 (1992).

[24] S. Henikoff and J. G. Henikoff, *Proc. Natl. Acad. Sci. U.S.A.* **89,** 10915 (1992).

directly by identifying blocks of conserved residues that are at least 45% (BLOSUM45), 50% (BLOSUM50), or 62% (BLOSUM62) identical. The BLOSUM matrices perform significantly better than the PAM120 or PAM250 matrices used in earlier versions of BLASTP and FASTA,[25] and perform very well with the rigorous Smith–Waterman algorithm.[9] Although extrapolated matrices can perform as well as the BLOSUM matrices, the BLOSUM matrices work very well with a wide range of comparison algorithms and gap penalties. The BLASTP program uses the BLOSUM62 matrix by default; FASTA and SSEARCH use BLOSUM50.

Modern scoring matrices based on either extrapolation (JTT160, Gonnet92) or direct transition frequencies (BLOSUM45–62) can provide very effective similarity searches when used with optimized gap penalties and length-normalized similarity scores.[9] Current statistical theory does not provide much guidance for selecting "optimal" gap penalties, but empirical studies suggest that a range of penalties, from -10, -2 to -14, -4 (-10 for the first residue and -2 for additional residues; these are equivalent to a penalty of the form $q + rk$ where $q = -8$ and $r = -2$), are appropriate for scoring matrices such as PAM250 and BLOSUM50 that are scaled in 1/3 bit units (the standard BLOSUM62 matrix is scaled in 1/2 bit units and thus works best with lower gap penalties[26]). Higher gap penalties (e.g., -14, -4) improve performance when partial sequences [e.g., expressed sequence tags (ESTs)] are compared. FASTA and SSEARCH use gap penalties of -12, -2 by default for protein sequence comparison; these values can be changed with a command line option. TFASTA and FASTA DNA sequence comparisons use higher gap penalties.

While it is often stated that differences in scoring matrices and gap penalties can cause large changes in sequence alignments, similarity scores for clearly related sequences are usually not very sensitive to changes in scoring matrix or gap penalty. In Table IV all of the similarity scores between the clearly related sequence pairs have statistical significance except for one of the PAM250 alignments. The similarity scores for the very distant relationships shown in Table IV are more dependent on the particular choice of scoring matrix and gap penalty. These latter relationships are at the border of detectable similarity and must be confirmed with additional comparisons. One should be very cautious when evaluating "significant" alignments that are seen only with one or two combinations of scoring matrices and gap penalties.

The default gap penalties used by FASTA, TFASTA, and SSEARCH are at the lower end of the useful range; searches can sometimes be im-

[25] S. Henikoff and J. G. Henikoff, *Proteins* **17,** 49 (1993).
[26] S. F. Altschul, *J. Mol. Biol.* **219,** 555 (1991).

TABLE IV
SCORING MATRICES AND GAP Penalties[a]

| Matrix | Gap penalty | Expectation value | |
		Clearly related	Distantly related
G-protein-coupled receptors			
BLOSUM50	−12/−2	2×10^{-4}	0.013
	−14/−2	1×10^{-4}	0.23
	−12/−1	7×10^{-4}	0.15
BLOSUM62[b]	−8/−2	4×10^{-4}	0.065
	−8/−1	0.009	0.27
	−6/−2	0.01	0.10
GONNET92	−10/−2	8×10^{-3}	1.52
	−12/−2	3×10^{-3}	1.89
	−14/−1	4×10^{-3}	5.98
PAM250	−12/−2	0.04	0.51
	−12/−1	0.20	1.16
	−14/−2	0.06	1.74
Glutathione transferases			
BLOSUM50	−12/−2	2×10^{-4}	0.02
	−14/−2	4×10^{-6}	0.009
	−12/−1	6×10^{-4}	0.27
BLOSUM62[b]	−8/−2	10^{-4}	0.11
	−8/−1	10^{-3}	0.40
	−6/−2	0.03	3.43
GONNET92	−10/−2	5×10^{-3}	1.79
	−12/−2	2×10^{-4}	0.45
	−14/−1	2×10^{-5}	0.016
PAM250	−12/−2	5×10^{-3}	0.48
	−12/−1	0.10	3.49
	−14/−2	8×10^{-3}	0.26

[a] Expectation values for moderately and distantly related sequences using different scoring matrices and gap penalties are given. G-protein-coupled receptors examined OPSD_HUMAN versus TA2R_MOUSE (clearly related) and CAR1_DICDI (distantly related). Glutathione transferases examined GTB1_MOUSE versus GT2_DROME, GTTR_RAT.

[b] BLOSUM62 is scaled in 1/2-bit units and thus uses lower gap penalties than the other matrices, which are scaled in 1/3-bit units.

proved by using a slightly greater penalty (e.g., -14, -2 instead of -12, -2). If many apparently unrelated sequences obtain expectation values below 0.2, gap penalties should be increased. Table V shows that changing the gap penalty from -12, -2 to -14, -2 can increase the expectation values for unrelated sequences 8-fold and reduce the number of unrelated sequences with expectation values below 1.0 by 4-fold.

Searching in Practice

In addition to selecting a sequence comparison algorithm and scoring matrix, investigators frequently encounter more practical problems such as which database should be searched and which computer is best. Although the answers to such questions will vary from site to site, this section provides a few guidelines.

TABLE V
INCREASING GAP PENALTIES TO IMPROVE SELECTIVITY[a]

Highest scoring unrelated sequences	Source	Score	Z score	$E()$
BLOSUM50, -12, -2				
CYB_CROLA	Cytochrome *b*	132	149.6	0.068
CYB_MONDO	Cytochrome *b*	126	142.4	0.18
CYB_ANGRO	Cytochrome *b*	122	141.8	0.19
CYB_CYPCA	Cytochrome *b*	125	141.2	0.20
CYB_XENLA	Cytochrome *b*	122	137.6	0.32
GABP_ECOLI	GABA permease	123	137.0	0.35
YGAK_HAEIN	Hypothetical 47.5 kDa protein	121	135.4	0.42
CYB_PARLI	Cytochrome *b*	115	129.2	0.94
VME1_CVPFS	E1 glycoprotein precursor	111	127.7	1.15
CYB_SPHLE	Cytochrome *b*	111	124.4	1.75
CYB_NEGBR	Cytochrome *b*	111	124.4	1.75
CYB_CARPL	Cytochrome *b*	110	123.2	2.04
BLOSUM50, -14, -2				
AAC2_DICDI	AAC-rich mRNA clone AAC11	108	133.7	0.53
ARP_PLAFA	Asparagine-rich protein	107	130.8	0.77
SUP2_PICPI	Omnipotent suppressor protein 2	106	126.6	1.32
CYB_CROLA	Cytochrome *b*	101	125.5	1.51
GABP_ECOLI	GABA permease	102	125.1	1.59
CYB_CYPCA	Cytochrome *b*	100	124.1	1.81
VME1_CVPPU	E1 glycoprotein precursor	97	123.2	2.02

[a] High-scoring unrelated sequences from a search of SWISS-PROT with CAR1_DICDI using different gap penalties are given.

Selecting a Database. The perfect database for similarity searching would have as few duplicate sequences as possible, yet be completely up to date. Many more sequence identification discoveries are missed by searching an out-of-date database than by using the wrong algorithm or scoring matrix. Unfortunately, the most up-to-date sequence database (GenPept with cumulative updates) is full of duplicate entries because it is translated automatically from GenBank. In practice, this means that GenPept is larger than necessary, so that expectation values calculated from this database may be artificially high[1] and thus less significant. Duplicate entries can also be misleading when they appear as the highest scoring results from a search. One may find that a query sequence has a marginally significant similarity with three or four sequences with different names, and then conclude that the similarity is genuine because a "match" has been found to several members of the family. If each of the family members is greater than 80% identical, then the additional matches provide little new information; they are only matches to the same sequence several times.

Well-curated nonredundant databases are time-consuming to maintain and are therefore less up-to-date. The different database producers use different strategies for updating the databases, so that the latest release of SWISS-PROT[3] is always missing some sequences in the PIR database,[27] and vice versa. To make certain that one does not miss a significant similarity because the sequence was missing from a database, it makes sense to search two or three databases: SWISS-PROT and/or PIR, and GenPept. The BLASTP E-mail server (blastp@ncbi.nlm.nih.gov) provides a nonredundant protein database that combines PIR, SWISS-PROT, and GenPept. However, the criterion for nonredundancy in this database is very strict, so many almost duplicate sequences (>90% identical) are included.

Search Speed and Sensitivity. The sequence comparison programs in the FASTA package, namely, FASTA, TFASTA, and SSEARCH, can be used on UNIX, DOS, and Macintosh systems with only minor differences in the implementations. A desktop Pentium or PowerPC can provide performance similar to a Sun workstation (Table III). In addition, the differences in performance between the BLASTP, FASTA, and Smith–Waterman algorithms can be modest (Table II). More important is the quality and timeliness of the database. Every characterization of an unknown protein should include a BLASTP search against the nonredundant protein database (send the E-mail message help to blast@ncbi.nlm.nih.gov). If available computer time is limited, it may be more effective to search several databases with a fast algorithm than to search only one database with a slow algorithm.

Currently, the Smith–Waterman algorithm with length-scaled similarity

[27] W. C. Barker, D. G. George, and L. T. Hunt, this series, Vol. 183, p. 31.

scores is the most effective method for identifying homologous proteins when full-length query sequences are used.[9] Smith–Waterman searches are slow; however, high-performance UNIX workstations can perform 20–50 searches per day against a comprehensive protein database like SWISS-PROT. Parallel versions of the FASTA programs can run on multiple UNIX workstations using the PVM parallel environment. PVM[28] runs on networked UNIX workstations from many vendors. A network of five DEC Alpha AXP 300 workstations can perform about 400 Smith–Waterman searches of SWISS-PROT per day. In general, most sequence comparison algorithms are available first in the UNIX environment.

Evaluating Sequence Similarities

Statistical Significance from Unrelated Similarity Scores

FASTA and SSEARCH estimate statistical significance by examining the relationship between the similarity score S and $\ln n$, the length of the library sequence. The first 10,000–20,000 sequences from a library search are used to calculate the regression line $S = a + b \ln n$ (Fig. 1) after very high scoring (presumably related) sequences are removed from the sample. The average variance of the normalized scores is also calculated, and the regression line and average variance are used to calculate

$$Z \text{ score} = [S - (a + b \ln n)]/var \qquad (3)$$

The distribution of Z scores should follow the extreme value distribution, so that

$$P(Z > x) = 1 - \exp(-e^{-1.282Z - 0.5772}) \qquad (4)$$

and, as before, $E(Z > x) = PD$. The Z' values in Fig. 5 are converted from Z scores using $Z' = 50 + 10Z$ so that they have about the same range as traditional similarity scores.

In general, the Z scores calculated during FASTA and SSEARCH searches match the extreme value distribution very closely (Fig. 3). The Kolmogorov–Smirnof statistic (Fig. 4) reports the maximum deviation of the score distribution from the extreme value distribution; however, this statistic is relatively insensitive. The best way to evaluate the accuracy of the statistical estimates is to examine the number of high-scoring unrelated sequences with expectations less than 1.0. If there are more than 5–10

[28] A. Geist, A. Beguelin, J. Dongarra, W. Jiang, R. Mancheck, and V. Sunderam, "PVM 3 User's Guide and Reference Manual," Technical Report ORNL/TM-12187, 1993.

unrelated sequences with low expectation values, the expectation values may be suspect.

Statistical Significance by Random Shuffling

Statistical estimates derived from database searches measure the difference between an observed similarity score and that expected for a sequence with the amino acid composition of the database. Such tests may overestimate significance in cases where the amino acid composition of the query sequence differs from that of the database. Thus, membrane proteins with their hydrophobic transmembrane domains may have high similarity scores with nonhomologous membrane proteins because of a restricted amino acid composition in the transmembrane domains. A more challenging test compares the similarity score between a query and library sequence with the distribution of scores obtained by comparing the query sequence to random sequences with the same length and amino acid composition as the library sequence. Such sequences are easily generated by randomly shuffling the library sequence, either globally by exchanging randomly each amino acid with any other position in the sequence or locally by performing the exchanges within a window of 10–20 residues. Because this Monte Carlo test measures the significance of the order of the two amino acid sequences, rather than the difference between the highest scoring sequences and the rest of the database, it tends to be more demanding.

As before, similarity scores for random sequences should follow the extreme value distribution, and a fit of the distribution of scores can be used to estimate the significance of an unshuffled score. However, to extrapolate an expectation value from shuffled sequences to that for a library search, the $E()$ value must be multiplied by the ratio of the number of sequences in the library to the number of shuffled sequences. Thus, in the example below, an $E()$ value from 500 shuffles must be multiplied by 80 to be comparable to an $E()$ value from the 40,000-entry SWISS-PROT database. As expected, the $E()$ value from the actual search, 2×10^{-4}, is slightly more significant than that from the shuffled distribution, 3×10^{-3}.

```
Comparison of OOHU (human opsin) with TA2R_MOUSE
  (thromboxane A2 receptor)
BLOSUM50 matrix, gap penalties: −12, −2
unshuffled s-w score: 160; shuffled score range: 38 − 92
Lambda: 0.15076 K: 0.017357; P(160) = 7.4282e-08
For 500 sequences, a score >=160 is expected 3.71e-05
  times
```

Table VI compares the statistical significance for clearly related, dis-

TABLE VI
SEARCH ALGORITHMS AND STATISTICAL SIGNIFICANCE[a]

Algorithm	Clearly related	Distantly related	Unrelated
OPSD_HUMAN versus	TA2R_MOUSE	CAR1_DICDI	APPC_ECOLI
Smith–Waterman	10^{-4}	0.01	0.57
PRSS[b]	10^{-4}	0.007	0.45
PRSS (window = 20)[b]	10^{-3}	0.23	3.0
FASTA, *ktup* = 1, *opt*	10^{-4}	0.02	0.39
FASTA, *ktup* = 2, *opt*	10^{-4}	2.2	0.36
BLASTP	0.07	$\geqslant 1.0$	$\geqslant 1.0$
GTB1_MOUSE versus	GT2_DROME	GTTR_RAT	MOD5_YEAST
Smith–Waterman	10^{-5}	0.02	4.4
PRSS[a]	10^{-7}	0.001	0.48
PRSS (window = 20)[a]	10^{-5}	0.001	1.56
FASTA, *ktup* = 1, *opt*	10^{-5}	0.025	1.9
FASTA, *ktup* = 2, *opt*	10^{-4}	1.5	>10
BLASTP	10^{-4}	10^{-4}	$\geqslant 1.0$

[a] Expected numbers of times that a similarity score as high or higher than that obtained by the indicated library sequence would be obtained by chance in a search of SWISS-PROT (43,000 entries) are given.
[b] Values given are expected times this score would be obtained after 1000 shuffles of the indicated library sequence with either global (PRSS) or local (window = 20) amino acid exchanges.

tantly related, and unrelated protein sequences in the G-protein-coupled receptor and glutathione transferase families using SSEARCH, FASTA, and BLASTP. For both families, FASTA *ktup* = 2 does not calculate significant similarity scores for the distant relationship, although FASTA *ktup* = 1 and SSEARCH do. BLASTP has mixed success: it does not detect the relationship between OPSD_HUMAN and CAR1_DICDI, but it does the best job of detecting the distant relationship between GTB1_MOUSE and GTTR_RAT.

The statistical significance calculated by the PRSS program differs in the two families as well. For the soluble glutathione transferase family, shuffled statistical estimates are similar to those determined from the database search, and there is little difference in expectation values calculated by the local and global shuffles. Presumably, this reflects the similarity between the amino acid composition of the soluble glutathione transferases and the SWISS-PROT database as a whole. In contrast, statistical estimates for the hydrophobic G-protein-coupled receptors are less significant when shuffling is done, and the local shuffling has more of an effect. G-protein-coupled receptors have highly localized amino acid composition bias in the seven transmembrane regions, and these regions are the most highly conserved. Thus, a window shuffle tends to reduce the statistical significance of the alignment, because high similarity scores can be produced by shuffling residues in the transmembrane domains.

Exploring Distant Relationships: Extending a Family

Although accurate statistical estimates for similarity scores can be very valuable in interpreting the results of similarity searches, they must be evaluated with caution. Distantly related homologous sequences often do not share statistically significant similarity. This is seen in Fig. 5, where many members of the glutathione transferase family have similarity scores between 0.1 and 10.

Proteins with high, but not statistically significant, similarity scores can often be separated from nonhomologous sequences by additional searches. For example, in Fig. 5, the similarity between the mouse class-mu glutathione transferase query sequence and GTTR_RAT is high, but not statistically significant. However, a search of the SWISS-PROT database using GT1_MUSDO (a housefly glutathione transferase, Fig. 7) shows clearly that GTTR_RAT, along with the plant auxin induced proteins (ARP1_TOBAC and others not shown), small heat-shock proteins (HS26_SOYBEAN), and elongation factors EF1γ, are also clearly related to GT1_MUSDO, and thus to the mouse class-mu enzyme. As expected for a family of homologous proteins, there is considerable overlap between the high-scoring sequences in Fig. 5 and Fig. 7. In contrast, when β-spectrin (the highest scoring unrelated sequence in Fig. 5) is used to search SWISS-PROT again, the highest scoring glutathione transferase has an expectation value of 5.5, whereas significant matches are seen with actinins and dystrophins.

Likewise, when the MOD5_YEAST tRNA isopentenyltransferase (the highest scoring unrelated sequence found using SSEARCH) is used to search the database again, 6 of the 32 unrelated sequences with expectation values ranging from 0.28 to 10.0 are glutathione transferases, but a variety of different proteins that are clearly unrelated to glutathione transferases have similar scores. Thus, one can identify high-scoring unrelated sequences simply by looking for high-scoring sequences that are unrelated to other high-scoring sequences. Re-searching a database with such sequences provides a simple confirmation of the quality of the statistical estimates, even when one is not certain whether a high-scoring sequence is related to the query sequence.

The best scores are:		initn	init1	opt	Z-opt	E(43267)
GT1_MUSDO	GST 1	1404	1404	1404	1728.6	0
GT1_DROME	GST 1-1	1229	1229	1230	1515.5	0
GT_PLEPL	GST A	174	117	274	344.7	10^{-12}
GTH1_DIACA	GST 1	189	157	263	331.4	10^{-11}
GTT1_RAT	GST 5	224	224	242	305.1	10^{-10}
GTTR_RAT	GST Yrs-Yrs	212	171	237	298.9	10^{-9}
DCMA_METSP	dichloromethane dehalogenase	220	150	210	264.7	10^{-8}
GTH2_DIACA	GST 2	153	153	205	263.2	10^{-7}
GT32_MAIZE	GST III	148	122	193	245.7	10^{-6}
GT1_WHEAT	GST 1	228	166	175	223.4	10^{-5}
URE2_YEAST	URE2 protein.	176	88	176	221.7	10^{-5}
GTB_TOBAC	GST	142	115	159	204.3	10^{-4}
EF1G_SCHPO	EF1γ	39	39	156	196.3	10^{-4}
GT28_SCHMA	GST 28 kd	66	41	146	188.5	10^{-3}
SSPA_ECOLI	stringent starvation protein A	116	116	145	187.2	10^{-3}
SSPA_HAESO	stringent starvation protein A	102	102	141	182.3	10^{-3}
GT28_SCHHA	GST 28 kd	66	41	140	181.1	0.0012
GTB_SILCU	GST	129	129	138	178.5	0.0017
ARP1_TOBAC	auxin-induced protein	61	39	138	178.3	0.0017
PRP1_SOLTU	pathogenesis-related protein	31	31	136	176.0	0.0023
HS26_SOYBN	heat shock protein 26A.	54	54	135	174.6	0.0028
GTY2_ISSOR	GST Y-2	48	48	134	174.5	0.0028
EF1G_HUMAN	EF1γ	29	29	138	173.8	0.0031
GT28_SCHBO	GST 28 kd	66	41	133	172.5	0.0036
EF1G_RABIT	EF1γ	29	29	134	168.9	0.0058
GTP_BOVIN	GST P	74	48	125	162.8	0.013
EF1G_XENLA	EF1γ	35	35	129	162.8	0.013
ARP2_TOBAC	auxin-induced protein	68	44	123	159.9	0.018
GTB1_MOUSE	GST GT8.7	88	52	122	158.9	0.021
GTAC_CHICK	GST, CL-3 SUBUNI	38	38	120	156.1	0.03
EF1X_CAEEL	EF1β/1δ	91	91	119	155.3	0.033
GT2_CHICK	GST 2	40	40	112	146.6	0.1
LIGF_PSEPA	LIGF protein.	57	57	107	139.4	0.25
GTP_HUMAN	GST P	42	42	102	134.7	0.47
GTH1_HUMAN	GST A1-1	40	40	102	134.3	0.49
GTM4_HUMAN	GST muscle	70	46	101	133.2	0.56
EF1B_WHEAT	EF1β'	48	48	96	127.1	1.2
PCPC_FLAS3	tetrachloro-p-HQ red. dehalogenase	37	37	96	126.2	1.4
EF1B_ORYSA	EF1β'	48	48	93	123.2	2.0
GTM1_HUMAN	GST HB4	47	47	89	118.5	3.7
CATE_ECOLI	**catalase HPII**	**74**	**47**	**92**	**113.8**	**6.8**
BLAB_AERHY	**β-lactamase**	**32**	**32**	**86**	**113.8**	**6.8**
GTM5_HUMAN	**GST testis/brain**	**43**	**43**	**84**	**112.1**	**8.4**
SYV_HUMAN	**Valyl-tRNA synthetase**	**41**	**41**	**93**	**111.6**	**8.9**
LGB_PSOTE	**leghemoglobin**	**58**	**58**	**81**	**111.4**	**9.2**

FIG. 7. Glutathione transferase extended family. High-scoring sequences from a search of SWISS-PROT with GT1_MUSDO (housefly glutathione transferase). Unrelated sequences are set in boldface type.

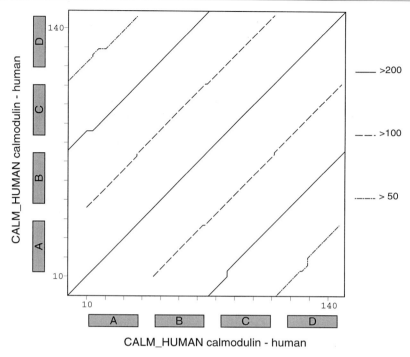

FIG. 8. Internal duplications in calmodulin. A PLALIGN plot of the local similarities in human calmodulin (CALM_HUMAN) is shown. The corresponding alignments are shown in Fig. 9.

Identifying Repeated Structures in Proteins

Thus far, we have focused on the identification and statistics of the single most significant similarity score shared by two sequences. As can be seen in Figs. 8 and 9, however, there are sometimes several nonoverlapping local alignments with high similarity scores. These suboptimal alignments can be used to infer the duplication events that gave rise to the protein sequence. An algorithm for the best N nonoverlapping local alignments has been described.[29]

The FASTA package also provides improved programs for identifying repeated domains in proteins (LALIGN and PLALIGN), which replace the earlier LFASTA and PLFASTA programs. All four programs find multiple local alignments between two sequences; if a sequence has an internal duplication, like the EF-hand calcium-binding domain in calmodu-

[29] M. S. Waterman and M. Eggert, *J. Mol. Biol.* **197,** 723 (1987).

```
Comparison of:
(A) >CALM_HUMAN Calmodulin                              - 148 aa
(B) >CALM_HUMAN Calmodulin                              - 148 aa
using matrix file: BLOSUM50, gap penalties: -14/-4

  47.7% identity in 65 aa overlap; score:  214

            20        30        40        50        60        70
CALM_H EFKEAFSLFDKDGDGTITTKELGTVMRSLGQNPTEAELQDMINEVDADGNGTIDFPEFLTMMARK
       :..::: .:::::.: :.. :: :: .::.. :. :....:: :.: ::.: ... ::. ::. :
CALM_H EIREAFRVFDKDGNGYISAAELRHVMTNLGEKLTDEEVDEMIREADIDGDGQVNYEEFVQMMTAK
            90        100       110       120       130       140
----------
  32.4% identity in 102 aa overlap; score:  177

       10        20        30        40        50        60        70
CALM_H AEFKEAFSLFDKDGDGTITTKELGTVM-RSLGQNPTEAELQDMINEVDADGNGTIDFPEFLTMMARKMKD
       ::... .. : ::.::: :. :.: :.. .. .: :... : : :::: :. :. .:. .
CALM_H AELQDMINEVDADGNGTIDFPEFLTMMARKMKDTDSEEEIREAFRVFDKDGNGYISAAELRHVMT-NLGE
       50        60        70        80        90        100       110

       80        90        100       110
CALM_H TDSEEEIREAFRVFDKDGNGYISAAELRHVMT
       ..::. : .: : ::.: .. :. ..::
CALM_H KLTDEEVDEMIREADIDGDGQVNYEEFVQMMT
       120       130       140
----------
  36.1% identity in 36 aa overlap; score:   55

             10        20        30
CALM_H DQLTEEQIAEF-KEAFSLFDKDGDGTITTKELGTVM
       ..::.:.. :. .::     : :::: .. .:. .:
CALM_H EKLTDEEVDEMIREA----DIDGDGQVNYEEFVQMM
       120       130       140
----------
  40.0% identity in 20 aa overlap; score:   53

       70        80
CALM_H LTMMARKMKDTDSEEEIREA
       .: ...:. : . .: ::::
CALM_H MTNLGEKLTDEEVDEMIREA
       110       120
----------
```

FIG. 9. Calmodulin internal alignments. Local alignments and similarity scores for calmodulin (CALM_HUMAN) produced with LALIGN are shown.

lin, there will be several alternative alignments of the two sequences that produce statistically significant similarity scores. LALIGN and PLALIGN both use the sim algorithm[30] to report the N best alignments between two

[30] X. Huang and W. Miller, *Adv. Appl. Math.* **12**, 337 (1991).

sequences. Unlike BLASTP, FASTA and SSEARCH report only the best alignment between a query sequence and a library sequence; LALIGN can be used to display multiple high-scoring alignments.

Because they provide rigorous nonoverlapping alignments, LALIGN (and the graphical version PLALIGN) perform more reliably with protein sequences than the earlier LFASTA and PLFASTA programs.[19] LFASTA and PLFASTA, or other programs developed by Miller and colleagues,[31] may be required to find local alignments in long DNA sequences.

Figure 8 shows a graphical plot of the local similarities within the calmodulin calcium binding protein. The same similarities are shown as alignments in Fig. 9. Calmodulin contains four EF-hand calcium-binding domains that are well conserved. The highest scoring alignment in Fig. 9 aligns domains A–B with C–D; the second highest aligns A–B–C with B–C–D; the third aligns A with D.

Summary: A Strategy for Identifying Homologous Proteins

Although there are several different comparison programs available (e.g., BLASTP, FASTA, SSEARCH, and BLITZ) that can be used with different scoring systems (e.g., PAM120, PAM250, BLOSUM50, BLOSUM62) and different databases (e.g., PIR, SWISS-PROT, GenPept), the following search protocol should identify homologous sequences whenever they can be found.

1. Always compare protein sequences if the genes encode proteins. Protein sequence comparison will typically double the evolutionary lookback time over DNA sequence comparison.

2. Search several sequence databases using a rapid sequence comparison program (e.g., BLASTP or FASTA, $ktup = 2$). Well-curated databases like PIR[27] or SWISS-PROT[3] tend to have fewer redundant sequences, which improves the statistical significance of a match,[1] but they are less comprehensive and up-to-date than GenPept.[32]

3. If there is good agreement between the distribution of scores and the theoretical distribution, and the alignments do not include "simple sequence" domains, accept sequences with FASTA $E()$ values or BLASTP $P()$ values[33] below 0.02 as homologous.

4. If no library sequences are found with E values below 0.02, perform additional searches with FASTA, $ktup = 1$, or SSEARCH. If library se-

---

[31] K.-M. Chao, R. C. Hardison, and W. Miller, *J. Comput. Biol.* **1,** 271 (1994).

[32] D. Benson, M. Boguski, D. J. Lipman, and J. Ostell, *Nucleic Acids Res.* **22,** 3441 (1994).

[33] P values are related to FASTA or SSEARCH $E()$ values according to the formula $P = 1 - e^{-E}$; for $E < 0.1$, $P \approx E$.

quences with E values less than 0.02 are found, the sequences are probably homologous, unless a low-complexity domain is aligned. However, sequences with similarity scores from 0.02 to 10.0 may be homologous as well. To characterize these more distantly related sequences, select "marginal" library sequences and use them to search the databases. Additional family members should have E values less than 0.05.

5. Homologous sequences share a common ancestor, and thus a common protein fold. Depending on the evolutionary distance and divergence path, two or more homologous sequences may have very few absolutely conserved residues. However, if homology has been inferred between A and B, between B and C, and between C and D, A and D must be homologous, even if they share no significant similarity.

6. Sequences with marginal E values should also be tested using the PRSS program. Compare the query and library sequences using at least 200 (and preferably 1000) shuffles. Shuffles using a window (-w) of 10–20 are more stringent than a uniform shuffle. Use the E value after 1000 shuffles to confirm an inference of homology.

7. Homologous sequences are usually similar over an entire sequence or domain, typically sharing 20–25% or greater identity for more than 200 residues. Matches that are more than 50% identical in a 20- to 40-amino acid region occur frequently by chance and do not indicate homology.

By following these steps, one will very rarely assert that two sequences are homologous when in fact they are not. However, these criteria are stringent; distantly related homologous sequences may fail to be detected because their similarity is not statistically significant. These tests are biased toward missing some distantly related sequences to avoid the possibility of misidentifying unrelated ones. In most database searches, the ratio of related to unrelated sequences is more than $4000:1$ (e.g., 10 related and 40,000 unrelated sequences). Thus, one is more likely to mistakenly identify two sequences as related than to overlook a genuine relationship, and our conservative evaluation criteria reflect that bias.

Appendix A: Programs in FASTA Package

Program	Description
Programs for database searching	
FASTA	Compares a protein sequence to a protein sequence database or a DNA sequence to a DNA sequence database using the FASTA algorithm
SSEARCH	Compares a protein sequence to a protein sequence database using the Smith–Waterman[5] algorithm

Program	Description
Programs for database searching	
TFASTA	Compares a protein sequence to a DNA sequence database, translating each DNA database sequence in all six frames
Statistical significance from random shuffles	
PRDF	Creates a library of 200–1000 (or more) sequences with the same length and amino acid composition as the library sequence examined. The program then calculates an optimized FASTA similarity score for the unshuffled sequence and for each of the shuffled sequences in the library. Probability estimates are calculated based on a fit to an extreme value distribution
PRSS	Performs the same shuffling and statistical calculations as PRDF but uses the Smith–Waterman algorithm in place of FASTA
RANDSEQ	Creates shuffled sequence(s)
Programs for finding duplicated and mosaic regions	
LALIGN	Implementation of the sim algorithm[30] to calculate multiple noninter-secting local alignments
PLALIGN	Version of LALIGN that plots the alignments in a dot-matrix format
LFASTA	Multiple local alignments using the FASTA algorithm. LALIGN is preferred for protein sequences but may be too slow for DNA sequences. PLFASTA is also available

Appendix B: FASTA/SSEARCH Program Options

-a	show complete sequences, rather than the similar region, in alignments
-b #	maximum number of best scores displayed
-c	(FASTA only) cutoff for calculating optimized score
-d #	the maximum number of alignments shown
-E #	expectation value cutoff for showing best scores and alignments (10.0 by default for proteins)
-f #	penalty for first residue in a gap (-12 by default for proteins)
-g #	penalty each additional residue in a gap (-2 by default for proteins)
-h	do not show histogram of scores
-i	search with DNA reverse complement (DNA only)
-l *file*	alternate file describing library selection
-m 0–4	alternative formats for alignments
-n	force sequence to be considered DNA
-o	(FASTA only) turn off default optimization step
-q	quiet mode; do not prompt for information
-r *file*	save all scores to a results file
-s *file*	alternate scoring matrix file or -s 250 for PAM250
-w #	line width for alignments (60 by default)
-x #	offset of alignment numbering
-z	do not calculate statistical estimates

[16] Discovering and Understanding Genes in Human DNA Sequence Using GRAIL

By Edward C. Uberbacher, Ying Xu, and Richard J. Mural

Introduction

Over the next few years a number of genome centers are each expecting to sequence about 100 megabases of human DNA. A number of techniques will be used to examine the contents of these sequences, ranging from database searches for homology with known proteins to pattern recognition methods for gene prediction. What is found will suggest experimental work with cDNAs from tissue-specific libraries and targeted mutagenesis to examine gene function. Although the currently existing collections of expressed sequence tags (ESTs) boast numbers in the hundreds of thousands, the number of complete cDNAs is much smaller, and approximately 50% of the new genes discovered have no detectable homologs in the protein sequence databases. This means that pattern recognition methods will play a crucial role in elucidating the locations and significance of genes throughout the genome.

Analysis of several key issues suggests that the methodology and quality of analysis and annotation can have a dramatic effect on the utility of the sequence in further research. One important consideration is that the more comprehensive and accurate the analysis during the initial computational pass through the sequence, the less time-consuming and costly experimental work will have to be done to clarify sequence function. Relatively speaking, computational analysis is very efficient and cost effective. Furthermore, computational analysis can be used to intelligently select and design experiments where they are most needed and most useful, and it can play a significant role in directing the path of experimental investigation. Care in the analysis now could save considerable time and expense later.

The goal of the GRAIL project[1-4] has been to create a comprehensive

[1] E. C. Uberbacher and R. J. Mural, *Proc. Natl. Acad. Sci. U.S.A.* **88,** 11261 (1991).

[2] R. J. Mural, J. R. Einstein, X. Guan, R. C. Mann, and E. C. Uberbacher, *Trends Biotechnol.* **10,** 66 (1992).

[3] Y. Xu, R. J. Mural, M. Shah, and E. C. Uberbacher, *in* "Genetic Engineering: Principles and Methods" (J. Setlow, ed.), Vol. 16, p. 241. Plenum, New York, 1994.

[4] E. C. Uberbacher, J. R. Einstein, X. Guan, and R. J. Mural, *in* "Proceedings of the Second International Conference on Bioinformatics, Supercomputing and Complex Genome Analysis" (H. A. Lim, J. W. Fickett, C. R. Cantor, and R. J. Robbins, eds.), p. 465. World Scientific, Singapore, 1993.

Copyright © 1996 by Academic Press, Inc.
All rights of reproduction in any form reserved.

analysis environment where a host of questions about genes and genome structure can be answered as quickly and accurately as possible. Constructing this system has entailed solving a number of significant technical challenges including (a) making coding recognition in sequences more sensitive and accurate, (b) compensating for isochore base compositional effects in coding predictions, (c) developing methods to determine which parts of each strand of a long genomic DNA are the coding strand, (d) improving the accuracy of splice site prediction and recognition of nonconsensus sites, and (e) recognizing variable regulatory structures such as polymerase II promoters. An additional challenge has been to construct algorithms which compensate for the deleterious effects of insertion or deletion (indel) errors in the coding region recognition process. This chapter addresses the progress made with regard to these technical issues and the current state of sequence feature recognition methods.

Constructing GRAIL has also required consideration of many practical issues related to users and useful analysis procedures. We describe and consider the merits of several basic analysis and annotation paradigms available to the user within the GRAIL and genQuest[5] tool suite, ranging from totally automated processing to interactive analysis of large genomic regions, and demonstrate what biological insights can be gained or missed in the analysis of several types of DNAs.

Methods for Recognition of DNA Sequence Features

Recognition of Coding Region Candidates

The GRAIL coding region recognition system has evolved considerably since its inception in 1991, while still retaining similar design principles.[1-4] The original system used a neural network to combine a number of coding indicators calculated within a fixed sequence window. This network effectively weighted the various coding indicators on the basis of empirical training data. One advantage of this approach, compared to simple weighting mechanisms[6,7] and the more recent use of linear discriminant analysis for combining coding indicators,[8] is that a neural network containing hidden layers can discover higher order correlations and relationships in the indicator data. The use of empirical data for training allows

[5] X. Guan, R. Mural, S. Petrov, and E. C. Uberbacher, "Proceedings of Genome Sequencing and Analysis Conference V." Hilton Head Island, South Carolina, 1993.
[6] J. W. Ficket, *Nucleic Acids Res.* **17,** 5303 (1982).
[7] J. W. Ficket and C. S. Tung, *Nucleic Acids Res.* **20,** 6441 (1992).
[8] V. V. Solovyev, A. A. Salamov, and C. B. Lawrence, *Nucleic Acids Res.* **22,** 5156 (1994).

the system to utilize each indicator optimally in the presence of all the others, without *a priori* assumptions about the independence of the indicators or their relative strengths.

The power of this approach is even more evident in versions of the coding region (CR) prediction system which include not only coding indicators, but measures of splice site strength and surrounding intron character, in the evaluation of a potential CR.[3] The relationships between the input sensors are very rich, and, for example, the fact that shorter exons have stronger splice sites can and is learned by the neural network system through the training process. The neural network can also compensate for differences in coding algorithm scores due to GC composition by being provided with a measure of the composition of the exon candidate or its isochore. These types of refinements are difficult to achieve using other methods for combining information about coding regions.

The GRAIL II coding system[3] considers discrete CR candidates (with specific edge signals), rather than using a fixed size sliding window in which to evaluate coding potential. The discrete candidate approach is designed to normalize the sensitivity of the system for both long and short coding regions. As one of the input features, the network is provided with a measure of the length of the CR candidate and can therefore consider the coding measures in light of the candidate.

A schematic for the current GRAIL neural net coding recognition system is shown in Fig. 1. The input to the neural net system contains the following: (1) a 6-mer in-frame score calculated over the candidate region and adjusted by its isochore GC composition (note that in GRAIL each candidate is generated with a fixed translation frame); (2) a 6-mer in-frame score calculated over the candidate region and adjusted by the candidate GC composition; (3) a 6-mer in-frame score calculated over a 60-base region right of the candidate region and adjusted by the isochore GC composition; (4) a 6-mer in-frame score calculated over a 60-base region right of the candidate region and adjusted by the candidate GC composition; (5) a 6-mer in-frame score calculated over a 60-base region left of the candidate region and adjusted by the isochore GC composition; (6) a 6-mer in-frame score calculated over a 60-base region left of the candidate region and adjusted by the candidate GC composition; (7) a frame-dependent Markov model score calculated over the candidate region (Markov scores are calculated and interpolated using two sets of separate tables, i.e., high AT and high GC tables); (8) the isochore GC composition; (9) the candidate GC composition; (10) a score derived from a length profile for CRs (candidates with lengths corresponding to more populated parts of the CR length histogram receive higher scores than those which are out of norm); (11) the candidate length; (12) a score measuring the strength

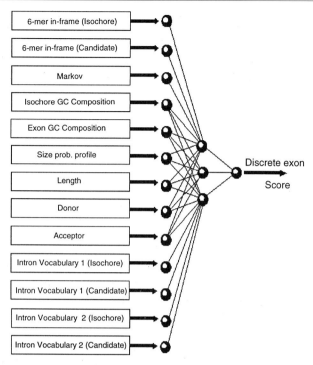

FIG. 1. Schematic of the neural network for evaluating internal protein coding exons in GRAIL.

of a donor junction candidate as determined by a separate neural net system described later; and (13) a score measuring the strength of a YAG or AA acceptor junction candidate as determined by a separate neural net system described later.

This system contains several coding indicators, but fewer than in the original system. The basic coding measures listed above include two frame-sensitive "vocabulary"-based algorithms. The 6-mer in-frame algorithm is basically identical to that in the first GRAIL version (except for updated statistical 6-mer tables), where the score of the algorithm for a candidate in a particular translation frame r, say $r = 1$, is

$$
\begin{aligned}
score = \log\left[\frac{f_1(a_1 \cdots a_6)}{f_n(a_1 \cdots a_6)}\right] + \log\left[\frac{f_2(a_2 \cdots a_7)}{f_n(a_2 \cdots a_7)}\right] \\
+ \log\left[\frac{f_0(a_3 \cdots a_8)}{f_n(a_3 \cdots a_8)}\right] + \log\left[\frac{f_1(a_4 \cdots a_9)}{f_n(a_4 \cdots a_9)}\right] + \cdots
\end{aligned}
\tag{1}
$$

where the frequencies f_i of 6-mers in frame i are calculated from a large set of coding regions, f_n is the frequency of the 6-mer in noncoding DNA, and $a_1 a_2 \ldots a_n$ represents the candidate. The 6-mer denoted by f_0, f_1, and f_2 are each calculated using separate tables. The Bayes formula is used, under the assumption of DNA being a fifth-order nonhomogeneous Markov chain, to measure coding potentials:

$$score = \frac{p_1(a_1 \cdots a_n)}{\sum_{r=0}^{2} p_r(a_1 \cdots a_n) + C p_n(a_1 \cdots a_n)} \tag{2}$$

where

$$
\begin{aligned}
p_r(a_1 \cdots a_n) &= p_r(a_1 \cdots a_5) p_r(a_6 | a_1 \cdots a_5) p_{N(r)}(a_7 | a_2 \cdots a_6) p_{N[N(r)]} \\
&\quad (a_8 | a_3 \cdots a_7) p_r(a_9 | a_4 \cdots a_8) \cdots \\
p_n(a_1 \cdots a_n) &= p_n(a_1 \cdots a_5) p_n(a_6 | a_1 \cdots a_5) p_n(a_7 | a_2 \cdots a_6) p_n \\
&\quad (a_8 | a_3 \cdots a_7) p_r(a_9 | a_4 \cdots a_8) \cdots
\end{aligned}
\tag{3}
$$

and C is the estimate of the ratio of coding versus noncoding bases in DNA, $p_r(a_1 \ldots a_5)$ and $p_n(a_1 \ldots a_5)$ are *a priori* probabilities of $a_1 \ldots a_5$ in frame r and in noncoding regions, $p_r(X|Y)$ and $p_n(X|Y)$ are the conditional probabilities of X in the presence of Y in translational frame r and noncoding region, respectively, and $N(i) = (i + 1) \bmod 3$.

As mentioned, in addition to measuring coding potential in the CR candidate, the coding potential in the neighborhood surrounding a coding region candidate is also measured (isochore vocabulary indicator). These two 60-base regions should have low coding scores as they are presumably intronic. If they do not, this score informs the neural network that the putative CR is probably larger than the current candidate (splice site may be wrong). This inclusion of scores from introns helps discriminate different edge options for a given coding region. For each putative CR, a number of candidates with different edge options and their scores are retained by the system for potential use in gene modeling (described later).

The prediction of coding regions using k-tuple methods is known to have strong dependence on isochore base composition and is more difficult in AT-rich domains. The composition-based behavior of the coding vocabulary is quite interesting. If we estimate the frequencies of frame-sensitive coding 6-mers and noncoding 6-mers in the high GC domain, and use these tables to calculate coding scores for a set of coding regions and their 60-base intronic flanking regions in all ranges of composition, an unexpected pattern results as shown in Fig. 2. The coding scores for both the coding regions and their flanks are much lower in the AT-rich domain compared to the GC-rich domain. A very similar behavior is seen if one uses the 6-mer frequency estimates from the AT-rich domain as the statistical set for

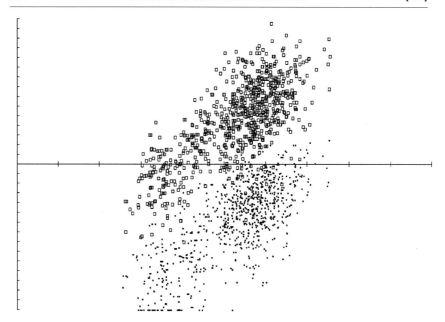

FIG. 2. The X axis represents the GC composition, and the Y axis represents the 6-mer scores. Each tick mark on the horizontal axis represents 10% in GC composition, with 0% on the left and 100% on the right. The large squares represent the coding regions, and the small dots represent the regions flanking coding regions.

the coding measure. Interestingly, though the relative separation between coding regions and their intronic flanking regions is similar at both ends of the composition range, many introns in high GC isochores have a higher coding score than many coding regions in AT-rich regions. This certainly highlights the necessity to treat compositional effects carefully.

GRAIL II corrects for the slope shown in Fig. 2 within the frame-sensitive 6-mer algorithm and in the Markov chain model. Earlier versions of GRAIL II used coding measures that were dependent on the GC composition of the candidate and its isochore. Basically the frequency for each 6-mer in the three coding frames and noncoding DNA was estimated from data sets in the high GC and high AT domains. Interpolation was used in each 6-mer instance to derive the estimate for each given candidate depending on the compositional circumstances.[4] A 2-kilobase neighborhood was used to estimate the isochore composition. In the current system, however, scores of a candidate derived from both the high GC tables and the high AT tables are presented to the network, which through training, and with the candidate and isochore composition as input, accomplishes the necessary fusion. For simplicity, this duplication of input indicators is

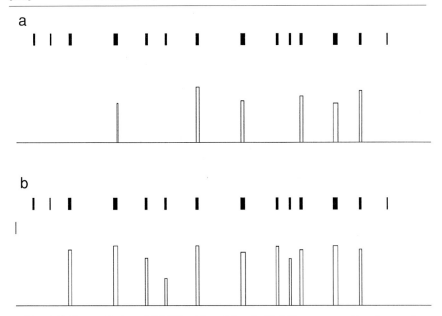

FIG. 3. Coding prediction of GRAIL without (a) and with (b) base compositional correction on the sequence HUMAFP (35% GC). The X axis represents the sequence axis, and the Y axis represents the value of neural net prediction scores. The solid bars on top represent the positions of the known exons, and the hollow rectangles are the predicted exons.

not represented in Fig. 1. A comparison of GRAIL II with and without correction for composition is shown in Fig. 3 for a gene in an AT-rich region.

Gene Modeling

The gene model construction algorithm takes as input the scored CR candidates generated by the coding region prediction neural networks and builds a single gene model in a specified region by appending a series of nonoverlapping CR candidates under the constraints that (1) the first CR candidate should start with an ATG and the last CR candidate should end with an in-frame stop codon, (2) adjacent CR candidates are translation-frame compatible, (3) no in-frame stop codons can be formed when appending two adjacent CR candidates, and (4) the distance between two adjacent candidates has to be larger than the minimum intron size. A dynamic programming algorithm[9] is used to construct a gene model that has the highest total CR scores. Certain constraints are allowed to be

[9] Y. Xu, R. Mural, and E. C. Uberbacher, *Comput. Appl. Biosci.* **10,** 613 (1994).

violated with a penalty when the dynamic programming algorithm constructs the highest scoring gene model. For example, a number of translation-frame compatibility violations are allowed in case an exon is missed or a splice site does not have the necessary consensus. This allows recovery of the remainder of the gene model.

Determining Coding Strand in Genomic DNA

Because a region of genomic DNA can have genes on both strands of the DNA, a method is needed to determine the coding strand in each portion of the sequence. We have designed a smooth filter which weighs initial coding region predictions on both strands and determines which strand is most likely to be coding at each location. The function has the form shown in Eq. (4):

$$
S(i) = \frac{100 \times \left[\sum_{j=-W}^{W} coding_f(i+j)(1 - |j|/W) + r\right]}{\sum_{j=-W}^{W} coding_f(i+j)(1 - |j|/W) + \sum_{j=-W}^{W} coding_r(i+j)(1 - |j|/W) + r} - 50 \quad (4)
$$

where $S(i)$ determines the coding strand at base i (with negative values for the reverse strand and positive values for the forward strand), $coding_f(i)$ and $coding_r(i)$ are the neural net predicted coding score at base i for the forward and reverse strands, respectively, W is one-half the measuring window size ($W = 500$), and r is a small positive number used to prevent the denominator from having a value of zero.

This function is used in the "shadow exons off" option within GRAIL which eliminates from view CR candidates that are perceived to be on the noncoding strand. This applies to CR candidates that are noncoding strand shadows of coding strand candidates and also false positives that are within but on the strand opposite an obvious gene. The function is sensitive enough to follow correctly the strand changes for genes embedded in introns. Figure 4 shows a plot of this function for a region with several genes.

Indel Error Detection and "Correction"

The goal of the GRAIL error detection system is to localize indel errors in coding regions which ordinarily may prevent recognition of a coding region and the discovery of sequence homology through database search. The problem of frameshift errors can be dealt with in several ways, and we have developed methods that recognize such errors both during

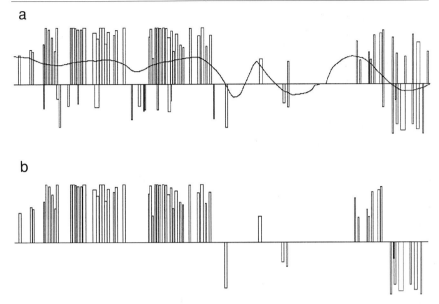

Fig. 4. Coding strand assignment. Each rectangle represents a predicted exon with width representing the size of the exon and height representing the predicted score. The curve in (a) is the strand calling function; (b) shows the result of strand calling.

the coding region recognition process[10] and during sequence comparison to the databases.[11] The former system is discussed here in more detail.

A basic property of coding recognition systems is the dependence on DNA vocabularies in the various frames. When the frame is disrupted, the recognition system becomes less sensitive and potentially fails. Few systems have considered the effects of errors on coding region prediction accuracy and noise level. GRAIL contains technology specifically for locating insertions and deletions of bases in coding regions and "correcting" these errors. This technology has proved to be very useful for single-pass EST/cDNA or genomic sequences. It can also improve the quality of assembled sequences from large genomic clones.

In GRAIL, a coding region candidate is recognized along with its preferred translation frame, the frame which has the highest coding probability. The error detection algorithm consists of two main steps. It first statistically finds points in the sequence where the preferred translation frame changes

[10] Y. Xu, R. Mural, and E. C. Uberbacher, *Comput. Appl. Biosci.* **11,** 117 (1995).
[11] X. Guan and E. C. Uberbacher, "Alignment of DNA and Protein Sequences Containing Frameshift Errors," ORNL/TM-12976 (1995).

and, second, evaluates the coding potentials on both sides of each transition point. If a transition point occurs between regions of high coding potential, the point is determined to be an indel, and one or two "C" bases are inserted to recover the frame consistency. "C" is used avoid the potential for creating stop codons on the coding strand.

To find the transition points, the algorithm divides a DNA sequence into segments in such a way that two adjacent segments have different preferred translation frames, each segment has a minimum length (to prevent short-range fluctuations), and the total coding potential along the preferred frames are maximized.

More formally, let $P_0(X)$, $P_1(X)$, and $P_2(X)$ denote the preference values* of a 6-mer X appearing in a coding region in translational frame 0, 1, and 2 versus appearing in noncoding regions, respectively. For a segment $a_j \ldots a_k$ of DNA $D = a_1 \ldots a_n$, we define

$$P_r(a_j \cdots a_k) = \sum_{i=j}^{k-5} P_{(i-r) \bmod 3}(a_i a_{i+1} \cdots a_{i+5}) \tag{5}$$

We call r the preferred reading frame of $a_j \ldots a_k$ if $P_r(a_j \ldots a_k)$ has the highest value among $P_0(a_j \ldots a_k)$, $P_1(a_j \ldots a_k)$, and $P_2(a_j \ldots a_k)$. We want to partition D into segments $D_1, D_2, \ldots, D_m$ such that the following objective function is maximized

$$\sum_{i=1}^{m} P_{r(i)}(D_i) \tag{6}$$

under the constraint that each D_i is at least K bases long and no two adjacent segments have the same preferred reading frame, where $r(i)$ denotes the preferred reading frame of segment D_i.

This problem is solved using dynamic programming[10] with $K = 30$, and we omit details here. The result of such a segmentation into preferred reading frame regions is shown in Fig. 5. The potential for each transition point to be within a coding region is evaluated using the fifth order Markov model in the GRAIL coding recognition system for 30-base regions before and after the transition point. The results of indel detection and "correction" are also shown in Fig. 5.

In applications of this technology for ESTs or cDNAs, the coding strand is assumedly known, and the method can be applied to just one strand using this option in GRAIL. In long genomic regions, genes may occur on both strands or even be embedded in introns and on the "other" strand.

---

* $P_r(X) = \log[p_r(X)/p_n(X)]$, where $p_r(X)$ is the probability of X appearing in a coding region in the translational frame r, and $p_n(X)$ is the probability of X appearing in a noncoding region.

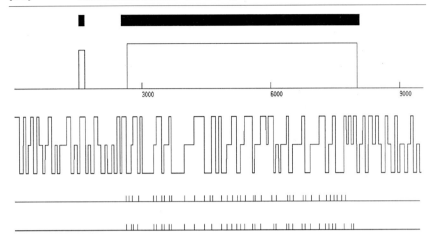

FIG. 5. Error detection. A sequence (HUMTRHYAL) is corrupted by deleting bases at the positions at the hash marks on the upper line at the bottom of the figure. The predicted indels are marked on the lower line. The center step function shows translation frame transitions. The solid bars represent actual exons, and hollow rectangles are the GRAIL predictions.

In these cases, the correction system is applied independently to each strand of the DNA after initial coding prediction, and the coding strand determination function is applied to determine the coding strand in each region. As described in the previous section on coding recognition, this function is a smooth filter that weighs initial coding region predictions on both strands to determine which strand is coding at each location as illustrated in Fig. 4. Indels that are detected on the proper strand are "corrected." Once this is done, the sequence is reprocessed for coding regions and processed for other features in the normal manner.

Table I shows a set of statistics for the performance of the error detection system.

TABLE I
STATISTICS ON INDEL ERROR DETECTION ALGORITHM

Error rate (%)	Total errors	Found errors	Av. dist.[a]
0.5	372	292 (78%)	9.7 bases
1.0	774	587 (76%)	9.4 bases
2.0	1339	856 (64%)	13.4 bases

[a] The average distance between a predicted indel and its corresponding actual indel.

Splice Junction Recognition

Recognition of donor and acceptor splice junctions remains an imprecise art, primarily due to a very significant background of nonfunctional sequences containing a splice consensus. Significant improvements in selectivity have been made in the last several years. The current GRAIL system recognizes acceptor junctions having the usual YAG consensus, as well as the nonstandard AAG consensus, with reasonable reliability. Donor splice sites containing the GT consensus are also recognized.

The basic method for locating splice junctions is based on a number of different "vocabulary"- or "word"-based measures in the neighborhood of true splice sites versus false sites (containing minimal splice consensus). There are seven measures used in the YAG acceptor neural network recognition system: (1) a 5-mer preference model calculated as a sum of log frequency ratios for positionally dependent 5-mers in true splice sites versus false sites, and covering a positional range of -23 to 0 where 0 is the "Y" in the YAG; (2) a single base log frequency ratio matrix over positions -27 to 4 (not including the AG) based on frequencies for bases in true and false splice sites; (3) a pyrimidine weighting system creates a weighted sum of the occurrences of C or T bases in positions -27 to 0, with the weight of each being the square root of the distance from position -28, and thus pyrimidines closer to the YAG have higher weight; (4) the distance to nearest 5' YAG is calculated from the position difference of the YAG under consideration and the nearest YAG in the 5' direction (the score is zero if the distance is less than 10 and linearly increases to 1.0 as the distance increases to 20 bases); (5, 6) coding potentials in adjacent regions, both 60 bases 5' and 35 bases 3' of the YAG, are used to indicate a transition between noncoding and coding sequences, with the frame-dependent 6-mer preference algorithm being used here; and (7) correlations between single bases across the site are compared using a frequency table of base–base occurrences obtained from true splice sites (the comparison is between position -27 and 4, except the consensus AG; in a given site, each base–base occurrence is given a score in the range of 0 to 1 based on its frequency in the table, and the individual scores are added to provide the overall score).

All algorithms are normalized so that their useful range falls between 0 and 1. A feed-forward neural network is trained to combine these measures based on feature vectors obtained from true splice sites and false sequence examples containing a CAG or TAG triplet. Figure 6 shows the performance statistics on an independent test set of YAG acceptor prediction system.

Acceptors with the nonstandard AAG consensus are recognized using basically the same measures, but with different statistical tables for the log

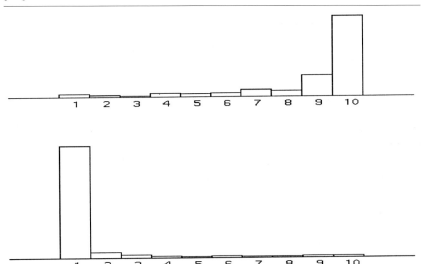

FIG. 6. YAG splice acceptor prediction. A total of 227 true (top) and 5127 false YAG acceptors (bottom) were tested. The height represents the percentage of acceptor candidates that were scored in the interval.

frequency ratio matrix, the base–base correlations, and the 5-mer prefer-ence model. In these acceptor junctions, perhaps to compensate for the poor consensus, the additional pattern elements are particularly strong. A neural network is trained to combine these measures on the basis of feature vectors from true AAG acceptors and false sequence examples containing the AAG triplet.

The donor splice junction recognition system contains four algorithms similar to some of those in the acceptor systems. A log frequency ratio matrix is used from positions −9 to 12, with position 0 being the "G" in the GT consensus. Coding potential is estimated 35 bases to the 5' side of position 0 and 60 bases to the 3' side. A base–base correlation system like that described above is used from position 2 to 33 (in the presumed intron). A neural network is trained to combine these measures on the basis of feature vectors obtained from true donor junctions with the GT consensus and false sequence examples containing a GT.

Repetitive DNA Recognition

GRAIL recognizes a number of families of complex repeats and both perfect and imperfect low-complexity repeat regions. Complex repeats are found using a combination of BLAST[12] and the Smith–Waterman algorithm[13] against a library of repetitive elements provided by Jerzy Jurka (Genetic Information Research Institute). This combination is used so that small isolated fragments of repeats can be recognized individually or potentially merged with larger repeat regions which may contain gaps and which can be effectively recognized by Smith–Waterman.

A fast linear time lookup algorithm is used to recognize both perfect and imperfect low-complexity repeats.[14] It utilizes indices calculated from noncontinuous overlapping k-tuples so that tandem repeats with insertions and deletions can be recognized. The approach is somewhat like that used in the IBM Flash sequence comparison server,[15] where a table of words is kept along with positional indices for the occurrences of words.

CpG Islands

The dinucleotide CpG is very underrepresented in the DNA sequences of vertebrates. CpG islands, regions with unusually high concentrations of CpG, are often considered to be gene markers in genomic DNA and are frequently found at the 5' end of genes.[16,17] The GRAIL system locates CpG islands as defined by Gardiner-Garden and Frommer.[18] Operationally these are regions greater than 200 bases in length which have more than 50% G+C and have a CpG content of at least 0.6 of that expected on the basis of the G+C content of the region.

Polymerase II Promoter Detection

The position of control elements relative to other parts of the gene and to each other is highly variable. A major class of RNA polymerase II promoters contain a TATAA sequence or variant approximately 30 bases upstream of the site of transcription initiation (cap site). Additional signals are often present at varying distances from the cap site. With current technology we are limited to recognizing promoters that fall into this category and cannot, for example, find promoters with initiator elements but not TATA boxes. The elements considered in GRAIL are the TATA, CAAT, and GC elements, the cap site, and the translation start site. Each

[12] S. F. Altschul, W. Gish, W. Miller, E. W. Myers, and D. J. Lipman, *J. Mol. Biol.* **215,** (1990).

[13] T. F. Smith and M. S. Waterman, *J. Mol. Biol.* **147,** 195 (1981).

[14] X. Guan and E. C. Uberbacher, "Proceedings of First Pacific Symposium on Biocomputing," pp. 718–719. World Scientific Publishing, Singapore, 1996.

[15] A. Califano and I. Rigoutsos, *CABIOS* in press.

[16] F. Antequera and A. Bird, *Proc. Natl. Acad. Sci. U.S.A.* **90,** 11995 (1993).

[17] F. Larsen, G. Gundersen, R. Lopez, and H. Prydz, *Genomics* **13,** 1095 (1992).

[18] M. Gardiner-Garden and M. Frommer, *J. Mol. Biol.* **196,** 261 (1987).

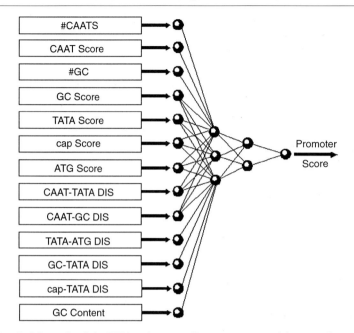

FIG. 7. Schematic of the RNA polymerase II promoter recognition neural network.

of these is initially scored using a log frequency ratio matrix, and then each potential combined group of elements is evaluated using a neural network. This network has been trained to consider the scores of the individual elements, the number of elements, and the distance relationships between elements. Although the distances between elements are variable, known promoter examples show preferred ranges of distance between many element pairs. Scores from this information are obtained from the normalized distance histograms of element pairs and used as neural net input. A schematic of the neural network system is shown in Fig. 7.

The neural network predictions of potential RNA polymerase II promoters are further refined by the application of a set of rules that examines the context of each promoter candidate relative to coding region predictions. Basically, the very high scoring candidates are retained regardless of location, while candidates with moderate scores are required to be within reasonable distance 5' of coding regions and cannot be inside or extremely close to the 3' end of a coding region. More details of the statistical matrices, neural network, and rules are described in Matis et al.[19]

[19] S. Matis, Y. Xu, M. Shah, X. Guan, J. R. Einstein, R. Mural, and E. C. Uberbacher, *Comput. Chem.* **20,** 135–140 (1996).

Recognition of Polyadenylation Sites

A log frequency ratio matrix is used to recognize potential polyadenylation sites. Because much experimental evidence exists for the participation of sequences downstream of the poly(A) site, we initially included significant flanking sequence in the matrix and then refined the matrix limits to 6 bases 5′ and 59 bases 3′ of the consensus AATAAA hexamer. Within the range of possible matrix scores, very high scores correspond to only real poly(A) sites and very low scores to only false sites. In the intermediate range a probabilistic score is provided for the element based on the Bayes formula. A more detailed description is provided in Matis *et al.*[19]

GRAIL Types: I, Ia, and II

There are three basic types of GRAIL systems: I, Ia, and II. In this chapter, our technical descriptions have focused on type II.

GRAIL I has been in place since June 1991 and uses a neural network described in Uberbacher and Mural,[1] which recognizes coding potential within a fixed-size window (100 bases). It evaluates coding potential without using additional context information such as splice junctions, etc. GRAIL Ia is an updated version of GRAIL I. It first uses a fixed-size window to locate potential coding regions and then evaluates a number of discrete candidates of different incremental lengths around each potential coding region, using information from the two 60-base regions adjacent to the coding region to find the best approximate boundaries for the coding region. The primary use of this system is for cDNAs where splice sites are not expected.

GRAIL II uses variable-size windows tailored to each potential CR candidate, defined as an open reading frame bounded by a pair of translation start/donor, acceptor/donor, acceptor/translation stop, or translation start/stop sites. This scheme facilitates the use of more genomic context information (splice junctions, translation starts, noncoding scores of 60-base regions on either side of a putative exon, candidate GC composition, and isochore) in the exon recognition process. GRAIL II is not designed for sequences without genomic context (cDNAs/ESTs).

Analysis of Genomic Sequence Using GRAIL

There are many points at which a large-scale (megabase) DNA sequencing project can benefit from computational analysis of the sequence. During the sequence determination phase, not only are computational tools critical to assembling sequences generated from a "shotgun" mode, but it is also possible to identify biologically meaningful features by computational means before the complete sequence of the region is known.[20]

The characteristics of the sequencing process itself affect decision points with respect to analysis of the sequence. In large-scale operations where assembly of cosmids or other large clones is very rapid, it makes sense to analyze the DNA sequence after the assembly process is complete. If assembly is slower or problematic, earlier analysis is useful for locating important genes and can be used to provide information that may aid in the sequence assembly process.

There are two primary ways of finding genes in DNA sequence using computational methods. Homology-based methods and feature recognition methods are complementary and need to be used together in order to assure complete analysis of a sequence.[20] Homology-based methods, for example, searching for related sequences in current database using tools like BLAST, are insufficient because it has been found that perhaps as many as 50% of newly discovered genes encode proteins which lack recognizable homologs in current databases.[21-24] Using tools like GRAIL to recognize the protein coding potential of a region of sequence prompts the investigator to design experiments to identify transcripts in order to confirm the existence of a potential new gene.

When the sequence of a genomic DNA region has been determined, an investigator can use GRAIL to identify and annotate features of biological importance in the sequence prior to submitting it to one of the public databases. GRAIL combines both the pattern-based and homology-based feature recognition. The sorts of features that can be examined (i.e., are accessible computationally) include potential protein coding regions, regulatory regions, relationships to known sequences, and the location of repetitive DNA elements.

There are two modes that can be used to annotate genomic sequence. (In an "automatic" mode, a number of computational tools examine the sequence and report their analysis (location and types of features found) to a local/public database without any need for human intervention. In an interactive mode, where the output from the various computational tools is examined by a knowledgeable individual, further analysis is carried out based on what is found. The GRAIL server can support automatic modes

[20] J.-M. Claverie, *Genomics* **23,** 575 (1994).
[21] C. Fields, M. D. Adams, O. White, and J. Craig Venter, *Nat. Genet.* **7,** 345 (1994).
[22] A. Bairoch and B. Boeckman, *Nucleic Acids Res.* **21,** 3093 (1993).
[23] D. Benson, D. J. Lipman, and J. Ostell, *Nucleic Acids Res.* **21,** 2963 (1993).
[24] H. S. Bilofsky and C. Burks, *Nucleic Acids Res.* **16,** 1861 (1988).

of analysis, and the XGRAIL graphical client–server interface can be used for interactive analysis.

GRAIL analysis is available to the user in graphic form in the X-window-based client–server system XGRAIL, through Mosaic interfaces, or by E-mail server. The E-mail version of GRAIL can be accessed at grail@ornl.gov, and the E-mail version of genQuest can be accessed at Q@ornl.gov. Instructions can be obtained by sending the word "help" to either address. The XGRAIL, Batch GRAIL, and XgenQuest client software is available by anonymous ftp from arthur.epm.ornl.gov (128.219.9.76). Both GRAIL and genQuest are accessible over the World Wide Web (URL http://avalon.epm.ornl.gov/). Communications with the GRAIL staff should be addressed to GRAILMAIL@ornl.gov.

Automatic Mode

The automatic mode of analysis is supported through a generic command line interface that allows the user to request a particular analysis of a sequence, for example, the locations and types of repetitive DNA elements, and returns an output in a standard format that can be parsed into whatever data structure the user has designed. The various types of analysis that are supported by the GRAIL command line interface are outlined below. A complete description of this interface and its uses can be found the current XGRAIL manual (available by anonymous ftp from arthur. epm.ornl.gov).

Coding. The GRAIL CR prediction neural networks provide an indication of coding region candidates on both strands of the DNA. Depending on the request, the server will return the following information: GRAIL I, predicted coding regions, standard scores, and translations of predicted coding regions; GRAIL Ia, predicted coding regions, "shadow" exons, and translations of predicted coding regions; GRAIL II, predicted coding regions, "shadow" exons, and translations of predicted coding regions. Each predicted CR has an associated translation frame called the "preferred frame" (one of the three possible frames) and the ORF (open reading frame) in that frame. Its frame is determined statistically (as described in previous section) and is virtually 100% reliable. Splice site prediction is inherent in the prediction of the candidate (GRAIL II) and is part of the normal output from the server.

Frameshift Error Detection/Compensation. The command line interface gives the user access to the frameshift error detection/compensation system described above.

Gene Modeling. The command line interface gives the user access to the gene modeling program. A table is returned with a predicted gene model for a user-specified region in the sequence.

CpG Islands. The CpG islands are calculated across the sequence and are returned as a list to the user.

Polymerase II Promoters. In response to a request to the GRAIL server, polymerase II promoter candidates are predicted across the sequence and available as a list. Each promoter prediction contains a number of possible elements and their scores including the CAAT box, GC box, TATA box, cap site, and translation start site. This system recognizes about 65% of TATA-containing promoters with a false-positive rate of about 1 per 20 kb. More detailed description is provided in the methods section.

Poly(A) Addition Sites. Poly(A) addition sites are automatically recognized using a large statistical matrix and constraints related to the presence of putative GRAIL CRs.

Repetitive DNAs. Low-complexity perfect and imperfect repeats are located throughout the sequence using a fast algorithm developed for the GRAIL project (see previous section). The list of low-complexity repeats is returned to the user. Complex repeats are found using a combination of BLAST and Smith–Waterman against a library of repetitive DNAs provided by Jerzy Jurka. A list is returned to the user.

Sequence Comparison Based on GRAIL Coding Regions. The genQuest server can be accessed through the command line interface allowing the user to automatically include a wide variety of homology-based methods in the automatic analysis of the sequence under study.

Batch GRAIL and Its Applications

For many applications the user is interested in obtaining the maximum information from a DNA sequence long before the "finished" sequence is completed. This might mean examining first-pass, shotgun sequences for the presence of coding regions or examining cDNA sequences for homologs in various databases. Batch GRAIL is designed for this purpose. It allows the user to analyze up to 1000 sequence fragments, usually first-pass sequence (300–500 bases long) from genomic or cDNA clones. The system finds potential coding regions, translates them in the preferred reading frame, and searches the translation against the SWISS-PROT database using BLAST. These results can be viewed in an XGRAIL-like environment that generates a list in which the sequences with GRAIL scores or BLAST scores greater than some user-specified threshold are highlighted. Clicking on a line in the list displays the result in a window and allows access to the database search information for the selected sequence.

Prior to assembly, the rate of errors and especially indels is likely to be considerably higher than in finished sequence. A useful option for this phase is the GRAIL error detection system that recognizes indels in coding regions and rectifies the reading frame. This options recovers about 75% of coding indels (misses those very near coding region edges) and largely restores coding region recognition processes.

A number of strategies have been described for gene discovery based on limited sequencing of genomic regions. The Batch GRAIL system is ideally suited for this kind of analysis. The following illustrates what one might expect to find in such an analysis. A 60-kb region of HSMHCAPG[25] was randomly segmented into 300 sequence fragments each 400 bases long in order to simulate a shotgun sequencing approach to gene discovery (this number of "clones" of this length represent 2× coverage). When these sequences were analyzed using Batch GRAIL, 36 fragments had high GRAIL scores, high BLAST scores, or both. This region has four genes, containing 30 exons that are annotated in GenBank. Of these 30 exons, 21 were found. For each gene, the number of exons found were as follows: PRC-7, 4 of 6; TAP-1, 6 of 7; LMP-7, 3 of 6; and TAP-2, 8 of 11. Thus, when only about 87% of the sequence has been sampled, all of the genes contained in the region have been identified.

Information from the analysis of first-pass sequence fragments can potentially be used to aid in the sequence assembly process. Most typically, if coding regions from several fragments are recognized and they have the same homologs in the protein sequence database, they are likely to belong to the same gene, and the sequence comparison can provide insight into the spatial relationship of fragments separated by gaps. We are aware, however, of no sequence assembly programs that make use of such information, and use of this information remains a manual operation for now.

Interactive Mode

The XGRAIL client–server system is designed to support fully the interactive mode of sequence analysis and annotation. This system provides a convenient interface for analyzing sequences up to 100 kb in length, provides a number of pattern-based and homology-based modes of analysis, and allows the user to generate a rich annotation report. It also allows the user to explore a number of "what if" scenarios through such means as altering the limits for assembling gene models or invoking algorithms that can detect and compensate for frameshift errors. Result of homology-based analyses are immediately available to the user through a transparent interface to the genQuest database search server.

All of the features that can be found automatically can be found and

---

[25] S. Beck, A. Kelly, E. Radley, F. Khurshid, R. P. Alderton, and J. Trowsdale, *J. Mol. Biol.* **228**, 433 (1992).

displayed in a user-friendly interface in the interactive mode. By actively interacting with the annotation process, a knowledgeable user can enhance the quality of the annotation in ways that are difficult to implement automatically. For example, in the interactive mode it is relatively easy to consider several gene models each having different constraints. This is very difficult to do in the automatic mode. A user can also pursue different levels of analysis depending on what is learned during the analysis. For example, searching motif databases can be informative if no close relatives of a predicted gene are found in various protein databases but is less useful when the gene under analysis can be placed within a well-understood gene family. Choosing or limiting various analysis paths is natural in the interactive mode.

In the interactive mode of analysis the user can pursue a number of paths to enhance the annotation of the sequence of interest. If a homolog is found when searching a database with a newly obtained sequence, the user can begin to get clues to potential gene function by using this homology information as an entry point into other rich data sources. These include a wide range of text and bibliographic data resources (MEDLINE, OMIM); mapping databases (GDB); biochemical, enzymological (Enzyme, LIGAND), and structural databases (PDB); as well as metabolic and physiological information. A skilled user can annotate hundreds of thousands of bases of sequence per day at a cost of well under U.S. $0.01 per base.

Figure 8 shows the XGRAIL interface with an analysis of a region of the sequence HSMHCAPG. The GenBank annotation of this region identifies parts of four genes. Gene modeling, and subsequent database searching, verifies that the genes are correctly identified. A number of other features that can be located and displayed in the XGRAIL system are shown in Fig. 8. These include repetitive DNA elements, shown as a light, arrowheaded box, and CpG island, represented by a box with no arrow. In the center of the figure, a gene, TAP-1, is modeled. All of the data seen on this screen, as well as a wide variety of user-supplied information, can be included in an XGRAIL generated annotation report.

Conclusion

Gaining a full understanding of the genome will, no doubt, prove to be even more challenging than initially generating its sequence. The tools and techniques we have outlined will aid in this analysis by identifying biologically relevant features in the sequence and providing some insight into their potential function. With modest effort, an investigator can greatly enrich the value of the sequence under study by including descriptions of

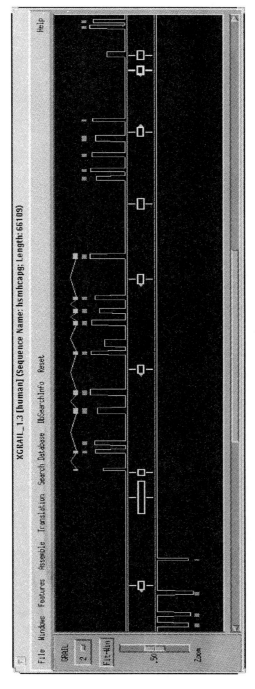

FIG. 8. An XGRAIL sequence analysis window of a portion of the sequence HSMCAPG. This window represents a region of approximately 31 kb. Bars above the upper line represent predicted coding regions on the forward strand, and those below the lower line are predicted coding regions on the complementary strand.

the genes, proteins, and regulatory regions that are present. Such analysis will provide a starting point to this most exciting phase of genome research.

Acknowledgments

This research was supported by the Office of Health and Environmental Research, United States Department of Energy, under contract DE-AC05-84OR21400 with Martin Marietta Energy Systems, Inc.

[17] Linguistic Analysis of Nucleotide Sequences: Algorithms for Pattern Recognition and Analysis of Codon Strategy

By Graziano Pesole, Marcella Attimonelli, and Cecilia Saccone

Introduction

In biology, genetic material and genomes are usually considered and treated as written texts which contain the information needed for replication and expression, as well as relevant regulatory signals. Most of the terms used in molecular biology that describe the properties of genetic material, such as transcription, translation, and editing, are actually linguistic metaphors.[1] Any written text is based on a language; knowledge of that language is required for its interpretation. In this context nucleotide sequences represent the four-letter language used by the genetic material.

Nucleotide and amino acid sequences have several features resembling modern written languages: (1) they consist of one-dimensional combinations of characters (letters for nucleotides or amino acids) with a considerable degree of repetitiveness, and (2) strings having nonrandom statistical properties may be identified. The lack of fixed ends as well as of spaces and punctuation in the text makes the definition of "word" difficult. Moreover, only a small number of the various possible combinations of the four nucleotides in a sequence can actually be found in nature. However, this number is large enough to allow given information to be expressed in multiple ways. In other words, the genetic language reflects one of the fundamental properties of the living matter: its capability to evolve.

The linguistic approach to the analysis of nucleotide sequences reveals a powerful tool for a number of purposes including: the identification of

[1] D. B. Searls, *Am. Sci.* **80**, 579 (1992).

Copyright © 1996 by Academic Press, Inc.
All rights of reproduction in any form reserved.

sequence motifs having a functional role; the establishment of functional correlations between strings; and the study of phylogenetic relationships between genetic texts (i.e., evolutionary analyses). Obviously linguistic approaches to the analysis of genetic material are numerous and differ according to the particular goal of the study. A survey of the literature in this field is not the aim of this chapter.

After a short introduction to the commonest aspects and treatments of nucleotide sequences as a language, we present two contributions of our group. The algorithm WORDUP[2] is aimed at the identification of statistically significant oligonucleotide motifs. Such a method is particularly suitable to the analysis of the huge number of sequences having unknown functions produced by automatic sequencing procedures. The algorithm CODONTREE[3] is aimed at the study of codon strategy in protein coding genes.

General Aspects of Linguistic Analysis

As in natural languages, in the biological language an alphabet may be defined by the number of elements determining its size, D. In the case of nucleic acids, the alphabet is made up of four elements (A, C, G, T; thus $D = 4$); for proteins the size (D) of the alphabet is 20, or it can be reduced if the amino acids are grouped into classes on the basis of chemical and functional features.

An alphabet is commonly used to compose words. Biological words are referred to in the literature with various terms such as strings, motifs, and patterns. For the sake of clarity, we shall try to give a novel definition of these terms in a slightly more accurate way, keeping in mind that they can always be used as synonyms and this normally does not affect linguistic analyses.

We shall call string any of the $L - w + 1$ subsequences w characters long contained in a nucleotide sequence L nucleotides long. In theory, the number of strings w nucleotides long obtainable with an alphabet having size D is equal to D^w. Hence, the occurrence probability of a string w nucleotides long should be equal to $1/D^w$ in a random text. Note also that in order to have a sequence that theoretically contains all possible D^w strings, its minimum length, L_{min}, must be $D^w + w - 1$. This implies that the significance of linguistic analysis hinges on the length of the sequence under examination. In particular, for $L \ll D^w$, it is possible to assume the

[2] G. Pesole, N. Prunella, S. Liuni, M. Attimonelli, and C. Saccone, *Nucleic Acids Res.* **20,** 2871 (1992).

[3] G. Pesole, M. Attimonelli, and S. Liuni, *Nucleic Acids Res.* **16,** 1715 (1988).

strings are Poisson distributed, given that the occurrence of any w-mer within a sequence of length L is a rare event.

The application of pattern recognition algorithms has demonstrated that not all strings have the same preference: some are overused whereas others are avoided.[4,5] We define pattern or motif as those strings whose usage is statistically nonrandom. Indeed, the statistical significance of a particular oligonucleotide string, in a given sequence collection, can be determined from the comparison of its observed and expected occurrence. The expected occurrence of a given string can be determined on the basis of probabilistic models that take into account the base composition of the sequence in which the relevant string is contained. In general, the occurrence probability of a given string is lower the longer the string length and the larger the alphabet used. Statistically significant patterns or motifs having different lengths make up the motifs vocabulary, where those patterns whose biological function is established may be classified as words. According to our definition a motif (or a pattern) will become a word when its function is clarified. Thus, only experimental studies such as site-specific mutagenesis and footprinting experiments can determine a motif to be a true word.

A nucleotide sequence of a given length can be more or less simple (or complex) depending on the richness of the vocabulary it actually uses as compared to the maximal possible vocabulary size. The elements of the vocabulary are here contiguous oligonucleotides of various sizes. The greater the number of strings used in a genetic text as compared to the total of possible strings, the more complex the genetic text. For example, if we consider a nucleotide sequence L nucleotides long, the greater the number of different oligonucleotide strings of length 1, 2, 3, ..., $L - 1$ occurring in the sequence, the higher the global complexity of the sequence. The linguistic complexity of a nucleotide sequence can be thus described according with the simple empirical formula[6] shown in Eq. (1):

$$C = \prod_{j=1}^{L-1} U_j \tag{1}$$

with

$$U_j = \left\{ \sum_{i=1}^{4^j} \frac{X_{ij}}{L - j + 1} \right\} \max\left(1, \frac{L}{4^j - j + 1} \right) \tag{2}$$

[4] V. Brendel, J. S. Beckmann, and E. N. Trifonov, *J. Biomol. Struct. Dyn.* **4**, 11 (1986).

[5] P. A. Pevzner, M. Y. Borodovsky, and A. A. Mironov, *J. Biomol. Struct. Dyn.* **6**, 1013 (1989).

[6] E. N. Trifonov, *in* "Structure and Methods: Human Genome Initiative and DNA Recombination" (R. H. Sazma and M. H. Sarme, eds.), p. 69. Adenine Press, Schenectady, New York, 1990.

where $X_{ij} = 1$ if the ith string of length j is present at least once in the sequence under examination, $X_{ij} = 0$ if it is absent. For example, if we consider the three tetramers AAAA, ACAC, and ACGT, their linguistic complexity is given by

$$C_{AAAA} = U_1 U_2 U_3 = \tfrac{1}{4} \times \tfrac{1}{3} \times \tfrac{1}{2} = 0.04$$
$$C_{ACAC} = U_1 U_2 U_3 = \tfrac{2}{4} \times \tfrac{2}{3} \times \tfrac{2}{2} = 0.33$$
$$C_{ACGT} = U_1 U_2 U_3 = \tfrac{4}{4} \times \tfrac{3}{3} \times \tfrac{2}{2} = 1.00$$

Obviously, the linguistic complexity of a genetic text also depends on its length. The global complexity of a nucleotide sequence can be also determined using various entropy measurements[7-10] where greater entropy corresponds to higher complexity.

Given that different regions of a nucleotide sequence may display different mean linguistic complexity values (e.g., coding versus noncoding regions) various methods have been developed which calculate local complexity in nucleotide and protein sequences.[11] As in natural languages, the rules according to which words are put together in a biological text set up its grammar.[1] The structural analysis of patterns allows the definition of words as well as the identification of possible long- or short-range relationships and hence the definition of biological grammar rules. Table I lists the most relevant algorithms developed for the identification of sequence patterns that could be defined as biological words. These algorithms are particularly useful for mapping functional domains in unannotated nucleotide sequences now produced in large amounts in megasequencing projects and for guiding and speeding up experimental investigations.

WORDUP: A Method to Discover Novel Motifs in Nucleic Acid Sequences

WORDUP algorithm has been specifically designed to identify statistically significant strings and motifs that are shared, or avoided, in sets of sequences functionally related but not evolutionarily homologous (e.g., promoter regions, introns, mRNA untranslated regions). The statistical significance of each string s_k ($k = 1, 4^w$) is determined by comparing, through a specific chi-square (χ^2) test, the actual number of different sequences in which s_k is present with the expected occurrences. Expectations are calculated on the basis of two assumptions: (1) oligonucleotides are Poisson

[7] G. Pesole, M. Attimonelli, and C. Saccone, *Trends Biotechnol.* **12**, 401 (1994).
[8] W. Ebeling and M. A. Jimenez-Montano, *Math. Biosci.* **52**, 53 (1980).
[9] L. L. Gatlin, *J. Theor. Biol.* **10**, 281 (1966).
[10] L. L. Gatlin, *J. Theor. Biol.* **18**, 181 (1968).
[11] P. Salamon and A. J. Konopka, *Comput. Chem.* **16**, 117 (1992).

distributed (see previous section), and (2) nucleotide sequences can be generated according to a first-order Markov chain.

The linguistic methods developed for the analysis of sequence information are mostly based on the assumption that the nucleotide sequence is a Markov chain.[9,12] In a Markov chain the nucleotide sequences can be regarded as a chain created by a "sequence generator." The generator operates according to a principle whereby the chain has finite memory h, that is, the probability of having nucleotide x at position n of string s whose $n - 1$ positions have already been generated depends exclusively on the $n - h$ immediately preceding nucleotides. This concept is represented by the probability condition in Eq. (3):

$$p(a_n | a_{n-1} \ldots a_1) = p(a_n | a_{n-1} \ldots a_{n-h}) \tag{3}$$

This rule allows us to calculate the occurrence probability for a given string according to different statistics depending on the value of h, which determines the order of the Markov chain. If $h = 0$, the string occurrence probability is based on the frequency of mononucleotides; if $h = 1$, it is based on the frequency of dinucleotides, and so on. It has been observed that a Markov chain of the first order gives a reliable description of actual nucleotide sequences that show the strongest bias at the level of dinucleotide composition.[12]

Algorithm Description

Consider a set of N nonhomologous sequences, $S_1, S_2, \ldots, S_N$, L_i nucleotides long, where $i = 1, \ldots, N$. Denote string s_k, of length w, by $a_1 a_2 \ldots a_j \ldots a_w$, where a_j is in $\{A, C, G, T\}$. If the oligomers are Poisson distributed, the probability $\pi_i(s_k)$ that s_k is found at least once in the sequence S_i is given by Eq. (4):

$$\pi_i(s_k) = 1 - e^{-\mu_{ik}} \tag{4}$$

with

$$\mu_{ik} = q_i(s_k)(L_i - w + 1) \tag{5}$$

and

$$q_i(s_k) = f(u_{1,2})f(u_{2,3}) \ldots f(u_{w-1,w})/f(u_2) \ldots f(u_{w-1}) \tag{6}$$

where μ_{ik} is the average number of the w-mer k contained in the sequence i and $q_i(s_k)$ is the occurrence probability of the oligomer s_k in the sequence

[12] H. Almagor, *J. Theor. Biol.* **104**, 633 (1983).

TABLE I

ALGORITHMS FOR SEARCHING SPECIFIC WORDS[a]

Word class	Algorithm description	Ref.
Codons	Codon preference plot for locating genes in sequenced DNA, predicting the relative level of expression, and detecting DNA sequencing errors	b
	Statistical method based on multinomial and Poisson distributions for measuring bias in codon usage table of a gene	c
	Method for measuring synonymous codon usage bias using reference set of highly expressed genes from a species, the Codon Adaptation Index	d
	GENEMARK: parallel gene recognition for both DNA strands (access genemark@ford.gatech.edu or genemark@embl.ebi.ac.uk)	e
	Statistical method for predicting protein coding regions in nucleic acid sequences	f
	GCWIND, a microcomputer (IBM-compatible) program for identification of protein-coding open reading frames	g
Splice junctions	Program GeneParser through dynamic programming (DP) is applied to problem of identifying internal exons and introns in genomic DNA sequences	h
	Using concept suggested by robotic environmental sensing, the method combines set of sensor algorithms and neural network to localize protein-coding portions of genes in anonymous DNA sequences (access[j] grail@ornl.gov)	i
	Splice-site prediction algorithm based on triplet frequencies of splice-site regions and preferences of oligonucleotides in exon and intron regions	k
	SITEVIDEO: computer system for functional site analysis and recognition; investigation of human splice sites	l
	Neural network predictions of splice sites in vertebrate genes (for GENIUS, access netserv@genius.embnet.dkfz-heidelberg.de, and for NETGENE, access netgene@virus.fki.dth.dk)	m
Palindromes, direct and inverted repeats	Text-editing algorithms that search for exact nucleotide sequence repetition and genome duplication	n
	Mobiles, modules, and motifs	o
	Method for fast database search for all k-nucleotide repeats	p
	UNIREP: microcomputer program to find unique and repetitive nucleotide sequences in genomes	q
	Method used to reconstruct the evolution of Alu's	r
Promoters	Method, of nonparametric statistical significance, selects important conserved single-base positions in combination with two-base coupling relations of identity and complementarity in DNA sequences	s
	Application of neural networks and information theory to the identification of *Escherichia coli* transcriptional promoters	t

TABLE I (*continued*)

Word class	Algorithm description	Ref.
Promoters	Application of perceptron algorithm for *E. coli* promoter searching	*u*
	Compilation and analysis of eukaryotic RNA polymerase II promoter sequences	*v*
	Rigorous analytical methods for finding unknown patterns that occur imperfectly in a set of several sequences allowing discovery of "consensus" sequences for −10 and −35 regions without any prior assumptions	*w*
Restriction sites	Dynamic programming algorithms for restriction map comparison	*x*
	Algorithm for searching restriction maps	*y*
DNA–protein interaction sites	Multialphabet consensus algorithm for identification of low specificity protein–DNA interactions	*z*
Ribosomal binding sites	Identification of ribosome binding sites in *Escherichia coli* using neural network models	*aa*

[a] Some of the algorithms are available at E-mail servers whose names and addresses are reported within parentheses.

[b] M. Gribskov, J. Devereux, and R. R. Burgess, *Nucleic Acids Res.* **12,** 539 (1984).

[c] A. D. McLachlan, R. Staden, and D. R. Boswell, *Nucleic Acids Res.* **12,** 9567 (1984).

[d] P. M. Sharp and W.-H. Li, *Nucleic Acids Res.* **15,** 1281 (1987).

[e] M. Y. Borodovsky and J. D. McIninch, *Comput. Chem.* **17,** 123 (1993).

[f] G. A. Fichant and C. Gautier, *Comput. Appl. Biosci.* **3,** 287 (1987).

[g] S. C. Schields, D. G. Higgins, and P. M. Sharp, *Comput. Appl. Biosci.* **8,** 521 (1992).

[h] E. E. Snyder and G. D. Stormo, *Nucleic Acids Res.* **21,** 607 (1993).

[i] E. C. Uberbacher and R. J. Mural, *Proc. Natl. Acad. Sci. U.S.A.* **88,** 11261 (1991).

[j] X. Guan, R. J. Mural, J. R. Einstein, R. C. Mann, and E. C. Uberbacher, *Proc. 8th IEEE Conf. AI Appl.* 9 (1992).

[k] V. V. Solovyev, A. A. Salamov, and C. B. Lawrence, *Nucleic Acids Res.* **22,** 5156 (1994).

[l] A. E. Kel, M. P. Ponomarenko, E. A. Likhachev, Y. L. Orlov, I. V. Ischenko, L. Milanesi, and N. A. Kolchanov, *Comput. Appl. Biosci.* **9,** 617 (1993).

[m] A. Milosavljevic and J. Jurka, *Mach. Learn.* **12,** 69 (1993).

[n] R. J. Nussinov, *J. Mol. Evol.* **19,** 283 (1983).

[o] P. Bork, *Curr. Opin. Struct. Biol.* **2,** 413 (1992).

[p] G. Bensoni and M. S. Waterman, *Nucleic Acids Res.* **22,** 4828 (1994).

[q] J. Mrazek and J. Kypr, *Comput. Appl. Biosci.* **9,** 355 (1993).

[r] S. Brunak, J. Engelbrecht, and S. Knudsen, *J. Mol. Biol.* **220,** 49 (1991).

[s] F. Rozkot, P. Sazelova, and L. Pivec, *Nucleic Acids Res.* **17,** 4799 (1989).

[t] K. Abremski, K. Sirotkin, and A. Lapedes, 8th International Conference on Mathematical and Computer Modeling (1991) *in* "Mathematical Modelling and Scientific Computing" (X. J. R. Arula, ed.), vol. 2, pp. 636–641. Principie Scientia, St. Louis (1993).

[u] N. N. Alexandrov and A. A. Mironov, *Nucleic Acids Res.* **18,** 1847 (1990).

[v] P. Bucher and E. N. Trifonov, *Nucleic Acids Res.* **14,** 10009 (1986).

[w] D. J. Galas, M. Eggert, and M. S. Waterman, *J. Mol. Biol.* **186,** 117 (1985).

[x] X. Huang and M. S. Waterman, *Comput. Appl. Biosci.* **8,** 511 (1992).

[y] W. Miller, J. Barr, and K. E. Rudd, *Comput. Appl. Biosci.* **7,** 447 (1991).

[z] A. V. Ulyanov and G. D. Stormo, *Nucleic Acids Res.* **23,** 1434 (1995).

[aa] D. Bisant and J. Maizel, *Nucleic Acids Res.* **23,** 1632 (1995).

i, calculated according to Stückle *et al.*[13] by assuming a first-order stationary Markov chain, with $f(u_{x,y})$ and $f(u_z)$ being the respective frequencies of dinucleotide xy and mononucleotide z ($x, y, z \in \{A, C, G, T\}$).

The observed number of occurrences of the w-mer k in the set of N sequences is given by Eq. (7):

$$P(s_k) = \Sigma_i \, p_i(s_k) \tag{7}$$

where $p_i(s_k)$ is 1 if s_k is found in the sequence i; otherwise $p_i(s_k)$ is 0. The expected number of occurrences is given by Eq. (8):

$$\Pi(s_k) = \Sigma_i \, \pi_i(s_k) \tag{8}$$

The statistical significance of the occurrences of the kth oligonucleotide can be established according to elementary χ^2 statistics:

$$\chi_k^2 = [P(s_k) - \Pi(s_k)]^2 / \Pi(s_k) \tag{9}$$

The w-mers having a χ^2 value above a suitable threshold (usually 20) form a vocabulary $V_m = \{w_j, j = 1, \ldots, d\}$ in which the d significant strings w_j are sorted according to the decreasing χ^2 value. Given that biologically significant strings can be of various lengths and also longer than the string size used for the statistical analysis reported above, we carry out the iterative procedure described in Fig. 1 to construct a new vocabulary, V_w, containing all significant patterns of length greater or equal to w.

Analysis of ATG Initiator Codon Oligonucleotide Context in Human Genes Using WORDUP Algorithm

To illustrate use of the WORDUP program we have applied the program to a set of human sequences spanning from position -20 to $+20$ with respect to the ATG initiator codon to investigate whether ATG needs to be located within specific contexts to start translation. A set of 8935 human sequences have been extracted by using the ACNUC retrieval program[14] from release 88 of the GenBank collection. This data set was reduced to 5184 entries after the removal of duplicated entries (i.e., 100% similar) by using the program CLEANUP.[14a] Indeed, the presence of redundant data makes the risk of assigning high significance to nonsignificant patterns very high.

Figure 2 shows a sample run of the program WORDUP, and a sample output is shown in Fig. 3. The "main searching file" (Fig. 3A) shows the w-mer patterns determined to be statistically significant, namely, above the

[13] E. E. Stückle, C. Emmrich, U. Grob, and P. J. Nielsen, *Nucleic Acids Res.* **18,** 6641 (1990).
[14] M. Gouy, C. Gautier, M. Attimonelli, C. Lanave, and G. Di Paola, *CABIOS* **1,** 167 (1985).
[14a] G. Grillo, M. Attimonelli, S. Liuni, and G. Pesole, *CABIOS* **12,** in press (1996).

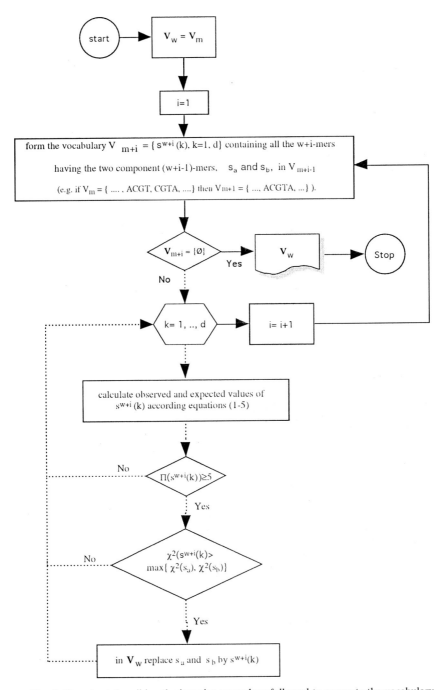

FIG. 1. Flowchart describing the iterative procedure followed to generate the vocabulary containing all statistically significant patterns of length greater or equal to *w*. Equation numbers corresond to equations in text.

```
% WORDUP

==============================================================================
                      >>>  W O R D U P  <<<
==============================================================================


Input Options
-------------
    Pattern reading from file     (Y/N)  :  N
    Dynamic pattern  computation  (Y/N)  :  Y
    Chi-square test               (Y/N)  :  Y

Input Data
----------
    Pattern length                  [6]:  6
    Chi-square threshold           [20]:  20

Input Files
-----------
    Sequence file                        :  input.seq

Output Files
------------
    Main searching file      [input.m]  :  <CR>
    Dynamic pattern file     [input.d]  :  <CR>
    Input data file          [input.i]  :  <CR>

Processing Sequences...
C.P.U. time [2.19 sec.]

Processing Patterns...
C.P.U. time [2.19 sec.]

Searching...
C.P.U. time [450.66 sec.]

Processing Dynamic Patterns...
C.P.U. time [3.24 sec.]


Execution Terminated!
```

FIG. 2. Sample run of WORDUP. Input data provided by the user are shown in bold-face type.

χ^2 fixed threshold. The "dynamic pattern file" (Fig. 3B) shows the statistically significant patterns obtained by the iterative word extension procedure described in Fig. 1.

The application of WORDUP determined a large number of statistically significant oligonucleotide patterns containing the ATG triplet (Fig. 3), thus clearly demonstrating the importance of favorable oligonucleotide contexts for initiating translation. The most significant motif turns out to be the heptamer ACCATGG (see Fig. 3B) formed by the dynamic word elongation procedure after merging of the significant hexamers ACCATG

A

```
================================================================
            W O R D U P   -   MAIN SEARCH FILE
================================================================
                        OCCURRENCES
WORDS           OBSERVED        EXPECTED        CHI-SQUARE
----------------------------------------------------------------
ACCATG          579             60.60064        4434.57210
CCATGG          745             103.03568       3999.76206
GCCATG          607             111.94065       2189.40811
CATGGA          430             77.54703        1601.90642
CACCAT          385             67.64057        1489.00284
CATGGC          462             108.49938       1151.73637
AAGATG          326             62.90758        1100.30648
CGCCAT          228             43.91259        771.71877
ATGGAG          312             78.38686        696.22760
ATGGCG          225             47.62966        660.51785
ATGGCC          292             77.11214        598.82648
AACATG          192             44.82686        483.19091
AGCCAT          219             60.90656        410.35872
ATGGAC          192             51.95716        377.46476
CAAGAT          164             41.23455        365.50312
CCATGA          268             89.13010        358.96340
.................................................................
.................................................................
================================================================
```

B

```
================================================================
            W O R D U P   -   DYNAMIC PATTERNS FILE
================================================================
                        OCCURRENCES
----------------------------------------------------------------
WORDS           OBSERVED        EXPECTED        CHI-SQUARE
----------------------------------------------------------------
[W = 6]
GCCATG          607             111.94065       2189.40820
CATGGA          430             77.54703        1601.90637
CACCAT          385             67.64057        1489.00281
CATGGC          462             108.49938       1151.73633
CGCCAT          228             43.91259        771.71875
.................................................................
.................................................................
[W=7]
ACCATGG         319             15.89421        5780.29025
CAAGATG         147             15.37392        1126.93577
CAGGATG         126             24.10293        430.77812
AGATGGC         105             18.30815        410.49901
GGAAGAT         68              12.34592        250.88268
.................................................................
.................................................................
[W=8]
CAGGACAC        37              6.65888         138.24894
[W=9]
GGCGGCGGC       40              6.63456         167.79591
CGGCGGCGG       31              5.49946         118.24404
================================================================
```

FIG. 3. Sample output of WORDUP application on the data set containing the 5184 human sequence regions spanning from position -20 to position $+20$ with respect to the ATG initiator codon. (A) Sample of the main searching file showing hexanucleotide patterns ranked according to their statistical significance. (B) Sample of the dynamic pattern file showing the vocabulary $V_{w=6}$ generated by the dynamic word extension procedure described in Fig. 1.

and CCATGG. Other oligonucleotide motifs have been found to be statistically significant (e.g., CAAGATG, GCCATG, and CAGGATG), most of them containing the translation initiator codon ATG. Despite the fact that some of the motifs fall into the GCCGCC(A/G)CCATGG initiation consensus proposed by Kozak[15] based on a survey of 699 vertebrate mRNAs, we found novel significant motifs such as CAAGATG and CAGGATG that are not referable to the above consensus. Further investigations are needed to determine if these novel signals are actually functional and if different motifs are related to specific gene classes.

A striking feature of the WORDUP algorithm is that its application does not define a consensus sequence, where it is known that not all degenerated patterns defined by a consensus have the same biological activity but, more accurately, provide a vocabulary of sequence motifs ranked according to their relative statistical significance as inferred from the χ^2 test. Indeed, the functional relatedness of the analyzed sequences (in this case the nucleotide context of the ATG starting codon) does not exclude the possibility that different motifs, for which a consensus representation is impracticable, can be found to be significant in different subsets at similar positions.

Indeed, if the above human sequence collection is searched for the canonical consensus motif described by Kozak,[15] namely, GCCGCC (A/G)CCATGG, only 17 matches are found in the 5184 sequences, of which 2 with GCCGCCACCATGG and 15 with GCCGCCGCCATGG show once more[16] that the consensus representation is inadequate and that the oligonucleotides generated from the consensus are not equivalent. The motif vocabularies constructed from collections of functionally equivalent sequences (e.g., promoters, mRNA untranslated regions, introns, ATG starting codon contexts) can be usefully used to scan the anonymous sequences that are now produced in large amounts in megasequencing projects with the aim of identifying their biological functions on the basis of significant motif matching with one or more specific vocabularies.

We have also implemented another simple program, LOCATEWORD, to locate the position of specific motifs, for example, those found statistically significant by WORDUP, within each sequence of the data set under investigation. By using LOCATEWORD, one can infer, besides the significant overrepresentation of a given sequence motif, if a motif is also located in preferential positions within the sequences.

CODONTREE: Linguistic Analysis of Codon Strategy

The classic triplet code responsible for the translation of a mRNA sequence into the amino acid sequence of a protein was the first deciphered

[15] M. Kozak, *J. Cell Biol.* **108,** 229 (1989).
[16] G. Stormo, this series, Vol. 183, p. 193.

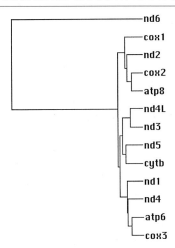

FIG. 4. Binary tree obtained by CODONTREE for the 13 genes coding for human mitochondrial DNA proteins (accession number V00662). Branch lengths are proportional to codon usage distance.

code of the genetic language. Owing to the degeneration of the genetic code, different triplets (codon families) can code for the same amino acid. As soon as the first nucleotide sequences were produced, it was immediately clear that for a given gene the usage of different synonymous codons was not random but rather species specific,[17] and also linked to its expression[18] and/or to the genetic compartment where it was located.

To investigate the linguistic properties of coding regions, that is, their codon strategy, we devised the algorithm CODONTREE that is able to determine gene relationships on the basis of codon usage similarity. Given a data set of N coding sequences, CODONTREE performs a hierarchical cluster analysis based on a codon usage similarity matrix.

The codon usage distance between sequence i and sequence j can be computed according to the χ^2 formula in Eq. (10):

$$\chi^2_{ij} = \sum_{k=1}^{61} \left[\frac{(n_i^k - E_{ij}^k)^2}{E_{ij}^k} + \frac{(n_j^k - E_{ji}^k)^2}{E_{ji}^k} \right] \tag{10}$$

where n_i^k and n_j^k are the total occurrences of codon k in sequences i and j, respectively, with

$$E_{ij}^k = (n_i^k + n_j^k) \frac{N_i^k}{(N_i^k + N_j^k)} \quad \text{and} \quad E_{ji}^k = (n_i^k + n_j^k) \frac{N_j^k}{(N_i^k + N_j^k)} \tag{11}$$

[17] R. Grantham, C. Gautier, M. Gouy, M. Jacobzone, and R. Mercier, *Nucleic Acids Res.* **9,** r42 (1981).
[18] H. Grosjean and W. Fiers, *Gene* **18,** 199 (1982).

that is, the expected values for the occurrence of codon k in sequences i and j under the null hypothesis of equivalent codon usage in both sequences weighted by the relative abundance, N_i^k and N_j^k, of the relevant amino acid in sequences i and j.

To calculate the χ^2 distance on the basis of the number of independent variables in both sequences, Eq. (10) is modified by dividing by the number of degrees of freedom, NDF_{ij}, where

$$NDF_{ij} = \sum_{h=1}^{20} [F_{ij}^h (DCF_h - 1)] \qquad (12)$$

F_{ij}^h is 1 or 0 depending on whether the hth codon family is present in both sequences and DCF_h is the number of degenerate codons in the hth family. So corrected, the distance formula to be employed in the similarity matrix for cluster analysis becomes

$$D_{ij} = \frac{\chi_{ij}^2}{NDF_{ij}} \qquad (13)$$

The result of clustering is graphically represented by a binary tree so that one or more groups of sequences using a similar codon strategy may be distinguished.

This method, besides being applicable to studying codon usage strategies in various organisms and/or different genome compartments, can be fruitfully used to construct reliable codon usage tables to be used in back-translation techniques aimed at the designing of suitable oligonucleotide probes. A simple example of the application of the CODONTREE program is given in Fig. 4, where the binary tree reporting the analysis carried out on the complete set of 13 genes coding for proteins of human mitochondrial (mt)DNA is shown. It clearly points out that the nd6 gene, the only one coded by the L strand of mtDNA, has a markedly different codon strategy with respect to the other 12 genes coded for by the H strand.

The programs WORDUP, CLEANUP, and CODONTREE are available free of charge by anonymous FTP at area.ba.cnr.it (IP 193.205.35.100). They have been written in C language and run on VMS and UNIX operating systems.

Acknowledgments

This work was partially financed by Ministero Universitá e Ricerca Scientifica, Italy and Progetto Finalizzato Ingegneria Genetica, Consiglio Nazionale delle Ricerche, Italy.

[18] Protein Sequence Comparison at Genome Scale

By Eugene V. Koonin, Roman L. Tatusov, and Kenneth E. Rudd

Introduction

Genome sequencing has become a reality. Adding to the numerous complete genome sequences of viruses and organelles,[1] the first bacterial genome sequence, *Haemophilus influenzae,* has been reported.[2] The first eukaryotic genome sequence, yeast *Saccharomyces cerevisiae,* is expected to become available in 1996. The utility of genome sequences critically depends on the availability of computer methods allowing researchers to extract the maximal possible information from them in a timely fashion. In this chapter we discuss a set of computer procedures for systematic analysis of a genome-scale set of protein sequences. The sequences of about 75% of the proteins encoded in the genome of *Escherichia coli,* arguably the best studied model organism,[3,4] were chosen for this study. A loosely organized *E. coli* genome sequencing project has resulted in several long contiguous sequences, with the longest (about 1.5 Mb) coming from the Laboratory of Genetics at the University of Wisconsin–Madison.[5,6] Therefore, the available *E. coli* sequence is representative of the entire genome rather than being a collection of the best studied genes. The requirements for an efficient computer approach and the problems arising in the analysis of this set of protein sequences are expected to be characteristic of genome-scale protein sequence analysis in general. Functional and evolutionary implications of the analysis of a somewhat smaller set of *E. coli* proteins are discussed elsewhere.[7,8]

[1] P. Bork, C. Ouzounis, and C. Sander, *Curr. Opin. Struct. Biol.* **4,** 393 (1994).

[2] R. D. Fleischmann, M. D. Adams, O. White, R. A. Clayton, E. F. Kirkness, A. R. Kerlavage, C. J. Bult, J.-F. Tomb, B. A. Dougherty, J. M. Merrick *et al., Science* **269,** 469.

[3] F. Neidhardt, J. L. Ingraham, K. B. Low, B. Magasanik, M. Schaechter, and H. E. Umbarger (eds.), *"Escherichia coli* and *Salmonella typhimurium:* Cellular and Molecular Biology." American Society for Microbiology, Washington, D.C., 1987.

[4] M. Riley, *Microbiol. Rev.* **57,** 862 (1993).

[5] D. L. Daniels, G. Plunkett, V. Burland, and F. R. Blattner, *Science* **257,** 771 (1992).

[6] V. Burland, G. Plunkett, H. J. Sofia, D. L. Daniels, and F. R. Blattner, *Nucleic Acids Res.* **23,** 2105 (1995).

[7] E. V. Koonin, R. L. Tatusov, and K. E. Rudd, *in "Escherichia coli* and *Salmonella typhimurium:* Cellular and Molecular Biology" (F. Neidhardt, R. Curtiss, III, J. L. Ingraham, E. C. C. Lin, K. B. Low, B. Magasanik, W. Reznikoff, M. Riley, M. Schaechter, and H. E. Umbarger, eds.), p. 2203, 2nd Ed., American Society for Microbiology, Washington, D.C., 1996.

[8] E. V. Koonin, R. L. Tatusov, and K. E. Rudd, *Proc. Natl. Acad. Sci. USA* **92,** 11921 (1995).

Generation of Nonredundant Set of *Escherichia coli*
  Protein Sequences

In addition to the general purpose public databases containing the *E. coli* sequences, at least four dedicated *E. coli* sequence databases are available.[9-12] The set of protein sequences analyzed in this study, EcoProt8, is a conceptual translation of EcoSeq8, a nonredundant *E. coli* DNA sequence collection.[9] Approximately 75% of the proteins encoded in the *E. coli* K12 genome are present in EcoProt8. About 45% of the protein sequences in EcoProt8 are putative products of open reading frames (ORFs) of unknown function. They are assigned provisional names beginning with the letter Y.[9,13] The data set includes many new genes identified in the unannotated regions of *E. coli* GenBank entries, as previously reported.[14,15] Some gene sequences seem to be frameshifted as revealed by comparison with homologs. These frameshifts are likely to represent sequencing errors and have been brought back into frame as accurately as possible on the basis of coding potential, as assessed using the GeneMark program,[16] and sequence alignment analysis. A review of start codons that have not been experimentally determined was done using GeneMark and inspection of putative ribosome-binding sites, and our starts may differ from those proposed by the original authors. These optimized protein sequences are present in the SWISS-PROT protein sequence database,[17] which includes cross-references to the EcoGene database (the protein-coding subset of EcoSeq).

Expectations Regarding Analysis of Genome-Scale
  Protein Sequence Set

Sequence Conservation

Analysis of the range of sequence conservation is the most straightforward and one of the most important aspects of any study of a large protein ensemble. A phylogenetic approach to sequence conservation appears to be the most logically consistent one. The availability of the majority of

[9] K. E. Rudd, *ASM News* **59,** 335 (1993).
[10] R. Wahl, P. Rice, C. M. Rice, and M. Kröger, *Nucleic Acids Res.* **22,** 3450 (1994).
[11] C. Medigue, A. Viari, A. Henaut, and A. Danchin, *Microbiol. Rev.* **57,** 623 (1993).
[12] T. Kunisawa, M. Nakamura, H. Watanabe, J. Otsuka, A. Tsugita, L. S. Yeh, D. G. George, and W. C. Barker, *Protein Sequences Data Anal.* **3,** 157 (1990).
[13] A. Stewart (ed.) "TIG Genetic Nomenclature Guide." Elsevier, Amsterdam, 1994.
[14] M. Borodovsky, E. V. Koonin, and K. E. Rudd, *Trends Biochem. Sci.* **19,** 309 (1994).
[15] M. Borodovsky, K. E. Rudd, and E. V. Koonin, *Nucleic Acids Res.* **22,** 4756 (1994).
[16] M. Borodovsky and J. McIninch, *Comput. Chem.* **17,** 123 (1993).
[17] A. Bairoch and B. Boeckmann, *Nucleic Acids Res.* **22,** 3578 (1994).

protein sequences encoded in a bacterial genome is expected to allow a robust estimate of the fraction of highly conserved proteins that contain regions with similarity to eukaryotic and/or archaeal proteins (ancient conserved regions or ACRs[18]), moderately conserved proteins that are similar to proteins from distantly related bacteria, and proteins that are unique to a single organism or to a group of closely related organisms. Another important facet of the comparative analysis is the establishment of correlations between protein sequence conservation and function.

Clusters of Paralogs: Relationships among Proteins Encoded in Same Genome

Homologous genes in the same organism whose products perform related but not identical functions are called paralogs.[19] Characterization of clusters of paralogs provides us with important information on the genetic repertoire of an organism. Of particular interest is the correlation between the number of paralogs and protein function. The existence of paralogs in E. coli has been recognized for some time.[20-22] Now that the sequence of a major part of the genome has been determined, construction of a nearly complete catalog of the paralogous clusters has become feasible.

Prediction of Protein Functions

Functional prediction is arguably the most important immediate outcome of sequence analysis. As a result of the genome-scale study of protein sequences, we obviously would like to know whether we can make experimentally testable functional predictions for a sizable number of proteins. More specifically, how many of the putative proteins that are only conceptual at this time can be assigned functions, even if defined only in general terms? Furthermore, it is important to evaluate the ability of increasingly sensitive methods for sequence analysis to boost significantly our ability to make functional predictions.

General Strategy of Analysis of Genome-Scale Protein Sequence Set

Figure 1 shows the general scheme employed in our analysis of the E. coli protein sequence set. The principal components of this strategy are

[18] P. Green, D. J. Lipman, L. Hillier, R. Waterston, D. States, and J.-M. Claverie, Science **259**, 1711 (1993).

[19] W. M. Fitch, Syst. Zool. **19**, 99 (1970).

[20] D. Zipkas and M. Riley, Proc. Natl. Acad. Sci. U.S.A. **72**, 1354 (1975).

[21] D. Zipkas, L. Solomon, and M. Riley, J. Mol. Evol. **11**, 47 (1978).

[22] B. Labedan and M. Riley, J. Bacteriol. **177**, 1585 (1995).

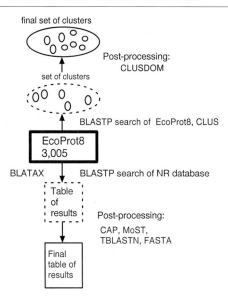

final set of clusters

Post-processing:
CLUSDOM

set of clusters

BLASTP search of EcoProt8, CLUS

EcoProt8
3,005

BLATAX   BLASTP search of NR database

Table
of
results   Post-processing:

CAP, MoST,
TBLASTN, FASTA

Final
table of
results

FIG. 1. Strategy for computer analysis of the *E. coli* protein sequences contained in the EcoSeq8 database.

(i) database searches using programs of the BLAST family, (ii) evaluation of the marginally significant alignments in the "twilight zone" using different types of motif analysis, (iii) clustering of paralogous proteins on the basis of the BLAST search results, and (iv) delineating distinct conserved domains among paralogous proteins. Table I lists the programs used to implement this scheme. All the original programs were written in C language and implemented under UNIX.

Protein Sequence Conservation

Scaling up BLAST Search

The BLA program[23] will use a file containing multiple protein sequences in the FASTA format[24] to screen a database with each sequence, one by one; a particular program of the BLAST family[25,26] is specified by the user. BLA accepts all the BLAST parameters. The BLOSUM62 amino acid

[23] R. L. Tatusov and E. V. Koonin, *Comput. Appl. Biosci.* **10,** 457 (1994).

[24] W. R. Pearson and D. J. Lipman, *Proc. Natl. Acad. Sci. U.S.A.* **85,** 2444 (1985).

[25] S. F. Altschul, W. Gish, W. Miller, E. W. Myers, and D. J. Lipman, *J. Mol. Biol.* **215,** 403 (1990).

[26] S. F. Altschul, M. S. Boguski, W. Gish, and J. C. Wootton, *Nat. Genet.* **6,** 119 (1994).

TABLE I
PROGRAMS USED FOR GENOME-SCALE ANALYSIS OF *Escherichia coli* PROTEIN SEQUENCES

Program	Purpose	Ref.
BLASTP	Similarity search of protein sequence databases with protein queries	25, 26
TBLASTN	Similarity search of conceptually translated sequence databases with protein queries	25, 26
BLA	Batchwise execution of BLAST with simultaneous search for PROSITE-type sequence patterns	23
SEG	Detection and masking of low-complexity regions in protein sequences; used as a parameter with BLAST	30–32
FASTA	Similarity search of protein sequence databases with protein queries	24, 44
CAP (Consistent Alignment Parser)	Construction of multiple alignment blocks from the BLAST output	38
MoST (Motif Search Tool)	Screening of protein sequence databases with position-dependent weight matrices constructed from alignment blocks produced by CAP	38
BLATAX (BLAST Taxonomy)	Classification of BLAST results according to taxonomic origin of sequences similar to the query; presentation of multiple BLAST outputs in table form; production of summary of multiple BLAST searches using table generate by BLATAX as the input	This chapter
CLUS	Single-linkage clustering of protein sequences on basis of exhaustive BLAST analysis of protein set	This chapter
CLUSDOM	Delineation of distinct, conserved domains using BLAST outputs for set of proteins comprising a cluster; splitting artificial clusters including multidomain proteins into single-domain clusters	This chapter

residue substitution matrix that has been shown to ensure the best combination of sensitivity and selectivity of sequence similarity detection in the BLOSUM and PAM series of matrices[27,28] is used by default. The BLAST searches are automatically combined with the search for patterns from

[27] S. Henikoff and J. Henikoff, *Proc. Natl. Acad. Sci. U.S.A.* **89,** 10915 (1992).
[28] S. Henikoff and J. Henikoff, *Proteins: Struct. Funct. Genet.* **17,** 49 (1993).

the PROSITE library[29] or another user-specified pattern library in the PROSITE format. By default, low-complexity regions that tend to produce spurious alignments in BLAST searches are masked in the query sequences using the SEG program.[30–32] A conserved pattern is defined as one that is conserved in at least 50% (or another, user-specified fraction) of the alignments with database sequences produced by BLAST for the given query sequence; the rationale for finding conserved patterns is that they are more likely to correspond to genuine functionally important sites in proteins.[23] A concise summary of the pattern search results is produced by the program. The following command line was used to perform the screening of the nonredundant (NR) protein sequence database maintained at the National Center for Biotechnology Information (NCBI):

bla ecp8 nr prosite.pat blast=blastp

Here, ecp8 is a FASTA library containing all the protein sequences from EcoProt8. The program creates directories BLAST, for the database search output files, and MOTIF, for the pattern search output files.

Phylogenetic Classification of BLAST Results

The program BLATAX (Table I) will use the output files in the BLAST directory produced by the BLA program as the input and will produce a table of "hits," classified according to the user-specified taxa, as the output. The taxonomy tree currently implemented in the GenBank database is used as the basis for this classification. To allow flexible taxonomic definitions, Boolean operations are incorporated. When analyzing the *E. coli* protein sequences, we were interested in (i) the most similar protein in the database, (ii) the most similar protein from distantly related bacteria, and (iii) the most similar protein from eukaryotes or Archaea (i.e., the most significant ACR). Accordingly, the following command was used in order to classify the BLAST results for the *E. coli* proteins:

blatax root eubacteria/proteobacteria eukaryotae+archaea fun =escherichia -i95%

Here, root indicates the best hit with any protein from the database (to eliminate the self-alignment, the -i95% option was used, disregarding alignments with greater than 95% identity); eubacteria/proteobacteria indicates the best hit with a protein from distantly related bacteria, which were

[29] A. Bairoch, *Nucleic Acids Res.* **22**, 3583 (1994).
[30] J. C. Wootton and S. Federhen, *Comput. Chem.* **17**, 149 (1993).
[31] J. C. Wootton, *Comput. Chem.* **18**, 269 (1994).
[32] J. C. Wootton and S. Federhen, this volume [33].

defined as bacteria outside the Proteobacteria domain[33]; eukaryotae + archaea indicates the most significant ACR; and fun=escherichia indicates that the apparent protein function is taken from the NR database definition line for the *E. coli* protein itself (if its function is known).

Postprocessing of Database Search Results: Evaluation of Sequence Similarities in Twilight Zone

It is unrealistic to expect in the foreseeable future that the process of evaluating, annotating and interpreting the results of the database search will be fully automatic. Manual postprocessing of the table produced by the BLATAX program using additional programs for sequence analysis is an essential component of our strategy for analyzing a genome-scale set of protein sequences. The goal of the sequence comparison is to reveal as many biologically relevant sequence similarities as possible, with a minimal admixture of false positives. Theoretically, a biologically relevant alignment most frequently includes homologous sequences, that is, sequences that share both common evolutionary origin and some degree of functional commonality. In some cases, however, an alignment may be biologically relevant and important for functional prediction but not represent true homology; examples of such similarities that perhaps may be due to functional convergence include membrane proteins and nucleotide-binding motifs.[34]

With any method for sequence analysis, it is impossible to choose a single cutoff that would discriminate between biologically relevant and spurious alignments. A twilight zone always exists that contains both relevant and irrelevant alignments. A biologically important sequence similarity is not necessarily highly statistically significant in itself. In many cases, the significance can be demonstrated by analyzing multiple sequences and delineating conserved motifs that may be presented as patterns or profiles. Motif analysis is the principal approach for verifying the significance of the twilight zone alignments; additional considerations such as the size of the aligned proteins and the position of the aligned segments in them, and the existence of known functional analogy, also have to be taken into account. As a whole, the exploration of the twilight zone, at least at this time, remains an area of expert analysis aided by various computer methods but not ready for complete automation.

Evidently, it is not possible to give a precise definition of the twilight zone; its boundaries will always remain fuzzy. The most appropriate defini-

[33] G. J. Olsen, C. R. Woese, and R. Overbeek, *J. Bacteriol.* **176,** 1 (1994).
[34] R. F. Doolittle, *Trends Biochem. Sci.* **19,** 15 (1994).

TABLE II
DATABASE SEARCH RESULTS WITH *Escherichia coli* PROTEINS
PRIOR TO POSTPROCESSING[a]

Score[b]	Best hit	Eubacteria–Proteobacteria	Eukaryotae
—	7 (0%)	383 (13%)	62 (2%)
<90	962 (32%)	1283 (43%)	2067 (69%)
>90	2041 (68%)	1344 (45%)	881 (29%)
Maximum:	6759	1922	2571
Average:	503	160	126

[a] Generated from the output of the BLATAX program.
[b] Dash indicates no hit.

tion is operational: the twilight zone consists of those alignments that require validation by additional methods after the initial database screening. Programs of the BLAST family calculate an alignment score for each of the high-scoring segment pairs (HSPs), using an appropriate substitution matrix, and the associated probability (P) that the alignment was produced by chance. In addition, the P values for multiple, compatible HSPs are calculated [$P(2)$, $P(3)$, etc.]. The P values for single HSPs are calculated on the basis of the extreme value distribution using the Karlin–Altschul statistics[35,36] and should be considered robust estimates, with significant deviations observed only for compositionally biased sequences that are mostly masked by the SEG program.[26] In contrast, $P(2)$, $P(3)$, etc., are calculated using an heuristic approach,[26] and artifactually low values are frequently observed. Therefore, the cutoff for database searches has to be defined as a score value rather than a P value. A score of 90 corresponds to the P value for an ungapped alignment of about 0.001 (for a protein consisting of ~300 amino acid residues, a typical bacterial protein size). We found that this is an appropriate threshold, below which lies the twilight zone. In retrospect, we found only two apparently spurious alignments with a score greater than 90 in our analysis of over 3000 *E. coli* proteins; these included database sequences with an obviously biased amino acid composition (E. V. Koonin, unpublished observations, 1995). The lower bound of the twilight zone is determined by the BLAST default, namely, the parameter E (expected number of HSPs) equal to 10, which approximately corresponds to a score of 30.

Table II shows the summary of the BLATAX output before postprocessing. For a significant fraction of the sequences, the best alignments are

[35] S. Karlin and S. F. Altschul, *Proc. Natl. Acad. Sci. U.S.A.* **87,** 2264 (1990).
[36] S. Karlin and S. F. Altschul, *Proc. Natl. Acad. Sci. U.S.A.* **90,** 5873 (1993).

```
          Overall Statistics
Sequences:                                3005
Sequences with BLAST hits:                3005
Sequences with motifs found:              1311
Sequences with motifs found in BLAST:     1281
Sequences with conserved motifs in BLAST:449

Total motifs: 477
     NOT conserved in any group:   266
     conserved in 1 group:         141
     conserved in several groups:   70

Finish:   Sun Jun 25 11:14:45 1995
```

FIG. 2. Summary of the PROSITE motifs detected in the *E. coli* protein sequences. In this summary file generated by the BLA program, the "Sequences with BLAST hits" line refers to the default BLASTP, without any cutoff.

in the twilight zone. Moreover, among the alignments with proteins from eukaryotes and Archaea, the majority belong to the twilight zone, making the additional analysis essential for constructing a robust list of the ACRs.

Generally, the evaluation of the twilight zone hits is based on the consistency of the alignments between the query and different sequences from the database. Consistent pairwise alignments may be combined in multiple alignment blocks. It is well known that such blocks contain more information and allow the demonstration of the statistical significance for more subtle sequence similarities than pairwise alignments.[37–39] The information contained in multiple alignment blocks is captured in different types of motifs, some of which are already known and some are new.

We used two approaches to motif analysis. The first approach includes the detection of motifs from the PROSITE library. Figure 2 shows the summary of the PROSITE motifs in *E. coli* proteins that was generated by the BLA program. Clearly, the motifs that are conserved in at least some of the alignments reported by BLAST output are most likely to be functionally relevant. Figure 3 shows an example of a conserved motif reported by BLA. In this example, the detection of the motif typical of N^6-adenine DNA methylases allows one to confidently predict this activity for the previously uncharacterized, putative protein YfiC. It has to be noticed that in some cases inspection of the alignments produced by BLAST allows one to detect motifs that are not included in PROSITE or are modifications of PROSITE signatures. The example in Fig. 4 shows the Zn protease signature in alignments of three *E. coli* proteins produced by BLASTP. Even though only one of these alignments is statistically signifi-

[37] M. Gribskov, A. D. McLachlan, and D. Eisenberg, *Proc. Natl. Acad. Sci. U.S.A.* **84**, 4355 (1987).

[38] R. L. Tatusov, S. F. Altschul, and E. V. Koonin, *Proc. Natl. Acad. Sci. U.S.A.* **91**, 12091 (1994).

[39] P. Bork and T. J. Gibson, this volume [11].

```
(#2089) >YfiC 220
  Length = 220,      12 hits

  === 1 (1) matches of (#303) [LIVMAC][LIVMFYA]1[DN]PP[FYW] -
  N-6 Adenine-specific DNA methylases signature

YfiC              115  rfdlIIsNPPY
>sp|P31825|       115  rfdlIIsNPPY
>gp|U28944|       184  lfdlIIaDPPW
>sp|P10835|       112  kfdfIVgNPPY
>sp|P39199|        42  qydlIVtNPPY
>pir|JN0797       136  kfdfVVgNPPY
>pir|S26851       105  kfdlIIgNPPY
>sp|P39200|        38  kydlIVsNPPY
```

FIG. 3. Conservation of the N^6-adenine DNA methylase motif in the BLASTP output for the putative *E. coli* protein YfiC. The multiple alignment block containing the signature (shown in uppercase letters) was automatically generated by the BLA program. The second sequence in the block is the YfiC sequence from SWISS-PROT (P31825).

cant, the conservation of the signature makes it very likely that all three proteins possess Zn-dependent protease activity. In fact, protease activity has already been demonstrated for the Lit protein.[40]

Under the second, more general approach, multiple alignment blocks are delineated in twilight zone BLAST outputs using the CAP (Consistent Alignment Parser) program.[38] The blocks are ranked by score calculated using the Dirichlet distribution mixture method.[41] A typical CAP command line is

cap aa.br aa.mot s=70 -l70%

Here, aa.br is a BLASTP output file, and aa.mot is the resulting multiple alignment block (motif) file; the parameters s and l indicate the BLASTP score cutoff and the percentage of the alignments with a score of s or greater that are required to contain a consistent segment more than eight amino acids in length for a block to be constructed. In practice, the parameters s and l have to be adjusted on a case-by-case basis, in order to derive alignment blocks that would be effective in a subsequent database search. Figure 5a shows an example of a multiple alignment block derived from the BLASTP output for the uncharacterized, putative *E. coli* protein YjgH. The consistence of different pairwise alignments revealed by CAP in itself suggests that these alignments represent a valid relationship, even if, individually, they are not statistically significant. To demonstrate the uniqueness

[40] Y.-T. N. Yu, and L. Snyder, *Proc. Natl. Acad. Sci. U.S.A.* **91,** 802 (1994).

[41] M. Brown, R. Hughey, A. Krogh, I. S. Mian, K. Sjolander, and D. Haussler, *in* "Proceedings of the First International Conference on Intelligent Systems in Molecular Biology" (L. Hunter, D. Searls, and J. Shavlik, eds.), p. 47. AAAI, Menlo Park, California, 1993.

of the motif represented by the alignment block and to search the sequence database for possible new members of the respective protein superfamily, the block may be used as the input for the MoST program.[38] Briefly, a position-dependent weight matrix is calculated using the Dirichlet mixture method and is used to scan the database. The segments that score above the cutoff, which is typically defined as the ratio of the expected number of retrieved segments to the observed number, are added to the block, the matrix is recalculated, and the search process is repeated iteratively until convergence. A typical MoST command line is

most nr aa.mot r.01 i80% >aa.mres

Here, the parameter r indicates the expected/observed ratio that is used as the cutoff, and the parameter i shows the percentage identity between sequence segments that is used to cluster them, with the segments within each cluster downweighted for the position-dependent matrix calculation, proportionally to the cluster size. As with CAP, the value of the parameter r has to be adjusted for individual alignment blocks. Figure 5b shows the MoST output for the block in Fig. 5a after the subsequent iterative database screening. The search converged after two iterations, detecting additional members of a new protein superfamily that is conserved in both bacteria and eukaryotes but still does not have a known function. The MoST analysis demonstrates that the multiple alignment block derived from the BLAST output indeed is a unique determinant of this superfamily.

Additional criteria that frequently help evaluate twilight zone alignments are the size of the query and the database sequence, and relative position of the aligned regions. When the sequences are of approximately the same size and are conserved through most of their length, this may suggest an authentic relationship, even when the statistical significance of the alignment is not very high. These considerations are particularly important for small proteins, for which the statistical significance of alignments produced by BLAST is frequently limited. Figure 6 shows an example of such an alignment between two small proteins from *E. coli* and *Bacillus subtilis* that have an identical size; in this case, the BLASTP search produced two consistent HSPs, resulting in a relatively low $P(2)$ value.

As a supplement or an alternative to CAP, tools for visualizing BLAST results in a compact form such as the MSPcrunch-BLIXEM workbench[42,43] and BLANCE[44] may be used for revealing mutual consistency between

[42] E. L. Sonnhammer and R. Durbin, *Comput. Appl. Biosci.* **10**, 301 (1994).
[43] T. L. Madden, R. L. Tatusov, and J. Zhang, this volume [9].
[44] W. R. Pearson, *Genomics* **11**, 635 (1991).

```
BLASTP 1.4.8MP [19-Dec-94] [Build 13:14:38 Apr 24 1995]

Reference: Altschul, Stephen F., Warren Gish, Webb Miller, Eugene W. Myers,
and David J. Lipman (1990).  Basic local alignment search tool.  J. Mol. Biol.
215:403-10.

Query=  yggG 252
        (252 letters)

Database:  Non-redundant PDB+SwissProt+SPupdate+PIR+GenPept+GPupdate
           147,297 sequences; 43,086,076 total letters.
Searching................................................done
```

```
                                                            Smallest
                                                              Sum
                                                   High   Probability
Sequences producing High-scoring Segment Pairs:    Score    P(N)      N

sp|P25894|YGGG_ECOLI HYPOTHETICAL 31.8 KD PROTEIN IN TKTA...  1173  2.7e-156   1

sp|P23894|HTPX_ECOLI HEAT SHOCK PROTEIN HTPX PRECURSOR. >...    75  4.9e-05    3

sp|P11072|LIT_ECOLI  BACTERIOPHAGE T4 LATE GENE EXPRESSIO...    43  0.57       3
```

```
>sp|P23894|HTPX_ECOLI HEAT SHOCK PROTEIN HTPX PRECURSOR. >pir|A43659|A43659
            heat shock protein htpX - Escherichia coli >gp|M58470|ECOHTPX_1
            htpX gene product [Escherichia coli]
            Length = 293

 Score = 75 (34.0 bits), Expect = 4.9e-05, Sum P(3) = 4.9e-05
 Identities = 14/30 (46%), Positives = 21/30 (70%)
                                     *
Query:   109 IRVYSGLMDMMTDNEVEAVIGHEMGHVALG 138
             + V +GL+  M+ +E EAVI HE+ H+A G
Sbjct:   118 VAVSTGLLQNMSPDEAEAVIAHEISHIANG 147

 Score = 40 (18.1 bits), Expect = 4.9e-05, Sum P(3) = 4.9e-05
 Identities = 7/11 (63%), Positives = 9/11 (81%)

Query:    93 VYMAKDVNAFA 103
             +Y A D+NAFA
Sbjct:    98 IYHAPDINAFA 108

 Score = 38 (17.2 bits), Expect = 4.9e-05, Sum P(3) = 4.9e-05
 Identities = 8/24 (33%), Positives = 13/24 (54%)

Query:     5 ALLVAMSVATVLTGCQNMDSNGLL 28
             A++V  +    LTG Q+   GL+
Sbjct:    13 AVMVVFGLVLSLTGIQSSSVQGLM 36


>sp|P11072|LIT_ECOLI BACTERIOPHAGE T4 LATE GENE EXPRESSION BLOCKING PROTEIN
            (GPLIT). >pir|A30386|BVECLT lit protein - Escherichia coli
            >gp|M19634|ECOLIT_1 E.coli lit gene encoding a bacteriophage T4
            late gene expression blocking protein (gplit), complete cds.
            [Escherichia coli]
            Length = 297

 Score = 43 (19.5 bits), Expect = 0.85, Sum P(3) = 0.57
 Identities = 9/22 (40%), Positives = 14/22 (63%)

Query:   175 LGEKLVNSQFSQRQEAEADDYS 196
             L  LV + FS ++E EAD ++
Sbjct:   167 LQHPLVTTAFSTQEEREADSHA 188

 Score = 40 (18.1 bits), Expect = 0.85, Sum P(3) = 0.57
 Identities = 10/30 (33%), Positives = 16/30 (53%)

Query:   209 GLATSFEKLAKLEEGRQSSMFDDHPASAER 238
             G+AT+   + LE    + + HPA+ ER
Sbjct:   209 GIATAVLCIQSLEVENYFCLQNTHPAAYER 238

 Score = 39 (17.7 bits), Expect = 0.85, Sum P(3) = 0.57
 Identities = 6/13 (46%), Positives = 9/13 (69%)
                       *
Query:   127 VIGHEMGHVALGH 139
             ++ HE+ HV L H
Sbjct:   157 ILHHEISHVVLQH 169
```

pairwise alignments produced by BLAST. These programs are particularly useful for analyzing large proteins that may have multiple domains.

The procedures outlined above partition the twilight zone proteins into two sets, those proteins for which the relevance of weak similarities was corroborated by the additional analysis and those for which the significance remained uncertain or was refuted. Table III is an excerpt of the final, edited table of the database search results with the *E. coli* proteins contained in EcoProt8. For illustration purposes, we included a range of proteins with varying degrees of sequence conservation, from one for which no information at all could be obtained (YgcJ) to those that show only low, not statistically significant similarity to other proteins, to those with limited but statistically significant conservation, to highly conserved proteins. Note that for proteins without statistically significant sequence conservation, the relevance of the included alignments was supported by detecting already known motifs (e.g., YieN, YihG, YghR) or by delineating a new motif (YjgH; see Fig. 5). Analysis of the twilight zone is the most labor-intensive but not the only component of the postprocessing of the table produced by BLATAX. Regions conserved between bacterial proteins and proteins encoded in the genomes of endosymbiotic organelles (mitochondria and chloroplasts) cannot be considered ACRs. Unfortunately, with the currently implemented taxonomy structure, organellar gene products are not easily recognized. Therefore, they had to be manually deleted or replaced by true ACRs. Also, it is clear that the "function" field had to be edited extensively as many functional descriptions in SWISS-PROT are incomplete and, for many proteins, the function is predicted anew.

Beyond the Twilight Zone

Over 600 *E. coli* proteins remained "in the dark," that is, without detectable sequence conservation, after the initial database searches and the twilight zone analysis described above. This set of proteins was subjected to further analysis using various additional computer methods to determine if they could reveal relationships that were not detected by BLASTP with SEG filtering, even as twilight zone alignments.

Effect of Low-Complexity Filtering. However important for eliminating spurious hits, low-complexity filtering may occasionally mask highly con-

---

FIG. 4. Local alignment of two *E. coli* protein sequences with a conserved metalloproteinase signature. A portion of the BLASTP output for the putative protein YggG, with the conserved signature highlighted by boldface type, is shown. A number of apparently irrelevant alignments with other proteins have been omitted.

a

```
Query= yjgH 131        (131 letters)Database:  Non-redundant
PDB+SwissProt+SPupdate+PIR+GenPept+GPupdate          147,297 sequences; 43,086,076 total
letters.Using cut-off score: 60


    Total number of segments: 8
    Limit height   6 of 15 seq
    Limit width    7
    Limit overlap  0
    CriteriON      Dirichlet mixture


    height: 7/15 (47%)  width: 36,    *216,    **1296  Score: 638
                            ....:....|....:....|....:....|....:.
    yjgH 131                48 DFQQQVRLAFDNLHATLAAAGCTFDDIIDVTSFHTD 83
    >sp|P39332|YJGH_ECOLI   48 DFQQQVRLAFDNLHATLAAAGCTFDDIIDVTSFHTD 83  S = 691,  P = 0
    >sp|P42631|YHAR_ECOLI   66 DVQDQARLSLENVKAIVVAAGLSVGDIIKMTVFITD 101 S = 81,   P = 4.3e-05
    >sp|P40431|YVN1_AZOVI   47 DFEAQTVRVFENLKAVVEAAGGSFADIVKLNIFLTD 82  S = 78,   P = 1.9e-08
    >sp|P37552|YABJ_BACSU   45 DIKEQTHQVFSNLKAVLEEAGASFETVVKATVFIAD 80  S = 71,   P = 0.0016
    >gp|U14003|ECOUW93_15   58 DVAAQARQSLDNVKAIVEAAGLKVGDIVKTTVFVKD 93  S = 61,   P = 0.14
    >sp|P39330|YJGF_ECOLI   79 DVAAQARQSLDNVKAIVEAAGLKVGDIVKTTVFVKD 79  S = 61,   P = 0.11
                            0.8 D...Q.R....N.KA...AAG....DI.K.T.F..D
                            0.6 .........FD.L...VE.....F...V...V..T.
                            0.4 .FAA.A.LALE.VH.ILA...CKVD..IDT.S.HK.
                            0.4 .VQQ.T.QS......T.....LS.G....V...I..
                            0.4 .....V..V......V......T..........V..
```

b

```
Searching command: moss ../../nr <block> r.02 i80%

moss ../../nr remost.res r.02 i80% > most.res
Method: Dirichlet mixture
Block (5x36)
   2 DVQDQARLSLENVKAIVVAAGLSVGDIIKMTVFITD      499  >sp|P42631|YHAR_ECOLI
   5 DVAAQARQSLDNVKAIVEAAGLKVGDIVKTTVFVKD      484  >gp|U14003|ECOUW93_15
   3 DFEAQTVRVFENLKAVVEAAGGSFADIVKLNIFLTD      483  >sp|P40431|YVN1_AZOVI
   4 DIKEQTHQVFSNLKAVLEEAGASFETVVKATVFIAD      472  >sp|P37552|YABJ_BACSU
   1 DFQQQVRLAFDNLHATLAAAGCTFDDIIDVTSFHTD      461  yjgH

Database:  Non-redundant PDB+SwissProt+SPupdate+PIR+GenPept+GPupdate, 5:13 AM
           EDT Jun 23, 1995
           147,297 sequences; 43,086,076 total letters.

1 iteration
Block of 5 items resulted in 9 matches - 3 new
                                    ....:....|....:....|....:....|....:.  Expected   Ratio
    >pir|S30349|S30349         12: gVAEeAkQALkNLgeILkAAGCdFtnVVKTTVlLAD  3.39e-08   4.29e-09
    >sp|P40185|YIF1_YEAST      64: sISEkAeQVFqNVKnILAEsnsS1DnIVKVNVFLAD  6.23e-07   6.89e-08
    >sp|P40037|YEO7_YEAST      48: sIADkAeQViqNiKnVLEAsnsS1DrVVKVNIFLAD  4.72e-06   4.65e-07
    >sp|P42631|YHAR_ECOLI   2  66: DVQDQARLSLENVKAIVVAAGLSVGDIIKMTVFITD         0   9.87e-28
    >sp|P39330|YJGF_ECOLI   5  44: DVAAQARQSLDNVKAIVEAAGLKVGDIVKTTVFVKD         0   3.34e-26
    >sp|P40431|YVN1_AZOVI   3  47: DFEAQTVRVFENLKAVVEAAGGSFADIVKLNIFLTD         0   3.34e-26
    >sp|P37552|YABJ_BACSU   4  45: DIKEQTHQVFSNLKAVLEEAGASFETVVKATVFIAD         0   5.64e-25
    >sp|P39332|YJGH_ECOLI   1  48: DFQQQVRLAFDNLHATLAAAGCTFDDIIDVTSFHTD         0   8.1e-24


2 iteration
Block of 8 items resulted in 10 matches - 1 new
                                    ....:....|....:....|....:....|....:.  Expected   Ratio
    >gi|699176|gp|U15181|      46: DmADQmgeVLrrIKsaLKqmGSaLTDgVhTrIhVTD  0.0676     0.0062
    >sp|P37552|YABJ_BACSU   7  45: DIKEQTHQVFSNLKAVLEEAGASFETVVKATVFIAD         0   3.92e-24
    >sp|P40185|YIF1_YEAST   2  64: SISEKAEQVFQNVKNILAESNSSLDNIVKVNVFLAD         0   1.07e-23
    >sp|P42631|YHAR_ECOLI   4  66: DVQDQARLSLENVKAIVVAAGLSVGDIIKMTVFITD         0   9.42e-24
    >sp|P39330|YJGF_ECOLI   5  44: DVAAQARQSLDNVKAIVEAAGLKVGDIVKTTVFVKD         0   7.44e-24
    >sp|P40037|YEO7_YEAST   3  48: SIADKAEQVIQNIKNVLEASNSSLDRVVKVNIFLAD         0   2.38e-23
    >sp|P40431|YVN1_AZOVI   6  47: DFEAQTVRVFENLKAVVEAAGGSFADIVKLNIFLTD         0   3.45e-23
    >sp|P39332|YJGH_ECOLI   8  48: DFQQQVRLAFDNLHATLAAAGCTFDDIIDVTSFHTD         0   2.15e-20
    >pir|S30349|S30349      1  12: GVAEEAKQALKNLGEILKAAGCDFTNVVKTTVLLAD         0   9.16e-20
```

FIG. 5. Multiple alignment blocks derived from BLAST outputs and their use for database screening. (a) Multiple alignment block derived from the BLASTP output for the putative protein YjgH using the CAP program. The first and last amino acid residue numbers are

```
BLASTP 1.4.8MP [19-Dec-94] [Build 13:14:38 Apr 24 1995]

Reference:  Altschul, Stephen F., Warren Gish, Webb Miller, Eugene W. Myers,
and David J. Lipman (1990).  Basic local alignment search tool.  J. Mol. Biol.
215:403-10.

Query=  yjbR 118
        (118 letters)

Database:  Non-redundant PDB+SwissProt+SPupdate+PIR+GenPept+GPupdate
           147,297 sequences; 43,086,076 total letters.
Searching................................................done

                                                    Smallest
                                                      Sum
                                             High  Probability
Sequences producing High-scoring Segment Pairs:  Score  P(N)      N

sp|P32699|YJBR_ECOLI HYPOTHETICAL 13.4 KD PROTEIN IN TYRB...  594  8.3e-77   1
sp|P37507|YYAQ_BACSU HYPOTHETICAL 13.9 KD PROTEIN IN COTF...   76  0.00030   2


>sp|P37507|YYAQ_BACSU HYPOTHETICAL 13.9 KD PROTEIN IN COTF-TETB INTERGENIC
           REGION. >gp|D26185|BAC180K_34 unknown [Bacillus subtilis]
           Length = 118

  Score = 76 (34.5 bits), Expect = 0.00030, Sum P(2) = 0.00030
  Identities = 11/32 (34%), Positives = 22/32 (68%)

Query:   69 PSRHLNKAHWSTVYLDGSLPDSQIYYLVDASY 100
            P  H++K HW ++ L+ + P+ +IY L++ S+
Sbjct:   83 PGYHMDKEHWISIVLERTDPEGEIYNLIEQSF 114

  Score = 31 (14.1 bits), Expect = 0.00030, Sum P(2) = 0.00030
  Identities = 6/14 (42%), Positives = 10/14 (71%)

Query:   48 VSLKTSPELAXXLR 61
            ++LK  PE++  LR
Sbjct:   63 LNLKCPPEISDRLR 76
```

FIG. 6. Local alignment between two small, identical size proteins from *E. coli* and *B. subtilis*. An excerpt of the BLASTP output for the putative protein YjbR is shown.

served, functionally important regions in proteins, thus precluding the detection of biologically relevant similarities. Therefore, we repeated the BLASTP search with the 637 protein sequences that did not show any sequence conservation in the first round of analysis. Predictably, high-scoring alignments including compositionally biased regions were observed for many of these proteins, confirming the importance of filtering in the initial search. Significant sequence similarities were detected for 10 proteins. Notably, in two of these proteins (GlpB and YhiN), the regions that have

indicated for each sequence segment as are the BLAST scores for the local alignments, from which the segments have been extracted. The program also derives a consensus amino acid residue pattern, with the frequencies of the included residues indicated in the leftmost column in the sequence alignment at bottom. (b) Result of iterative database search using a position-dependent weight matrix derived from the alignment block in (a). The two rightmost columns include the expected number of retrieved sequence segments with a given score and the ratio of the expected to the actually retrieved number of segments.

TABLE III

PHYLOGENETICALLY CLASSIFIED RESULTS OF NONREDUNDANT SEQUENCE DATABASE SCREENING WITH SEQUENCES OF *Escherichia coli* PROTEINS[a]

Protein	Size	Best hit	Score P[b]	Eubacteria–Proteobacteria	Score P	Eukaryotae + Archaea	Score P	Function[c]
YgcJ	363		—		—		—	
Lit	297	YGGG_ECOLI	43 0.83	—		—		Zn protease
YieN	440	CC46_YEAST	58 1	—		CC46_YEAST	58 1	ATPase
YihG	310	YBP2_YEAST	59 1.4e-07	—		YBP2_YEAST	59 1.4e-07	Acyltransferase
YghR	252	HUMCHSBS_1	60 0.26	S38764	50 0.30	HUMCHSBS_1	60 0.26	ATPase
YfiC	220	CELC18A3_5	75 0.045	JN0797	61 0.99	CELC18A3_5	75 0.045	Adenine-specific DNA methylase
YhjQ	242	B42465	77 0.002	CVU09867_1	76 0.042	YOU5_CAEEL	71 0.19	ATPase
YjgH	131	YVN1_AZOVI	78 1.9e-0.8	YABJ_BACSU	71 0.0016	YIF1_YEAST	51 0.79	
YgcB	888	DNU17138_2	80 0.049	—		YKB7_YEAST	66 0.61	ATPase

YhdO	481	S44955	80 / 0.022	S44955	80 / 0.022	—		Zn protease
YehL	384	SYOATPBP_2	81 / 0.02	SYOATPBP2	81 / 0.02	CC48_YEAST	69 / 0.86	ATPase
YbiF	295	YWFM_BACSU	91 / 0.00032	YWFM_BACSU	91 / 0.00032	—		Permease
YgfP	439	RERTRZA_2	94 / 7e-09	RERTRZA_2	94 / 7e-09	CELT12A2_8	59 / 0.97	Hydrolase
YrfG	237	JC2075	103 / 3.8e-07	YFGS_LACCA	50 / 0.94	YCR5_YEAST	65 / 0.69	Hydrolase
YgjR	221	S43165	115 / 2.1e-10	SPU10405_6	80 / 1.4e-05	DARAX110_1	104 / 3.4e-06	Dehydrogenase
AtoS	608	HYDH_SALTY	324 / 3.3e-41	KINC_BACSU	188 / 1.7e-34	SLN1_YEAST	89 / 4.4e-12	Sensor protein
DnaE	1160	DP3A_SALTY	5751 / 0	MC129_1	268 / 9.5e-40	—		DNA polymerase
RpoB	1342	RPOB_SALTY	6759 / 0	MSGRPOB_1	741 / 0	RPB2_HUMAN	193 / 0	RNA polymerase

[a] An excerpt of the table produced by the BLATAX program from the BLAST results and modified at the postprocessing step as described in the text is shown. Horizontal lines separate a protein for which no similarity could be detected, proteins belonging to the twilight zone, and proteins with highly significant similarities.

[b] Dash indicates no hit; $e\text{-}n = 10^{-n}$.

[c] Either a predicted function (for all the "Y" proteins) or the actually known function is indicated.

been masked included glycine-rich dinucleotide-binding motifs (P loops). Thus, even though the chance of masking a conserved region is apparently quite small, a control database search without low-complexity filtering is an important precaution.

Alternative Matrices for Amino Acid Residue Comparison. Matrices with a low BLOSUM (or PAM) number have been constructed so as to reflect distant relationships between protein sequences[27] and, supposedly, might be useful for increasing the search sensitivity. However, a database search using the BLOSUM50 matrix failed to reveal any new, significant relationships for the set of nonconserved *E. coli* proteins.

Using FASTA with k-tuple 1. The FASTA algorithm detects high-scoring local alignments with gaps.[24] It has been reported that, when used with the *k*-tuple 1, FASTA is significantly more sensitive than BLAST and approximately as sensitive as the complete Smith–Waterman algorithm.[44] Using FASTA 2.0, we obtained relevant alignments that could be confirmed by additional analysis for 26 of the *E. coli* proteins for which BLAST failed to detect any sequence similarities.

Searching Nucleotide Databases. A considerable number of genes, particularly those for which only partial sequences are available, have never been annotated in databases. Therefore, a search of the nucleotide sequence database translated in all six reading frames frequently reveals important protein sequence similarities.[12,13,45,46] An increasingly useful resource for detecting sequence conservation is the rapidly growing database of expressed sequence tags (ESTs), dBEST.[47] Screening of the nucleotide version of the NR database and the dBEST using the TBLASTN[26] program revealed significant sequence similarities for 73 *E. coli* proteins that did not show such similarities with sequences contained in protein databases. This will become increasingly important with the rapid growth of dBEST and should be considered an essential postprocessing step.

Summary of Sequence Conservation and Functional Prediction for Escherichia coli Proteins

Table IV shows the relative contribution of different methods of database screening to the detection of biologically relevant sequence conservation in *E. coli* proteins. Clearly, the majority of these similarities were identified using BLASTP with subsequent motif analysis. To reach the goal of detecting as many relationships between proteins as possible using the

[45] K. Robison, W. Gilbert, and G. M. Church, *Nat. Genet.* **7**, 205 (1994).
[46] A. Krogh, I. S. Mian, and D. Haussler, *Nucleic Acids Res.* **22**, 4768 (1994).
[47] M. S. Boguski, T. M. Lowe, and C. M. Tolstoshev, *Nat. Genet.* **4**, 332 (1993).

TABLE IV
CONTRIBUTION OF DIFFERENT DATABASE SEARCH PROCEDURES TO
DETECTION OF SEQUENCE CONSERVATION IN *Escherichia coli* PROTEINS

Method	Number of proteins with detectable similarity	% of total (3005 sequences)
BLASTP/SEG	2057	68.5
BLASTP/SEG + motif analysis	2368	78.8
BLASTP without SEG	2378	79.1
TBLASTN	2451	81.6
FASTA (k tup = 1)	2477	82.4

sequence-based approach, the other tested methods are also essential, even though relatively few additional similarities are identified.

Table V summarizes the results of the sequence similarity searches from the phylogenetic prospective. First, it is important to emphasize that, generally, *E. coli* proteins should be characterized as highly conserved (44% contain ACRs, and 60% contain regions conserved in proteins from distantly related bacteria). Altogether, almost 70% of the *E. coli* proteins contain regions conserved in proteins from phylogenetically distant organisms. Second, it is apparent that, predictably, the fraction of weak similarities increases with the phylogenetic distance. In particular, about one-third of the ACRs are in the twilight zone, and, accordingly, their detection critically depends on the methods for motif analysis and other approaches complementary to the straightforward database screening.

TABLE V
DATABASE SEARCH RESULTS WITH *Escherichia coli* PROTEINS
AFTER POSTPROCESSING[a]

Score[b]	Best hit	Eubacteria–Proteobacteria	Eukaryotae
—	500 (17%)	1202 (40%)	1713 (56%)
<90	392 (13%)	462 (15%)	468 (16%)
>90	2122 (70%)	1351 (45%)	833 (28%)
Maximum:	6759	1922	1675
Average:	421	145	83

[a] Generated from the output of the BLATAX program that was post-processed as described in the text.
[b] Dash indicates no hit.

Almost one-half of the protein sequences in the EcoProt8 database have been defined only as ORF products (with names starting with the letter Y as described above), in contrast to products of known, named genes. This gives an estimate of the number of functionally uncharacterized proteins (this estimate is only approximate as some of the Y proteins have been experimentally studied, whereas several genes have been named arbitrarily and have not been actually characterized in terms of expression and function). For the majority of these uncharacterized, putative proteins, sequence comparison allows functional predictions; altogether, there were over 650 such putative proteins with a predicted function, a sizable, 22% fraction of the *E. coli* proteins (Fig. 7). Clearly, these predictions provide ample starting material for directed experimentation. It has to be stressed, however, that in many cases the observed level of sequence conservation is sufficient only to make a general prediction, such as "a permease," "an ATPase," or "a transcription regulator"; more detailed predictions in these cases would be unreliable and should be avoided. Interestingly, some of the putative proteins, for which no prediction was possible, belong to phylogenetically highly conserved families (e.g., Fig. 5) and may represent yet unknown, essential cellular functions.

Identifying Clusters of Paralogs among
  Escherichia coli Proteins

Characterization of clusters of paralogs is an important aspect of the computer-assisted genome analysis which may provide both functional and

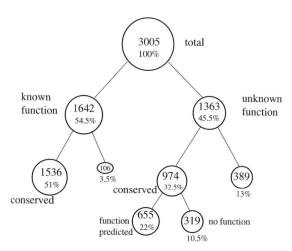

Fig. 7. The *E. coli* protein sequence set: sequence conservation and functional characterization.

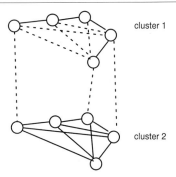

FIG. 8. Protein sequence clusters produced by the single-linkage algorithm. Solid lines indicate scores above the cutoff chosen for clustering, and dotted lines indicate lower scores. Two hypothetical clusters are shown, one that is fully connected (all sequences are connected by scores above the cutoff) and another that is only minimally connected.

evolutionary insights. Numerous methods for clustering related objects are available.[48] Because the existing methods for sequence comparison are not transitive, we choose a "greedy," single-linkage clustering procedure (a similar approach to protein clustering using a smaller set of *E. coli* proteins has also been described[49]). This method appears to be the most useful one for protein sequence clustering as it captures the existing relationships more completely than other clustering methods. The single-linkage algorithm defines a cluster within a set of protein sequences as a subset comprising a connected component of the self-comparison graph, with each edge corresponding to an alignment with a score greater than the chosen cutoff; it is not required, however, that each pair of sequences within a cluster score above the cutoff (Fig. 8). We found that when the *E. coli* protein sequence set was compared to itself using BLAST and SEG, only a very small fraction of alignments with a score greater than 70 were artifactual. Therefore, this relatively low score was chosen as the threshold for clustering, in order to characterize paralogous proteins as fully as possible. The clustering is performed using the CLUS program. A typical command line is

clus ecp8 70 >ec8.clu

By default, the program will cluster all the sequences for which it finds BLAST outputs (in the BLAST directory created by the BLA program); ecp8 is the FASTA library including all the sequences from EcoProt8. In

[48] H. C. Romesburg, "Cluster Analysis for Researchers." Krieger, Melbourne, Florida, 1989.
[49] H. Watanabe and J. Otsuka, *Comput. Appl. Biosci.* **11,** 159 (1995).

addition to the file ec8.clu containing the results of clustering, CLUS will create FASTA libraries corresponding to each cluster and a file called lonely which includes sequences not belonging to any of the clusters. Figure 9 shows a portion of the ec8.clu file; brief descriptions of each cluster were added manually.

The single-linkage clustering procedure may run into serious difficulty with clusters including multidomain proteins. When distinct domains of such a protein are homologous to sets of unrelated proteins, this results in

```
1458 entries involved in 7450 pairs with S = 70
  Score: 2619
        RHS repeat core proteins
1 (  6):                      rhsA 1377
                              rhsB 1411
                              rhsC 1397
                              rhsD 1426
                              rhsE' 681
                              yibJ 233

        Molybdopterin dehydrogenases
  Score: 244
  2 ( 10):                    bisC 739
                              dmsA 785
                              fdhF 715
                              fdnG 1016
                              fdoG 1016
                              narG 1247
                              narZ 1246
                              nuoG 820
                              torA 848
                              napA 809

        Pyruvate-formate lyases
  Score: 652
  3 (  4):                    pflA 760
                              pflB 765
                              yfiD 127
                              yhaS 746

        Permeases
  Score: 1242
  4 (  5):                    acrB 1049
                              acrD 1038
                              acrF 1034
                              ybdE' 177
                              yhiV 1037

        Fumarate dehydratases
  Score: 2585
  5 (  2):                    fumA 548
                              fumB 548
```

FIG. 9. Single-linkage clustering of *E. coli* protein sequences. A portion of the CLUS program output is shown. Indicated are the cluster number, the number of sequences in each cluster (in parentheses), and the mean score for each cluster.

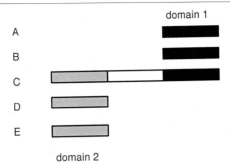

FIG. 10. Multidomain organization of proteins may result in artificial clustering. A hypothetical example of an artificial clustering of five proteins, one of which (C) contains two conserved domains, is shown. Homologous domains in the hypothetical proteins are shown by identical shading. Clearly, the cluster has to be split into two single-domain clusters, namely, {A, B, C} and {C, D, E}.

artifactual lumping of the unrelated proteins into a single cluster (Fig. 10). In fact, a cluster of homologous proteins should be more accurately defined as a set of conserved domains that contain at least one alignment block represented in all sequences. To split artifactual clusters including multidomain proteins into internally consistent, single-domain clusters, the program CLUSDOM was written. CLUSDOM will analyze the BLAST outputs for proteins belonging to a given cluster and delineate distinct, nonoverlapping conserved regions presumably corresponding to domains; a list of such apparent domains, along with a pseudographical output, is produced. A typical CLUSDOM command line is

clusdom seq low=70 >domfile

Here, seq is the name of the first sequence in the respective cluster produced by CLUS. The list of the cluster members is read directly from the CLUS output (Fig. 9); as with CLUS, the default location of the BLAST outputs is the BLAST directory. The parameter low=70 indicates that only alignments with scores of 70 or greater are included. Figure 11 shows the CLUSDOM output for a set of proteins that have been identified as a cluster by CLUS. This cluster included several aminoacyl-tRNA synthetases and an uncharacterized, putative protein YgjH. The CLUSDOM analysis showed that this cluster is in fact artificial and should be split into two single-domain clusters, one of them corresponding to the aminoacyl-tRNA synthetase catalytic domain and the other one to a new, small domain with an unknown function that comprises the whole YgjH sequence and the N- or C-terminal regions of two aminoacyl-tRNA synthetases.

Postprocessing of the paralogous cluster table produced by CLUS in-

```
IleS 939
5 sequences (10 segments)
                                                       length: 939
ooooooooooooooooooooooooooooooooooooooooooooooooooooo    >IleS
--oooooooooo------ooooooooooo-----ooo----------------    >ValS
---oooo-----------ooo--------------------------------    >LeuS
---oooo----------------------------------------------    >MetG
-----------------------------oooooo------------------    >CysS

MetG 680
6 sequences (7 segments)
                                                       length: 680
ooooooooooooooooooooooooooooooooooooooooooooooooooooo    >MetG
-ooooo-----------------------------------------------    >IleS
ooooo------------------------------------------------    >LeuS
-ooooooo-----------------oooo-------------------------    >CysS
-------------------------------------------ooooo----    >YgjH
-ooo-------------------------------------------------    >ValS

PheT 795
2 sequences (2 segments)
                                                       length: 795
ooooooooooooooooooooooooooooooooooooooooooooooooooooo    >PheT
---ooo-----------------------------------------------    >YgjH

YgjH 110
3 sequences (3 segments)
                                                       length: 110
ooooooooooooooooooooooooooooooooooooooooooooooooooooo    >ygjH
-------ooooooooooooooooooo----------------------------   >pheT
ooooooooooooooooooooooooooooo-------------------------   >metG

Total  2 domains
       5   LeuS 860
           IleS 939
           MetG 680
           ValS 951
           CysS 461

       3   YgjH 110
           PheT 795
           MetG 680
```

FIG. 11. Delineation of distinct conserved domains in an artificial protein cluster including multidomain proteins. A portion of the CLUSDOM program output for a cluster of seven proteins, six of which are aminoacyl-tRNA synthetases of different specificity (except for the putative protein YgjH for which no function has been reported), is shown. The program produces a schematic map of local alignments with a score above the cutoff (in this case, a score of 70) for each sequence in the cluster (four such maps are shown) and then combines them to delineate distinct conserved domains.

cluded resolution of the artificial clusters including multidomain proteins as well as elimination of the few artifactual links due to spurious alignment with scores above the cutoff. The initial list produced by CLUS consisted of 339 clusters including 1458 proteins. The final list included 1419 proteins (39 proteins apparently were initially included in clusters based on unreliable alignments and were removed at the postprocessing stage) belonging to 361 clusters. The increase in the number of clusters resulting from splitting the artificial groups is relatively small, as most of the paralogous clusters include only 2 to 4 members (Fig. 12). However, certain functional categories of proteins, notably those involved in metabolite transport and transcription regulation, tend to form large clusters.

Future Directions: Different Levels of Genome Analysis

The combination of several stand-alone programs for protein sequences analysis described allows one to perform a relatively in-depth analysis of a genome-scale protein sequence set in a timely fashion. The analysis of more than 3000 *E. coli* proteins described in this chapter took about 3 weeks of full-time work of one researcher. There is no doubt that integration of these and additional programs in a single, interactive workbench linked to standard and special-purpose databases will provide for further acceleration and improvement of genome-scale protein sequence analysis. The hypertext approach opens previously unimaginable possibilities for linking different data structures. GeneQuiz, a prototype workbench for large-scale protein analysis implemented as a relational database, has been described.[50]

Genome sequences may be analyzed at several different levels. Primary genome sequencing papers typically include only the first level of analysis, with strong similarities to database sequences, PROSITE motifs (usually without a critical evaluation of their relevance), and most obvious functional predictions being reported. The approach described here represents a second level of analysis, which involves the revaluation of relatively weak sequence similarities using various computer methods, phylogenetic classification of the conserved regions, and delineation of clusters of paralogous proteins. Clearly, a higher level, more sophisticated analysis is both feasible and necessary, even if its automatic implementation presents a challenge. This next-step analysis, currently underway in our group with the *E. coli* protein sequence set, includes derivation of signature motifs for each conserved protein and the use of these motifs to delineate superclusters combining the original clusters of paralogs with some of the "lonely" sequences; analysis of the chromosomal locations of paralogous genes that may allow us to detect large-scale duplications; and, further, a more detailed exploration of the correlations between protein function, sequence conservation, and the tendency to form clusters of paralogs. A further significant increase in the sensitivity of detection of relationships between proteins may be achieved by comparing three-dimensional structures that are becoming

[50] M. Scharf, R. Schneider, G. Casari, P. Bork, A. Valencia, C. Ouzounis, and C. Sander, *in* "Proceedings of the First International Conference on Intelligent Systems in Molecular Biology," p. 348. San Francisco, 1994.

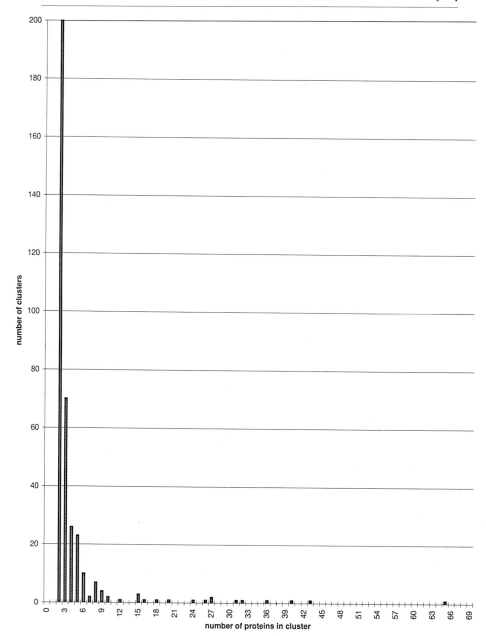

FIG. 12. Size distribution of the clusters of paralogs among *E. coli* proteins.

increasingly available[51] and by "threading" sequences through databases of known tertiary structures or structural motifs.[52] Implementing these methods at a genome scale is a major goal for the near future. An extremely important direction that will require the development of a new group of methods is the comparative analysis of protein sets encoded in several completely sequenced genomes which should become available soon. We believe that as these studies proceed, new questions that are not yet foreseen will emerge.

Summary

An adequate set of computer procedures tailored to address the task of genome-scale analysis of protein sequences will greatly increase the beneficial impact of the genome sequencing projects on the progress of biological research. This is especially pertinent given the fact that, for model organisms, one-half or more of the putative gene products have not been functionally characterized. Here we described several programs that may comprise the core of such a set and their application to the analysis of about 3000 proteins comprising 75% of the *E. coli* gene products. We find that the protein sequences encoded in this model genome are a rich source of information, with biologically relevant similarities detected for more than 80% of them. In the majority of cases, these similarities become evident directly from the results of BLAST searches. However, methods for motif analysis provide for a significant increase in search sensitivity and are particularly important for the detection of ancient conserved regions. As a result of sequence similarity analysis, generalized functional predictions can be made for the majority of uncharacterized ORF products, allowing efficient focusing of experimental effort. Clustering of the *E. coli* proteins on the basis of sequence similarity shows that almost one-half of the bacterial proteins have at least one paralog and that the likelihood that a protein belongs to a small or a large cluster depends on the function of this particular protein.

Availability of Programs and Data

The source code and executable commands for the programs described in this chapter are available via anonymous ftp at ncbi.nlm.nih.gov from

---

[51] C. Orengo, *Curr. Opin. Struct. Biol.* **4**, 429 (1994).
[52] M. S. Johnson, N. Srinivasan, R. Sowdhamini, and T. L. Blundell, *Crit. Rev. Biochem. Mol. Biol.* **29**, 1 (1994).

directories pub/bla (BLA, BLATAX, CLUS, and CLUSDOM) and pub/most (CAP and MoST). The EcoSeq8 protein sequence set, the processed table of phylogenetically classified database search results, and the table of paralogous protein clusters are available in the repository/Eco directory.

Acknowledgments

We are grateful to Vladimir Soussov for generating a taxonomy table that is used by the BLATAX program and to Tatiana Tatusov for writing procedures for database access. We thank Peer Bork and David Lipman for helpful discussions.

[19] Iterative Template Refinement: Protein-Fold Prediction Using Iterative Search and Hybrid Sequence/Structure Templates

By Tau-Mu Yi and Eric S. Lander

Introduction

Rather than predicting the detailed, atomic resolution structure of a protein from its sequence using computer-intensive simulations, many researchers have turned their attention to the less daunting task of predicting the general structural geometry or fold of a protein. This approach is motivated by the popular assumption that the universe of protein folds is limited.[1] The challenge of fold recognition is to encode sufficient information about the structure into a template so that sequences adopting that fold can be distinguished from proteins with a different topology. Individual fold templates can be used to search the sequence database for compatible sequences, or a given sequence can be aligned against a library of templates.

Roughly speaking, one can divide the approaches to fold prediction into sequence-based methods and structure-based methods. For years scientists have employed sequence searching techniques to identify structurally related proteins,[2] and the sensitivity of these methods has been enhanced by exploiting the information in templates containing multiple aligned se-

---

[1] C. Chothia, *Nature (London)* **357,** 543 (1992).
[2] W. R. Pearson and D. J. Lipman, *Proc. Natl. Acad. Sci. U.S.A.* **85,** 2444 (1988).

Copyright © 1996 by Academic Press, Inc.
All rights of reproduction in any form reserved.

quences or consensus sequence motifs.[3-8] The structure-based methods can be further divided into two camps: those that incorporate structural information into a local structure–environment template and those that encode structural information into a residue–residue contact potential. Bowie et al.[9] have pioneered the former strategy by defining a set of 18 structural environment classes based on secondary structure, solvent accessibility, and polarity. Jones et al.,[10] Sippl and Weitckus,[11] and Bryant and Lawrence[12] were among the first groups to adopt the latter approach of transforming the structural data into a contact potential that measured the propensity for two residues to be in close proximity. Both strategies have achieved excellent results.

We have developed an automated procedure for fold prediction that combines sequence-based and structure-based (local environment) information into a set of dynamic templates. The technique, termed iterative template refinement (ITR), employs an iterative search scheme that detects related proteins and sequentially adds them to the template. In this fashion, the initial seed template, constructed from a protein of known structure, gives rise to a tree of descendent templates containing both a structure–environment component as well as an expanding multiple sequence component.

Below, we outline the overall design and implementation of ITR. The description of the method is divided into five subtopics: (1) template, (2) profile, (3) database search, (4) test of significance, and (5) iteration. Particular attention is paid to the measures adopted to improve computational efficiency and to the selection of several important parameters including the gap penalty, sequence/structure scoring ratio, and the threshold for significance. Finally, we evaluate the performance of ITR in terms of both the correct identification of target proteins structurally related to the seed protein and the accuracy of the resulting alignments. In particular, we review the results from the application of ITR to six proteins: arabinose-binding protein (ABP), plastocyanin, cytochrome c, chymotrypsin, lactate

[3] W. R. Taylor, J. Mol. Biol. 188, 233 (1986).
[4] D. Bashford, C. Chothia, and A. M. Lesk, J. Mol. Biol. 196, 199 (1987).
[5] M. Gribskov, A. D. McLachlan, and D. Eisenberg, Proc. Natl. Acad. Sci. U.S.A. 84, 4355 (1987).
[6] G. J. Barton and M. J. E. Sternberg, J. Mol. Biol. 212, 389 (1991).
[7] S. Henikoff and J. G. Henikoff, Nucleic Acids Res. 19, 6565 (1991).
[8] A. F. Neuwald and P. Green, J. Mol. Biol. 239, 698 (1994).
[9] J. U. Bowie, R. Luthy, and D. Eisenberg, Science 253, 164 (1991).
[10] D. T. Jones, W. R. Taylor, and J. M. Thornton, Nature (London) 358, 86 (1992).
[11] M. J. Sippl and S. Weitckus, Proteins 13, 258 (1992).
[12] S. H. Bryant and C. E. Lawrence, Proteins 16, 92 (1993).

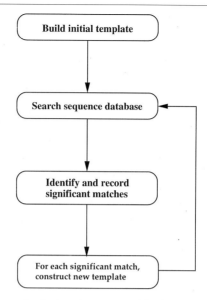

Fig. 1. Flowchart illustrating the iterative process of database search and template construction. (Adapted from Yi and Lander.[13])

dehydrogenase (Rossmann domain), and the α subunit of tryptophan synthase.

Overview of Method

The ITR approach consists of repeated cycles of database search and new template generation (see Fig. 1[13]). At the start, a single seed protein of known structure is converted to a Level 1 template comprised of a sequence component and a structure–environment component. During the search phase, the template is compared against all sequences in the database, and target sequences that register a significant alignment score are recorded. During the new template generation phase, the target sequences representing significant hits are added individually to the sequence component of the template, generating a set of Level 2 templates. These new templates are then used to start the next (second) cycle. In theory, the process can be repeated indefinitely; in practice, the search is terminated after the fifth cycle, and a list of significant hits is compiled.

[13] T.-M. Yi and E. S. Lander, *Protein Sci.* **3,** 1315 (1994).

Template

The ITR templates are distinctive because they contain both a sequence component and a structure-derived component, and because they can be modified by the addition of new sequences. The sequence component of each template is dynamic, growing through the alignment of sequences with significant similarity scores; the structure component, on the other hand, remains fixed. More formally, we define a Level k template rooted at P_1 to be the structure–environment string derived from the seed protein P_1 together with a multiple subsequence alignment of P_1 and $(k - 1)$ other sequences $P_2, \ldots, P_k$. A target sequence possessing a significant matching score is aligned to the sequence portion of the template, creating a Level $(k + 1)$ template (see Fig. 2[14]). The structure component consists of a local environment string constructed by assigning each position in the structure to one of 15 structure–environment classes defined in Yi and Lander[14] using the Bowie–Eisenberg methodology.[9]

Profile

A profile is a $L \times 20$ matrix (L is the profile length) in which the entry $PRF(j, x)$ specifies the score for matching the jth position in the profile with amino acid x. The sequence and structure components of each template were combined into a single profile using an extension of the procedure developed by Gribskov and colleagues.[5] First, separate profiles for the sequence component (SEQ_PRF) and the structure component (STR_PRF) were constructed:

$$SEQ\_PRF(j, x) = \sum_{i=1}^{k} A(a_{ij}, x) \tag{1a}$$

$$STR\_PRF(j, x) = B(e_j, x) \tag{1b}$$

where A denotes the amino acid by amino acid scoring matrix calculated by Benner and colleagues,[15] and B denotes an environment class by amino acid scoring matrix as originally specified by Yi and Lander.[14] The variable a_{ij} represents the amino acid at position j in the protein P_i, and e_j represents the environment symbol at the jth position in protein P_1.

To combine the two profiles in the appropriate proportion, it was necessary to calculate the magnitude of each profile. We defined the magnitude to be the sum of the positive terms in the profile divided by the profile length (i.e., the average sum of the positive terms in each row):

[14] T.-M. Yi and E. S. Lander, *J. Mol. Biol.* **232**, 117 (1993).
[15] G. H. Gonnet, M. A. Cohen, and S. A. Benner, *Science* **256**, 1443 (1992).

Level 3 Template

```
1                                                                    80
DAA555BCDE BDB1400232 1340244CCA EBC978ACE2 -440241034 024DEADAAA ABACCDECAE 201421443E    3D structure-environment string (protein 1)

KLGFLVKQPE EPWFQTEWKF ADKAGKDLGF EVIKIAVPDG -EKTLNAIDS LAASGAKGFV ICTPDPKLGS AIVAKARGYD    sequence (protein 1)
TIALVVSTLN NPFFVSLKDG AQKEADKLGY NLVVLDSQNN PAKELANVQD LTVRGTKILL INPTDSDAVG NAVKMANQAN    sequence (protein 2)
TILVIVPDIC DFFFSEIIRG IEVTAANHGY LVLIGDCAHQ NQQEKTFIDL IITKQIDGML L--LGSRLPF DA-SIEEQRN    sequence (protein 3)
```

```
...AIGLVTPEND VPFNSGVFMD MVSCISRELA YHDIDLLLIA DDEHADCHSY MRLVESRRID ALIIAHTLDD DPRITH...    target sequence
```

Level 4 Template

```
1                                                                    80
DAA555BCDE BDB----140 0232134024 4CCAEBC978 ACE2-44024 1034024DEA DRAAAB-ACC DECAE20142    3D structure-environment string (protein 1)

KLGFLVKQPE EPW----FQT EWKFADKAGK DLGFEVIKIA VPDG-EKTLN AIDSLAASGA KGFVIC-TPD PKLGSAIVAK    sequence (protein 1)
TIALVVSTLN NPF----FVS LKDGAQKEAD KLGYNLVVLD SQNNPAKELA NVQDLTVRGT KILLIN-PTD SDAVGNAVKM    sequence (protein 2)
TILVIVPDIC DPF----FSE IIRGIEVTAA NHGYLVLIGD CAHQNQQEKT FIDLIITKQI DGMLL---LG SRLPFDA-SI    sequence (protein 3)
AIGLVTPEND VPFNSGVFMD MVSCISRELA YHDIDLLLIA DDEH-ADCHS YMRLVESRRI DALIAHTLD DDPRITH---    sequence (protein 4)
```

FIG. 2. Dynamic nature of ITR templates. A Level 3 template rooted at arabinose-binding protein (1ABP) is shown spawning a Level 4 template. Each template consists of a sequence component and a structure–environment component (represented using hexadecimal notation as in Yi and Lander[14]). Only positions 1–80 are presented. (Adapted from Yi and Lander.[13])

$$m_{\text{SEQ}} = \frac{1}{L} \times \sum_{j=1}^{L} \sum_{x} \max[\text{SEQ\_PRF}(j, x), 0] \tag{2a}$$

$$m_{\text{STR}} = \frac{1}{L} \times \sum_{j=1}^{L} \sum_{x} \max[\text{STR\_PRF}(j, x), 0] \tag{2b}$$

For a given ratio R of sequence to structure information, the final combined profile was created using Eq. (3):

$$\text{PRF}(j, x) = \left(\frac{R}{(R + 1)m_{\text{SEQ}}} \right) \text{SEQ\_PRF}(j, x)$$
$$+ \left(\frac{1}{(R + 1)m_{\text{STR}}} \right) \text{STR\_PRF}(j, x) \tag{3}$$

A ratio of $R = 2:1$ was used in these studies, placing twice as much weight on the sequence component as the structure component. R corresponds to the ratio of the weighting parameters (α, β) as defined in Yi and Lander.[13]

From the six ITR searches carried out by Yi and Lander,[13] we selected a sample of 100 template/target pairs in which the target sequence was known to be structurally related to the seed protein in order to investigate the effect of altering the sequence/structure scoring ratio R. Five values of R were examined: $1:0$, $2:1$, $1:1$, $1:2$, and $0:1$. For each comparison, a set of Z scores (see Test of Significance section for details) was calculated using the different ratios, and then the scores were normalized with respect to the $R = 2$ case ($Z_{\text{norm}} = Z - Z_{2:1}$). The average of these data are presented in Fig. 3. Interestingly, the $R = 2:1$ ratio resulted in a gain of 2 standard deviation (SD) units over the sequence alone ($1:0$) and 5 SD units over the structure alone ($0:1$), thus confirming the benefits of combining sequence and structure information. Moreover, the sequence component made a greater contribution to the strength of a match than the structure component. Finally, the $2:1$ and $1:1$ ratios did not produce significantly different Z scores, suggesting that the scoring system is not overly dependent on a particular choice of R.

Database Search

The large number of profiles being run against the sequence database necessitated the development of an efficient search procedure. Because dynamic programming is relatively time-consuming, we implemented a fast prescreening phase using a hashing method to identify the 2000 top scoring proteins. These sequences were then reanalyzed by local dynamic programming. Below, we describe the hash search method which is based on ideas

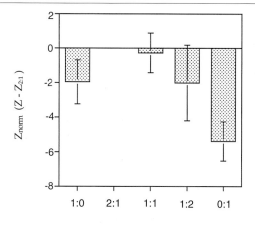

R (Sequence/Structure Ratio)

FIG. 3. Effect of varying the sequence/structure scoring ratio R on the strength of a match. Five values of R are displayed on the x axis, and the y axis depicts the Z scores normalized to the 2:1 case ($Z_{\mathrm{norm}} = Z - Z_{2:1}$). The mean and standard deviation for 100 template/target pairs are presented.

developed by Altschul *et al.*[16] for the program BLAST. We also describe the calculation of a normalized gap penalty for the dynamic programming part of the search.

Hash Search

The first step was to create a ghost sequence for each profile by identifying the amino acid at each row in the profile possessing the highest value [i.e., GHOST_SEQ(j) = arg max$_x$ {PRF(j, x)}]. Both the ghost sequence and the sequence database was divided into 2-mer words. By assembling a hash table of 2-mer words in the database, exact matches between the ghost sequence and the database sequences could be quickly located. These matches were extended in both directions by calculating a similarity score based on the alignment of the target sequence with the appropriate positions in the profile (not the ghost sequence). A list of the 2000 proteins with the highest segment scores was compiled, and these sequences were then recompared to the profile using the Smith–Waterman local dynamic programming algorithm.[17]

---

[16] S. F. Altschul, W. Gish, W. Miller, E. W. Myers, and D. J. Lipman, *J. Mol. Biol.* **215**, 403 (1990).

[17] T. F. Smith and M. S. Waterman, *J. Mol. Biol.* **147**, 195 (1981).

Gap Penalty

Given a profile and a target sequence, one of the most important parameters for the calculation of the alignment score by dynamic programming is the gap penalty. Our gapping strategy incorporated several elements: (1) we used both a gap initiation and gap extension term ($g_{total} = g_{init} + L_{gap}g_{ext}$) in a 5:1 ratio, (2) the gap penalties were normalized to the magnitude of the profile, and (3) we used a modified form of the "pay once" gapping protocol described by Smith and Smith[18] for multiple sequence alignments.

It is important to calibrate the gap penalties to the size of the values in the profile. The normalization was accomplished using Eq. (4):

$$(g_{open}, g_{ext}) = m_{PRF}(5g, g) \tag{4}$$

where m_{PRF} is the magnitude of the profile. In this work, g was set to 0.15. For comparison, the traditional gap penalties used with the Dayhoff PAM250 matrix, namely, $(g_{open}, g_{ext}) = (6, 2)$, correspond to normalized gap penalty values of $(0.53, 0.18)$ for the typical sequence [$m_{PRF} = 1$ when $R = 1:0$, see Eq. (3)]. Likewise, the gap penalties recommended by Gonnet et al.[15] for use with the Benner matrix, $(g_{open}, g_{ext}) = (20.63, 1.65)$, correspond to the normalized values of $(1.91, 0.15)$. Thus, the gap penalties used in ITR approximate the gap parameters employed in standard sequence comparison systems.

When adding new sequences to a multiple sequence alignment, Smith and Smith[18] suggested that the presence of gaps in specific positions of the alignment should influence the gap penalty. They defined a pay once rule which makes new gaps less costly when inserted near a previous gap. In ITR, positions in the profile in which at least one-third of the sequences in the sequence component contained a gap were assigned a special gap character. Then, adopting an abridged version of this rule, (i) new gaps in the target sequence across from a gap character incur no gap penalty, and (ii) gaps in the profile adjacent to a gap character are subjected to only the gap extension penalty. The net effect of this provision was that gaps could be made more easily at specific locations in the profile.

To characterize this system for assigning gap penalties, we examined the alignment length of various profiles against random sequences (i.e., sequences composed by randomly selecting amino acids from a given distribution) as a function of the gap penalty. The idea was that there should be an inverse relationship between the magnitude of the gap penalties and the length of the alignments. Indeed, as the gap penalty was varied around the normalized value $g_0^{ITR} = (0.75, 0.15)$ for a set of ITR profiles, from

---

[18] R. F. Smith and T. F. Smith, *Proc. Natl. Acad. Sci. U.S.A.* **87,** 118 (1990).

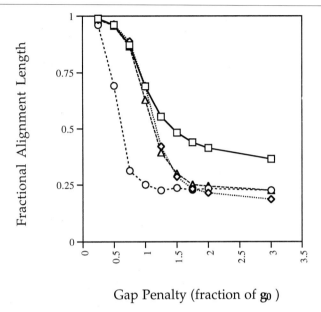

Gap Penalty (fraction of g_0)

FIG. 4. Alignment length for comparisons against random sequences as a function of the gap penalty. Random sequences were twice the length of the profile. Alignment length was measured in terms of fractional coverage of the seed protein. The gap penalty spanned a range of values around g_0, the standard normalized penalty. Four scoring matrix/gap penalty combinations were examined: ($\square$) ITR scoring system, $g_0 = (0.75, 0.15)$; ($\diamond$) Dayhoff matrix, $g_0 = (0.53, 0.18)$; ($\bigcirc$) Benner matrix, $g_0 = (1.91, 0.15)$; ($\triangle$) Benner matrix, $g_0 = (0.75, 0.15)$.

$(g_0^{\text{ITR}}/4)$ to $(3g_0^{\text{ITR}})$, there was a transition from longer to shorter alignments (see Fig. 4), with the most dramatic change occurring in the interval between $(3g_0^{\text{ITR}}/4)$ and $(3g_0^{\text{ITR}}/2)$. For comparison, the Dayhoff matrix and gap penalties produced a curve that was also centered at approximately $g_0^{\text{Dayhoff}} = (0.53, 0.18)$, whereas the transition midpoint for the Benner scoring system was around $(g_0^{\text{Benner}}/2)$. Reducing the gap initiation term for the Benner matrix from 1.91 to 0.75, however, shifted the graph to the right, closer to the Dayhoff and ITR curves. Preliminary data on a sample of profiles and target sequences suggest that the Z scores are maximized when the gap penalty lies in the transition region.

Test of Significance

The significance of a match was calculated by converting the raw alignment score $Y = Y(\text{PRF}, P_{\text{target}})$ into a Z score using a Monte Carlo shuffling technique.[19] The profile was compared to a collection of random sequences of the same length n as P_{target}, but with the amino acids chosen randomly according to the frequencies in the database. The mean $\overline{Y}_n$ and standard deviation σ_n from 1800 trials were used to determine the final Z-score: $Z = (Y - \overline{Y}_n)/\sigma_n$.

To avoid calculating Z scores for all 2000 target proteins that passed the prescreening stage in this laborious fashion, we first calculated an estimated Z score for each sequence. For each profile, we determined values for $\overline{Y}_n$ and σ_n for a set of random sequences of length $n = 100, 200, 400, 600, 800$. Then, for any length n and score Y, an estimated Z score was linearly interpolated or extrapolated from these reference values. Tests showed that the majority ($\geq 90\%$) of estimated values were within 1 SD unit of the final Z score. For proteins with an estimated Z score of greater than 6.0, a final Z score was calculated. Matches that did not span at least two-thirds of the length of the starting protein were discarded in order to prevent short alignments.

Many programs calculate Z scores by permuting the sequence of the target protein so that the composition of the random sequences is the same as the original sequence (Z^{SEQ}), whereas ITR randomly selects amino acids according to the distribution observed in the database (Z^{DB}). To examine any possible discrepancy, we calculated Z scores using both approaches for a set of 100 template/target pairs. We found that Z^{DB} was approximately 1.1 SD units larger than Z^{SEQ}. However, in all but six cases, Z^{DB} was within 2 SD units of Z^{SEQ}. Even for the compositionally biased protein hypothetical 125K protein (JQ0316 from the arabinose-binding protein search) in which the chi-square (χ^2) statistic measuring the deviation from the average database composition was 308.3 ($df = 19$), Z^{DB} (9.8) was relatively similar to Z^{SEQ} (8.5) for the highest-scoring match to the protein in the search. With ITR, one has the option to determine Z^{SEQ} along with Z^{DB}, but the Z^{SEQ} values should be evaluated with respect to a lower significance scale.

We have established a Z score of 7.5 as the threshold for significance. Sequence comparison methods using the sequence permutation shuffling technique typically judge a Z score of 6.0 or greater as significant.[20] The value of 7.5 represents a compromise between maximizing the number of true positives while minimizing the number of false positives. Using this threshold, there were 18 true positives and 2 false positives from the six searches. Increasing the threshold to 8.0 eliminates both false positives, but it also removes 2 of the 18 true positives. Furthermore, 14 of the 52 protein classes identified in all the searches had Z scores between 7.5 and 8.0.

[19] S. Karlin, P. Bucher, and V. Brendel, *Annu. Rev. Biophys. Biophys. Chem.* **20**, 175 (1991).
[20] W. R. Pearson, this series, Vol. 183, p. 63.

Decreasing the threshold to 7.0 would increase the number of false positives by 2, while adding 3 true positives. Thus, one would expect the majority of matches with Z scores above 7.5 to be legitimate, with an even higher level of confidence being accorded to hits with scores above 8.0.

Iteration

Through iteration, a multiple sequence alignment is built up in the sequence component of the upper level templates. The consensus pattern that emerges in these alignments facilitates the detection of more distant relationships, especially when coupled to the structure–environment component of the template. In addition, from a more global perspective, one can view the iteration process as the breadth-first exploration of template space. Thus, the results of a completed search can be depicted as a tree of templates rooted at the seed protein.

Example

To illustrate the gradual refinement of the templates, we provide selected snapshots from a search using arabinose-binding protein (1ABP). In this example, many of the less obvious matches did not occur until later in the search process with the higher level templates. The Level 1 template identified only galactose-binding protein and ribose-binding protein (see Fig. 5a). The two Level 2 templates turned up many members of the *lac* repressor family. By the fourth level, however, there were hits to more distantly related proteins such as leucine-binding protein and phosphofructokinase. New protein classes such as the two-component transcriptional regulators and atrial natriuretic receptor also emerged in the later levels.

Focusing on a single target protein, leucine-binding protein (2LBP), and a single lineage through the tree of templates, it is possible to observe how refinement of the sequence portion of the template eventually results in the detection of a distant relative. From the Level 1 template 1ABP to the Level 4 template 1ABP_JGECR_RPECG_JV0031 (the name of a template is derived from the PDB/PIR codes of the member proteins), the Z score of the match to 2LBP increased from 3.05 to 8.07 (see Fig. 5b). The bulk of this increase can be attributed to the enhanced signal in the sequence component; the Z score for sequence alone ($R = 1:0$) rose from 1.44 to 5.41, whereas the Z score for the structure component remained constant. The importance of casting a broad net by employing the breadth-first exploration of possible templates was underscored by the observation that only 3 of the 40 Level 4 templates possessed a significant match with 2LBP. Pursuing one or two paths in the tree by depth-first search may have easily missed these templates.

Tree of Templates

The results from an ITR search can be represented by a tree of templates derived from the seed template (root node). In Fig. 6, we present a path in the 1ABP template tree taken from the example in the previous section. Each node corresponds to a template, and the edge connecting a Level k template to a descendent Level $k + 1$ template represents the strength of the match (Z score) between the Level k template and the sequence of $P_{(k+1)}$. In theory, each node can generate as many children as the number of significant hits. The templates are labeled with the minimum-valued edge on the path from the root node to that template. At the end of the search, all the sequences belonging to one or more of the templates in the tree are collected, and the maximum Z score associated with each sequence is recorded.

Because the growth of the template tree is exponential, several measures were adopted to prevent excessive branching. First, the tree was not extended beyond six levels. Second, the list of high-scoring proteins from each database search was pruned for sequences possessing greater than 40% sequence identity with either a sequence in the template or another member of the list. Third, the total number of Level k templates allowed to spawn active children (i.e., Level $k + 1$ templates that were run against the database) was limited to 6 on Level 2, 30 on Level 3, and 48 on Levels 4 and 5. When the number of templates exceeded these bounds, the most promising templates were selected according to (i) the total number of significant matches and (ii) the number of significant matches with a higher Z score than previously recorded. Finally, the number of active children per template was restricted to 6 on Level 1, 5 on Level 2, 4 on Level 3, and 3 on Level 4. If the number of significant matches surpassed these limits, then the sequences were clustered into the appropriate number of classes using a hierarchical clustering algorithm in which successive pairs of proteins with the greatest similarity were grouped together. A single sequence was chosen to represent each cluster. Thus, the maximum number of database searches for a single ITR run is $1 + 6 + (6 \times 5) + (30 \times 4) + (48 \times 3) = 301$.

Analysis of Output

In this section, we describe the results from six ITR searches using the following initial seed proteins: arabinose-binding protein (1ABP), plastocyanin (1PCY), cytochrome c (1CCR), chymotrypsin (2CGA), the dinucleotide-binding domain of lactate dehydrogenase (5LDH, domain 1), and the α subunit of tryptophan synthase (1WSY_A).[13] Among the 20 proteins of

a

LEVEL 1

Template = *1ABP*

	Z-score	CODE	TITLE
1	>15.00	JGECR	D-Ribose-binding protein precursor - Escherichia coli
2	13.99	JGECG	D-Galactose-binding protein - Escherichia coli

LEVEL 2

Template = *1ABP_JGECR*

	Z-score	CODE	TITLE
1	>15.00	JGECG	D-Galactose-binding protein - - Escherichia coli
2	>15.00	RPECDU	pur repressor - Escherichis coli
3	>15.00	RPECL	lac represor - Escherichia coli
4	13.95	RPECCT	cyt repressor - Escherichia coli
5	12.56	RPECG	gal repressor - Escherichia coli
6	11.51	RPECEG	ebg repressor - Escherichia coli
7	10.41	JV0031	MalI protein - Escherichia coli
8	9.10	A35160	Repressor protein RafR -Escherichia coli
9	9.04	B24925	lac repressor - Klebsiella pneumoniae

LEVEL 4

Template = *1ABP_JGECR_RPECCT_A35160*

	Z-score	CODE	TITLE
1	>15.00	RPECDU	pur repressor - Escherichis coli
2	>15.00	RPECG	gal repressor - Escherichia coli
3	>15.00	RPECL	lac represor - Escherichia coli
4	>15.00	JGECG	D-Galactose-binding protein - Escherichia coli
5	>15.00	B24925	lac repressor - Klebsiella pneumoniae
6	>15.00	JV0031	MalI protein - Escherichia coli
7	>15.00	RPECEG	ebg repressor - Escherichia coli
8	8.03	KIRBF	6-Phosphofructokinase (EC 2.7.1.11) - Rabbit
9	7.88	S03321	Regulatory protein nifR1 - Rhodobacter capsulatus

Fɪɢ. 5. Iteration leads to the detection of distantly related proteins. (a) Three snapshots from the arabinose-binding protein (1ABP) search. The high-scoring matches to the Level 1 template, to a Level 2 template, and to a Level 4 template are depicted. Only significant Z scores are shown, and sequences closely related to one of the template sequences or to another

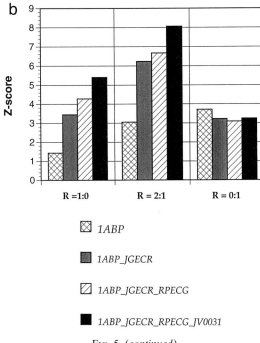

FIG. 5. (*continued*)

known structure detected by these searches that did not possess significant sequence similarity with the seed sequence, 18 were true positives (i.e., shared the same structural fold) and 2 were false positives. Many of the valid hits were to distant relatives of the starting protein. For example, the arabinose-binding protein search identified phosphofructokinase, the plastocyanin search detected the immunoglobulins, and the chymotrypsin search picked up several bacterial serine proteases. The two false positives, arabinose-binding protein and p21 Ras, arose from the tryptophan synthase run. All three proteins possess parallel α/β topologies consisting of repeated $\beta\alpha$ units. Although there were many false negatives in each of the searches, we did not expect a single ITR search to identify all the members of a structural family. Rather, the concern was with limiting the number of false positives.

member of the list have been removed. (Adapted from Yi and Lander.[13]) (b) The Z scores from comparison of leucine-binding protein (2LBP) against the Level 4 template 1ABP_JGECR_RPECG_JV0031 (black bars) and its three ancestors (1ABP, cross-hatched bars; 1ABP_JGECR, gray bars; 1ABP_JGECR_RPECG, hatched bars). Three values of R were examined: 1:0, 2:1, and 0:1.

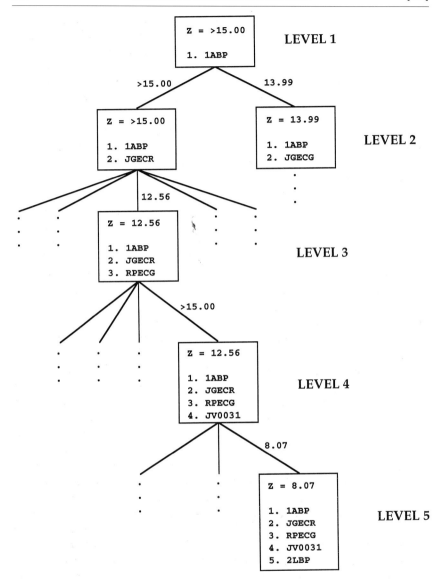

Fig. 6. Graphical representation of a single path in the 1ABP search tree. An expanding tree of templates grows from the starting protein during the course of the search. Each square (node) represents a distinct template. The template and the sequences contained in the template are assigned a Z score based on the minimum-valued edge on the path from the root protein to the template.

Since the publication of the original description of ITR,[13] the structures of two more proteins identified in the searches have been solved. In both cases, the structural data confirmed the prediction of structural similarity between the seed protein and target protein. First, the corepressor-binding domain of the PurR repressor (*lac* repressor family) displayed remarkable structural homology to the members of the periplasmic-binding protein family including 1ABP.[21] Second, Holm *et al.*[22] demonstrated that $3\alpha,20\beta$-hydroxysteroid dehydrogenase, a member of the short-chain dehydrogenase family, is structurally related to the dinucleotide-binding domain of the classic dehydrogenases including lactate dehydrogenase. Better statistics on ITR will be obtained in the future as more searches are performed and as more structures are solved.

A good fold-prediction algorithm should not only identify proteins that possess the same overall fold as the starting protein, but should also produce an alignment that corresponds closely to the structural (true) alignment. We determined the accuracy of alignment for a series of template/target pairs in which the structure of the target was known to be homologous to the starting protein. The quality of the alignment was assessed in terms of the percentage of positions in the ITR alignment that agreed to within δ positions of the alignment generated by a structural comparison of the seed and target proteins (see Table I). Only positions found in both alignments were counted. Three values of δ were used: 0, 2, and 5. The template/target pairs were divided into three categories on the basis of the Z score of the match: (i) $Z \geq 10.0$, (ii) $7.5 \leq Z < 10.0$, and (iii) $2.0 \leq Z < 6.0$. Also presented in Table I are data from the alignment of the seed sequence and the structure–environment string against the target sequence.

The best alignments occurred when the Z score of the match was greater than 10.0 (82% agreement, $\delta = 0$). In all of these cases, the seed sequence possessed significant sequence similarity with the target sequence, and, as a result, the seed sequence alignment was also quite good (80%, $\delta = 0$). There was a decline in alignment quality for the template/target pairs in which the Z score was between 7.5 and 10.0 (36%, $\delta = 0$; 67%, $\delta \leq 2$). Thus, despite the significance of the match, approximately 25% of the positions were seriously misaligned ($\delta > 5$). Not surprisingly for $Z < 6.0$, the alignment quality deteriorated further. Finally, comparing the results from using only the seed sequence or only the structure–environment string, it appears that the sequence does a better job of finding the correct alignment.

[21] M. A. Schumacher, K. Y. Choi, H. Zalkin, and R. G. Brennan, *Science* **266,** 763 (1994).
[22] L. Holm, C. Sander, and A. Murzin, *Nat. Struct. Biol.* **1,** 146 (1994).

TABLE I
ACCURACY OF ALIGNMENTS[a]

Z group	$\delta = 0$	$\delta \leq 2$	$\delta \leq 5$	Average Z score
$Z \geq 10.0$				
Full template	0.82	0.93	0.99	17.36
P_1 ($R = 1:0$)	0.80	0.93	1.00	14.69
P_1 ($R = 0:1$)	0.44	0.62	0.71	4.27
$7.5 \leq Z < 10.0$				
Full template	0.36	0.67	0.76	8.57
P_1 ($R = 1:0$)	0.20	0.43	0.58	2.86
P_1 ($R = 0:1$)	0.19	0.32	0.36	2.42
$2.0 \leq Z < 6.0$				
Full template	0.21	0.51	0.68	3.62
P_1 ($R = 1:0$)	0.19	0.42	0.60	2.38
P_1 ($R = 0:1$)	0.19	0.30	0.38	2.03

[a] Accuracy of alignments for three groups of template/target pairs. Accuracy is measured by the proportion of positions in the alignment that deviate δ positions or less from the structural alignment. The last column presents the average Z score of the matches in each group. The alignment quality for the comparison of the seed sequence against the target protein, and for aligning the structure–environment string against the target, is also shown.

Conclusion

In summary, ITR attempts to exploit the vast amount of information in the sequence database. Iterative searching of the database for related sequences leads to the refinement of the templates. Coupling sequence information with structure information increases the sensitivity of the search. The enhanced signal in templates containing multiple sequence alignments has been reported,[3-8] but in these previous studies the sequences comprising the multiple sequence templates were already known to belong to the same family. Pickett et al.[23] explored the merging of sequence and structure components on hybrid templates, but for each search, they used a single static template with no iteration. Thus, despite the domination of the fold-prediction field by structure-based methods, one is advised not to ignore sequence information.

The conceptual simplicity of ITR stands in contrast to the technical complexity of its implementation, especially the details regarding the selection of parameters and the application of specialized methods to decrease the running time. ITR employs three important timesaving measures: (1)

[23] S. D. Pickett, M. A. S. Saqi, and M. J. E. Sternberg, *J. Mol. Biol.* **228,** 170 (1992).

prescreening the database for high-scoring proteins using a hashing technique; (2) limiting the branching of the template tree; and (3) calculating an estimated Z score in lieu of a final Z score for many of the comparisons. Because of these provisions, it is possible to complete a five-level ITR search within 1 week. For the specification of key parameters such as the gap penalty and the sequence/structure scoring ratio, we have determined normalized values based on the magnitude of each profile. Furthermore, we have found that the performance of ITR was relatively robust to small variations in the value of these parameters.

Surprisingly, a significant Z score did not guarantee a good alignment. Indeed, we found that only when the Z score was above 10.0 and the seed protein possessed sequence similarity with the target could one be confident about the accuracy of the alignment. One possible remedy .is to include additional types of structural information into the scoring system. Jones *et al.*,[10] Sippl and Weitckus,[11] and Bryant and Lawrence[12] have all had good success using a residue–residue contact potential to evaluate the threading of sequences on structures. Likewise, it may be possible to incorporate predictions of secondary structure and solvent accessibility into the overall scoring scheme. This additional information may also enhance the sensitivity of the templates.

Availability of Program

The ITR package consists of seven core subprograms that carry out the following functions during each cycle: (1) construct a profile for each template, (2) run each profile against the database using hashing technique, (3) reanalyze the highest scoring 2000 proteins using local dynamic programming, (4) calculate estimated Z scores for comparisons, (5) calculate final Z scores for selected matches, (6) select active templates for next round, and (7) create next level templates.

The input for the program is the sequence and structure–environment string of the seed protein. If the structure of the seed protein is not known, then the sequence alone serves as the input. Otherwise, a separate program assigns each position in the structure to a local environment class based on information derived from the PDB coordinate file. The raw data from the search, initially represented as a tree of templates, are converted to an ordered list of target sequences and their associated maximum final Z scores.

The whole package is written in the C language and can be compiled using a standard C compiler. We have successfully run ITR under the UNIX operating system on DEC Alpha, Sun SPARC 10, and Silicon Graphics Indigo workstations. On a DEC Alpha computer, the typical ITR search takes about 1 week. The program can be obtained by anonymous FTP to the Internet address genome.wi.mit.edu.

Section III

Multiple Alignment and Phylogenetic Trees

[20] Multiple Protein Sequence Alignment: Algorithms and Gap Insertion

By WILLIAM R. TAYLOR

Introduction

The basic method of sequence alignment has remained unchanged for over 20 years. In that time the most significant development has been to add a termination condition to display local alignments; all other algorithms and modifications have been concerned with increasing speed, the effect of various gap penalties and relatedness matrices, or extensions for multiple sequence data. This status quo has remained despite results in which it is clear to the eye that the alignments produced sometimes do not reflect the expected biological pattern of regions of weak matches (including gaps) and regions of good matches. This characteristic of the data has not been ignored, but has been dealt with by methods other than conventional dynamic programming (typically pattern matching). Within a dynamic programming approach, failure to attain the desired alignment has usually been attributed to an incorrect combination of gap penalty and relatedness matrix; however, the underlying problem is more fundamental and some novel (and revived) algorithms are described below that have been developed by the author in an attempt to overcome these defects.

Outline of Algorithms

Score Run-Length Enhancement

Consider the following simple example: the sequence ABQDDEFHR-SKKLMO, when matched in phase with the alphabet, gives rise to six scattered matches. However, if displaced back by one position two segments (DEF and KLM) align, also giving six matches. With real sequences, the trained eye, of one who is familiar with remotely related protein sequence alignments, might suspect that a scattered score distribution is less biologically significant than a clumped distribution. This quality in a sequence alignment that makes it look correct has been quantified and turned into a method to improve basic pairwise and multiple sequence alignment.[1]

---

[1] W. R. Taylor, *J. Comput. Biol.* **1,** 297 (1994).

Copyright © 1996 by Academic Press, Inc.
All rights of reproduction in any form reserved.

Profile Gap Weighting

In the development of a multiple sequence alignment profile, gaps can accumulate and make the profile longer than the average sequence length and even longer than any sequence that it contains. This elongation creates a bias for the profile to recognize longer members of the family as these require less insertions to be aligned and hence incur less penalty. An algorithm is described that prevents this undesirable behavior by comparing the average distance (excluding gaps) between pairs of positions and using this difference to modify the gap penalty.[2]

Gap Bias by Structure Prediction

The use of known and predicted secondary structure in sequence matching, while frequently examined, has not been fully exploited. For the insertion of gaps, previous methods considered only the opening of a gap at a single point and not what is contained in the gap once opened or what terminates the broken sequence ends. Incorporating all these aspects is computationally more difficult (and slower) but makes a difference both to the stability of the correct alignment and to the detailed placing of gaps.[3]

Score Run-Length Enhancement

Background

Sequence alignment has revealed many unexpected and often important similarities that have given insight into previously obscure systems. Underlying much of this success is the widely used dynamic programming (DP) algorithm[4–6] which finds the optimal alignment of the sequences under a given scoring scheme. Although the algorithm is rigorous, the parameters are poorly characterized. These include the model relatedness between sequence elements (either residues or nucleotides) and the penalty for gaps. Despite much analysis, there is still little to guide the best choice of these.

Multiple sequence alignment typically identifies strongly scoring regions (where gaps are less frequent), which, in situations where the structure is known, are often found to correspond to core secondary structures (or motifs). With only a pair of sequences, introducing correlation between

[2] W. R. Taylor, *Bull. Math. Biol.* in press (1996).
[3] W. R. Taylor, *Gene* **165**, GC27 (1995).
[4] S. B. Needleman and C. D. Wunsch, *J. Mol. Biol.* **48**, 443 (1970).
[5] P. H. Sellers, *J. Combinator. Theor.* **16**, 253 (1974).
[6] T. F. Smith and M. S. Waterman, *J. Mol. Biol.* **147**, 195 (1981).

adjacent matches should emulate this uneven distribution of score, enhancing the score for well-matched regions, not only for the matching elements but also for the surrounding region to an extent expected for the size of a typical core structure. The following method describes how this can be achieved, dealing with the problems of correlation length and score normalization. For simplicity, the method is sometimes referred to below as the motif-bias method.

Algorithm

Basic Dynamic Programming. Dynamic programming compares sequences by finding the best path through a matrix of scores from all pairwise matches of elements between the two sequences. This is achieved by incrementally extending each path with a locally optimal step. Element $d_{i,j}$ in the score matrix can extend any path from the preceding row or column to produce the highest path score. Applying this condition to each matrix element transforms the pairwise score matrix into a matrix of path scores. This can be represented recursively:

$$s_{i,j} = d_{i,j} + \max \begin{cases} s_{i-1,j-1} \\ s_{i-1,m} - g & (m < j - 1) \\ s_{n,j-1} - g & (n < i - 1) \end{cases} \tag{1}$$

where g is a gap penalty. Path connectivity is recorded in a matrix of pointers, and if all d_{ij} are positive, then the path from the highest score is the optimal (global) alignment.

Modified Algorithm. The modified DP algorithm should give higher scores for runs of good matches. This can be achieved by accumulating a running product of match scores and giving a higher score to long runs of matches relative to more scattered matches.

Running-product score. The effect of the running-product score depends on the match score values: allowing a zero score, the product will disappear and not recover afterward; whereas if the minimum is one, the product will grow continually. It is therefore necessary to rescale the match scores into a reasonable range, which is equivalent to continually damping by a constant value (a):

$$r_i = (r_{i-1} + 1)(d_i + 1)/a \tag{2}$$

where d_i is ith in the series $\{d_1 \ldots d_N\}$ (with $d_i \geq 0$, $\forall_i$) and r_i is the current product.

Automatic damping. Real (as distinct from random) sequences allow the possibility that very similar sequences will give astronomic products unless the value of a is chosen suitably large in anticipation. This problem was avoided by defining a in terms of the final score:

$$a^n = r_N/N + \bar{d} \tag{3}$$

where r_N is the terminal product in a sequence length N and $\bar{d}$ the average score (n is a parameter). Defining a as a function of itself means that a solution for its value must be found by iteration, and to give stability to this convergence, a further constant c was introduced:

$$r_i = (d_i + 1)(r_{i-1} + 1)/(ac) \tag{4}$$

A good starting estimate of a was $a^n = \bar{d} + 1$.

Dynamic programming formulation. To include gaps, the above scoring function must be incorporated into the DP algorithm, and, in keeping with the overall goal, the running product [Eq. (4)] should not jump across gaps:

$$r_{i,j} = \begin{cases} (d_{i,j} + 1)(r_{i-1,j-1} + 1)/(ac) \\ 0, \quad \text{after gaps} \end{cases} \tag{5}$$

In Eq. (5), match scores are now elements of a matrix.

To produce the desired property that the score for good sequence matches tends toward the unbiased score, the unmodified sum [Eq. (1)] was combined with the running product into a sum t:

$$t_{i,j} = d_{i,j} + \max$$

$$\begin{cases} t_{i-1,j-1} + (d_{i,j} + 1)(r_{i-1,j-1} + 1)/(ac) &: r_{i,j} = (d_{i,j} + 1)(r_{i-1,j-1} + 1)/(ac) \\ t_{i-1,m} - g & (m < j - 1) &: r_{i,j} = 0 \\ t_{n,j-1} - g & (n < i - 1) &: r_{i,j} = 0 \end{cases}$$

$$\tag{6}$$

As r_{ij} depends on the maximum in Eq. (6), its conditional assignment is shown on each line of the equation. Designating the expression left of the ":" as *left* and that right of the "=" as *right,* this construct reads: "**if** the |*left*| is maximum **then** assign r_{ij} the |*right*|" (for |*expression*|, read "the value of the *expression*").

Application

All pairwise alignments of 12 remotely related aspartyl proteases were calculated and assessed by the percentage of correctly aligned motifs for different gap penalty g and similarity matrix m. (See Ref. 1 for details.)

Unbiased Alignment. The results for the unbiased method showed little sensitivity to the form of the similarity matrix but were strongly dependent on the gap penalty. Above $g = 10$, accuracy dropped from around 70–75% toward 40–50% at $g = 30$ (Fig. 1a).

Motif-Biased Alignment. Using the motif-biased algorithm [Eq. (6)], different parameter values for n [Eq. (3)] and c [Eq. (5)] produced different behavior. However, almost all combinations extended the region of good accuracy (better than 65%) into higher gap penalties, and with $n = 3$, $c = 5$ this accuracy was obtained up to $g = 30$. In addition, the region of highly gapped alignments (over 10 gaps) contracted slightly (Fig. 1b).

Gap-Normalized Controls. The scores with the motif-biased method will always be greater than those with the unbiased method. To allow for this, the scores were normalized by the number of gaps. The line of 10 gaps is roughly the same between Fig. 1a and Fig. 1b, implying that they are comparable. If fewer gaps were taken then the plots would need to be rescaled. For five gaps this factor is 1.4, but even after this rescaling, the accuracy of the motif-biased method at $g = 30$ still exceeds the unbiased results by 5–10%.

Multiple Alignment. The method was applied to multiple sequence alignment as implemented in the program MULTAL.[7,8] On the basis of results for the pairwise alignments, parameters $n = 3$, $c = 5$, $m = 3$ (Fig. 1b) were fixed and just the gap penalty (g) varied. Using motif bias on aspartyl protease sequences, the correct alignment was obtained over the range $g = 14$–16; by contrast, no correct alignment was found with the unbiased method.

From experience with hierarchic profile clustering, better results were expected from profile/profile matching using a greater weight on identity matches in the later alignment cycles. This was implemented by gradually reducing m with each cycle (moving toward identity matching). With the unbiased method a stable range (gap penalty 11–15) was now found. Using the motif-bias method (with $n = 3$), however, this was extended to 10–22. (Table I).

Alignment Sensitivity. The sensitivity of the modified algorithm was evaluated by comparing all pairwise scores within a family to those obtained between families. For this the aspartyl proteases (Dp) were compared to an equivalent selection of immunoglobulin (Ig) sequences. An average alignment score was obtained within each family (Dp/Dp, Ig/Ig) and expressed relative to the average interfamily (Dp/Ig) score (Table II). With protease data the unbiased method achieved 15% separation of averages but managed only half this with immunoglobulin sequences. Using the biased method, continual improvement in separation was found with in-

[7] W. R. Taylor, *J. Mol. Evol.* **28,** 161 (1988).

[8] W. R. Taylor, this series, Vol. 183, p. 456.

a

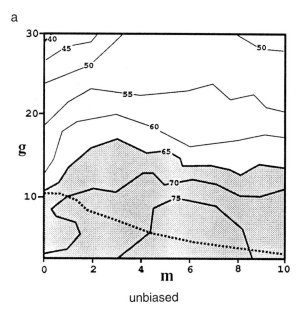

unbiased

b

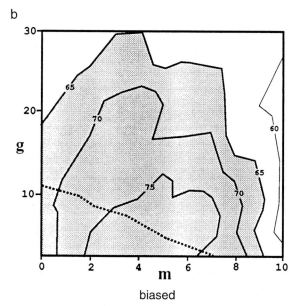

biased

FIG. 1. Accuracy plot varying gap penalty and matrix. The percentage of correctly aligned motifs is plotted for varying gap penalty g against matrix composition m ($m = 0$ = identity matrix, $m = 10$ = Dayhoff matrix) for the unbiased method (a) and for the biased methods (b) with $n = 3$ and $c = 5$. Both plots were sampled at intervals of $g = 3$ and $m = 1$ (excluding the region $g < 3$). The area below the dotted line contains alignments with an excessive number of gaps (more than 10).

TABLE I

MULTIPLE ALIGNMENT STABILITY UNDER GAP PENALTY VARIATION[a]

	Gap penalty						
	- - - -5 - - - 10 - - -15 - - -20 - - - 25 - - -30 - - -35
	. . . . : . . . . | . . . . . : . . . . | . . . . : . . . . | . . . . :
Unbiased	- - - - - - -+++|||||++ - - - - - - - - - - - - - - - -
Biased ($n = 3$)	- - - - - - - -+||||||| | |||||++ - - - - - - - - - - -

[a] At each value of the gap penalty a "|" indicates that an alignment was obtained that correctly aligned the motifs across all sequences. A "+" indicates an incorrect alignment (even if only by one motif in one sequence), whereas a "-" indicates that no test was made.

creasing n, reaching maximum at $n = 5$ with 13% for the immunoglobulins and 20% for the proteases.

Relationship to Other Approaches

Method of Vingron and Argos. Segment matching has been used by Argos and co-workers,[9,10] allowing high-scoring segment combinations to be selected manually or semiautomatically.[11] Their segments have a limited range of fixed lengths, and although this is sufficient for practical purposes, the approach has the theoretical problem that the segments must be scored prior to concatenation, which does not fully exploit the synergism in a number of consecutive weak segments. An equivalent approach has been used for predicting transmembrane segments.[12]

Method of Boswell and McLachlan. The approach of Boswell and McLachlan[13] is similar to the current method in its use of exponential damping to enhance local features. They used a running sum (as distinct from a product in the current method), resulting in less extreme values and so avoiding the complexities of score normalization. They also used a second reverse pass across the matrix to balance lag effects in the score response. Using a product, this is less critical since a product has a sharper response and allows a single pass to be used in the current method. Boswell and McLachlan used their approach mainly to look for local alignments, again allowing them to be less concerned about the score normalization that must be considered with a global method.

[9] P. Argos, *J. Mol. Biol.* **193**, 385 (1987).
[10] P. Argos and M. Vingron, this series, Vol. 183, p. 352.
[11] R. Rechid, M. Vingron, and P. Argos, *CABIOS* **5**, 107 (1989).
[12] D. T. Jones, W. R. Taylor, and J. M. Thornton, *Biochemistry* **33**, 3038 (1994).
[13] D. R. Boswell and A. D. McLachlan, *Nucleic Acids Res.* **12**, 457 (1984).

TABLE II

ALIGNMENT SENSITIVITY UNDER GAP PENALTY (g) VARIATION FOR
PROTEASE AND IMMUNOGLOBULIN SEQUENCES[a]

	g for protease				g for immunoglobulin			
n	15	20	25	30	15	20	25	30
0	15.0	15.1	14.7	14.3	7.5	7.5	7.1	6.5
2	16.7	16.9	16.8	16.5	6.9	7.2	7.3	7.3
3	18.7	18.8	19.0	18.8	9.3	9.7	9.9	10.0
4	19.4	19.6	19.5	19.3	10.6	11.1	11.3	11.3
5	20.0	20.4	20.4	20.1	12.4	13.0	13.0	13.3

[a] With and without the motif bias ($n > 1$ and $n = 0$, respectively).
The parameter n which controls the emphasis on the run bias was
introduced in Eq. (3), and the associated parameter c [Eq. (4)] was
assigned the value $n + 2$. The table values are the relative percentage
separations of the mean intrafamily score over the mean interfamily score.

Method of Huang. Huang[14] has described a simple enhancement for
matched elements in ungapped runs. The sequences on which this method
was tested, however, were all more closely related than those for which
the current method was developed. Consequently, the main difference in
the methods is the degree of bias imposed, which, being relatively mild in
the Huang method, avoided the complexities of score normalization.

Profile Gap Weighting

Background

Accumulation of gaps is a problem with profiles, both when aligning a
profile with another profile and when aligning with a single sequence. Gaps
make the profile longer than the mean sequence length, even leading to a
profile longer than any sequence in the alignment. Containing the problem
with strong gap penalties can lead to incorrect alignments and is therefore
not an ideal solution. Previously, pragmatic steps have been taken to avoid
this problem, including excision of variable regions,[15] and altering the gap

[14] X. Huang, *in* "Combinatorial Pattern Matching, Volume 807 of Lecture Notes in Computer
Science" (M. Crochemore and D. Gusfield, eds.), p. 54. Proceedings, 5th Annual Symposium,
CPM. Springer-Verlag, Berlin, 1994.

[15] J. D. Thompson, D. G. Higgins, and T. J. Gibson, *CABIOS* **10**, 19 (1994).

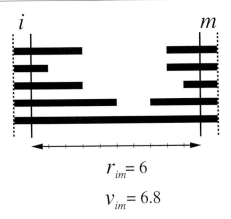

$$r_{im} = 6$$

$$v_{im} = 6.8$$

FIG. 2. Sequence profile position separation. Five protein sequences are shown schematically (black bars) aligned into a profile containing gaps (white space). Ten positions separate positions i from m in the profile, but the mean number of intervening residues (r_{im}) is only six (with a variance, v_{im}, of 6.8).

penalty locally.[16-19] Although the former approach obviously solves the problem, it does so with the undesirable loss of data. The second approach is not a solution but only moderates the problem by reducing the penalties: clearly, a sequence that is greater than average length but equal in length to the profile will experience no penalties.

The method described in this section is a generalization of an approach used in matching sequence templates.[20] In that application, discrete patterns embodied a consensus sequence description, and gaps were constrained not only between adjacent templates but also between all pairs (with a weight that reflected the degree to which the ideal separation would be attained). Reapplied at the residue level, the ideal separation becomes the mean separation (excluding gaps) and the target weight, the variance (Fig. 2).

Implementation with combinatorial matching was simple since each template location was determined before the pairwise gap function was evaluated. Within dynamic programming, however, the alignment cannot be calculated independently of the gap penalties. This circularity could be overcome iteratively; however, a single-pass approximation is presented

[16] A. M. Lesk, M. Levitt, and C. Chothia, *Protein Eng.* **1**, 77 (1986).
[17] G. J. Barton and M. J. E. Sternberg, *Protein Eng.* **1**, 89 (1987).
[18] K. Masaharu, F. Kishimoto, Y. Ueki, and H. Umeyama, *Protein Eng.* **2**, 347 (1989).
[19] J. D. Thompson, D. G. Higgins, and T. J. Gibson, *Nucleic Acid Res.* **22**, 4673 (1994).
[20] W. R. Taylor, *Prog. Biophys. Mol. Biol.* **54**, 159 (1989).

below and the implied danger of an asymmetric solution ignored. This potential problem has been fully evaluated[2] and is discussed further in the concluding discussion.

Algorithm

Length-Difference Function. In two aligned profiles **a** $(a_1 \ldots a_M)$ and **b** $(b_1 \ldots b_N)$, the pairs of positions a_i, b_j and a_m, b_n $(i < m, j < n)$ have been matched. If the mean number of residues (excluding gaps) between i and m is ${}^a r_{im}$ and that between j and n is ${}^b r_{jn}$, then the length difference $({}^{ab}d)$ is

$$ {}^{ab}d_{ij,mn} = {}^a r_{im} - {}^b r_{jn} \tag{7} $$

A negative difference means ${}^b r_{jn}$ is too long and further gaps in **b** should be discouraged; similarly, with positive d, gaps in **a** should be avoided. This could be implemented using d to alter the gap penalty (g) and can be inversely weighted by the variance (v), giving a modified gap penalty (g') as

$$ g'_{ij} = g + {}^{ab}d_{ij,mn} / ({}^a v_{im} + {}^b v_{jn} + c) \tag{8} $$

In Eq. (8), c moderates damping sensitivity and must be greater than zero.

Alignment Path Sum. The DP algorithm requires a decision for every element in the score matrix on whether to match, insert, or delete. This should be influenced not only by a single separation (as above) but by all separations over the current alignment. If the current matrix position is $\{i,j\}$ and q designates an alignment, then the path of q can be specified as ${}^q\mathbf{p}_{ij} = \{{}^q(n,m)_1, {}^q(n,m)_2, \ldots, {}^q(n,m)_K\}$. A gap score $({}^q s_{ij})$ based on the separation differences (d) over the path ${}^q\mathbf{p}_{ij}$ can then be calculated by modifying Eq. (8) to

$$ {}^q s_{ij} = \frac{t}{10} \sum_{k=1}^{K} \frac{|{}^{ab}d_{ij,{}^q m_k {}^q n_k}|}{k({}^a v_{i,{}^q m_k} + {}^b v_{j,{}^q n_k} + c)} \tag{9} $$

with ${}^q m_k, {}^q n_k$ locating the path at node k. The parameters t and c control the magnitude and sensitivity of the bias. Each contribution is normalized by its path length (k).

Dynamic Programming Formulation. Use of the gap scores to modify the gap penalty directly [suggested by Eq. (8)] is not ideal, as this neglects the fact that extension with no indel also has a gap score. To allow for this, the score for the straight (nonindel) extension was subtracted from those involving an indel. This leaves the score of any path without indels equal to the score obtained with the unmodified algorithm.

From the current matrix element, let x be a path with an immediate insertion, y a path with an immediate deletion, and z the straight (diagonal)

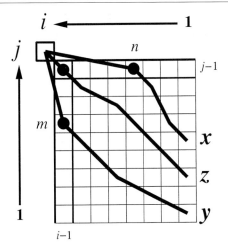

FIG. 3. Dynamic programming with trace back. The calculation of the $\{i,j\}$th element of the score matrix $(e_{i,j})$ requires the values of all elements $\{n < i, m < j\}$ (represented by the grid). Conventionally, the path to $\{i,j\}$ is extended from the cell with the maximum value of $\{e_{i-1,j-1}, e_{i-1,m} - g, e_{n,j-1} - g\}$ (g is the gap penalty); however, in the modified algorithm, the optimal alignments from each of these positions (designated z, y, and x, respectively) are also considered. These alignments are specified as a series of paired integers, and in the graph (where $i = 9$ and $j = 10$) the path of alignment y, for example, is ${}^y\mathbf{p}_{9,10} = \{(1,1),(3,5),(8,6)\}$.

path with no immediate indel (Fig. 3). The gap scores for each can be used in the DP algorithm as follows:

$$e_{i,j} = f_{i,j} + \max \begin{cases} e_{i-1,j-1} \\ e_{i-1,m} - g - {}^y s_{ij} + {}^z s_{ij} & (m < j - 1) \\ e_{n,j-1} - g - {}^x s_{ij} + {}^z s_{ij} & (n < i - 1) \end{cases} \quad (10)$$

where, e_{ij} is the element being calculated, f_{ij} the contribution from sequence positions i,j (typically from a relatedness matrix), and g the gap penalty.

The additional requirement over the conventional DP algorithm is the extraction of all alignments for each element (this normally happens only once at the end). This step only requires tracing chains of precalculated pointers, but, compared to just finding the largest of three numbers, it imposes considerable additional burden. To reduce this, a so-called greedy algorithm was implemented in which m and n are preselected to correspond with maximum e:

$$e_{i,j} = f_{i,j} + \max$$

$$\begin{cases} e_{i-1,j-1} \\ e_{i-1,w} - g - {}^y s_{ij} + {}^z s_{ij} & (e_{i-1,w} = \max\{e_{i-1,m}, m < j - 1\}) \\ e_{u,j-1} - g - {}^x s_{ij} + {}^z s_{ij} & (e_{u,j-1} = \max\{e_{n,j-1}, n < i - 1\}) \end{cases} \quad (11)$$

This requires evaluation of only three alignments per element (compared to $i + j - 1$) and so retains overall quadratic time dependence (given equal-length profiles). Most importantly, it has been shown to make no significant difference in the results.[2]

Application

Data and Scoring. Three families, aspartyl protease, globin, and SH3 domains, were taken as test data. Each included representatives of known structure, and all members within each family were very distantly related. As previously, alignments were assessed by aligned characters in predefined motifs (see Ref. 2 for details). These were summed pairwise over all aligned sequences and normalized by the number of pairs.

Globins. Globin sequences were aligned initially with the unmodified algorithm using a variety of values for the parameters g [gap penalty, Eq. (9)] and m (matrix softness). Of 110 alignments, 11 gave perfect results. When the modified algorithm was similarly tested no improvement was found (measured by the area containing the correct alignment), despite tests over a variety of values for the parameters t and c [Eq. (9)] (Table III). This lack of improvement might be attributed to the few number of gaps required to align the globins. Detailed examination of the alignments indicated that much of the error derived from the misalignment of one very remotely related (bacterial) sequence.

Acid Proteases. The area of the correct alignment for the proteases (Table IV) in $\{g,m\}$ space was similarly investigated for a variety of combinations of t and c values. In contrast to the globins, improvements were found

TABLE III
GLOBIN CORRECT ALIGNMENT AREAS[a]

			t		
c	1	3	5	7	9
1	4	1	0	0	0
5	9	6	4	3	2
10	10	9	7	4	4
15	11	8	7	7	1

[a] The area of the correct globin alignment in the parameter space of m (relatedness matrix) and g (gap penalty) is shown for combinations of the parameters t and c. The ummodified algorithm had an area of 11.

TABLE IV
PROTEASE CORRECT ALIGNMENT AREAS[a]

c	\multicolumn{8}{c}{t}							
	1	3	5	7	9	11	13	15
1	12	15	15	12	9	8	3	7
5	14	12	15	15	16	20	15	16
10	14	15	13	14	13	13	16	14
15	14	14	9	12	12	11	12	15

[a] The area of the correct protease alignment in the parameter space of m (relatedness matrix) and g (gap penalty) is shown for combinations of the parameters t and c. The unmodified algorithm had an area of 13.

over half the $\{t,c\}$ parameter space, with the largest occurring with high trace-weight (t) (Fig. 4).

SH3 Domains. With SH3 sequences, the correct alignment (which is highly gapped) was never obtained with the unmodified algorithm, whereas with the modified algorithm some correct alignments were found.

Structural Bias for Gaps

Background

Alignment has greatest power when applied to a family of related sequences, particularly, if structural data are available. With such data, the method becomes less sensitive to both the choice of gap penalty and amino acid relatedness matrix (the two weaknesses of the method, whatever the alignment algorithm). With good multiple sequence data most practical methods develop a consensus for each position in the alignment, giving, effectively, a specific local relatedness matrix. As for the case of gaps, as the alignment grows (aligning the most similar sequences first), early gaps (inserted with greatest confidence) provide preferred sites for further insertion.

Gap insertion has been thoroughly investigated.[16-19,21] However, these methods have been concerned mainly with the source of the local bias (typically, based on hydrophobicity or known structure) and not its implementation in the alignment algorithm. In the basic DP algorithm, the gap penalty is modified locally depending on known (or predicted) structure,

[21] R. F. Smith and T. F. Smith, *Protein Eng.* **5,** 35 (1992).

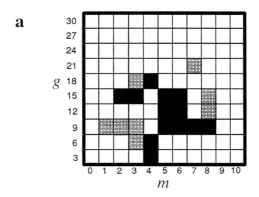

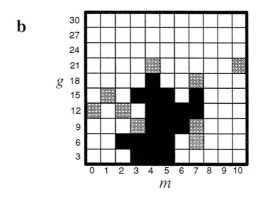

FIG. 4. Parameter-space plots of the acid protease family. The correct multiple alignment of the acid protease motifs is shown as filled cells for combinations of the gap penalty (g) and the matrix softness (m) parameters. The shaded cells indicate almost correct alignments in which only one motif in one sequence is misaligned. Plot (a) shows the results with the unmodified algorithm, whereas (b) shows the largest area found with the gap-bias algorithm using the parameter combination $t = 11$, $c = 5$ (see Table IV).

but this ignores the nature of the inserted sequence and also the opposing (broken) sequence ends. However, in an extended algorithm for structure comparison, Zhu et al.[22] combine these components (using observed solvent accessibilities), and, in the method described in this section, their approach is adopted but accessibility is substituted by a propensity of the sequence to be unstructured.

[22] Z.-Y. Zhu, A. Šali, and T. L. Blundell, *Protein Eng.* **5,** 43 (1992).

Algorithm

Variable Gap Penalty Function. Aligning residue i with j and $i - k + 1$ with $j - 1$, the gap penalty function of Zhu *et al.*[22] can be written

$$s = k(p_j + p_{j-1} + v) + \sum_{m=1}^{k} q_{i-m} \tag{12}$$

where p is the propensity to be broken and q the propensity to be inserted, summed over k residues (v is a constant). Note that Eq. (12) is linearly dependent on length insert; this feature was not adopted, and instead a length-independent function was developed.

As predicted gap sites are less reliable than solvent exposure measurements, a range ($n = 3$) around the broken ends was considered, giving the three sums:

$$a_i = \sum_{m=1}^{k} q_{i-m}/k \tag{13}$$

$$b_j = \sum_{m=0}^{n-1} p_{j+m}/(m + 1) \tag{14}$$

$$c_j = \sum_{m=1}^{n} p_{j-m}/m \tag{15}$$

In addition, the flanking sums (b, c) were weighted by their distance from the break. The function allows the inserted segment and the flanking ends to have different propensities (p and q); however, little justification was seen for this, and a common propensity (p) was used (see below).

A score for insertion (t) can be formed as a function of the three sums (a, b, c) as

$$t = u[da + e(bc)^{1/2}]^f \tag{16}$$

with d, e, and u being weights. A product of ends (bc) was taken to balance the contributions on either side of the break and the square root then taken to moderate the resulting large contribution. Values of the parameter $f = 1$ and $f = \frac{1}{2}$ were investigated. When $f = 1$, $u = 0.01$; when $f = \frac{1}{2}$, $u = 0.1$. For use as a penalty, the function must be inverted, giving

$$r = g/(1 + t) \tag{17}$$

In this form the factor g is equivalent to the conventional gap penalty, which becomes reduced by t when it is favorable to insert a gap.

Empirical and Predicted Gap Propensities

Conservation and Variation. Sequence variation can be measured either by similarity, using one of the many relatedness matrices,[23-25] or by difference. The latter is less well quantified (except using simple identity), but the two approaches can usually be interconverted with little information loss.[26]

Given a matrix of differences (**D**), a score for sequence variation at a given position (k) is

$$v_k = \frac{2}{N(N-1)} \sum_{i=1}^{N-1} \sum_{j=i+1}^{N} D_{a,b} \tag{18}$$

where a,b are indices of **D** for residues found in sequences i,j, respectively, in N aligned sequences. The parameter v was taken directly as the propensity p for insertion. Similarly, using a similarity matrix **M**,

$$w_k = \frac{2}{N(N+1)} \sum_{i=1}^{N} \sum_{j=i}^{N} M_{a,b} \tag{19}$$

and the inverted measure $100/w$ was taken as the propensity p.

Gaps do not appear in any of the above matrices and so must be treated separately. Relative to a matched identity score of 10, an acid/gap match scored $10 + h$ and a gap/gap match scored $10 + 2h$. When using a similarity matrix, gaps were taken to have no similarity to any acid or themselves: their effect was therefore dependent on the absolute values in the relatedness matrix.

Pascarella and Argos Scales. Preference parameters for the likelihood of gap insertion have been calculated by Pascarella and Argos[27] from their database of alignments based on known structures. Ideally for the current application, the preferences were calculated separately for the inserted region (subdivided by size) and the residues on the broken ends of the other sequence. The results of this analysis are summarized in Table V by ranking the amino acids according to their propensity.

The preference scales of Pascarella and Argos[27] span a range that is roughly an order of magnitude larger than their mean standard deviation. While this is sufficient for all the scales to exhibit the expected rough partition of hydrophobic and hydrophilic properties, there is little significant

[23] M. O. Dayhoff, R. M. Schwartz, and B. C. Orcutt, *in* "Atlas of Protein Sequence and Structure" (M. O. Dayhoff, ed.), Vol. 5, Suppl. 3, p. 345. National Biomedical Research Foundation. Washington, D.C., 1978.

[24] S. Henikoff and J. G. Henikoff, *Proc. Natl. Acad. Sci. U.S.A.* **89,** 10915 (1992).

[25] D. T. Jones, W. R. Taylor, and J. M. Thornton, *CABIOS* **8,** 275 (1992).

[26] W. R. Taylor and D. T. Jones, *J. Theor. Biol.* **164,** 65 (1993).

[27] S. Pascarella and P. Argos, *J. Mol. Biol.* **224,** 461 (1992).

TABLE V
RANKED PASCARELLA–ARGOS PREFERENCES[a]

Ends	Inserts			
	1–5	6–10	11–15	16–56
G	N	S	T	W
N	S	G	P	Y
R	G	T	N	G
P	P	Q	Y	E
S	T	D	D	H
T	D	N	Q	R
K	Q	P	E	L
D	K	A	G	N
H	H	F	K	S
Y	A	R	S	D
Q	E	Y	V	T
A	L	C	H	K
C	R	K	L	Q
F	Y	V	M	A
L	F	L	I	I
W	V	W	A	P
V	W	E	R	M
M	M	H	F	F
E	I	I	C	V
I	C	M	W	C

[a] The preferences calculated by Pascarella and Argos[27] have been simplified to a rank ordering. These were calculated for the residue at the broken ends of the insert and the residues found in the insert (classed by the length ranges shown). In the ends and shorter inserts there is a clear preference for G, P, and the small hydrophilic amino acids (D, N, S, and T).

separation between neighbors in the rank order of the acids (Table V), and much of the variation between scales can be interpreted as sampling error, especially for the less numerous larger sized insertions. Within this noise level, there is also little to distinguish the preferences within insertions from those for the broken ends. This can be seen more graphically by combining the results for the insertions as a weighted mean and plotting these values against the scale for the broken ends (Fig. 5).

The broken-end scale of Pascarella and Argos[27] has been used to good effect by Thompson et al.,[15] and although it would be possible to combine this with the values for the inserted segment,[27] they used instead a simple

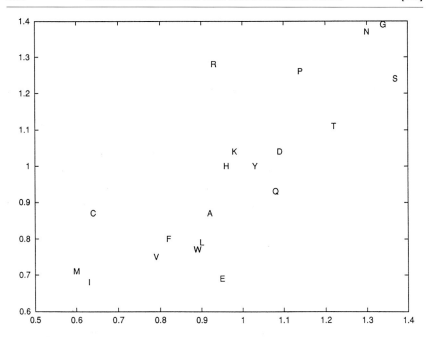

Fig. 5. Comparison of Pascarella and Argos end and insert preferences. The preferences of Pascarella and Argos[27] for a residue to be adjacent to a gap are plotted against the weighted mean of the preferences to be in an insert of any size. The average error in the y axis (ends) is 0.06, and that on the x axis (inserts) is roughly twice this value. Therefore, only arginine (R), cysteine (C), and glutamate (E) deviate significantly.

hydrophilic bias for insertion. It seems doubtful whether there is sufficient distinction among the scales to justify their full implementation.

Chou and Fasman Preferences. A simpler source of preferences can be found in the propensities for residues to be in secondary structure. This source is attractive as it avoids the inherent circularity of deriving parameters to control alignments, from alignments. The simplest source of preferences was taken from the method of Chou and Fasman[28] rather than the more complex methods, such as that of Garnier *et al.,*[29] which consider a local region of sequence.

Concentrating on gaps, the type of secondary structure is not important, only whether there is secondary structure present. For this two-state prediction, preference parameters were not recalculated, but instead a simple

[28] P. Y. Chou and G. D. Fasman, *Adv. Enzymol.* **47,** 45 (1978).
[29] J. Garnier, D. J. Osguthorpe, and B. Robson, *J. Mol. Biol.* **120,** 97 (1978).

score (p) was based on the propensities for turn (P_t), α helix (P_α) and β sheet (P_β) as

$$p = P_t/(P_\alpha + P_\beta) \tag{20}$$

The order of the amino acids ranked on p (Table VI), conforms to expectations: those with unique main-chain stereochemistry (G, P) are highest, followed by those with small, polar side chains (S, N, D), with all the hydrophobic amino acids at the bottom (A, W, F, L, I, M, V). In absolute terms, the scale expands at the hydrophilic end; however, if the logarithm of the values is taken, the scale becomes comparable to those of Pascarella and Argos[27] (Fig. 6). The outlying points on this plot correspond to arginine (R), cysteine (C), proline (P), and glutamate (E), with

TABLE VI
SECONDARY STRUCTURE-BASED GAP
PREFERENCE SCALES[a]

Amino acid	P_α	P_β	P_t	$p - p_{min}$	$\ln p$
P	0.57	0.55	1.59	11.95	2.65
G	0.57	0.75	1.50	9.12	2.43
S	0.77	0.75	1.32	6.44	2.16
N	0.67	0.89	1.35	6.41	2.16
D	1.01	0.54	1.20	5.50	2.05
C	0.70	1.19	1.18	4.00	1.83
H	1.00	0.87	1.06	3.42	1.73
R	0.98	0.93	1.04	3.20	1.69
T	0.83	1.19	1.07	3.05	1.67
K	1.16	0.74	0.98	2.91	1.64
Y	0.69	1.47	1.06	2.66	1.59
E	1.51	0.37	0.84	2.22	1.50
Q	1.11	1.10	0.86	1.65	1.36
A	1.42	0.83	0.70	0.86	1.13
W	1.08	1.37	0.75	0.81	1.12
F	1.13	1.38	0.71	0.58	1.04
L	1.21	1.30	0.68	0.46	1.00
I	1.08	1.60	0.66	0.22	0.90
M	1.45	1.05	0.58	0.07	0.84
V	1.06	1.70	0.62	0.00	0.81

[a] The Chou and Fasman[28] propensities for amino acids to be in α, β, and turn are tabulated with the derivative measure p and its logarithm (on which the entries are ranked). WIth few exceptions this latter measure corresponds well with the Pascarella and Argos[27] scales.

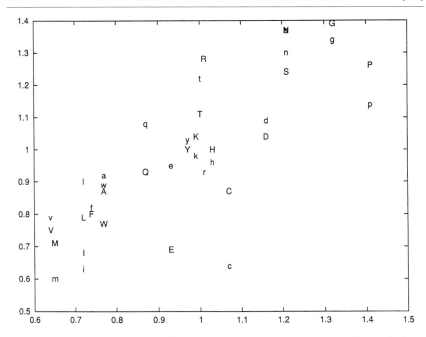

F$_{IG}$. 6. Comparison of Chou–Fasman and Pascarella–Argos preferences. The preferences of Pascarella and Argos[27] for a residue to be associated with a gap (Fig. 5) are plotted on the y axis against the scale (p) derived from the Chou–Fasman propensities for secondary structure (Table VI). The linear scaling of the p values was found that gave the minimum sum-of-squares deviation from both Pascarella–Argos scales (ends and inserts). Points are plotted using the one-letter amino acid code in uppercase for the Pascarella–Argos ends values and lowercase for the Pascarella–Argos insert values.

the three latter acids having a greater propensity to associate with gaps by the p measure.

Application

Calculation of Phase Overlap. As in the previous studies described above, the results of the algorithm were assessed by the size of the phase corresponding to the correct alignment in the parameter space. The parameter space of the two weights {d,e} (corresponding to the inserted segment and the flanking ends) was investigated over a square at 132 points from d = e = 0 to d = e = 10. For each point the area of the correct alignment in the parameter space defined by gap penalty g and matrix softness m was found.

Areas were found for two protein families and their overlap expressed as the percentage size of the union of the two areas. (Perfect overlap would thus score 100 and no overlap 0.) This comparative use of two proteins greatly reduces the possibility of achieving trivial improvements simply by the indirect rescaling of one of the parameters defining the observation space and its adoption as a generally robust protocol will be discussed further below.

The globins and aspartyl proteases were again employed as test data.

Unbiased Method. Without any modification, the basic DP algorithm, as implemented in MULTAL,[7,8] did not result in any overlap of the correct alignment phase for the two proteins. All the overlaps described below are therefore an improvement over the basic algorithm.

Conservation-Based Measures. Using the Dayhoff PAM120 matrix and $f = 1$ [Eq. (16)] produced a band of slight improvement in overlap running antidiagonal across the matrix; similarly, with the square-root variation [$f = \frac{1}{2}$, Eq. (16)], an equivalent stripe across the matrix was observed (Table VII).

Difference-Based Measures

No special gap treatment. The difference measure based only on amino acid identity (with $f = 1$) gave substantial increases in overlap over a broad range of weight combinations, with the best results obtained when both

TABLE VII
PERCENTAGE PHASE-AREA OVERLAP WITH AMINO ACID SIMILARITY
$(f = \frac{1}{2})^a$

e	\multicolumn{11}{c}{d}
	0	1	2	3	4	5	6	7	8	9	10
0								10	10	9	9
1						9	20	10	8	10	8
2						8	11	10	9	10	8
3					8	20	11	9	11	8	9
4			13		8	11	11	10	11	8	8
5			10		10	11	11	14	8	8	8
6		10		8	11	11	10	11	8	8	
7		10	10	20	11	10	11	10			
8	13		9	9	11	10					
9	13	10	20	10							
10	13	9									

[a] Only nonzero values are entered. The weights d and e were applied to the inserted component and the broken-end components.

TABLE VIII

PERCENTAGE PHASE-AREA OVERLAP WITH AMINO ACID DIFFERENCES
$(f = \frac{1}{2})^a$

e	d										
	0	1	2	3	4	5	6	7	8	9	10
0		14					4				
1		18	21	14	18	8	10	3	7	7	13
2		11	26	21	21	15	9	3	15	6	12
3		6	9	16	15	15	12	11	10	6	5
4		14	5	12	14	11	14	21	17	11	9
5				3	15	14	14	17	13	10	10
6				4	17	6	18	17	14	9	16
7		15	5	9		8	18	14	12	11	10
8			11			12	19	14	6	9	8
9		12	4	4			6	10	14	12	12
10		13	5			7	3	8	13	14	14

^{a} Only nonzero values are entered. The parameters d and e are the insert and end weights.

components were weighted. This synergism between the components was even more marked with $f = \frac{1}{2}$ (Table VIII).

Special gap treatment. Setting the gap bonus parameter $h = 5$ with $f = \frac{1}{2}$, made little difference, giving similar results to the unweighted gaps (Table VIII) and exhibiting a wedge of good improvement expanding from the origin (Table IX). Increasing the gap bonus weight to $h = 10$ produced only slight improvement around the diagonal.

Analysis of Results. Cooperativity between the two components (broken ends and insert) was found only with the difference-based measure. This was unexpected and is difficult to rationalize but may be associated with the different softness of the similarity and difference matrices. A position conserving, say, I, L, V, would appear conserved by amino similarity but appear divergent based on a count of different identities. Further studies will be necessary to see if this is a general phenomenon.

Conclusions

Theoretical Considerations

Strictly, the dynamic programming algorithm should be applied only with a metric scoring function; that is, the optimal path can be cut anywhere and the resulting fragments are themselves optimal alignments. The algo-

TABLE IX

PERCENTAGE PHASE-AREA OVERLAP WITH AMINO ACID DIFFERENCES
($f = \frac{1}{2}$) AND FURTHER ENHANCED GAP SIGNIFICANCE ($h = 5$)[a]

e	d										
	0	1	2	3	4	5	6	7	8	9	10
0		20									
1		28	26	19	16						10
2			22	25	19	17	17		10	10	10
3			14	24	19	19	15	13	16	10	11
4		13		15	18	16	20	24	20	14	13
5					21	13	19	17	15	14	14
6		13			18		23	24	23	16	19
7						13	20	17	16	16	17
8						13	18	13	13	12	12
9		13						12	19	18	15
10		10					10	12	15	14	15

[a] Only values over 9 are entered. The parameters d and e are the insert and end weights.

rithms presented in this work, to varying degrees, violate that condition as the maximal path is determined at one end of the sequence by considering information obtained directly from the other end (and not through a locally accumulating sum). The alignment at one end of the sequences (or one end of a gap) thus has a view of the sequence that was not available when the other end was aligned, so breaking the symmetry of the calculation. This implies that if the algorithm were applied to the sequences in reverse, the same answer (alignment and score) might not be obtained. The very existence of such a possibility means that the algorithm cannot guarantee to find the optimal solution.

The related problems of nonoptimality and directional asymmetry would normally be seen as good reasons to abandon the DP algorithm in favor of an algorithm that can guarantee the optimum soluton. However, problems of the type reviewed in the current work (that consider nonlocal interactions) correspond in complexity to the related problems of structure alignment[30,31] and optimal sequence threading,[32] which have been shown to be NP-complete in computational complexity.[33] The only algorithm then available to guarantee an optimal solution is combinatoric enumeration or,

[30] W. R. Taylor and C. A. Orengo, *J. Mol. Biol.* **208,** 1 (1989).
[31] A. Šali and T. L. Blundell, *J. Mol. Biol.* **212,** 403 (1990).
[32] D. T. Jones, W. R. Taylor, and J. M. Thornton, *Nature* (*London*) **358,** 86 (1992).
[33] R. H. Lathrop, *Protein Eng.* **7,** 1095 (1994).

with less certainty, stochastic searching (e.g., simulated annealing). As both these approaches are computationally very expensive on sequences of sufficient length to be of interest, there is strong motivation to retain the dynamic programming approach and either quantify the probable error sizes[2] or deal with them by iteration.[34]

Assessing Alignment Quality

A general problem that has recurred throughout the preceding descriptions is the measurement of alignment accuracy. With known structures, alignment can be based on structural geometry,[30] giving a criterion of truth against which pure sequence alignment can be compared and assessed. With remotely related sequences however, comparison even with known structures can become ambiguous or systematically differ from expectation. For example: the structural alignment of a β hairpin in two proteins might give

$$\text{B B B B B} - - -**- - -\text{B B B B B}$$
$$\text{B B B B B B B B} **\text{B B B B B B B B}$$

where B is a residue in a β strand and * indicates equivalent positions in the turn (– is a gap). Unless the turn is strongly conserved in the sequence, it is unlikely that this alignment would be reproduced with only sequence data: more probable would be the insertion of a single gap.

A measure of alignment quality is needed for remotely related sequences that does not depend on the unreliable exact location of gaps. Previously, motifs have been used to circumvent this problem,[35,36] and, while effective, this approach brings its own problem in the choice of motifs to monitor. If these are too few they will not give a full reflection of alignment quality, but if too many are used, the problems associated with single residue matching reemerge. A balance has been found in the work described above by using relatively few motifs (between two and ten) associated with well-spaced core secondary structures.

The use of relatively few motifs, however, can result in the situation where all motifs are easily aligned, creating a lack of discriminating power (if two parameter sets both get the motifs aligned correctly, it is not possible to decide which is better). This problem was overcome through monitoring the area in parameter space corresponding to the correct alignment. This approach is particularly useful as it quantifies the stability of the correct

---

[34] C. A. Orengo and W. R. Taylor, *J. Theor. Biol.* **147,** 517 (1990).

[35] W. R. Taylor, *J. Mol. Biol.* **188,** 233 (1986).

[36] M. A. McClure, T. K. Vasi, and W. M. Fitch, *Mol. Biol. Evol.* **11,** 571 (1994).

alignment allowing parameter sets to be discriminated by their distance from any incorrect alignment.

Unfortunately, the use of phase areas in parameter space to monitor performance leads to the complication that the area might be trivially changed by modifying a parameter that correlates with a dimension of the space being monitored. For example; if one dimension was the gap penalty, then a new method that simply halved the gap penalty would increase the area of the correct alignment phase by 50%. In the first study described above, where such correlation was possible, it was monitored by taking the number of gaps as an invariant control.[1]

Although the number of gaps is satisfactory, it is not an ideal control, and it was avoided in the later study of structure-biased gap penalties where the coupling between area and parameter is direct. In this work, a better control was found by simultaneously monitoring the phase space of two protein families and using the percentage overlap as a measure of improvement. This is exactly the requirement of practical studies, such as multiple alignment or databank scanning, in which an investigator takes a default parameter set and expects it to be optimal over as wide a range of proteins as possible.

Having established the protocol many further applications await. Of great practical importance would be to search for the parameter combination that gives the correct alignment for the maximum number of sequence families in the data banks. In a more theoretical direction, the phase space of the correct alignments could be investigated at increasing sample densities to delineate their outlines. With just two sequences these areas are polygons,[37] but given the much greater complexity of multiple alignment, irregular (possibly fractal) shapes would be expected.

Integration with Iteration

The algorithms described above have been coded into separate modified versions of MULTAL.[7,8] Each method still requires considerable work to fully explore their individual behavior over a larger number of proteins, however, and work is currently underway to combine the methods into a common implementation that uses iteration to overcome directional asymmetry in the calculations and simultaneously provide a stopping condition.

---

[37] M. Vingron and M. S. Waterman, *J. Mol. Biol.* **235,** 1 (1994).

[21] Progressive Alignment of Amino Acid Sequences and Construction of Phylogenetic Trees from Them

By DA-FEI FENG and RUSSELL F. DOOLITTLE

Introduction

In 1970, Needleman and Wunsch published an elegant algorithm for optimally aligning two protein sequences.[1] Moreover, the scheme could be used with weighting scales that assigned scores to each set of paired residues and assessed penalties for unpaired ones (gaps). Arguably, the most popular of these amino acid substitution matrices was the PAM scale devised by Dayhoff and co-workers, which was based on observed mutations for sets of closely related protein sequences.[2,3] During the 1970s and early 1980s, numerous investigators used the Needleman–Wunsch algorithm in conjunction with various versions of the Dayhoff PAM matrix to generate binary (pairwise) alignments. On the other hand, multisequence alignments were usually made manually, the binary alignments serving as a guide.

Although in principle the Needleman and Wunsch approach could be extended to multiple sequences, in practice the computational time for such a task proved prohibitively long, the memory required for storing the necessary arrays being enormous even for sets of very short sequences.[4,5] Although global methods were devised that yielded reasonable alignments on longer sequences, they were still restricted to five or fewer sequences.[6] An even more disappointing aspect was that the pattern of gaps in multiple alignments was often inconsistent with what was observed in binary alignments; thus, the optimal alignment between the two closest sequences as indicated by a binary alignment was often altered in the presence of a third or fourth sequence.[6]

[1] S. B. Needleman and C. D. Wunsch, *J. Mol. Biol.* **48,** 443 (1970).

[2] M. O. Dayhoff, R. V. Eck, and C. M. Park, *in* "Atlas of Protein Sequence and Structure" (M. O. Dayhoff, ed.), Vol. 5, p 89. National Biomedical Research Foundation, Washington, D.C., 1972.

[3] R. M. Schwartz and M. O. Dayhoff, *in* "Atlas of Protein Sequence and Structure" (M. O. Dayhoff, ed.), Vol. 5, Suppl. 3, p. 353. National Biomedical Research Foundation, Washington, D.C., 1978.

[4] R. A. Jue, N. W. Woodbury, and R. F. Doolittle, *J. Mol. Evol.* **15,** 129 (1979).

[5] M. Murata, J. S. Richardson, and J. L. Sussman, *Proc. Natl. Acad. Sci. U.S.A.* **82,** 3073 (1985).

[6] M. S. Johnson and R. F. Doolittle, *J. Mol. Evol.* **23,** 267 (1986).

Copyright © 1996 by Academic Press, Inc.
All rights of reproduction in any form reserved.

In 1987, we reported a simple progressive alignment procedure that circumvented the need for aligning sequences exhaustively.[7] Like other methods introduced at about the same time,[8,9] it actually used a binary alignment algorithm[1] iteratively, first to align the two most closely related sequences and then to align the next most similar one to those, etc. The strategy was further guided by the simple rule "once a gap, always a gap," the rationale being that the positions and lengths of gaps introduced between the more similar pairs of sequences should not be affected by distantly related ones.[7] As such, sequences were compared in an approximately descending order of similarity, each new overall alignment score being determined by comparisons of residues of the last added sequence against the averaged scores of all previously aligned residues. The program was written in such a way that the order could be refined automatically by checking the next two sequences at each round, priority being determined on the basis of the higher similarity score. This operation was continued until all the sequences were aligned. We referred to this procedure as progressive alignment. Because progressive alignment is not exhaustive, the required computational time is short compared with global alignment procedures.

Progressive Alignment and Phylogenetic Trees

Although many investigators are only interested in alignments per se, our interest has been driven by a need for the automatic construction of sequence-based phylogenetic trees. Historically, the first step in constructing a quantitative phylogenetic tree has always been to find an objective procedure for obtaining a multiple alignment. In our 1987 paper[7] we presented a suite of programs that allowed users to go directly from the aligned sequences to a phylogenetic tree, a number of different programs being used sequentially. Some steps had to be performed by hand, however, and we later described a version in which many of the steps were tied together in a more user-friendly fashion.[10] Even then, the system still had some annoying limitations, however. In this chapter we describe a much improved, easier to use, set of programs, which we refer to as ProPack: a packet of programs centering around progressive alignment (Table I). The changes pertain mainly to speeding up the calculations and reducing the number of manual operations; the basic idea of progressive alignment remains the same.

[7] D. F. Feng and R. F. Doolittle, *J. Mol. Evol.* **25,** 351 (1987).
[8] W. R. Taylor, *CABIOS* **3,** 81 (1987).
[9] G. J. Barton and M. J. E. Sternberg, *J. Mol. Biol.* **198,** 327 (1987).
[10] D. F. Feng and R. F. Doolittle, this series, Vol. 183, p. 375.

TABLE I

PROPACK SUITE OF PROGRAMS FOR PROGRESSIVE ALIGNMENT AND MAKING TREES

Program	Objective
FORMAT	Convert sequences to Old Atlas format
INSPECT	Find matching boundaries
CROP	Cut sequences to order
ARRANGE	Determine an approximate branching order
PREALIGN	Prealign cluster of sequences
ALIGN	Determine multiple alignment only
MULPUB	Change aligned sequences to compact format
TREE	Make a multiple alignment and construct a phylogenetic tree from it
BLEN	Determine branch lengths based on distance matrix
NONEG	Find best branching order with no negative branch lengths
SETREE	Convert information to a format recognized by tree drawing program
BTREE	Sample multiple alignments and generate new trees for bootstrapping procedure
CLUS	Analyze clusters (nodes) from BTREE for agreement with initial tree

Improvements

The principal new features are (a) introduction of an option for choosing an alternative amino acid substitution matrix, (b) automatic elimination of negative branch lengths in the calculation of phylogenetic trees, (c) addition of a bootstrap analysis option, and (d) inclusion of a simple program for converting output to a form that can be used to draw trees on a microcomputer with standard software. Additionally, experience has shown that the prealignment of subclusters is seldom necessary, and as a result the program PREALIGN is downplayed in the new ensemble. Instead, sequences can be input in any order, with the program ARRANGE automatically sorting them into an approximate order for processing by the main program TREE. A summary of these programs is presented in Table I, and a flowchart of the new scheme is depicted in Fig. 1.

Programs

The computer programs described here are for protein sequence comparisons (not DNA or RNA). They are written in the C language[11] and run under the UNIX operating environment, although they have also been used successfully by others with a VMS system after minor modifications.

[11] B. W. Kernighan and D. M. Ritchie, "The C Programming Language." Prentice-Hall, Englewood Cliffs, New Jersey, 1978.

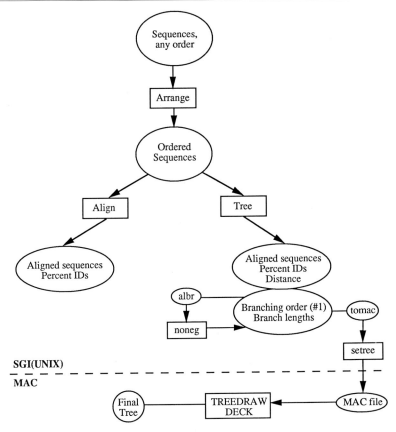

FIG. 1. Flow diagram of progressive alignment and phylogenetic tree construction. The names in ovals denote files; square boxes represent programs. All programs shown above the dashed line run under the UNIX operating system via a command line. A description of the command line format for each of the programs can be obtained simply by typing the name of the program after these programs have been compiled on the home computer.

In their present form they are exactly adapted for running on Silicon Graphics machines (IRIX 4.0.5 IOP).

The primary program TREE, which produces the multiple alignment, the branching order, and the branch lengths, can handle as many as 52 sequences (a number that is incidental to the number of lowercase and uppercase letters that are used as designation for entries). The maximum lengths of the sequences are strictly a function of the available computer memory. At the present time, sequences of 1000 residues can be aligned easily. The sequences must be provided in the "Old Atlas" format (Fig.

```
TDEHU  L-lactate dehydrogenase (EC 1.1.1.27) chain M - human
PDEHU    1 K I T V V G V G A V G M A C A I S I L M K D L A D E L A L V
PDEHU   31 D V I E D K L K G E M M D L Q H G S L F L R T P K I V S G K
PDEHU   61 D Y N V T A N S K L V I I T A G A R Q Q E G E S R L N L V Q
PDEHU   91 R N V N I F K F I I P N V V K Y S P N C K L L I V S N P V D
PDEHU  121 I L T Y V A W K I S G F P K N R V I G S G C N L D S A R F R
PDEHU  151 Y L M G E R L G V H P L S C H G W V L G E H G D S S V P V W
PDEHU  181 S G M N V A G V S L K T L H P D L G T D K D K E Q W K E V H
PDEHU  211 K Q V V E S A Y E V I K L K G Y T S W A I G L S V A D L A E
PDEHU  241 S I M K N L R R V H P V S T M I K G L Y G I K D D V F L S V
PDEHU  271 P C I L G Q N G I S D L V K V T L T S E E E A R L K K S A D
PDEHU  301 T L W G I Q *
```

FIG. 2. Example of Old Atlas format.

2). The program FORMAT changes sequences into the proper form so that they are recognized by all the programs.

Although program names are capitalized throughout this chapter in order to emphasize what is actually typed, on a day-to-day basis we ordinarily use lowercase. It must be emphasized that UNIX is rigorously case sensitive.

Converting Similarity to Distance

Similarity scores for pairs of aligned sequences are converted to difference scores by the following version of the Poisson equation:[12]

$$D = -\ln S \tag{1}$$

where

$$S = [S_{real}(ij) - S_{rand}(ij)]/[S_{iden}(ij) - S_{rand}(ij)] \times 100 \tag{2}$$

$S_{real}(ij)$ is the observed similarity score for the two sequences being aligned, and $S_{iden}(ij)$ is the average of the two scores for the two sequences compared with themselves. $S_{rand}(ij)$ provides a measure of the background noise between sequences i and j. Previously $S_{rand}(ij)$ was determined by a computer-intensive shuffling operation; now it is calculated by a formula, as discussed below.

S_{rand}

The random score between two optimally aligned sequences, $S_{rand}(ij)$, is a function of the composition of the sequences, the number of internal gaps in the multiple alignment, the gap penalty, and the scoring matrix used. To put this on a mathematical basis, the following conditions obtained:

[12] D. F. Feng, M. S. Johnson, and R. F. Doolittle, J. Mol. Evol. 21, 112 (1985).

a_i, b_j represent the residue type in sequence i and j, $N_a(i)$, $N_b(j)$ are the number of times a_i, b_j appear in sequences i and j, N_g is the number of internal gaps unique to either sequence i or j, irrespective of their lengths, *pen* is the gap penalty used in the calculation, and L is the overall length of the sequences after they are aligned. Now, if we avail ourselves of a suitable amino acid substitution matrix M, then

$$S_{rand} = (1/L) \Sigma \Sigma M(a_i, a_j)N_a(i)N_b(j) - N_g * pen \qquad (3)$$

Equation (3) allows S_{rand} to be calculated independently of a random number generator; not only is the process speeded up, but it is more consistent.

Outline of Operations

In essence, there are 10 steps to making a sequence-based tree with our programs: (a) gather the sequences, (b) format and otherwise edit the sequences, (c) catenate the sequences into a single file, (d) arrange the sequences into an appropriate order, (e) align the sequences, (f) calculate the similarity scores and pairwise distances, (g) make a distance (difference) matrix, (h) determine the branching order and branch lengths, (i) if necessary, find an alternate solution without negative branch lengths, and (j) draw a tree. One also has the option of performing a bootstrap analysis, and we comment further on this subsequently.

Input

Gathering the sequences from appropriate databases is a straightforward matter. To put them into a suitable format for our programs, one must first strip out all nonsequence alphabetic characters, including the title (numbers will be ignored).

FORMAT temp >seq1

The program will then ask for an appropriate title and a four-character identifier. Once a sequence is in the "old Atlas" format, it can also be readily edited with a program called CROP so that only specified segments are called on.

CROP seq1 23 487 >seq1X

In this case an edited sequence composed of residues 23 to 487 is put into a separate file. CROP is a very useful program because good alignments of distantly related sequences cannot be obtained if the lengths of the sequences are too different. In this regard, the program INSPECT[13] is

---

[13] R. F. Doolittle, "URFs and ORFs: A Primer on How to Analyze Derived Amino Acid Sequences," p. 21. University Science Books, Mill Valley, California, 1986.

helpful for identifying appropriate cut points in sequences of different lengths or mosaic composition.

Once the sequences have been gathered, formatted, and cropped to specification, they need to be catenated into a single file. As suggested above, the order is important. Often it is sufficient to input the sequences in an approximate biological order. A more rigorous way is to find an approximate branching order based on pairwise alignments only. The program called ARRANGE makes an alignment of every pair and uses the scores to find an approximate branching order by the method of Fitch–Margoliash.[14] This approximate branching order can then be used as a guide to catenating the sequences into a starting file for the multiple alignment.

Alignment Strategy

Once the sequences are in a single file in an approximate order, they can be aligned with a program called ALIGN. Consider what goes on in this program. Let A, B, C, D, E, ..., be a set of sequences in descending similarity order as determined by ARRANGE. The ALIGN program first aligns A and B. If (AB) represents an optimal alignment between these two sequences, then C(AB) and (AB)C are tried next, the set with the higher similarity score determining whether C(AB) or (AB)C will be the three-way alignment. Assuming that (AB)C has a higher score, the next step involves trying the two alignments, ((AB)C)D and ((AB)D)C. Again, the higher similarity score determines whether ((AB)C)D or ((AB)D)C will be the four-way alignment. This procedure is repeated until all the sequences are aligned. Because of this iterative process, the order of the sequences in the resultfile alignment may be different from that generated by ARRANGE as presented in seqfile.

$$\text{ALIGN seqfile seqfile1 } M > \text{resultfile}$$

where M is the amino acid mutation matrix which will be discussed in more detail below. Seqfile1 is a file written during execution that contains the sequences with X's inserted as neutral elements at all gap positions. These elements are neutral in the sense that the matching of an X with a residue in the other sequence results in a value of zero. The number of X's tends to increase as more sequences are brought into the alignment, the direct result of the rule "once a gap, always a gap." In addition to its use in scoring within the ALIGN program itself, the file is also used as the input

---

[14] W. M. Fitch and E. Margoliash, *Science* **155**, 279 (1967).

file for MULPUB, a program that converts the arrangement to a close-pack form of any specified size (in column lengths). Thus,

MULPUB seqfile1 80 > seqfileM

rearranges the aligned sequences in a close-packed array 80 columns wide (Fig. 3).

Choice of Matrices

In addition to the Dayhoff (PAM250) matrix,[3] which has been used commonly in the past, a number of other scoring matrices have been published, including one by Gonnet et al.[15] and another based on block alignments by Henikoff and Henikoff.[16] Either of these matrices can be used with the ARRANGE, ALIGN, or TREE programs. In an extensive study involving enzyme sequences, we found that the differences between the PAM250 and the Gonnet matrices are small, and both produce reliable results for sequences that are closely related to each other. The BLOSUM62 matrix, on the other hand, appears to give better alignments for more distantly related sequences. For the PAM250 and Gonnet matrices the gap penalty should be set at 8; in the case of the BLOSUM62 matrix, we have found a gap penalty of 6 to be optimum. These values were initially determined by a consideration of the distribution of the scores used in these matrices and then verified empirically by inspection of the alignments produced by them. The Dayhoff PAM250, Gonnet, or BLOSUM matrices can be called simply by adding a D, G, or B to the command line, respectively.

ALIGN seqfile seqfile1 D > resultfile   (use Dayhoff PAM250, gap penalty = 8)

or

     ALIGN seqfile seqfile1 G > resultfile   (use Gonnet, gap penalty = 8)

or

     ALIGN seqfile seqfile1 B > resultfile   (use BLOSUM62, gap penalty = 6)

Making a Tree

Alternatively, the sequences can be aligned and a tree constructed directly by the program called TREE. The early operations of the ALIGN

[15] G. H. Gonnet, M. A. Cohen, and S. A. Benner, *Science* **256,** 1443 (1992).
[16] S. Henikoff and J. G. Henikoff, *Proteins: Struct. Funct. Genet.* **89,** 10915 (1992).

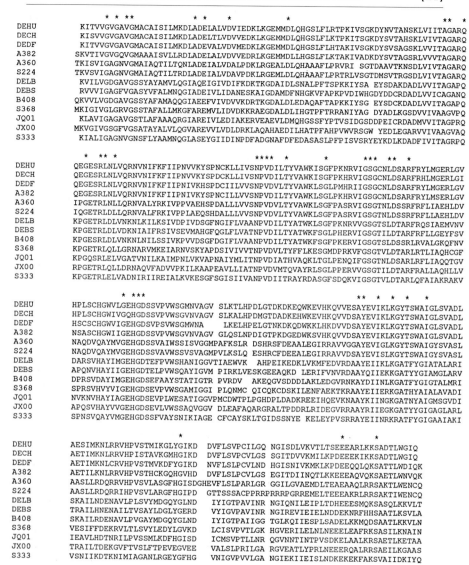

Fig. 3. Example of close-pack format that emerges from MULPUB set at 80 columns across. Thirteen L-lactate dehydrogenase sequences were aligned. The four-letter designations are as follows: DEHU, human; DECH, chicken; DEDF, spiny dogfish; A382, sea lamprey; A360, barley; S224, maize; DELB, *Lactobacillus casei*; DEBS, *Bacillus stearothermophilus*; B408, *Lactobacillus plantarum*; S368, *Thermotoga maritima*; JQ01, *Bifidobacterium*; JX00, *Thermus aquaticus*; S333, *Mycoplasma*.

and TREE programs are identical; the only reason to use ALIGN is the time saving if a tree is not the goal. The TREE program goes on to calculate the branching order and branch lengths.

$$\text{TREE seqfile seqfile1 } M > \text{seqfile2}$$

Moreover, the ARRANGE and TREE programs can be called in a single step by use of the shell file OVERALL, which consists of the following two lines,

$$\text{ARRANGE \$1 \$4} > \text{\$2}$$
$$\text{TREE \$1 \$3 \$4} > \text{\$5}$$

The user must specify the starting file with the collected sequences and designate three new files for the storage of outputs:

$$\text{OVERALL seqfile seqfile1 seqfile2 } M > \text{seqfile3}$$

In this case, the final alignment, percent identities, distance, branching order, and branch lengths appear in seqfile3.

Removing Negative Branch Lengths

There are a number of different ways of reducing a difference matrix composed of m intergroup distances to a phylogenetic tree with n branch lengths. One can solve all the simultaneous equations, for example, or one can use a least squares approach, as suggested by Klotz and Blanken[17] and Li.[18] Although we favor the least squares approach, it does have the flaw that often the best mathematical solution contains one or more negative values, always associated with the inner branch segments of the tree. Most often, the problem can be remedied by simply switching the two taxa (or groups of taxa) that arise at either end of the offending segment (Fig. 4). Many times, however, finding the best arrangement free of negative values is challenging to the point of frustration.

Accordingly, we devised a program that looks for the best matrix-derived tree with no negative segments. It is called NONEG. It works by switching taxa in an iterative manner and testing the result both for the presence of negative segments and the quality of the tree. The latter is measured by a comparison of the initially provided intergroup distances with the final intergroup distances obtained from summing the appropriate branchlengths and is given as a percent standard deviation. The input data

[17] L. C. Klotz and R. L. Blanken, *J. Theor. Biol.* **91**, 261 (1981).
[18] W.-H. Li, *Proc. Natl. Acad. Sci. U.S.A.* **78**, 1085 (1981).

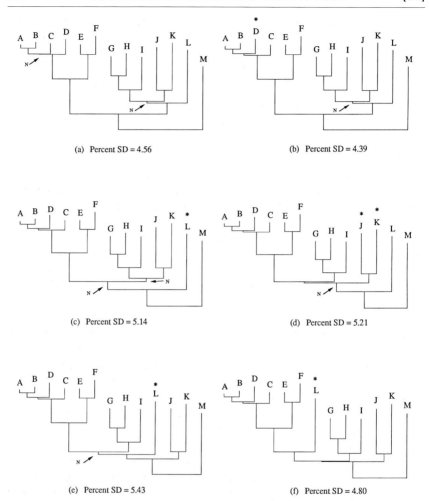

FIG. 4. Illustration of branch swapping that occurs during the elimination of negative branch lengths. In the original tree (a) the arrows indicate two negative branch lengths (for drawing purposes, they are shown as having positive lengths). Asterisks (*) denote taxa involved in switching at various steps. For example, switching the taxa C and D eliminates the first of the negative branch lengths (b) and at the same time improves the overall tree as evidenced by the percent standard deviation dropping from 4.56 to 4.39. Subsequent switchings (c, d, or e) actually make the tree worse without eliminating the final negative value, until finally (f) a solution is reached.

for NONEG appear automatically in a file called ALBR (for alternate branching) every time the program TREE is run. The protocol is simply

NONEG albr > newfile

The newfile contains the initial, intermediate, and final branching orders, along with their respective percent standard deviations.

Bootstrap Analysis

The purpose of making a phylogenetic tree is to provide an estimate of the underlying biological relationships. Because the tree is derived from a multiple sequence alignment, which in turn depends on fluctuations in the real biological world, and to a lesser extent on systematic variables such as the gap penalty and the scoring matrix, the information is expected to have inherent variability. Felsenstein applied the bootstrapping method to phylogenetic trees in an effort to estimate these statistical fluctuations.[19] The method has become a popular way of expressing confidence about the branching order on a node by node basis.

In our version, bootstrapping is invoked by substituting the program BTREE for TREE. The bootstrap itself is performed as follows: the alignment and branching order are found as described above. Then, columns of residues are sampled randomly from the multiple alignment, with replacement. Columns in which gaps (represented as neutral element X) occur in more than half of the positions, and erratic overhangs, if present, are ignored. This sample selection process continues until the lengths of the sequences are the same as the originally aligned sequences. Equations (1), (2), and (3) are used to determine the distance matrix. The best branching order with all positive branch lengths is then saved. The procedure is repeated at least 100 times. Finally, all the branching orders are compared with the original branching order on a node by node basis. The number of times a cluster (around a node) is the same as in the original tree is tabulated; the results are expressed in percentages and placed at the nodes.

Tree Draw Interface

The program TREE produces two hidden files, ALBR and TOMAC. Both contain a branching order and a set of branch lengths, but their formats are different. As noted above, ALBR (alternate branch lengths) is used as an input file for NONEG whenever one or more negative branch

[19] J. Felsenstein, *Evolution* **39,** 783 (1985).

```
(((((AB)C)D) (EF)) ((((GH)I) (JK))L))M
     % s.d. (obs. vs calc.) =    4.56

(((((AB)D)C) (EF)) ((((GH)I) (JK))L))M
     % s.d. (obs. vs calc.) =    4.39

((((((AB)D)C) (EF)) (((GH)I) (JK)))L)M
     % s.d. (obs. vs calc.) =    5.14

(((((((AB)D)C) (EF)) ((GH)I)) (JK))L)M
     % s.d. (obs. vs calc.) =    5.21

(((((((AB)D)C) (EF)) ((GH)I))L) (JK))M
     % s.d. (obs. vs calc.) =    5.43

((((((((AB)D)C) (EF))L) ((GH)I)) (JK))M

Branch lengths are  (r1,r2,r3,r4,r5):
                     (r6,r7,........):

      1.96      3.74      3.05      9.65      0.99
      6.91     15.76      6.24     11.94     16.38
     14.84     33.56      5.91     12.79     14.85
      9.28     21.67      4.94      0.44     28.33
     31.22      5.01     50.53

      % s.d. (obs. vs calc.) =    4.80
```

FIG. 5. Output of NONEG using the L-lactate dehydrogenase tree example.

lengths appear. It depicts the branching order with a set of consecutive lines that describe clusters:

ABCDEFG

DE

*

The asterisk denotes the end of branching order information and signifies the beginning of the array of branch lengths.

TOMAC, on the other hand, is the input file to use on the way to tree drawing; it is automatically updated by NONEG. The file begins with the convenient one-line notation of the branching order as described by Fitch[20] and is followed directly by the branch lengths, data that can be read by the program SETREE. SETREE is an interactive program in which the user has a choice of using the default four-letter code or of specifying a new name for each taxon (limited to 20 characters; no spaces or parentheses allowed). The output file, whatever it is named, will contain the phylogenetic tree information in a format that can be read by a hypercard tree drawing program called TREEDRAW DECK that has been adapted from the

[20] W. M. Fitch, *Am. Nat.* **111**, 223 (1977).

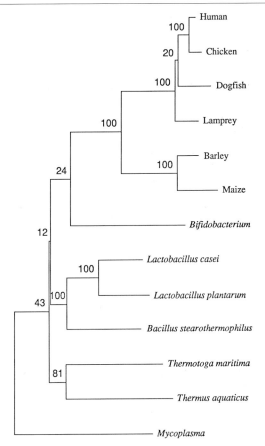

FIG. 6. Bootstrap analysis (100 trials) of 13 L-lactate dehydrogenase sequences. The analysis was restricted to trees with all positive branch lengths.

Felsenstein PHYLIP programs[21] by D. G. Gilbert of Indiana University. This program can be downloaded from the Internet (dgilbert@iubio.bio.indiana.edu) and installed on a suitable microcomputer (our version happens to interface with a MacIntosh). All of the above steps are summarized in Fig. 1.

Example

An illustration of these programs can be afforded by a consideration of a set of L-lactate dehydrogenase (LDH) sequences. From the PIR data-

[21] J. Felsenstein's PHYLIP programs can be obtained by anonymous ftp from 128.95.12.41.

base, 13 LDH sequences were taken, run through the program FORMAT, and then catenated into a file called TEST1, without regard to order. The program ARRANGE was used to calculate first all the binary alignments and put the sequences in an approximate order, and then to proceed directly to making the multiple alignment and finding a branching order.

OVERALL TEST1 TEST2 TEST3 D > TEST4

The output file TEST4 contains the aligned sequences, tables of similarity and distance, a branching order, and a set of branch lengths determined by the method of least squares. As it happens, in this instance, the table of branch lengths contains two negative values (Fig. 4a). Accordingly, the program NONEG was used to find a valid branching order.

NONEG ALBR > TEST5

The output (in TEST5) shows the beginning, intermediate, and final branching orders and their respective percent standard deviations (Fig. 5).

The SETREE program is then used to generate a file for tree draw.

SETREE tomac > tomac1

Finally, the bootstrap version of the program was conducted separately:

BTREE seqfile seqfile1 D > resultfile
CLUS tomac1 rboot > rboot1

It must be underscored that the bootstrapping is restricted to solutions with no negative branch lengths, a necessary condition if the likelihoods are to have any meaning. This condition is often ignored by other bootstrapping procedures involving amino acid alignments made by matrix methods. The results are stored in a hidden file named rboot. A full calculation generally consists of 100 bootstrap trials. When rboot is analyzed by the program CLUS, the number of times (for 100, it is the percentage) a node is in agreement with that in the starting tree is shown on the phylogenetic tree (Fig. 6).

Program Availability

All the programs described in this chapter are available by anonymous ftp from juno.ucsd.edu. After logging in with any identifying name as a password, claimants should go to the directory progs/.

[22] Using CLUSTAL for Multiple Sequence Alignments

By Desmond G. Higgins, Julie D. Thompson,
and Toby J. Gibson

Introduction

The simultaneous alignment of many nucleotide or amino acid sequences is now one of the commonest tasks in computational molecular biology. It forms a common prelude to phylogenetic analysis of sequences, the prediction of secondary structure (DNA or protein), the detection of homology between newly sequenced genes and existing sequence families, the demonstration of homology in multigene families, and the finding of candidate primers for PCR (polymerase chain reaction). For such a widely required analysis, it is surprising how long it took for practical methods to appear. Up until 1987, it was standard practice to construct multiple alignments manually. This is very tedious and error prone. The basic problem was that direct extensions of the standard dynamic programming approach for the alignment of two sequences were computationally impossible for more than three real sequences. Some methods had been invented[1–4] based on trying to find alignment blocks or on iterating toward a consensus sequence, for example, but these were not very widely used. The first practical methods for the sensitive multiple alignment of many protein sequences appeared in 1987 and 1988[5–10] and were based on an original idea by David Sankoff.[11] The idea is to exploit the fact that groups of sequences are phylogenetically related (if they can be aligned, there is usually an underlying phylogenetic tree). This approach is commonly referred to as progressive alignment.[6] Most of the automatic multiple alignments that appear in the current literature are carried out using this approach.

[1] W. Bains, Nucleic Acids Res. 14, 159 (1986).
[2] E. Sobel and H. M. Martinez, Nucleic Acids Res. 14, 363 (1986).
[3] D. J. Bacon and W. F. Anderson, J. Mol. Biol. 191, 153 (1986).
[4] M. S. Johnson and R. F. Doolittle, J. Mol. Evol. 23, 267 (1986).
[5] W. R. Taylor, CABIOS 3, 81 (1987).
[6] D.-F. Feng and R. F. Doolittle, J. Mol. Evol. 25, 351 (1987).
[7] G. J. Barton and M. J. E. Sternberg, J. Mol. Biol. 198, 327 (1987).
[8] W. R. Taylor, J. Mol. Evol. 28, 161 (1988).
[9] D. G. Higgins and P. M. Sharp, Gene 73, 237 (1988).
[10] F. Corpet, Nucleic Acids Res. 16, 10881 (1988).
[11] D. Sankoff, SIAM J. Appl. Math. 78, 35 (1975).

METHODS IN ENZYMOLOGY, VOL. 266
Copyright © 1996 by Academic Press, Inc.
All rights of reproduction in any form reserved.

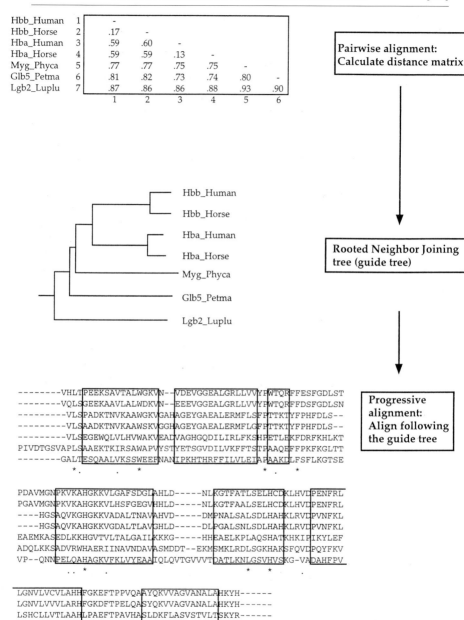

Hbb_Human	1	-					
Hbb_Horse	2	.17	-				
Hba_Human	3	.59	.60	-			
Hba_Horse	4	.59	.59	.13	-		
Myg_Phyca	5	.77	.77	.75	.75	-	
Glb5_Petma	6	.81	.82	.73	.74	.80	-
Lgb2_Luplu	7	.87	.86	.86	.88	.93	.90
		1	2	3	4	5	6

Pairwise alignment:
Calculate distance matrix

Hbb_Human
Hbb_Horse
Hba_Human
Hba_Horse
Myg_Phyca
Glb5_Petma
Lgb2_Luplu

Rooted Neighbor Joining
tree (guide tree)

Progressive
alignment:
Align following
the guide tree

```
--------VHLTPEEKSAVTALWGKVN--VDEVGGEALGRLLVVYPWTQRFFESFGDLST
--------VQLSGEEKAAVLALWDKVN--EEEVGGEALGRLLVVYPWTQRFFDSFGDLSN
---------VLSPADKTNVKAAWGKVGAHAGEYGAEALERMFLSFPTTKTYFPHFDLS--
---------VLSAADKTNVKAAWSKVGGHAGEYGAEALERMFLGFPTTKTYFPHFDLS--
---------VLSEGEWQLVLHVWAKVEADVAGHGQDILIRLFKSHPETLEKFDRFKHLKT
PIVDTGSVAPLSAAEKTKIRSAWAPVYSTYETSGVDILVKFFTSTPAAQEFFPKFKGLTT
--------GALTESQAALVKSSWEEFNANIPKHTHRFFILVLEIAPAAKDLFSFLKGTSE
        *.           .        *                    *  .    *
```

```
PDAVMGNPKVKAHGKKVLGAFSDGLAHLD-----NLKGTFATLSELHCDKLHVDPENFRL
PGAVMGNPKVKAHGKKVLHSFGEGVHHLD-----NLKGTFAALSELHCDKLHVDPENFRL
----HGSAQVKGHGKKVADALTNAVAHVD-----DMPNALSALSDLHAHKLRVDPVNFKL
----HGSAQVKAHGKKVGDALTLAVGHLD-----DLPGALSNLSDLHAHKLRVDPVNFKL
EAEMKASEDLKKHGVTVLTALGAILKKKG-----HHEAELKPLAQSHATKHKIPIKYLEF
ADQLKKSADVRWHAERIINAVNDAVASMDDT--EKMSMKLRDLSGKHAKSFQVDPQYFKV
VP--QNNPELQAHAGKVFKLVYEAAIQLQVTGVVVTDATLKNLGSVHVSKG-VADAHFPV
        ..  *       .                      *    *       .
```

```
LGNVLVCVLAHHFGKEFTPPVQAAYQKVVAGVANALAHKYH------
LGNVLVVVLARHFGKDFTPELQASYQKVVAGVANALAHKYH------
LSHCLLVTLAAHLPAEFTPAVHASLDKFLASVSTVLTSKYR------
LSHCLLSTLAVHLPNDFTPAVHASLDKFLSSVSTVLTSKYR------
ISEAIIHVLHSRHPGDFGADAQGAMNKALELFRKDIAAKYKELGYQG
LAAVIADTVAAG--------DAGFEKLMSMICILLRSAY-------
VKEAILKTIKEVVGAKWSEELNSAWTIAYDELAIVIKKEMNDAA---
     .     .        .                 .
```

Progressive alignment involves making initial guesses as to the phylogenetic relatedness of the sequences and using the branching order in an initial phylogenetic tree to align larger and larger groups of sequences. Start by aligning the most closely related pairs of sequences using dynamic programming and gradually align these groups together, keeping gaps that appear in early alignments fixed. At each stage, align two sequences or one sequence to an existing subalignment or align two subalignments. There are now many variations on the approach. The initial tree may be modified as the procedure progresses[12]; alternative branching orders may be tested at each stage[6]; the initial tree may be replaced by a tree calculated from the fully aligned sequences and the procedure iterated until the alignment (or tree) converges[10]; sequences may be aligned together in a nonphylogenetic manner but rather by adding sequences one at a time to a growing alignment.[7] Further, one can use different strategies for aligning the groups of prealigned sequences. For example, one can base the alignment only on the alignment of the two most closely related sequences (one from each alignment).[5] In most current implementations, the subalignments are aligned together using information from all of the constituent sequences with an extension of the profile alignment approach.[13] An example of the approach is shown in Fig. 1 for the alignment of seven globin sequences of known tertiary structure.

Algorithmically, the progressive approach is considered to be "heuristic" insofar as the method is not guaranteed to produce alignments with any particular mathematical property such as maximum alignment score. The method is, however, soundly based biologically and has the great

[12] J. Hein, *Mol. Biol. Evol.* **6,** 649 (1989).
[13] M. Gribskov, A. D. McLachlan, and D. Eisenberg, *Proc. Natl. Acad. Sci. U.S.A.* **84,** 4355 (1987).

FIG. 1. Outline of the progressive multiple alignment approach in CLUSTAL W. Seven globin sequences of known tertiary structure are used (the sequence names are identifiers from the SWISS-PROT sequence database). The alignment was carried out using default parameters except that the Dayhoff PAM weight matrices [M. O. Dayhoff, R. M. Schwartz, and B. C. Orcutt, *in* "Atlas of Protein Sequence and Structure" (M. O. Dayhoff, ed.), Vol. 5, Suppl. 3, p. 345. National Biomedical Research Foundation, Washington, D.C., 1978] were used instead of BLOSUM [S. Henikoff and J. G. Henikoff, *Proc. Natl. Acad. Sci. U.S.A.* **89,** 10915 (1992)]. The approximate positions of the seven α helices, common to all sequences, are shown as boxes. The alignment is carried out by aligning the two α-globins (Hba_Human and Hba_Horse) together; then the two β-globins are aligned (Hbb_Human and Hbb_Horse); then the two aligned α-globins are aligned to the two aligned β-globins; finally, the whale myoglobin (Myg_Phyca), the lamprey cyanohemoglobin (Glb5_Petma), and the lupine leghemoglobin (Lgb2_Luplu) are aligned one at a time to the growing alignment.

advantage of speed and simplicity. One can align hundreds of sequences, even on personal computers. More importantly, the sensitivity of the approach, as judged by the ability to align distantly related sequences, is very high. In simple cases, it is common to derive alignments which are impossible to improve by eye. In these cases, the method can be said to be a satisfactory replacement for manual alignment. In more difficult cases, the method usually gives a useful starting point for further refinement. In this chapter, we describe some of the problems of progressive multiple alignment and some simple modifications which, we believe, greatly improve the sensitivity for difficult protein alignments. All of the methods described here are freely available in a computer program called CLUSTAL W which can be run under a wide variety of operating systems.

Problems with Progressive Alignment

There are two obvious and interrelated problems inherent in the progressive alignment approach: (1) the local minimum problem and (2) the parameter choice problem. The local minimum problem stems from the "greedy" nature of the algorithm. Every time an alignment is carried out, some proportion of the residues will be misaligned. This proportion will be very small (or nonexistent) for very closely related proteins (e.g., at least 50% identical) but will increase as more and more divergent sequences are used. Any mistakes that appear during early alignments in a progressive multiple alignment cannot be corrected later as new sequence information is added. If the data set contains sequences of different degrees of divergence, the first alignments may be very accurate, and by the time the most diverged sequences are aligned, some information about gap frequency and residue conservation at each position will be available. In this case, the progressive approach may work very well. In other cases, however, if the first alignments are not correct they cannot be corrected later.

It is commonly argued that the local minimum problems stems largely from errors in the branching order of the initial phylogenetic tree. Consequently, many authors have devoted great effort to investigating alternative topologies or advocating particular methods of phylogenetic analysis. For example, previous versions of CLUSTAL[9,14,15] have been criticized[16] for using UPGMA[17] to generate initial trees as UPGMA is notorious for giving

[14] D. G. Higgins and P. M. Sharp, *CABIOS* **5,** 151 (1989).
[15] D. G. Higgins, A. J. Bleasby, and R. Fuchs, *CABIOS* **8,** 189 (1992).
[16] C.-B. Stewart, *Nature* (*London*) **367,** 26 (1994).
[17] P. H. A. Sneath and R. R. Sokal, "Numerical Taxonomy." Freeman, San Francisco, 1973.

incorrect branching orders when rates of substitution vary greatly in different lineages. This criticism is erroneous. If the alignment is simple enough, almost any tree will give the correct alignment. If the alignment is sufficiently difficult, almost any tree will give the wrong alignment. There is no one-to-one correspondence between having the correct tree topology and getting the right alignment. The better the tree, then the better the chances of getting a good alignment, but there are no guarantees. Even if UPGMA is not ideal for obtaining correct tree topologies, it can be argued that it is still very useful for alignment purposes. At each stage in the alignment process, align the most similar remaining sequences or subalignments so as to minimize the alignment errors at that step. UPGMA gives this property automatically. Nonetheless, we now provide the neighbor-joining method[18] for making initial trees because it seems to provide more reliable tree topologies and gives better estimates of tree branch lengths which we use to weight sequences and adjust the alignment parameters dynamically.

The local minimum problem is intrinsic to progressive alignment, and we do not provide any direct solutions. The only way to correct it is to use an overall measure of multiple alignment quality and find the alignment which maximizes this measure. This can be done directly for small numbers of sequences using the program MSA[19] but is, for now, uncomputable for more than about seven sequences. MSA computes approximate bounds for the location of the best pathway through the N-dimensional (for N sequences) dynamic programming array and then carries out full dynamic programming in this restricted area. It may be possible to use stochastic or iterative optimization procedures[20,21] to optimize multiple alignments of many sequences in the future. Even if practical solutions are found, we believe that the parameter choice problem, described below, is just as important and will still preclude high accuracy alignments if it is not addressed.

The parameter choice problem stems from using just one set of parameters (normally an amino acid substitution matrix and two gap penalties) and hoping that these will be appropriate over all parts of all the sequences to be aligned. If the sequences are very similar, almost any reasonable parameters will give a good alignment. For highly divergent sequences, however, the exact choice of parameters may have a great effect on align-

[18] N. Saitou and M. Nei, *Mol. Biol. Evol.* **4,** 406 (1987).

[19] D. Lipman, S. F. Altschul, and Kececioglu, *J. Proc. Natl. Acad. Sci. U.S.A.* **86,** 4412 (1989).

[20] O. Gotoh, *CABIOS* **9,** 361 (1993).

[21] C. E. Lawrence, S. F. Altschul, M. S. Boguski, J. S. Liu, A. F. Neuwald, and J. C. Wooton, *Science* **262,** 208 (1993).

ment quality. Different weight matrices from the well-known PAM[22] or BLOSUM[23] series are appropriate for aligning sequences of different evolutionary distances. For very similar sequences, even an identity matrix will provide sensible alignments, but for sequences in the so-called twilight zone[24] (sequences of roughly 20–25% identity), the exact values given to each type of substitution may be critical. Further, there is no particular reason why the same gap penalties should work equally well at all positions in an alignment. Gaps tend to occur far more often between the main secondary structure elements of α helices and β strands than within.[25] With the latest CLUSTAL program (CLUSTAL W), we attempt to attack this problem by computing position-specific gap opening and extension penalties as the alignment proceeds. We also use different amino acid weight matrices; "hard" ones for closely related sequences and "softer" ones for more divergent sequences.

We believe that the correct use of parameters at different positions is important. We provide simple heuristic methods for doing this which seem to work well in difficult test cases. Other authors have attacked the problem using more systematic methods such as hidden Markov models[26] or have exploited structural information when the structure of one or more of the sequences is known.[27]

CLUSTAL W

CLUSTAL W[28] is derived directly from the CLUSTAL[9,14] and CLUSTAL V series of programs. The "W" in the name stands for "weighting" as we now give different weights to sequences and parameters at different positions in alignments. The main feature of the old programs was the ability to align many sequences quickly, even on a personal computer, with a minimal sacrifice in sensitivity. The new program offers similar speed as before (CLUSTAL W is in fact slower than the older programs, but, fortunately, advances in the power of personal computers and workstations have canceled this out) but provides a number of new features and appears

[22] M. O. Dayhoff, R. M. Schwartz, and B. C. Orcutt, in "Atlas of Protein Sequence and Structure" (M. O. Dayhoff, ed.), Vol. 5, Suppl. 3, p. 345. National Biomedical Research Foundation, Washington, D.C., 1978.

[23] S. Henikoff and J. G. Henikoff, Proc. Natl. Acad. Sci. U.S.A. **89,** 10915 (1992).

[24] R. F. Doolittle, "URFs and ORFs: A Primer on How to Analyze Derived Amino Acid Sequences." University Science Books, Mill Valley, California, 1987.

[25] S. Pascarella and P. Argos, J. Mol. Biol. **224,** 461 (1992).

[26] A. Krogh, M. Brown, S. Mian, K. Sjölander, and D. Haussler, J. Mol. Biol. **235,** 1501 (1994).

[27] G. J. Barton and M. J. E. Sternberg, Protein Eng. **1,** 89 (1987).

[28] J. D. Thompson, D. G. Higgins, and T. J. Gibson, Nucleic Acids Res. **22,** 4673 (1994).

to be more sensitive for difficult protein alignments. The main new features are (1) support for more file formats for trees, sequence data sets, and alignments; (2) optional, full dynamic programming alignments for estimating the initial pairwise distances between all the sequences; (3) neighbor-joining[18] trees for the initial guide trees, used to guide the progressive alignments; (4) sequence weighting to correct for unequal sampling of sequences at different evolutionary distances; (5) dynamic calculation of sequence- and position-specific gap penalties as the alignment procedes; (6) the use of different weight matrices for different alignments; and (7) improved facilities for adding new sequences to an existing alignment.

The source code of CLUSTAL W, version 1.5 (April 1995), is available free of charge from the EMBL E-mail file server or by anonymous ftp to the EMBL ftp server. Compiled versions are also available for MS-DOS and Macintosh computers. With the Macintosh version, we also supply the NJplot program of Manolo Gouy (University of Lyon) which allows for the graphical display and manipulation of phylogenetic trees.

Use ftp to connect to ftp.ebi.ac.uk and give user name anonymous and full E-mail address as password. The four versions are as follows:

/pub/software/vax/clusalw.uue	uuencoded ZIP archive for VMS
/pub/software/unix/clustalw.tar.Z	compressed tar archive for UNIX
/pub/software/dos/clustal$.exe	self-extracting archive for MS-DOS
/pub/software/mac/clustalw.sea.hqx	Binhex encoded self-extracting archive

The MS-DOS and UNIX versions should be transferred in binary mode; the VMS and Mac versions in ASCII.

Position-Specific Gap Penalties

Here we give a brief summary of the methods used to calculate position-specific gap penalties for protein alignments. Two gap penalties are used initially: a gap opening penalty (GOP) and a gap extension penalty (GEP). Traditionally, one will choose values for these parameters before alignment and use the same values for all sequences. The exact values used are usually the default values offered by the software, which are often chosen empirically by the software authors by trial and error for a given amino acid weight matrix. In a simple world where one knew the positions of all secondary structure elements (α helices and β strands) in all or some of the sequences, one could increase the GOP (and GEP) at each position inside a helix or strand and decrease it between them.[27] This would force gaps to occur most often in loop regions, which is what is observed in practice with test cases from protein structure superposition.[25] One could be more sophisticated and make the GOP highest at the center of helices and strands and reduce it at the edge, allowing some gaps to occur at the ends of secondary structure elements. If one does not know the secondary structure of the sequences, as is normally the case, this is not possible. In CLUSTAL W, we use a set of very simple rules to help modify the GOP and GEP at each position in a sequence or prealigned group of sequences, depending on the residues that occur at each position and the frequency of gaps at each position. These rules are simple heuristics that seem to work very well in practice, although it should be possible to derive similar rules with greater statistical or mathematical validity.

Before any two sequences or prealigned groups of sequences are aligned, we calculate initial values for the GOP and GEP as functions of the amino acid weight matrix to be used, the sequence (or alignment) lengths, and the divergence between the sequences. The values for GOP and GEP are set from a user-controlled menu (defaults are offered) and then modified as follows:

$$GOP \rightarrow A * B * \{GOP + \log[\min(N, M)]\} \tag{1}$$

where N and M are the lengths of the sequences to be aligned, A is the average value for a mismatch in the amino acid weight matrix, and B is the percent identity of the two sequences. The GEP is then modified using the following formula:

$$GEP \rightarrow GEP * [1.0 + |\log(N/M)|] \tag{2}$$

where N and M are, again, the lengths of the two sequences.

The overall effect of these transformations is to allow for the use of different weight matrices with sequences of different degrees of divergence and to try to correct for some side effects of using sequences of different lengths. If the sequences are greatly different in length (as measured by the ratio N/M), the GEP is increased to try to inhibit the appearance of too many long gaps in the shorter sequence.

Next, tables of GOP and GEP values are calculated for each of the two sequences or groups of sequences to be aligned, one GOP and GEP for each position. Initially, these values are all the same, as calculated using Eqs. (1) and (2) above. These are then modified at each position using four rules. The overall aim is to encourage gaps to occur in likely loop regions. Informally, the rules are (1) use lower gap penalties at positions where gaps already occur; (2) increase gap penalties adjacent to positions where gaps already occur; (3) reduce gap penalties where stretches of hydrophilic residues occur; and (4) increase or decrease gap penalties using tables of the observed frequencies of gaps adjacent to each of the 20 amino acids.[25]

If there are gaps at a position in a group of prealigned sequences (this rule and the following one do not apply to single sequences), then the GOP is reduced in proportion to the number of sequences with a gap at that position and the GEP is lowered by one-half. The new GOP is calculated as

$$GOP \rightarrow GOP * 0.3 * (W/N) \tag{3}$$

where W is the number of sequences without a gap at the position and N is the number of sequences.

If a position contains no gaps but is within eight residues of an existing gap (this value of 8 can be changed from a menu), the GOP is increased as follows:

$$GOP \rightarrow GOP * \{2 + [8 - (D) *2]/8\} \tag{4}$$

where D is the distance from the gap.

A run of five (this number can be changed from a menu) consecutive, hydrophilic residues is considered to be a hydrophilic stretch. The residues that are considered to be hydrophilic are conservatively set to D, E, G, K, N, Q, P, R, and S by default but can be changed by the user. Any positions with no gaps that are spanned by such a stretch of residues get the GOP reduced by one-third.

In Table I, we list 20 residue-specific gap propensity values. These are derived from the observed frequencies of gaps adjacent to each residue in

TABLE I
RESIDUE-SPECIFIC GAP OPENING PENALTY FACTORS[a]

Residue	Penalty	Residue	Penalty
A	1.13	M	1.29
C	1.13	N	0.63
D	0.96	P	0.74
E	1.31	Q	1.07
F	1.20	R	0.72
G	0.61	S	0.76
H	1.00	T	0.89
I	1.32	V	1.25
K	0.96	Y	1.00
L	1.21	W	1.23

[a] These values are derived from the observed frequencies of gaps adjacent to each residue in alignments of sequences of known tertiary structure.[25] The values were transformed from the published values such that the bigger the number, the less likely a gap is to occur adjacent to that residue. The numbers are then used as simple multiplication factors to modify gap opening penalties, normalized around a value of 1.0 for histidine.

alignments of sequences with known three-dimensional structure.[25] If a position does not contain a gap or a hydrophilic stretch, then the values in Table I are used as simple multiplication factors to increase or decrease the GOP; for example, a position with only glycine will get a reduced GOP (multiplied by 0.61), whereas a position with only methionine will get an increased GOP (multiplied by 1.29). If there is a mixture of residues at a position, then the multiplication factor is the average of those for each residue, one from each sequence.

The overall effect of the four rules can be seen in Fig. 2 on a small stretch of alignment from four globin sequences. The GOP is highest adjacent to a gap and lowest at a gap position or where hydrophilic runs occur. These rules are most useful when there are already some sequences correctly aligned. Then, new gaps will tend to concentrate in areas where gaps already occur and will promote a blocklike appearance in the final alignment. For the first alignments, before any sequences are aligned, there are no gaps yet and the first two rules above cannot be used; only the residue-specific gap frequency rule and hydrophilic stretch rules can be used.

Sequence Weighting

In most real data sets of protein sequences, it is common to have unequal sampling of sequences at different evolutionary distances. Frequently, there

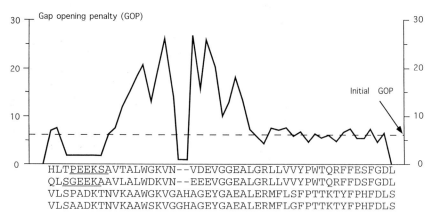

FIG. 2. Illustration of the effect of modifying the gap opening penalty (GOP) on a stretch of globin alignment from Fig. 1. The initial GOP is shown as a dotted line, and the position-specific GOP values are plotted along the alignment. The GOP is lowest at positions with gaps and where hydrophilic stretches occur (two such stretches are underlined). The GOP is highest within eight residues of gaps. The rest of the variation is caused by the residue-specific factors from Table I.

are clusters of closely related sequences and small numbers of highly diverged ones. For example, in the data set illustrated in Fig. 1, there are two mammalian α-globins and two β-globins. These pairs are almost identical and provide little extra information for alignment purposes. Traditionally, during multiple alignment or when using aligned sequences as profiles for database searching, one gives equal weight to all sequences. In the globin example, this means that the two α-globins are given as much weight each as the more diverged leghemoglobin. Several authors have addressed the problem of sequence weighting in order to correct for this effect[29]; closely related sequences are down-weighted, while relatively more distant ones receive greater weight. In profile searches, this has been shown to increase the sensitivity of the search as measured by the ability to detect distant relatives of the sequences in the profile.[30-32] Sequence weights were first used for multiple alignments by Vingron and Argos[33] and are also used in the MSA[19] program.

There are several methods available for sequence weighting. For our purposes, it must be possible to derive weights for unaligned sequences as the weights will be used to help arrive at the multiple alignment itself. This prevents the use of methods that use multiple alignments as input. We use a simple tree-based method that only requires a tree describing the rough relatedness of the sequences.[30] The guide trees, used to guide the multiple alignment, provide this. The method requires a rooted tree with branch lengths. As a first approximation, the weights are calculated as the distance of each sequence from the root of the tree. This gives increased weight to the most diverged sequences and less to the more conserved ones. Second, if two sequences (or groups of sequences) share a common internal branch, the length of the internal branch is shared when deriving the weights. This automatically downweights related sequences in proportion to the degree of relatedness. An example is shown in Fig. 3, using the guide tree from Fig. 1. Here the leghemoglobin (Lgb2_Luplu) gets a weight of 0.442, which is equal to the length of the branch from the root to it. The human α-globin (Hba_Human) receives a weight equal to the length of branch leading to it that is not shared by any other sequences (0.055) plus one-half the length of the branch shared with the horse α-globin (0.219/2) plus one-quarter the length of the branch shared by all four hemoglobins (0.061/4) plus one-fifth the branch shared between the hemoglobins and the myo-

[29] M. Vingron and P. R. Sibbald, *Proc. Natl. Acad. Sci. U.S.A.* **90,** 8777 (1993).
[30] J. D. Thompson, D. G. Higgins, and T. J. Gibson, *CABIOS* **10,** 19 (1994).
[31] R. Lüthy, I. Xenarios, and P. Bucher, *Protein Sci.* **3,** 139 (1994).
[32] S. Henikoff and J. G. Henikoff, *J. Mol. Biol.* **243,** 574 (1994).
[33] M. Vingron and P. Argos, *CABIOS* **5,** 115 (1989).

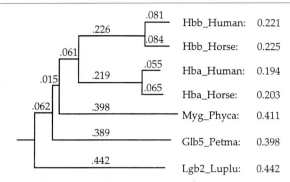

FIG. 3. Sequence weights for the seven globin sequences from Fig. 1. A rooted neighbor-joining tree is shown with branch lengths. The weights are shown for each sequence before normalization (the weights are normalized so as to make the largest equal to 1.0).

globin (0.015/5) plus one-sixth the branch shared by all the vertebrate globins (0.062/6). This gives a total weight of 0.194 before normalization. All weights are normalized to make the largest weight equal to one.

These weights are intuitive, simple to calculate, and have been shown to be useful in increasing the sensitivity of profile searches.[30,32] With multiple alignment, they are used to give different weight to the contributions of different sequences to the alignment scores when aligning two subalignments or when aligning a sequence to a subalignment. The weights are used as simple multiplication factors when calculating the alignment score between two positions.

Weights for Adding New Sequences to Existing Alignment

Sequence weights are also useful when adding new sequences to an existing alignment. In CLUSTAL W, we provide facilities to do this in three ways: (1) add a single sequence to an alignment; (2) add a set of new sequences one at a time to an alignment; (3) align two existing alignments. With methods 1 and 2, there is a further use of sequence weighting. If the sequence to be aligned is much closer to some of the sequences in the alignment (the new sequence can be said to go "inside" the underlying tree), then one can exploit this to give extra weight to the most closely related examples (most closely related to the new sequence). This will help to make sure that the placement of new gaps is most influenced by the close relatives rather than distantly related sequences. The weights that achieve this effect are, again, very simple to derive. Pairwise alignments and resulting simple alignment distances (mean number of differences per site, ignoring gaps and not corrected for multiple hits) are calculated between the new sequence and each sequence in the old alignment. Then, new weights are calculated for each sequence in the alignment as 1.0 minus the distance from the new sequence. This has the effect of giving a weight of 1.0 to identical sequences (identical to the new sequence) and a low weight to distant relatives. Finally, these weights are multiplied with the original tree weights for each sequence in order to combine the properties of the two types of weights, and the weights are normalized so as to sum to 1.

An example is shown in Fig. 4 where two globins are added to the previous globin alignment. The two sequences are compared to each of the sequences in the alignment. The sequences are added to the alignment such that the most similar ones (most similar, on average, to the sequences in the alignment) are added first. In this case, the trout β-globin (Hbb1_Salir) is added first using the weights shown in Fig. 4. These weights give increased weight to the most closely related sequences (closest to the new sequence) but also downweight sequences that are closely related to each other. Then, a new set of weights are calculated for the addition of the new leghemoglobin sequence (Lgba_Phavu), including a weight for the trout β-globin.

We have not examined the properties of these new weights for adding sequences to an alignment in any detail or carried out extensive empirical

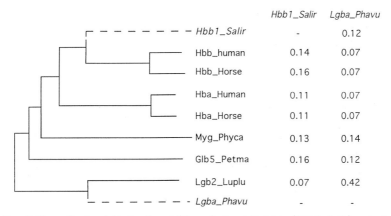

FIG. 4. Normalized weights for the addition of a trout β-globin (Hbb1_Salir) and a bean leghemoglobin (Lgba_Phavu) to the alignment of seven globin sequences in Fig. 1. The positions in the tree for the new sequences are shown with dashed lines. First, the trout β-globin is added, and the seven weights needed for this are shown. Then the bean globin is added, and eight weights are needed for this (one weight for each of the original seven globins and one for the trout globin). The weights have the effect of simultaneously upweighting sequences that are similar to the new sequence and downweighting sequences that have other close relatives in the tree. The weights are normalized to sum to 1.0.

evaluations. They are, however, intuitive and simple to calculate. It seems to us to be of great importance to make use of such weights or to develop and evaluate new weighting schemes as more and more databases of large alignments are created that need to be updated automatically. These weights help to exploit the information contained in large alignments. An alignment of a few hundred or more sequences that has been carefully created by experts, sometimes incorporating structural information, is an important and useful resource which contains much information to help characterize new sequences.

Phylogenetic Trees

All trees in CLUSTAL W are calculated using the neighbor-joining method.[18] These are used as guide trees to guide the multiple alignments or can be produced after multiple alignment with a bootstrap[34] option. This is a distance matrix approach which is fast to calculate but which gives the correct tree topology in a wide variety of situations. The method produces unrooted trees with estimates of branch lengths for each branch.

For the calculation of guide trees, the user can choose between fast approximate pairwise alignment of sequences using a k-tuple approach[35] based only on identities or full dynamic programming[36] with a weight matrix and two gap penalties. The former are extremely fast to calculate and are useful if the user has hundreds of sequences, but the latter are more accurate. Distances are calculated as the number of exact matches in the best alignment between the pair of sequences divided by the number of positions considered, ignoring positions with gaps. With the guide trees, there is no correction for multiple substitutions. The distances are given to the neighbor-joining method, and an unrooted guide tree is produced. The tree is rooted by a "midpoint" approach.[30] Root placement involves making the mean distance from the root to the tips of the tree equal on both sides of the root. The biological validity of this method of placing the root depends on the quality of the "molecular clock" in the data set. For the current application, however, it does not matter if the root is incorrectly placed. This is because placing the root using the midpoint method is the equivalent of always aligning the next most closely related sequences or groups of sequences on the tree, ending when all sequences are aligned. This is the desired behavior.

After multiple alignment (or after reading a full multiple alignment

[34] J. Felsenstein, *Evolution* **39,** 783 (1985).
[35] W. J. Wilbur and D. J. Lipman, *Proc. Natl. Acad. Sci. U.S.A.* **80,** 726 (1983).
[36] S. B. Needleman and C. Wunsch, *J. Mol. Biol.* **48,** 444 (1970).

from a file), more accurate trees can be calculated using distances calculated from the fully aligned sequences. For each distance calculation, gaps are not considered, but there is an option to ignore all sites where a gap occurs in any of the sequences. Although this sounds wasteful of data (it removes any site where a gap occurs) it does have the advantage of basing all distance calculations on the same number of sites. For sequences of similar length, this makes little difference. It has the added advantage of automatically removing the most ambiguous sites from the alignment (those that are hardest to align and where the exact alignment may be arbitrary or an artifact of the alignment process). There is a second option to correct the distances for multiple substitutions. This has little effect on small distances (e.g., sequences more than 80% identical) but will stretch large distances considerably to compensate for the great number of hidden substitutions that are estimated to have occurred. For example, using the Dayhoff model of protein evolution,[22] if one observes two amino acid sequences that are 20% identical (distance of 0.80 differences per site), one estimates that 250 substitutions have occurred per 100 sites (distance of 2.5 substitutions per site). Gaps are not considered in these distance calculations for two reasons. First, it is problematic how to score gaps, and, second, it is not known whether gaps appear and disappear in a clocklike manner. For many protein sequences, it appears that substitutions occur in a reasonably regular manner over time (the so-called molecular clock hypothesis), and this allows one to use sequences to derive phylogenetic information. There may not be an equivalent "gap clock," and if there is it is not known how to calibrate it.

To correct simple protein distances (number of observed differences per site) for multiple hits, it is common practice to use a formula from Kimura[37]:

$$K_{aa} = -\ln(1 - D - D^2/5) \tag{5}$$

where K_{aa} is the estimated, corrected distance and D is the observed distance. This formula is a curve-fitting approximation to a table of corrected distances derived from the Dayhoff model of protein evolution.[22] The formula is simple to calculate and is an extremely accurate fit to the Dayhoff model in the range 0.0 to 0.75 observed distance. Above this, however, the approximation becomes inaccurate, and at D above 0.85 or so, the formula cannot be evaluated as it requires finding the logarithm of a negative number. With CLUSTAL W, we use the Kimura formula for D up to 0.75 and use precalculated tables of corrected distances for all D above this in intervals of 0.001. These tables were calculated using the Dayhoff model

[37] M. Kimura, "The Neutral Theory of Molecular Evolution." Cambridge Univ. Press, Cambridge, 1983.

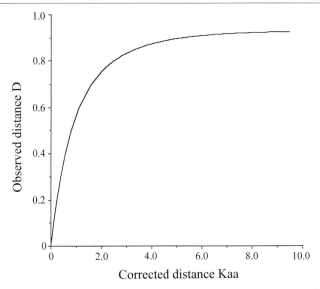

FIG. 5. Relationship between the simple observed distance between two protein sequences (mean number of differences per site) and the estimated corrected distance (the number of substitutions per site that have occurred, including multiple hits). These values were calculated using the Dayhoff model of amino acid substitution[22] and are hard coded in CLUSTAL W.

of amino acid substitution, from which the well-known PAM tables of amino acid similarity are derived. The graph in Fig. 5 shows the relationship between K_{aa} and D. These values were calculated using the Dayhoff model and hard coded into CLUSTAL W.

The Dayhoff model cannot correct any observed distances greater than 0.93 or so as the model reaches equilibrium at this point. For idealized proteins that have reached a divergence of 0.93 observed differences per site, substitutions are just as likely to restore an identity at a site as remove one. Therefore, we correct any observed distances above 0.93 to the arbitrary level of 10.0. This is crude, but such distances are rare with real sequences. If the sequences are this divergent, then alignment is very difficult to begin with and any tree making methods will have great problems making sensible trees. During bootstrapping, however, such distances can occur randomly, even if no sequences are as diverged as this. With CLUSTAL W, the user is warned if this occurs. Provided it occurs rarely, it seems to have no detrimental effect on the bootstrap results. If it becomes critical, however, users are advised to use the more sophisticated distance calculation method provided in the PROTDIST program of the PHYLIP package.[38] This method uses a Dayhoff model to estimate numbers of substitutions per site, but does so taking all mismatches into account as well as identities, and will be appropriate for sequences of any amino acid composition.

For graphical display of trees, we recommend that users use the DRAWGRAM and DRAWTREE programs of the PHYLIP package[38] or TREETOOL,[39] if the user has access to a SUN computer. For Macintosh computers, we provide the extremely useful and simple to use NJPLOT program of Manolo Gouy.[40] NJPLOT reads trees in the widely used New Hampshire format and allows the user to make simple manipulations and save the tree as PICT format, which can then be used in many Macintosh drawing and graphics packages. The bootstrap trees produced by CLUSTAL W include the bootstrap support levels as extra labels. These are displayed by both TREETOOL and NJPLOT.

Summary

We have tested CLUSTAL W in a wide variety of situations, and it is capable of handling some very difficult protein alignment problems. If the data set consists of enough closely related sequences so that the first alignments are accurate, then CLUSTAL W will usually find an alignment that is very close to ideal. Problems can still occur if the data set includes sequences of greatly different lengths or if some sequences include long regions that are impossible to align with the rest of the data set. Trying to balance the need for long insertions and deletions in some alignments with the need to avoid them in others is still a problem. The default values for our parameters were tested empirically using test cases of sets of globular proteins where some information as to the correct alignment was available. The parameter values may not be very appropriate with nonglobular proteins.

We have argued that using one weight matrix and two gap penalties is too simplistic to be of general use in the most difficult cases. We have replaced these parameters with a large number of new parameters designed primarily to help encourage gaps in loop regions. Although these new parameters are largely heuristic in nature, they perform surprisingly well and are simple to implement. The underlying speed of the progressive

[38] J. Felsenstein, *Cladistics* **5,** 164 (1989).
[39] M. Maciukenas, University of Illinois, USA, unpublished. Available by anonymous ftp from rdp.life.uiuc.edu (/pub/RDP/programs/TreeTool).
[40] M. Gouy, University of Lyon, France, unpublished. Available for Apple Macintosh computers by anonymous ftp frpm ftp.ebi.ac.uk (/pub/software/mac/NJplot.sea.hqx).

alignment approach is not adversely affected. The disadvantage is that the parameter space is now huge; the number of possible combinations of parameters is more than can easily be examined by hand. We justify this by asking the user to treat CLUSTAL W as a data exploration tool rather than as a definitive analysis method. It is not sensible to automatically derive multiple alignments and to trust particular algorithms as being capable of always getting the correct answer. One must examine the alignments closely, especially in conjunction with the underlying phylogenetic tree (or estimate of it) and try varying some of the parameters. Outliers (sequences that have no close relatives) should be aligned carefully, as should fragments of sequences. The program will automatically delay the alignment of any sequences that are less than 40% identical to any others until all other sequences are aligned, but this can be set from a menu by the user. It may be useful to build up an alignment of closely related sequences first and to then add in the more distant relatives one at a time or in batches, using the profile alignments and weighting scheme described earlier and perhaps using a variety of parameter settings.

We give one example using SH2 domains. SH2 domains are widespread in eukaryotic signalling proteins where they function in the recognition of phosphotyrosine-containing peptides.[41] In the chapter by Bork and Gibson ([11], this volume), Blast and pattern/profile searches were used to extract the set of known SH2 domains and to search for new members. (Profiles used in database searches are conceptually very similar to the profiles used in CLUSTAL W: see the chapters [11] and [13] for profile search methods.) The profile searches detected SH2 domains in the JAK family of protein tyrosine kinases,[42] which were thought not to contain SH2 domains. Although the JAK family SH2 domains are rather divergent, they have the necessary core structural residues as well as the critical positively charged residue that binds phosphotyrosine, leaving no doubt that they are bona fide SH2 domains.

The five new JAK family SH2 domains were added sequentially to the existing alignment of 65 SH2 domains using the CLUSTAL W profile alignment option. Figure 6 shows part of the resulting alignment. Despite their divergent sequences, the new SH2 domains have been aligned nearly perfectly with the old set. No insertions were placed in the original SH2 domains. In this example, the profile alignment procedure has produced better results than a one-step full alignment of all 70 SH2 domains, and in

[41] I. Sadowski, J. C. Stone, and T. Pawson, *Mol. Cell. Biol.* **6**, 4396 (1986).

[42] A. F. Wilks, A. G. Harpur, R. R. Kurban, S. J. Ralph, G. Zuercher, and A. Ziemiecki, *Mol. Cell. Biol.* **11**, 2057 (1991).

```
CLUSTAL W(1.5) multiple sequence alignment


H_Src         EWYFGKI-----TRRESERLLLNA----ENPRGTFLVRESET---TKGAYCLSVSDFD--
Ce_B0523.1    AYFHGLI-----QREDVFQLLDN--------NGDYVVRLSDPKPGEPRSYILSVMFNN--
H_Crk         SWYWGRL-----SRQEAVALLQG------QRHGVFLVRDSST---SPGDYVLSVSENS--
H_SLP_76      EWYVSYI-----TRPEAEAALRK-----INQDGTFLVRDSSK-KTTTNPYVLMVLYKD--
M_3bp2        SVFVNTT-----ESCEVERLFKATDPRGEPQDGLYCIRNSST---KSGKVLVVWDESS--
H_PlcG1/1     KWFHGKLGAGRDGRHIAERLLTEYCIETGAPDGSFLVRESET---FVGDYTLSFWRNG--
H_PlcG1/2     EWYHASL-----TRAQAEHMLMR-----VPRDGAFLVRKRN----EPNSYAISFRAEG--
H_Ptp1c/1     RWFHRDL-----SGLDAETLLKG-----RGVHGSFLARPSRK---NQGDFSLSVRVGD--
H_Ptp1c/2     RWYHGHM-----SGGQAETLLQA-----KGEPWTFLVRESLS---QPGDFVLSVLSDQ--
Gg_Tensin     YWYKPDI-----SREQAIALLKD------REPGAFIIRDSHS---FRGAYGLAMKVASPP
H_Jak1        NGCHGPIC----TEYAINKLRQE-----GSEEGMYVLRWSCT-DFDNILMTVTCFEKS--
H_Tyk2        DGIHGPL-----LEPFVQAKLRP-------EDGLYLIHWSTS---HPYRLILTVAQRS--
R_Jak2        SNCHGPI-----SMDFAISKLKKAG----NQTGLYVLRCSPK---DFNKYFLTFAVER--
R_Jak3        ELCHGPI-----TLDFAIHKLKAAG----SLPGSYILRRSPQ---DYDSFLLTACVQTPL
Dm_Hop        LHCHGPI-----GGAYSLMKLHEN----GDKCGSYIVRECDR---EYNIYYIDINTKIMA


H_Src         ----------NAKGLNVKHYKIRKLD----------------SGGFYITS----RTQFN
Ce_B0523.1    ---------KLDENSSVKHFVINSVE---------------NKYFVNN----NMSFN
H_Crk         --------------RVSHYIINSSGPRPPVPPSPAQPPPGVSPSRLRIGD-----QEFD
H_SLP_76      --------------KVYNIQIRYQK---------------ESQVYLLGTGLRGKEDFL
M_3bp2        --------------NKVRNYRIFEKD----------------SKFYLEG----EVLFA
H_PlcG1/1     --------------KVQHCRIHSRQ-------------DAGTPKFFLTD----NLVFD
H_PlcG1/2     --------------KIKHCRVQQEG---------------QTVMLGN-----SEFD
H_Ptp1c/1     --------------QVTHIRIQNSG----------------DFYDLYG----GEKFA
H_Ptp1c/2     ------PKAGPGSPLRVTHIKVMCEG----------------GRYTVGG----LETFD
Gg_Tensin     PTVMQQNKKGDITNELVRHFLIETSP---------------RGVKLKG-CPNEPNFG
H_Jak1        -------EQVQGAQKQFKNFQIEVQK----------------GRYSLHG---SDRSFP
H_Tyk2        ------QAPDGMQSLRLRKFPIEQQD---------------GAFVLEG---WGRSFP
R_Jak2        ----------ENVIEYKHCLITKNE----------------NGEYNLSG---TKRNFS
R_Jak3        G-------------PDYKGCLIRQDP---------------SGAFSLVG--LSQLHR
Dm_Hop        KKTD-------QERCKTETFRIVRKD---------------SQWKLSYNN---GEHVLN


H_Src         SLQQLVAYYSKH
Ce_B0523.1    TIQQMLSHYQKS
H_Crk         SLPALLEFYKIH
H_SLP_76      SVSDIIDYFRKM
M_3bp2        SVGSMVEHYHTH          Representative pre-aligned SH2 domains
H_PlcG1/1     SLYDLITHYQQV
H_PlcG1/2     SLVDLISYYEKH
H_Ptp1c/1     TLTELVEYYTQQ
H_Ptp1c/2     SLTDLVEHFKKT
Gg_Tensin     CLSALVYQHSIM
H_Jak1        SLGDLMSHLKKQ
H_Tyk2        SVRELGAALQGC          JAK SH2 domains
R_Jak2        SLKDLLNCYQME
R_Jak3        SLQELLTACWHS
Dm_Hop        SLHEVAHIIQAD
```

FIG. 6. Profile alignment adding sequentially five newly detected JAK SH2 domains to an existing SH2 alignment. Because of space restrictions, just 10 of the 65 original SH2 domains are shown. All 65 domains were actually used for the profile alignment. The sole error occurs in H_Jak1 block 2 (shown in boldface italics) which is misaligned one residue leftward.

considerably less time. In this example, it is roughly five times faster to add the new sequences one at a time to the existing SH2 alignment than it is to recalculate the full alignment. It is also more accurate and gives the user greater control.

[23] Combined DNA and Protein Alignment

By Jotun Hein and Jens Støvlbæk

Introduction

Most long DNA sequences contain coding regions, and thus it is optimal to use information from both the DNA sequence and the coded protein when comparing such a sequence to a homologous variant. In this chapter we present a heuristic algorithm that can compare DNA with both coding and noncoding regions, but also multiple reading frames, and determine which exons are homologous. A program, GenAl (genomic alignment), has been developed that implements the algorithm. It is demonstrated by comparing HIV2 (human immunodeficiency virus type 2) with HIV1. A stochastic model of the evolution of the complete virus has also been developed that allows estimation of the amount of selective constraint on different genes (including overlapping regions), the equilibrium base composition, and lastly the transversion and transition distances. This method has been applied to two HIV2's.

Comparisons of longer genomic DNA sequences will typically contain both coding and noncoding regions, which cannot be analyzed by the traditional dynamical programming algorithm.[1] As a protein evolves slower than its coding DNA, it will be more reliable to align the protein than the underlying DNA, and an algorithm that compares genomic DNA should incorporate the information from the protein. Presently, this problem is solved by separating the sequences into coding and noncoding parts, then analyzing them separately, and, finally, patching the resulting alignments into a global alignment. This is laborious and cannot be done if the DNA has overlapping reading frames. In this chapter we present an algorithm that solves these problems.

It should be noted that all evolutionary events happen at the DNA level, as proteins do not replicate. The basic events are (1) substitutions, which in coding regions can have the additional consequence of changing

[1] S. B. Needlemann and C. D. Wunsch, *J. Mol. Biol.* **48**, 444 (1970).

Copyright © 1996 by Academic Press, Inc.
All rights of reproduction in any form reserved.

the coded protein; and (2) insertions–deletions (indels) in coding and non-coding regions. In coding regions, indels will normally have lengths that are a multiple of 3. Other evolutionary events are possible but are not allowed in a standard alignment.

Heuristic DNA–Protein Alignment Algorithm

Peltola *et al.*[2] introduced an algorithm that allowed a long DNA string to be searched for reading frames coding for a given protein sequence, that is, one protein versus one DNA string. Hein[3] considered algorithms aligning DNA sequences, but measuring the distances in terms of the induced change at the protein level. These algorithms were slow. The algorithm presented here is a fast but heuristic algorithm solving the same problem and is an elaboration on the algorithm first presented in Hein and Støvlbæk.[4]

The central approximation in the heuristic algorithm is to represent the coding DNA as if only the middle nucleotide in a codon coded for an amino acid. This allows a heuristic algorithm that is very similar to the simple DNA (or protein, but not both) comparing algorithm. This representation will create a new string with the same length as the original DNA string, but a position can now contain not only a nucleotide, but also an amino acid (potentially two amino acids, if there were a reading frame in the opposite direction). If two matched reading frames code for homologous proteins, there must have been a reading frame in the DNA all the way back to the most recent common ancestor and all indels should have lengths that are multiples of three nucleotides. In regions where no homologous proteins are matched, indels can have any length, as there could have been a period in its recent history when the DNA was noncoding. If the homology relationships between proteins in the two sequences are known beforehand, this could be part of the data and help in solving the problem. This option in the program is called guiding and is advantageous if the method could deduce the homology relationships itself. Because homologous proteins are usually more similar (less distant), there should be an inherent tendency to match homologous instead of nonhomologous proteins.

A general algorithm for aligning DNA with reading frames can now be formulated as follows: Let $s1$ and $s2$ be two sequences of length l_1 and l_2, respectively. The substring consisting of the first i elements of sk ($k = 1$, 2) is denoted sk_i, and $sk[i]$ refers to the ith element of sk. Let $D_{i,j}$ be the minimal distance between $s1_i$ and $s2_j$, when all insertion–deletions have

[2] H. Peltola, H. Söderlund, and E. Ukkonen, *Nucleic Acids Res.* **14,** 1.99 (1986).
[3] J. J. Hein, *J. Theor. Biol.* **167,** 169 (1994).
[4] J. J. Hein and J. Støvlbæk, *J. Mol. Evol.* **38,** 310 (1994).

lengths that are multiples of 3 and the differences in DNA are weighted by the amino acids that they invoke. Let the indices i' and j' be such that $i - i'$ (or $j - j'$) is 3, when only indels of length three are allowed, otherwise the value is 1. The distance between two elements in the sequences is $d[s1(i), s2(j)] = dn\{nuc[s1(i)], nuc[s2(j)]\} + da\{aa[s1(i)], aa[s2(j)]\}$, where $dn(,)$ is the distance between two nucleotides (nuc) and $da\{aa[s1(i)]$, $aa(s2)\}$ is the distance between the amino acids (aa) associated with that nucleotide. If there is one amino acid at each nucleotide, then it is the traditional amino acid distance; if there is only one amino acid as opposed to no amino acid, then it will be g_a, the cost of deleting one amino acid (g_n is the cost of deleting one nucleotide). In contrast to ordinary alignment, the matching term can also contain a weight that corresponds to the insertion–deletion of amino acids. The algorithm can then be written as

Initialization, $D_{0,j}$ (analogous for $D_{i,0}$):
$$D_{0,0} = 0$$
$$D_{0,j} = D_{0,j-1} + d[-, s2(j)]$$
$$D_{i,0} = D_{i-1,0} + d[s1(i), -]$$

The $D_{i,j}$ values then obey the recursion:

$D_{i,j}$ = minimum of following three quantities, when $i > 0$ and $j > 0$
1. Insertion in s1: $\{D_{i',j} + g_a*[$number of aa with codons overlapping $(i, i') + g_n]\}$
2. Insertion in s2: $\{D_{i,j'} + g_a*[$number of aa with codons overlapping $(j, j') + g_n]\}$
3. Matching nucleotides: $\{D_{i-1,j-1} + d[s1(i), s2(j)]\}$

The algorithm needs two sets of parameters, one set for proteins and one for DNA. Each set has a gap penalty and a distance function on the elements of the sequences: amino acids in the case of proteins and nucleotides in the case of DNA. The guiding principle for the parameters to be used is that very frequent events should have a low weight, whereas relatively rare events should have a higher weight. Many schemes have been devised to accomplish this.[5] A new problem in this analysis is how to weight events at the protein level relative to events at the DNA level. As proteins are more conserved than DNA and can retain observable similarity over much longer periods of time, the protein events should have higher costs than the DNA events. In principle it would be possible to let the protein level have complete precedence over the DNA level, corresponding to weighting proteins infinitely higher than DNA. DNA would still be a determining

---

[5] R. F. Doolittle, "Of URFs and ORFs." University Science Books, Mill Valley, California, 1986.

factor in the alignment, deciding the precise position of indels within a codon and choosing between equally good protein alignments. This algorithm can also handle overlapping reading frames.

The use of this algorithm is illustrated in two very simple cases in Fig. 1. Figure 2 shows the cost of different alignments, assuming different configurations of reading frames. It illustrates that the cheapest alignment for a given configuration of reading frames is the one that matches those reading frames. In general, it is more expensive to assume the presence of reading frames than their absence, because this avoids the cost associated with events for the corresponding reading frames. The information about the reading frames is essential to the algorithm, and it does not seem easy to modify the algorithm to also find the reading frames.

A program has been written, GenAl, that implemented this algorithm. GenAl can be obtained from the Netserver at EMBL. The algorithm was extended in several relevant ways that enhances its applicability to real data. (1) Frameshift indels do occur, but at very low frequency. This can be incorporated by allowing indels of length 1 and 2, but assigning them a very high cost. (2) Early in the history of sequence analysis it was realized that long stretches of gap signs in an alignment were likely to be due to one long indel and not to a series of adjacent small indels. This should be reflected in the gap cost function. The most used gap penalty function has the form $a + bk$ (where k is the length of the indel), as this allows for a fast algorithm).[6] This has been incorporated into GenAl and will improve the alignment involving medium size indels (<100 bp). For much larger indels a concave weighting would be superior).[7] (3) The program uses the Hirschberg linear memory algorithm,[8] so it can analyze very long sequences. (4) Alignment algorithms can be sped up and memory requirements reduced, without serious loss in reliability, by finding long stretches of common segments to the two sequences. This was implemented using the DAWG (directed acyclic word graph) as defined by Blumer et al.[9] This will also allow terminal indels to have a weight of zero, which is realistic if the sequences being compared have been sequenced to different extents. (5) The GenAl program can also search for close-to-optimum alignments, which can be useful in determining which regions are well aligned. Well-aligned regions will also have close-to-optimum alignments, while arbitrarily aligned regions will not.

[6] O. Gotoh, *J. Mol. Biol.* **162,** 705 (1981).
[7] W. Miller and E. W. Myers, *Bull. Math. Biol.* **50,** 2.97 (1988).
[8] D. S. Hirschberg, *Commun. ACM* **18,** 341 (1975).
[9] A. Blumer, J. Blumer, D. Hausler, and M. McConnell, *J. Assoc. Comput. Mach.* **34,** 3.578 (1987).

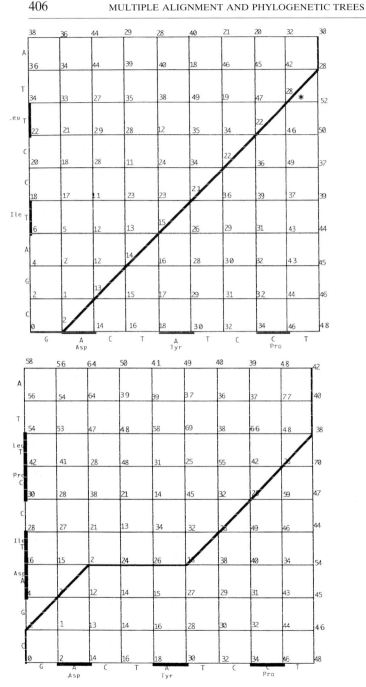

Total Alignment

```
Asp  Tyr  Pro
G A C T A T C C T -
- C G A T C C T T A
  -   Ile  Leu
```

Protein Alignment

```
Asp Tyr Pro
 -  Ile Leu
```

Total Alignment

```
  Asp  Tyr  Pro
- G A C T A T C C T - -
C G A - - - T C C T T A
  Asp       Pro
        Ile  Leu
```

Protein Alignment

```
        -    -
Asp  Tyr  Pro
Asp   -   Pro
     Ile  Leu
```

		Configuration		
		0-0	1-1	1-2
	0-0	<u>8</u>	58 (8+50)	78 (8+70)
Alignment	1-1	11	<u>30</u> (11+19)	50 (11+39)
	1-2	12	62 (12+50)	<u>42</u> (12+30)

FIG. 2. If the two sequences shown in Fig. 1 were aligned without reading frames, the optimal alignment would be gapless and have a cost of 8. So now there are three different alignments, depending on which reading frames are assigned to the DNA sequences. The cost of the gapless alignment would be 58 if one reading frame were present in each sequence and 78 if the second sequence had two reading frames. The cheapest alignment (underlined) for a given configuration is the alignment that uses the information from the reading frames of that configuration.

FIG. 1. Alignment path, HIV2ST vs HIVJRFL. Two DNA sequences, with one reading frame each, are aligned in the upper matrix. The first sequence is at the x axis of the matrix and the second at the y axis. The weights used include (1) substitutions between nucleotides is 1, (2) a mutation of an amino acid is 5, (3) an indel of a nucleotide is 2, and (4) an indel of an amino acid is 10. In other words, events at the protein level cost five times the events at the DNA level. However, a protein indel within a singly coding region corresponds to three nucleotide indels. The distance between $s1_i$ and $s2_j$ is found at the node (i, j). $d(s1i, -)$ (the cost of deleting the first i elements of $s1$) is $10 \times$ (number of amino acids in first i elements) $+ 2i$. The path up through the matrix corresponding to a minimal alignment is shown in thick lines. The total weight of the alignment is 30, which corresponds to the deletion of two nucleotides (cost 4), deletion of amino acids (cost 10), the cost of two amino acid replacements (cost 10), and last the substitution of six nucleotides. To calculate the value of the node (8,7) with one asterisk and value 28, three previous cells must be considered: (7,6) and, since (8,7) is within both reading frames, indels must come in groups of three and the relevant nodes will be (5,7) and (8,4). The value coming from (5,7) is 49 plus the cost for deleting three nucleotides (6) and one amino acid (10), which will be 65. The value coming from (8,4) is calculated analogously to be 53. The value coming from (7,6) is 22 plus the cost for both mutating a nucleotide (1) and an amino acid (5), which is 28. Since 28 is the smallest of the three, this value is assigned to (8,7). If a node is not within exons in both sequences, then indels do not have to come in groups of three. To calculate the value associated with node (5,2), which is outside the reading frame in sequence 2, the nodes that must be considered are (4,1), (5,1), and (4,2). The resulting total alignment and protein alignment is shown to the right of the matrix.

In the lower matrix an additional reading frame has been introduced into $s2$, so the algorithm has to choose which reading frame the protein in $s1$ is to be matched with in $s2$. It chooses the new reading frame and deletes the old reading frame. The cost of the alignment found is 42, which corresponds to the deletion of six nucleotides (12), the deletion of one reading frame of two amino acids (20), and the deletion of one additional amino acid (10).

```
                                        -----pol---------pol---------pol-----
                              -----gag---------gag---------gag-----
      1           303         548        1717        1840        2113        2122
      +-----------+-----------+-----------+-----------+-----------+-----------+
      0            1          113        1279        1402        1600        1609
                              -----gag---------gag---------gag---------gag-----
                                        -----pol---------pol---------pol-----


      -----pol---------pol---------pol-----              -----vpx---------vpx-----
                -----vif---------vif---------vif-----              -----vif-----
   2122         4868        4930        4938        5343        5515        5552
   +-----------+-----------+-----------+-----------+-----------+-----------+
   1609         4358        4413        4425        4807        4864        4876
                -----vif---------vif---------vif---------vif---------vif-----
   -----pol---------pol-----


                    -----vpr---------vpr---------vpr---------vpr---------vpr-----
      -----vpx-----
                              -----tat2---------tat2---------tat2----
   5552         5681        5756        5844        5958        5980        5998
   +-----------+-----------+-----------+-----------+-----------+-----------+
   4876         4876        4936        5024        5147        5166        5214
   -----vif---------vif-----              -----tat2---------tat2----

   -----vpr---------vpr---------vpr---------vpr---------vpr-----


                                                              -----env---------env-----
                    -----rev2----
      -----tat2---------tat2----
   5998         6070        6139        6148        6149        6210        8291
   +-----------+-----------+-----------+-----------+-----------+-----------+
   5214         5286        5361        5378        5541        5623        7668
   -----tat2---------tat2----              -----vpu---------vpu-----
                    -----rev2----
                                                              -----env---------env-----


      -----env---------env---------env---------env---------env-----
      -----rev3---------rev3----
      -----tat3----                         -----nef---------nef---------nef-----
   8291         8387        8544        8562        8584        8728        8729
   +-----------+-----------+-----------+-----------+-----------+-----------+
   7668         7758        7900        7918        7942        8084        8086
   -----tat3----
   -----rev3---------rev3---------rev3---------rev3----
   -----env---------env---------env---------env---------env-----


      -----nef-----
   8729         9329        9541        9672
   +-----------+-----------+-----------+
   8086         8736        8896        8896
   -----nef-----
```

FIG. 3. Compact overview of the total alignment of a variant of HIV2 (HIV2ST) with a variant of HIV1 (HIVJRFL). The reading frames above the center line, which shows the sequence position, belong to HIV2 and the ones below to HIV1. Homologous reading frames are shown equidistant from the sequence line.

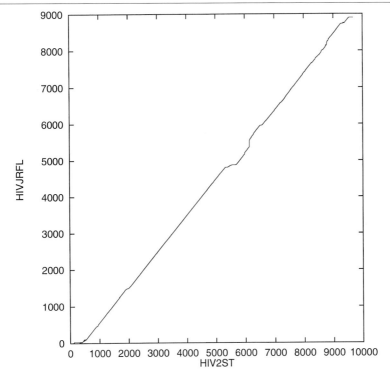

FIG. 4. Optimal alignment of two viruses (HIV2ST and HIVJRFL). The parameters used were 10 for the initiation of an indel, 1 for mutating a nucleotide, 5 for mutating an amino acid, and a nucleotide indel of length 1 cost 2 and an amino acid indel cost 10. Indels of length 1 or 2 in coding regions cost 80 (frameshifts). All these parameters are user options and could be changed.

Analysis of Human Immunodeficiency Virus

The input needed for GenAl is, then, the complete nucleotide sequences, a list of the genes in the two sequences, and the parameters used in the alignment. Each gene consists of a list of exons. An overview of the resulting alignment of HIV1 and HIV2 is shown in Fig. 3. Homologous exons are shown the same number of lines above and below the center line which represents the sequence. The optimal path up through the matrix is shown in Fig. 4. The HIV1 variant has not been fully sequenced at the ends.

Close to optimal alignments have been used to give a nonstatistical confidence estimate on the alignment by, for instance, Zuker.[10] If alignment paths that are within d of the optimum are also drawn, certain areas will

[10] M. Zuker, *J. Mol. Biol.* **221,** 403 (1991).

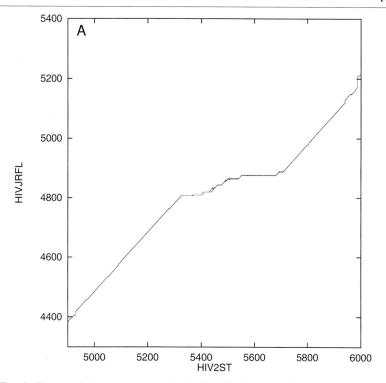

F𝐼G. 5. Closeup of the alignment paths of HIV2 (in the region from the end of the *pol* gene to the end of the *vpr* gene) and HIV1 (in the region from the end of the *pol* gene to the beginning of the *vpu* gene). (A) Optimal paths. (B) Set of paths with a cost within 25 of the optimum for the same region. It can be seen, that the position of the large indel is badly determined.

contain very different paths, while other areas will only contain paths that are very similar to the optimal path. The first seems to be arbitrarily aligned, while the latter is aligned with greater certainty. Figure 5 gives an example of an optimal and a set of suboptimal (within 25 of the optimum) HIV2–HIV1 alignments. A small part of the total alignment, which is about 40 pages of text, is shown in Fig. 6.

Prospects

In longer sequences nonstandard events, like inversions, duplications, and transversions, are to be expected, and in such situations the present method will fail, as it can only interpret sequence evolution in terms of

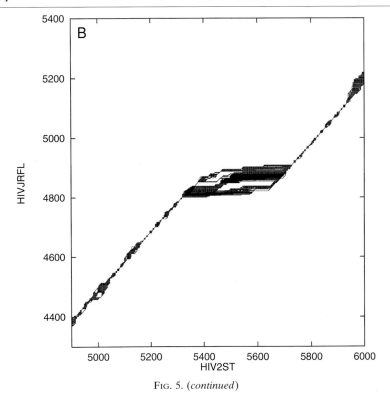

FIG. 5. (*continued*)

substitutions and indels. The present method could be combined with a method such as Schoeninger and Waterman[11] for analyzing such situations.

It would also be useful to have a multiple alignment version of this method. There are numerous methods that derive multiple alignments from pairwise alignments. However, the complexity of combining DNA and protein alignment, and the increased length of the sequences analyzed by this method, have so far prevented a satisfactory implementation of such a method. The presence of recombination in viruses would also undermine the rationale of using tree-based multiple alignment methods.

Evolutionary Model of Complete Virus

The events proposed by a minimal alignment of two homologous sequences give an underestimate of the amount of evolution that has occurred

---

[11] M. Schoeniger and M. S. Waterman, *Bull. Math. Biol.* **54,** 4.521 (1992).

```
      Asp  Arg  Gly  Leu  Pro  Ala  Ala  Arg  Glu  Thr  Arg  Asp
        Thr  Glu  Asp  Phe  Leu  Gln  Leu  Glu  Lys  Gln  Glu  Thr
2014 G A C A G A G G A C T T C C T G C A G C T C G A G A A A C A A G A G A C
1510 G A C A G C A A C T C C C T C T C A G - - - - - - A A G C A G G A G C C
        Thr  Ala  Thr  Pro  Ser  Gln              Lys  Gln  Glu  Pro
      Asp  Ser  Asn  Ser  Leu  Ser              Glu  Ala  Gly  Ala


      Thr  Met  Gln  Arg  Asp  Asp  Arg  Gly  Leu  Ala  Ala  Pro
        Pro  Cys  Arg  Glu  Thr  Thr  Glu  Asp  Leu  Leu  His  Leu
2050 A C C A T G C A G A G A G A G A C G A C A G A G G A C T T G C T G C A C C T
1540 G A T A G A C A A G G A A A T G T - - - A T C C T T T A A C T T C C C T
        Ile  Asp  Lys  Glu  Met              Tyr  Pro  Leu  Thr  Ser  Leu
      Asp  Arg  Gln  Gly  Asn  Val              Ser  Phe  Asn  Phe  Pro


      Gln  Phe  Ser  Leu  Trp  Lys  Arg  Pro  Val  Val  Thr  Ala
        Asn  Ser  Leu  Phe  Gly  Lys  Asp  Gln  ! ! !
2086 C A A T T C T C T C T T T G G A A A A G A C C A G T A G T C A C A G C A
1573 C A G A T C A C T C T T T G G C A A C G A C C C C T C G T C A C A A T A
        Arg  Ser  Leu  Phe  Gly  Asn  Asp  Pro  Ser  Ser  Gln  ! ! !
      Gln  Ile  Thr  Leu  Trp  Gln  Arg  Pro  Leu  Val  Thr  Ile
```

FIG. 6. Alignment showing part of the small region involved in the coding of both the *pol* gene and the end of the *gag* gene. The amino acids just above and below the DNA belong to the gag protein, and the uppermost and lowermost amino acids belong to the *pol* protein. There are two indels in this region, one of length 3 and one of length 6.

since their most recent common ancestor. To correct for this and to analyze the evolutionary dynamics, an evolutionary model is needed.

A model is presented for sequence evolution that can analyze combinations of noncoding, singly coding, and multiply coding regions of aligned homologous DNA sequences. It combines features of the Hasegawa, Kishino, and Yano (HKY) substitution model,[12] allowing for bias both in nucleotide frequencies and in transition/transversion rates with those of the Li *et al.*[13] transition–transversion model, with selection against replacement substitutions. In addition, it is generalized to apply to any combination of overlapping reading frames.

Hasegawa, Kishino, and Yano

The HKY model is a stationary continuous time Markov chain. The rate matrix has three parameters for the equilibrium base frequencies, π_A, π_C, and π_G [π_T is not a free parameter because $\pi_T = (1.0 - \pi_A - \pi_C - \pi_G)$] and two parameters for the transition rate, α, and the transversion rate, β. The transition probability will also have a time parameter, t. The

[12] M. Hasegawa, H. Kishino, and T. Yano, *J. Mol. Evol.* **22,** 160 (1985).
[13] W.-S. Li, C.-I. Wu, and C.-C. Luo, *Mol. Biol. Evol.* **22,** 150 (1985).

$Q_{i,j}$	A	C	G	T
A	$-$	$\beta*\pi C$	$\alpha*\pi G$	$\beta*\pi T$
C	$\beta*\pi A$	$-$	$\beta*\pi G$	$\alpha*\pi T$
G	$\alpha*\pi A$	$\beta*\pi C$	$-$	$\beta*\pi T$
T	$\beta*\pi A$	$\alpha*\pi C$	$\beta*\pi G$	$-$

Equilibrium frequencies of nucleotides: $(\pi A, \pi C, \pi G, \pi T)$.

Transition Probabilities: $P(t) = e^{tQ}$

Distance between sequence: $2*t*[\beta*\pi Y*\pi R + \alpha*(\pi T*\pi C + \pi G*\pi A)]$

$\pi R = (\pi A + \pi G)$ (purines), $\pi Y = (\pi C + \pi T)$ (pyrimidines)

FIG. 7. Hasegawa, Kishino, and Yano substitution model. The rate matrix, **Q**, of the HKY model assumes a substitution rate toward a nucleotide (N) that is proportional to its equilibrium frequency, π_N, multiplied by the transition rate (α) if the substitution is a transition or by the transversion rate (β) if the substitution is a transversion. The element in a row has sum 0.0, so the diagonal entry ($-$) will be minus the sum of the off-diagonal elements in a row. The transition probability matrix, **P**, can be calculated by matrix exponentiation of $t\mathbf{Q}$ ($\mathbf{P} = e^{t\mathbf{Q}}$), where t is the amount of time elapsed. The entries of P have five parameters π_A, π_C, π_G, a, and b, where $a = \alpha t$ and $b = \beta t$. For instance, the probability for changing an A to a C can be written as $P_{A,C}(\pi_A, \pi_C, \pi_G, a, b)$. The expected number of events per site is used as a distance measure between the two sequences and is readily calculated from the parameters.

time parameter will always appear as αt and βt, and the rates and time cannot be estimated separately and are formulated as a distance. Figure 7 illustrates the rate matrix of this model.

Selection against Replacements

The analysis of sequences coding for proteins is complicated by selection against substitutions that cause amino acid replacements. For single coding sequences this causes substitutions to be partitioned into synonymous (also called silent) and replacement (also called nonsynonymous) substitutions.

Li et al.[13] modeled the evolution of coding regions by assuming that the genetic code was more regular than it actually is. It was assumed that varying any nucleotide in a codon among the four possible nucleotides would give the same amino acid (4-fold degenerate) (4), two amino acids, each pair differing by a transition (2-fold degenerate) (2:2), or four different amino acids (nondegenerate) $(1:1:1:1)$. It was also assumed that the nucleotide at different positions evolved independently.

Each position is modeled by the two-parameter model of Kimura,[14] which is the special case of the HKY model where $\pi_T = \pi_A = \pi_C = \pi_G$. It is also assumed that any replacement substitution is accepted with a probability f (selection factor), expected for purifying selection to lie between 0.0 and 1.0. This can readily be incorporated into the scheme of Kimura by defining α' and β' for the three site types. In a nondegenerate site, both transitions and transversions will change the amino acid, and therefore we will have $\alpha' = f\alpha$ and $\beta' = f\beta$; in a 2-fold degenerate site only transversions will be amino acid changing, yielding $\alpha' = \alpha$ and $\beta' = f\beta$; and a 4-fold degenerate site will have no selection factor incorporated, with $\alpha' = \alpha$ and $\beta' = \beta$ (Fig. 8A).

Purifying selection will slow down all events in nondegenerate sites and only transversions in 2-fold degenerate sites; it should have no effect in 4-fold degenerate sites. Properly quantified, this slowdown can be used to estimate the strength of selection and then to recover the underlying biochemical substitution rate. It has been shown in many data sets that the HKY model is superior to the Kimura two-parameter model),[15] so this underlying model was used.

Generalization

This scheme is easily carried over to overlapping reading frames (Fig. 8B) and will be illustrated in the case of two overlapping reading frames. Now, there will be a selection factor for each protein, f_1 and f_2, and one for both $f_{1,2}$, and each site will have to be classified according to its kind in each reading frame. If selection works independently on the two genes, $f_{1,2} = f_1 f_2$.

For the model of the complete virus each gene has its own selection intensity that must be the same for different regions. The selection induced by different genes in areas of overlap are assumed to work independently. This gives a model with five parameters from the HKY model plus nine selection factors from nine genes. The parameters were then estimated by maximum likelihood.[16] Figure 9 shows the overview of an HIV1–HIV1 alignment, and the estimated parameters are shown in Fig. 10. The estimated distance is 50% larger in this model than a previous model by Hein

[14] M. Kimura, *J. Mol. Evol.* **16,** 111 (1980).
[15] N. Goldman, *J. Mol. Evol.* **36,** 182 (1993).
[16] A. W. F. Edwards, "Likelihood." Cambridge Univ. Press, Cambridge, 1972.

A

Sites types

1-1-1-1:	$\alpha' =$	$f*\alpha$	$\beta' =$	$f*\beta$
2-2 :	$\alpha' =$	α	$\beta' =$	$f*\beta$
4 :	$\alpha' =$	α	$\beta' =$	β

B

2nd RF\1st RF	1-1-1-1	2-2	4
1-1-1-1:	$(f1*f2*\alpha, f1*f2*\beta)$	$(f2*\alpha, f1*f2*\beta)$	$(f2*\alpha, f2*\beta)$
2-2 :	$\underline{(f1*\alpha, f1*f2*\beta)}$	$(\alpha, f1*f2*\beta)$	$(\alpha, f2*\beta)$
4 :	$(f1*\alpha, f1*\beta)$	$(\alpha, f1*\beta)$	(α, β)

C

2. pol:	Arg	Gly
1. gag:	Glu	Asp
	A G $\underline{A}$ G G A	
	A G $\underline{C}$ A A C	
1. gag:	Ala	Thr
2. pol:	Ser	Asn

FIG. 8. (A) Single reading frame selection. With one reading frame all amino acid changing substitutions will be assumed to be reduced with a factor, f, relative to if they had not changed the amino acid. Ignoring a few irregularities of the genetic code allows this f to be incorporated directly into the Kimura model. (B) Two independent overlapping reading frame selection. If two reading frames, 1 and 1, are present, there will be two distinct selection factors, f_1 and f_2, that can be incorporated as well. If there were interaction between the joint effects of replacements in both reading frames, then $f_1 f_2$ should be replaced with $f_{1,2}$. This is readily generalized to more than two reading frames. (C) Likelihood for $\underline{AC}$. Using the upper sequence to classify sites, then it is a 1-1-1-1 site with respect to the second reading frame and a 2-2 site with respect to the first reading frame. Thus, $a' = f_2 a$ and $b' = f_1 f_2 b$. The probability for observing $\underline{AC}$ is $pA*P_A, c(p_A, p_C, p_G, f_2 a, f_1 f_2 a)$, that is, the probability of picking A according to the equilibrium distribution and then having A evolve into C, subject to the selective slowdown from the two reading frames of which it is a part.

and Støvlbæk[17] that used the Kimura two-parameter model, not HKY, as underlying model.

As a statistical model for the evolution of the virus, the present model is lacking in biological realism. A more complete model should take the

[17] J. J. Hein and J. Støvlbæk, *J. Mol. Evol.* **40**, 181 (1995).

```
              -----gag---------gag-----              -----vif---------vif-----
                        -----pol---------pol---------pol-----
     1          336       1634      1838      4587      4642      5105
     +-----------+-----------+-----------+-----------+-----------+-----------+
     1          336       1631      1874      4623      4678      5141
                        -----pol---------pol---------pol-----
              -----gag---------gag-----              -----vif---------vif-----


       -----vpr---------vpr---------vpr-----
       -----vif-----                                 -----rev2----
                        -----tat2--------tat2--------tat2--------tat2----
     5105       5165      5376      5395      5515      5590      5607
     +-----------+-----------+-----------+-----------+-----------+-----------+
     5141       5201      5412      5431      5551      5626      5643
                        -----tat2--------tat2--------tat2----
       -----vif-----                                 -----rev2----
       -----vpr---------vpr---------vpr-----


              -----env---------env---------env---------env---------env-----
                        -----rev3-------rev3-------rev3----
       -----vpu---------vpu-----         -----tat3--------tat3----
     5607       5770      5852      7915      7960      7999      8189
     +-----------+-----------+-----------+-----------+-----------+-----------+
     5643       5803      5888      7972      8017      8056      8246
       -----vpu---------vpu-----         -----tat3----
                        -----rev3-------rev3-------rev3----
              -----env---------env---------env---------env---------env-----


       -----env---------env-----
       -----rev3----
                                        -----nef-----
     8189       8195      8331      8333      8953      8956      9175
     +-----------+-----------+-----------+-----------+-----------+-----------+
     8246       8252      8388      8390      9007      9010      9229
                                        -----nef---------nef-----

       -----env---------env-----


     9175       9176
     +-----------+
     9229       9229
```

Fig. 9. GenAl was used to align two HIV1's (HIVELI and HIVLAI) with the same parameters as in Fig. 3.

following into account: varying rates at different positions, codon bias, and uneven selection along the gene and recombination. Recombination will have an effect because the time back until the most recent common ancestor for the two sequences can differ from region to region. More importantly,

Hasegawa, Kishino & Yano Substitution Model Parameters:

$\alpha * t$	$\beta * t$	πA	πC	πG	πT
0.350	0.105	0.361	0.181	0.236	0.222
0.015	0.005	0.004	0.003	0.003	

Selection Factors

GAG	0.385	(s.d. 0.030)
POL	0.220	(s.d. 0.017)
VIF	0.407	(s.d. 0.035)
VPR	0.494	(s.d. 0.044)
TAT	1.229	(s.d. 0.104)
REV	0.596	(s.d. 0.052)
VPU	0.902	(s.d. 0.079)
ENV	0.889	(s.d. 0.051)
NEF	0.928	(s.d. 0.073)

Estimated Distance per Site: 0.194

FIG. 10. HIV1 analysis, showing parameters estimated for the model describing the evolution of the aligned HIV1's in Fig. 9. Transitions happen more than three times as frequently as transversions. The sequences also seem very A rich. All estimated parameters are accompanied by a standard deviation. It is seen that longer genes (*gag, pol, env*) have their selection factors better determined than shorter genes (i.e., they have smaller standard deviations). It is also seen that the *pol* gene is the slowest evolving. The *tat* gene has an *f* value higher than 1.0, and the three last small genes have *f* values that are very high. The large envelope gene (*env*) has evolved considerably faster than the other two large genes (*pol, gag*). The expected number of events per site is 0.195.

a model should allow not only the analysis of two sequences but of many, which means that the phylogeny with recombination problem would have to be addressed.

Conclusion

The basic idea is to combine the protein alignment problem with the DNA alignment problem and then solve them simultaneously. The alignment will then also align homologous exons with homologous exons, because this is more parsimonious.

The solution of the combined alignment problem is a mosaic of noncoding, singly coding, and possibly multiply coding regions. It is of interest to measure rates of evolution and selective effects of different coding regions and to estimate the expected number of events per site in the absence of selection as a distance measure between the sequences. A model has been proposed that combines one of the better models for the evolution of nucleotides with one of the most widely used methods of incorporating selection against amino acid replacements to yield a model for complete viruses, including regions with overlapping reading frames.

Acknowledgments

J. H. was supported by the Carlsberg Foundation and the Danish Research Council, Grants 11-8916-1 and 11-9639-1. J. S. was supported by the Danish Research Council, Grants 11-8916-1 and 11-9639-1, and by the Carlsberg Foundation. We thank Bernt Guldbrandtsen for comments on the manuscript and for help with the figures, and Dr. G. Myers for making the HIV database available.

[24] Inferring Phylogenies from Protein Sequences by Parsimony, Distance, and Likelihood Methods

By Joseph Felsenstein

Introduction

The first molecular sequences available were protein sequences, so it is not surprising that the first papers on inferring phylogenies from molecular sequences described methods designed for proteins. Eck and Dayhoff[1] described the first molecular parsimony method, with amino acids as the character states. Fitch and Margoliash[2] initiated distance matrix phylogeny methods with analysis of cytochrome sequences. Neyman[3] presented the first likelihood method for molecular sequences, using a model of symmetric change among all amino acids.

After a long period in which attention shifted to nucleotide sequences, attention is again being paid to models in which the amino acid sequences explicitly appear. This is not only because of the increased availability of protein sequence data, but also because the conservation of amino acid sequence and protein structure allows us to bring more information to bear on ancient origins of lineages and of genes. In this chapter I briefly review

---

[1] R. V. Eck and M. O. Dayhoff, "Atlas of Protein Sequence and Structure 1966." National Biomedical Research Foundation, Silver Spring, Maryland, 1966.

[2] W. M. Fitch and E. Margoliash, *Science* **155**, 279 (1967).

[3] J. Neyman, *in* "Statistical Decision Theory and Related Topics" (S. S. Gupta and J. Yackel, eds.), p. 1. Academic Press, New York, 1971.

Copyright © 1996 by Academic Press, Inc.
All rights of reproduction in any form reserved.

the work on using protein sequence and structure to infer phylogeny, in the process describing some methods of my own.

Parsimony

Eck and Dayhoff[1] did not describe their algorithms in enough detail to reproduce them, but it is apparent that the model of amino acid sequence evolution they used did not take the genetic code into account. It simply considered the amino acids as 20 states, with any change of state able to result in any of the other 19 amino acids. The realization that more information could be extracted by explicitly considering the code shortly led to more complex models. Fitch and Farris[4] gave an approximate algorithm to calculate for any set of amino acid sequences, on a given tree, how many nucleotide substitutions must, at a minimum, have occurred. As certain amino acid replacements would then require two or three base substitutions, this would differentially weight amino acid replacements. Moore[5,6] had already presented an exact, though more tedious, algorithm to count the minimum number of nucleotide substitutions needed, and he pointed out the approximate nature of the Fitch and Farris method.[7]

These papers might have settled the matter for the parsimony criterion, except that they count as equally serious those nucleotide substitutions that do and do not change the amino acid. For example, we might have a phenylalanine that is coded for by a UUU, which ultimately becomes a glutamine that is coded for by a CAA. This requires three nucleotide substitutions. It is possible for one of these to be silent, as we can go from UUU (Phe) → CUU (Leu) → CUA (Leu) → CAA (Glu). Presumably the second of these changes will not be as improbable as the others, as it will not have to occur in the face of natural selection against change in the amino acid, nor wait for a change of environment or genetic background that favors the amino acid replacement.

In the PROTPARS program of my PHYLIP package of phylogeny programs, I have introduced (in 1983) a parsimony method that attempts to reflect this. In the above sequence it counts only two changes, allowing the silent substitutions to take place without penalty. In effect the method uses the genetic code to designate which pairs of amino acids are adjacent, and it allows change only among adjacent states. Sankoff[8] and Sankoff and

---

[4] W. M. Fitch and J. S. Farris, *J. Mol. Evol.* **3**, 263 (1974).

[5] G. W. Moore, J. Barnabas, and M. Goodman, *J. Theor. Biol.* **38**, 459 (1973).

[6] G. W. Moore, *J. Theor. Biol.* **66**, 95 (1977).

[7] G. W. Moore, *in* "Genetic Distance" (J. F. Crow and C. Denniston, eds.), p. 105. Plenum, New York, 1974.

[8] D. Sankoff, *SIAM J. Appl. Math.* **28**, 35 (1975).

Rosseau[9] have presented a generalized parsimony algorithm that allows us to count on a given tree topology how many changes of state are necessary, where we can use an arbitrary matrix of penalties for changes from one state to another. The PROTPARS algorithm is equivalent to the Sankoff algorithm, being quicker but less general.

The set of possible amino acid states in the PROTPARS algorithm has 23 members, these being the 20 amino acids plus the possibilities of a gap and a stop codon. Serine is counted not as one amino acid but as two, corresponding to the two islands of serine codons in the genetic code. These are {UCA, UCG, UCC, UCU} and {AGU, AGC}, which make serine the only amino acid whose codons fall into two groups that cannot be reached from each other by a single mutation. PROTPARS copes with this by regarding them as two amino acid states (ser1 and ser2) and treats an observation of serine as an ambiguity between these two.

Imagine that we know, for a node in the tree, the set of amino acid states that are possible at this node. If the node is a terminal (tip) species, these are just the observed amino acid, there being more than one if serine is observed or if any of Asn, Gln, or Glx are observed. There is also the possibility that the amino acid is unknown, but known not to be a gap, and the possibility that the amino acid could be any one including a gap. More complex ambiguities are also possible and can arise in the process of reconstruction of the states at interior nodes in the tree. Any of these can be represented by designating the members of the set S_0 of possible states.

Given the particular version of the genetic code that we are using, we can also precompute, for each amino acid a, the set N_a of amino acid states that are one or fewer steps away. In PROTPARS, gaps are counted as being three steps away from all the amino acids and from stop codons. Having these precomputed sets allows us to take the sets S_0 at the tips of the tree, and compute for them S_1 and S_2, the sets of amino acid states one or fewer steps away, and two or fewer steps away. In our program, all states, including gaps, are three or fewer steps away, so that we do not need a set S_3. In PROTPARS the three sets S_0, S_1, and S_2 are updated down the tree, and the number of steps needed for the tree counted, in the following way.

Imagine that there is an internal node in the tree with two descendants, and whose sets of possible states are the L_i and the R_i. We are computing the sets S_i for the internal node. First, L_0 and R_0 are compared. If they are the same then the L_i must be identical to the R_i, and the S_i are simply

[9] D. Sankoff and P. Rousseau, *Math. Prog.* **9**, 240 (1975).

set to be the L_i, and no steps are counted. Otherwise, we compute the four sets

$$
\begin{aligned}
T_0 &= L_0 \cap R_0 \\
T_1 &= (L_1 \cap R_0) \cup (L_0 \cap R_1) \\
T_2 &= (L_2 \cap R_0) \cup (L_1 \cap R_1) \cup (L_0 \cap R_2) \\
T_3 &= R_0 \cup (L_2 \cap R_1) \cup (L_1 \cap R_2) \cup L_0
\end{aligned}
\tag{1}
$$

They are computed one after the other. Their interpretation is straightforward. For example, T_1 is the set of amino acid states that, if present at the internal node, requires one step to give rise to L_0 and none to give rise to R_0, or else one step to give rise to R_0 and none to give rise to L_0. Thus it is the set of states which, if present at the interior node, require one extra step in the subtree that is above that node. As soon as one of these, say T_k, turns out to be nonempty, we know that a minimum of k more steps will be needed at this node, and that the set S_0 for that node will be T_k. In addition, T_3 at least must be nonempty, as it contains the union of R_0 and L_0.

Now, having found S_0, all we need to do is to compute S_1 and S_2 for the internal node. The formulas for doing so are

$$
S_k = \bigcup_{a \in S_{k-1}} N_a, \qquad k = 1, 2
\tag{2}
$$

This of course does not need to be done if $L_0 = R_0$, as the sets L_1 and L_2 (or R_1 and R_2) can then be used directly.

This method of calculation using sets is equivalent to having a vector of numbers, one for each amino acid state, which are 0, 1, 2, or 3. The Sankoff algorithm asks us to specify for each state the number of extra steps that would be required above that point in the tree if that state existed in that internal node. In our model the possible values for the number of extra steps are 0, 1, 2, and 3. The sets S_i are just the amino acids that would have the number of extra steps less than or equal to i. The algorithm is then equivalent to the appropriate application of the Sankoff algorithm. It could probably be speeded up further, as most of the time the set S_2 is the set of all amino acids, and that could be used as the basis for some further economies.

Figure 1 shows the sets that would be stored on a small sample tree for one amino acid position, and the counting of steps. At each node the three sets S_0, S_1, and S_2 are shown, and at interior nodes the number of steps that are counted are also shown in circles. There are four different amino acids at the tips of the tree. If any amino acid could change to any other the tree would require only three steps, but in my protein parsimony model it requires five.

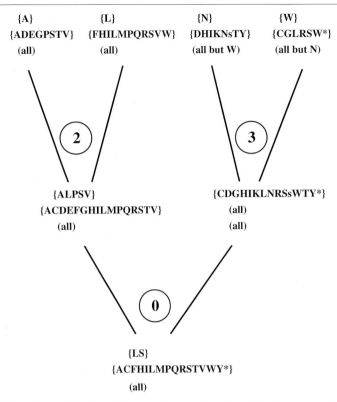

{A}   {L}   {N}   {W}

{ADEGPSTV} {FHILMPQRSVW} {DHIKNsTY} {CGLRSW*}

(all)   (all)   (all but W)   (all but N)

2   3

{ALPSV}   {CDGHIKLNRSsWTY*}

{ACDEFGHILMPQRSTV}   (all)

(all)   (all)

0

{LS}

{ACFHILMPQRSTVWY*}

(all)

Fig. 1. Small tree with the calculation of the sets S_0, S_1, and S_2 shown at each node, for a site where the tips have amino acid states alanine, leucine, asparagine, and tryptophan, respectively. The sets are shown as sets of one-letter amino acid representations. S and s are the two codon "islands" of serine, and an asterisk (*) represents stop codons. The number of steps counted at each fork is shown in a circle.

Protein parsimony methods exactly equivalent to PROTPARS are also available in the programs PAUP and MacClade, using predefined matrices of costs of substitution between amino acid states, with the costs being taken into account by the Sankoff algorithm.

Distances

Distance matrix methods calculate for every pair of sequences an estimate of the branch length separating them, where branch length is the product of time and rate of evolution. That tree is then chosen that, by some criterion, makes the best prediction of these pairwise distances. For protein sequences we need to specify a probabilistic model of evolution. Jukes and Cantor[10] were the first to do this for protein sequences (see also Farris[11]). This model was highly oversimplified, as it had equal probabilities of change between all pairs of amino acids. Dayhoff and Eck[12] and Dayhoff *et al.*[13] empirically tabulated probabilities of change between amino acids over short evolutionary times, producing a table of transition probabilities between amino acids. This model does not take explicit account of the genetic code, and it is subject to errors from the limited sample size on which it was based. Nevertheless, the genetic code should affect its transition probabilities, and so should the biochemical properties of the amino acids. A more recent empirical model of amino acid change is that of Jones *et al.*[14] They have also produced models for specific subclasses of proteins that may be more useful in those contexts.[15] Other compilations of scoring matrices for evaluating the similarity of amino acid sequences[16,17] are not in the form of transition probability tables. For this reason they cannot be used to compute the branch length estimates that we require here.

A naive alternative to these empirical matrices is to divide the amino acids into a number of categories, based on their chemical properties. Suppose that we imagine mutations occurring in the genetic code table, with the starting points being codons generated at random from a given base composition. Now imagine single base substitutions. If these do not change the biochemical class of the amino acid, they are accepted; if they do, they are only accepted with probability p. We omit the stop codons from consideration: if either the starting point or the destination of a change is a stop codon, the change is not made. This model, once given the amino acid categories, the base frequencies, and the probability p, generates a transition probability table between all pairs of amino acids.

Version 3.5 of PHYLIP contains a program, PROTDIST, which computes distances based on the PAM001 model[13] and the transition probability matrix generated by the categories model. It also can compute distances

[10] T. H. Jukes and C. Cantor, *in* "Mammalian Protein Metabolism" (M. N. Munro, ed.), p. 21. Academic Press, New York, 1969.

[11] J. S. Farris, *Am. Nat.* **107,** 531 (1973).

[12] M. O. Dayhoff and R. V. Eck, "Atlas of Protein Sequence and Structure 1967–1968," National Biomedical Research Foundation, Silver Spring, Maryland, 1968.

[13] M. O. Dayhoff, R. M. Schwartz, and B. C. Orcutt, *in* "Atlas of Protein Sequence and Structure." (M. O. Dayhoff, ed.), Vol. 5, Suppl. 3, p. 345. National Biomedical Research Foundation, Washington, D.C., 1978.

[14] D. T. Jones, W. R. Taylor, and J. M. Thornton, *Comput. Appl. Biosci.* **8,** 275 (1992).

[15] D. T. Jones, W. R. Taylor, and J. M. Thornton, *FEBS Lett.* **339,** 269 (1994).

[16] G. H. Gonnet, M. A. Cohen, and S. A. Benner, *Science* **256,** 1443 (1992).

[17] S. Henikoff and J. G. Henikoff, *Proc. Natl. Acad. Sci. U.S.A.* **89,** 10915 (1992).

using the formula of Kimura[18] which bases the distance on the fraction of amino acids shared between the sequences, without regard to which amino acids they are. The categories model, as implemented in PROTDIST, can use several different genetic codes (the universal code and several kinds of mitochondrial codes). Three categorizations of the amino acids are used, one the categories given by George et al.,[19] one from a categorization in a "baby biochemistry" text, and one the opinion of a colleague. Interestingly, all three of these turn out to be subdivisions of one linear order of amino acids. We have found that a value of $p = 0.45$ brings the ratio of between- to within-category change in the category model of George et al.[19] close to that in the Dayhoff model. In the next release (4.0) of PHYLIP, we hope to expand the range of models by including the model of Jones et al.[14] and allowing for a gamma distribution of evolutionary rates among sites, in the manner of Jin and Nei[20] and Nei et al.[21]

Given the evolutionary model, we use maximum likelihood estimation to compute the distances. In effect we are specifying a two-species tree, with but one branch, between the pair of species, and estimating that branch length by maximum likelihood. If we observe n_{ij} changes between amino acids i and j, and if the model we are using has equilibrium frequency f_i for amino acid i and transition probability $P_{ij}(t)$ over time t, the expected fraction of sites which will have amino acid i in one species and j in the other is $f_iP_{ij}(t)$. The PAM001 matrix gives the conditional probabilities P_{ij}, but they are not reversible. To make a reversible model that is as close as possible to PAM001, we have used instead

$$Q_{ij} = (f_iP_{ij} + f_jP_{ji})/2 \tag{3}$$

This gives us symmetric joint probabilities of observing i and j in two closely related sequences. Suppose that the **M** are transition probabilities that would lead to the joint probabilities **Q**, and that π is the vector of equilibrium frequencies which is implied by **M**. We start out knowing **Q** but not **M** or π. It is not hard to show that the eigenvalues of $\pi'\mathbf{M}$ are the same as the eigenvalues of **Q**, and the eigenvalues of **M** can also be directly derived from those of **Q**. The eigenvalues and eigenvectors of **M** are computed in this way (they are precomputed in the PAM001 case and computed by the program in the categories cases).

[18] M. Kimura, "The Neutral Theory of Molecular Evolution." Cambridge Univ. Press, Cambridge, 1983.
[19] D. G. George, W. C. Barker, and L. T. Hunt, this series, Vol. 183, p. 333.
[20] L. Jin and M. Nei, Mol. Biol. Evol. **7**, 82 (1990).
[21] M. Nei, R. Chakraborty, and P. A. Fuerst, Proc. Natl. Acad. Sci. U.S.A. **73**, 4164 (1976).

From the eigenvalues and eigenvectors of **M** we can readily compute the transition probabilities $M_{ij}(t)$, and their derivatives with respect to t. The likelihood which we must maximize is

$$L = \prod_i \prod_j [\pi_i M_{ij}(t)]^{n_{ij}} \tag{4}$$

The log-likelihood is maximized over values of t by Newton–Raphson iteration.

The resulting distance computation is not fast, but it seems adequate. However, it makes one assumption that is quite severe. All amino acid positions are assumed to change at the same rate. This is unrealistic. To some extent we can compensate for this by correcting the distances by using the approach of Jin and Nei.[20] However there is information that is being lost by doing this. We would like to be able to use the variation in an amino acid position in one part of the data set to infer whether that position allowed change to occur at a high rate, and thus to help us evaluate other parts of the same data set. But no distance matrix method can do this, as they consider only pairs of sequences.

Likelihood Methods

Neyman[3] and Kashyap and Subas[22] developed maximum likelihood methods for inferring phylogenies from protein data. They used the highly oversimplified Jukes–Cantor[10] model of symmetric change among amino acids, and they could not handle more than three or four sequences in the tree in a reasonably exact way. I have shown[23] how to make the likelihood computations practical for larger numbers of species. Likelihood methods for proteins have not been developed further until more recently, because of the computational burden. Where nucleotide sequence likelihood methods use a 4×4 transition probability matrix, in protein models these must be either 20×20 or 64×64, thus requiring either 25 or 256 times as much computation. With increased speed of desktop and laboratory computers, developing a reasonable likelihood method for protein sequences has become more of a priority.

Adachi and Hasegawa[24] and Adachi et al.[25] have developed such a method, using the Dayhoff PAM matrix[13] as the transition probability

[22] R. L. Kashyap and S. Subas, *J. Theor. Biol.* **47,** 75 (1974).
[23] J. Felsenstein, *J. Mol. Evol.* **17,** 368 (1981).
[24] J. Adachi and M. Hasegawa, *Jpn. J. Genet.* **67,** 187 (1992).
[25] J. Adachi, Y. Cao, and M. Hasegawa, *J. Mol. Evol.* **36,** 270 (1993).

matrix among amino acid states, but without any direct use of the genetic code. Their program, which is similar to existing DNA likelihood programs but has some effort put into requiring fewer evaluations of the likelihood, is available in their MOLPHY package from their ftp site at sunmh.ism.ac.jp.

It is tempting to develop a method that takes the genetic code explicitly into account. In principle one could have 64 states, one for each codon, and regard the amino acids as ambiguous observations (e.g., alanine would be regarded as an observation of "either TCA or TCG or TCC or TCT"). The computational difficulties would be severe. One could also hope to take into account both observed protein sequence and the underlying DNA sequence, which is often known. Hein[26] and Hein and Støvbæk[27,28] have made a start on such models.

A more serious limitation of existing protein maximum likelihood models is that they assume that all positions change at the same expected rate. This assumption has been removed from nucleotide sequence likelihood models, using hidden Markov model techniques.[29-32] Its extension to proteins is straightforward and badly needed, but it does promise to slow down the computer programs severalfold.

Structure, Alignment, and Phylogeny

Beyond any of these complications is the challenge of taking protein structure into account. Researchers on analysis of RNA sequences have found that there is a synergism between inferences of phylogeny, alignment, and structure. It is just beginning to become widely recognized that the same will be true with proteins, the advantages being probably greater. Structure-based hidden Markov models (HMMs) have been used to improve sequence alignment of proteins, although without taking phylogeny into account.[33,34] Three-dimensional protein structures can be used to infer phylogenies.[35] Structural context affects not only amino acid composition,

[26] J. Hein, *J. Theor. Biol.* **167**, 169 (1994).

[27] J. Hein and J. Støvlbæk, *J. Mol. Evol.* **38**, 310 (1994).

[28] J. Hein and J. Støvlbæk, *J. Mol. Evol.* **40**, 181 (1995).

[29] J. Felsenstein and G. A. Churchill, *Mol. Biol. Evol.* **13**, 93 (1996).

[30] Z. Yang, *Mol. Biol. Evol.* **10**, 1396 (1994).

[31] Z. Yang, *J. Mol. Evol.* **39**, 306 (1994).

[32] Z. Yang, *Genetics* **139**, 993 (1995).

[33] P. Baldi, Y. Chauvin, T. Hunkapiller, and M. A. McClure, *Proc. Natl. Acad. Sci. U.S.A.* **91**, 1059 (1994).

[34] A. Krogh, M. Brown, I. S. Mian, K. Sjölander, and D. Haussler, *J. Mol. Biol.* **235**, 1501 (1994).

[35] M. S. Johnson, A. Šali, and T. L. Blundell, this series, Vol. 183, p. 670.

but the substitution process itself.[36] When residues interact, there may result patterns of compensating substitutions. This has begun to be examined for proteins.[37]

In RNAs, phylogenies and inferences of structure are increasingly important to one another. Patterns of compensating substitutions are strong and have led to mathematical models of this substitution process.[38,39] One can imagine a unified process of inference for proteins and protein-coding regions that simultaneously infers phylogeny, alignment, secondary structure, and three-dimensional structure. The computational problems will be severe, but many of the components needed are already being worked on. Having coordinated our inferences of structure and evolutionary history, we will then be free to dream of considering function as well.

Acknowledgments

This work has been supported by grants from the National Science Foundation (DEB-9207558) and the National Institutes of Health (1 R01 GM 51929-01).

[36] J. Overington, D. Donnelly, M. S. Johnson, A. Šali, and T. L. Blundell, *Protein Sci.* **1,** 216 (1992).
[37] W. R. Taylor and K. Hatrick, *Protein Eng.* **7,** 341 (1994).
[38] E. R. M. Tillier, *J. Mol. Evol.* **39,** 409 (1994).
[39] E. R. M. Tillier and R. A. Collins, *Mol. Biol. Evol.* **12,** 7 (1995).

[25] Reconstruction of Gene Trees from Sequence Data

By Naruya Saitou

Properties of Gene Tree

Reconstruction of the phylogeny of genes is essential not only for the study of evolution but also for biology in general because replication of nucleotide sequences automatically produces a bifurcating tree of genes. It should be emphasized that the phylogenetic relationship of genes is different from the mutation process. The former always exists, whereas mutations may or may not happen within a certain time period and DNA region. Therefore, even if several nucleotide sequences happen to be identical, there must be a genealogical relationship for those sequences.

However, it is impossible to reconstruct the genealogical relationship without the occurrence of mutational events. In this respect, the extraction

Copyright © 1996 by Academic Press, Inc.
All rights of reproduction in any form reserved.

of mutations from genes and their products is important for reconstructing phylogenetic trees of genes. The advancement of molecular biotechnology has made it possible to produce nucleotide sequences routinely. We therefore focus on the analysis of nucleotide sequences. However, a substantial part of this chapter also applies to other molecular data.

Formal Characteristics of Trees

A tree is a kind of graph. A graph is composed of node(s) and branch(es). There should be only one path between any two nodes on a tree. In evolutionary studies, a node represents a gene, species, or population, depending on the purpose, and a branch represents the topological relationship between nodes (often including information on lengths that represent mutational changes or evolutionary time). Nodes are divided into external and internal ones. The former are also called operational taxonomic units (OTUs). There are five OTUs (1–5) and four internal nodes (X, Y, Z, and R1) in the tree in Fig. 1a. Branches are also divided into external and internal ones. An external branch connects an external node and an internal node (e.g., branch 1-X of Fig. 1), whereas an internal branch connects two internal nodes (e.g., branch X-Z of Fig. 1).

A tree can be either rooted or unrooted. A rooted tree has a special node called the root which is defined as the position of the common ancestor

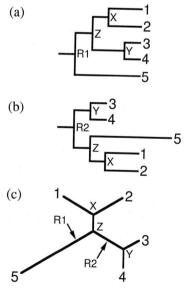

Fig. 1. Examples of rooted trees (a, b) and an unrooted tree (c) for five OTUs.

(see Fig. 1a,b). There will be a unique path from the root to any other node, and the direction of this is, of course, that of time. A phylogenetic tree in an ordinary sense is a rooted tree. Unfortunately, however, many methods for building phylogenetic trees produce unrooted trees, such as the tree of Fig. 1c. An unrooted tree can be converted to a rooted tree if the position of the root is specified. The trees in Fig. 1a,b were produced from the unrooted tree of Fig. 1c.

This relation between rooted and unrooted trees is used for the "outgroup" method of rooting as follows. When we are interested in determining the phylogenetic relationship among n sequences, we will add one (or more) sequence that is known to be an outgroup to the n sequences. The unrooted tree obtained for the $n + 1$ sequences can easily be converted to a rooted tree of n sequences. Sequence 5 corresponds to the outgroup in the tree in Fig. 1c when the root is R1, and the tree of Fig. 1a is then obtained. When the root is R2, sequences 3 and 4 are considered to be the outgroup to sequences 1, 2, and 5, and we obtain the tree of Fig. 1b.

The number of possible tree topologies rapidly increases with an increase in the number of OTUs. The general equation for the possible number of topologies for bifurcating unrooted trees (Tn) for n (≥ 3) OTUs is given by[1]

$$Tn = (2n - 5)!/[2^{n-3}(n - 3)!] \tag{1}$$

If we apply Eq. (1), there are 221,643,095,476,699,771,875 possible tree topologies for 20 OTUs. It is clear that the search for the true phylogenetic tree of many sequences is a very difficult problem. This is why so many methods have been proposed for building phylogenetic trees.

Gene Trees and Species Trees

Phylogenetic trees of genes and species are called gene trees and species trees, respectively, and there are several important differences between them. One such difference is illustrated in Fig. 2. Because a gene duplication occurred before the speciation of species A and B in Fig. 2a, both species have two homologous genes in their genomes. In this situation, we should distinguish orthology, which is homology of genes reflecting the phylogenetic relationship of species, from paralogy, which is homology of genes caused by gene duplication(s).[2] Thus, genes 1 and 3 (and 2 and 4) are orthologous, whereas genes 1 and 4 (and 2 and 3) are paralogous. If one is not aware of the gene duplication event, the gene tree for 1 and 4

[1] L. L. Cavalli-Sforza and A. W. F. Edwards, *Am. J. Hum. Genet.* **19**, 233 (1967).
[2] W. M. Fitch and E. Margoliash, *Evol. Biol.* **4**, 67 (1970).

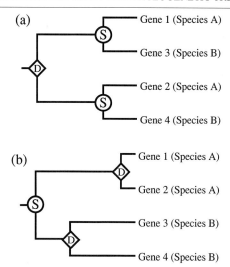

FIG. 2. Two possibilities of a gene tree for four genes sampled from two species. (a) Gene duplication (denoted by D) occurred before speciation (denoted by S). (b) Speciation occurred before two gene duplications.

may be misrepresented as the species tree of A and B, and thus a gross overestimation of the divergence time may occur. It should also be noted that the divergence time between genes 1 and 3 is identical with that between genes 2 and 4, since both times correspond to the same speciation event.

When two homologous gene copies are found in species A and B, another situation is possible, as shown in Fig. 2b. Now two gene duplications occurred after the speciation of species A and B, and two gene copies in the genome of each species are more closely related with each other than the corresponding homologous genes at different species. Because two duplication events occurred independently, the divergence time between genes 1 and 2 is different from that between genes 3 and 4.

Even when orthologous genes are used, a gene tree may be different from the corresponding species tree. This difference comes from the existence of gene genealogy in the ancestral species. A simple example is illustrated in Fig. 3a. A gene sampled from species A has its direct ancestor at the speciation time T_1 generations ago, and so does a gene sampled from species B. Thus, the divergence time between the two genes sampled from the different species always overestimates that of the species. The amount of overestimation corresponds to the coalescence time in the ancestral species, and its expectation is $2N$ for neutrally evolving nuclear genes of

(a)   (b)

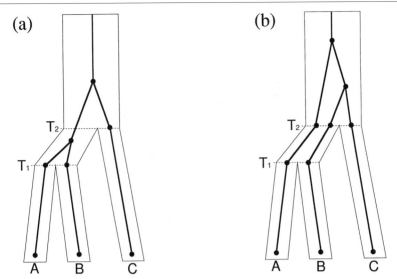

FIG. 3. Difference between a gene genealogy and species tree. (a) Topology of the gene genealogy is the same as that for the species tree. (b) Topology of the gene genealogy is different from that for the species tree. Full circles and thick lines denote the genealogical relationship, whereas thin lines (outlining the gene tree) denote the species tree. A, B, and C denote three genes each sampled from extant species, whereas X and Y denote ancestral genes. T_1 and T_2 denote the two speciation times.

diploid organism, where N is the population size of the ancestral species.[3] Therefore, if the two speciation events (T_1 and T_2) are close enough, the topological relationship of the gene tree may become different from that of the species tree, as shown in Fig. 3b. Although species A and B are more closely related than to C, genes sampled from species B and C happen to be more closely related with each other than to that sampled from species A. The probability (P_{error}) of obtaining an erroneous tree topology is given by[4]

$$P_{error} = (2/3) \, e^{-T/2N} \qquad (2)$$

where $T = T_2 - T_1$ generations. For example, P_{error} is 0.404 when $T = 50,000$ and $N = 50,000$. Therefore, a species tree estimated from a single gene may not be correct even if the gene tree was correctly estimated. In this case, we should use more than one gene.

[3] F. Tajima, *Genetics* **105**, 437 (1983).
[4] M. Nei, "Molecular Evolutionary Genetics." Columbia Univ. Press, New York, 1987.

(a)

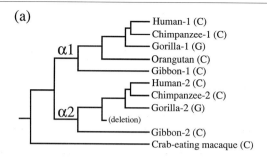

(b)

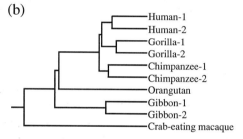

Fɪɢ. 4. Alteration of an estimated gene tree caused by gene conversion. (a) The most presumable gene tree for the primate immunoglobulin α1 and α2 genes. C or G in parentheses after species names indicate one nucleotide configuration possibly caused by gene conversion in the gorilla genome. (b) A spurious gene tree (modified from Kawamura et al.[5]).

When gene conversion and/or recombination has occurred within the gene region under consideration, a gene tree may be different from the species tree. Kawamura et al.[5] examined primate immunoglobulin α genes 1 and 2. Figure 4a shows the plausible gene tree; the gene duplication clearly preceded speciation of hominoids, followed by deletion of the α2 gene from the orangutan genome. However, there are many nucleotide sites that possibly experienced gene conversion. One such example is shown in Fig. 4a; two gorilla genes were both G at a particular nucleotide site, while the remaining genes were C. This suggests either parallel substitution in the gorilla lineage or gene conversion between two gorilla genes. If this kind of nucleotide configuration is contiguous, gene conversion is suspected. The resulting spurious gene tree (Fig. 4b) is distorted from the tree of Fig. 4a because of the strong effect of gene conversion.

Ideally, branch lengths of a phylogenetic tree are proportional to the

[5] S. Kawamura, N. Saitou, and S. Ueda, J. Biol. Chem. **267**, 7359 (1992).

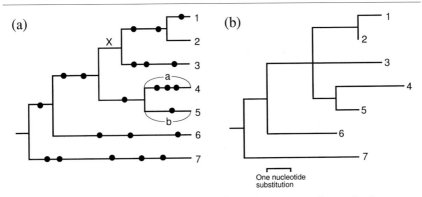

FIG. 5. Examples of the expected gene tree (a) and the corresponding realized gene trees (b). Filled circles on the expected gene tree denote nucleotide substitutions. Because no substitution occurred at branch X of the expected gene tree (a), the corresponding branch does not exist in the realized gene tree (b).

physical time since divergence. Thus the branch a and b of Fig. 5a should be the same length. We call this type of rooted tree the expected tree.[4] Both species and gene trees have their expected trees, but their properties are somewhat different from each other. An expected gene tree directly reflects the history of DNA replications, whereas an expected species tree is a gross simplification of the course of differentiation of populations. Therefore, the speciation time is always unclear.

As mentioned earlier, the genealogical relationship of genes, or expected gene tree, is independent from the mutation process. However, mutation events are essential for the reconstruction of phylogenetic trees. Thus, we can at best estimate a gene tree according to the mutation events realized on its expected gene tree. We call this ideal reconstruction of the gene tree the realized gene tree (Fig. 5b), whereas the reconstructed one from observed data is called the estimated gene tree.[6] Branch lengths of realized and estimated gene trees are proportional to mutational events. These mutational events are not necessarily proportional to physical time. By definition, expected gene trees are strictly bifurcating, while realized and estimated gene trees may be multifurcating. This is because of the possibility of no mutation at a certain branch, such as branch X of Fig. 5a.

A species tree reconstructed from observed data is called an estimated species tree, but there is no realized species tree. It should also be noted that both expected and realized trees are rooted, while estimated trees are often unrooted due to the limitations of available information.

[6] N. Saitou, in "Molecular Biology: Current Innovations and Future Trends Part 2" (H. G. Griffin and A. M. Griffin, eds.), p. 115. Horizon Scientific Press, Norfolk, England, 1995.

TABLE I
CLASSIFICATION OF TREE-MAKING METHODS

Method	Stepwise clustering	Exhaustive search
Distance matrix	UPGMA	KITCH
	Distance Wagner	Fitch–Margoliash
	Neighbor joining	Minimum evolution
Character state		Maximum parsimony
		Compatibility
		Maximum likelihood

Methods for Building Phylogenetic Trees of Genes

Classification of Tree-Building Methods

Many methods have been proposed for building a phylogenetic tree from observed data. To clarify the nature of each method, it is useful to classify these methods from various aspects. Tree-building methods can be divided into two types in terms of the type of data they use: distance matrix methods and character-state methods. A distance matrix consists of all the possible pairwise distances, whereas an array of character states is used for the character-state methods. UPGMA (unweighted pairgroup method using arithmatic mean),[7] the Fitch and Margoliash method,[8] the distance Wagner method,[9] the neighbor-joining method,[10] and the minimum evolution methods[1,11,12] are distance matrix methods, whereas the maximum parsimony method,[13] the compatibility method,[14] and the maximum likelihood method[15] are character-state methods (Table I).

Another classification is by the strategy of a method to find the best tree. One way is to examine all or a large number of possible tree topologies and choose the best one according to a certain criterion. We call this the exhaustive search method. The other strategy is to examine a local topological relationship of OTUs and find the best tree. These types of methods are called stepwise clustering methods. Both strategies are used for the distance matrix methods, while the exhaustive search strategy is usually used for character-state methods (Table I).

[7] P. H. P. Sneath and R. Sokal, "Numerical Taxonomy." Freeman, San Francisco, 1977.
[8] W. M. Fitch and E. Margoliash, *Science* **155,** 279 (1967).
[9] J. S. Farris, *Am. Nat.* **106,** 645 (1972).
[10] N. Saitou and M. Nei, *Mol. Biol. Evol.* **4,** 406 (1987).
[11] N. Saitou and T. Imanishi, *Mol. Biol. Evol.* **6,** 514 (1989).
[12] A. Rzhetsky and M. Nei, *Mol. Biol. Evol.* **9,** 945 (1992).
[13] W. M. Fitch, *Am. Nat.* **111,** 223 (1977).
[14] W. J. Le Quesne, *Syst. Zool.* **18,** 201 (1969).
[15] J. Felsenstein, *J. Mol. Evol.* **17,** 368 (1981).

TABLE II
ESTIMATED NUMBER OF NUCLEOTIDE SUBSTITUTIONS PER SITE BETWEEN EVERY
PAIR OF 10 SEQUENCES[a]

	1	2	3	4	5	6	7	8	9
2	0.0516								
3	0.0550	0.0031							
4	0.0483	0.0221	0.0253						
5	0.0582	0.0651	0.0685	0.0549					
6	0.0094	0.0416	0.0450	0.0384	0.0549				
7	0.0125	0.0584	0.0619	0.0551	0.0651	0.0157			
8	0.0284	0.0687	0.0722	0.0654	0.0754	0.0317	0.0285		
9	0.0925	0.1221	0.1259	0.1185	0.1370	0.0820	0.0786	0.0927	
10	0.1921	0.2183	0.2228	0.2054	0.2309	0.1798	0.1795	0.1833	0.1860

[a] Gaps were eliminated from the comparison, and a total of 323 nucleotide sites were compared. Kimura's two-parameter method was used [M. Kimura, *J. Mol. Evol.* **16**, 111 (1980)]. Sequence identifications: 1, *Mus mus domesticus* functional gene; 2, *M. mus domesticus* pseudogene; 3, *M. mus castaneus* pseudogene; 4, *M. spicilegus* pseudogene; 5, *M. leggada* pseudogene; 6, *M. mus domesticus* cDNA; 7, *M. leggada* functional gene; 8, *M. platythrix* functional gene; 9, *Rattus norvegicus* cDNA; 10, *Homo sapiens* cDNA.

In distance matrix methods, a phylogenetic tree is constructed by considering the relationship among the distance values D_{ij} (distance between OTUs i and j). An example of a distance matrix is presented in Table II. The distances were computed from the nucleotide sequences for ten p53 functional genes and pseudogenes.[16] These sequence data will be used consistently in this chapter for worked-out examples. There are many methods for estimating evolutionary distances from nucleotide sequences.

Because there are already many reviews on tree-building methods,[4,6,17,18] we describe only the following six methods; UPGMA, the neighbor-joining method, the minimum evolution method, the maximum parsimony method, the maximum likelihood method, and network methods.

Methods Assuming Molecular Clock

When constancy of the evolutionary rate, or a molecular clock, is assumed, we can reconstruct rooted trees. This is because sequences should

[16] H. Ohtsuka, M. Oyanagi, Y. Mafune, N. Miyashita, T. Shiroishi, K. Moriwaki, R. Kominami, and N. Saitou, *Mol. Phylogenet. Evol.* in press.

[17] N. Saitou, *in* "Handbook of Statistics, Volume 8: Statistical Methods for Biological and Medical Sciences" (C. R. Rao and R. Chakraburty, eds.), p. 317. Elsevier, Amsterdam, 1990.

[18] D. L. Swofford and G. J. Olsen, *in* "Molecular Systematics" (D. M. Hillis and C. Moritz, eds.), p. 411. Sinauer Associates, Sunderland, Massachusetts, 1990.

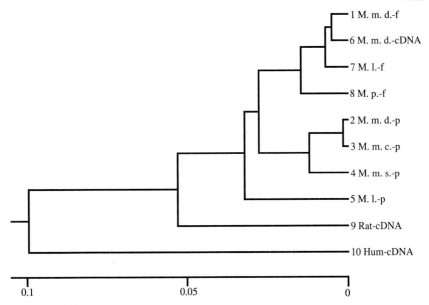

FIG. 6. UPGMA tree for the distance matrix in Table II. Sequence identifiers correspond to those of Table II.

be clustered in the order of their mutational difference if the amount of mutational changes is strictly proportional to evolutionary time. There are many ways to obtain such rooted trees from a distance matrix.[7] In this section, UPGMA and KITCH are discussed.

Let us briefly explain the UPGMA algorithm using the distance matrix shown in Table II. We first choose the smallest distance, D_{23} (=0.0031). Then OTUs 2 and 3 are combined and the distances between the combined OTU [2–3] and the remaining eight OTUs are computed by taking arithmetic means. At the next step, again the smallest distance ($D_{16} = 0.0094$) is chosen from the distance matrix. Then the OTUs 1 and 6 are combined into OTU [1–6]. This process is continued until all the OTUs are finally clustered into a single one. The resultant rooted tree is shown in Fig. 6. Although *Mus* functional genes (sequences 1, 6, 7, and 8) formed a monophyletic cluster, the corresponding *Mus* pseudogenes (sequences 2–5) did not form a monophyletic one.

Because of the long history of UPGMA (originally proposed by Sokal and Michener[19]), there are many computer programs available for UP-GMA, and these are not specified. It should be noted that there are several

[19] R. Sokal and C. D. Michener, *Univ. Kansas Sci. Bull.* **28**, 1409 (1958).

synonyms for UPGMA, such as the simple linkage method, the clustering method, and the nearest neighbor method.

KITCH is a computer program in the PHYLIP package[20] and is related to the Fitch and Margoliash method,[8] but constancy of the evolutionary rate is assumed. Because of this restriction, the result of KITCH is usually quite close to that of UPGMA. In fact, when KITCH was applied to the distance matrix of Table II, a result (not shown) identical to that of UPGMA was obtained. It seems that there is no use for this exhaustive search program if one already has the result using a UPGMA program.

A simulation study[10] has shown that UPGMA is not efficient in reconstructing the true topological relationship when the constancy of evolutionary rate is not assumed. Therefore, it is not advisable to use methods assuming a molecular clock for estimating realized trees. However, those are still useful for estimating expected trees, where all the branch lengths are proportional to physical time.

When we have only an unrooted tree with no outgroup, there is a way of rooting it if we assume a rough constancy of the evolutionary rate. Given the unrooted tree topology, we successively cluster OTU pairs starting from the smallest distance similar to UPGMA.[21] If we apply this algorithm to the unrooted tree of Fig. 1c, we will obtain the tree of Fig. 1a.

Neighbor-Joining Method

A pair of OTUs are called neighbors when these are connected through a single internal node in an unrooted bifurcating tree. For example, OTUs 1 and 2 of Fig. 1c are a pair of neighbors. If we combine these OTUs, this combined OTU [1–2] and OTU 5 become a new pair of neighbors. It is thus possible to define the topology of a tree by successively joining pairs of neighbors and producing new pairs of neighbors. In general, $n - 3$ pairs of neighbors are necessary to define the topology of an unrooted tree with n OTUs.

The neighbor-joining method[10] produces a unique final unrooted tree by sequentially finding pairs of neighbors by examining a distance matrix. Thus the neighbor-joining method is a distance matrix method as well as a stepwise clustering method. The principle of minimum evolution is used in the neighbor-joining method, and it has been proved that the expected value of the sum of branch lengths is smallest for the tree with the true branching pattern.[22] Because of the simple algorithm, more than 100 OTUs

[20] J. Felsenstein, "PHYLIP: Phylogeny Inference Package, Version 3.5c." Univ. of Washington, Seattle, 1993.

[21] N. Ishida, T. Oyunsuren, S. Mashima, H. Mukoyama, and N. Saitou, *J. Mol. Evol.* **41,** 180 (1995).

[22] A. Rzhetsky and M. Nei, *Mol. Biol. Evol.* **10,** 1073 (1993).

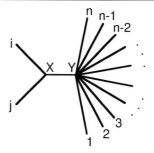

Fig. 7. Tree of N OTUs in which OTUs i and j are neighbors.

can be handled within a relatively short computer time by using the neighbor-joining method. For example, Horai et al.[23] produced a neighbor-joining tree for 193 human mitochondrial DNA sequences.

The following explanation of the neighbor-joining algorithm is based on Saitou.[6] We start from a starlike tree, which is produced under the assumption of no clustering among all the n OTUs compared. Under this tree, the sum (S_0) of n branch lengths can be shown to be

$$S_0 = Q/(n - 1) \tag{3}$$

where

$$Q = \Sigma_{i<j} \, D_{ij} \tag{4}$$

In practice, some pairs of OTUs are more closely related to one another than other pairs are. Among all the possible pairs of OTUs [$n(n - 1)$]/2 pairs for n OTUs], we choose the one that gives the smallest sum of branch lengths. Let us consider the tree of Fig. 7, where OTUs i and j are assumed to be neighbors. The sum of branch lengths is defined by

$$S_{ij} = (B_{iX} + B_{jX}) + B_{XY} + \Sigma_{k \neq i,j} \, B_{kY} \tag{5}$$

where $B_{\alpha\beta}$ is branch length between nodes α and β. There are the following relationships between distances and branch lengths:

$$D_{ij} = B_{iX} + B_{jX} \tag{6a}$$
$$D_{ik} = B_{iX} + B_{XY} + B_{kY} \quad (k \neq i, j) \tag{6b}$$
$$D_{jk} = B_{jX} + B_{XY} + B_{kY} \quad (k \neq i, j) \tag{6c}$$
$$D_{kl} = B_{iY} + B_{jY} \quad (k, l \neq i, j) \tag{6d}$$

[23] S. Horai, R. Kondo, Y. Nakagawa-Hattori, S. Hayashi, S. Sonoda, and K. Tajima, Mol. Biol. Evol. **10**, 23 (1993).

With the tree shown in Fig. 7, it can be shown by applying the above relationship that

$$B_{XY} = [Q - (n - 1)D_{ij} - (n - 1) \Sigma_{k,l \neq i,j} D_{kl}/(n - 3)]/2(n - 2) \quad (7)$$

If we neglect OTUs i and j in Fig. 7, the remaining $n - 2$ OTUs form a starlike tree, as is clear from Eq. (6d). Thus we apply Eq. (3) and obtain

$$\Sigma_{k \neq i,j} B_{kY} = \Sigma_{k,l \neq i,j} D_{kl}/(n - 3) \quad (8)$$

We also note that

$$\Sigma_{k,l \neq i,j} D_{kl} = Q - (R_i + R_j - D_{ij}) \quad (9)$$

where $R_i = \Sigma_j D_{ij}$ and $R_j = \Sigma_i D_{ij}$. Putting Eqs. (6a), (7), and (8) into Eq. (5) with consideration of Eq. (9), we obtain

$$S_{ij} = D_{ij}/2 + [2Q - R_i - R_j]/2(n - 2) \quad (10)$$

Equation (10) was first proposed by Studier and Keppler.[24]

This S_{ij} value is computed for all $n(n - 1)/2$ pairs of OTUs, and the pair that has the smallest S_{ij} value is chosen as neighbors. This pair of OTUs is then regarded as a single OTU, and the new distances between the combined OTU and the remaining ones are computed by averaging. This procedure is continued until all pairs of neighbors are found.

If OTUs i and j are chosen as neighbors as shown in Fig. 7, the branch lengths are estimated using the Fitch and Margoliash procedure[8] as

$$B_{iX} = D_{ij}/2 + (R_i - R_j)/2(n - 2) \quad (11a)$$

and

$$B_{jX} = D_{ij} - B_{iX} \quad (11b)$$

Therefore, all the branch lengths as well as the tree topology will be determined after $n - 2$ steps for n OTUs.

Table III shows the output of the computer program NJNUC, and Fig. 8 shows the neighbor-joining tree. Human p53 cDNA sequence (OTU 10) was assumed to be the outgroup. Branch lengths are estimated numbers of nucleotide substitutions that occurred in this p53 sequence, and all of them are integer values. To obtain those numbers, estimated numbers of nucleotide substitutions per site (numbers in parentheses in Table III) were multiplied with the number of compared nucleotide sites, then the resulting values were rounded. If a branch length turned out to be zero,

---

[24] J. A. Studier and K. J. Keppler, *Mol. Biol. Evol.* **5**, 729 (1988).

TABLE III
OUTPUT OF PROGRAM NJNUC FOR p53 SEQUENCE DATA[a]

Node 11	OTU 9 = 14.632 (4.530E-02)	OTU 10 = 45.440 (1.407E-01)
Node 12	OTU 2 = −0.065 (−2.011E-04)	OTU 3 = 1.069 (3.311E-03)
Node 13	Node 12 = 4.463 (1.382E-02)	OTU 4 = 2.693 (8.339E-03)
Node 14	Node 13 = 4.281 (1.325E-02)	OTU 5 = 11.547 (3.575E-02)
Node 15	OTU 8 = 5.427 (1.680E-02)	Node 11 = 9.108 (2.820E-02)
Node 16	OTU 7 = 1.913 (5.923E-03)	Node 15 = 1.235 (3.822E-03)
Node 17	Node 14 = 5.102 (1.580E-02)	OTU 6 = 0.532 (1.646E-03)
Node 18	(Last node)	
OTU 1	1.675 (5.186E-03)	
Node 17	0.907 (2.808E-03)	
Node 16	1.410 (4.364E-03)	

[a] The distance matrix of Table II was used. Numbers after the OTU or node designation are branch lengths in terms of nucleotide substitutions that occurred at the compared sequence region between that node/OTU and the node written at the top of each row. Numbers in parentheses are branch lengths in terms of nucleotide substitutions per site.

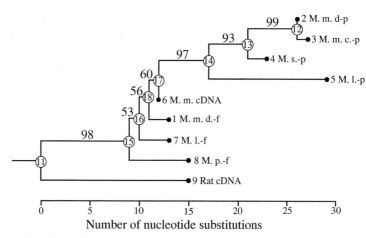

FIG. 8. Neighbor-joining tree constructed from the distance matrix of Table III. This tree was drawn on the basis of the output shown in Table IV. Branch lengths are proportional to the number of nucleotide substitutions per branch. Sequence identifiers correspond to those of Table II, and numbers in circles are internal node identifications. Numbers above internal branches are bootstrap probabilities (%). Human cDNA sequence was assumed to be the out-group.

such as the branch 2–12, that branch was truncated. A tree obtained by applying this procedure, first proposed by Nerurker et al.,[25] is an estimation of the realized tree. The topology of the neighbor-joining tree (Fig. 8) is somewhat different from that of the UPGMA tree (Fig. 6). Now the *Mus* pseudogenes (sequences 2–5) form a monophyletic cluster, while *Mus* functional counterparts (sequences 1, 6, 7, and 8) do not form a monophyletic cluster.

Numbers above internal nodes in the tree of Fig. 8 are bootstrap probabilities (percentages) based on 1000 replications (program NJBOOT2 was used for obtaining the bootstrap probabilities). For example, all the four pseudogene sequences are clustered with a high bootstrap probability (97%) at the internal node C. The bootstrap method was proposed for estimating variances from unknown probability distributions[26] and was introduced into phylogenetic study.[27] Character-state data are necessary to use the bootstrap method, but trees built using any distance matrix method can be tested using this technique. We first randomly resample n nucleotide sites from the given sequence data of n nucleotides with replacement. This resampling is replicated at least 1000 times. For example, one replication may have n nucleotide sites with the positions $1, 2, 2, 4, 5, \ldots, n-2, n,$ and n. Resampling is usually done by generating pseudorandom numbers. Each replicated sequence data set is then used as the input data to build phylogenetic trees. A bootstrap probability of a certain internal branch is simply the number of trees that realize this branch divided by the total number of replications. These probabilities are often summarized on the phylogenetic tree estimated by using the original sequence data. The bootstrap method is currently widely used, but its influence on phylogenetic inference is not thoroughly known; theoretical studies are still going on.

A series of programs (NJ, NJNUC, etc.) are included in the TreeTree package developed by the author. This method is also available in packages PHYLIP,[20] MEGA,[28] CLUSTAL W,[29] and MOLPHY.[30] Programs NJBOOT2 and TREEVIEW run on MS-DOS developed by K. Tamura (E-mail: Koichiro-Tamura@c.metro-u.ac.jp) are also available.

[25] V. R. Nerurkar, K.-J. Song, N. Saitou, R. R. Mallan, and R. Yanagihara, *Virology* **196,** 506 (1993).

[26] B. Efron, *Ann. Stat.* **7,** 1 (1979).

[27] J. Felsenstein, *Evolution* **39,** 783 (1985).

[28] S. Kumar, K. Tamura, and M. Nei, "MEGA: Molecular Evolutionary Genetics Analysis, Version 1.0." The Pennsylvania State Univ., University Park, 1993.

[29] J. D. Thompson, D. G. Higgins, and T. J. Gibson, *Nucleic Acids Res.* **22,** 4673 (1994).

[30] J. Adachi and M. Hasegawa, "MOLPHY: Programs for Molecular Phylogenetics, Version 2.2." Institute of Statistical Mathematics, Tokyo, 1994.

Minimum Evolution Methods

The concept of minimum evolution was used in the neighbor-joining method, and this concept was first used by Cavalli-Sforza and Edwards.[1] Saitou and Imanishi[11] proposed a simple method applying the principle of minimum evolution. In this method, branch lengths of a given tree are estimated by applying the procedure of Fitch and Margoliash,[8] and the tree with the smallest sum of branch lengths is chosen as the best tree. Rzhetsky and Nei[12] proposed a minimum evolution method in which branch lengths with their standard errors are computed by applying the least squares method. A neighbor-joining tree is first constructed as the candidate tree, and the related trees with only small topological differences are then searched. A simplified algorithm for computing least squares estimates of branch lengths has been proposed to reduce the computation time.[22]

Table IV shows an output of a minimum evolution program ME_TREE. There is one topological difference between the neighbor-joining tree (see Fig. 8) and this minimum evolution tree, regarding the clustering of sequences 1 and 7. The sum of branch lengths are 0.346188 and 0.346526 for the minimum evolution and neighbor-joining trees, respectively. Standard errors of branch lengths are small when the bootstrap values of the corresponding branches (see Fig. 8) are high. For example, the internal branch

TABLE IV
OUTPUT OF PROGRAM ME_TREE FOR p53 SEQUENCE DATA[a]

Branch	Branch length ± SE	Significance level (%)
1 and 16	0.005795 ± 0.003568	89.48
2 and 12	−0.000243 ± 0.000247	67.30
3 and 12	0.003349 ± 0.003354	67.78
4 and 13	0.008107 ± 0.005287	87.14
5 and 14	0.036693 ± 0.011213	99.90
6 and 17	0.001429 ± 0.002244	47.14
7 and 16	0.006695 ± 0.003946	90.90
8 and 15	0.017395 ± 0.006561	99.20
9 and 11	0.045297 ± 0.012698	99.96
10 and 11	0.140679 ± 0.023459	99.96
11 and 15	0.027608 ± 0.011227	98.58
12 and 13	0.014052 ± 0.006792	96.06
13 and 14	0.013044 ± 0.007296	92.50
14 and 17	0.014755 ± 0.006999	96.42
15 and 18	0.005024 ± 0.004819	70.16
16 and 18	0.002454 ± 0.004777	39.00
17 and 18	0.004056 ± 0.002930	83.24

[a] The distance matrix of Table II was used.

TABLE V

CLASSIFICATION OF NUCLEOTIDE CONFIGURATIONS OF p53 SEQUENCE DATA OF 323
NUCLEOTIDES FOR MAXIMUM PARSIMONY METHOD

Category	Observed number	Minimum number of changes
Noninformative configuration		
Invariant	236	0
Variant with 2 nucleotides	50	50
Variant with 3 nucleotides	9	18
Variant with 4 nucleotides	0	0
Informative configuration		
Variant with 2 nucleotides	26	26
Variant with 3 nucleotides	2	4
Variant with 4 nucleotides	0	0
Total	323	98

(0.014755 ± 0.006999) connecting nodes 14 and 18 is significantly larger than zero (significance level of 96.42%), and the corresponding bootstrap probability for the neighbor-joining tree is 97%.

There is a computer program (ME_TREE) run on MS-DOS.[31] Another program run on SUN workstations has been developed by Igor Belyi [WWW (World Wide Web) home page is http://www.cse.psu.edu/~belyi].

Maximum Parsimony Methods

The principle of maximum parsimony was first used for morphological data,[32] but it was independently proposed also for molecular data.[33] There are several kinds of maximum parsimony methods based on various assumptions, but the one that produces unrooted trees as in the case of the neighbor-joining method is mainly used for nucleotide sequence data.[13] The maximum parsimony principle is the minimization of the character-state changes (tree length) on the given tree topology, and is related to the principle used in minimum evolution methods. However, the performance of the two methods in choosing the best topology can be quite different.

Let us consider the example sequence data. We first classify the 323 nucleotide sites into different configurations (Table V). A nucleotide configuration is a distribution pattern of nucleotides for a given number of

[31] A. Rzhetsky and M. Nei, *Comput. Appl. Biosci.* **10,** 409 (1994).
[32] J. H. Camin and R. Sokal, *Evolution* **19,** 311 (1965).
[33] R. V. Eck and M. O. Dayhoff, *in* "Atlas of Protein Sequence and Structure" (M. O. Dayhoff ed.). National Biomedical Research Foundation, Silver Spring, Maryland, 1966.

sequences. The possible number (C_n) of configurations for n sequences is given by[34]

$$C_n = (4^{n-1} + 3 \times 2^{n-1} + 2)/6 \tag{12}$$

For example, there are 51 possible nucleotide configurations for five sequences. It should be noted that the number of possible configuration increases if we distinguish transitional differences from transversional ones.

Those configurations are first divided into noninformative and informative ones (Table V). Configurations that do not contribute to the selection of tree topology are called noninformative for the maximum parsimony method. All the sequences have the same nucleotide at the invariant configuration. There were 236 sites that fell into this category. We do not need any nucleotide substitution for this configuration under the maximum parsimony principle. One and two substitutions are necessary for any topology for variant with two and three nucleotides of the noninformative configuration, respectively. An informative nucleotide configuration should have more than one kind of nucleotide, and at least two of these should be observed in more than one of the sequences.[13] Only 28 of 323 sites had informative configurations. In total, we need at least 98 nucleotide substitutions for this data set.

Because there already exist several descriptions of the maximum parsimony method,[4,6,17,18] we will skip the explanation of the method in this chapter and show only the worked-out example. The result of the maximum parsimony analysis using PAUP is presented in Table VI. Nine equally parsimonious trees that require 112 substitutions were found by using branch-and-bound as well as heuristic options. However, two of them turned out to be identical with each other if we truncate an internal branch with zero length. Thus, the real number of equally parsimonious trees was eight (trees 1–8 of Table VI). Tree 6 had the same topology with the neighbor-joining tree (Fig. 8), and tree 4 was the maximum likelihood tree. Trees 9–12 required 113 substitutions and thus are subparsimonious. Biologically, however, tree 10 or 11 seems to be more reasonable.[16] It is also interesting to note that the minimum evolution tree (tree 12 of Table VI) was not a maximum parsimonious tree.

The principle of maximum parsimony attracted many because of its simplicity and logical clarity. However, there are some problems with this method when molecular data are used. Felsenstein[35] showed analytically that the maximum parsimony method may be positively misleading when the rate of evolution is grossly different among lineages of four sequences.

[34] N. Saitou and M. Nei, *J. Mol. Evol.* **24,** 189 (1986).
[35] J. Felsenstein, *Syst. Zool.* **27,** 401 (1978).

TABLE VI
MAXIMUM PARSIMONY AND MAXIMUM LIKELIHOOD ANALYSES

Tree ID[a]	Tree topology[b]	RNM[c]	Differences of log L[d]
1	((((((((2,3),4),5),8),6),1),7,(9,10))	112	−4.07
2	((((7,8),1),6),(((2,3),4),5),(9,10))	112	−0.29
3	(1,6,(((((2,3),4),5),8),((9,10),7)))	112	−4.09
4	((((((((2,3),4),5),6),1),8),7,(9,10))	112	Best
5	(((((2,3),4),5),(6,1)),8),7,(9,10))	112	−5.39
6	((((((((2,3),4),5),6),1),7),8,(9,10))	112	−2.55
7	(((((2,3),4),5),(6,(1,7))),8,(9,10))	112	−4.31
8	(((((2,3),4),5),6),((1,7),8),(9,10))	112	−6.10
9	(((1,6),7),((((2,3),4),5),8),(9,10))	113	−9.02
10	((((1,6),7),(((2,3),4),5)),8,(9,10))	113	−8.80
11	((((1,6),7),8),(((2,3),4),5),(9,10))	113	−8.99
12	(((((2,3),4),5),6),(1,7)),8,(9,10))	113	−7.37

[a] Trees 6 and 12 are the neighbor-joining tree (Fig. 8) and the minimum evolution tree (Table IV), respectively.
[b] Sequence identifications are the same as those of Table II.
[c] Required number of mutations when the maximum parsimony method was applied.
[d] Differences of log likelihood values from that of the best tree (tree 4; its log likelihood was −1016.75).

When the expected number of required substitutions for the true tree is larger than that for a wrong one, the maximum parsimony method will give more and more wrong answers as the number of compared nucleotides is increased (problem of efficiency). The same problem was found even when constancy of the evolutionary rate is assumed.[36,37] Saitou[38] showed that the gross underestimation of the branch lengths occurred when the divergence (number of nucleotide substitutions per site) among sequences was larger than 0.2. Therefore, we should be careful when using the maximum parsimony method.

After the tree topology is determined, however, the principle of maximum parsimony can be useful for estimating the location of mutational events. For example, Gojobori et al.[39] estimated the direction of nucleotide substitutions, and Saitou and Ueda[40] mapped the insertions and deletions on the assumed phylogenetic tree of primates. Jermann et al.[41] have recon-

[36] A. Zharkikh and W.-H. Li, Syst. Biol. 42, 113 (1993).
[37] N. Takezaki and M. Nei, J. Mol. Evol. 39, 210 (1994).
[38] N. Saitou, Syst. Zool. 38, 1 (1989).
[39] T. Gojobori, W.-H. Li, and D. Graur, J. Mol. Evol. 18, 360 (1982).
[40] N. Saitou and S. Ueda, Mol. Biol. Evol. 11, 504 (1994).
[41] T. M. Jermann, J. G. Opitz, J. Stachkouse, and S. A. Benner, Nature (London) 374, 57 (1995).

structed ancestral ribonuclease proteins from the estimated tree for the artiodactyls. It should be noted that the maximum parsimony principle can be applied to any tree irrespective of the methods used for constructing it.

PAUP 3.1.1 (a commercial product distributed from Illinois Natural History Servey) is run on a Macintosh with many user-friendly options. Maximum parsimony analysis is possible also for PHYLIP[20] and MEGA.[28] MacClade[42] has various useful features for molecular data, although it does not search the topology space.

Maximum Likelihood Methods

The maximum likelihood method is often used for parameter estimation in statistics, and it was first applied to building phylogenetic trees for allele frequency data.[1] Later, various maximum likelihood methods and computer programs were developed for sequence data.

The core algorithm of the maximum likelihood method is as follows. We first define the probability $P_{\alpha\beta} \equiv Pr(N_\alpha, N_\beta, B_{\alpha\beta})$ for observing nucleotides N_α and N_β at a particular nucleotide site at nodes α and β, respectively, when branch length is $B_{\alpha\beta}$. It is necessary to define the nucleotide transition matrix to compute $P_{\alpha\beta}$. Because each nucleotide site is assumed to evolve independently, the likelihood values for all the nucleotide sites are multiplied to obtain the overall likelihood. As is usually done in maximum likelihood techniques, the logarithm of the likelihood ($\log L$) is computed by changing branch lengths, and the maximum likelihood solution is determined for this tree topology. This maximum likelihood solution is ideally obtained for all the possible topologies, and the one that shows the highest value is chosen. Interested readers may refer to more detailed descriptions of this method.[4,6,15,43]

Table VI shows the result of the DNAML computation (user tree option was used). Tree 4, one of 8 equally parsimonious trees, was found to have the highest likelihood among the 12 trees compared. Likelihood values for subparsimonious trees (with 113 required substitutions) are somewhat lower than those for equally parsimonious trees.

Because the maximum likelihood method requires massive computer time, there are several searching methods other than the exhaustive search. The default method of DNAML[15,20] is the sequential addition of sequences. Saitou[44] proposed a stepwise clustering of sequences for the maximum likelihood method, and this searching method is the same as that of the

[42] W. P. Maddison and D. R. Maddison, "MacClade Version 3." Sinauer Associates, Sunderland, Massachusetts, 1992.

[43] N. Saitou, this series, Vol. 183, p. 584.

[44] N. Saitou, *J. Mol. Evol.* **27**, 261 (1988).

neighbor-joining method. NucML of MOLPHY[30] has several options for topology searches, and one of them (star decomposition) is similar to that of the Saitou[44] method.

DNAML and DNAMLK (molecular clock is assumed) are included in PHYLIP.[20] There is also a modified version of DNAML called fast-DNAML.[45] NucML for nucleotide sequences and ProtML for amino acid sequences are included in MOLPHY.[30]

Methods Producing Networks, Not Trees

The evolutionary history of a gene should be presented as a tree. When we analyze real sequence data, however, this tree structure may not be clearly observed. Bandelt and Dress[46] proposed the split decomposition method for distance matrix data. Unlike most tree-building methods, it usually produces a network, not a tree. A relaxed condition is used for estimating splitting patterns among OTUs, and both the signal (suggesting the tree structure) and the noise (suggesting patterns inconsistent with the tree structure) can be presented simultaneously. The resultant network is shown in Fig. 9. Four parallelograms suggest the existence of some parallel nucleotide changes, though the overall structure is quite close to an unrooted tree.

This network construction can also be applied to sequence data directly. When two nucleotide positions show an incongruent partition pattern, a discordancy diagram[13] appears. Bandelt[47] has extended this idea and proposed the phylogenetic network method. A network structure is useful for delineating anomaly in the history of gene trees. For example, when two regions of a gene experienced recombination(s), we may obtain a network, not a tree, if we analyze the sequence data by combining the two regions.

Regarding program availability, Daniel Huson (huson@mathematik. uni-bielefeld.de) and Rainer Wetzel have developed a shareware called SplitsTree run on Macintosh. A program for the phylogenetic network method is under development by H.-J. Bandelt.

Freely Distributed Computer Packages

PHYLIP[20] contains many programs in the form of both source code and executable files. Various kinds of maximum likelihood methods, maximum parsimony methods, and distance matrix methods can be used. It can be retrieved from evolution.genetics.washington.edu (128.95.12.41) or from

[45] G. J. Olsen, H. Matsuda, R. Hagstrom, and R. Overbeek, *Comput. Appl. Biosci.* **10,** 41 (1994).
[46] H.-J. Bandelt and A. Dress, *Adv. Math.* **92,** 47 (1992).
[47] H.-J. Bandelt, *Verhandlungen des Naturwissenschaftlichen Vereins in Hambrug* **34,** 51 (1994).

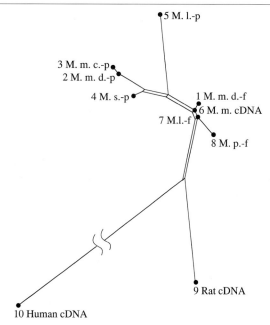

FIG. 9. Network constructed by using the SplitsTree program for the distance matrix of Table II. The length of an external branch to sequence 10 (human cDNA) was truncated, but other branch lengths were drawn proportional to the estimated lengths.

the PHYLIP WWW home page (http://evolution.genetics.washington. edu/phylip.html).

MEGA[28] is run on MS-DOS under a user-friendly environment. Many kinds of evolutionary distance estimation methods can be used, including synonymous and nonsynonymous substitutions. For further information, contact the following E-mail address: imeg@psuvm.psu.edu.

CLUSTAL W[29] is capable of doing multiple sequence alignment. After the alignment, it constructs neighbor-joining trees with bootstrapping. It can be retrieved from ftp.ebi.ac.uk (193.62.196.6) or from the EBI WWW home page (http://www.ebi.ac.uk/software/software.html).

MOLPHY[30] includes programs for maximum likelihood methods for both nucleotide and amino acid sequences. It can be retrieved via ftp from sunmh.ism.ac.jp (133.58.12.20).

Dendro-Maker (developed by Tadashi Imanishi) is run on Macintosh, and draws UPGMA and neighbor-joining trees. It can be retrieved via ftp from ftp.nig.ac.jp:/pub/mac/bio/dendromaker/. Treetool (developed by Mike Maciukenas) is run on Sun Sparc workstations and works with Newick format tree files for drawing trees. It can be retrieved through the RDP WWW homepage (http://rdp.life.uiuc.edu/).

TreeTree is a package of various programs mainly related to the neighbor-joining method developed by the author (E-mail address: nsaitou@genes.nig.ac.jp). Program NJ requires a distance matrix, whereas NJNUC requires nucleotide sequences. It can be retrieved through the author's WWW home page (http://smiler.nig.ac.jp/).

Acknowledgment

This chapter was partly supported by grants-in-aid for scientific researches from the Ministry of Education, Science and Culture, Japan.

[26] Estimating Evolutionary Distances between DNA Sequences

By WEN-HSIUNG LI and XUN GU

Introduction

Estimation of the evolutionary distance between two DNA sequences requires a stochastic model for nucleotide substitution. Most models for DNA evolution can be regarded as a time-continuous Markovian process, which can be characterized by the rate matrix $\mathbf{R}$, or, equivalently, the nucleotide substitution pattern. The most general model for $\mathbf{R}$ has 12 parameters to be estimated, but its application in practice is difficult. Indeed, one usually uses a simpler model (i.e., a simplified substitution pattern) to derive an analytical formula for estimating the distance (see, e.g., Refs. 1–4).

However, if some assumptions of a simple substitution model are violated, the estimate of a distance will be biased and will not increase linearly with time, that is, it will be nonadditive.[5] Additivity is a highly desirable property for evolutionary distances, because if it does not hold, all distance matrix methods of tree reconstruction may become statistically inconsistent, that is, may lead to an erroneous tree with a probability approaching 1 as

[1] T. H. Jukes and C. R. Cantor, in "Mammalian Protein Metabolism" (H. N. Munro, ed.), p. 21. Academic Press, New York, 1969.
[2] M. Kimura, *J. Mol. Evol.* **16,** 111 (1980).
[3] F. Tajima and M. Nei, *Mol. Biol. Evol.* **1,** 269 (1984).
[4] K. Tamura and M. Nei, *Mol. Biol. Evol.* **10,** 512 (1993).
[5] J. Felsenstein, *Annu. Rev. Genet.* **22,** 521 (1988).

Copyright © 1996 by Academic Press, Inc.
All rights of reproduction in any form reserved.

the sequence length increases to infinity.[6] To avoid this problem, it is necessary to develop a method for estimating distance under a general model of nucleotide substitution. Moreover, nonadditivity can also be generated if the assumption of a uniform rate among nucleotide sites is violated.[7] Because rate variation among sites is a common phenomenon, it should be taken into account in estimating evolutionary distances. In this chapter, we discuss an additive distance measure under a general substitution model and extend the method to the case where the substitution rate varies among sites. We first present the theory and then the estimation procedure. We also discuss the dissimilarities between our method and that of Lanave *et al.*,[8] which is one of the most general models available.

Stationary and Time-Reversible Model

Let **R** be the rate matrix whose ijth element r_{ij} is the substitution rate from nucleotide i to nucleotide j, $i \neq j$ ($i, j = 1, 2, 3$, and 4); the diagonal elements are given by $r_{ii} = -\Sigma_{j \neq i} r_{ij}$. By the Markovian process, the matrix of transition probabilities for t time units is generally given by

$$\mathbf{P}(t) = e^{\mathbf{R}t} \tag{1}$$

where the ijth element of $\mathbf{P}(t)$ is $P_{ij}(t)$, the probability of transition from nucleotide i to nucleotide j after t evolutionary time units.

The nucleotide substitution model to be used is called the SR model, which assumes that the substitution process is stationary and reversible in time. Stationarity means that the expected nucleotide frequencies in the sequences do not change in time. We denote the equilibrium frequencies by $\pi_i, i = 1, 2, 3$, and 4 for A, T, C, and G, respectively. The time reversibility means that the relation

$$\pi_i r_{ij} = \pi_j r_{ji} \tag{2}$$

holds for any i and j. Thus, the SR model can be presented by the following matrix.

	A	T	C	G
A	r_{11}	$\pi_2 v_1$	$\pi_3 v_2$	$\pi_4 s_1$
T	$\pi_1 v_1$	r_{22}	$\pi_3 s_2$	$\pi_4 v_3$
C	$\pi_1 v_2$	$\pi_2 s_2$	r_{33}	$\pi_4 v_4$
G	$\pi_1 s_1$	$\pi_2 v_3$	$\pi_3 v_4$	r_{44}

[6] R. W. DeBry, *Mol. Biol. Evol.* **9,** 537 (1992).
[7] Y. Tateno, N. Takezaki, and M. Nei, *Mol. Biol. Evol.* **11,** 261 (1994).
[8] C. Lanave, G. Preparata, C. Saccone, and G. Serio, *J. Mol. Evol.* **20,** 86 (1984).

TABLE I
MODELS OF NUCLEOTIDE SUBSTITUTION

JC model

	A	T	C	G
A	$-3v$	v	v	v
T	v	$-3v$	v	v
C	v	v	$-3v$	v
G	v	v	v	$-3v$

K2P model

	A	T	C	G
A	$-(2v + s)$	v	v	s
T	v	$-(2v + s)$	s	v
C	v	s	$-(2v + s)$	v
G	s	v	v	$-(2v + s)$

TN model

	A	T	C	G
A	r_{11}	$\pi_2 v$	$\pi_3 v$	$\pi_4 v$
T	$\pi_1 v$	r_{22}	$\pi_3 v$	$\pi_4 v$
C	$\pi_1 v$	$\pi_2 v$	r_{33}	$\pi_4 v$
G	$\pi_1 v$	$\pi_2 v$	$\pi_3 v$	r_{44}

TmN model

	A	T	C	G
A	r_{11}	$\pi_2 v$	$\pi_3 v$	$\pi_4 s_1$
T	$\pi_1 v$	r_{22}	$\pi_3 s_2$	$\pi_4 v$
C	$\pi_1 v$	$\pi_2 s_2$	r_{33}	$\pi_4 v$
G	$\pi_1 s_1$	$\pi_2 v$	$\pi_3 v$	r_{44}

Because $r_{ii} = -\Sigma_{j \neq i} r_{ij}$, for example, $r_{11} = -(\pi_2 v_1 + \pi_3 v_2 + \pi_4 s_1)$, the SR model is a nine-parameter model. The SR model includes many models as special cases, for example, the Jukes and Cantor model (JC),[1] the Kimura two-parameter model (K2P),[2] the Tajima and Nei model (TN),[3] and the Tamura and Nei model (TmN)[4] (see Table I).

General Additive Distance Measure

Consider two sequences X and Y that have evolved from O, a common ancestor, t time units ago. The transition probability matrix in each of the two lineages is given by Eq. (1). By time reversibility, the substitution process from the common ancestor O to sequences X and Y is equivalent to that from X through O to Y (or from Y through O to X). The transition probability matrix from X to Y is given by

$$\mathbf{P}(2t) = e^{2t\mathbf{R}} \tag{3}$$

The number of substitutions per site (K) between two DNA sequences, which is a standard measure of evolutionary distance, is defined by

$$K = 2t \sum_{i=1}^{4} \pi_i \sum_{j \neq i} r_{ij} = -2t \sum_{i=1}^{4} \pi_i r_{ii} \tag{4}$$

because $r_{ii} = -\sum_{j \neq i} r_{ij}$. By spectral decomposition, the diagonal elements of $\mathbf{R}$ can be expressed as

$$r_{ii} = \sum_{k=1}^{4} u_{ik} v_{ki} \lambda_k \tag{5}$$

where λ_k $(k = 1, 2, 3, \text{ and } 4)$ is the ith eigenvalue of $\mathbf{R}$, one of which is zero, say $\lambda_4 = 0$; u_{ik} is the ikth element of the eigenmatrix $\mathbf{U}$; and v_{ki} is the kith element of matrix $\mathbf{V} = \mathbf{U}^{-1}$. By putting Eq. (5) into Eq. (4), we have

$$K = -2t \sum_{k=1}^{3} b_k \lambda_k \tag{6}$$

where the constants b_k $(k = 1, 2, \text{ and } 3)$ are defined by

$$b_k = \sum_{i=1}^{4} \pi_i u_{ik} v_{ki} \tag{7}$$

Under the SR model, it is difficult to derive an analytical formula for K because the eigenvalues of $\mathbf{R}$ cannot be expressed in analytical forms. However, we can solve this problem as follows. Let z_k be the kth eigenvalue of $\mathbf{P}(2t)$. By matrix theory, Eq. (3) implies that

$$z_k = e^{2t\lambda_k} \tag{8}$$

Because $\lambda_4 = 0$ and $z_4 = 1$, there are only three nontrivial equations in Eq. (8). Equation (3) also implies that $\mathbf{R}$ and $\mathbf{P}(2t)$ have the same eigenmatrix. Thus, the number of substitutions per site can be computed by

$$K = - \sum_{k=1}^{3} b_k \ln z_k \tag{9}$$

where z_k and b_k can be estimated from sequence data (see below).

The method for estimating K under the SR model can be extended to

TABLE II
CONSTANTS c_k IN GENERAL ADDITIVE DISTANCE
UNDER SR [EQ. (10)] OR SRV [EQ. (24)] MODEL[a]

Distance	c_k ($k = 1, 2,$ and 3)
K	$\sum_{i=1}^{4} \pi_i u_{ik} v_{ki}$
A	$\sum_{i=1}^{4} \sum_{j \neq i \in Ts} \pi_i u_{ik} v_{kj}$
B	$\sum_{i=1}^{4} \sum_{j \neq i \in Tv} \pi_i u_{ik} v_{kj}$
D_{ij}	$\pi_i u_{ik} v_{kj}$
d_m	1/4, if $z_k = \max(z_1, z_2, z_3)$; 0, otherwise

[a] K is the number of substitutions per site; A is the number of transitional substitutions per site; B is the number of transversional substitutions per site; D_{ij} is the number of substitutions from nucleotides i to j per site; and d_m is the minimum distance defined by Eq. (26). The subscripts $j \neq i \in Ts$ and $j \neq i \in Tv$ mean that the differences between nucleotides i and j are transitional (Ts) and transversional (Tv), respectively.

a general additive (time-linear) distance d, which is defined by the linear combination of $\ln z_k$, namely,

$$d = - \sum_{k=1}^{3} c_k \ln z_k \tag{10}$$

for some constants c_k. For example, the number of transitional substitutions per site (A), the number of transversional substitutions per sites (B), and the number of substitutions from nucleotides i to j (D_{ij}) are the special cases of d by choosing appropriate constants c_k (see Table II).

However, the general additive distance defined by Eq. (10) requires the condition that all eigenvalues z_k (or λ_k) are real. We[9] have shown that this is the case.

Estimation of Distances and Sampling Variances

As all the above quantities (K, A, B, and D_{ij}) can be expressed in the same form (Table II), we can treat all of them in the same way. Let J_{ij} be

[9] X. Gu and W. H. Li, *Proc. Natl. Acad. Sci. U.S.A.*, in press (1996).

the expected frequency of sites where the nucleotide is i in sequence X and j in sequence Y, which, by Markovian properties, is given by

$$J_{ij} = \sum_{k=1}^{4} \pi_k P_{ki}(t) P_{kj}(t), \qquad i,j = 1, \ldots, 4 \qquad (11)$$

By time reversibility, that is, $\pi_i P_{ij}(t) = \pi_j P_{ji}(t)$, we have

$$J_{ij} = \sum_{k=1}^{4} \pi_i P_{ik}(t) P_{kj}(t) = \pi_i \sum_{k=1}^{4} P_{ik}(t) P_{kj}(t) = \pi_i P_{ij}(2t) \qquad (12)$$

where $\sum_{k=1}^{4} P_{ik}(t) P_{kj}(t) = P_{ij}(2t)$ is a basic property of transition probabilities. Obviously, Eq. (12) gives a simple method for estimating the transition probability $P_{ij}(t)$ directly from sequence data J_{ij}.

Let matrix $\mathbf{J}$ consist of J_{ij}. It can be shown that, if the substitution process is stationary and reversible, $\mathbf{J}$ is symmetric, that is, $J_{ij} = J_{ji}$. To test whether the observed data deviate significantly from this condition, we suggest the following χ^2 test. Let N_{ij} be the observed number of sites at which the nucleotide is i in sequence X and j in sequence Y. Note that there are six independent equations if $\mathbf{J}$ is symmetric, namely, $E[N_{12}] = E[N_{21}]$, $E[N_{13}] = E[N_{31}]$ and so forth, where E means taking expectation. For each pair, say $E[N_{12}]$ versus $E[N_{21}]$, we can construct the statistic $(N_{12} - N_{21})^2/(N_{12} + N_{21})$ to test whether the observed data deviate significantly from the null hypothesis of $E[N_{12}] = E[N_{21}]$, which follows a χ^2 distribution with $df = 1$. Therefore, the following statistic

$$S = \sum_{i=1}^{3} \sum_{j=i+1}^{4} \frac{(N_{ij} - N_{ji})^2}{N_{ij} + N_{ji}} \qquad (13)$$

follows a χ^2 distribution with $df = 6$. Thus, if $S > 12.59$, the null hypothesis that $N_{ij} = N_{ji}$ for all $i \neq j$ is rejected at the 5% significance level; in this case, the SR model cannot be applied.

In summary, the procedure for estimating distances can be outlined as follows (a computer program for the entire procedure is available on request).

Step 1. For each pair of sequences, count N_{ij}, the number of sites at which the nucleotide is i in the first sequence and is j in the second sequence.

Step 2. Compute the statistic S given by Eq. (13) to test whether matrix $\mathbf{J}$ is symmetric. Stop if $\mathbf{J}$ is nonsymmetric, that is, if $S \geq 12.59$.

Step 3. Estimate matrix $\mathbf{J}$ by

$$\hat{J}_{ij} = \frac{N_{ij} + N_{ji}}{2L}, \qquad i,j = 1, \ldots, 4 \qquad (14)$$

where L is the sequence length.

Step 4. Estimate the transition probability matrix $\mathbf{P}(2t)$ as

$$\hat{P}_{ij} = \frac{\hat{J}_{ij}}{\hat{\pi}_i}, \qquad i, \ldots, 4 \tag{15}$$

where $\hat{\pi}_i$ is the frequency of nucleotide i estimated by taking (simple) average between the two sequences.

Step 5. Compute eigenvalues $\hat{z}_k$ ($k = 1, \ldots, 4$) by solving the characteristic equation $\det(\hat{\mathbf{P}} - z\mathbf{I}) = 0$, where $\hat{\mathbf{P}}$ consists of $\hat{P}_{ij}$ and $\mathbf{I}$ is the identity matrix; the corresponding eigenmatrix $\mathbf{U}$ and its inverse matrix $\mathbf{V}$ are also obtained simultaneously by a standard algorithm.

Step 6. Compute the evolutionary distance d from Eq. (10); the constants c_i depend on the specified distance measure (see Table II).

Step 7. Finally, compute the sampling variance of d approximately given by

$$\mathrm{Var}(d) = \sum_{i=1}^{3} \frac{c_i^2}{z_i^2} \mathrm{Var}(z_i) + \sum_{i=1}^{3} \sum_{j \neq i} \frac{c_i c_j}{z_i z_j} \mathrm{Cov}(z_i, z_j) \tag{16}$$

where $\mathrm{Var}(z_i)$ is the sampling variance of z_i and $\mathrm{Cov}(z_i, z_j)$ is the sampling covariance between z_i and z_j. Let $\mathbf{W}$ be the variance–covariance matrix of z_i values, which is defined by $w_{ii} = \mathrm{Var}(z_i)$ and $w_{ij} = \mathrm{Cov}(z_i, z_j)$ ($i \neq j$). We[9] have developed a simple method to compute $\mathbf{W}$ by inverting $\mathbf{I}_f$, that is, $\mathbf{W} = \mathbf{I}_f^{-1}$, where the klth element of matrix $\mathbf{I}_f$ is given by

$$I_{kl} = \sum_{i=1}^{4} \sum_{j=1}^{4} \frac{N_{ij}}{P_{ij}^2} u_{ik} u_{il} v_{kj} v_{lj}, \qquad k, l = 1, 2, 3 \tag{17}$$

General Additive Distance Under Variable Rates

The general additive distance defined by Eq. (10) requires the assumption of the same substitution rate for all sites. If this assumption is violated, Eq. (10) may no longer be additive.[7] Thus, we need to extend the new method to the case where the substitution rate varies among sites (the SRV model, where V stands for variable rates).

We assume that the rate variation among sites follows a gamma distribution. The model is as follows. The ijth element of the rate matrix $\mathbf{R}$ is expressed by $r_{ij} = h_{ij}u$, where h_{ij} is a constant and u varies randomly according to the following gamma distribution

$$\phi(u) = \frac{\beta^\alpha}{\Gamma(\alpha)} u^{\alpha-1} e^{-\beta u} \tag{18}$$

where the mean of u is given by $\bar{u} = \alpha/\beta$.

First, we consider the expected number of substitutions per site K, which, under the SRV model, is defined by

$$K = -2t \sum_{i=1}^{4} \pi_i \bar{r}_{ii} \tag{19}$$

where $\bar{r}_{ij} = h_{ij}\bar{u}$. Let $\mathbf{\bar{R}}$ be the matrix of expected rates, that is, $\mathbf{\bar{R}}$ consists of $\bar{r}_{ii}$. By spectral decomposition, Eq. (19) can be written as

$$K = -2t \sum_{k=1}^{3} b_k \bar{\lambda}_k \tag{20}$$

where constants b_k are given by Eq. (7); $\bar{\lambda}_k$ ($k = 1, \ldots, 4$) is the kth eigenvalue of matrix $\mathbf{\bar{R}}$ ($\bar{\lambda}_4 = 0$).

Note that, when the rate varies among sites, only the expectation (over all sites) can be observed from sequence data. That is, the $\hat{P}_{ij}$ of Eq. (15) is actually an estimate of the average transition probability $\bar{P}_{ij}(2t)$, which is defined by

$$\bar{P}_{ij}(2t) = \int_0^\infty P_{ij}(2t)\phi(u)\,du \tag{21}$$

Let $\mathbf{\bar{P}}(2t)$ be the matrix of average transition probabilities, and $\bar{z}_k$ ($k = 1, \ldots, 4$) be the kth eigenvalues of matrix $\mathbf{\bar{P}}(2t)$. To derive an estimation formula of K under the SRV model, one must find a relation between $\bar{\lambda}_k$ and $\bar{z}_k$. We[9] have shown that, when the substitution rate varies among sites according to a gamma distribution, $\bar{z}_k$ and $\bar{\lambda}_k$ have the following relation

$$-2\bar{\lambda}_k t = \alpha(\bar{z}_k^{-1/\alpha} - 1), \qquad i = 1, \ldots, 4 \tag{22}$$

The expected number of substitutions per site under the SRV model is given by

$$K = \alpha \sum_{k=1}^{3} b_k(\bar{z}_k^{-1/\alpha} - 1) \tag{23}$$

and the general additive distance under the SRV model is given by

$$\bar{d} = \alpha \sum_{k=1}^{3} c_k(\bar{z}_k^{-1/\alpha} - 1) \tag{24}$$

for some constants c_k (see Table II). It can be shown that $\bar{d} \to d$ of Eq. (10) when $\alpha \to \infty$, that is, the substitution rate is uniform among sites.

The procedures for estimating distance $\bar{d}$ under the SRV model are the same as that of the SR model except the last two steps, that is, after step 5, go to the following steps:

Step 6'. Compute the evolutionary distance $\bar{d}$ by Eq. (24) if the gamma distribution parameter α is known; it can be estimated by the parsimony method[4] or the maximum likelihood method[10]; the constants c_i depend on the specified distance measure (see Table II).

Step 7'. Finally, compute the sampling variance of $\bar{d}$ approximately given by

$$\text{Var}(\bar{d}) = \sum_{i=1}^{3} \frac{c_i^2}{z_i^{2(1+1/\alpha)}} \text{Var}(z_i) + \sum_{i=1}^{3} \sum_{j \neq i} \frac{c_i c_j}{(z_i z_j)^{1+1/\alpha}} \text{Cov}(z_i, z_j) \qquad (25)$$

Obviously, $\text{Var}(\bar{d}) \to \text{Var}(d)$ as $\alpha \to \infty$. $\text{Var}(\bar{d})$ in Eq. (25) should be interpreted as a lower bound of the sampling variance because it assumes that α is estimated precisely without error.

Minimum Distance and Inapplicable Cases

When the sequence length is short, the sampling effect can become serious. The large sampling variance can not only nullify the additivity of a distance, but can also make the SR method inapplicable. An inapplicable case occurs if any estimated z_k is negative. Therefore, a distance measure that minimizes the sampling effects may be useful in practice. Let $z_m = \max(z_1, z_2, z_3)$. Then the minimum distance is defined by

$$d_m = -\frac{1}{4} \ln z_m \qquad (26)$$

The sampling variance of d_m is given by

$$\text{Var}(d_m) = \left(\frac{1}{4 z_m}\right)^2 \text{Var}(z_m) \qquad (27)$$

where $\text{Var}(z_m)$ can be computed as above. Note that d_m is always positive because $0 < z_1, z_2, z_3 < 1$ under the SR model. The minimum distance is additive and less affected by saturation so that it has few inapplicable cases. Note that d_m is more general than the transversion distance. Indeed, under the Tamura and Nei model,[4] which is a special case of the SR model, the

---

[10] X. Gu, X. Y. Fu, and W. H. Li, *Mol. Biol. Evol.* **12**, 546 (1995).

transversional distance is the minimum distance d_m if the rates of transversion are smaller than those of transition.

However, as shown by Kelly,[11] negative eigenvalues can arise if the nucleotide substitution does not follow a Markov model. We can assess whether the substitution process fits the general Markovian model, using a test that the smallest eigenvalue is positive. We suggest a Z test by computing the following statistic

$$Z = \frac{z_0}{\text{Var}(z_0)^{1/2}} \qquad (28)$$

where $z_0 = \min(z_1, z_2, z_3, z_4)$. Note that in the case of $z_0 < 0$, the SR method is inapplicable. If z_0 is not significantly smaller than zero, the inapplicability may be regarded as due to sampling effects, and the minimum distance or a simpler method may be used instead. However, if z_0 is significantly smaller than 0, the substitution process may not be Markovian; in this case, a more complex model is needed.

Discussion

The model of Lanave et al.[8] (LA for short) has been shown to be a nine-parameter rather than twelve-parameter model.[12] So, it is in fact equivalent to the SR model. However, their method and ours differ in several aspects. First, our additive distance is general and includes many distance measures as special cases; the LA method considers only the number of substitutions per site. Second, our method uses a χ^2 test to assess whether the SR model is suitable; the test in the LA method is for testing the stationarity but not time reversibility. Third, and most important, we extended the general additive distance under the SR model to the case where the substitution rate varies among sites (i.e., the SRV model); there is no similar extension in the LA method. Therefore, our method has several advantages over the LA method.

Rodriguez et al.[13] proposed a general formula for estimating the number of substitutions per site as

$$K = -tr[\mathbf{F} \ln(\mathbf{F}^{-1}\mathbf{J})] \qquad (29)$$

[11] C. Kelly, Biometrics 50, 653 (1994).
[12] A. Zharkikh, J. Mol. Evol. 39, 315 (1994).
[13] F. Rodriguez, J. F. Oliver, A. Marin, and J. R. Medina, J. Theor. Biol. 142, 485 (1990).

where tr is the trace of a matrix and $\mathbf{F}$ is the diagonal matrix of nucleotide frequencies, that is, $\mathbf{F} = \text{diag}(\pi_1, \pi_2, \pi_3, \pi_4)$. It can be shown that Eq. (29) is mathematically equivalent to Eq. (9). However, Eq. (29) is not convenient for extension to include the effect of rate heterogeneity. The general additive distance defined by Eq. (10) is more general and easier for conducting statistical analysis of sequence data.

It is worth mentioning that if $c_i = 1/4$, Eq. (10) is the paralinear or LogDet distance.[14-16] Let matrix $\mathbf{J}$ consist of J_{ij}. Then, Eq. (12) can be expressed as $\mathbf{J} = \mathbf{FP}$, or $\mathbf{P} = \mathbf{F}^{-1}\mathbf{J}$. Because $z_1 z_2 z_3 z_4 = \det(\mathbf{P}) = \det(\mathbf{F}^{-1}\mathbf{J}) = \det(\mathbf{J})/\det(\mathbf{F})$, we have

$$d = -\frac{1}{4}\ln(z_1 z_2 z_3 z_4) = -\frac{1}{4}\ln\frac{\det(\mathbf{J})}{\det(\mathbf{F})} \tag{30}$$

The paralinear distance was recommended for phylogenetic reconstruction because its additivity holds even when the process is not reversible in time, that is, it is based on the general (12-parameter) substitution model. However, the paralinear distance is an unbiased estimate of the number of substitutions per site (K) if and only if the equilibrium frequencies of the four nucleotides are 1/4. More seriously, the additivity does not hold if the substitution rate varies among sites. To our knowledge, the SRV method is to date the most general that includes the effect of rate variation among sites. Since rate variation is a common phenomenon, the SRV method is preferable over the paralinear distance.

Acknowledgments

We thank Dr. A. Zharkikh for valuable discussion. This study was supported by National Institutes of Health Grant GM30998.

[14] D. Barry and J. A. Hartigan, *Biometrics* **43,** 261 (1987).
[15] J. A. Lake, *Proc. Natl. Acad. Sci. U.S.A.* **91,** 1455 (1994).
[16] P. J. Lockhart, M. A. Steel, M. D. Hendy, and D. Penny, *Mol. Biol. Evol.* **11,** 605 (1994).

[27] Local Alignment Statistics

By Stephen F. Altschul and Warren Gish

Introduction

Because protein and DNA sequences frequently share only isolated regions of similarity, the most widely used sequence alignment algorithms seek subalignments[1-3] as opposed to global alignments.[4-7] A subalignment (also called a local alignment) places into correspondence a segment from each of the two sequences being compared. Null characters (often represented by dashes) may be added to either segment to represent insertions or deletions. Subalignment quality generally is measured by a score calculated by adding substitution scores for each aligned pair of letters and gap scores for each run of nulls in one segment aligned with letters in the other. An ungapped subalignment is one in which no nulls or gaps are allowed.

Given a particular scoring system, a central question has always been how high a score may be expected to occur purely by chance. Through studying the comparison of random sequences, much progress has been made on this question. The optimal scores from ungapped subalignments are known to approach an extreme value distribution, and explicit formulas are available for the parameters of the distribution.[8-11] The distribution for the sum of the r best ungapped subalignment scores is also known.[12] Some rougher asymptotic results are available for the scores of subalignments with gaps,[13] and there is much empirical evidence that the statistical theory for ungapped subalignments generalizes in outline to this case.[14-18]

[1] T. F. Smith and M. S. Waterman, *J. Mol. Biol.* **147,** 195 (1981).

[2] W. R. Pearson and D. J. Lipman, *Proc. Natl. Acad. Sci. U.S.A.* **85,** 2444 (1988).

[3] S. F. Altschul, W. Gish, W. Miller, E. W. Myers, and D. J. Lipman, *J. Mol. Biol.* **215,** 403 (1990).

[4] S. B. Needleman and C. D. Wunsch, *J. Mol. Biol.* **48,** 443 (1970).

[5] D. Sankoff, *Proc. Natl. Acad. Sci. U.S.A.* **69,** 4 (1972).

[6] P. H. Sellers, *SIAM J. Appl. Math.* **26,** 787 (1974).

[7] D. Sankoff and J. B. Kruskal, "Time Warps, String Edits and Macromolecules: The Theory and Practice of Sequence Comparison." Addison-Wesley, Reading, Massachusetts, 1983.

[8] S. Karlin and S. F. Altschul, *Proc. Natl. Acad. Sci. U.S.A.* **87,** 2264 (1990).

[9] A. Dembo, S. Karlin, and O. Zeitouni, *Ann. Prob.* **22,** 2022 (1994).

[10] A. Dembo and S. Karlin, *Ann. Prob.* **19,** 1737 (1991).

[11] A. Dembo and S. Karlin, *Ann. Prob.* **19,** 1756 (1991).

[12] S. Karlin and S. F. Altschul, *Proc. Natl. Acad. Sci. U.S.A.* **90,** 5873 (1993).

[13] R. Arratia and M. S. Waterman, *Ann. Appl. Prob.* **4,** 200 (1994).

[14] T. F. Smith, M. S. Waterman, and C. Burks, *Nucleic Acids Res.* **13,** 645 (1985).

Copyright © 1996 by Academic Press, Inc.
All rights of reproduction in any form reserved.

In this chapter, we study further the distribution of optimal gapped subalignment scores. We provide evidence that two parameters are sufficient to describe both the form of this distribution and its dependence on sequence length. Using a random protein model, the relevant statistical parameters are calculated for a variety of substitution matrices and gap costs. An analysis of these parameters elucidates the relative effectiveness of affine as opposed to length-proportional gap costs. We show empirically that the theory for the sum of the r best ungapped subalignment scores generalizes to gapped subalignments. Thus sum statistics provide a method for evaluating sequence similarity that treats short and long gaps differently. By example, we show how this method has the potential to increase search sensitivity. The statistics described can be applied to the results of FASTA[2] searches, or to those from a variation of the BLAST programs, described here, that permits gaps.

Statistics of Ungapped Subalignment Scores

The scores of ungapped subalignments are inherited completely from scores for aligning pairs of residues. The score for aligning residues i and j we will call s_{ij}, and we assume generally that $s_{ij} = s_{ji}$. To approach analytically the question of what subalignment scores are statistically significant, a model of random sequences is needed. The simplest model assumes that all residues are drawn independently, with respective probabilities $p_1, \ldots, p_r$ for the various residue types; r is 20 for proteins and 4 for nucleic acids. For the statistical theory described here to hold, a central requirement is that the expected score $\Sigma_{i,j=1}^{r} p_i p_j s_{ij}$ for a pair of randomly chosen residues be negative. This is, in fact, a desirable restriction, for a positive expected score implies that long subalignments will tend to have high scores, even when the constituent segments bear no biological relationship.

Let S be the optimal ungapped subalignment score from the comparison of two random sequences of lengths m and n. When m and n are sufficiently large, it can be shown that the distribution of S is well approximated by an extreme value distribution,[19] whose cumulative distribution function is given by

$$P(S < x) = \exp[-e^{-\lambda(x-u)}] \tag{1}$$

[15] J. F. Collins, A. F. W. Coulson, and A. Lyall, *CABIOS* **4,** 67 (1988).

[16] R. Mott, *Bull. Math. Biol.* **54,** 59 (1992).

[17] M. S. Waterman and M. Vingron, *Proc. Natl. Acad. Sci. U.S.A.* **91,** 4625 (1994).

[18] M. S. Waterman and M. Vingron, *Stat. Sci.* **9,** 367 (1994).

[19] E. J. Gumbel, "Statistics of Extremes." Columbia Univ. Press, New York, 1958.

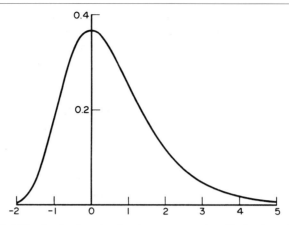

FIG. 1. Probability density function for the extreme value distribution with $u = 0$ and $\lambda = 1$.

This distribution has two parameters: the characteristic value u, which can be thought of as the center of the distribution, and the decay constant or scale parameter λ. The probability density function for the standard extreme value distribution with $u = 0$ and $\lambda = 1$ is shown in Fig. 1.

Analytic formulas are available for λ and u in the case considered here.[8-10] λ is the unique positive solution for x in Eq. (2):

$$\sum_{i,j=1}^{r} p_i p_j e^{s_{ij}x} = 1 \tag{2}$$

u is dependent on the lengths of the sequences compared and is given by

$$u = (\ln Kmn)/\lambda \tag{3}$$

where K is a constant given by a geometrically convergent series dependent on the p_i and s_{ij}.[8-10] Combining Eqs. (1) and (3), we can eliminate u and write simply that

$$P(S \geq x) = 1 - \exp(-Kmn\, e^{-\lambda x}) \tag{4}$$

This is an expression for the probability that the optimal ungapped subalignment attains a score of at least x. More generally, it may be shown that the number of distinct ungapped subalignments with score at least x is approximately Poisson distributed,[8] with parameter $Kmn\, e^{-\lambda x}$. The meaning of "distinct" will be discussed below at greater length. These results may be generalized to Markov-dependent sequences[11] and to ungapped subalignments of multiple sequences.

We note that when a set of scores s_{ij} is multiplied by a positive scalar, the effect is to divide λ by that same constant. Because the new scores differ from the original ones in no important way (they both imply the same optimal subalignments), it is useful to define normalized scores, which are λ times the nominal scores, that remain fixed. These scores are said to be expressed in nats,[20] and we use this terminology below. When P of Eq. (4) is less than 0.1, increasing the score of a subalignment by 1 nat decreases its probability of chance occurrence by a factor of about e.

Dependence of Statistical Parameters on Search Space Size

An extreme value distribution has been shown to hold in the asymptotic limit for other measures of local similarity, such as the longest run of matching letters between two sequences, allowing for k mismatches.[21] As above, explicit formulas for λ and u are available. If p is the probability that two random letters match, $\lambda = \ln 1/p$, and u is a function of $(1 - p)mn$ involving constant, log, and log–log terms.[21]

The measure of sequence similarity S of interest to us here is the optimal subalignment score, allowing gaps, frequently called a Smith–Waterman score.[1] We will consider only affine gap costs,[22,23] parameterized by a gap-opening penalty a, and a gap-extension penalty b, so that a gap of length k receives a score of $-a - (k - 1)b$. It has been shown that for a range of substitution and gap scores, S grows asymptotically as a logarithmic function of the search space size mn.[13] The scores for which this is true are analogous to scores with negative expected value in the no-gap case, and it is only in this local scoring regime that we will be interested.

Although the distribution of S has not been shown analytically to approach an extreme value distribution, a number of empirical studies strongly suggest that it does.[14–18] To obtain estimates of statistical significance under this assumption, one need only estimate the relevant parameters λ and u for use in Eq. (1). These parameters will, in general, depend on the particular scoring system employed, and on the compositions and lengths of the sequences being compared.

If a particular protein scoring system has been chosen, and proteins can be assumed to have a standard amino acid composition, then λ and u will be a function only of the lengths of the sequences being compared. There is much empirical evidence that λ remains unchanged with increasing search

[20] S. F. Altschul, *J. Mol. Biol.* **219,** 555 (1991).
[21] R. Arratia, L. Gordon, and M. S. Waterman, *Ann. Stat.* **14,** 971 (1986).
[22] O. Gotoh, *J. Mol. Biol.* **162,** 705 (1982).
[23] W. M. Fitch and T. F. Smith, *Proc. Natl. Acad. Sci. U.S.A.* **80,** 1382 (1983).

space size, and this agrees with the analytic results for simpler measures of local similarity. It would be convenient if, as hypothesized by Waterman and Vingron,[18] the dependence of u on search space size takes the simple form of Eq. (3) that it does when gaps are disallowed: a simulation for one search space size could then be generalized to all sizes. It is not immediately clear, however, that this should be the case. We know, for instance, that for the longest matching run with k mismatches, the formula for u contains a log–log term in mn,[21] and there is empirical evidence that such a term arises as well for various other local similarity functions.[24–26] An earlier empirical study finds evidence of a log–log term for Smith–Waterman scores,[16] and Waterman and Vingron find some deviation from their estimated probability curves when mn is varied.[18] We show in the next section that a naive data analysis indeed suggests the presence of a log–log term in the formula for u, but that a simple "edge effect" correction eliminates the evidence for such a term. Thus it appears that the simple form of the theory for ungapped subalignments does extend to gapped subalignments.

Estimation of Statistical Parameters

The most straightforward way to estimate λ and u for the comparison of sequences of lengths m and n is to generate many pairs of random sequences and calculate for each the optimal subalignment score S. The parameters for an extreme value distribution may be fit to these data using the method of moments,[24] maximum likelihood estimation,[16] or linear regression on a transform of the data[15,17,18]; all these methods give virtually identical results. A rapid method for estimating the parameters, which involves collecting scores from the r locally best subalignments, has also been described.[17,18] In the present study, λ and u are estimated by the method of moments.

Using the background amino acid frequencies described by Robinson and Robinson,[27] we generated 10,000 pairs of random protein sequences for each of a large range of sequence lengths $m = n$. For each pair, we calculated the optimal subalignment score based on the BLOSUM62 substitution scores,[28] coupled with ($a = 12$, $b = 1$) affine gap costs. For reasons discussed below, we collected as well the length of each optimal subalignment. As in other studies,[16–18] the resulting scores fit well an extreme value

[24] S. F. Altschul and B. W. Erickson, *Bull. Math. Biol.* **48,** 617 (1986).

[25] M. S. Waterman and L. Gordon, *in* "Computers and DNA" (G. I. Bell and T. G. Marr, eds.), p. 127. Addison-Wesley, Reading, Massachusetts, 1990.

[26] S. F. Altschul, *J. Mol. Evol.* **36,** 290 (1993).

[27] A. B. Robinson and L. R. Robinson, *Proc. Natl. Acad. Sci. U.S.A.* **88,** 8880 (1991).

[28] S. Henikoff and J. G. Henikoff, *Proc. Natl. Acad. Sci. U.S.A.* **89,** 10915 (1992).

TABLE I
EMPIRICAL VALUES FOR l, u, λ, AND K AS FUNCTION OF SEARCH SPACE SIZE[a]

n, m	l	$\ln nm$	$\ln n'm'$	u	λ	K
191	22.6	10.5	10.25	26.45	0.298	0.073
245	25.8	11.0	10.78	28.31	0.286	0.055
314	29.4	11.5	11.30	30.21	0.282	0.051
403	32.4	12.0	11.83	32.04	0.275	0.041
518	36.3	12.5	12.35	33.92	0.279	0.048
665	40.3	13.0	12.87	35.94	0.273	0.041
854	43.9	13.5	13.39	37.84	0.272	0.040
1097	48.1	14.0	13.91	39.75	0.275	0.046
1408	51.6	14.5	14.43	41.71	0.268	0.036
1808	55.1	15.0	14.94	43.54	0.271	0.041
2322	59.1	15.5	15.45	45.53	0.267	0.035
2981	63.5	16.0	15.96	47.32	0.270	0.040

[a] Ten thousand random protein sequence pairs, using amino acid frequencies from Robinson and Robinson,[27] were generated for each value of n and m. Optimal subalignment scores were calculated using (12, 1) affine gap costs and the BLOSUM62 amino acid substitution matrix.[28] Standard errors for u and λ are 0.05 and 0.003, respectively.

distribution. We confine our attention here to the attendant estimates of λ and u, shown in Table I. For further analysis the table also provides estimates of the parameter K, under the assumption that Eq. (3) is valid.

As expected,[14–18] the estimates of λ remain essentially constant for m and n sufficiently large (>350). We turn thus to the question of whether the parameter u can in fact be written in the form of Eq. (3), or whether a log–log term is suggested as well. Equation (3) implies that plotting u against $\ln mn$ should yield a straight line with slope $1/\lambda$ and y intercept $(\ln K)/\lambda$. Linear regression of u versus $\ln mn$ indeed yields an almost perfectly straight line ($r > 0.9999$), but the implied value of 0.261 for λ is significantly lower than the average value of 0.272 found for those points with m and n greater than 350. The greater than anticipated slope could be explained by a log–log term, as invoked by Mott.[16] Although the 4% discrepancy in λ may seem small, a similar experiment using PAM250[29,30]

[29] M. O. Dayhoff, R. M. Schwartz, and B. C. Orcutt, in "Atlas of Protein Sequence and Structure" (M. O. Dayhoff, ed.), Vol. 5, Suppl. 3, p. 345. National Biomedical Research Foundation, Washington, D.C., 1978.

[30] R. M. Schwartz and M. O. Dayhoff, in "Atlas of Protein Sequence and Structure" (M. O. Dayhoff, ed.), Vol. 5, Suppl. 3, p. 353. National Biomedical Research Foundation, Washington, D.C., 1978.

TABLE II
STATISTICAL PARAMETERS CALCULATED BY VARIOUS METHODS

Scoring system	Method for calculating parameters	λ	K
BLOSUM62 matrix,	Average case (n, $m > 350$)	0.272	0.041
(12, 1) gap costs	Linear regression	0.261	0.026
	Edge-corrected linear regression	0.270	0.041
PAM250 matrix,	Average case (n, $m > 350$)	0.202	0.037
(15, 3) gap costs	Linear regression	0.189	0.019
	Edge-corrected linear regression	0.199	0.037
BLOSUM62 matrix,	Average case (n, $m > 200$)	0.321	0.12
infinite gap costs	Linear regression	0.311	0.08
	Edge-corrected linear regression	0.319	0.12
	Analytic theory	0.318	0.13

substitution scores and (15, 3) gap costs yielded an even larger discrepancy of 6% (see Table II).

Before postulating the necessity for a more complicated formula than Eq. (3), we tested an edge effect correction of a type that has been used for many years by the BLAST programs.[3] The basic idea is that a subalignment does not exist at a point, but has a certain length. Any subalignment starting near the end of either sequence is thus likely to run out of sequence before it can attain a score sufficient to become optimal. A sequence might therefore be considered to have an effective length shorter than its actual one by the length l of a typical optimal subalignment. Accordingly, we provide in Table I a column l representing the mean of the observed lengths of the optimal subalignments, and a column ln $m'n'$, where $m' = m - l$ and $n' = n - l$. Performing a linear regression using these edge-corrected values for search space size yields a value of 0.270 for λ, in very close agreement with the average value of λ estimated from individual points. The same improved agreement was found for the PAM250-based scoring system described above (Table II).

To test the validity of this approach, we performed a similar computational experiment for a case in which the asymptotic values for λ and K are known analytically, that is, one in which no gaps are allowed. As shown in Table II, the parameter estimates both from edge-corrected linear regression and from averaging the estimates from multiple values of m and n agree quite well with the values of λ and K derived from theory.

Although no computational experiments are sufficient to establish the asymptotic validity of a specific formula for u, or indeed the validity of the extreme value distribution, our results certainly are consistent with the hypothesis of Waterman and Vingron[18] that u can be well approximated by Eq. (3). This is fortunate, for it allows a simulation for a single pair of sequence lengths to determine K, and thereby to determine u for all other search space sizes. Our one suggestion is that, in addition to parameters λ and K, the relative entropy H[20] of the scoring system should be estimated. Basically, H is the expected score, per aligned pair of residues, of an optimal random subalignment. By estimating the mean length l described above, H expressed in nats[28] may be written as

$$H \approx \lambda u/l \approx (\ln Kmn)/l \qquad (5)$$

This formula may then be inverted to calculate the edge correction described above. Specifically, $m' \approx m - (\ln Kmn)/H$, and similarly for n'. When $(\ln Kmn)/H$ is a large fraction of either m or n, it suggests that edge effects are substantial, and that asymptotic formulas are likely to lose some accuracy.

Local Optimality

Two sequences may share more than a single region of similarity, and it is therefore desirable to seek not just the optimal subalignment, but other high-scoring subalignments as well. The immediate problem is that the second, third, and fourth highest scoring subalignments are all likely to be slight variations of the optimal one. A definition is needed of when two subalignments are distinct.

The first such definition is due to Sellers,[31] who described a subalignment as locally optimal if it has a better score than any subalignment it intersects within a path graph. He provided an algorithm to find all and only the locally optimal subalignments,[31] but it required an undetermined number of sweeps of a path graph, and therefore was not provably $O(n^2)$. An $O(n^2)$ time algorithm was subsequently described.[32]

Altschul and Erickson introduced a definition of weak local optimality, somewhat more inclusive than the Sellers original definition.[32] A subalignment is weakly locally optimal if it intersects within a path graph no better subalignment that is weakly locally optimal. This definition is not circular but recursive; it is anchored by the optimal subalignment. One problem with seeking only locally optimal subalignments is that a very strong subalignment A can suppress a good subalignment C it never intersects by means of a subalignment B, with intermediate score, that intersects both.[24,32] By the Sellers definition, only subalignment A could be locally optimal; by the definition of Altschul and Erickson, both A and C could be weakly

[31] P. H. Sellers, *Bull. Math. Biol.* **46,** 501 (1984).
[32] S. F. Altschul and B. W. Erickson, *Bull. Math. Biol.* **48,** 633 (1986).

locally optimal. From a mathematical standpoint local optimality is the more natural definition, but weak local optimality generally captures better most people's intuitive notion of when two subalignments are distinct.

Algorithms for finding all weakly locally optimal subalignments using nonlinear similarity functions were described by Altschul and Erickson.[24,32] Waterman and Eggert described a more efficient algorithm for the standard linear similarity function under consideration here.[33] For real biological sequences the Waterman–Eggert algorithm generally requires only $O(n^2)$ time, but its worst case behavior has not been established. A linear-space variation of this algorithm has been described by Huang *et al.*[34] Barton has described an algorithm claimed to be more time efficient than that of Waterman and Eggert.[35] Although the algorithm is more efficient, it does not in fact always succeed in finding all and only the locally optimal subalignments by either definition above,[36,37] and it must therefore be regarded as a heuristic.

In the next section we consider the random behavior of the r best locally optimal subalignments. We use the Sellers definition of local optimality because it permits a provably $O(n^2)$ algorithm[32] that is simpler than that of Waterman and Eggert for weakly locally optimal subalignments. We note that for random sequences, where the optimal subalignment never has a very large score, and where locally optimal subalignments tend to be far removed from one another, the two definitions usually are congruent. Thus our results should apply both to locally optimal subalignments and to weakly locally optimal subalignments. For the reasons discussed above, however, when real sequences are to be compared we recommend the weak local optimality definition,[32] and thus the algorithms of Waterman and Eggert[33] or Huang *et al.*[34]

Sum Statistics

When several distinct regions of similarity are found within a pair of sequences, it is sometimes desirable to report a combined assessment of their statistical significance. This is more involved, however, than it might first appear. Suppose, for example, that using a particular scoring system the two best distinct subalignments have scores of 65 and 40. It is tempting to ask for the probability that the highest score is 65 or greater and that

[33] M. S. Waterman and M. Eggert, *J. Mol. Biol.* **197,** 723 (1987).
[34] X. Huang, R. C. Hardison, and W. Miller, *CABIOS* **6,** 373 (1990).
[35] G. J. Barton, *CABIOS* **9,** 729 (1993).
[36] O. Gotoh, *CABIOS* **3,** 17 (1987).
[37] W. Miller and M. Boguski, *CABIOS* **10,** 455 (1994).

the second highest score is 40 or greater. This, however, is inappropriate, because a legitimate p value represents the probability of attaining a given result or a better one, and thus requires a linear ranking of all possible results. Call the above result A, and consider a different result B in which the highest score is 52 and the second highest 45. Implicitly, result B is not considered better than A, because its high score is lower than that of A. But neither is result A considered superior, because its second high score is lower than that of B.

One way to rank all results is to consider only the rth highest score and calculate, for instance, the probability that the second best score is S or greater. In the above example, result B would be considered superior to A because its second best score is the greater (see Fig. 2). In the no-gap case, the probability that the rth highest scoring locally optimal subalignment has score at least S is derived from the Poisson distribution,[8] and it is well approximated by

$$1 - e^{-a} \sum_{i=0}^{r-1} \frac{a^i}{i!} \tag{6}$$

where $a = Kmn\ e^{-\lambda S}$. For subalignments allowing gaps, simulation has shown this formula to hold extremely well, once K and λ are estimated.[17,18]

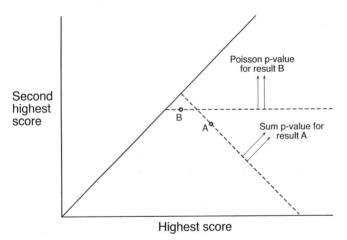

FIG. 2. The calculation of Poisson and sum p values. The region on or below the main diagonal represents feasible joint values of the first and second highest locally optimal subalignment scores. A probability density may be imagined to cover this space.[12] The probability that the second highest score is greater than or equal to 45 is that lying on or above the horizontal line through point B. The probability that the sum of the two highest scores is greater than or equal to 105 is that lying on or above the antidiagonal line through point A. Which result is considered the more significant depends on which criterion is employed.

An alternative and perhaps more natural way to assess a collection of the r highest distinct subalignment scores is by means of their sum. In the above example, result A would be considered better than B because 105, the sum of its two scores, is greater than 97 (see Fig. 2). When gaps are disallowed, the asymptotic distribution for the sum of the r highest scores from locally optimal subalignments has been derived.[12] If S_r is the rth highest score, let $T_r = \sum_{i=1}^{r} (\lambda S_i - \ln Kmn)$. Then for large mn the probability density function for T_r approaches

$$f(t) = \frac{e^{-t}}{r!(r-2)!} \int_0^{\infty} y^{r-2} \exp(-e^{(y-t)/r}) \, dy \qquad (7)$$

To obtain the tail probability that $T_r \geq x$ one must integrate $f(t)$ from Eq. (7) for t from x to infinity.[12] This double integral is easily calculated numerically, and a program for the purpose in the C programming language is available from the authors.

Here, we wish to investigate whether the sum statistics described by Eq. (7) are applicable as well to subalignments with gaps. To test this hypothesis, we generated 10,000 pairs of random protein sequences of length 600. Using BLOSUM62 substitution scores and (12, 1) gap costs, we collected for each sequence pair the 10 best locally optimal subalignment scores. We estimated λ and K from the optimal scores by the method of moments. We then tabulated, for r from 1 to 10, the frequency with which the sum of the r highest scores achieved various values, and calculated from Eq. (7) the expected number of times these values should have been attained. The results for $r = 5$ are shown in Fig. 3 and Table III, and Table III also shows a χ^2 goodness-of-fit calculation. Because λ and K were estimated from the data, the number of degrees of freedom in this example is 71 and the χ^2 statistic is 89.7, not surprisingly high under the null hypothesis that the theory is exactly valid. Table IV gives χ^2 values for all r tested, along with the scores characterizing several calculated and observed significance levels. It is evident that the sum statistics perform quite well: although the χ^2 goodness-of-fit test evinces imperfect conformity with the theory, the 95, 99, and 99.9% significance levels are predicted within an error of ± 1 in almost every case. Similar or better results were obtained for other substitution matrices and gap costs. We note that the gap costs were small enough, in this example, for over one-half the optimal subalignments to contain gaps, so that we are not simply verifying the theory for the no-gap case, where it is known to hold.

We conclude that the statistical theory for ungapped alignments carries over essentially unchanged to gapped alignments. All that is required for its application are good estimates of λ and K and, if edge corrections are desired, H as well. The statistical significance of optimal subalignment

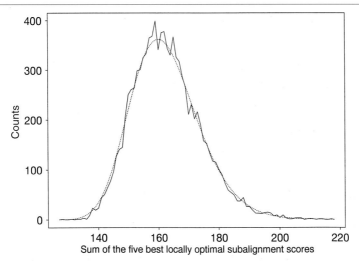

Fig. 3. Expected and observed counts, from 10,000 random trials, for the sum of the five highest locally optimal subalignment scores. The dotted line represents the theoretical distribution, based on values of λ and K estimated from the data.[12] The solid line connects data points from the observed distribution (see Table III). The χ^2 goodness-of-fit statistic (with 71 degrees of freedom) is 89.7. Even were the theory precisely valid, a worse fit would be obtained in about 7% of all stochastic experiments. Curves with this or a better fit to theory were obtained for five of the ten values of r tested (see Table IV).

scores, and of the sum of locally optimal subalignment scores, is then readily calculated.

On a practical note, one rarely knows *a priori* whether the criterion one uses for assessing a given comparison of two sequences should be the highest subalignment score, or the sum of the two highest, or the sum of the three highest, etc. One way to decide is to choose the collection of subalignments that yields by Eq. (7) the smallest p value. A difficulty is that one has then performed multiple tests, and the p value one had calculated is accordingly no longer valid. A simple and conservative remedy has been suggested by P. Green and implemented in the BLAST programs.[3] The basic idea is to use a geometric series in r to discount the significance of results involving r scores.[38] The same strategy may be applied when the Poisson statistics of Eq. (6) are used in place of sum statistics.

It is interesting to compare the joint evaluation of multiple independent subalignments by means of sum statistics, and the consolidation of these subalignments using gaps and their attendant costs. Sum statistics allow gaps of arbitrary length, in either or both sequences, between the subalignments

[38] S. F. Altschul, M. S. Boguski, W. Gish, and J. C. Wootton, *Nat. Genet.* **6,** 119 (1994).

TABLE III
STOCHASTIC EXPERIMENT FOR SUM OF FIVE BEST LOCALLY OPTIMAL SUBALIGNMENT SCORES[a]

Sum of scores	Expected number	Observed number	χ^2	Sum of scores	Expected number	Observed number	χ^2
<133	4.0	2	1.00	170	250.3	211	6.16
133, 134	6.9	2	3.48	171	233.0	231	0.02
135	6.4	3	1.78	172	215.7	202	0.87
136	9.4	1	7.50	173	198.7	216	1.51
137	13.5	8	2.23	174	182.1	191	0.43
138	18.9	23	0.90	175	166.1	158	0.40
139	25.8	19	1.80	176	150.9	153	0.03
140	34.4	23	3.79	177	136.4	133	0.09
141	45.1	44	0.03	178	122.8	117	0.28
142	57.7	50	1.03	179	110.2	115	0.21
143	72.5	64	0.99	180	98.4	101	0.07
144	89.3	77	1.70	181	87.6	90	0.07
145	108.1	96	1.36	182	77.6	73	0.28
146	128.7	127	0.02	183	68.6	59	1.35
147	150.7	141	0.62	184	60.4	54	0.68
148	173.7	145	4.75	185	53.0	49	0.31
149	197.4	206	0.38	186	46.4	37	1.91
150	221.1	251	4.03	187	40.5	37	0.30
151	244.5	261	1.12	188	35.2	44	2.19
152	266.9	264	0.03	189	30.6	26	0.68
153	287.8	299	0.44	190	26.4	26	0.01
154	306.7	301	0.11	191	22.8	19	0.63
155	323.3	325	0.01	192	19.6	13	2.23
156	337.2	333	0.05	193	16.8	13	0.88
157	348.2	365	0.81	194	14.4	12	0.41
158	356.0	368	0.40	195	12.3	14	0.23
159	360.7	399	4.07	196	10.5	15	1.93
160	362.2	341	1.24	197	8.9	13	1.85
161	360.6	376	0.65	198	7.6	7	0.04
162	356.2	378	1.34	199	6.4	9	1.03
163	349.1	344	0.07	200	5.4	4	0.38
164	339.6	330	0.27	201	4.6	9	4.25
165	327.8	367	4.69	202	3.9	3	0.19
166	314.8	326	0.40	203, 204	6.0	2	2.67
167	299.9	317	0.98	205, 206	4.2	6	0.77
168	284.0	286	0.01	207–210	4.9	6	0.25
169	267.4	266	0.01	>210	4.4	4	0.04

[a] Ten thousand random protein sequence pairs of length 600 were generated, using amino acid frequencies from Robinson and Robinson.[27] Subalignment scores were calculated using (12, 1) gap costs and the BLOSUM62 matrix.[28]

TABLE IV
SIGNIFICANCE LEVELS AND χ^2 STATISTICS FOR SUM OF THE r BEST LOCALLY OPTIMAL
SUBALIGNMENT SCORES

| r | Estimated/observed significance levels | | | χ^2 statistic | Degrees of freedom | P value (%) |
	95%	99%	99.9%			
1	47/47	53/54	62/63	42.3	32	10
2	85/84	92/92	102/103	59.1	45	8
3	119/119	128/128	139/140	61.9	54	20
4	152/152	162/162	174/175	62.3	63	50
5	184/183	194/195	207/207	89.7	71	7
6	214/214	226/226	239/239	117.2	77	0.2
7	244/244	256/256	271/270	111.7	83	2
8	274/273	286/286	301/301	111.5	88	5
9	302/301	316/314	332/332	121.4	94	3
10	331/330	344/343	361/362	135.7	99	0.8

involved, and in fact do not require that these subalignments be combinable into a consistent larger subalignment. (The statistics may be sharpened by requiring consistent ordering of the subalignments.)[12] Each such gap is paid for by an increase in r, and thus an extra subtraction of ln Kmn in the calculation of T_r. The net effect is, to first order, an effective cost of ln Kmn nats for combining subalignments. This is substantially greater than the costs normally employed for short gaps,[39] as will be seen below, but it does permit great flexibility in the combination of subalignments.

Statistical Parameters for Frequently Used Substitution Matrices

For reference purposes, we have compiled in Tables V–VII statistical parameters for several frequently used amino acid substitution matrices. Several warnings should be given concerning the accuracy of these tables. First, Tables V–VII were constructed using a specific amino acid frequency model,[27] which may be poor for any particular pair of proteins. Mott has attempted to identify systematic dependencies of statistical parameters on amino acid composition,[16] and this may prove a fruitful direction for further research. Second, except those for infinite gap costs, which were derived from theory, all parameter values are stochastically determined and therefore involve random error. Finally, gap costs which involve relative entropies less than about 0.15 nats have optimal subalignments with average

[39] W. R. Pearson, *Protein Sci.* **4**, 1145 (1995).

TABLE V

STATISTICAL PARAMETERS FOR (a, b) AFFINE GAP COSTS IN CONJUNCTION WITH BLOSUM62
SUBSTITUTION SCORES[a]

a	b	λ	K	H (nats)	a	b	λ	K	H (nats)
∞	0–∞	0.318	0.13	0.40	8	7, 8	0.270	0.06	0.25
					8	4–6	0.262	0.05	0.23
12	3–12	0.305	0.10	0.38	8	3	0.243	0.035	0.18
12	2	0.300	0.09	0.34	8	2	0.215	0.021	0.12
12	1	0.275	0.05	0.25	8	1		Borderline	
11	3–11	0.301	0.09	0.36	7	6, 7	0.247	0.05	0.18
11	2	0.286	0.07	0.29	7	4, 5	0.230	0.030	0.15
11	1	0.255	0.035	0.19	7	3	0.208	0.021	0.11
					7	2	0.164	0.009	0.06
10	4–10	0.293	0.08	0.33	7	1		Linear	
10	3	0.281	0.06	0.29					
10	2	0.266	0.04	0.24	6	5, 6	0.200	0.021	0.10
10	1	0.216	0.014	0.12	6	4	0.179	0.014	0.08
					6	3	0.153	0.010	0.05
9	5–9	0.286	0.08	0.29	6	1, 2		Borderline or linear	
9	3, 4	0.273	0.06	0.25					
9	2	0.244	0.030	0.18	5	5	0.131	0.009	0.04
9	1	0.176	0.008	0.06	1–5	1–4		Borderline or linear	

[a] Parameters for infinite gap costs were calculated using the analytic theory.[8] Linear designates scoring systems in the linear as opposed to the logarithmic domain, where the
extreme value theory does not apply.[13] Borderline designates scoring systems near the
logarithmic–linear phase transition.[13] From Ref. 28.

length greater than 50. This is a large fraction of the length of the random
sequences generated, and thus edge effects begin to be substantial. The
most important effect, suggested by the trends seen in Table I, may be an
overestimate of the asymptotic value of λ.

It should furthermore be understood that both the PAM[29,30]and
BLOSUM[28] matrices consist of log-odds scores constructed explicitly from
target frequencies for aligned amino acid pairs. The theory that supports
the use of such scores[20] breaks down once gaps are allowed. Thus, although
nothing prevents these matrices from being used in conjunction with gap
scores, substitution scores constructed in some other manner may be preferable. The practice of subtracting some value from all matrix elements,[16]
for example, has sometimes been advocated. However, except in broad
outline, theory remains silent on the best way to choose not only gap scores,
but also the substitution scores that are used with them.

There are several lessons to be drawn from the numbers in Tables
V–VII. For a given gap-opening cost a, the statistical parameters do not

TABLE VI
STATISTICAL PARAMETERS FOR (a, b) AFFINE GAP COSTS IN CONJUNCTION WITH BLOSUM50 SUBSTITUTION SCORES[a]

a	b	λ	K	H (nats)	a	b	λ	K	H (nats)
∞	0–∞	0.232	0.11	0.34	11	8–11	0.197	0.05	0.21
					11	6, 7	0.190	0.04	0.19
16	4–16	0.222	0.08	0.31	11	5	0.184	0.04	0.17
16	3	0.213	0.06	0.27	11	4	0.177	0.031	0.15
16	2	0.207	0.05	0.24	11	3	0.167	0.028	0.11
16	1	0.180	0.024	0.15	11	2	0.130	0.009	0.06
					11	1		Linear	
15	8–15	0.222	0.09	0.31					
15	6, 7	0.219	0.08	0.29	10	8–10	0.183	0.04	0.17
15	4, 5	0.216	0.07	0.28	10	6, 7	0.178	0.035	0.16
15	3	0.210	0.06	0.25	10	5	0.168	0.026	0.13
15	2	0.202	0.05	0.22	10	4	0.156	0.020	0.10
15	1	0.166	0.018	0.11	10	3	0.139	0.013	0.07
					10	2	0.099	0.007	0.03
14	8–14	0.218	0.08	0.29	10	1		Linear	
14	5–7	0.214	0.07	0.27					
14	4	0.205	0.05	0.24	9	7–9	0.164	0.029	0.13
14	3	0.201	0.05	0.22	9	5, 6	0.152	0.021	0.10
14	2	0.188	0.034	0.17	9	4	0.134	0.014	0.07
14	1	0.140	0.009	0.07	9	3	0.107	0.008	0.04
					9	1, 2		Linear	
13	8–13	0.211	0.06	0.27					
13	5–7	0.205	0.05	0.24	8	8	0.139	0.017	0.08
13	4	0.202	0.05	0.22	8	7	0.134	0.015	0.07
13	3	0.188	0.034	0.18	8	6	0.127	0.013	0.06
13	2	0.174	0.025	0.13	8	5	0.117	0.011	0.05
13	1	0.114	0.006	0.04	8	4	0.101	0.009	0.03
					8	1–3		Borderline or linear	
12	7–12	0.205	0.06	0.24					
12	5, 6	0.197	0.05	0.21	7	7	0.100	0.010	0.04
12	4	0.192	0.04	0.18	7	6	0.094	0.010	0.03
12	3	0.178	0.028	0.15	7	1–5		Borderline or linear	
12	2	0.158	0.019	0.10					
12	1		Borderline		1–6	1–6		Linear	

[a] From Ref. 28.

change appreciably for gap-extension costs b greater than some value b'. This may at first seem surprising but it indicates that, with these costs, very few locally optimal subalignments that arise by chance contain gaps of length greater than one. One conclusion is that for a given a it is unrewarding to use any gap extension penalty greater than b'. The noise from high-

TABLE VII

STATISTICAL PARAMETERS FOR (a, b) AFFINE GAP COSTS IN CONJUNCTION WITH PAM250 SUBSTITUTION SCORES[a]

a	b	λ	K	H (nats)	a	b	λ	K	H (nats)
∞	$0-\infty$	0.229	0.09	0.23	11	7–11	0.186	0.04	0.13
					11	5, 6	0.180	0.034	0.11
16	4–16	0.217	0.07	0.21	11	4	0.165	0.021	0.09
16	3	0.208	0.05	0.18	11	3	0.153	0.017	0.07
16	2	0.200	0.04	0.16	11	2	0.122	0.009	0.04
16	1	0.172	0.018	0.09	11	1		Linear	
15	5–15	0.215	0.06	0.20	10	8–10	0.175	0.031	0.11
15	4	0.208	0.05	0.18	10	7	0.171	0.029	0.10
15	3	0.203	0.04	0.16	10	6	0.165	0.024	0.09
15	2	0.193	0.035	0.14	10	5	0.158	0.020	0.08
15	1	0.154	0.012	0.07	10	4	0.148	0.017	0.07
					10	3	0.129	0.012	0.05
14	6–14	0.212	0.06	0.19	10	1, 2	Borderline or linear		
14	4, 5	0.204	0.05	0.17					
14	3	0.194	0.035	0.14	9	7–9	0.151	0.020	0.07
14	2	0.180	0.025	0.11	9	6	0.146	0.019	0.06
14	1	0.131	0.008	0.04	9	5	0.137	0.015	0.05
					9	4	0.121	0.011	0.04
13	6–13	0.206	0.06	0.17	9	3	0.102	0.010	0.03
13	4, 5	0.196	0.04	0.14	9	1, 2		Linear	
13	3	0.184	0.029	0.12					
13	2	0.163	0.016	0.08	8	7, 8	0.123	0.014	0.05
13	1	0.110	0.008	0.03	8	6	0.115	0.012	0.04
					8	5	0.107	0.011	0.03
12	7–12	0.199	0.05	0.15	8	1–4	Borderline or linear		
12	5, 6	0.191	0.04	0.13					
12	4	0.181	0.029	0.12	7	7	0.090	0.014	0.02
12	3	0.170	0.022	0.10	7	1–6	Borderline or linear		
12	2	0.145	0.012	0.06					
12	1		Borderline		1–6	1–6		Linear	

[a] From Refs. 29 and 30.

scoring random subalignments will not decrease appreciably, but the occasional true subalignment with a long gap will be penalized.

For a given set of gap costs, it is also useful to consider the ratio of λ to λ_∞ for infinite gap costs. This ratio indicates the proportion of information in ungapped alignments that must be sacrificed in the hope of extending the alignments using gaps. For example, (12, 1) gap costs used in conjunction with BLOSUM62 scores sacrifice about 14% of the information in ungapped alignments, reducing, for example, an alignment with a score of 30 nats to one of about 26 nats. However, gap costs frequently allow ungapped alignments to be extended, recouping in the process more than the amount of information lost. The best gap costs will be those that on average maximize the difference between information gained and information lost.

Fairly low gap costs are sometimes employed so that subalignments involving long insertions or deletions will not be too severely penalized. This, however, decreases the value of any score obtained by lowering the value of λ. (An alternative view is that it raises the noise level by permitting the appearance of higher scoring random alignments.) One possible compromise is to charge fairly high gap penalties, but to evaluate the appearance of multiple high-scoring locally optimal subalignments using the sum statistics described above. This allows small gaps to be incorporated without unduly deflating the currency of alignment score, but it also permits gaps of arbitrary size to contribute to the significance of a result. Although such ideas may inform the search, the choice of appropriate gap costs still relies on empiricism.[39,40]

Empirical Statistics from Database Searches

In many situations, searching protein or DNA sequence databases for local alignments using the Smith–Waterman algorithm[1] or its variations[31–34] requires too much time to be practical. Accordingly, parallel architecture computers,[41–45] specialized VLSI chips,[46–48] and heuristic algorithms[2,3,49] have been emplyed for the purpose. In the examples below, we use a heuristic algorithm BLASTGP, which is a new version of the BLAST programs that permits gaps in protein sequence comparisons. This experi-

[40] G. Vogt, T. Etzold, and P. Argos, *J. Mol. Biol.* **249,** 816 (1995).

[41] A. F. W. Coulson, J. F. Collins, and A. Lyall, *Comput. J.* **30,** 420 (1987).

[42] R. Jones, *CABIOS* **8,** 377 (1992).

[43] G. Vogt and P. Argos, *CABIOS* **8,** 49 (1992).

[44] D. L. Brutlag, J.-P. Dautricourt, R. Diaz, J. Fier, B. Moxon, and R. Stamm, *Comput. Chem.* **17,** 203 (1993).

[45] S. S. Sturrock and J. F. Collins, "MPsrch Version 1.3." Biocomputing Research Unit, University of Edinburgh, 1993.

[46] E. T. Chow, T. Hunkapiller, J. C. Peterson, B. A. Zimmerman, and M. S. Waterman, *in* "Proceedings of the 1991 International Conference on Supercomputing," p. 216. ACM Press, New York, 1991.

[47] R. P. Hughey, Ph.D. Thesis, Brown University, Providence, Rhode Island (1991).

[48] C. T. White, R. K. Singh, P. B. Reintjes, J. Lampe, B. W. Erickson, W. D. Dettloff, V. L. Chi, and S. F. Altschul, *in* "Proceedings of the 1991 IEEE International Conference Computer Design: VLSI in Computers and Processors," p. 504. IEEE Computer Society Press, Los Alamitos, California, 1991.

[49] A. Califano and I. Rigoutsos, *in* "Proceedings of the First International Conference on Intelligent Systems for Molecular Biology" (L. Hunter, D. Searls, and J. Shavlik, eds.), p. 56. AAAI Press, Menlo Park, California, 1993.

mental program is available from the authors on request.[50] Like the program LFASTA, its basic strategy is to use a banded Smith–Waterman algorithm[51] centered on high-scoring segment pairs (HSPs) identified by the BLAST strategy.[3] When all HSPs found are processed in this way, the search time is about 50% greater than that of BLASTP; when only HSPs significant in the context of the pairwise comparison in which they occur are processed, the search time is increased by about 10%. Using BLASTGP, we can study the use of sum statistics on gapped subalignments involving real sequences.

To test whether the statistics developed above are approximately valid for protein database searches, we compared shuffled versions of 1000 randomly selected proteins to the SWISS-PROT protein sequence database, Release 31.0.[52] Before searching, each shuffled sequence was filtered using the SEG program[38,53] to remove regions of highly biased amino acid composition; such regions may arise even in shuffled sequences when the source sequence begins with a biased composition. The database searches were performed using the BLOSUM62 substitution matrix[28] and (12, 2) affine gap costs. The corresponding values of $\lambda = 0.300$ and $K = 0.09$ from Table V were used to calculate sum statistic p values for any set of locally optimal subalignments found for a given database sequence. A p value cutoff was imposed so that 10 sequences were expected to be found in each database search.[38] For the 1000 trials, the mean number of sequences reported was 13.9, and the median number 8, in good agreement with prediction. For a comparable 1000 BLASTP searches, in which the statistical parameters were calculated analytically, the mean and median number of sequences found were 10.6 and 7.

Biological Example

To illustrate several potential effects of gap costs and sum statistics, we consider a database search of SWISS-PROT Release 31[52] (43,470 sequences; 15.3 million residues) with a hypothetical yeast protein of length 234 (SWISS-PROT accession number P40582)[54] as query. Using the BLASTGP program in conjunction with BLOSUM62 substitution scores and (12, 2) affine gap costs, two statistically significant similarities are found. The highest score for a single subalignment (Fig. 4a) is 74 (32.0 bits[20]) and involves a *Silene cucubalus* glutathione *S*-transferase (SWISS-PROT accession num-

[50] Write Warren Gish (gish@watson.wustl.edu).
[51] K.-M. Chao, W. R. Pearson, and W. Miller, *CABIOS* **8,** 481 (1992).
[52] A. Bairoch and B. Boeckmann, *Nucleic Acids Res.* **22,** 3578 (1994).
[53] J. C. Wootton and S. Federhen, *Comput. Chem.* **17,** 149 (1993).
[54] B. G. Barrel *et al.,* unpublished.

(a)

```
P40582    6 IKVHWLDHSRAF-RLLWLLDHLNLEYEIVPYKRDANFRAPPELKKIHPLGRSPLLE 60
            IKVH    S A  R+L  L   +LE+E VP    A       P    ++P G+ P LE
Q04522    2 IKVHGNPRSTATQRVLVALYEKHLEFEFVPIDMGAGGHKQPSYLALNPFGQVPALE 57
            ***********************************************
```

```
         61 VQDRETGKKKILAESGFIFQYVLQHFDHSH  90
            G+ K+  ES  I +Y+    DH +
         58 -----DGEIKLF-ESRAITKYLAYTHDHQN  81
```

(b)

```
P40582   35 YKRDANFRAPPELKKIHPLGRSPLLEVQDR  64
            Y+  A+ R  P LK   P+G+ P+LEV  +
P41043   74 YEDVAHPRRVPALKPTMPMGQMPVLEVDGK 103
            ******************************
```

```
P40582  118 LMIEFILSKVKDSGMPFPI-SYLAR-KVADKISQAYSSGEVKNQFDFVEGEISKNN 171
            L I+ ++  + D  +   + SY   ++ +K    ++  +    + +E  + N+
P41043  130 LQIDIVVDTINDFRLKIAVVSYEPEDEIKEKKLVTLNAEVIPFYLEKLEQTVKDND 185
                                                      *********
```

```
        172 GYLVDGKLSGADILMS-FPLQMAFERKFAAPEDYPAISKWLKTITSEESYAASKEK 226
            G+L  GKL+ AD+  +      M + K    E YPA+  +  + + E   A  EK
        186 GHLALGKLTWADVYFAGITDYMNYMVKRDLLEPYPAVRGVVDAVNALEPIKAWIEK 241
            **************
```

(c)

```
P40582   18 RLLWLLDHLNLEYEIVPYKRDANFRAPPELKKIHPLGRSPLLEVQDRETGKKKILA 73
            R+  +L+  L++EIVP         P+   ++P G+ P L    D        ++L
P04907   16 RVATVLNEKGLDFEIVPVDLTTGAHKQPDFLALNPFGQIPALVDGD------EVLF 65
            ******************************************
```

```
         74 ESGFIFQYVLQHF 86
            ES  I +Y+   +
         66 ESRAINRYIASKY 78
```

FIG. 4. Locally optimal subalignments from the comparison of a hypothetical yeast protein (SWISS-PROT accession number P40582),[54] with various sequences in SWISS-PROT Release 31.[52] Alignments were found using the program BLASTGP, with BLOSUM62[28] amino acid substitution costs and (12, 2) affine gap costs. Central lines of alignments echo identities, and plus symbols (+) indicate positions with positive substitution score. Sections of alignments underlined with asterisks are ungapped alignments found by BLASTP[3] using the same substitution costs. (a) Locally optimal subalignment with score 74 and E value 0.045 involving an *S. cucubalus* glutathione *S*-transferase sequence (accession number Q04522).[55] When gaps are disallowed, the underlined region has score 62 and E value 0.90. (b) Two locally optimal subalignments, with respective scores 54 and 57, involving a *Drosophila* glutathione *S*-transferase sequence (accession number P41043).[56] The sum statistic E value[12,38] for the two subalignments is 0.014. When gaps are disallowed, the score of the underlined region in the second subalignment is 47, and the combined E value is 0.096. (c) Locally optimal subalignment with score 59 and E value 4.8 involving a maize glutathione *S*-transferase III sequence (accession number P04907).[57] When gaps are disallowed, the underlined region has score 58 and E value 3.5.

ber Q04522).[55] In the context of the search,[38] the expected number of occurrences (E value)[38] of a subalignment at least as good as this one is 0.045. If no gaps are allowed, the subalignment shrinks to the segment pair with score 62 (28.6 bits) indicated in Fig. 4a, whose E value of 0.9 is twenty times larger.

An even more significant result involves the two locally optimal subalignments with a *Drosophila* glutathione *S*-transferase sequence (SWISS-PROT accession number P41043)[56] shown in Fig. 4b. Neither subalignment, the first with score 54 (23.4 bits) and the second with score 57 (24.7 bits), would by itself be significant in the context of a database search.[38] However, using the sum statistics described above,[12] the two alignments in conjunction are equivalent to a single alignment of 33.7 bits, which here has an E value of 0.014. When no gaps are allowed, the second subalignment is trimmed to the indicated segment pair, with score 47 (21.7 bits), shown in Fig. 4b; the E value for the combined segment pairs increases by a factor of seven. The subalignments shown would be joined were lower gap costs used, but this would decrease λ and thereby dilute the value of the nominal score attained.

Not every alignment, of course, is rendered more significant by allowing gaps. A locally optimally subalignment of the query with a maize glutathione *S*-transferase III sequence (SWISS-PROT accession number P04907)[57] is shown in Fig. 4c. This alignment has score 59 (25.5 bits), whereas the one to which it shrinks when gaps are disallowed has score 58 (26.7 bits). Although the nominal score increases with the introduction of gaps, the 6% decrease in λ (Table V) renders the score less surprising.

The substitution matrix and gap costs employed in this example are not necessarily the best for recognizing distant similarities[39,40] and are used only for illustration. The program BLASTGP is similar in many ways to more recent experimental versions of FASTA, and a thorough comparison[39] of their relative speeds and sensitivities awaits further study.

Acknowledgments

We thank Dr. Martin Vingron for helpful conversions, and Dr. Gregory Schuler for assistance in the production of Figure 2.

[55] T. M. Kutchan and A. Hochberger, *Plant Physiol.* **99,** 789 (1992).
[56] C. Beall, C. Fyrberg, S. Song, and E. Fyrberg, *Biochem. Genet.* **30,** 515 (1992).
[57] R. E. Moore, M. S. Davies, K. M. O'Connell, E. I. Harding, R. C. Wiegand, and D. C. Tiemeier, *Nucleic Acids Res.* **14,** 7227 (1986).

[28] Parametric and Inverse-Parametric Sequence Alignment with XPARAL

By D. GUSFIELD and P. STELLING

Introduction

When aligning DNA or amino acid sequences using numerical-based optimization, there is often considerable disagreement about how to weight matches, mismatches, insertions and deletions (indels), and gaps. Most alignment methods require the user to specify fixed values for those parameters, and it is widely observed that the quality of the resulting alignment can be greatly affected by the choice of parameter settings.

Parametric alignment attempts to avoid the problem of choosing fixed parameter settings by computing the optimal alignment as a function of variable parameters for weights and penalties. The goal is to partition the parameter space into regions (which are necessarily convex) such that in each region one alignment is optimal throughout and such that each region is maximal for this property. Thus parametric alignment allows one to see explicitly, and completely, the effect of parameter choices on the optimal alignment. Parametric sequence alignment was first used in a paper by Fitch and Smith[1] and was studied more extensively in papers by Gusfield *et al.*,[2,3] Waterman *et al.*,[4] and Vingron and Waterman.[5] The last four papers concern the number, shape, pattern, and interpretation of the regions. The basic algorithmic ideas for computing two-dimensional parametric decompositions were first developed in contexts other than sequence alignment.[6]

In this chapter we first describe a publicly available, user-friendly interactive software package, XPARAL, that solves the parametric alignment problem; we emphasize newer features in XPARAL. We next illustrate the use of XPARAL by reexamining a study done by Barton and Sternberg[7] on gap weights used in aligning protein secondary structure. Finally, we discuss the empirical and theoretical efficiency of XPARAL. We use stan-

---

[1] W. Fitch and T. Smith, *Proc. Natl. Acad. Sci. U.S.A.* **80,** 1382 (1983).

[2] D. Gusfield, K. Balasubramonian, and D. Naor, "Proceedings of the Third Annual Symposium on Discrete Algorithms," p. 432. Orlando, FL. 1992.

[3] D. Gusfield, K. Balasubramonian, and D. Naor, *Algorithmica* **12,** 312 (1994).

[4] M. Waterman, M. Eggert, and E. Lander, *Proc. Natl. Acad. Sci. U.S.A.* **89,** 6090 (1992).

[5] M. Vingron and M. Waterman, *J. Mol. Biol.* **235,** 1 (1994).

[6] D. Gusfield, *J. ACM* **30,** 551 (1983).

[7] G. Barton and M. Sternberg, *Protein Eng.* **1,** 89 (1987).

Copyright © 1996 by Academic Press, Inc.
All rights of reproduction in any form reserved.

dard terminology for sequence alignment, but note that the term gap refers to a maximal contiguous run of spaces (possibly only a single space).

Parametric Alignment and XPARAL Features

The XPARAL package allows the user to specify alignment objective functions with or without character-specific scoring matrices (such as PAM matrices). The treatment (mathematical and algorithmic) of these two cases is somewhat different and is discussed separately.

Any alignment $\mathscr{A}$ of two strings contains a specific number of matches, mismatches, indels, and gaps. We denote these numbers by $mt_{\mathscr{A}}$, $ms_{\mathscr{A}}$, $id_{\mathscr{A}}$, and $gp_{\mathscr{A}}$, respectively. Without the use of character-specific scoring matrices, the value of alignment $\mathscr{A}$ is therefore $v_{\mathscr{A}}(\alpha, \beta, \gamma, \delta) \equiv \alpha mt_{\mathscr{A}} - \beta ms_{\mathscr{A}} - \gamma id_{\mathscr{A}} - \delta gp_{\mathscr{A}}$, where α, β, γ, and δ are variables that adjust the relative importance of matches, mismatches, indels, and gaps. Note that the value of the alignment is a linear function of the four parameters. When these four parameters have fixed values α_0, β_0, γ_0, and δ_0, then the fixed-parameter problem is to find an alignment $\mathscr{A}$ maximizing the objective function: $\alpha_0 mt_{\mathscr{A}} - \beta_0 ms_{\mathscr{A}} - \gamma_0 id_{\mathscr{A}} - \delta_0 gp_{\mathscr{A}}$.

The XPARAL package allows the user to select two of the parameters α, β, γ, and δ to be variable, and to choose constant settings for the other two parameters. For illustration, suppose γ and δ are chosen to be variable, while α and β are fixed at one. This is a choice that is typical for aligning amino acid sequences. As a function of two variable parameters, the value of alignment $\mathscr{A}$ specifies a plane in three-dimensional space. Thus,[2–4] for any pair of input sequences, the γ, δ parameter space decomposes into convex polygons such that any alignment which is optimal for some α, γ point in the interior of a polygon $\mathscr{P}$ is optimal for all points in $\mathscr{P}$ and nowhere else. XPARAL computes and displays this polygonal decomposition. The user can then select any particular polygon $\mathscr{P}$ to see an alignment that is optimal for all points in $\mathscr{P}$, and to display the number of matches, mismatches, indels, and gaps in that alignment. There may be alternative alignments that are optimal throughout $\mathscr{P}$ (and nowhere else), but all those cooptimal alignments have exactly the same number of matches, mismatches, indels, and gaps. XPARAL can count the number of cooptimal alignments in polygon $\mathscr{P}$ and display each one in turn.

Figure 1 shows the XPARAL display of one such polygonal decomposition of the γ, δ space. The grid points are superimposed and will be explained later. The topmost windows display menu buttons. The two horizontal windows below the menu buttons display the input sequences (which must be scrolled to see the complete sequences); the main window shows the polygonal decomposition and contains one dark polygon; the horizontal

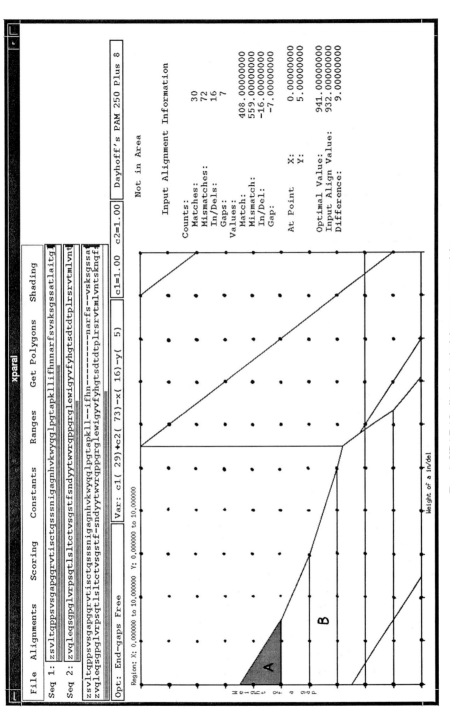

FIG. 1. XPARAL display for indels versus gap weights.

window below the input windows shows an alignment that is optimal for all the γ, δ points in the dark polygon. The smaller horizontal windows below the alignment window give information on the settings the user has selected as well as on the number of matches, mismatches, spaces, and gaps in the displayed alignment. Those numbers remain the same throughout the entire polygon, but the value of the alignment changes throughout the polygon. When the user selects a particular point, the value of the optimal alignment at that point is displayed to the right of the main window.

Use of Scoring Matrices

The XPARAL package allows the user to define and use any character-specific scoring matrix, such as a PAM matrix, to assign a weight to each possible pair of characters (either a match or a mismatch). With the use of scoring matrices we let $smt_{\mathscr{A}}$ denote the total weight given to the matches in alignment $\mathscr{A}$ and let $sms_{\mathscr{A}}$ denote the total weight given to the mismatches in $\mathscr{A}$. Then the value of $\mathscr{A}$ is $v_{\mathscr{A}}(\alpha, \beta, \gamma, \delta) \equiv \alpha smt_{\mathscr{A}} + \beta sms_{\mathscr{A}} - \gamma id_{\mathscr{A}} - \delta gp_{\mathscr{A}}$. The term $\beta sms_{\mathscr{A}}$ is added, rather than subtracted, because the scoring matrices already incorporate the appropriate sign of the mismatch penalty. When selecting two parameters to be variable, the value of alignment $\mathscr{A}$ is again a function of those two variable parameters and defines a plane in three-space.

Choice of Alignment Model and Input Mode

Using XPARAL, the user can specify whether the alignment should be global (Needleman–Wunsch), global with end gaps free, local (Smith–Waterman), or internal, where end spaces in the longer string are penalized but end spaces in the shorter string are ignored. Sequences are input to XPARAL either from files or at the input window. XPARAL also allows the input sequences to be taken from a file that holds a previously determined reference alignment $\mathscr{A}$. Then, when the user selects a point p in the decomposition, XPARAL displays the value of the input alignment at p, the value of the optimal alignment at p, and the difference between the two values. In this mode, XPARAL can also solve the inverse-parametric problem: For what points in the γ, δ space does the optimal alignment (computed by dynamic programming) have a value which is closest to the value of the reference alignment $\mathscr{A}$? That is, if we let $v(\alpha, \beta, \gamma, \delta)$ denote the value of the optimal alignment at point $(\alpha, \beta, \gamma, \delta)$, then we seek the γ, δ points where $v(\alpha_0, \beta_0, \gamma, \delta) - v_{\mathscr{A}}(\alpha_0, \beta_0, \gamma, \delta)$ is minimum. These points are called inverse-optimal.

By convexity, the inverse-optimal point(s) must occur either at a single vertex, at a single edge (line segment between two vertices), or at a single complete polygon of the polygonal decomposition. The latter case happens when the plane representing the value of $\mathscr{A}$ is parallel (or identical) to a plane in the decomposition. For efficiency, the inverse parametric problem is solved using gradient-descent rather than by first finding the complete decomposition.

Inverse-parametric computation is useful for trying to deduce parameter settings where the optimal alignment (with respect to a chosen objective function) might likely reconstruct correct alignments that have been determined by other methods. Of course, a parameter setting that minimizes the numerical difference between the value of the optimal alignment and the value of the reference alignment is not necessarily a parameter setting where the optimal alignment is most similar in form (an undefined concept) to the reference alignment. In experiments we have done, however (see the next section), we see a good correlation between this numerical distance and the ability to recapture significant features of the reference alignment. Inverse-parametric computations also allow an efficient, but rough, test of the validity of both the alignment model and the reference alignment. For if a numerical-based alignment model is valid in some biological setting, then a known alignment that is correct in that setting should have a numerical value that is close to optimal, at least for some points in the parameter space.

XPARAL is written in C++ and compiled for the DECstation 5000 and the DEC α. Compiled versions can be obtained at the following web site: http://wwwcsif.cs.ucdavis.edu/~gusfield/strpgms. Source code is available from the authors on request.

Using XPARAL to Study Gap Models for Secondary Structure

We illustrate one use of XPARAL by reexamining a study done by Barton and Sternberg[7] on the effectiveness of global and end-gap free sequence alignment to correctly align protein secondary structures. Their study examined sequences whose three-dimensional structure was known. Superimposing the structures for two proteins gives a reference alignment against which they evaluated alignments computed from sequence data alone. They scored a sequence alignment $\mathscr{A}$ by identifying the residues in regions of known secondary structure and counting the number of those residue pairs which align the same in $\mathscr{A}$ as in the reference alignment.

The sequence alignment objective function they used was the following: Maximize $smt_{\mathscr{A}} + sms_{\mathscr{A}} - \gamma id_{\mathscr{A}} - \delta gp_{\mathscr{A}}$ over all alignments $\mathscr{A}$, where the scoring matrix was integer-rounded PAM250 with a constant of 8 added to every entry.[8] They considered every integer combination of γ and δ from 0 to 10 and found alignments with optimal values (using both global and end-gap free models) at each of those 121 integer γ, δ combinations. Each optimal was then scored, as above, to evaluate it against the reference alignment. They considered five such pairs of sequences, but full details were published for only one pair (immunoglobulins FABVL versus FABVH). In that case, the best of the 121 optimal sequence alignments they found received a score of 32 of a possible maximum of 41. Optimal global and end-gap free alignments at the 121 integer points were next computed using a new alignment model that penalizes gaps and indels in the known regions of secondary structure more heavily than gaps and indels outside those regions. With this differential gap model, they found several integer points where the optimal alignments had scores of 36.

One of their main conclusions[7] is that standard alignment models are not adequate for aligning secondary structure, but the differential gap model described above is effective. A similar conclusion was reached by Lesk *et al.*[9] The final recommendation, to use secondary structure information when available, is not in dispute, but such information is not always available. Moreover, since secondary structure is often predicted from sequence alone, one would like an independent alignment method to test those predictions, rather than simply incorporate them. So the effectiveness of standard alignment models remains of interest.

Vingron and Waterman[5] have also examined the immunoglobulins FABVL and FABVH using their parametric alignment program. In contrast to the previous study, they found polygons in the γ, δ space where the resulting alignment agreed with the main features of the known structural alignment. This gives a different picture of the effectiveness of standard alignment models to align secondary structure. However, Vingron and Waterman[5] used corrected sequences compared to those used by Barton and Sternberg,[7] modified PAM250 differently, used local alignment, and did not use a numerical score to evaluate alignment quality. So those differences alone might account for the differing results. Alternatively, the different results might be due to chance, since only one optimal alignment was computed at each sample point, even if several cooptimal alignments exist, or it may be that complete parametric decomposition yields a wider, more informative range of alignments than is obtained from the (grid) sampling approach.[7]

---

[8] For clarity, the score of an alignment $\mathscr{A}$ is the number given by the Barton and Sternberg evaluation criteria, while the value of $\mathscr{A}$ is the number given by the dynamic programming objective function. The term optimal refers to the value.

[9] A. Lesk, M. Levitt, and C. Chotia, *Protein Eng.* **1,** 77 (1986).

We used XPARAL to examine why the two studies obtained differing results. We first kept the exact alignment conditions of Barton and Sternberg[7] (using their exact sequences, using integer-rounded PAM250 plus 8, and computing global and end-gap free alignments), but we used XPARAL to completely decompose the γ, δ parameter space. We considered three questions. First, with these conditions, are there polygons where an optimal alignment has a larger score than 32? Second, if there are such alignments, were they missed[7] because of a limited choice of test points, or because the polygons are small, or because only a single optimal alignment was computed at each test point? Third, how well does the inverse-parametric feature of XPARAL find parameter settings where high-scoring alignments are obtained?

Results from Empirical Study

The polygonal decomposition obtained is shown in Fig. 1, with the 121 integer points superimposed. The striking feature is that although there are only seventeen polygons in the decomposition, the 121 integer points miss the interior of all but four of the seventeen polygons. Moreover, several of the polygons have multiple (up to 12) sample points on their boundaries but none in their interiors, so polygon size is not the reason that interior points were missed. Conversely, more than 40% of the test points fall into the three largest polygons, and so give redundant information.

Sampling at interior points gives robust alignments since any alignment that is optimal at an interior point is optimal throughout the polygon, whereas an alignment that is optimal on an edge or vertex may only be optimal on that edge or vertex. In particular, with only seventeen alignments, no mater what γ, δ point p is selected (in Fig. 1), one of those seventeen alignments will be optimal at p. The larger set of 121 alignments computed at integer sample points need not have that property. Interior points are further desirable because the number of cooptimal alignments in the interior of a polygon is always less than the number on any of its boundaries, except at borders of the parameter space. So although the integer points provide a dense, regular, sampling of the parameter space, they may only provide a biased representation of the alignment space and miss important alignments. We believe this happened.

Without using differential gap penalties, XPARAL found a polygon containing optimal alignments with score as high as the best alignments found when Barton and Sternberg[7] used their differential gap penalties. The interior of the polygon labeled A contains 24 cooptimal alignments, 18 with a score of 36 and 6 with a score of 32. The alignment XPARAL displayed for polygon A had a score of 36. The differences among the 18 high-scoring alignments are outside of the regions of secondary structure, as are the differences among the 6 lower scoring alignments. So with respect to secondary structure, there are effectively two different optimal alignments in polygon A. Three integer points fall on the boundary of A, but none falls in the interior. The polygon bordering A from below has six cooptimal alignments, all with score 32. So the three alignments computed[7] at the grid points on the boundary of A missed the high scoring alignments, essentially by chance. However, the chance that a deterministically computed alignment is one of the high-scoring alignments is improved by sampling in the interior of A. For example, when an alignment is computed on the boundary of A and the polygon below it, the algorithm chooses a single alignment (and always the same one no matter how many points on the boundary are sampled) from among 18 with a high score and 12 with a low score. Inside A, the algorithm choses from among 18 with a high score and 6 with a low score. The way, then, to optimize the chances that a deterministic algorithm will find a high-scoring alignment, without enumerating cooptimals, is to compute one alignment in the interior of each polygon and one on each boundary and vertex.

We have no theorem that integer points miss polygon interiors with high probability, but it is not surprising that they do (especially near parameter settings where biologically reasonable alignments occur), since one should expect those boundaries to preferentially contain integer points. To see this, consider the equations for the values of two neighboring alignments $\mathscr{A}_1$ and $\mathscr{A}_2$, and let id_i and gs_i denote the number of indels and gaps in alignment $\mathscr{A}_i$. Each equation describes a plane, and every polygon boundary lies on the intersection of two such planes. A boundary therefore lies on a line

$$\delta = \frac{C}{gs_2 - gs_1} + \gamma \frac{id_1 - id_2}{gs_2 - gs_1}$$

where C is the total of the match and mismatch weights for $\mathscr{A}_2$ minus the total of the match and mismatch weights for $\mathscr{A}_1$. With an integer PAM matrix, C is an integer, and so the intercept of the line will be an integer if C is a multiple of $gs_2 - gs_1$. That is likely because $gs_2 - gs_1$ is an integer and tends to be small, often one. The reason is that the number of gaps tends to be small for real protein data, and the number of gaps must monotonically fall along any vertical line of increasing δ, whenever the line crosses a boundary. For example, in Fig. 1, on the vertical line starting at $(0.25, 0)$ and ending at $(0.25, 10)$, the alignments encountered have 14, 11, 10, 8, 7, 6, 5, and 3 gaps in that order. Those two facts (small gap number an monotonic change) suggest that $gs_2 - gs_1$ will tend to be a small integer for alignments in neighboring polygons. In fact, 17 of the 27 interior boundary edges in Fig. 1 lie on lines with integer intercepts, and 5 more have half-integer intercepts. The argument for integer (or small denominator) slope is similar, and 21 of the 27 boundary edges have integer slopes. Integer (or half-integer) slopes and/or intercepts imply that the edges will preferentially contain integer points.

Results from Inverse-Parametric Alignment

We used the same immunoglobulin sequences to test the inverse-parametric feature of XPARAL. For reference alignment we used the alignment of score 36 that was obtained[7] with differential gap and indel penalties. XPARAL found that all the points in the polygon labeled B (Fig. 1) are inverse-optimal, and hence the plane for the reference alignment is parallel to the plane for polygon labeled B. The reference alignment has 30 matches, 72 mismatches, 16 indels, and 7 gaps, whereas the three optimal alignments for polygon B have 29 matches, 73 mismatches, 16 indels, and 7 gaps. It is interesting how similar the two vectors are. Moreover, the optimal alignment value (at every point in B) differs from the reference alignment value by less than 1%. So the reference alignment is very close to optimal in that polygon, although each of the three cooptimals in B have a score of only 32. This provides a rough confirmation of the validity of the standard alignment model (applied to FABVL and FABVH sequences). There are over 2^{200} alignments of these two sequences, and yet there are only 120 alignments that are (co-)optimal anywhere in the entire γ, δ parameter space, and only 110 (co-)optimal alignments inside the 17 polygons found in the bounded 10 by 10 region.

Related Results

All of the above results remain essentially the same when global alignment is used in place of end-gap free alignment. However, there are alignment models in which the results are stronger. For example, when PAM250 is used instead of PAM250 plus 8, polygon A essentially expands to contain three interior integer points, and XPARAL again returns a displayed alignment of score 36. Departing completely from these conditions, using global alignment, choosing α and γ to be variable, and setting $\beta = \delta = 0$ (as is sometimes suggested for DNA), but using the PAM250 matrix, then the decomposition contains 11 polygons, including a large one where the reference alignment[7] is optimal. In other words, the standard alignment model without differential gap weights not only returns an alignment with a score as good as the reference alignment, it returns the reference alignment exactly.

Conclusion from Empirical Study

To study properties of particular alignment models, parametric alignment can more selectively and effectively home in on critical parameter regions than can a grid sampling of the parameter space. Cooptimal or near-optimal alignments in those critical regions can then be generated and studied in more depth. Generating a range of alignments, such as all of the optimal alignments from each polygon that neighbors the polygon which contains some initial parameter setting, also provides an alternative to the recommended practice of generating suboptimal alignments for a single fixed parameter setting.

Efficiency of XPARAL

We have put a great deal of effort into making XPARAL practical. The decomposition shown in Fig. 1 took less than 6 sec to compute using a DEC α, and 14 sec on a slower DECstation 5000/25. As another empirical measure, XPARAL computed only 160 fixed-parameter alignments to determine that decomposition. With additional programming effort, that number can be reduced by about one-half. We will now prove that for most alignment models, the number of fixed-parameter alignments that XPARAL must compute to find the decomposition is proportional to the number of polygons in the decomposition. This issue was stated as an open question by Pevzner and Waterman.[10] The result to be established here was claimed earlier[3] without a proof.

Basic Algorithm for XPARAL

We need to describe the inner workings of XPARAL at a high level. Our approach is different in some important ways from the Waterman *et al.*[4] method, and does not use infinitesimals. For illustration, we again assume that γ and δ are the variable parameters. The workhorse of XPARAL is the following problem called the ray-search problem: Given an alignment $\mathscr{A}$, a point p where $\mathscr{A}$ is optimal, and a ray h in γ, δ space starting at p, find the furthest point (call it r^*) from p on ray h where $\mathscr{A}$ remains optimal. If $\mathscr{A}$ remains optimal until h reaches a border of the

[10] P. Pevzner and M. Waterman, *Proc. Israel Symposium on Theory of Computing and Systems*, 158. (1995).

parameter space, then r^* is that border point on h. The ray-search problem is solved as follows:

Set r to the (γ, δ) point where h intersects a border of the parameter space. While $\mathcal{A}$ is not an optimal alignment at point r do
begin
Find an optimal alignment $\mathcal{A}^*$ at point r.
Set r to be the unique point on h where the value of $\mathcal{A}$ equals the value of $\mathcal{A}^*$.
end;
Set r^* to r.

This algorithm is Newton's classic zero finding method specialized to a piecewise linear function. The following three facts will be needed in the analysis and are easy to establish: Newton's method finds r^* exactly; unless $\mathcal{A}$ is optimal at the initial setting of r, the last computed alignment $\mathcal{A}^*$ is cooptimal with $\mathcal{A}$ at r^* and yet is also optimal on h for some nonzero distance beyond r^*; and, when it computes an alignment at a point r on h, none of the alignments computed previously (in this execution of Newton's algorithm) are optimal at r.

We now explain how to find the edges of polygon $\mathcal{P}(\mathcal{A})$, given an alignment $\mathcal{A}$ that is optimal at a point p, and known to be optimal for an (unknown) polygon $\mathcal{P}(\mathcal{A})$. First pick any ray h from p and solve the ray-search problem along h. There are two degenerate cases that can occur: one is that r^* lies on a border of the parameter space, and the other is that r^* is a vertex of the decomposition. We will consider those degenerate cases later and assume for now that they do not occur. Therefore, the ray search along h will find a point r^* that lies on an edge e of polygon $\mathcal{P}(\mathcal{A})$. By Newton's second fact, the ray search will also return an alignment $\mathcal{A}^*$ that is optimal in the interior of the polygon bordering edge e. The intersection of the two planes for $\mathcal{A}$ and $\mathcal{A}^*$ describes a line l^* that contains edge e, so the full extent of e can be found by solving two more ray-search problems using $\mathcal{A}$. In one problem, ray h is the half-line of l^* starting at r^* and running in one direction along l^*, and in the other problem ray h is the remaining half-line of l^* in the other direction. These two ray searches find the opposite endpoints of edge e. Once edge e is fully described, we select another ray h from p that does not intersect edge e, and find a second edge of $\mathcal{P}(\mathcal{A})$. By linking identical endpoints of edges of $\mathcal{P}(\mathcal{A})$ that have been found, it is easy to continue selecting rays from p that do not intersect previously discovered edges or vertices of $\mathcal{P}(\mathcal{A})$. In this way, we find all the edges of $\mathcal{P}(\mathcal{A})$, stopping when the discovered edges of $\mathcal{P}(\mathcal{A})$ link together to form a closed cycle.

Consider now the two degenerate cases that may occur when trying to find an edge of $\mathscr{P}(\mathscr{A})$. In the case that r^* is on a border of the parameter space, then that border line is used in place of l^*. In the other case, when r^* is a vertex, the algorithm will realize this because $\mathscr{A}$ will not be optimal past r^* on at least one of the two rays on l^* from r^*. When this occurs, the algorithm simply begins a new ray search from p using a ray that avoids r^* and all other previously discovered vertices and edges.

To compute a full polygonal decomposition, one first finds an alignment that is sure to be optimal for some (unknown) polygon. This is easy to do with a constant number of ray searches, and we omit details. Now we explain how the algorithm finds successive polygons. When finding the first polygon $\mathscr{P}$ (and for each additional polygon it finds), the algorithm inserts into a list, L, one distinct vector ($smt_{\mathscr{A}*}$, $sms_{\mathscr{A}*}$, $id_{\mathscr{A}*}$, $gp_{\mathscr{A}*}$) for each alignment $\mathscr{A}^*$ found to be optimal in the interior of a polygon bordering $\mathscr{P}$. When $\mathscr{P}$ is finished, the algorithm finds and marks one of the unmarked vectors from L, say for $\mathscr{A}'$, and then finds the polygon $\mathscr{P}(\mathscr{A}')$ where $\mathscr{A}'$ is optimal. The parameter space will be fully decomposed when all vectors in L are marked. Since the algorithm never chooses a marked vector, nor inserts two equal vectors, and since when a polygon is found the algorithm learns one alignment optimal at the interior or each neighboring polygon, each polygon in the full decomposition is found exactly once.

Time Analysis and New Idea

The above details lead to the following time analysis. Let R, E, and V be the number of polygons, edges, and vertices, respectively, in a decomposition, and let $O(nm)$ be the time to compute a single fixed-parameter alignment for sequence of lengths n and $m > n$. How many ray searches are executed to find a polygon $\mathscr{P}(\mathscr{A})$, given $\mathscr{A}$ and p? Let d be the number of edges of $\mathscr{P}(\mathscr{A})$. Then $3d$ ray searches are done to find the edges of $\mathscr{P}(\mathscr{A})$, and, in the highly degenerate case that selected rays from p intersect all the vertices of $\mathscr{P}(\mathscr{A})$, then another $3d$ (wasted) ray searches may be done as well. Hence at most $6d$ ray searches suffice to describe $\mathscr{P}(\mathscr{A})$. Each edge lies on at most two polygons, so the algorithm does at most $12E$ ray searches to find the complete decomposition. Further, from Newton's third fact each ray search requires at most R fixed-parameter alignment computations, so the complete decomposition requires at most $12RE$ fixed-parameter alignments which can be done in $O(ERnm)$ time. This unsatisfactory bound will be improved with one additional idea.

The new idea is to modify the Newton algorithm (given $\mathscr{A}$) to pick the initial point r far enough on h to be at or beyond the (unknown) point r^*, yet as close to r^* as present information allows. Consider any alignment

$\mathscr{A}'$ computed before the present execution of Newton's method. Compute the intersection of the planes for $\mathscr{A}$ and $\mathscr{A}'$ and project that line onto the γ, δ plane. If the projection intersects h, then the initial r need not be any further from p than r', since $\mathscr{A}'$ has greater value than $\mathscr{A}$ beyond r'. If the projection misses h, then $\mathscr{A}$ has greater value than $\mathscr{A}'$ at every point on h. Any intersection and projection can be done in constant time. Repeating this for each previously computed alignment $\mathscr{A}'$, we set the initial r to the point closest on h to p among all the computed r' points.

The modified Newton method clearly reduces the number of needed alignments, but by how much, and how much added time is needed to implement it? During the entire algorithm we will keep a list L' that is a superset of list L. Whenever any alignment is computed, the vector for that alignment is placed into L', if it is not already there. When we begin any ray search, we use L' as explained above to find the initial point r. It takes constant time to compute each point r', so the added cost of using L' in a single ray search is proportional to the size of L'. We claim that size is at most $V + E + R$, that is, over the entire running of the algorithm only $V + E + R$ distinct vectors will be computed. This follows because XPARAL is a deterministic algorithm so that no matter how many times it might compute an optimal alignment at the same vertex, it always returns the same alignment (hence, same vector). Similarly, even when XPARAL computes alignments at different points on the same polygon edge, it will return the same alignment each time, and it is true for alignments inside a polygon, since all optimal alignments inside a polygon have the same vector. So the added bookkeeping time for using L' is just $O(V + E + R)$ per ray search or $O[12E(V + E + R)]$ overall.

For the analysis of the number of needed fixed-parameter alignments, call an alignment computation redundant if it returns a vector that is already in L'. We claim that in any single ray search (using modified Newton with alignment $\mathscr{A}$) only the last alignment computation in that search could be redundant. To see this, note that if $\mathscr{A}$ is optimal at the initial r, then only one alignment is computed in that ray search; otherwise, the redundant alignment, $\mathscr{A}'$, is computed at a point closer to p than the initial r. Since, by Newton's third fact, each vector computed during the ray search is distinct, $\mathscr{A}'$ must have been in L' before the present ray search. But that would contradict the choice of the initial r.

We can now analyze the time to find a complete polygonal decomposition. There are still at most $12E$ ray searches, and, in each, at most one computed alignment is redundant. Each other alignment computation finds a new vector to add to L' (which has size at most $V + E + R$), so the complete polygonal decomposition is computed using at most $V + 13E + R$ fixed-parameter alignments. This leads to an overall time bound of

$O[12E(V + E + R) + (V + 13E + R)nm]$. Now a polygonal decomposition can be viewed as a connected planar graph, and when each vertex is incident with at least three edges (as in the case of a polygonal decomposition), then $V \leq E \leq 3R$. This is easy to show using Euler's classic theorem, but is not true for general planar graphs. So the terms $12E$ and $V + E + R$ and $V + 13E + R$ are each proportional to R, that is, each are $O(R)$. Hence, the above time bound becomes $O(R^2 + Rnm)$, which is $O(R + nm)$ per polygon.

The bound of $O(R + nm)$ per polygon holds no matter what choices are made in XPARAL. However, it was shown[2,3] that when no character-specific scoring matrices are used, and two variable parameters are picked, then $R = O(nm)$. In fact,[2,3] for global alignment, when no scoring matrices are used then $R < n^{2/3}$. When scoring matrices are used, but γ and δ are the chosen variable parameters, then again $R = O(nm)$. This establishes the claim that for most of the (important) parameter choices, a full polygonal decomposition can be found in $O(nm)$ time per polygon, proportional to the time needed to compute just a single fixed-parameter alignment. It is unknown how big R can be when scoring matrices are used and α, β are the variable parameters. The algorithm description and analysis above is much cruder than what is actually implemented in XPARAL but suffices for the main result.

Acknowledgments

We thank John Nguyen for helpful efforts and insights in working with XPARAL. Research was partially supported by Grant DE-FG3-9ER6999 from the U.S. Department of Energy.

Section IV

Secondary Structure Considerations

[29] Identification of Functional Residues and Secondary Structure from Protein Multiple Sequence Alignment

By CRAIG D. LIVINGSTONE and GEOFFREY J. BARTON

Introduction

There are now many techniques for the automatic alignment of multiple protein sequences. Most of the practical approaches follow the work of Feng and Doolittle[1] and Barton and Sternberg[2] by hierarchically combining pairwise comparisons of two sequences, either a sequence and a preexisting alignment or two preexisting alignments (see Ref. 3 or 4 for an introduction to multiple alignment). Hierarchical methods of alignment cope with large numbers of sequences and give reasonably accurate alignments. Having generated the alignment, the problem is to find out what it can tell us about the protein family. Interpretation of alignments can be particularly difficult when there are large numbers of sequences to examine.

Exactly how useful an alignment will be depends on both the biological context and the overall similarity between the aligned sequences. For example, sequences that show just enough similarity to be aligned reliably provide the most information about the alignment positions important to all members of the protein family. These may be active site residues in an enzyme or, more generally, residues critical to the conserved secondary structural core of the protein. However, when trying to pin down the residues important to functional differences between proteins, even a group of sequences that show high overall similarity may reveal interesting patterns.

In this chapter we describe a strategy for hierarchical analysis of residue conservation.[5] In summary, the method allows the residue-specific similarities and differences in physicochemical properties between groups of sequences to be identified quickly. The method also highlights conserved positions across a complete alignment and thus can help to identify patterns characteristic of regular secondary structures (α helices and β strands). We summarize a procedure for applying these patterns in secondary structure

[1] D. F. Feng and R. F. Doolittle, *J. Mol. Evol.* **25**, 351 (1987).

[2] G. J. Barton and M. J. E. Sternberg, *J. Mol. Biol.* **198**, 327 (1987).

[3] G. J. Barton, this series, Vol. 183, p. 403.

[4] G. J. Barton, *in* "Protein Structure Prediction: A Practical Approach" (M. J. E. Sternberg, ed.), in press. IRL Press at Oxford Univ. Press, Oxford, 1995. (A preprint is available at the URL http://geoff.biop.ox.ac.uk/)

[5] C. D. Livingstone and G. J. Barton, *Comput. Appl. Biosci.* **9**, 745 (1993).

Copyright © 1996 by Academic Press, Inc.
All rights of reproduction in any form reserved.

prediction, and we evaluate their predictive power in six blind secondary structure predictions.

Amino Acid Conservation

Many different methods of summarizing the amino acid composition at an alignment position have been described. For example, Kabat calculates variability,[6] Sander and Schneider entropy or variation,[7] Smith and Smith information,[8] and Brouillet et al. evolutionary divergence.[9] Techniques to color or shade alignments by various criteria, for example, hydrophobicity or charge, can also help highlight conserved features (e.g., Refs. 10, 11, and 12).

An alternative approach developed by Taylor[13] defines physicochemical properties that represent groups of amino acids. The method is based on the set representation of the amino acid properties shown in Fig. 1. The strengths of this representation are that it is conceptually simple and that it allows an amino acid to have more than one property. For example, lysine (K) is not only positive, charged, and polar, reflecting its terminal NH_3^+ group, but also hydrophobic on account of the four CH_2 groups in its side chain. In addition, lysine is not a member of the sets: small, proline, tiny, aliphatic, aromatic, or negative. Given the Venn diagram, an alignment position may be described by the minimal set of properties that are present at a position.[13]

It is also convenient to represent each position by a single number that encapsulates the conservation of physicochemical properties. These conservation numbers (C_n) were introduced by Zvelebil et al.[14] and extended by Livingstone and Barton.[5] The numbers are determined by calculating the total number of set boundaries (P) that must be crossed in order to visit all amino acids at the alignment position. This number is then subtracted from the total number of properties (N). For example, in Fig. 1 there are 10 properties, and a position that had only L and R would

[6] E. A. Kabat, "Structural Concepts in Immunology and Immunochemistry," 2nd Ed. Holt, Rinehart, and Winston, New York, 1976.

[7] C. Sander and R. Schneider, Proteins: Struct. Funct. Genet. 9, 56 (1991).

[8] R. F. Smith and T. F. Smith, Proc. Natl. Acad. Sci. U.S.A. 87, 118 (1990).

[9] S. Brouillet, J. L. Risler, and P. P. Slonimski, Biochimie 74, 571 (1992).

[10] D. J. Parry-Smith and T. K. Attwood, Comput. Appl. Biosci. 7, 233 (1991).

[11] J. Devereux, P. Haeberli, and O. Smithies, Nucleic Acids Res. 12, 387 (1984).

[12] G. J. Barton, Protein Eng. 6, 37 (1993).

[13] W. R. Taylor, J. Theor. Biol. 119, 205 (1986).

[14] M. J. J. M. Zvelebil, G. J. Barton, W. R. Taylor, and M. J. E. Sternberg, J. Mol. Biol. 195, 957 (1987).

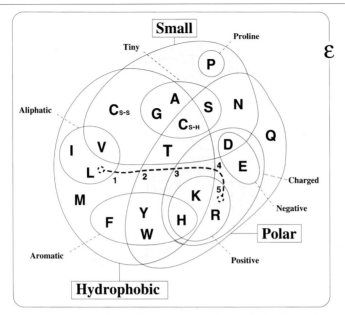

FIG. 1. The 20 common amino acids are shown with ten physicochemical properties.[5,13] Gray-filled areas define sets of properties possessed by none of the amino acids.

give a value of $P = 5$ and $C_n = (10 - 5) = 5$. Because gaps are normally associated with variable parts of a protein structure and are thus less likely to occur where there is a conserved functional residue, they are given all properties. This has the effect of reducing the conservation number at any position that includes a gap.

Figure 2 shows a segment of a multiple alignment of annexin domain sequences shaded where $C_n \geq 5$ (see Ref. 15 for the full alignment). The conserved pattern of amino acids at $i, i + 3, i + 4, i + 7$ is characteristic of an α helix where the conserved residues are important for packing against the core of the protein. Conservation patterns and their use in secondary structure prediction are discussed further below.

Hierarchical Analysis of Residue Conservation

When calculating variabilities, frequencies, or conservation numbers across a complete alignment, or when deriving a consensus, one is throwing away information about the relative similarities of each sequence. It is

[15] G. J. Barton, R. H. Newman, P. F. Freemont, and M. J. Crumpton, *Eur. J. Biochem.* **198,** 749 (1991).

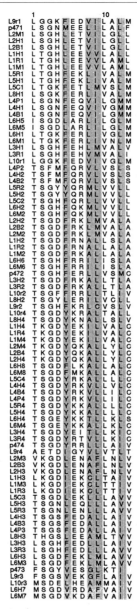

FIG. 2. Section of an alignment of 88 annexin sequences. A pattern of conservation characteristic of an α helix is revealed by gray-shaded positions that give conservation number values of ≥7.

normally more useful first to subgroup the sequences by similarity or other criteria (e.g., function), then compare subgroup to subgroup. In this way, position-specific amino acid preferences in each subgroup can be identified, and these preferences may suggest the location of residues that are important in defining the different functions of the proteins.

If a tree or dendrogram is calculated for the sequences, the overall similarity between the sequences will normally be easy to see. For example, Fig. 3 shows a dendrogram calculated from pairwise comparison of 17 flavodoxin sequences. The SWISS-PROT sequence identifier codes are shown at the right-hand side of Fig. 3. Sequences that are joined toward the right of Fig. 3 show greater similarity than those that are joined further to the left. Thus, there are three clear groupings (working up from the bottom of Fig. 3): {FLAV_MEGEL, FLAV_CLOBE}, {FLAW_DESGI, FLAV_DESGI, FLAV_DESSA, FLAV_DESVH, FLAV_DESDE, FLAV_DESDE}, and {FLAV_SYNP2, FLAV_ANASP, FLAV_SYNP7, FLAV_SYNY3, FLAW_ECOLI, FLAV_ECOLI, FLAV_CHOCR}; the two outliers are FLAV_CLOAB and FLAV_RHOCA.

The dendrogram is derived from the overall similarity between each sequence pair and hence reflects the total conservation of physicochemical properties at each position in the alignment. What we would like to do is identify the positions in the alignment that are making the most significant or "interesting" contribution to this clustering. What constitutes interesting is problem dependent, but we find the following strategy (see Fig. 4) useful for narrowing down the search.

1. First divide the sequence alignment into subgroups. In the flavodoxin example, we can choose five subgroups (the three groups summarized above and two separate sequences), but other groupings are possible. To simplify the explanation, in Fig. 4, only three subgroups have been identified and labeled A–C.

2. At each position in the alignment calculate the conservation number for each subgroup as if it were isolated from the rest of the alignment. For the alignment shown in Fig. 4 there are three conservation numbers at each position, C_A, C_B, and C_C. We set a threshold for conservation T. Subgroups at a position that score $\geq T$ are regarded as conserved, those that score $< T$ are unconserved.

3. We now compare all pairs of subgroups at each position and for each pair calculate the conservation number for the combined pair (e.g., C_{A+B}). The most interesting positions are those that are conserved in one or more subgroups. For example, for subgroups A and B, then if $C_A \geq T$ and $C_B \geq T$ the position in both subgroups is conserved. When the groups are

flavodoxin

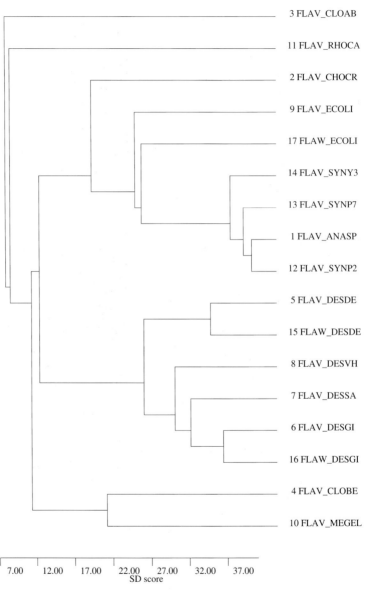

FIG. 3. Dendrogram or tree calculated by single linkage cluster analysis from pairwise comparisons of 17 flavodoxin sequences. The horizontal axis is calibrated in standard deviations (SD) from the mean of scores for randomly shuffled sequences of the same length and composition (see Ref. 3 for an explanation of SD scores).

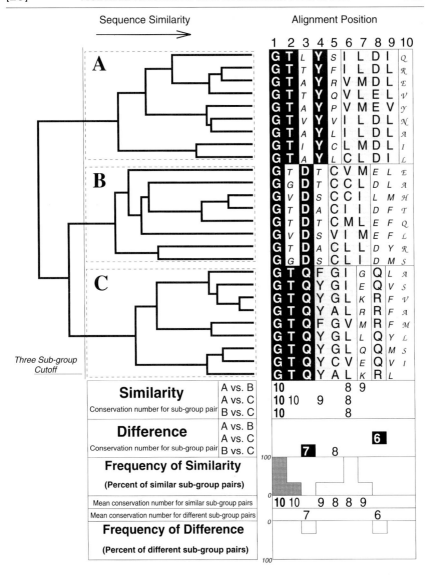

FIG. 4. Illustration of hierarchical conservation analysis. See text for explanation.

joined, the conservation number C_{A+B} may (a) stay the same, (b) be $< min(C_A, C_B)$, or (c) go below the threshold T. If it stays the same (a), then the pair is said to be similar; if (b) then the subgroups are different but conserved when together, and if (c) the subgroups are different but unconserved when together. Of these possibilities (c) is perhaps the most interesting since it points to a position that shows conserved physicochemical properties in each subgroup, but the properties conserved are different. Such differences may indicate a site of functional importance.

Figure 4 shows a stylized output from the program AMAS (analysis of multiply aligned sequences) that implements these ideas.[5] The sequence alignment has been shaded to illustrate similarities within each subgroup. Below the alignment, the lines for similarity show the conservation values obtained when each pair of subgroups is combined and the combined conservation number is not $< T$. For example, at position 7 subgroups A and B combine with a conservation number of 9 ($T = 8$). The lines for difference illustrate positions at which combination of subgroups lowers the conservation number $< T$. For example, at position 3, there is an identity in subgroup B and one in C, but when the groups are combined, the identity is lost and the conservation drops below T to a value of 7.

A summary of the similarities and differences is given as a frequency histogram. Each upward bar represents the proportion of subgroup pairs that preserve conservation, while each downward bar shows the percentage of differences. For example, at position 6 3/3 pairs are conserved (100%), whereas at positions 3 and 8 1/3 pairs show differences (33%). With a large alignment, the histogram can quickly draw the eye to regions that are highly conserved, or where there are differences in conserved physicochemical properties.

Figure 5 illustrates an alignment for the 17 flavodoxin sequences compared pairwise in Fig. 3. Although tertiary structures are known for at least four of the proteins, this information was not used when generating the alignment. Following from Fig. 3, five subgroups have been separated by horizontal lines. The shading on the alignment shows identities within each subgroup (white on black), conserved positions with each subgroup (black on gray), and unconserved positions (black on white). The AMAS program used to generate Fig. 5 also permits color highlighting. Color gives a clearer representation of the subgroup similarities and also allows identities across the complete alignment to be highlighted in a different color. The similarity/ difference histogram is below the alignment, as in Fig. 4, except that average conservation values are also printed below and above the bars. The four

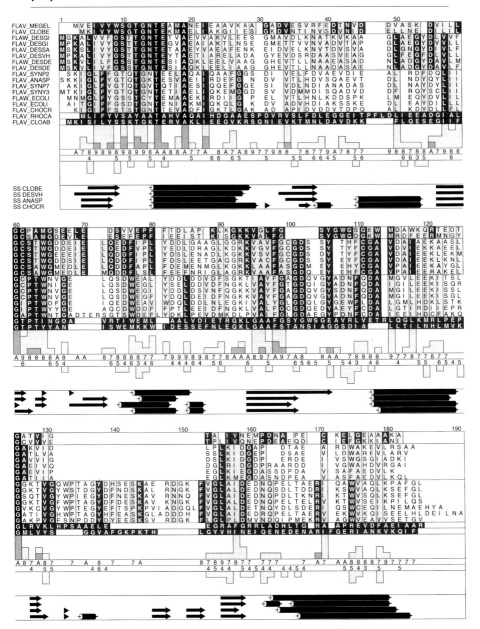

FIG. 5. Conservation analysis of the 17 flavodoxin sequences clustered in Fig. 3. The Taylor Venn diagram was used (Fig. 1) with a threshold of $T = 7$. See text for details.

lines of cylinders and arrows illustrate the known secondary structure for four of the protein sequences (taken from the SWISS-PROT feature tables).

Similarity Histogram Interpretation

The similarity histogram gives qualitatively similar information to conservation calculated across all sequences in the alignment (e.g., Fig. 2). However, unlike conservation across the alignment, the histogram will tolerate one subgroup having very different properties from the rest. Examination of the similarity histogram for the flavodoxin alignment shows a number of regions with conservation patterns characteristic of regular secondary structure. The classical patterns one would expect are (1) short runs of conserved hydrophobics–buried β strand; (2) $i, i + 2, i + 4$ pattern of conserved hydrophobics–surface β strand (alternating pattern); and (3) $i, i + 3, i + 4, i + 7$ pattern of conserved residues–α helix. Variations on the basic α-helix pattern are common, for example, $i, i + 4, i + 7$ or similar.

Examination of the similarity histogram in Fig. 5 shows conservation patterns characteristic of buried β strand at positions 6–8, 57–64, and 95–99. A helical conservation pattern is seen at positions 15–26, where conserved residues at positions 15, 19, 22, and 26 would lie on the same face of the helix. A further pattern is seen at positions 115–122 with clear conservation at 115 and 118 and weaker but still predominantly hydrophobic conservation at 122.

Examination of conservation patterns can often suggest where small errors in the sequence alignment exist. For example, there is a helical pattern starting at conserved position 77. Position $77 + 3$ is also conserved, but continuation of the pattern is masked by the insertion of Leu in one sequence at position 81. Closing this gap would allow the conservation pattern to continue at 82 and 85. A further error is seen at position 172 where one sequence has an inserted Phe. Closing this gap would clarify the $i, i + 3, i + 4, i + 7$ pattern at positions 171 (which would become 172), 175, 176, and 179.

Interpretation of Differences Histogram

The differences histogram in Fig. 5 highlights positions that are conserved within subgroups but where the conserved properties in each subgroup are different. For example, at position 9 Trp is conserved in subgroup 1, Gly in subgroup 2, and Gly/Ser in subgroup 3. Subgroups 4 and 5 have Val and Ser, respectively. This is an interesting difference in size as Trp is the largest amino acid and Gly the smallest. Inspection of protein three-dimensional structures representative of the two subgroups reveals that the protein secondary structures in subgroup 1 have shifted relative to subgroup 2 to accommodate the larger Trp residue.

A further difference is seen at position 157. In subgroup 2 K/R are conserved, while in subgroup 3 A/P is conserved. The protein three-dimensional structures show that in proteins in subgroups 1 and 2 this residue (K/R) points out into solvent. In subgroup 3 the residue packs against the inserted loop from positions 132 to 153. In this alignment there are many other positions at which differences in properties are observed between subgroups.

In general, if no tertiary structures are known for the protein family, then differences identified in this way may point to important structural or functional sites that can be investigated further by mutagenesis. If the structures of one or more members of the family are known, then the differences analysis may highlight positions that can be explained by their structural context.

Alternative Amino Acid Classifications: AMAS Program

The Taylor Venn diagram (Fig. 1) is a convenient general classification of amino acid properties. However, other classifications may exploit the same set formalism. Figure 6 illustrates four alternative set descriptions.

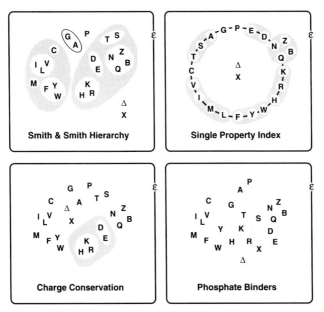

FIG. 6. Alternatives to the Taylor Venn diagram shown in Fig. 1.

The Smith and Smith diagram is quite similar to the Taylor Venn diagram, but derived from an amino acid hierarchy.[8] The single property index, again from the work of Taylor[13] permits each amino acid to have only one property (hydrophobic, charge, etc). Sometimes it is useful to examine specific subsets of properties. For example, one can highlight similarities and differences in charge by considering only the properties charge, positive, and negative. This approach is illustrated in Fig. 7 for the annexins. Annexins have four sequence repeats in a single chain.[15] Analysis of an alignment of repeats using the charge diagram reveals a conserved Glu in repeats 2 changing to conserved Arg in repeats 4, at position 31, but no conservation at the equivalent position in repeats 1 or 3. This suggests an important salt bridge between repeats 2 and 4. The X-ray structure of annexin V[16] confirms this prediction.

Analysis of phosphate binding sites in proteins has shown which residues are most frequently found binding to phosphate.[17] The phosphate binders set may be used to identify potential phosphate binding residues from multiple alignment.

Protein Secondary Structure Prediction from Multiple Alignment

One of the most successful applications of multiple sequence alignment has been to improve the accuracy of secondary structure prediction. In 1987, Zvelebil *et al.*[14] showed a 9% improvement in three-state (α helix, β strand, and coil) prediction accuracy through the use of multiple alignments. A number of groups have exploited multiple alignments either for blind predictions of structure for specific protein families (e.g., Refs. 15 and 18–21) or to develop general methods for secondary structure prediction on proteins of known structure (e.g., Refs. 22–24). In most examples, the overall accuracy of prediction has been around 70% which is close to the minimum level of agreement observed for secondary structures of two members of the same protein family when their three-dimensional structures are both known.[25] Together these results show that secondary struc-

[16] R. Huber, J. Romsich, and E.-P. Paques, *EMBO J.* **9**, 3867 (1990).
[17] R. R. Copley and G. J. Barton, *J. Mol. Biol.* **242**, 321 (1994).
[18] I. P. Crawford, T. Niermann, and K. Kirchner, *Proteins: Struct. Funct. Genet.* **2**, 118 (1987).
[19] R. B. Russell, J. Breed, and G. J. Barton, *FEBS Lett.* **304**, 15 (1992).
[20] S. A. Benner and D. Gerloff, *Adv. Enzyme Regul.* **31**, 121 (1990).
[21] C. D. Livingstone and G. J. Barton, *Int. J. Pept. Protein Res.* **44**, 239 (1994).
[22] J. M. Levin, S. P. A. Pascarella, P. Argos, and J. Garnier, *Protein Eng.* **6**, 849 (1993).
[23] A. A. Salamov and V. V. Solovyev, *J. Mol. Biol.* **247**, 11 (1995).
[24] B. Rost and C. Sander, *J. Mol. Biol.* **232**, 584 (1993).
[25] R. B. Russell and G. J. Barton, *J. Mol. Biol.* **234**, 951 (1993).

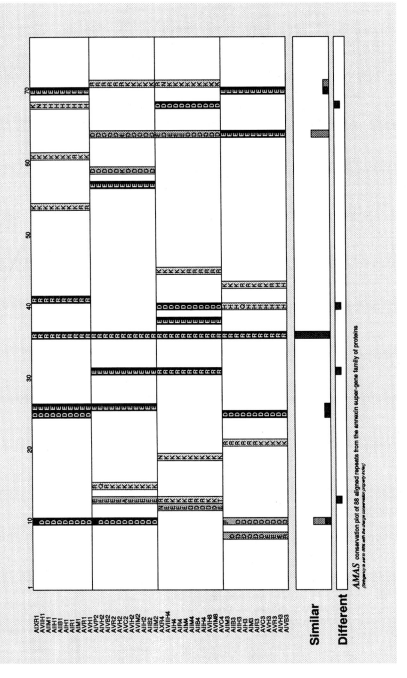

FIG. 7. Conservation analysis of annexin repeats using the charge diagram shown in Fig. 6. See text for discussion.

ture prediction from multiple alignment is significantly better than from a single sequence. The predictions are better, not only in raw percentage accuracy terms, but also in terms of the usefulness of the predictions since reliability can be assigned to each region. As shown in the next section, simply by considering conserved patterns in the alignment, it is possible accurately to assign confidence to each predicted region.

Experiences of Six Blind Predictions

Here we summarize the results of six secondary structure predictions that were made prior to any knowledge of the protein three-dimensional structure. The structures have now been determined for representatives of each family, so it is possible for us to evaluate the success of the predictions. The predictions are for the protein tyrosine phosphatase family, blood coagulation factor XIII, glucose-6-phosphate dehydrogenase, urease β and γ subunits, and synaptotagmin C_2. Having obtained all available sequences for each family, each prediction was performed by applying the following protocol.

1. Predict secondary structure for each sequence by the methods of Lim,[26] Garnier et al.,[27] and Chou and Fasman.[28] Predict turns by the methods of Rose[29] and Wilmot and Thornton.[30] Combine the results of the predictions at each position in the alignment. In addition, predict the secondary structure by the method of Zvelebil et al.[14]

2. Calculate conservation values for all positions in the alignment and identify patterns characteristic of α helix and β strand (see section on similarity histogram interpretation).

3. Predict secondary structure by combining the conservation pattern analysis and the results of the classic predictions. First, potential loops are assigned as regions where insertions/deletions are common in the alignment. The remainder of the alignment is predicted as potential core secondary structure. Accordingly, each region is examined for the presence of strong agreement between the secondary structure prediction methods and clear presence of a conservation pattern.

4. Rank the assignments of secondary structure according to the following confidence rules: predicted loop with insertions/deletions > loop containing conserved Gly or Pro and/or high polar amino acid content >

[26] V. Lim, J. Mol. Biol. **88,** 873 (1974).

[27] J. Garnier, D. J. Osguthorpe, and B. Robson, J. Mol. Biol. **120,** 97 (1978).

[28] P. Y. Chou and G. D. Fasman, Adv. Enzymol. **47,** 45 (1978).

[29] G. D. Rose, Nature (London) **272,** 586 (1978).

[30] A. C. M. Wilmot and J. M. Thornton, J. Mol. Biol. **203,** 221 (1988).

TABLE I
SECONDARY STRUCTURE PREDICTION ACCURACY[a]

Confidence	Number of residues	Mean residue-by-residue accuracy
High	569 (32.9%)	87.2%
Medium	748 (43.3%)	70.8%
Low	411 (23.8%)	49.0%
	1728	Mean: 71.4%

[a] For regions assigned with high, medium, and low confidence in six blind secondary structure predictions.

surface helix with characteristic patterns of conservation > surface strand with clear amphipathicity > buried strand (short, highly conserved hydrophobic run) > helix/strand with no associated conservation pattern. Each predicted secondary structure was assigned confidence of high, medium, or low by considering the local environment of each potential secondary structure.

The results over the six protein predictions are summarized in Table I. Approximately 33% of the residues were assigned to a secondary structure with high confidence, and of these the residue-by-residue accuracy was 87.2%. Those regions assigned with low confidence gave 49% accuracy, which is better than the random expected value of about 40%.

Evaluation of Conservation Patterns for Prediction

Although the sample size is small, it is useful to examine how frequently the conservation patterns expected to be associated with helix and strand

TABLE II
LEVEL OF AGREEMENT BETWEEN CONSERVATION PATTERNS[a]

	Number observed in:		
Predicted secondary structure type	Helix	Strand	Coil
Strand with clear alternating pattern of conserved and unconserved residues	0	17	0
Strand in continuous run of conserved hydrophobic residues	4	11	2
Helix with characteristic conservation patterns	22	1	1

[a] The predicted secondary structures are compared to the actual secondary structures observed in regions displaying the described patterns.

actually were found in the appropriate secondary structure. A summary of the patterns that contributed to the most confidently predicted regions is given in Table II. Clearly, if an alternating pattern of conservation is seen, then it is strong evidence for a β strand. Similarly, the helical patterns, when clear, are strong indicators of α helix. However, runs of conserved hydrophobic amino acids can be misleading. Of the 17 examples predicted as strand, 4 were actually in helix and 2 in coil.

To see whether these observations hold in general, we have analyzed a set of 78 multiple alignments for proteins where at least one is of known three-dimensional structure. The data confirm that clear helical conservation patterns in the set are diagnostic of helix; likewise, alternating conservation patterns are nearly always associated with strand. Interestingly, the more problematic runs of conserved hydrophobics are also well correlated with strand except when the conservation extends over more than seven residues. These longer runs of conserved residues are often associated with α helix.[31]

Implementation and Availability of Programs

The set-based analysis of multiple sequence alignments has been coded in the program AMAS. AMAS is written in C language, runs under UNIX, and is available by ftp from geoff.biop.ox.ac.uk (see the README file on the server for instructions). Alternatively, AMAS may be run via the World Wide Web by opening the URL http://geoff.biop.ox.ac.uk/servers/amas_server.html. The program produces a textual summary of the subgroups and physicochemical properties at each position in the alignment as well as a boxed and shaded (or colored) alignment in PostScript that may be printed.

Acknowledgment

We thank Dr. Rob. Russell for numerous helpful comments on secondary structure prediction and programming, and Prof. L. N. Johnson for support. G.J.B. thanks the Royal Society for the support of a University Research Fellowship. C.D.L. thanks the Medical Research Council for the support of a Research Studentship.

[31] C. D. Livingstone, Ph.D. Thesis, University of Oxford, Oxford (1995).

[30] Prediction and Analysis of Coiled-Coil Structures

By ANDREI LUPAS

Introduction

A coiled coil is a bundle of several helices (typically two or three) that interact at an angle of approximately $+20°$, and assume a side-chain packing geometry termed knobs-into-holes[1] in which, at the helix interface, a side chain of one helix (knob) packs into a space surrounded by four side chains of the facing helix (hole). The precise register of knobs and holes is achieved by giving the right-handed coiled-coil helices a left-handed twist (or supercoil) and thus reducing their periodicity from approximately 3.6 residues per turn to 3.5. The net result of this supercoiling is that every seventh residue occupies an equivalent position on the helix surface, and therefore the sequence shows a heptad periodicity in the chemical nature of the side chains. In a schematic notation, in which the seven different structural positions are named a–g, positions a and d are occupied by hydrophobic residues and form the helix interface, while positions b, c, e, f, and g are hydrophilic and form the solvent-exposed part of the helix surface.

This sequence periodicity allows the detection of coiled-coil segments by simple inspection,[2] although such an approach generally requires several successive heptads (5–10) in order to establish that the observed pattern is actually significant. Even then, the interpretation is strongly influenced by individual bias, particularly when the sequence contains skips or stutters in the heptad pattern. More recently, the approach by inspection has been further complicated by the discovery that many coiled-coil segments in otherwise globular proteins are very short (typically four or less heptads). This has led to a large number of coiled-coil sightings, many of which have not held up to closer scrutiny. The need for a quantitative method, independent of individual bias and giving clearer results than mere sequence inspection, led to the development of COILS, a statistically controlled, profile-based method for the prediction and analysis of coiled coils.[3,4]

[1] F. H. C. Crick, *Acta Crystallogr.* **6**, 689 (1953).
[2] C. Cohen and D. A. D. Parry, *Trends Biochem. Sci.* **11**, 245 (1986).
[3] D. A. D. Parry, *Biosci. Rep.* **2**, 1017 (1982).
[4] A. Lupas, M. Van Dyke, and J. Stock, *Science* **252**, 1162 (1991).

METHODS IN ENZYMOLOGY, VOL. 266

Copyright © 1996 by Academic Press, Inc.
All rights of reproduction in any form reserved.

COILS Program

In 1982, Parry[3] derived a coiled-coil profile from the sequences of nematode myosin, rabbit-skeletal tropomyosin, influenza hemagglutinin, and sheep keratin (1725 residues total) and presented a scheme whereby the compatibility of a sequence with the profile could be calculated by geometric averaging over the region of putative heptads. His article represents one of the first applications of a sequence profile to structure prediction and, although the scoring scheme was modified substantially in developing COILS, its main concepts form the basis of the program.

Calculating Scores

At the center of the method lies the profile, or scoring matrix, with which a given sequence is compared. It is generated by tabulating the residue frequency of each residue at each of the seven heptad positions in a database of coiled-coil sequences,[3–8] and it is either normalized so that the sum of the frequencies at any of the seven positions is unity[3,5–7] or is divided by the frequency of occurrence of residues in GenBank to yield the preference of each residue for each position.[4,8] Such matrices have been compiled for two-[4,5,8] and three-stranded coiled coils[6] and for four-helix bundles.[7] COILS currently utilizes two matrices, both of which have been derived from databases of two-stranded coiled coils. The first matrix[4] is based on a database of 16,968 residues from myosins, tropomyosins, and keratins (type I and II intermediate filaments); we refer to it as the MTK matrix. The second, more recent matrix[8] is based on 26,965 residues and also includes other intermediate filaments (types III to V), desmosomal proteins, and kinesins; we refer to it as the MTIDK matrix. The possibility of including scoring matrices specific for three- and four-stranded structures into COILS is discussed in the section on COILS applications.

The similarity between a protein sequence and the scoring matrix is computed residue by residue using a gliding window. For a given residue, the window is placed such that the residue is in the last position and then moved progressively until the residue is in the first position. For each position of the window, the sequence can be read in seven different frames, and therefore seven different scores must be calculated. Each score is obtained by assigning the residues in the window their frequency from the scoring matrix (e.g., using the MTK matrix, Leu in frame $b = 0.297$) and

[5] J. F. Conway and D. A. D. Parry, *Int. J. Biol. Macromol.* **12,** 328 (1990).
[6] J. F. Conway and D. A. D. Parry, *Int. J. Biol. Macromol.* **13,** 14 (1991).
[7] C. D. Paliakasis and M. Kokkinidis, *Protein Eng.* **5,** 739 (1992).
[8] A. Lupas, COILS documentation, Coils/vms folder at FTP.BIOCHEM.MPG.DE.

taking the geometric average over the window (i.e., for a window of size N, multiplying N frequencies and taking the Nth root). After all possible scores for a residue have been calculated by assuming seven frames for all positions of the window, the maximum score and the frame it was obtained in are assigned to the residue. The algorithm then moves on to the next residue and repeats the process. The gliding window ensures that even comparatively short coiled-coil segments will be detected in the larger context of globular proteins, while the score maximization for each residue allows a fairly accurate determination of the ends of such segments. The geometric rather than arithmetic average (summing over N frequencies and dividing by N) is necessary in order to give each residue the same weight. For example, when averaging a residue that is 10 times as frequent as expected with another residue 10 times less frequent than expected, the result should be 1 (geometric average) rather than 5.05 (arithmetic average).

In this scoring scheme, each of the seven heptad positions is given equal weight. Because coiled coils are generally fibrous, solvent-exposed structures and all but the internal a and d positions have a high likelihood of being occupied by hydrophilic residues, the program is biased toward hydrophilic, charge-rich sequences. This occasionally leads to highly charged sequences (e.g., polyglutamate) obtaining high coiled-coil scores in the obvious absence of a heptad periodicity. For this reason, COILS offers the option of giving the two hydrophobic positions a and d the same weight as the five hydrophilic positions b, c, e, f, and g. This is achieved by raising the frequencies of residues in a and d to the power 2.5 and by taking the tenth root for each heptad. This feature permits the rapid identification of segments whose high scores are not due to a heptad periodicity but to a high incidence of charged residues, since after increasing the weight of positions a and d, the scores for these segments decrease strongly.[8]

From Scores to Probabilities

The significance of scores is established by comparison with the score distributions in globular and coiled-coil proteins. To this end, score distributions were compiled for globular proteins of known structure and for a database of coiled coils, were approximated by Gaussian curves, and were scaled against GenBank.[4,8] From these data, the probability $P(S)$ that a residue with score S is part of a coiled coil can be calculated using the formula

$$P(S) = \frac{G_{cc}(S)}{[RG_{g}(S) + G_{cc}(S)]} \tag{1}$$

where $G_{cc}(S)$ and $G_g(S)$ are the values of the Gaussian curves for coiled coils (cc) and globular proteins (g) at score S, and R is the estimated ratio of globular to coiled-coil residues in GenBank. Because the distribution of scores changes with choice of the matrix, of window size, and of weighting, statistics were performed separately for each option using window sizes of 14, 21, and 28 residues. The statistical parameters of the resulting score distributions are given in the COILS documentation.[8] A comparison of these parameters yields the following conclusions. (i) The MTIDK matrix provides for a somewhat better resolution between the score distributions of globular and coiled-coil proteins than the MTK matrix. Also, the MTIDK matrix provides for a more consistent evaluation of the different protein families forming the coiled-coil database. (ii) Nevertheless, the differences between the two matrices are small, indicating that little further progress can be expected from the compilation of even larger databases of two-stranded coiled coils. (iii) With both matrices, weighting slightly decreases the resolution between globular proteins and coiled coils. (iv) For all scoring methods, the resolution between globular proteins and coiled coils decreases strongly with decreasing size of the scanning window.

A scan of coiled-coil proteins whose structures are known to atomic resolution (Fig. 1) shows that COILS is accurate in the analysis of parallel and antiparallel two-stranded structures and of parallel three-stranded structures, but generally does not detect all helices in antiparallel structures containing three or more helices (helical bundles). (Note that only a small subset of helical bundles can be considered to form coiled coils. Most helical bundles do not show knobs-into-holes packing and have helical crossing angles other than $+20°$.) Several reasons are responsible for this decrease in accuracy. (i) The helical surface buried in the core of coiled coils increases with the number of helices, leading to a broadening of the hydrophobic stripe. For this reason, helices from helical bundles do not fit as well the sequence consensus of two-stranded coiled coils. (ii) Helices in helical bundles are generally not supercoiled, leading to distortions in the knobs-into-holes packing (termed x and da layers) that are accompanied by frame breaks in the pattern of hydrophobic residues.[9] These frame breaks lead to a drop in scores, because COILS looks for a continuous frame. (iii) Helices in helical bundles occasionally assume the ridges-into-grooves packing[10] normally seen in globular proteins (e.g., spectrin). This packing mode leads to changes in the hydrophobic pattern that are not always consistent with the heptad pattern scored by COILS.

[9] A. Lupas, S. Müller, K. Goldie, A. M. Engel, A. Engel, and W. Baumeister, *J. Mol. Biol.* **248**, 180 (1995).
[10] C. Chothia, *Annu. Rev. Biochem.* **53**, 537 (1984).

parallel two-stranded coiled coils

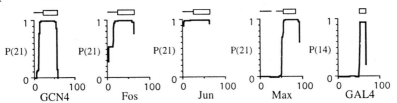

antiparallel two-stranded coiled coils

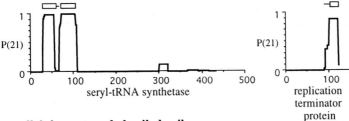

parallel three-stranded coiled coils

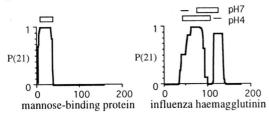

antiparallel helical bundles

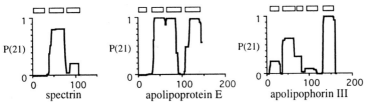

FIG. 1. COILS probabilities for coiled-coil proteins of known structure, obtained with the unweighted MTIDK matrix at a window size of 21 residues (14 residues for GAL4 as this coiled coil has a length of only 15 residues). Above the plots is shown the loctation of coiled-coil helices (open boxes) and of helical segments that interact with or extend the coiled-coil helices (lines).

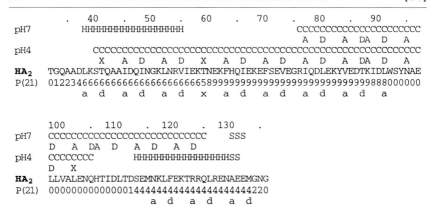

```
           .    40    .    50    .    60    .    70    .    80    .    90     .
pH7             HHHHHHHHHHHHHHHHHHH                   CCCCCCCCCCCCCCCCCCCCCCCCC
                                                      A   D    A   DA   D    A
pH4             CCCCCCCCCCCCCCCCCCCCCCCCCCCCCCCCCCCCCCCCCCCCCCCCCCCCCCCCCCCCCCCCC
                X    A   D    A   D    X    A   D    A   D    A   D    A   DA   D    A
HA₂    TGQAADLKSTQAAIDQINGKLNRVIEKTNEKFHQIEKEFSEVEGRIQDLEKYVEDTKIDLWSYNAE
P(21)  0122346666666666666666666666665899999999999999999999999999999888000000
                a    d    a    d    a    d    x    a    d    a    d    a    d    a
```

```
       100     .    110    .    120    .    130    .
pH7    CCCCCCCCCCCCCCCCCCCCCCCCCCCCCCC      SSS
       D    A   DA   D    A    D    A   D
pH4    CCCCCCCC           HHHHHHHHHHHHHHHHHHHSS
       D        X
HA₂    LLVALENQHTIDLTDSEMNKLFEKTRRQLRENAEEMGNG
P(21)  00000000000000014444444444444444444444220
                     a    d    a    d    a    d
```

Fig. 2. Summary of the COILS analysis for the influenza hemagglutinin rod segment (HA₂). Above the sequence are shown the structures at pH 7 and pH 4 (H, helix; C, coiled-coil helix). Beneath the sequence are shown the COILS probabilities abbreviated to the first digit as computed with an unweighted MTK matrix and a window size of 21 residues.

A scan of non-coiled-coil proteins from PDB (database of Fall 1993) shows that, occasionally, helices that do not fold into a coiled coil can obtain high coiled-coil probabilities. At probabilities exceeding 90% in a 28-residue scan, these are the N-terminal helix of λ repressor for the MTK matrix, a helix from pig lactate dehydrogenase for the weighted MTK matrix, helices from Hsc70, yeast triose phosphate isomerase, and *Bacillus stearothermophilus* tyrosyl-tRNA synthetase for the MTIDK matrix, and *B. stearothermophilus* tyrosyl-tRNA synthetase for the weighted MTIDK matrix. The occurrence of such high-scoring segments in globular proteins is entirely expected since, for example, at a probability of 90%, one of ten sequences should not fold into a coiled coil if COILS operates correctly. In addition, it cannot be excluded that these high-scoring segments in fact have coiled-coil potential and would form coiled coils under different ionic conditions or in a different protein context. As discussed below for influenza hemagglutinin, even judgments based on crystal structures can be misleading if the potential of these structures for dynamic rearrangement is ignored (Fig. 2).

Applications

COILS was initially developed for the analysis of chemotaxis receptors,[11] whose methylation sites show a heptad spacing. The discovery that the two regions of methylated residues have a high coiled-coil forming

[11] A. Lupas, Ph.D. Thesis, Princeton University, Princeton, New Jersey (1990).

potential[4,11] prompted a model whereby adaptation of the receptors to ligand binding involves an equilibrium between a four-helix bundle and two antiparallel coiled coils in the receptor dimer.[11,12] In this model, the transition between the two structures is regulated by the masking and unmasking of negative charges near the helix interface via reversible carboxymethylation. Since the proposal of this model, Krylov et al.[13] showed that the oligomer state of a coiled-coil structure (the leucine zipper of vitellogenin binding protein (VBP)) can indeed change from dimer to tetramer by an increase in hydrophobicity at a position flanking the helix interface.

The COILS program has been widely used both for predicting new coiled coils and for analyzing known structures. Its most noted success was the prediction that a loop connecting two coiled-coil helices in influenza hemagglutinin at pH 7 would itself assume a coiled-coil structure at pH 4,[14] uniting the two helices into a continuous rod that would propel the viral fusion peptide from close to the viral membrane into the cellular membrane. The crystal structure of influenza hemagglutinin at pH 4[15] demonstrated the correctness of the prediction but revealed an even more fundamental rearrangement of the structure, including the formation of a turn in the previously continuous coiled-coil rod. As can be seen from Fig. 2, the COILS analysis of hemagglutinin correctly anticipated many features of the pH 4 structure, including the newly formed turn and an x layer in the core of the coiled coil.

In its current stage of development, COILS does not give any direct information on the oligomer structure of the potential coiled-coil segments it identifies. In principle, such information could be obtained by comparing the sequence of these segments to matrices specific for two-, three-, or four-stranded structures. The matrix yielding the highest score would implicitly be the one most specific for the analyzed sequence. An implementation of this scheme[8] using available matrices[6–8] has yielded disappointing results, indicating that the available matrices for three- and four-stranded structures (which were compiled from comparatively small databases) may not be sufficiently accurate.

Recommendations for Application of COILS to Sequences of Unknown Structure

The COILS program is specific for solvent-exposed, left-handed coiled coils. Other types of coiled-coil structures, such as buried coiled coils (e.g.,

[12] J. B. Stock, G. S. Lukat, and A. M. Stock, *Annu. Rev. Biophys. Biophys. Chem.* **20,** 109 (1991).
[13] D. Krylov, I. Michailenko, and C. Vinson, *EMBO J.* **13,** 2849 (1994).
[14] C. M. Carr and P. S. Kim, *Cell (Cambridge, Mass.)* **73,** 823 (1993).
[15] P. A. Bullough, F. M. Hughson, J. J. Skehel, and D. C. Wiley, *Nature (London)* **371,** 37 (1994).

the central coiled coil in catabolite repressor protein, or some transmembrane domains) and right-handed coiled coils, are not detected by the program.

The program does not reach yes-or-no decisions based on a threshold value. Rather, it yields a set of probabilities that presumably reflect the coiled-coil forming potential of a sequence. This means that even at high probabilities (e.g., >90%), there will be (and should be) sequences that in fact do not form a coiled coil, though they may have the potential to do so in a different context.

The COILS program is biased toward hydrophilic, highly charged sequences. For this reason, all scans should be performed with a weighted and an unweighted matrix, and the results compared. Differences of more than 20–30 percentage points in the probabilities should be taken to indicate that a coiled-coil structure is unlikely, the elevated scores being mainly due to the high incidence of charged residues.

The MTK and MTIDK matrices both assign high probabilities to known coiled-coil segments, but they identify different helices at high probability in a database of globular proteins. This is a surprising feature, the reason for which is as yet unclear, but it can be exploited for predictive purposes. It is therefore useful to compare the results of scans made with the two matrices. Again, differences of more than 20–30 percentage points in the probabilities should be taken to indicate that a coiled-coil structure is unlikely.

The resolution between globular and coiled-coil score distributions decreases strongly with a decreasing size of the scanning window. The prediction of new coiled-coil segments should therefore be made using a 28-residue window, or in special cases a 21-residue window. Smaller 14-residue windows should normally be reserved for the analysis of local parameters (such as the frame) in known or predicted coiled coils.

The ends of coiled-coil segments appear to be most accurately identified in a 21-residue window. In general, we assume that residues with probabilities greater than 50% are part of the coiled-coil segment. In addition, a search for the most likely helix ends using the residue frequencies compiled by Richardson and Richardson[16] is recommended. The VMS release of COILS contains an auxiliary program, CAPS, that was written for this purpose.

Sequences with high coiled-coil probability from globular proteins rarely exceed a length of 30 residues. None is longer than 35 residues. Sequences with probabilities exceeding 80–90% that extend for more than 35 residues

---

[16] J. S. Richardson and D. C. Richardson, *Science* **240,** 1648 (1988).

are therefore more likely to assume a coiled-coil structure than is indicated by the obtained probabilities.

Where possible, sequences related to the protein of interest should also be analyzed for predicted coiled-coil segments (see section on the ALIGNED programs). It should be kept in mind, though, that the sequences must be related in the region of high scores in order for the comparison to be significant. In addition, comparison of the coiled-coil prediction with predictions of the secondary structure are generally useful, particularly if multiple related sequences are available.

Example: Cartilage Matrix Protein

As an example for the application of COILS, we discuss here briefly the coiled-coil prediction for cartilage matrix protein (CMP) (Fig. 3). This protein is a major component of the extracellular matrix of nonarticular cartilage, forms homotrimers, and has been sequenced in chicken and humans (SWISS-PROT accession names CAMA_CHICK and CAMA_HUMAN). The sequence contains two domains related to von Willebrand factor type A repeats, separated by a domain related to epidermal growth factor (EGF), and ends in a region of 40 residues without obvious similarity to other proteins. A COILS scan of human CMP shows that this region has an elevated coiled-coil forming potential. Coiled-coil probabilities in excess of 90% are obtained in a 28-residue window with both scoring matrices, weighted and unweighted, indicating strongly that the region does in fact form a coiled coil. The same results are obtained

cartilage matrix protein (CMP) domain structure

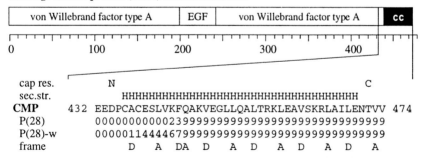

FIG. 3. Analysis of a putative coiled-coil region in cartilage matrix protein. The most likely N- and C-terminal cap residues were identified using CAPS. The secondary structure prediction was obtained from the PHD server. The COILS probabilities (abbreviated to the first digit) were computed with the MTIDK matrix, unweighted and weighted, at a window size of 28 residues.

with the chicken sequence, which is 60% identical to the human sequence in this region.

A secondary structure prediction for the aligned chicken and human sequences by the PHD server[17] shows that the C-terminal region is likely to form an α helix, but that the helix is probably longer by another heptad at the N-terminal end, a prediction which is supported by a search for the most likely N- and C-cap residues of the helix using the frequencies compiled by Richardson and Richardson. The 14-residue scans of the human sequence show a frame break at the beginning of the region of high scores that corresponds to a *da* layer,[9] such as is also found at the ends of two other trimeric coiled coils, namely, human mannose-binding protein and influenza hemagglutinin. This *da* layer gives a plausible reason for why the high scores do not extend over the entire length of the predicted helix.

In conclusion, CMP most likely contains a three-stranded coiled coil extending from Asp^{434} to the C-terminal end of the protein and containing a *da* layer at Phe^{444}–Gln^{445}. The presence of two closely spaced cysteines at the beginning of the coiled coil may indicate formation of a disulfide ring, as in the trimeric coiled-coil proteins fibrinogen, laminin, and thrombospondin.

ALIGNED Programs

As mentioned in the example above, it is useful to check whether coiled-coil regions predicted in a protein also appear in the same place in related proteins. For larger protein families, such prediction matching is tedious. We have therefore developed two programs, ALIGNED20 and ALIGNED80, that scan multiple alignments for predicted coiled-coil regions. The programs differ only in the size of the accepted alignments (up to 20 sequences in ALIGNED20 and up to 80 sequences in ALIGNED80) for reasons related to the output format.

Calculating Consensus Probabilities

The ALIGNED programs analyze each of the sequences in an alignment using the COILS algorithm and write an output file summarizing the predictions. In addition, they compute a consensus probability for each position in the alignment by arithmetic averaging over individual probabilities, after these have been weighted for the divergence of the respective sequence from the other sequences in the alignment. The weighting occurs by multiplication of probability and divergence. The divergence is calculated as the

[17] B. Rost, C. Sander, and R. Schneider, *CABIOS* **10**, 53 (1994).

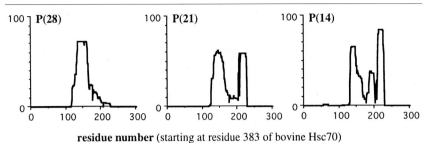

residue number (starting at residue 383 of bovine Hsc70)

FIG. 4. ALIGNED scan of the C-terminal domain of Hsp70 proteins. The scans were made at window sizes of 28, 21, and 14 residues using the unweighted MTK matrix.

inverse of the arithmetic average of the pairwise identities between the respective sequence and the other sequences in the alignment. Although this weighting counterbalances the influence of highly similar sequences within the alignment, our experience has been that consensus probabilities can still be biased if a large number of nearly identical sequences are present. For this reason we recommend that, for the calculation of consensus scores, alignments be purged of sequences that are more than 90% identical to a sequence already present in the alignment.

As has been discussed, COILS is accurate in detecting solvent-exposed two- and three-stranded coiled coils but is less accurate in the analysis of helical bundles. In such cases, it often only identifies one of several helices, but frequently not the same one in related sequences. This effect makes ALIGNED a very useful tool in analyzing protein families in which predicted coiled-coil segments are in fact most likely part of a helical bundle. For such families, the consensus prediction yields multiple, closely spaced, short peaks with probabilities in the 30 to 90% range.

Example: Heat-Shock Protein Hsp70

As an example for the application of ALIGNED, we discuss briefly the prediction for Hsp70 proteins (Fig. 4), for which we reported previously the presence of a C-terminal coiled-coil region.[4] Hsp70 heat-shock proteins have been sequenced in a large number of organisms and are therefore particularly well suited for a consensus prediction. For the purpose of this analysis we generated an alignment largely by automated means but would like to point out that, as in secondary structure predictions, the accuracy of ALIGNED is dependent on the correctness of the input alignment and that it is therefore useful to optimize alignments prior to analysis, for example, using MACAW.[18] For the present analysis, we extracted all Hs7*,

[18] G. D. Schuler, S. F. Altschul, and D. J. Lipman, *Proteins* **9,** 180 (1991).

Gr78, and Dnak entries from SWISS-PROT, eliminated the divergent Hsp110 subfamily, and kept only one member of each group of sequences that were more than 90% identical, ending with a list of 68 sequences. These we aligned in PileUp[19] using an increased gap penalty of 5. Analysis of the alignment with ALIGNED80 showed that, as the scanning window was decreased from 28 to 14, the initially detected first peak was joined by a second and then by a third, indicating the presence of a helical bundle (Fig. 4). Note that if these peaks were part of a segmented coiled coil, such as in intermediate filaments, the probabilities would be in the 90% range and the peaks would be longer. The location of the three peaks is matched by three helices predicted using the PHD server.[17] In conclusion, the analysis indicates the presence of a three-helix bundle at the C-terminal end of Hsp70 proteins.

Input File Formats

COILS accepts files in GCG (Genetics Computer Group) format and in Pearson (FASTA) format. In addition, users can adapt any sequence to be read by COILS by marking its beginning (by ">" or "[space][space]..") and end (by "*" or "//"). An input file may contain multiple sequences as long as they are delimited by markers.

ALIGNED accepts files created by PileUp and CLUSTAL V from SWISS-PROT entries (this limitation is connected to the space allocated in the alignment for the sequence names). An expanded range of input formats is planned.

Program Availability

All programs, source codes, and documentation can be downloaded from the Coils/vms folder of the anonymous ftp server FTP.BIOCHEM. MPG.DE. The programs are written in VAX Pascal and operate equally under VAX/VMS and OpenVMS. In addition, the coils folder contains C and c++ source codes for COILS that can be compiled under UNIX, as well as a compiled version of the c++ code for PC/DOS. Macstripe, a Macintosh adaptation of COILS by Alex Knight (knight@wi.mit.edu), is available on the World Wide Web at http://www.wi.mit.edu/matsudaira/ coilcoil.html.

A World Wide Web (WWW) server for COILS has become available at the Swiss Institute for Experimental Cancer Research (http://ulrec3.unil.ch/

---

[19] Genetics Computer Group, Madison, WI.

software/COILS_form.html) courtesy of Kay Hofmann (khofmann@ isrec-sun1.unil.ch).

Acknowledgments

I thank Janice Lupas for programming the algorithms described in this chapter and Jeff Stock for critically reading the manuscript.

[31] PHD: Predicting One-Dimensional Protein Structure by Profile-Based Neural Networks

By Burkhard Rost

Introduction

We still cannot predict protein three-dimensional (3D) structure from sequence alone, but we can predict 3D structure for one-fourth of the known protein sequences (SWISS-PROT[1]) by homology modeling based on significant sequence identity (>25%) to known 3D structures (Protein Data Bank, PDB[2]).[3] For the remaining, about 30,000 known sequences, the prediction problem has to be simplified. An extreme simplification is to try to predict projections of 3D structure, for example, one-dimensional (1D) secondary structure, solvent accessibility, or transmembrane location assignments for each residue. Despite the extreme simplification, the success of 1D predictions has been limited as segments from single sequences (used as input) do not contain sufficient global information about 3D structures.[4,5] Patterns of amino acid substitutions within sequence families are highly specific for the 3D structure of that family. Using such evolutionary information is the key to a significant improvement of 1D predictions.

In this chapter we describe three prediction methods that use evolutionary information as input to neural network systems to predict secondary

[1] A. Bairoch and B. Boeckmann, *Nucleic Acids Res.* **22,** 3578 (1994).
[2] F. C. Bernstein, T. F. Koetzle, G. J. B. Williams, E. F. Meyer, M. D. Brice, J. R. Rodgers, O. Kennard, T. Shimanouchi, and M. Tasumi, *J. Mol. Biol.* **112,** 535 (1977).
[3] C. Sander and R. Schneider, *Nucleic Acids Res.* **22,** 3597 (1994).
[4] W. Kabsch and C. Sander, *FEBS Lett.* **155,** 179 (1983).
[5] B. Rost, C. Sander, and R. Schneider, *Trends Biochem. Sci.* **18,** 120 (1993).

Copyright © 1996 by Academic Press, Inc.
All rights of reproduction in any form reserved.

structure (PHDsec[6–8]), relative solvent accessibility (PHDacc[9]), and trans-membrane helices (PHDhtm[10]) are described. Also illustrated are the possibilities and limitations in practical applications of these methods with results from careful cross-validation experiments on large sets of unique protein structures. All predictions are made available by an automatic E-mail prediction service (see section on availability). The baseline conclusion after some 30,000 requests to the service[11] is that 1D predictions have become accurate enough to be used as a starting point for expert-driven modeling of protein structure.[12–14]

Methods

Generating Multiple Sequence Alignment

The first step in a PHD prediction is generating a multiple sequence alignment. The second step involves feeding the alignment into a neural network system. Correctness of the multiple sequence alignment is as crucial for prediction accuracy as is the fact that the alignment contains a broad spectrum of homologous sequences. By default, PHD uses the program MaxHom (Fig. 1) that generates a pairwise profile-based multiple align-ment.[15] A key feature of MaxHom is the compilation of a length-dependent cutoff for significant pairwise sequence identity (Fig. 1).[15]

Multiple Levels of Computations

The PHD methods process the input information on multiple levels (Fig. 2). The first level is a feed-forward neural network with three layers of units (input, hidden, and output). Input to this first level sequence-to-structure network consists of two contributions: one from the local se-quence, that is, taken from a window of 13 adjacent residues, and another from the global sequence (Fig. 2). Output of the first level network is the 1D structural state of the residue at the center of the input window. For

[6] B. Rost and C. Sander, *J. Mol. Biol.* **232**, 584 (1993).

[7] B. Rost and C. Sander, *Proc. Natl. Acad. Sci. U.S.A.* **90**, 7558 (1993).

[8] B. Rost and C. Sander, *Proteins* **19**, 55 (1994).

[9] B. Rost and C. Sander, *Proteins* **20**, 216 (1994).

[10] B. Rost, R. Casadio, P. Fariselli, and C. Sander, *Protein Sci.* **4**, 521 (1995).

[11] B. Rost, C. Sander, and R. Schneider, *CABIOS* **10**, 53 (1994).

[12] T. J. P. Hubbard and J. Park, *Proteins* **23**, 398 (1995).

[13] B. Rost, "TOPITS: Threading One-Dimensional Predictions into Three-Dimensional Struc-tures." AAAI Press, Cambridge, July 16–19, 1995.

[14] B. Rost and C. Sander, *Proteins* **23**, 295 (1995).

[15] C. Sander and R. Schneider, *Proteins* **9**, 56 (1991).

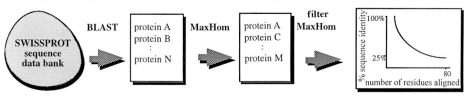

FIG. 1. First, for each protein, the SWISS-PROT database is searched for sequence homologs with a fast alignment method [BLAST, S. F. Altschul, W. Gish, W. Miller, E. W. Myers, and D. J. Lipman, *J. Mol. Biol.* **215**, 403 (1990)]. Second, the list of putative homologs found is reexamined with a more sensitive profile-based multiple alignment method [MaxHom, C. Sander and R. Schneider, *Proteins* **9**, 56 (1991)]. Third, a length-dependent cutoff for significant pairwise sequence identity is applied [25% + 5%, where +5% reflects a safety margin in the twilight zone (R. F. Doolittle, "Of URFs and ORFs: A Primer on How to Analyze Derived Amino Acid Sequences." University Science Books, Mill Valley, California, 1986)].

PHDsec and PHDhtm the second level is a structure-to-structure network (see below). The next level consists of an arithmetic average over independently trained networks (jury decision). The final level is a simple filter.

Number of Output Units Determined by Task

Secondary structure is coded by three units: helix, H (*H, G*, and *I* in DSSP, the database containing the secondary structure and solvent accessibility for proteins of known 3D structure[16]); strand, E (*E* and *B* in DSSP[16]); and none of the above, denoted loop, L. Transmembrane locations are coded by two units, one for residues being in a transmembrane helix, the other for non-membrane-bound residues (assignments from SWISS-PROT[1]). For solvent accessibility the output coding is not so straightforward. First, the value for accessibility is normalized to a relative accessibility (observed accessibility taken from DSSP[16] divided by maximal accessibility of a given residue type[9,17]) to enable a comparison between residues of different sizes. Second, the relative accessibility is projected onto ten states (for technical reasons; Fig. 2).[9]

Better Segment Prediction by Structure-to-Structure Networks

The output coding for the second level network is identical to the one for the first (Fig. 2). The dominant input contribution to the second level structure-to-structure network is the output of the first level sequence-to-structure network. The reason for introducing a second level is the follow-

---

[16] W. Kabsch and C. Sander, *Biopolymers* **22**, 2577 (1983).
[17] G. D. Rose, A. R. Geselowitz, G. J. Lesser, R. H. Lee, and M. H. Zehfus, *Science* **229**, 834 (1985).

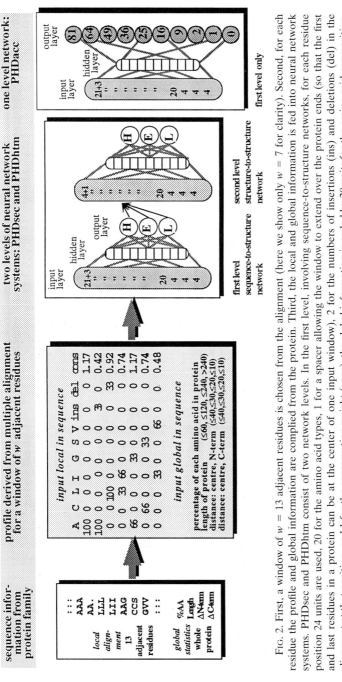

Fig. 2. First, a window of $w = 13$ adjacent residues is chosen from the alignment (here we show only $w = 7$ for clarity). Second, for each residue the profile and global information are complied from the protein. Third, the local and global information is fed into neural network systems. PHDsec and PHDhtm consist of two network levels. In the first level, involving sequence-to-structure networks, for each residue position 24 units are used, 20 for the amino acid types, 1 for a spacer allowing the window to extend over the protein ends (so that the first and last residues in a protein can be at the center of one input window), 2 for the numbers of insertions (ins) and deletions (del) in the alignment at that position, and 1 for the conservation weight (cons); the global information is coded by 20 units for the amino acid composition, 4 for the protein length, and 8 for the distances of the window with respect to the protein ends. The output units code for the 1D structural state of the central residue. For PHDsec, three output units code for helix, strand, and rest; for PHDacc, ten units code for ten levels of relative solvent accessibility (e.g., if the fourth unit has the maximal value, then the prediction is a relative solvent accessibility ≥9% and <16%); and for PHDhtm, two units code for transmembrane or not transmembrane helix. In the second level, involving structure-to-structure networks, the output of the first level is fed into a second level of structure-to-structure network, which additionally uses global information and the conservation weight as input; for example, for PHDsec, first level output = 3 units → local input to second level = 3 + 1 (spacer) + 1 (cons). The output of the second level is the same as that of the first level.

ing. Networks are trained by changing the connections between the units such that the error is reduced for each of the examples successively presented to the network during training. The examples are chosen at random. Therefore, the examples taken at time step t and at time step $t + 1$ are usually not adjacent in sequence. This implies that the network cannot learn that, for example, helices contain at least three residues. The second level structure-to-structure network introduces a correlation between adjacent residues with the effect that predicted secondary structure segments or transmembrane helices have length distributions similar to the ones observed.[6,7]

Balanced Predictions by Balanced Training

For the prediction of secondary structure and transmembrane helices, the distribution of the examples is rather uneven: about 32% of the residues are observed in helix, 21% in strand, and 47% in loop; about 18% of the residues in integral transmembrane proteins are located in transmembrane helices. Choosing the training examples proportional to the occurrence in the data set (unbalanced training) results in a prediction accuracy that mirrors this distribution; for example, strands are predicted inferior to helix or loop.[18–20] A simple way around the database bias is a balanced training: at each time step one example is chosen from each class, that is, one window with the central residue in a helix, one with the central residue in a strand, and one representing the loop class. This training results is a prediction accuracy well balanced between the output states.[6,7]

Compromise between Overprediction and Underprediction by Jury Decision

Balanced training results in improved predictions for the less populated output states (e.g., strand). However, this is associated with less accurate predictions for more populated states (loop). Consequently, the overall accuracy is lower for the balanced than for the unbalanced prediction. To find a compromise between networks with balanced and those with unbalanced training, a final jury decision is performed (effectively a compromise between over- and underprediction). The jury decision is a simple arithmetic average over, typically, four differently trained networks: all

---

[18] O. Gascuel and J. L. Golmard, *CABIOS* **4,** 357 (1988).

[19] B. Rost and C. Sander, *in* "1D Secondary Structure Prediction through Evolutionary Profiles" (H. Bohr and S. Brunak, eds.), p. 257. IOS Press, Amsterdam, Oxford, and Washington, D.C., 1994.

[20] A. A. Salamov and V. V. Solovyev, *J. Mol. Biol.* **247,** 11 (1995).

combinations of first level networks with balanced or unbalanced training, and with balanced or unbalanced training of second level networks (2 × 2). The final prediction is assigned to the unit with maximal output value (winner takes all).

Correcting Obvious Errors by Final Filter

For secondary structure prediction (PHDsec), the filter affects only drastic, unrealistic predictions (e.g., HEH → HHH; EHE → EEE; and LHL → LLL). For accessibility prediction (PHDacc), the filter performs an average over neighboring output units (i.e., not over adjacent residues). Only the filter used for predicting transmembrane helices (PHDhtm) is crucial for the performance. The currently implemented filter has been guided by previous experience.[21-24] Predicted transmembrane helices which are too long are either split or shortened. Predicted transmembrane helices which are too short are either elongated or deleted. All these decisions (split or shorten; elongate or delete) are based on the strength of the prediction and on the length of the transmembrane helix predicted.[10]

Avoiding Overestimating Prediction Accuracy

The three necessary conditions for an appropriate evaluation of prediction accuracy are first, that training and testing set are distinct; second, that the testing set is representative; and, third, that free parameters are not optimized on the test set which is used for the final evaluation. In more detail, first, the criterion for distinct sets is that no protein in one set has more than 25% pairwise sequence identity to any protein in the other.[15] Second, the test set has to be representative for the database (ideally for all existing proteins), that is, all known sequence families should be included, and they should be included only once. Third, no free parameter should be optimized with respect to the test set. A simple protocol for correct testing would be the following. (1) Choose a small test set (pretest, some 10 proteins) and adjust free parameters; (2) keeping the network fixed, compile the accuracy for all test proteins (real test, >100 proteins by cross-validation experiments; note that the number of splits between test and training sets for cross-validation is of no interest for the user); (3) apply the same network to another test set never used before (prerelease test, e.g., protein structures experimentally determined after the project

[21] G. von Heijne, *Nucleic Acids Res.* **14,** 4683 (1986).
[22] G. von Heijne and Y. Gavel, *Eur. J. Biochem.* **174,** 671 (1988).
[23] G. von Heijne, *J. Mol. Biol.* **225,** 487 (1992).
[24] L. Sipos and G. von Heijne, *Eur. J. Biochem.* **213,** 1333 (1993).

had started). A lower level of accuracy for the pre-release test than for the real test indicates an overfitting of free parameters. Step three should be reapplied whenever a considerable number of new structures have been added to the database (Table I).

Results

Values for Expected Prediction Accuracy Are Distributions

Statements such as secondary structure is about 90% conserved within sequence families,[25] or solvent accessibility is about 85% conserved within sequence families,[9] refer to averages of distributions. The same holds for the expected prediction accuracy (Fig. 3). Such distributions explain why some developers have overestimated the performance of their tools using data sets of only tens of proteins (or even fewer).[5] For the user interested in a certain protein, the distributions imply a rather unfortunate message: for that protein, the accuracy could be lower than 40%, or it could be higher than 90% (Fig. 3). For some of the worst predicted proteins, the low level of accuracy could be anticipated from their unusual features, for example, for crambin or the antifreeze glycoprotein type III. However, for others the reasons for the failure of PHDsec are not obvious; for example, both the phosphatidylinositol 3-kinase[26] and the Src homology domain of cytoskeletal spectrin have homologous structure,[27] but prediction accuracy varies between less than 40% (kinase) and more than 70% (spectrin). Another possible reason for a bad prediction is a bad alignment. In general, single sequences yield accuracy values about ten percentage points lower than multiple alignments.[6] Indeed, the worst case for a prediction so far is pheromone (1erp), a short protein structurally dominated by a disulfide bridge, for which there is no sequence alignment available: only 32% of the residues are predicted correctly.

Reliability of Prediction Correlating with Accuracy

An estimate where in the distributions (Fig. 3) a given prediction is to be expected is given by the prediction strength, that is, the difference between the output unit with highest value (winner unit) and the output unit with the next highest value. This difference is used to define a reliability

[25] B. Rost, C. Sander, and R. Schneider, *J. Mol. Biol.* **235,** 13 (1994).
[26] S. Koyama, H. Yu, D. C. Dalgarno, T. B. Shin, L. D. Zydowsky, and S. L. Schreiber, *Cell* (*Cambridge, Mass.*) **72,** 945 (1993).
[27] A. Musacchio, M. Noble, R. Pauptit, R. Wierenga, and M. Saraste, *Nature* (*London*) **359,** 851 (1992).

TABLE I

ACCURACY OF SECONDARY STRUCTURE AND ACCESSIBILITY PREDICTION

Method[a]	Set[b]	N[c]	Secondary structure			Solvent accessibility			Date[j]
			Q_3[d]	I[e]	Sov_3[f]	Q_3[g]	Q_2[h]	Corr[i]	
HM: SeqAli	1	80	88.4	0.62	89.7	71.6	83.8	0.68	
HM: StrAli	1	80				73.6	84.8	0.77	
RAN	1	80	35.2	0.01	30.6	33.9	52.0	0.01	
PHD	2	126	71.6	0.27	72.8	57.9	75.0	0.54	06 92
PHD	3	124	72.5	0.28	75.6				07 93
SIMPA	3	124	60.7	0.12	61.7				
PHD	3a	112				57.9	74.7	0.54	03 94
PHD	7	60	74.8	0.34	76.8				
LPAG	7	60	68.5[k]	—	—				
PHD	8	13				60.8	79.2	0.61	
Wako & Blundell	8	13				—	76.5[k]	—	
PHD	4	27	72.0	0.28	72.4	57.6	73.4	0.55	05 94
PHD	5	59	73.0	0.30	75.7	57.0	74.0	0.54	11 94
PHD	6	9	72.1	0.27	72.8		63.0	0.38	12 94
PHD	2–6	**337**	**72.3**	**0.28**	**73.8**				03 95
		318				**57.3**	**74.2**	**0.54**	

[a] HM: SeqAli, homology modeling based on sequence alignments within sequence families [B. C. Sander, and R. Schneider, *J. Mol. Biol.* **235**, 13 (1994); B. Rost and C. Sander, *Proteins* **20**, 216 (1994)]; HM: StrAli, homology modeling based on structural alignments [B. Rost and C. Sander, *Proteins* **20**, 216 (1994)]; *RAN*, random alignments, that is, worst prediction; PHD, neural network predictions; SIMPA, statistical prediction method [J. M. Levin, *et al.*, *FEBS Lett.* **205**, 303 (1986); note that SIMPA is not reported as the best method but scored better than others (GORIII, COMBINE) on set 7]; Wako & Blundell, statistical prediction method based on alignments [Wako and Blundell, *J. Mol. Biol.* **238**, 682 (1994)]; *LPAG*, statistical prediction method based on alignments [J. M. Levin, S. P. A. Pascarella, P. Argos, and J. Garnier, *Protein Eng.* **6**, 849 (1993)].

[b] Different tests sets are numbered to indicate identical sets. 1, B. Rost, C. Sander, and R. Schneider, *J. Mol. Biol.* **235**, 13 (1994); 2, 3, B. Rost and C. Sander, *Proteins* **19**, 55 (1994); 3a, subset of set 3, B. Rost and C. Sander, *Proteins* **20**, 216 (1994); 4 and 5, recently determined structures; 6, proteins from Asilomar prediction context, B. Rost and C. Sander, *Proteins* **23**, 295 (1995); 7, J. M. Levin, S. P. A. Pascarella, P. Argos, and J. Garnier, *Protein Eng.* **6**, 849 (1993); 8, H. Wako and T. L. Blundell, *J. Mol. Biol.* **238**, 682 (1994); 2–6, results for proteins from sets 2–6, 337 unique protein chains with a total of 74,901 residues for PHDsec, and 318 unique proteins with a total of 79,588 residues for PHDacc.

[c] *N*, Number of proteins used for testing (all results for test proteins with less than 25% sequence identity to proteins used for training).

[d] Q_3, Three-state overall pre-residue accuracy for secondary structure, that is, number of residues predicted correctly in helix, strand, or rest.

[e] *I*, Information, entropy measure for accuracy [B. Rost, C. Sander, and R. Schneider, *J. Mol. Biol.* **235**, 13 (1994); B. Rost and C. Sander, *J. Mol. Biol.* **232**, 584 (1993)].

index for the prediction of each residue [normalized to a scale from 0 (low) to 9 (high)]. Residues with higher reliability index are predicted with higher accuracy (Fig. 4). In practice, the reliability index offers an excellent tool to focus on some key regions predicted at high levels of expected accuracy. (Note however, that the reliability indices tend to be unusually high for poor alignments.)

Prediction of Secondary Structure at Better than 72% Accuracy

PHDsec was the first secondary structure prediction method to surpass a level of 70% overall three-state per-residue accuracy.[6] The last test set with more than 300 unique protein chains and a total of more than 70,000 residues compiled for this chapter yielded a three-state per-residue accuracy better than 72% (Table I). Besides the high level of overall accuracy, predictions are well balanced (high value for I in Table I). Furthermore, PHDsec meets the demands for a reasonable prediction tool in that the accuracy measured in segment-based scores[25] is higher than per-residue scores: about 74% of the segments are correctly predicted (Sov_3 in Table I).

Structural Class Prediction Comparable to Experimental Accuracy

Proteins can be sorted roughly into four structural classes based on secondary structure content: all-α (helix $\geq$ 45%, strand $<$ 5%), all-β (strand $\geq$ 45%, helix $<$ 5%), α/β (helix $\geq$ 30%, strand $\geq$ 20%), and all others.[28,29] An experimental way to measure secondary structure content is circular dichroism spectroscopy.[30,31] A simple alternative is to use the

[28] M. Levitt and C. Chothia, *Nature* (*London*) **261**, 552 (1976).
[29] C.-T. Zhang and K.-C. Chou, *Protein Sci.* **1**, 401 (1992).
[30] C. W. J. Johnson, *Proteins* **7**, 205 (1990).
[31] A. Perczel, K. Park, and G. D. Fasman, *Proteins* **13**, 57 (1992).

---

[f] Sov_3, Three-state overall per-segment accuracy, that is, overlap of predicted and observed secondary structure segments [B. Rost, C. Sander, and R. Schneider, *J. Mol. Biol.* **235**, 13 (1994)].

[g] Q_3, Three-state overall per-residue accuracy for accessibility, that is, percentage of residues correctly predicted as buried, intermediate, or exposed [B. Rost and C. Sander, *Proteins* **20**, 216 (1994)].

[h] Q_2, Two-state overall accuracy for accessibility, that is, percentage of residues correctly predicted as buried or exposed.

[i] Corr, Correlation between predicted and observed relative solvent accessibility [B. Rost and C. Sander, *Proteins* **20**, 216 (1994)].

[j] Date of collecting the data set.

[k] Result taken from literature.

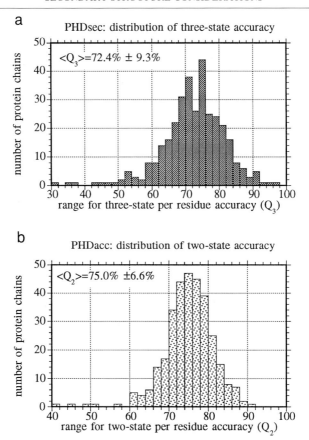

Fɪɢ. 3. Expected variation of prediction accuracy with protein chain. (a) Three-state per-residue overall accuracy for PHDsec (total of 337 chains). (b) Two-state per-residue overall accuracy for PHDacc (total of 318 chains). Given are the distributions, averages, and one standard deviation. (c) Cumulative percentage of protein chains predicted at an error level lower than the value given (error = 100 − accuracy). Error values are percentages of falsely predicted residues (PHDsec, PHDacc) and falsely predicted segments (PHDhtm). For example, for one-half of all chains, PHDhtm predicts all segments correctly (note that the total set for evaluating PHDhtm comprises only 69 chains), whereas PHDsec and PHDacc rate at about 25% falsely predicted residues.

predictions of PHDsec to compile the overall prediction of secondary structure content. Based on the predicted content, proteins are sorted into either of the four structural classes. The result is that for about 74% of all protein chains, the class is correctly predicted (Table II). The correlation between observed and predicted content is 0.88 for helix and 0.75 for strand. These values are comparable to results from circular dichroism spectroscopy (he-

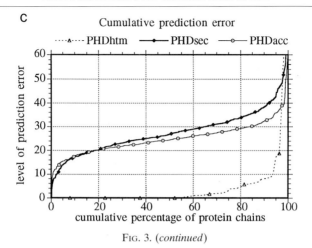

FIG. 3. (*continued*)

lix; 0.84; strand, 0.37–0.41[31]).[6] Of course, this does not imply that PHDsec can replace experiments. However, the high level of accuracy suggests using PHDsec prediction as a complement to experiments.

Prediction of Buried or Exposed Residues at 74% Accuracy

Comparing the conservation of secondary structure with that of solvent accessibility (measured in three states), we find that solvent accessibility is less conserved (Table I). Consequently, PHDacc is less accurate than PHDsec. However, the accessibility prediction is relatively close to the optimum given by homology modeling: the correlation between predicted and observed relative accessibility is 0.54 for PHDacc and would be 0.68 for sequence alignments if homology modeling were possible (Table I). More than 74% of the residues are predicted correctly in either of the two states, buried or exposed. Entirely buried residues (<4% accessible) are predicted best (data not shown).[9] PHDacc is, so far, superior to other methods (Table I). (*Note:* When a subset of 99 monomers was tested, the two-state accuracy rose to over 77%.[9])

Transmembrane Helices Predicted at 95% Accuracy

The problem in evaluating the performance of PHDhtm is the small set of proteins for which the locations of transmembrane helices have been determined reliably. Consequently, the results ought to be viewed with caution. The overall two-state per-residue accuracy of PHDhtm is 95% (Fig. 3), and the per-segment accuracy is about 96% (only 15 of 380 trans-

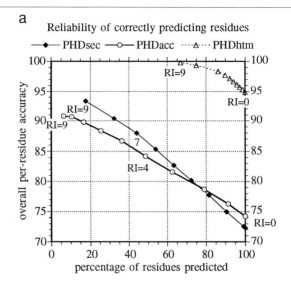

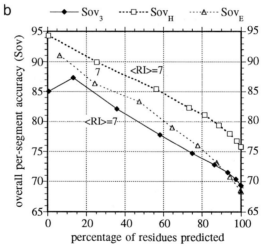

FIG. 4. (a) Expected per-residue accuracy for residues with a reliability index (*RI*) above a given cutoff; A level of accuracy comparable to homology modeling is reached for 49% of all residues by PHDacc (*RI* ≥ 4) and for 44% of all residues by PHDsec (*RI* ≥ 7). The small region covered by the reliability index of PHDhtm is dominated by strong predictions for nontransmembrane residues; the most accurately predicted residues in transmembrane helices reach a level of only 96% accuracy. (b) Expected per-segment accuracy for secondary structure segments with an average reliability index (⟨*RI*⟩) above a given cutoff. For example, an average reliability of ⟨*RI*⟩ ≥ 7 is reached for (i) 36% of all segments (*Sov₃* > 82%), (ii) 56% of all helices (*Sov_E* > 85%), and (iii) 24 of all strands (*Sov_H* > 86%). [Definitions of segment overlap are given in B. Rost, C. Sander, and R. Schneider, *J. Mol. Biol.* **235**, 13 (1994).].

TABLE II
PREDICTING SECONDARY STRUCTURE CONTENT AND STRUCTURAL CLASS

Method[a]	Set[a]	N_{prot} [a]	ΔHelix[b]	ΔStrand[b]	All-α[c]	All-β[c]	α/β[c]	Rest[c]	Q_{class} [c]
HM:SeqAli	Set 1	80	2.8 ± 3.8	2.7 ± 3.2	94.1	86.7	100.0	89.7	90.0
RAN	Set 1	80	32.1 ± 20.8	21.3 ± 14.5	0.0	0.0	0.0	71.2	44.7
PHD	Set 2	126	8.5 ± 8.0	7.5 ± 8.0	85.7	50.0	50.0	74.1	74.6
PHD	Set 3	124	7.8 ± 6.8	7.3 ± 7.9	94.1	0.0	55.6	74.5	75.8
PHD	Set 2–6	337	8.1 ± 7.9	7.1 ± 7.6	85.0	55.6	45.5	75.6	74.2

[a] See Table I, footnotes a–c.
[b] Error in predicting the content of helix or strand averaged over all protein chains in the data set. The error is computed as the difference between the percentage of helix (Δhelix) or strand (Δstrand) between observed and predicted. (Values are given ± one standard deviation.)
[c] Percentage of protein chains correctly predicted in either of the four classes: all-α, all-β, α/β, and all others. Q_{class} gives the percentage of protein chains correctly predicted in any of the four classes.

membrane helices in a set of 69 proteins were wrongly predicted).[10] Of further practical importance is the low level of false positives for PHDhtm. Of a set of 278 globular water-soluble proteins with unique sequences, PHDhtm predicts only 14 incorrect transmembrane helices; these errors occur mostly for proteins with highly hydrophobic β strands in the core.[10]

Availability

PHD predictions (and MaxHom alignments) are available on request by the automatic prediction service PredictProtein.[11] For detailed information send the word help as subject to the Internet address PredictProtein@EMBL-Heidelberg.DE or ventured through the World Wide Web (WWW) site: http://www.embl-heidelberg.de/predictprotein/predictprotein.html (Fig. 5). Because we sometimes have over 100 requests per day, returning a prediction may take a day or more. If no answer is received after two days, something has gone wrong (typical reasons: corrupted E-mail connection of sender or hardware problems at EMBL). In such a case, simply resubmit the request. Should the answer not appear after another two days, send a note to Predict-Help@EMBL-Heidelberg.DE. For further services (e.g., database) provided by the EMBL Protein Design Group, see http://www.sander.embl-heidelberg.de/descr/ or connect by anonymous ftp to ftp.embl-heidelberg.de.

Comments

Accuracy of Predictions. The expected levels of accuracy [PHDsec 72 ± 9% (three states), PHDacc 75 ± 7% (two states), and PHDhtm 94 ±

File Header	(optional statements in italics)

```
Joe Sequencer, Department of Advanced Protein Research,
National Univeristy, Timbuktu
joe@amino.churn.edu
    predict secondary structure, predict accessibility;
    predict transmembrane
    return MSF; return no alignment; return HSSP profiles;
    return graph
```

File Body - request = single sequence
(anything in line(s) after '#' is interpreted as one-letter amino acid sequence)

```
# incredulase from paracoccus dementiae, translated from cDNA
KELVLALYDYQEKSPREVTMKKGDILTLLNSTNKD
WWKVEVNDRQGFVPAAYVKKLD
```

FIG. 5. Example of a request to the automatic protein structure prediction server PredictPro-tein. All network methods (PHDsec, PHDacc, PHDhtm) are available. The file to be submitted consists of two parts: a head (optional key words are shown in italics) and the main body starting with a hash (#) in the first line and the one-letter code amino acid sequence in the following lines. Alternative options include submission of a list of sequences or a complete alignment. Details are given in the PredictProtein help file (see text).

6%] are valid for typical globular, water-soluble (PHDsec, PHDacc), or helical transmembrane proteins (PHDhtm) when the multiple alignment contains many and diverse sequences. High values for the reliability indices indicate more accurate predictions. (*Note:* For alignments with little varia-tion in the sequences, the reliability indices adopt misleadingly high values.)

Usefulness of Predictions. The prediction of secondary structure can be accurate enough to assist chain tracing. Furthermore, predictions can be used as a starting point for modeling 3D structure and predicting function.[13,32–34]

Confusion between Strand and Helix. PHDsec focuses on predicting hydrogen bonds. Consequently, occasionally strongly predicted (high relia-bility index) helices are observed as strands and vice versa.

Strong Signal from Secondary Structure Caps. The ends of helices and strands contain a strong signal. However, on average PHDsec predicts the core of helices and strands more accurately than the caps.[19]

Accessibility Useful to Provide Upper Limits for Contacts. The predicted solvent accessibility (PHDacc) can be translated into a prediction of the number of water atoms around a given residue. Consequently, PHDacc

[32] T. J. P. Hubbard, "Use of β-Strand Interaction Pseudo-Potential in Protein Structure Prediction and Modeling." IEEE Society Press, Los Alamitos, CA, 1994.

[33] B. Rost, *in* "Fitting 1D Predictions into 3D Structures" (H. Bohr and S. Brunak, eds.), in press. CRC Press, Boca Raton, Florida, 1995.

[34] T. Meitinger, A. Meindl, P. Bork, B. Rost, C. Sander, M. Haasemann, and J. Murken, *Nat. Genet.* **5,** 376 (1993).

can be used to derive upper and lower limits for the number of interresidue contacts of a certain residue (such an estimate could improve predictions of interresidue contacts[35]).

Protein Design and Synthesized Peptides. The PHD networks are trained on naturally evolved proteins. However, the predictions have proved to be useful in some cases to investigate the influence of single mutations. For short polypeptides, the following should be taken into account: the network input consists of 17 adjacent residues, and thus shorter sequences may be dominated by the ends (which are treated as solvent).

Prediction of Porins. PHDhtm predicts only transmembrane helices, and PHDsec has been trained on globular, water-soluble proteins. How does one predict the 1D structure for porins then? As porins are partly accessible to solvent, the prediction accuracy of PHDsec was relatively high (70%) for the known structures. Thus, PHDsec appears to be applicable.

Using Prediction of Transmembrane Helices. One possible application of PHDhtm is to scan, for example, entire chromosomes for possible transmembrane proteins. The classification as transmembrane protein is not sufficient to have knowledge about function, but it may shed some light on the puzzle on genome analyses. When using PHDhtm for this purpose, the user should keep in mind that on average about 5% of the globular proteins are falsely predicted to have transmembrane helices.

Acknowledgments

I express my gratitude to the colleagues from the European Molecular Biology Laboratories who help(ed) in developing PHD. First of all, thanks to Chris Sander for intellectual, emotional, and financial support. Second, thanks to Reinhard Schneider for valuable ideas, important discussions, and for help in setting up the prediction server. Third, thanks to Antoine de Daruvar for having rewritten the server software and for now maintaining the server. Fourth, thanks to Gerrit Vriend whose ideas paved the way for the first prediction above 70% accuracy. Fifth, thanks to Séan O'Donoghue for a thorough correction of the manuscript. Finally, thanks to all those who deposit data about protein structure in public databases and thus enable the development of tools such as PHD.

[35] U. Goebel, C. Sander, R. Schneider, and A. Valencia, *Proteins* **18,** 309 (1994).

[32] GOR Method for Predicting Protein Secondary Structure from Amino Acid Sequence

By Jean Garnier, Jean-François Gibrat, and Barry Robson

Introduction

How to extract properties from the amino acid sequence of a protein for understanding its function is one of the most timely and competitive areas of the biological sciences. It is related to a fundamental aspect of biology. A linear and ordered sequence of amino acids, coded and conserved by a linear and ordered sequence of nucleotide bases, codes for all that is characteristic of living organisms: specific and organized interactions in space and time between proteins, lipids, nucleic acids, and the cell metabolites. These characteristics depend on how a protein can fold in a unique active three-dimensional structure. This process is spontaneous under given environmental conditions, even though the living cell can add efficiency and control by use of some protein complexes, called chaperones, to catalyze this process.

Much effort has been devoted to the calculation of the spatial structure of a polypeptide chain from its amino acid sequence alone, with only limited but nevertheless encouraging success[1] when a polypeptide is longer than 10–20 amino acids. However, attempts to reduce the problem to simpler features of the protein fold such as α helix, β strands, and aperiodic or coil structure have yielded interesting results (see Refs. 2 and 3). These results have been an aid for designing new proteins, predicting the effect of point mutations, identifying the protein class, for instance, all-α or all-β proteins, predicting epitopes, etc. It is hoped that this information will be increasingly useful to molecular biologists and protein modelers. Usually the computing time is short, and many programs of secondary structure predictions are available on-line to the biologist.

The GOR method is one of the most popular of the secondary structure prediction schemes. This method is theoretically well founded in a series of earlier papers, and it has been the real first prediction of secondary structure implemented as a computer program. The three letters stand for

---

[1] Protein Structure Prediction Issue. *Proteins* **23,** 3 (1995).
[2] J. Garnier and J. M. Levin, *CABIOS* **7,** 133 (1991).
[3] B. Rost and C. Sander, *Trends Biochem. Sci.* **18,** 120 (1993).

METHODS IN ENZYMOLOGY, VOL. 266

Copyright © 1996 by Academic Press, Inc.
All rights of reproduction in any form reserved.

the first letter of the names of the authors of the original publication.[4] This method remains remarkably popular,[5] but users have overlooked the fact that improved versions of the method have since been published, and a full description of them can be found in the book edited by Fasman.[6] The addition of homologous sequence information through multiple alignments has given a significant boost to the accuracy of secondary structure predictions.[7–9] In this chapter, after presenting the major principles used by the GOR method, we give some results obtained with an updated version of this method.

Principles of Method

In a series of articles, Robson et al.[10,11] used the formalism of information theory and Baysian statistics to establish the code relating the amino acid sequence and the secondary structures of a protein. This led later to the development of the GOR method.[4]

Information theory was developed in the 1950–1960s[12,13] and the GOR method made use of an information function described by Fano,[14] $I(S; R)$, which is defined as

$$I(S; R) = \log[P(S|R)/P(S)] \tag{1}$$

Originally this formulation was concerned mainly with electronic transmission of information. In the present application, S is one of the three conformations, R is one of the 20 amino acid residues, $P(S|R)$ is the conditional probability for observing a conformation S when a residue R is present, and $P(S)$ is the probability of observing S. According to the definition of conditional probabilities, $P(S|R) = P(S, R)/P(R)$ where $P(S, R)$ is the joint probability of observing the events S and R and $P(R)$ is the probability of observing a residue R. It is easy to have an estimation of $I(S; R)$ from

[4] J. Garnier, D. Osguthorpe, and B. Robson, *J. Mol. Biol.* **120,** 97 (1978).

[5] L. B. M. Ellis and R. P. Milius, *CABIOS* **10,** 341 (1994).

[6] J. Garnier and B. Robson, *in* "Prediction of Protein Structure and the Principles of Protein Conformation" (G. D. Fasman, ed.), Chap. 10, p. 417. Plenum Press, New York, 1989.

[7] J. M. Levin, S. Pascarella, P. Argos, and J. Garnier, *Protein Eng.* **6,** 849 (1993).

[8] B. Rost and C. Sander, *J. Mol. Biol.* **232,** 584 (1993).

[9] V. di Francesco, P. J. Munson, and J. Garnier, *28th Annual Hawaii International Conference on System Sciences* (L. Hunter, ed.), p. 285. IEEE Computer Society Press, Los Alamos, 1995.

[10] B. Robson and R. H. Pain, *J. Mol. Biol.* **58,** 237 (1971).

[11] B. Robson, *Biochem. J.* **141,** 853 (1974).

[12] C. E. Shannon and W. Weaver, "The Mathematical Theory of Communication." Univ. of Illinois Press, Urbana, Illinois, 1949.

[13] L. Brillouin, "Science and Information Theory." Academic Press, New York, 1956.

[14] R. Fano, "Transmission of Information." Wiley, New York, 1961.

a database of known sequences and corresponding observed secondary structures since $P(S, R) = f_{S,R}/N$, $P(R) = f_R/N$ and $P(S) = f_S/N$ with N being the total number of amino acids in the database, $f_{S,R}$ the number of residues R observed in the conformation S in the same database, f_R the total number of residues R, and f_S the total number of residues observed in the conformation S in the same database. Then

$$I(S; R) = \log[(f_{S,R}/f_R)/(f_S/N)] \tag{2}$$

This quantity can be obtained easily from the database provided it is large enough.

A more general treatment requires corrections for levels of data (see Robson[11] and below). Robson[11] introduced the information difference,

$$I(\Delta S; R) = I(S; R) - I(n\text{-}S; R) = \log(f_{S,R}/f_{n\text{-}S,R}) + \log(f_{n\text{-}S}/f_S) \tag{3}$$

where n-S stands for the conformations other than S (non-S); for instance, if S is α helix (H), n-S will be β strand (E) and coil (C) for a three-state prediction. It gives the extra information for S on the two others. It represents a kind of normalization where the total number of amino acids, N, and residues, R, in the database have disappeared from the equation. In effect the positive hypothesis $(S; R)$ and the complementary negative hypothesis $(n$-$S; R)$ are treated in concert. This quantity also corresponds to one-residue information or single-residue information or self-information. Calculated for the three conformations, the highest value of Eq. (3) for one of the conformations S will be the predicted conformation and will be the propensity for that residue to be in that conformation, usually expressed in centinat units when natural logarithms are used. This underlines one of the differences with the Chou–Fasman propensities which correspond approximately to the mantissa of the log of Eq. (2).

Equations (1) to (3) can be extended to a local sequence along the polypeptide chain of n consecutive residues R:

$$\begin{aligned} I(\Delta S_j; R_1, \ldots, R_n) &= \log[P(S_j, R_1, \ldots, R_n)/P(n\text{-}S_j, R_1, \ldots, R_n)] \\ &\quad + \log[P(n\text{-}S)/P(S)] \end{aligned} \tag{4}$$

where $P(S_j, R_1, \ldots, R_n)$ is the joint probability of the conformation S at position j in the sequence and the local sequence $R_1, \ldots, R_n$. One may remark that

$$P(S_j, R_1, \ldots, R_n) + P(n\text{-}S_j, R_1, \ldots, R_n) = 1 \tag{5}$$

and that

$$P(S_j, R_1, \ldots, R_n)/P(n\text{-}S_j, R_1, \ldots, R_n) = P(S)/P(n\text{-}S)e^{I(\Delta S_j; R_1, \ldots, R_n)} \tag{6}$$

In predicting a residue to be in one conformation, one can predict either the one having the highest value of the information with Eq. (4) or the highest probability value taken from Eqs. (5) and (6). Probability values have been used for the prediction of Ramachandran zones.[15] They are more precise than confidence scales developed for other methods[8,16] and underline the fact that the decision to predict the conformation of the highest probability leaves the possibility that the other conformations have a definite probability to occur which can be close to the highest, and thus should not necessarily be ruled out.

One faces a fundamental problem when calculating information values. One needs to estimate terms such as $P(S_j, R_1, \ldots, R_n)$ involving N residues. It is impossible to evaluate such terms directly from the database, so one must resort to various approximations. The different versions of the GOR method correspond to various types of approximations we have tried in an effort to improve the accuracy of the method.

Approximations Involved in GOR Method

The first GOR version,[4] named GOR I, added to the single-residue information the so-called directional information of eight residues on each side of the residue to be predicted in the sequence. This limit of eight was not arbitrary but was based on studies of information content at increasing separations. To obtain the information measure, one starts by calculating from the database the frequency of each of the 20 amino acids residues at different positions, up to eight residues on the N-terminal and C-terminal side, when the central residue is observed in a given conformation but independently of the nature of that residue. In fact, in this approximation one assumes that there is no correlation between residues occurring at different positions in the window of 17 residues so defined. Then

$$I(\Delta S_j; R_1, \ldots, R_n) \approx I(\Delta S_j; R_j) + \Sigma_{m,m \neq 0} I(\Delta S_j; R_{j+m}) \qquad (7)$$

where j stands for any position j in the amino acid sequence of which the conformation ought to be predicted and m is between -8 (N-terminal side) and $+8$ (C-terminal side). The same version, GOR II, was updated with a new database in 1989.[6] Both versions predicted four conformations, H, E, C, and T, with T for turns. Subsequent versions predicted only three conformations H, E, and C, although the method has no intrinsic limitation in the number and nature of conformations. The different turn types and the relative difficulty of distinguishing between them using the DSSP program[17] led us to limit the prediction for the time being to three conforma-

[15] J. F. Gibrat, B. Robson, and J. Garnier, *Biochemistry* **30**, 1578 (1991).
[16] V. Biou, J. F. Gibrat, J. M. Levin, B. Robson, and J. Garnier, *Protein Eng.* **2**, 185 (1988).
[17] W. Kabsch and C. Sander, *Biopolymers* **22**, 2577 (1983).

tions so that helices and strands represent the major architectural structures of the conserved core of homologous proteins.

The next level of approximation, introduced in the GOR III version,[18] considered the correlation between the type of residues in the window and the type of the residue to be predicted. This version uses the so-called pair information:

$$I(\Delta S_j; R_1, \ldots, R_n) \approx I(\Delta S_j; R_j) + \Sigma_{m,m \neq 0} I(\Delta S_j; R_{j+m}|R_j) \qquad (8)$$

The second term on the right-hand side of Eq. (8) is a conditional information.[15] It involves the calculation from the database of pair frequencies of residues R_j and R_{j+m} with R_j having the observed conformations S_j and n-S_j at position j, with a frequency of f_{S_j,R_{j+m},R_j} and $f_{n\text{-}S_j,R_{j+m},R_j}$, respectively, but the conformation of residue R_{j+m} is not taken into consideration. We have

$$I(\Delta S_j; R_{j+m}|R_j) = \log(f_{S_j,R_{j+m},R_j}/f_{n\text{-}S_j,R_{j+m},R_j}) + \log(f_{n\text{-}S_j,R_j}/f_{S_j,R_j}) \qquad (9)$$

When this approximation was used, the database (at that time containing roughly 12,000 residues) was barely large enough to allow an easy calculation of terms for Eq. (9). Each of these terms involves two amino acids and a secondary structure conformation, so there are 1200 entries in the table. The average number of observations per entry was therefore 10. However, the amino acids are not all equiprobable; some like Trp or Met are rarer, and the number of observations for entries involving such amino acids were less than 10. As a consequence, the probabilities estimated using the ratio of such sparse frequencies were unreliable and were responsible for a decrease of the prediction accuracy. To circumvent this problem, we introduced so-called dummy frequencies. Readers interested in the precise definition of these dummy frequencies are referred to Refs. 15 and 17.

Dummy frequencies amount to the following considerations. Let us assume that we observe in the database two occurrences of a Met at position $j - 1$ when the residue at j is a Trp in helical conformation. We can calculate easily the expected number we would observe if the two events were uncorrelated, namely, this is the frequency of Met at position $j - 1$ multiplied by the frequency of Trp at position j having a helical conformation divided by the total number of residues. This number is relatively reliable since it is calculated using frequencies that involve only one residue (which thus are greater than frequencies for pairs by a factor of 20, on average). Now we can ask the question, How much do we trust the frequencies for pairs we observed in the database? If we trust them 100%, we just use these frequencies in the calculations of information values. If we mistrust them 100%, we can always use the numbers calculated assuming that

[18] J. F. Gibrat, J. Garnier, and B. Robson, *J. Mol. Biol.* **198**, 425 (1987).

the events are independent, but then we are back to the approximation of Eq. (7). In fact, empirically, to improve the accuracy prediction we need to consider an intermediary stage when we bias the observed frequencies toward the calculated (uncorrelated) ones by a given amount.

The database available (see Table I) now contains about 63,000 residues, so the average number of observations for pairs per entry is 50. This is large enough for us to compute terms involving pairs of residues without the need for introducing dummy frequencies. Note, however, that this new database does not allow the calculation of terms involving triplets of amino acids. We thus decided to include more pairs in our description of the window of 17 residues. Instead of considering only the 16 pairs R_{j+m}, R_j, with m varying from -8 to $+8$ and $m \neq 0$, that is, the pair formed by each residue in the window and the central one, we consider all the possible pairs in the window [there are $(17 \times 16)/2$ such pairs]. We thus have used for the results presented below the following approximation (GOR IV version):

$$\log \frac{P(S_j, LocSeq)}{P(n\text{-}S_j, LocSeq)} = \frac{2}{17} \sum_{\substack{m=-8, \\ n>m}}^{+8} \log \frac{P(S_j, R_{j+m}, R_{j+n})}{P(n\text{-}S_j, R_{j+m}, R_{j+n})}$$

$$- \frac{15}{17} \sum_{m=-8}^{+8} \log \frac{P(S_j, R_{j+m})}{P(n\text{-}S_j, R_{j+m})} \qquad (10)$$

where *LocSeq* stands for the local sequence $R_1, \ldots, R_n$ of 17 residues around the residue to be predicted. The values of $P(S_j, LocSeq)$ from Eq. (10) for the three conformations are then used directly for the predictions instead of using probabilities calculated from Eqs. (5) and (6) with the information value calculated from Eq. (9).

Database and Results

We used a database of 267 protein structures having a resolution better than 2.5 Å with an R factor less than 25% and whose length is greater than 50 residues (see Table I). There is no pair of proteins with an identity above 30%. The prediction is carried out using a jackknife: the protein to be predicted is removed from the database, the parameters are estimated using the 266 remaining proteins, and the prediction is done using these parameters. As mentioned above, this database is large enough that we do not need to use dummy frequencies anymore. Moreover, we do not use decision constants to adjust the predicted number of secondary structures to the observed number in the database. In fact, there is no optimization of any sort; we just estimate the probabilities according to Eq. (10) from the frequencies observed in the database.

TABLE I
Database Proteins[a]

1aaj.x	1aak.x	1aap.a	1aba.x	1abk.x	1abm.a	1add.x
1ads.x	1alk.a	1aoz.a	1apa.x	1apm.e	1arb.x	1atr.x
1avh.a	1ayh.x	1bab.a	1bbh.a	1bbp.a	1bet.x	1bge.a
1bll.e	1bmd.a	1bov.a	1bpb.x	1brs.d	1btc.x	1c2r.a
1caj.x	1cau.a	1cau.b	1cde.x	1cdt.a	1cew.i	1cgt.x
1chm.a	1cmb.a	1cob.a	1col.a	1cpc.a	1cpc.b	1cpt.x
1crl.x	1cse.i	1ctf.x	1ctm.x	1cus.x	1ddt.x	1dhr.x
1dog.x	1dsb.a	1eaf.x	1eco.x	1ede.x	1end.x	1epa.a
1fba.a	1fdd.x	1fha.x	1fia.a	1fkb.x	1fna.x	1fnr.x
1fxi.a	1gal.x	1gd1.o	1gdh.a	1gky.x	1glt.x	1gmf.a
1gof.x	1gox.x	1gp1.a	1gpb.x	1gpr.x	1gsr.a	1hbq.x
1hdx.a	1hiv.a	1hlb.x	1hle.a	1hmy.x	1hoe.x	1hpl.a
1hrh.a	1hsl.a	1huw.x	1ifc.x	1ipd.x	1isu.a	1ith.a
1l29.x	1le4.x	1len.a	1lga.a	1lis.x	1lla.x	1lmb.3
1lts.a	1lts.d	1mdc.x	1mgn.x	1min.a	1min.b	1mjc.x
1mpp.x	1mup.x	1nar.x	1nba.a	1ndk.x	1noa.x	1nsb.a
1nxb.x	1ofv.x	1olb.a	1omf.x	1omp.x	1onc.x	1osa.x
1pda.x	1pfk.a	1pgb.x	1pgd.x	1phh.x	1php.x	1pii.x
1plf.a	1poc.x	1poh.x	1pox.a	1ppa.x	1ppf.e	1ppf.i
1ppn.x	1prc.c	1prc.h	1prc.l	1prc.m	1pts.a	1pya.a
1pya.b	1pyd.a	1rcb.x	1rec.x	1rib.a	1rnd.x	1rop.a
1rve.a	1s01.x	1sac.a	1sbp.x	1ses.a	1sgt.x	1sha.a
1shf.a	1sim.x	1slt.b	1snc.x	1spa.x	1stf.i	1tbe.a
1tca.x	1tie.x	1tml.x	1tnd.a	1tpl.a	1trb.x	1trk.a
1tro.a	1ttb.a	1utg.x	1vaa.a	1vaa.b	1vmo.a	1wht.a
1wht.b	1wsy.a	1wsy.b	1yhb.x	1zaa.c	256b.a	2aai.b
2aza.a	2bop.a	2ccy.a	2cdv.x	2chs.a	2cmd.x	2cp4.x
2cpl.x	2cro.a	2ctc.x	2cts.x	2cyp.x	2dnj.a	2er7.e
2hbg.x	2hhm.a	2hip.a	2hpd.a	2ihl.x	2lh2.x	2liv.x
2mhr.x	2mnr.x	2msb.a	2mta.c	2mta.h	2mta.l	2pf1.x
2pia.x	2pol.a	2por.x	2reb.x	2rn2.x	2rsl.a	2sar.a
2sas.x	2scp.a	2sga.x	2sn3.x	2spc.a	2tgi.x	2tmd.a
2tpr.a	2tsc.a	3aah.a	3aah.b	3adk.x	3b5c.x	3cd4.x
3chy.x	3cla.x	3cox.x	3dfr.x	3eca.a	3gap.a	3gbp.x
3ink.c	3rub.l	3rub.s	3sdh.a	3tgl.x	451c.x	4blm.a
4enl.x	4fgf.x	4gcr.x	4ts1.a	4xis.x	5fbp.a	5p21.x
5tim.a	6fab.h	6fab.l	6taa.x	8abp.x	8acn.x	8atc.a
8atc.b	8cat.a	8i1b.x	8rxn.a	8tln.e	9ldt.a	9rnt.x
9wga.a						

[a] The database was prepared by J. M. Levin and checked for homologous sequences with the help of V. Di Francesco. This database has been modified to restore the total length of the sequences as defined in the SEQRES field of the Protein Data Bank (PDB) file (the DSSP program omits residues whose coordinates are missing in the PDB file, and thus if this occurs in the middle of the polypeptide chain it is split into two or more chains). Residues having no coordinates were assigned the conformation X and were not taken into account for the prediction accuracy although the prediction was done with the whole sequence length. The PDB code is followed by the chain name a, b, c, d, h (heavy), l (light), x (one chain only), e (enzyme), or i (inhibitor).

TABLE II
GLOBAL RESULTS FOR DATABASE PREDICTION

	Observed			
	H	E	C	Total
Predicted				
H	14,460	3094	4790	22,344
E	1124	4965	2089	8178
C	6002	5546	21,496	33,044
Total	21,586	13,605	28,375	63,566
Q_{prd} [a]	64.7	60.7	65.1	
Q_{obs} [b]	67.0	36.5	75.8	
Q_3 [c] = 64.4%				

[a] Number of correctly predicted residues/number of predicted residues.
[b] Number of correctly predicted residues/number of observed residues
[c] Total number of correctly predicted residues/total number of residues.

However, this sometimes leads to predictions that are not physically meaningful, for example, helices having only two residues, or mixtures of strand (E) and helix (H) residues. Several attempts have been made to solve that problem (see Rost and Sander[8] and Zimmermann[19]). Here we added a simple filter after the prediction which requires helices to be at least four residues and strands to be at least two residues. For instance, let us assume that we predict two isolated H residues. We then look for all the possibilities of extension of the two H residues (in this case, there are three possibilities: adding two H's before, adding one H before and one H after, and adding two H's after the predicted H's). We then calculate the product of the probabilities of the different secondary structures for the three segments so defined. The segment that is the most probable is selected leading either to an extension of the helix to four residues or to the suppression of the two isolated residues. Although this filter affects the prediction of particular proteins, on average for the whole database it has no effect on the prediction accuracy; it neither improves nor decreases the percentage of correctly predicted residues.

The global results for the database are shown in Table II. The percentage of correctly predicted residues is 64.4%, and an individual prediction output of the program is given in Table III with an extra column to compare with

[19] K. Zimmermann, *Protein Eng.* **7**, 1197 (1994).

TABLE III
PREDICTION OF EGLIN[a]

Seq	Obs	Prd	pH	pE	pC
T	X	C	0.00	0.00	1.00
E	X	C	0.00	0.02	0.98
F	X	C	0.00	0.08	0.92
G	X	C	0.01	0.13	0.87
S	X	C	0.02	0.14	0.84
E	X	C	0.04	0.24	0.72
L	X	C	0.09	0.33	0.59
K	C	C	0.15	0.24	0.61
S	C	C	0.19	0.16	0.65
F	C	C	0.12	0.12	0.77
P	C	C	0.29	0.12	0.59
E	C	C	0.35	0.26	0.39
V	C	C	0.37	0.30	0.34
V	C	C	0.35	0.24	0.41
G	C	C	0.25	0.20	0.55
K	C	C	0.26	0.27	0.47
T	C	C	0.24	0.34	0.42
V	H	C	0.37	0.19	0.44
D	H	H	0.54	0.08	0.38
Q	H	H	0.58	0.11	0.30
A	H	H	0.61	0.11	0.28
R	H	H	0.56	0.19	0.25
E	H	H	0.50	0.27	0.24
Y	H	H	0.46	0.35	0.19
F	H	H	0.34	0.44	0.22
T	H	H	0.29	0.38	0.32
L	H	C	0.20	0.35	0.44
H	H	C	0.09	0.22	0.69
Y	C	C	0.03	0.11	0.86
P	C	C	0.05	0.06	0.89
Q	C	C	0.09	0.15	0.76
Y	C	C	0.08	0.29	0.63
N	E	C	0.07	0.31	0.61
V	E	E	0.06	0.65	0.30
Y	E	E	0.04	0.75	0.21
F	E	E	0.02	0.76	0.22
L	E	C	0.01	0.30	0.69
P	E	C	0.02	0.09	0.88
E	C	C	0.02	0.03	0.95
G	C	C	0.01	0.01	0.98
S	C	C	0.01	0.02	0.97
P	C	C	0.09	0.12	0.79
V	E	C	0.18	0.33	0.49
T	E	H	0.23	0.51	0.26

TABLE III (continued)

Seq	Obs	Prd	pH	pE	pC
L	C	H	0.36	0.46	0.18
D	C	H	0.38	0.33	0.29
L	C	H	0.51	0.17	0.32
R	C	C	0.39	0.16	0.46
Y	C	C	0.33	0.22	0.46
N	C	C	0.24	0.18	0.57
R	E	C	0.20	0.31	0.49
V	E	E	0.17	0.57	0.26
R	E	E	0.12	0.71	0.17
V	E	E	0.07	0.80	0.13
F	E	E	0.05	0.71	0.25
Y	E	E	0.03	0.49	0.48
N	E	C	0.01	0.12	0.87
P	C	C	0.01	0.03	0.96
G	C	C	0.01	0.03	0.96
T	C	C	0.02	0.10	0.88
N	C	C	0.03	0.29	0.68
V	E	E	0.04	0.54	0.41
V	E	E	0.05	0.68	0.27
N	C	E	0.02	0.71	0.27
H	C	E	0.01	0.58	0.41
V	C	C	0.01	0.22	0.78
P	C	C	0.01	0.09	0.91
H	E	C	0.00	0.02	0.98
V	E	C	0.00	0.00	1.00
G	C	C	0.00	0.00	1.00

[a] Amino acid sequence (Seq) of eglin, a subtilisin inhibitor (1cse), with observed conformations (Obs), predicted conformations with filter (Prd), and the probability values pH, pE, and pC for the predicted α helix (H), β strands (E), and coil (C), respectively. The conformation X corresponds to residues for which the crystallographer gave no coordinates. For some residues, for instance, V-13, although the probability for H is higher, the filter assigned a coil (see text). The accuracy of prediction for the three conformations (Q_3) is 73%.

the observed conformations. The result when considering the prediction of individual proteins is 64.7% with a standard deviation of 9.3%. Figure 1a shows the number of proteins as a function of the percentage of correctly predicted residues. Figure 1b is just a check that this distribution does not depart significantly from a Gaussian distribution.

As suggested by Levin,[20] Fig. 2 shows a scatter plot of the percentage of correctly predicted residues as a function of the size of the protein

[20] J. M. Levin, to be published.

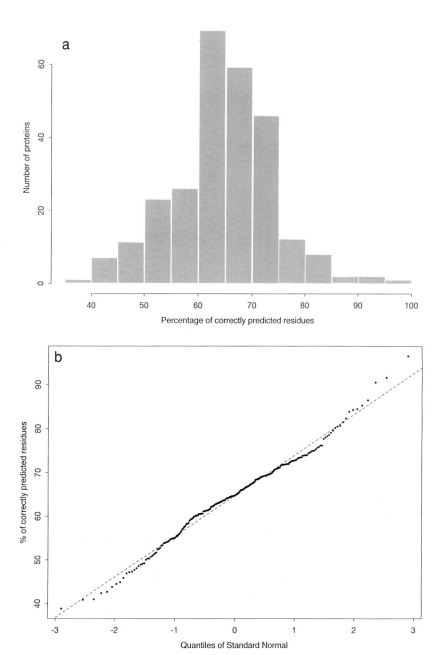

Fig. 1. (a) Histogram of secondary structure prediction accuracies (Q_3) for all the proteins of the database. (b) Normal quantile–quantile plot of the results showing the agreement of the distribution with a normal distribution. The individual values of Q_3 for each protein of the database are sorted according to the quantiles of a normal distribution with a mean of 64.4% and a standard deviation of 9.3%. The dashed line is the least-squares fit of the points.

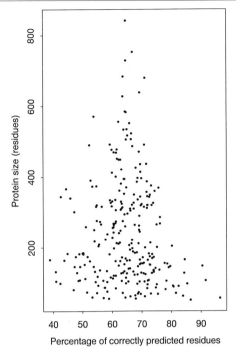

FIG. 2. Distribution of the sequence length (number of amino acid) of the database proteins as a function of the number of correctly predicted residues.

(number of residues). There seems to be no apparent effect of the length of the protein on the accuracy of prediction, except that the longer the protein, the closer the accuracy comes to the average value. For proteins of less than 200 residues the accuracy can lie anywhere between 40 and 90%. A consequence of this observation is that no protein is easier to predict than any other, but rather there are segments of the sequence easier to predict; thus the shorter is the protein, the more likely its accuracy of prediction will be different from the mean, and inversely for the longest ones.

Table IV shows results for the prediction of secondary structure segments, namely, helices and strands. The percentage of correctly predicted segments is given according to the minimum percentage of overlap that is allowed between the predicted and observed segments. For instance, in Table IV, for the row 75% a segment is considered as being correctly predicted if the predicted segment overlaps with at least 75% of the observed segment (counted as the number of residues).

Conclusion

Through the successive incorporation of observed frequencies of single, then pairs of residues on a local sequence of 17 residues, the accuracy of

TABLE IV
RESULTS FOR SEGMENTS: HELICES AND STRANDS

	Number of segments			Average length	
	H	E	Total	H	E
Observed	1989	2587	4576	10.9	5.9
Predicted	2148	2043	4191	10.6	4.1

Overlap	H segments	E segments
75%	51.1	23.7
50%	70.0	42.0
25%	75.7	50.2

the GOR method has been improved from about 55% (GOR I using a jackknife[21]) up to 64.4%. The increase of the database size from 67 proteins to the present database of 267 proteins and the use of a more detailed description of the local sequence resulted in an improvement of about 1% (GOR III, Q_3 = 63.3%; GOR IV, Q_3 = 64.4%; the corresponding standard deviations of Q_3 are 0.8% and 0.6%, respectively). However, the result of 63.3% for GOR III was reached using dummy frequencies and adding decision constants that we now believe resulted in a slight bias toward the database we were then using. This causes the overall accuracy of the method to be overestimated by a percent or so, as became apparent when parameters derived from the original database were used to predict new sets of proteins. This is the reason why, here, we avoided the use of decision constants (e.g., to adjust the number of predicted secondary structures to what is observed in the database) in order to obtain a more robust estimation of the accuracy of the method. This small increase in the prediction accuracy is consistent with a previous assumption we made.[15] We estimated that we were able to extract more or less all the information available in the local sequence. Other published methods including neural net methods, using only the protein sequence, are of similar or lower accuracy (see Refs. 2 and 3).

We attributed the limitation in the accuracy of the prediction[15] to the lack of long-distance effects. This appears to be confirmed here by the poor quality of the β-strand prediction. Because β sheets require the pairing of residues that may be distant along the sequence, this secondary structure is presumably more dependent on long-range interactions than are α helices or coils. Clearly the method does not fare well with the prediction of β strands. Although Q_{prd} for β strands is only slightly lower than Q_{prd} for the

[21] W. Kabsch and C. Sander, *FEBS Lett.* **155,** 179 (1983).

two other secondary structures, there is a chronic underprediction of this structure, Q_{obs} for β strands is thus significantly lower compared to Q_{obs} for helices and coils (even taking into account the overestimation of the corresponding Q_{obs} values, which is a consequence of the overprediction of helices and coils). This is also noticeable in the fact that, whereas the average length of predicted and observed helices corresponds closely, the average length of predicted strands is shorter by about one-third compared to the average length of the observed ones.

It is thus very important to consider the possibilities of including long-range interactions in our method (or other methods using only short-range information, that is, local sequence only, for that matter). One way to introduce long-distance effects is to use specific nonlocal pairs to improve β-strand prediction.[22] In other words, the prediction of β strands could be done using the local window as usual to first select putative β-strand segments and then one could slide a window along the sequence to look whether complementary segments to these putative β strand segments can be found. Another possibility is to use multiple alignments. This is based on the assumption that corresponding residues in the alignment, provided the alignment is correct, will have the same secondary structure, being at the same location in the fold. The use of multiple alignments has recently been the source of an improvement of the accuracy ranging from 5, for the GOR[7] and the quadratic logistic[9] methods up to 10 percentage points for a neural network method.[8]

The GOR method has the advantage over neural network-based methods or nearest-neighbor methods in that it clearly identifies what is taken into account for the prediction and what is neglected. Moreover, the method provides estimates of probabilities for the three secondary structures at each residue position, which can be useful for further application of the method.[15] It relies only on observed frequencies in the database; thus, the calculation of the parameters is straightforward and easy to update.

Availability

The corresponding program has been written in C language and currently runs on a platform with a UNIX operating system (but it will run equally well on other operating systems). It can be obtained by anonymous ftp at NCBI (National Center for Biotechnology Information) using the following procedure: ftp ncbi.nlm.nih.gov, move to the directory gibrat/ GOR. It is also available at INRA-Jouy-en-Josas: ftp locus.jouy.inra.fr, move to directory/pub/protein/GOR.

---

[22] T. J. P. Hubbard, *27th Annual Hawaii International Conference on System Sciences,* (L. Hunter, ed.) IEEE Computer Society Press, Los Alamos, 336 (1994).

[33] Analysis of Compositionally Biased Regions in Sequence Databases

By John C. Wootton and Scott Federhen

Introduction

Sequences of natural macromolecules are very different from random polymers, most strikingly in the numerous interspersed "simple" sequence regions that have significant biases in amino acid or nucleotide composition. From systematic analyses that use a conservative definition of significant bias, such regions account for approximately one-quarter of the amino acids in current sequence databases.[1,2] These include segments usually described by terms like glutamine-rich or glycine–arginine-rich, and also more weakly repetitive nonglobular domains. More than one-half of the proteins have at least one such region. Genomic DNA sequences are considerably more variegated than amino acid sequences, exhibiting many types of bias in different functional subclasses of sequences.[3]

Collectively, these regions exhibit a very broad range of compositional properties and lengths, and most of them have unknown structures, dynamics, and interactions.[4] Unprecedented classes keep on appearing in new genomic sequences and their coding sequence translations. The sequence simplicity varies from extreme, as in homopolymeric tracts, to very subtle as in some nonglobular domains of proteins. Locally abundant residues may be contiguous or loosely clustered, irregularly spaced or periodic. They tend to evolve rapidly, reflecting mutational processes such as replication slippage, unequal crossing-over, and biased nucleotide substitution. The relative roles of functional selection and mutational drive are generally unknown.

Therefore, lacking a consistent conceptual framework based on structure and evolution, computer analysis of low-complexity regions presents very different challenges than pairwise or multiple sequence alignment. In most cases it does not make sense from either structural or mutational

[1] J. C. Wootton and S. Federhen, *Comput. Chem.* **17**, 149 (1993).
[2] J. C. Wootton, *Comput. Chem.* **18**, 269 (1994).
[3] P. Salamon and A. K. Konopka, *Comput. Chem.* **16**, 117 (1992).
[4] J. C. Wootton, *Curr. Opin. Struct. Biol.* **4**, 413 (1994).

Copyright © 1996 by Academic Press, Inc.
All rights of reproduction in any form reserved.

viewpoints to attempt to align low-complexity sequences position by position. Instead, general methods are required to explore and analyze non-random compositional heterogeneity in sequence databases. We have developed algorithms for these purposes, which are implemented in the well-tested SEG family of programs.[1,2] We provide here a practical guide to using these programs, and also a brief outline of their basic, but relatively unfamiliar, underlying principles.

Programs SEG and PSEG are tuned for amino acid sequences and NSEG for nucleotide sequences. The programs can be applied to either (1) individual sequences, including whole chromosomes if appropriate, or (2) entire sequence databases. The low-complexity regions can be defined for further study in their own right,[2,4] or, alternatively, can be masked from query sequences for sequence similarity searches in order to focus attention on alignments of high-complexity regions.[5]

The SEG philosophy is to use unbiased inference in an exploratory spirit. A sequence or database is treated initially as a heterogeneous mixture with unknown statistical properties, and then attempts may be made to infer these properties. An initial assumption of equal uniform probabilities for the appearance of residues places all possible low-complexity segments on an equal footing. For example, polypeptide segments rich in generally common amino acids such as leucine and alanine are treated as no more or no less surprising than segments rich in histidine, methionine, or tryptophan. This approach is justified by the results: in current databases, clusters of leucine are relatively rare, whereas segments rich in, for example, histidine or methionine are relatively abundant.

Having identified low-complexity regions at a given level of significance, we would then like to answer questions like, What does the segment resemble, and do any similar segments have known functions? This research requires database-oriented methods for segment comparison and classification, based on compositional attributes. These methods are under development[6] and are omitted from this chapter, as is a large body of relevant and fascinating mathematical theory, much of which has been reviewed by Konopka[7] under the name Biomolecular Cryptology. At the time of writing, only a few of these possible theoretical approaches have been implemented in robust software that is generally available via Internet. In addition to

[5] S. F. Altschul, M. Boguski, W. Gish, and J. C. Wootton, *Nat. Genet.* **6,** 119 (1994).
[6] J. C. Wootton and S. Federhen, *in* "Bioinformatics and Genome Research" (H. A. Lim and C. R. Cantor, eds.), p. 159. World Scientific Publishing, Singapore, 1995.
[7] A. K. Konopka, *in* "BIOCOMPUTING: Informatics and Genome Projects" (D. Smith, ed.), p. 119–174. Academic Press, San Diego, 1994.

the SEG family, these include SAPS,[8] XNU,[9] PYTHIA,[10] and SIMPLE34.[11] A comparative review of these and other methods has been published separately.[12]

Definitions

The SEG programs use general measures[1,3,13] of the combinatorial complexity of sequences and their compositions. Terms like "low-complexity" and "local compositional complexity" are used here only in this specific sense, distinct from the complexity theory of physicists, from algorithmic complexity, and from sequence complexity of classical reassociation kinetic experiments. We try to avoid the terms "entropy" and "information content" which have suffered from inconsistent usage in the biological literature.

Compositional complexity is based only on residue composition, regardless of the patterns or periodicity of sequence repetitiveness. This contrasts with some alternative methods[11,14,15] that use counts of k-grams (k-letter words; e.g., TAGG is a 4-gram) to define residue patterns and clustering. Complexity, pattern, and periodicity are distinct abstract attributes of simple sequences. For example, the following sequences have identical (low) compositional complexity because of their identical compositions (A_8T_8), but differ in patterns and periodicity:

TATATAAATTTAATTA has neither significant pattern nor periodicity
TAATTAATTTTAATAA has notable k-gram patterns (TAA, AAT, and
TAAT) but these are irregularly spaced and do not show periodicity
TATATATATATATATA has periodicity, modulo-2, and hence significant
k-gram patterns as a consequence

In the SEG programs, the complexity state of a sequence or subsequence is represented by a list of numbers, a complexity state vector,[1,3,13,16] which summarizes the composition. For example, a 5-nucleotide window has six

[8] V. Brendel, P. Bucher, I. R. Nourbakhsh, B. E. Blaisdell, and S. Karlin, *Proc. Natl. Acad. Sci. U.S.A.* **89,** 2002 (1992).

[9] J.-M. Claverie and D. J. States, *Comput. Chem.* **17,** 191 (1993).

[10] A. Milosavljevic and J. Jurka, *Comput. Appl. Biosci.* **9,** 407 (1993).

[11] J. M. Hancock and J. S. Armstrong, *Comput. Appl. Biosci.* **10,** 67 (1994).

[12] J. C. Wootton, *in* "Nucleic Acid and Protein Sequence Analysis: A Practical Approach" (M. J. Bishop and C. J. Rawlings, eds.), in press. IRL Press, Oxford, 1995.

[13] P. Salamon, J. C. Wootton, A. K. Konopka, and L. K. Hansen, *Comput. Chem.* **17,** 135 (1993).

[14] P. A. Pevsner, M. Y. Borodovsky, and A. A. Mironov, *J. Biomol. Struct. Dyn.* **6,** 1013 (1989).

[15] S. Pietrokovski, J. Hirshon, and E. N. Trifonov, *J. Biomol. Struct. Dyn.* **7,** 1251 (1990).

[16] A. K. Konopka and J. Owens, *Genet. Anal. Tech. Appl.* **7,** 35 (1990).

possible complexity state vectors. In order of increasing compositional complexity, these are {5,0,0,0}, {4,1,0,0}, {3,2,0,0}, {3,1,1,0}, {2,2,1,0}, and {2,1,1,1}. These vectors are ranked lists of the numbers of each nucleotide, irrespective of which letter corresponds to each number. The complexity state {3,1,1,0}, for example, has 12 different possible compositions, which include (T_3,C,A) and (G_3,T,C). Each of these compositions has 20 possible sequences, or permutations, of the letters in the composition.

Each complexity state has the same number of sequences per composition. The possibility of 20 sequences per composition makes {3,1,1,0} a more complex state than, for example, {3,2,0,0} which has only 10 possible sequences per composition. The theoretical number of complexity state vectors, which is computable from well-established principles of number theory, becomes very large at longer windows.[1,2] For example, an amino acid window of length 40 generates 35,251 complexity states and a rounded total of 1.1×10^{52} sequences.

Methods

We first describe the SEG program and its various applications in analyzing interspersed low-complexity amino acid sequences. Then we consider additional methods implemented in PSEG and NSEG to analyze periodic compositional complexity and nucleotide sequences.

Segment Representation

Low-complexity segments in natural protein sequences are most commonly interspersed, so that compositional bias may be considered to be a local property of contiguous residues.[1] These local contrasts in complexity may be represented by optimized segments, that is, nonoverlapping subsequences with precise boundaries that are defined as either high complexity or low complexity. The number of such segments and their lengths are determined automatically by the optimization algorithm in SEG, subject to parameters (which may be user-specified) that control the granularity and stringency of segmentation.

The optimized segments produced by SEG are represented, in either human-readable or machine-readable forms, as digitized text strings, together with their sequence position coordinates and other data. This approach is distinct from graphical methods such as dot-matrix plots or complexity profiles, which also have their uses in representing compositional bias (see below) but are not readily applied to large sequence databases. SEG can analyze sequences of any length, and also entire sequence databases. The segmented sequence strings, or their sequence identifiers and coordinates, may be used as input for further computer analysis or stored as a specialized minidatabase.

Using SEG Program

The SEG program is run from a UNIX command line that specifies the input file. This file is an amino acid sequence, formatted as for the FASTA[17] or BLAST[18] programs, or a database file containing many such sequences. Using the example of the human prion protein (file "prion," Fig. 1A), the minimal command line for SEG is "seg prion" or, to direct the standard output to a file, "seg prion > prion.segout." This command implies the default parameters and options, as discussed below. The program runs in less than a second on typical UNIX workstations with sequences of up to approximately 1000 amino acids.

The default output is tree form (Fig. 1B), which is designed to be human-readable and to focus attention on segments of interest. Regions of contrasting complexity are displayed on the two sides of a central column of residue numbers, with low-complexity segments in lowercase letters on the left-hand side and high-complexity segments in uppercase letters on the right. The sequences of all segments read from left to right, and their position in the sequence is ordered top to bottom, as the residue numbers indicate. The sequence identifiers and definition line are also presented.

Some options of SEG give forms of output that are designed primarily for input to other computer programs. Commonly used examples are shown in Fig. 1C,D; a complete list of options is given in the documentation file available with the source code (see section on Availability below). The "-x" option (Fig. 1C) produces a masked FASTA-formatted file, ready for input as a query sequence for database search programs such as BLAST or FASTA. The amino acids in low-complexity regions are replaced with "x" characters. This option is implemented as a filter in the BLAST series of programs when the "-filter seg" option is specified.[19] Masking removes the potential confusion caused by low-complexity segments in database search methods such as BLAST or FASTA. Compositional bias is not encompassed by the random model used in these methods to evaluate local alignment statistics, and it can result in spuriously high scores and overwhelmingly large output lists.[5] One essential application of masking

[17] W. R. Pearson and D. J. Lipman, *Proc. Natl. Acad. Sci. U.S.A.* **85,** 2444 (1988).

[18] S. F. Altschul, W. Gish, W. Miller, E. W. Myers, and D. J. Lipman, *J. Mol. Biol.* **215,** 403 (1990).

[19] T. L. Madden, R. L. Tatusov, and J. Zhang, this volume [9].

(A) Input File (FASTA Format)

```
>PRIO_HUMAN MAJOR PRION PROTEIN PRECURSOR (PRP) (PRP27-30) (PRP33-35C).
MANLGCWMLVLFVATWSDLGLCKKRPKPGGWNTGGSRYPGQGSPGGNRYPPQGGGGWGQP
HGGGWGQPHGGGWGQPHGGGWGQPHGGGWGQGGGTHSQWNKPSKPKTNMKHMAGAAAAGA
VVGGLGGYMLGSAMSRPIIHFGSDYEDRYYRENMHRYPNQVYYRPMDEYSNQNNFVHDCV
NITIKQHTVTTTTKGENFTETDVKMMERVVEQMCITQYERESQAYYQRGSSMVLFSSPPV
ILLISFLIFLIVG
```

(B) seg prion

```
>PRIO_HUMAN MAJOR PRION PROTEIN PRECURSOR (PRP) (PRP27-30) (PRP33-35C).
                                    1-49    MANLGCWMLVLFVATWSDLGLCKKRPKPGG
                                            WNTGGSRYPGQGSPGGNRY
ppqggggwgqphgggwgqphgggwgqphgg      50-94
            gwgqphgggwgqggg
                                    95-112   THSQWNKPSKPKTNMKHM
         agaaaagavvgglggymlgsams    113-135
                                    136-187  RPIIHFGSDYEDRYYRENMHRYPNQVYYRP
                                             MDEYSNQNNFVHDCVNITIKQH
                  tvttttkgenftet    188-201
                                    202-236  DVKMMERVVEQMCITQYERESQAYYQRGSS
                                             MVLFS
            sppvillisflifliv       237-252
                                    253-253  G
```

(C) seg prion -x

```
>PRIO_HUMAN MAJOR PRION PROTEIN PRECURSOR (PRP) (PRP27-30) (PRP33-35C).
MANLGCWMLVLFVATWSDLGLCKKRPKPGGWNTGGSRYPGQGSPGGNRYxxxxxxxxxxxx
xxxxxxxxxxxxxxxxxxxxxxxxxxxxxxxxxxxxxxxxTHSQWNKPSKPKTNMKHMxxxxxxxxx
xxxxxxxxxxxxxxxxxRPIIHFGSDYEDRYYRENMHRYPNQVYYRPMDEYSNQNNFVHDCV
NITIKQHxxxxxxxxxxxxxxxxDVKMMERVVEQMCITQYERESQAYYQRGSSMVLFSxxxx
xxxxxxxxxxxxxG
```

(D) seg prion -l

```
>PRIO_HUMAN(50-94) complexity=1.92 (12/2.20/2.50)
ppqggggwgqphgggwgqphgggwgqphgggwgqphgggwgqggg

>PRIO_HUMAN(113-135) complexity=2.47 (12/2.20/2.50)
agaaaagavvgglggymlgsams

>PRIO_HUMAN(188-201) complexity=2.26 (12/2.20/2.50)
tvttttkgenftet

>PRIO_HUMAN(237-252) complexity=2.50 (12/2.20/2.50)
sppvillisflifliv
```

Fig. 1. SEG input and outputs. The sequence is the unprocessed human prion protein, SWISS-PROT accession number P04156, name PRIO_HUMAN [H. A. Kretzschmar, L. E. Stowring, D. Westaway, W. H. Stubblebine, S. B. Prusiner, and S. J. Dearmond, *DNA* **5,** 315 (1986)]. (A) FASTA-formatted input file. (B, C, and D) Different forms of output given by the command lines shown and described in the text. The structural and functional significance of the low-complexity segments of the prion protein is unknown.

has been for the database–database comparisons used to precompute neighbors for the *Entrez* retrieval system.[20]

Some SEG output options write the optimized segments as a multisequence, FASTA-formatted "minidatabase" file, suitable for a wide range of further computer analyses. These may contain only the low-complexity segments ("-l" option, Fig. 1D) or only high-complexity segments ("-h" option), or both ("-a" option). Each segment is a separate sequence entry with an informative header line. Complexity data (K_2 in units of bits, see below) and other program parameters are included in the header line. The "-l" option is particularly useful for research on the low-complexity segments of whole sequence databases such as SWISS-PROT.[21] The SEG outputs from the "-l" option at different stringencies then become input for further searches, classification, and statistical analyses.[2,4,6]

Parameters

The number of low-complexity segments and their lengths are determined automatically by the optimization algorithm in SEG. However, the granularity and stringency of the search for low-complexity segments are controlled by three numeric parameters, which may be specified by the user on the command line. These are, in obligatory order after the sequence file name and before the options, W (trigger window length, an integer greater than zero, default 12 residues), $K_2(1)$ (trigger complexity, default 2.2 bits), and $K_2(2)$ (extension complexity, default 2.5 bits). The latter complexity values, for 20-letter amino acid sequences, may be greater than or equal to zero up to a maximum of 4.322 bits (the rounded value of $\log_2 20$).

The roles of these parameters are described in detail below in the section on the SEG algorithm. It is not necessary to understand the meaning of these parameters to use SEG creatively for many purposes, although knowledge of the algorithm is useful for more flexible parameterization and to help interpret unexpected results from database analyses. The default parameters are a good compromise for most proteins for the purpose of masking low-complexity regions, with the "-x" option, in query sequences for the BLAST programs. Thus, "seg prion -x" has the same meaning as "seg prion 12 2.2 2.5 -x." Other recommended standard parameter sets are as follows (for sequence file "myseq"): (1) for homopolymer analysis, to examine all homopolymeric subsequences of length 7 or greater, for example, use "seg myseq 7 0 0" (the complexity values of zero force maximum possible stringency, which identifies only homopolymers), and (2) for long

[20] G. D. Schuler, J. A. Epstein, H. Ohkawa, and J. A. Kans, this volume [10].
[21] A. Bairoch and B. Boeckmann, *Nucleic Acids Res.* **22**, 3578 (1994).

nonglobular domains of protein sequences, diagnose at longer window lengths,[2] for example, as discussed below, it is valuable to use two different granularities: "seg myseq 25 3.0 3.3" and "seg myseq 45 3.4 3.75."

Granularity

Figure 2 illustrates the concept of granularity, using both graphical complexity profiles and optimal segments produced by the SEG algorithm from the human prion protein sequence. In this example, the W (trigger window length) parameter was varied (5, 10, 20, or 30 residues). Longer trigger windows define more sustained regions of low complexity (compare 30-residue windows with those of 10 or 5, Fig. 2), but they overlook some

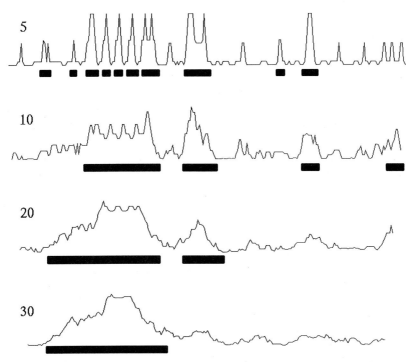

Fig. 2. Complexity profiles and optimized segments obtained from the human prion protein sequence (see Fig. 1 legend) at four different window lengths. The horizontal axis of the complexity profiles represents the 253 positions of the amino acid sequence. The vertical axis of each plot is the complexity, with low complexity at the top and high complexity at the bottom of each plot. The complexity measure is K_1 defined in the text section on the SEG algorithm, which is in the range 1–0. The solid bars below each plot show the extent of each optimal segment obtained in the corresponding four runs of SEG with the W parameter set to the same values as the window lengths of the profiles.

very short biased subsequences detected by the shorter windows. Thus, varying *W* changes the granularity, or resolution, of the search for low-complexity segments, but not necessarily the lengths of the optimized segments produced by SEG, which may be identical at different trigger window lengths.

Exploratory Strategy

The following is recommended as a routine strategy, to investigate a new protein sequence for possible low-complexity segments and nonglobular domains prior to database searches: (1) visual inspection of a self–self dot-matrix plot (many implementations are available), using a very low threshold, is recommended, so that low-complexity and repetitive regions appear as diagonal blotches and stripes (Fig. 3), providing a graphic, intu-

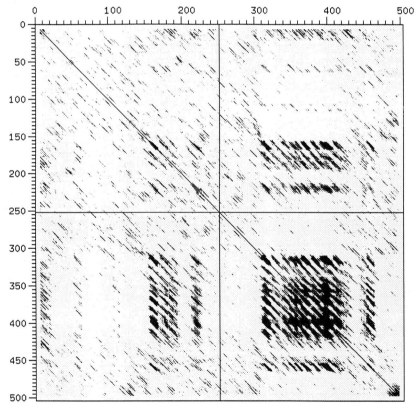

Fig. 3. A low-threshold, self–self dot-matrix plot of the human Wiskott–Aldrich syndrome protein (translated from GenBank accession number U12707, NCBI gi number 695151).

itive preliminary to more rigorous analysis using SEG; (2) SEG analysis with the default parameters and also the two sets (45 3.4 3.75 and 25 3.0 3.3) that target different length ranges of "nonglobular" regions is advised, as the default tree form output may focus attention on interesting regions and act as a guide to the use of masking in subsequent BLAST searches.

This strategy is illustrated in Figs. 3 and 4 for the human WASP gene product, which is mutationally modified in Wiskott–Aldrich syndrome, a hereditary immunodeficiency.[22] The protein is strikingly proline-rich, but the SEG results at different granularities show that a weakly repetitive pattern and possible nonglobular domains extend well beyond the obvious proline-rich segments. After masking of these potential nonglobular regions, a BLASTP output list then readily reveals possible sequence motifs in the high-complexity "globular" regions that weakly resemble parts of nucleolar transcription factor UBF-1. These features, and the potential nonglobular domains themselves, are candidates for further investigation by computer and experimental methods.

Nonglobular Domains

The advantage of using SEG to predict candidates for nonglobular domains of proteins is its generality.[2] Other, pattern-based methods may be used to classify specific classes of nonglobular structure, for example, types of coiled coil,[23] but the SEG method is sensitive to any regular or irregular weak repetitiveness of sequence, regardless of the actual polypeptide structure likely to be present.

Sequences of physicochemically defined nonglobular regions are partitioned accurately using the "nonglobular" SEG parameters recommended above, for example[2] (at $W = 45$), those of collagens, myosins, other coiled-coil proteins, elastins, mucins, proteoglycan core proteins, and long solvent-exposed α helices such as caldesmon, and (at $W = 25$) histones H1/H5 and nonhistone proteins. Candidates for new classes of unknown structure have also been predicted.[2] In the great majority of cases, these parameters appear to partition elongated domains, although intriguing exceptions are found in some regions of large heat-shock protein sequences that have subtle low-compositional complexity although they are known to possess a folded structure.

Periodic Compositional Complexity

Compositional bias commonly affects residues that are spaced at regular intervals rather than contiguous.[2,4,7] The concept of compositional complex-

[22] J. M. Derry, H. D. Ochs, and U. Francke, *Cell (Cambridge, Mass.)* **78,** 635 (1994).
[23] A. Lupas, M. Van Dyke and J. Stock, *Science* **252,** 1162 (1991).

(A) seg wiskott 25 3.0 3.3

>gp|U12707|HSU12707_1 Wiskott-Aldrich syndrome protein [Homo sapiens]

```
                                 1-147    MSGGPMGGRPGGRGAPAVQQNIPSTLLQDH
                                          ENQRLFEMLGRKCLTLATAVVQLYLALPPG
                                          AEHWTKEHCGAVCFVKDNPQKSYFIRLYGL
                                          QAGRLLWEQELYSQLVYSTPTPFFHTFAGD
                                          DCQAGLNFADEDEAQAFRALVQEKIQK
rnqrqsgdrrqlppppptpaneerrgglppl  148-200
   plhpggdqggppvgplslglatv
                                 201-204  DIQN
pditssryrglpapgpspadkkrsgkkkis   205-243
              kadigapsg
                                 244-311  FKHVSHVGWDPQNGFDVNNLDPDLRSLFSR
                                          AGISEAQLTDAETSKLIYDFIEDQGGLEAV
                                          RQEMRRQE
plpppppppsrggnqlprapivggnkgrsgp  312-434
lppvplgiapppptprgppppgrgglhhhp
lqlldvldhcplhplelvghpchhhrhhrh
rrpapgmdqplphslllwclpgawpgggrg
              all
                                 435-483  DQIRQGIQLNKTPGAPESSALQPPPQSSEG
                                          LVGALMHVMQKRSRAIHSS
       degedqagdededdewdd        484-501
```

(B) seg wiskott 45 3.4 3.75

>gp|U12707|HSU12707_1 Wiskott-Aldrich syndrome protein [Homo sapiens]

```
                                 1-116    MSGGPMGGRPGGRGAPAVQQNIPSTLLQDH
                                          ENQRLFEMLGRKCLTLATAVVQLYLALPPG
                                          AEHWTKEHCGAVCFVKDNPQKSYFIRLYGL
                                          QAGRLLWEQELYSQLVYSTPTPFFHT
fagddcqaglnfadedeaqafralvqekiq   117-259
krnqrqsgdrrqlppppptpaneerrgglpp
lplhpggdqggppvgplslglatvdiqnpd
itssryrglpapgpspadkkrsgkkkiska
       digapsgfkhvshvgwdpqngfd
                                 260-296  VNNLDPDLRSLFSRAGISEAQLTDAETSKL
                                          IYDFIED
qqgleavrqemrrqeplpppppppsrggnql  297-469
prapivggnkgrsgplppvplgiappppptp
rgppppgrgglhhhplqlldvldhcplhpl
elvghpchhhrhhrhrrpapgmdqplphsl
llwclpgawpgggrgalldqirqgiqlnkt
       pgapessalqpppqsseglvgal
                                 470-501  MHVMQKRSRAIHSSDEGEDQAGDEDEDDEW
                                          DD
```

FIG. 4. SEG results using searches of the human Wiskott–Aldrich syndrome protein (see Fig. 3) with two different granularities of "nonglobular" parameters. The SEG command lines are shown, with the parameters following the sequence file name "wiskott." (A) Low-complexity segments correspond approximately to the repetitive regions that are clear in Fig. 3. (B) Parameters with $W = 45$ reveal additional weakly biased sequences flanking the more striking proline-rich repeats, and these additional regions are recruited into consolidated low-complexity segments. For BLASTP searches, it is appropriate for this protein to mask these longer, consolidated segments, in order to focus attention on any weak similarities in the high-complexity, probably globular, regions.

ity is readily extended to these periodic cases, as implemented in programs PSEG for proteins and NSEG for nucleic acids.

Periodicity is a formal sequence attribute that is independent of compositional complexity and is defined as repetition of residue types or k-grams at a constant interval (period, modulo, or distance). It is useful to distinguish true periodicity, which is tandem repetition, exact or with variations, of a sequence pattern of constant length, from quasi-periodicity[7] in which repetition arises as a secondary consequence of different compositional biases in different phases. Quasi-periodic regions are abundant in DNA sequences, for example, modulo-3 in codon-biased mRNA coding sequences, and also occur in some helical polypeptides, for example, modulo-3 in collagen rods. Nonintegral periods may also arise from helical secondary structures in polynucleotide or polypeptide chains.

To generalize the concept of complexity state vectors to noncontiguous positions in a sequence, periodic compositional complexity is calculated from complexity state vectors whose numbers are counts of residues spaced at any defined interval. For example, a modulo-3 window of length 12 counts only the residues marked "1", and not the residues marked "0", in a 36-residue subsequence:

$$(1,0,0,1,0,0,1,0,0,1,0,0,1,0,0,1,0,0,1,0,0,1,0,0,1,0,0,1,0,0,1,0,0,1,0,0)$$

Sliding this window along a sequence in steps of single residues defines three overlapping phases. The period in PSEG is set by the "-z 3" option (for period 3). PSEG output uses lowercase characters, or "x" characters for masking with the "-x" option, to represent only those individual residues that are defined as low-complexity in a phase-specific manner (Fig. 5). With PSEG and NSEG, tree form output is specified by the "-p" command line option.

Figure 5 illustrates the phase specificity of PSEG results, using a typical collagen sequence. As with the "nonglobular" parameters of SEG, the sequence is partitioned accurately and automatically by PSEG into its globular domains (Fig. 5A, right-hand blocks) and nonglobular rods (Fig. 5A, left-hand blocks). However, PSEG has the important advantage that high-complexity information is retained within the rod sequences (shown by uppercase characters among the lowercase low-complexity phases). Masking with PSEG using the "-x" option (Fig. 5B) enables BLASTP to be used to investigate the subclasses and phylogeny of collagens from their rod sequences alone. The glycine phase of the rod segments and the proline-rich (including hydroxyproline) parts of the other two phases are defined by PSEG as low complexity, so that large output lists of spurious matches to glycine-rich or proline-rich noncollagen proteins are completely eliminated.

(A) **pseg collagen -z 3 -p**

>CA19_HUMAN COLLAGEN ALPHA 1(IX) CHAIN PRECURSOR.

```
                                  1-266    MKTCWKIPVFFFVCSFLEPWASAAVKRRPR
                                           FPVNSNSNGGNELCPKIRIGQDDLPGFDLI
                                           SQFQVDKAASRRAIQRVVGSATLQVAYKLG
                                           NNVDFRIPTRNLYPSGLPEEYSFLTTFRMT
                                           GSTLKKNWNIWQIQDSSGKEQVGIKINGQT
                                           QSVVFSYKGLDGSLQTAAFSNLSSLFDSQW
                                           HKIMIGVERSSATLFVDCNRIESLPIKPRG
                                           PIDIDGFAVLGKLADNPQVSVPFELQWMLI
                                           HCDPLRPRRETCHELPARITPSQTTD
ergppgeqgppgasgppgvpgidgidgdrg    267-356
pkgppgppgpagepgKpgApgKpgTpgAdg
LtgPdgSpgSigSkgQkgEpgVpgSrgFpg
rgIpgppgppgtaglpgelgrvgpvgdpgr    357-754
rgppgppgppgprgtIgFHdGDpLCpNAcP
PgRSgYPgLPgMRgHKgAKgEIgEPgRQgH
KgEEgDQgELgEVgAQgPPgAQgLRgITgL
VgDKgEKgARgLDgEPgPQgLPgAPgDQgQ
RgPPgEAgPKgDRgAEgARgIpgLpgPkgD
tgLpgVdgRdgIpgMpgTkgEpgKpgPpgD
agLqgLpgVpgIpgAkgVagEkgStgApgK
pgQmgNsgKpgQqgppgevgprgpqglpgs
rgelgpvgspglpgklgslgspglpglpgp
pglpgmkgdrgvvgepgpkgeqgaSgeEge
AgeRgeLgdIglPgpKgsAgnpgepglrgp
egsrglpgVegPrgPpgPrgVqgEqgAtgL
                        pgVqgPpg
                                  755-786  RAPTDQHIKQVCMRVIQEHFAEMAASLKRP
                                           DS
gATgLpgRpgPpgPpgPpgEngFpgQmgIr    787-899
gLpgIkgPpgAlgLrgPkgDlgEkgErgPp
gRgPNgLPgAIgLPgDPgPAsYgkNgrDge
           RgppglagIpgVpgPpgPpgLpg
                                  900-931  FCEPASCTMQLVSEHLTKGLTLERLTAAWL
                                           SA
```

(B) **pseg collagen -z 3 -x**

>CA19_HUMAN COLLAGEN ALPHA 1(IX) CHAIN PRECURSOR.

```
MKTCWKIPVFFFVCSFLEPWASAAVKRRPRFPVNSNSNGGNELCPKIRIGQDDLPGFDLI
SQFQVDKAASRRAIQRVVGSATLQVAYKLGNNVDFRIPTRNLYPSGLPEEYSFLTTFRMT
GSTLKKNWNIWQIQDSSGKEQVGIKINGQTQSVVFSYKGLDGSLQTAAFSNLSSLFDSQW
HKIMIGVERSSATLFVDCNRIESLPIKPRGPIDIDGFAVLGKLADNPQVSVPFELQWMLI
HCDPLRPRRETCHELPARITPSQTTDxxxxxxxxxxxxxxxxxxxxxxxxxxxxxxxxxxxx
xxxxxxxxxxxxKxxAxxKxxTxxAxxLxxPxxSxxSxxSxxQxxExxVxxSxxFxxxxIx
xxxxxxxxxxxxxxxxxxxxxxxxxxxxxxxxxxxxIxFHxGDxLCxNAxPPxRS
xYPxLPxMRxHKxAKxEIxEPxRQxHKxEExDQxELxEVxAQxPPxAQxLRxITxLVxDK
xEKxARxLDxEPxPQxLPxAPxDQxQRxPPxEAxPKxDRxAExARxIxxLxxPxxDxxLx
xVxxRxxIxxMxxTxxExxKxxPxxDxxLxxLxxVxxIxxAxxVxxExxSxxAxxKxxQx
xNxxKxxQxxxxxxxxxxxxxxxxxxxxxxxxxxxxxxxxxxxxxxxxxxxxxxxxxxxxx
xxxxxxxxxxxxxxxxxxxxxxxSxxExxAxxRxxLxxIxxPxxKxxAxxxxxxxxxxxxxx
xxxxVxxPxxPxxPxxVxxExxAxxLxxVxxPxxRAPTDQHIKQVCMRVIQEHFAEMAAS
LKRPDSxATxLxxRxxPxxPxxExxFxxQxxIxxLxxIxxPxxAxxLxxPxxDxxEx
xExxPxxRxPNxLPxAIxLPxDPxPAxYxxNxxDxxRxxxxxxxIxxVxxPxxPxxLxxF
CEPASCTMQLVSEHLTKGLTLERLTAAWLSA
```

FIG. 5. Tree form (A) and masked (B) output from PSEG using the command line options shown. Other parameters are the defaults, namely, $W = 12$ residues, $K_2(1) = 2.2$ bits, $K_2(2) = 2.5$ bits. The sequence is human collagen alpha I type IX, SWISS-PROT accession number P20849, name CA19_HUMAN [Y. Muragaki, T. Kimura, Y. Ninomiya, and B. R. Olsen, *Eur. J. Biochem.* **192,** 703 (1990)].

"Simple" Sequences in DNA

Compositional bias in genomic DNA is much more intricate than in proteins, and it differs markedly in different organisms. DNA sequences are mosaics of different patches such as microsatellites, variable dinucleotide, trinucleotide, and other tandem repeats, dispersed repeats, telomeric sequences, recombinational hot spots, CpG islands, longer domains of different compositional bias, and many quasi-periodicities of different degrees of subtlety in both coding and noncoding sequences. In some cases, statistically distinct domains may overlap one another. This variegation cannot be adequately represented by simple random or Markov statistical models based on k-grams, and local low complexity is only one of the linguistic characteristics.[7,24]

The NSEG program is a useful discovery tool for exploring this difficult territory. The principles of NSEG are exactly as described above for SEG and PSEG, but tuned to the four-letter alphabet. The "-z" option specifies the periodicity. For example, "-z 1" defines a period of 1 and thus restricts the analysis to continuous segments of residues (as described above in the section on using the SEG program). Values of "-z" greater than 1 give behavior as described above for PSEG (see section on periodic compositional complexity).

The NSEG program is also effective for masking residues in biased regions prior to database searches by programs such as BLASTN that compare nucleotide sequences. The "-x" option in NSEG replaces masked nucleotides with "n" characters. The results of masking at several different periods can be integrated, using the utility program NMERGE to combine the "n" characters obtained from different NSEG output files and produce a single FASTA-formatted input file for BLASTN. A corresponding utility, PMERGE, combines different sets of "x" characters from amino acid sequences masked at different periods.

As with protein sequences, if the DNA sequence is not too long, a valuable preliminary step involves visual inspection of a self–self dot-matrix plot. For DNA, this should include matches of the sequence to its complementary strand. This plot may indicate the approximate positions and extent of repeats and compositional bias. For more rigorous identification of biased regions in DNA, which vary greatly in length, periodicity, and complexity, an interactive, exploratory approach using NSEG is recommended. This may be achieved by varying the command line parameters and investigating a range of periods, for example, from 1 to at least 6.

Possible parameters for an initial investigation might be, for period 1

[24] S. Karlin and V. Brendel, *Science* **259**, 677 (1993).

(contiguous residues), "nseg mychromosome 21 1.3 1.3 -z 1 -p" and, for periods of 2 or greater, more stringent parameters, for example, "nseg mychromosome 21 1.1 1.1 -z 3 -p." Shorter windows, for example 15 or 17 instead of 21, are likely to reveal additional short segments. Figure 6 shows a sample of NSEG results at periods 1 and 6 from part of a large intron of the human ABL gene, illustrating a purine-rich feature and a very subtle modulo-6 quasi-periodicity.

Following a few exploratory runs with different parameters, NSEG may then readily be used with the "-l" option to report, for example, all trinucleotide repeat segments in a whole genome or database at a required level of stringency.

SEG Algorithm

SEG is a two-pass algorithm. The first stage identifies approximate raw segments of low complexity. The second stage is local optimization within each raw segment. At the first stage, the search for low-complexity raw segments is determined by W, $K_2(1)$, and $K_2(2)$, described in the section on parameters above. $K_2(2)$ must not be less than $K_2(1)$ and is usually set slightly greater in order to permit the local optimization at the second stage of SEG to operate within a relatively liberal range.

The compositional complexity, K_1, of a complexity state vector (as used in the Fig. 2 profiles) is a measure of the information needed per position, given the composition of the window, to specify a particular residue sequence.[1,3] For an N-residue alphabet (usually, $N = 4$ or 20) and a window of length L:

$$K_1 = \frac{1}{L} \log_N \Omega \tag{1}$$

where Ω is the multinomial coefficient ($L!/\prod_{i=1}^{N} n_i!$), which gives the number of sequences per composition characteristic of the complexity state vector. The n_i are the N numbers in the complexity state vector. The logarithm is taken to base N to place K_1 in the range 0 to 1. To express complexity in frequently used information units, logarithms may be taken to base 2, giving bits, or to base e, giving nats.

Another informational measure of compositional complexity, K_2, expressed in bits, is used instead of K_1 for computational efficiency at the first stage of the SEG program:

$$K_2 = -\sum_{i=1}^{N} \frac{n_i}{L} \left(\log_2 \frac{n_i}{L} \right) \tag{2}$$

(A) <u>nseg abl.1b 21 1.1 1.1 -z 1 -p</u>

>U07562 Human ABL gene, intron 1b, subsequence 43040-43749.

	1-28	CCTGGGTGACAAAGTGAGACCCTATCTC
aaaaaagaaagaaagaaaggaag	29-51	
	52-60	TTAGAAAGT
gggggagggagggaaggaggaag	61-83	
	84-84	T
gaggaaggaaagaaaggaaagaaaacaaag agaaagaaa	85-123	
	124-710	CTTGTCTGATTTGGTCTGCACTTTGAGTGA GAGGAAACAAAAACCTCCCGAGATAACTTT TATTCTAAGCATGCCTTCTTTTAAACTTTG TTTGAATTTACCTGGCCTCTGTAGTTTGTG AAATTCCAAGAAGAGACATTATTTTAGAGA GATACTAGGTTGCTAGAGTTCCTAGATTAC TACAGTTCCTAGTTAGCTGGCAGAAGTCAA TGCATATCCTCTTGGACTATATCTATCTTT TTCTCAGTCTTAACACCAGATTCCCAAAGA TAAAGCTTCTCCAAACATGATCTCACAATC CAAAACTAAGAACATGAGAAAACAAACAAC CTTGCATCACAGAGCAGGAAAATGAACAGT AGAACTATAGTACTTTATAGCTTTAGATAT AAAAACTGTCAGATTCAGAATATAAAATAC TGGTTTAAAAACATTTAAAGAAATAAAATG TGAAATAGAAAACAAGTAAGGACTAAAATC CTATGAAGAAAAATCAAATAATTGAAAAAG AACTAAAAAAGAAATGAAAACTACGATCAG TGAAATTAAAACTTCAGTGGATGGGTTTAA CAAGTTAGACACAACTG

(B) <u>nseg abl.1b 21 1.1 1.1 -z 6 -p</u>

>U07562 Human ABL gene, intron 1b, subsequence 43040-43749.

	1-441	CCTGGGTGACAAAGTGAGACCCTATCTCAA AAAAGAAAGAAAGAAAGGAAGTTAGAAAGT GGGGGAGGGAGGGAAGGAGGAAGTGAGGAA GGAAAGAAAGGAAAGAAAACAAAGAGAAAG AAACTTGTCTGATTTGGTCTGCACTTTGAG TGAGAGGAAACAAAAACCTCCCGAGATAAC TTTTATTCTAAGCATGCCTTCTTTTAAACT TTGTTTGAATTTACCTGGCCTCTGTAGTTT GTGAAATTCCAAGAAGAGACATTATTTTAG AGAGATACTAGGTTGCTAGAGTTCCTAGAT TACTACAGTTCCTAGTTAGCTGGCAGAAGT CAATGCATATCCTCTTGGACTATATCTATC TTTTTCTCAGTCTTAACACCAGATTCCCAA AGATAAAGCTTCTCCAAACATGATCTCACA ATCCAAAACTAAGAACATGAG
aAAACAaACAACcTTGCAtCACAGaGCAGG aAAATGaACAGTaGAACTaTAGTAcTTTAT aGCTTTaGATATaAAAACtGTCAGaTTCAG aATATAaAATACtGGTTTaAAAACaTTTAA aGAAATaAAATGtGAAATaGAAAAcAAGTA aGGACTaAAATCcTATGAaGAAAAaTCAAA tAATTGaAAAAGaACTAAaAAAGAaATGAA a	442-652	
	653-710	ACTACGATCAGTGAAATTAAAACTTCAGTG GATGGGTTTAACAAGTTAGACACAACTG

FIG. 6. Part of intron 1b of the human ABL gene, analyzed for contiguous-residue low-complexity segments (A) and quasi-periodicities of period 6 (B) using NSEG with the command lines shown. The subsequence indicated is from GenBank accession number U07562 (unpublished cosmid sequence submitted to GenBank by B. R. Roe, 1994).

K_2 is an approximation that converges toward K_1 at large window lengths.[1] K_2 is an adequate estimate of complexity for the first, noncritical, stage of SEG. K_2 is analogous to "entropy" in Shannon's information theory, whereas K_1, being based on combinatorial enumerations of states, resembles "entropy" in the sense of Boltzmann.

A sliding window of length W is moved in single-residue steps along the sequence, and complexity K_2 is computed at each step. Trigger windows are those of complexity less than or equal to $K_2(1)$ bits. A raw segment may consist of a single trigger window or a contig of overlapping windows. The contig is constructed by merging each trigger window in both directions with any overlapping trigger windows, and also with any overlapping extension windows of length W and complexity less than or equal to $K_2(2)$ bits. Each nonoverlapping contig is a raw segment.

At the second stage of SEG, each raw segment is reduced to a single optimal low-complexity segment, which may be the entire raw segment but is usually a shorter subsequence. The optimal subsequence is that with the most improbable composition, calculated on the assumption of equal (uniform) probabilities for the appearance of the 4 nucleotides or 20 amino acids. The probability of occurrence, P_0, of each complexity state is minimized:

$$P_0 = \frac{1}{N^L} \Omega F \tag{3}$$

where F is the combinatorial expression $N!/\Pi_{k=0}^{L} r_k!$, which is the number of compositions that have this complexity state vector.[1,3] Here, r_k is the count of the number of times the number k occurs among the n_i of the complexity state vector. Other symbols are as for formula Eq. (1). N^L is the total number of possible sequences for a window of length L and alphabet size N.

The segment of minimal P_0 is found by exhaustive search over all subsequences of each raw segment, using precomputed lookup tables of log factorials for efficiency. P_0 is particularly suitable for this optimization because it gives closely similar expected values at all window lengths.[1] The use of uniform prior probabilities of residues, as embodied in this probability function, is critical to ensure that low-complexity segments are defined in a consistent manner throughout a sequence or database.

Availability

The programs SEG, PSEG, NSEG, PMERGE, and NMERGE are available as standard C language source code that compiles on Sun, Solaris, or Silicon Graphics systems and has been compiled on other UNIX workstations with trivial modifications. Executables are also provided for Sun and Silicon Graphics systems. All these programs are available by anonymous ftp from ncbi.nlm.nih.gov in subdirectory /pub/seg together with documentation. A default version of SEG for masking purposes is also available as the "-filter seg" option within the BLAST programs.[19]

Future Developments

It is increasingly important to have robust methods for comparison and classification of compositionally biased segments of sequences. For a growing number, still a minority, of low-complexity protein segments, molecular interactions and functional assignments have been made experimentally. Given the involvement of such segments in biological functions such as morphogenesis and human molecular diseases, we would like to understand their crucial molecular characteristics that may be compositionally determined and generally evolve rapidly. Software to support such research is under development.[6]

For genomic studies, it is essential to view compositional bias in the context of many types of other features such as recognizable functional sites, transcripts, coding sequences, and homologies. For this purpose, the SEG family of programs is being integrated into software packages, or workbenches, that have graphic multilevel browsing facilities and include zoom functions. For example, graphical complexity profiles and SEG optimal segments can be viewed, at different granularities, in conjunction with sequence matches from BLAST programs in an enhancement of the BLIXEM viewer[25] that is under test at the time of writing.[26]

[25] E. L. Sonnhammer and R. Durbin, *Comput. Appl. Biosci.* **10,** 301 (1994).
[26] E. L. Sonnhammer and J. C. Wootton, unpublished (1995).

Section V

Three-Dimensional Considerations

[34] Discrimination of Common Protein Folds: Application of Protein Structure to Sequence/Structure Comparisons

By MARK S. JOHNSON, ALEX C. W. MAY, MICHAEL A. RODIONOV, and JOHN P. OVERINGTON

Introduction

The comparison of protein sequences and structures is a critical first step for a wide range of tasks that include, for example, the identification of related proteins—finding common protein folds during data bank searches (fold recognition),[1-4] phylogenetic reconstruction—based on data from aligned macromolecular sequences,[3] and protein modeling. In protein modeling, an estimate of the atomic coordinates for a protein is based on accurate alignment of its sequence to one or more tertiary structures from related proteins.[5]

For very similar sets of proteins (sequence identity >45%), there is usually no difficulty in identifying whether a relationship truly exists or in obtaining an alignment that is highly accurate [accurate in the sense that matched positions in the alignment occupy equivalent spatial positions within the three-dimensional (3D) structures]. As the dissimilarity between proteins increases, however, it becomes increasingly difficult to achieve either of these goals. Such difficulties in resolving relationships among proteins typically result from the influences of low sequence similarity on the alignment: there are few conservative changes and identical residues shared at equivalent positions in the proteins. In addition, with more dissimilar proteins there are usually more insertions/deletions (indels or gaps), and their true boundaries within the alignment are often obscured by low surrounding sequence similarity. Thus, one can be left with an alignment that is largely incorrect or with a data bank search score for a true match that is indistinguishable from the mass of scores from comparisons with unrelated proteins.

[1] M. S. Johnson, *Mol. Med. Today* **1**, 188 (1995).
[2] F. Eisenhaber, B. Persson, and P. Argos, *Crit. Rev. Biochem. Mol. Biol.* **30**, 1 (1995).
[3] R. F. Doolittle (ed.), this series, Vol. 183.
[4] T. L. Blundell and M. S. Johnson, *Protein Sci.* **2**, 877 (1993).
[5] M. S. Johnson, N. Srinivasan, R. Sowdhamini, and T. L. Blundell, *Crit. Rev. Biochem. Mol. Biol.* **29**, 1 (1994).

METHODS IN ENZYMOLOGY, VOL. 266

Copyright © 1996 by Academic Press, Inc.
All rights of reproduction in any form reserved.

Dynamic programming[6] is still the method of choice for the alignment of proteins. It can also be used to search a protein against a collection of proteins within a data bank. In dynamic programming[6] there are two competing factors that influence the resulting alignment: the scores attributed to residue matches and the penalties assessed to reduce excessive gaps. Both of these factors contribute toward the overall alignment score and the alignment that results; they are therefore obvious targets to address when seeking to improve the resolution of protein relationships.

For sequence comparisons, 20 by 20 matrices of values suitable for scoring amino acid matches have most recently been produced from large sets of aligned sequences[7-9] and sets of aligned 3D structures.[10] Although they are an improvement[10,11] over the original Dayhoff comparison matrix[12] and others based on physicochemical properties of amino acids, they are all limited by their generic nature since the same set of scores applies to different locations within the 3D fold and for proteins from different families. Typically, a constant gap penalty or one that employs a length limiting component is also used.

If a 3D structure is known, then a sequence can be compared with the structure using one of two basic strategies.[1,2] In the first approach, one-dimensional (1D) sequences are compared with 1D templates or profiles derived from known 3D structures.[13,14] The template provides a set of scores for matching a position within the structure with those in the sequence. Scores have been based on local residue environments,[14] the environmental dependence of residue replacements,[15,16] or the contact preferences of the amino acids.[17] Sippl[18] introduced the use of pseudo-energy potentials[18-20]

[6] S. B. Needleman and C. Wunsch, *J. Mol. Biol.* **48,** 444 (1970).

[7] S. Henikoff and J. G. Henikoff, *Proc. Natl. Acad. Sci. U.S.A.* **89,** 10915 (1992).

[8] G. H. Gonnet, M. A. Cohen, and S. A. Benner, *Science* **256,** 1443 (1992).

[9] D. T. Jones, W. R. Taylor, and J. M. Thornton, *CABIOS* **8,** 275 (1992).

[10] M. S. Johnson and J. P. Overington, *J. Mol. Biol.* **233,** 716 (1993).

[11] S. Henikoff and J. G. Henikoff, *Proteins* **17,** 49 (1993).

[12] M. O. Dayhoff, R. M. Schwartz, and B. C. Orcutt, *in* "Atlas of Protein Sequence and Structure" (M. O. Dayhoff, ed.), Vol. 5. Suppl. 3, p. 345. National Biomedical Research Foundation, Washington, D.C., 1978.

[13] A. Šali, J. P. Overington, M. S. Johnson, and T. L. Blundell, *Trends Biochem. Sci.* **15,** 235 (1990).

[14] R. Lüthy, A. D. McLachlan, and D. Eisenberg, *Proteins: Struct. Funct. Genet.* **10,** 229 (1991).

[15] J. P. Overington, M. S. Johnson, A. Šali, and T. L. Blundell, *Proc. R. Soc. London Ser. B* **241,** 132 (1990).

[16] M. S. Johnson, J. P. Overington, and T. L. Blundell, *J. Mol. Biol.* **231,** 735 (1993).

[17] C. Ouzounis, C. Sander, M. Scharf, and R. Schneider, *J. Mol. Biol.* **232,** 805 (1993).

[18] M. J. Sippl, *J. Mol. Biol.* **213,** 859 (1990).

[19] D. Jones, W. R. Taylor, and J. M. Thornton, *Nature (London)* **358,** 86 (1992).

[20] A. Godzik, A. Kolinski, and J. Skolnick, *J. Mol. Biol.* **226,** 227 (1992).

based on the distances between residues in known 3D structures. Hence, the second strategy has evolved from evaluation of the total energy of a sequence given that it is folded similarly to the known 3D structure. This of course requires an alignment prior to evaluation, and these methods typically repetitively examine alternative alignments or folds, since the optimization procedure is far more complex.

Functions for the assignment of gap penalties have also been proposed based on examination of gaps in aligned 3D structures[21] or sequences.[22] Neither of these penalty regimes is dependent on the location of the gap within the protein. Indeed, Barton and Sternberg[23] showed that if gap penalties were set according to secondary structure assignments, where penalties are higher within helices and strands than without, the accuracy of alignments is significantly improved.

Our approach to increasing the sensitivity of comparisons has been to examine families of aligned 3D structures and the residue replacements that have occurred since divergence from a common ancestor. Amino acid interchanges that appear in aligned homologous proteins are constrained to varying degrees by the local environment in the folded protein structure as well as the obvious functional constraints that apply. Each amino acid has distinctive attributes, such as hydrophobicity, hydrogen bonding capacity, and conformational preferences, that allow it to fulfill a relatively unique role in its contribution toward a protein fold. Yet, surprisingly few positions within a 3D structure are completely protected from replacement during evolution. A key to improvements within protein sequence–structure comparisons may be in considering not only the residues involved but also their local environmental details where known.

This approach has several advantages. First, we consider structural alignments that are clearly more accurate than sequence comparisons, even when the sequence identity has dipped to as low as 10% in the realm of alignments between unrelated protein sequences. Second, knowledge about the local physicochemical environment of every residue within each 3D structure is directly obtainable. Thus, the fold can be interpreted in terms of amino acid "flavors" that extend beyond the 20 naturally occurring residues that generally make up a protein. For example, a hydrogen bonded polar amino acid such as aspartate, glutamate, threonine, or serine, buried in the protein core, is highly conserved, whereas these same residues at the surface of a protein and exposed to solvent are not.[15,16] It is also clear that gap penalties should not be the same at all positions along a fold:

[21] S. Pascarella and P. Argos, *J. Mol. Biol.* **224,** 461 (1992).
[22] S. A. Benner, M. A. Cohen, and G. H. Gonnet, *J. Mol. Biol.* **229,** 1065 (1993).
[23] G. J. Barton and M. J. E. Sternberg, *Protein Eng.* **1,** 89 (1987).

indels within the protein core are seldom seen. Thus, our approach is to make an informed assignment of gap penalties that vary at each position along the sequence or 3D fold according to environmental preferences.

Environment-Specific Amino Acid Substitutions

Ninety-six families of 3D structures (443 total structures) were structurally aligned[15,24,25] by rigid-body superposition using a dynamic programming version (P. Thomas and M. S. Johnson, 1990, unpublished results; Ref. 26) of the computer program MNYFIT[27] and COMPARER.[26,28] Residues at each aligned position in all pairs of structures contributing to each family were classified by the computer program JOY[15,24,25] according to the following criteria: (1) the type of amino acid replacement observed at matched positions in the structural alignments (cysteine was segregated from the half-cystine of disulfide bonds, giving in effect 21 amino acid types); (2) the secondary structure of the residue according to the DSSP assignment scheme of Kabsch and Sander[29] (as implemented by D. Smith, 1989, unpublished results), namely, helices (α helices and 3_{10} helices), β strand, irregular structure, and residues with a positive ϕ main-chain torsion angle (primarily the forbidden α_L region of the Ramachandran map) (leads to four classes of secondary structure); (3) the solvent accessibility of the residue,[30] where a 7% cutoff is employed to distinguish between residues of the solvent-inaccessible core ($\leq 7\%$) and those with a higher exposure to solvent[31] (leads to two classes: accessible or inaccessible residues at the 7% level); and (4) the potential of the residue for hydrogen bonding, based on a simply heavy atom donor–acceptor distance of at most 3.5 Å (angular criteria were not considered): (i) side-chain to side-chain hydrogen bonds (two classes: residue has or has not the potential to form a hydrogen bond), (ii) hydrogen bonds between a side chain and main-chain NH (two classes), and (iii) hydrogen bonds between a side chain and main-chain CO (two classes).

In total, classification according to the above features leads to 64 (4 $\times$ 2 $\times$ 2 $\times$ 2 $\times$ 2) 21 by 21 substitution matrices, although some amino acid

[24] J. P. Overington, D. Donnelly, A. Šali, M. S. Johnson, and T. L. Blundell, *Protein Sci.* **1,** 216 (1992).

[25] A. Šali and J. P. Overington, *Protein Sci.* **3,** 1582 (1994).

[26] M. S. Johnson, A. Šali, and T. L. Blundell, this series, Vol. 183, p. 670.

[27] M. J. Sutcliffe, I. Haneef, D. Carney, and T. L. Blundell, *Protein Eng.* **1,** 377 (1987).

[28] A. Šali and T. L. Blundell, *J. Mol. Biol.* **212,** 403 (1990).

[29] W. Kabsch and C. Sander, *Biopolymers* **22,** 2577 (1983).

[30] B. Lee and F. M. Richards, *J. Mol. Biol.* **55,** 379 (1971).

[31] T. J. P. Hubbard and T. L. Blundell, *Protein Eng.* **1,** 155 (1987).

interchanges will not be observed for residues that either cannot form hydrogen bonds or cannot act as both hydrogen bond acceptors and donors: 766 amino acid "flavors" result. These tables can also be coalesced to represent fewer features but with better statistics at the expense of some specificity. The variation that occurs in substitution patterns for serine exposed to 16 different local environments is shown in Fig. 1.[15,16]

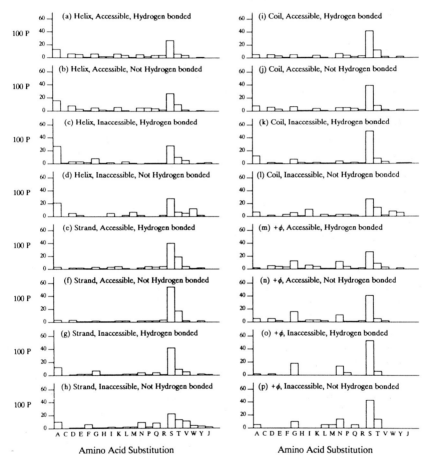

FIG. 1. Sixteen environment-dependent distributions for one of the 21 amino acid types, namely, serine. The probabilities (100*P*) for matching a serine in a particular local environment with each of the other amino acids (half-cystine, C, cysteine, J), regardless of their environment, are shown. Structural features considered here include the residue secondary structure, residue solvent accessibility, and whether a hydrogen bond can involve the hydroxyl of the serine side chain. (If the hydrogen bonding class was differentiated into the three categories described in the text, 64 distributions would result.) Used with permission.[16]

Thus, residues at each position within a 3D structure or aligned set of structures can be recorded as a template of scores that reflects not only the structural environment at that position in the fold, but also the pattern of residue replacement likely for that residue type.[16] Once the 3D fold is encoded in this way, matches can be made against 1D sequences whose structures are not known (Fig. 2).[16] Two strategies are available[16,32]: one can search a sequence against all known 3D structures encoded as templates, or a structure template can be compared against all of the sequences in a data bank (see below). From the comparisons within these search procedures alignment scores are obtained, and these values, normalized by the length of the alignment excluding terminal overhangs, can be used to discriminate between authentic matches and alignments between likely unrelated proteins.

If we integrate over each of the 64 environment-dependent tables, a residue exchange matrix is produced that is suitable for comparisons of sequences when no information from a 3D structure is available for one of a pair of proteins.[10] Such a matrix, updated to include contributions from 96 aligned families of structures, is shown in Table I. An earlier version based on 65 aligned families has proved to be among the more sensitive matrices for comparing distantly related protein sequences.[10,11]

Residue–Residue Contact Substitutions

Another source of information that can be applied to scoring matches between a structure and a sequence is reflected in the contacts that occur among side chains within the protein structure and how the conservation of these contacts varies among related proteins.[33] Contacts provide details on the local 3D fold, are often contributed by residues distant along the protein sequence, and are highly conserved even when the sequence similarity between the proteins is low. For example, compare the distribution of pairwise percentage of conserved contacts from 1261 aligned pairs of related protein tertiary structures (Fig. 3a) with the distribution of percentage sequence identity for each of these same comparisons (Fig. 3b). Elsewhere we have considered contacts between side chains and main-chain polar groups.[32,34]

Phillips introduced the concept of the contact matrix.[35] To construct a

---

[32] M. S. Johnson, J. P. Overington, Y. Edwards, A. C. W. May, and M. A Rodionov, *Proc. 27th Hawaii Int. Conf. Syst. Sci.* **5,** 296 (1994).

[33] M. A. Rodionov and M. S. Johnson, *Protein Sci.* **3,** 2366 (1994).

[34] Y. J. Edwards, M. S. Johnson, D. S. Moss, and T. L. Blundell, *in* "Techniques in Protein Chemistry V" (J. W. Crabb, ed.), p. 405. Academic Press, New York, 1994.

[35] D. S. Phillips, *Biochem. Soc. Symp.* **N31,** 11 (1970).

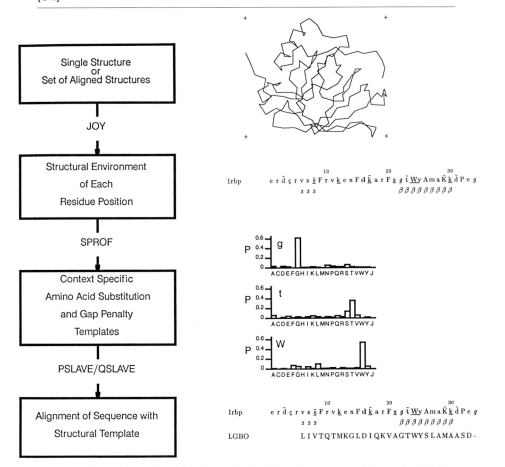

Fig. 2. Comparison illustrating the analysis of the 3D structure of the retinol-binding protein (1rbp) and the alignment with the sequence of bovine β-lactoglobulin (LGBO). In the first step, structural features such as hydrogen bonding, secondary structure, and residue solvent accessibility are calculated for 1rbp with the program JOY. In the next step, these structural features of 1rbp are used by a program SPROF to specify context-specific values from the structural substitution matrices and produce amino acid replacement scores and gap-penalty information according to the environment of each residue position in this structure. Scores (P) for amino acid replacement at positions 22–24 of the 1rbp structure (glycine–threonine–tryptophan) are shown (the standard one-letter amino acid code is used, but C refers to half-cystine and J to cystine). In the final step, either of two programs PSLAVE (a template against a sequence data bank) or QSLAVE (a sequence or multiple alignment against a template data bank) is used to produce an alignment of the structural template with the sequence of β-lactoglobulin. The compatibility of the contacts within the sequence with those in the known structure can subsequently be evaluated. Used with permission.[16]

TABLE I
STRUCTURE-BASED SEQUENCE COMPARISON MATRIX[a]

Diagonal cells are shown as *scoring value / all-positive value*. Upper triangle = structure-based amino acid scoring table; lower triangle = all-positive table.

	A	C	D	E	F	G	H	I	K	L	M	N	P	Q	R	S	T	V	W	Y
A	5.9 / 20.1	-11.0	-1.5	-0.8	-3.5	-0.7	-3.7	-2.4	-1.0	-2.7	-1.1	-1.5	-1.1	-1.1	-2.3	0.3	-1.0	-0.5	-4.8	-3.6
C	3.1	19.7 / 33.9	-11.8	-8.9	-6.4	-10.8	-9.2	-8.9	-10.5	-8.6	-7.4	-7.0	-14.1	-7.3	-5.1	-10.0	-9.0	-11.0	-5.9	-10.2
D	12.6	2.3	8.3 / 22.4	1.7	-7.1	-1.8	-1.2	-6.2	-0.8	-7.1	-6.9	2.3	-1.4	-0.4	-2.7	-0.4	-1.7	-5.5	-7.8	-5.0
E	13.3	5.2	15.8	7.8 / 22.0	-5.7	-2.2	-1.8	-4.7	0.9	-5.0	-3.4	-0.6	-1.1	2.1	-0.3	-0.9	-1.0	-3.7	-6.5	-3.6
F	10.6	7.7	7.0	8.4	9.9 / 24.0	-7.1	-1.9	0.3	-5.1	1.9	0.7	-5.1	-5.6	-4.3	-5.6	-4.6	-4.8	-1.4	2.2	4.3
G	13.4	3.3	12.3	11.9	7.0	8.2 / 22.3	-3.3	-7.0	-3.1	-6.6	-5.3	-1.0	-3.3	-2.8	-4.2	-1.4	-3.8	-6.1	-5.4	-5.6
H	10.4	4.9	12.9	12.3	12.2	10.8	11.9 / 26.1	-5.6	-0.2	-3.8	-3.4	0.7	-3.0	1.5	-0.2	-2.4	-2.6	-4.1	-4.8	0.4
I	11.7	5.2	7.9	9.4	14.4	7.1	8.5	7.9 / 22.0	-4.4	2.6	1.5	-5.6	-4.9	-5.0	-4.2	-5.5	-2.6	3.8	-2.6	-2.2
K	13.1	3.6	13.3	15.0	9.0	11.0	13.9	9.7	6.8 / 21.0	-3.5	-2.1	0.1	-1.4	1.4	3.1	-1.1	-0.4	-3.3	-4.3	-3.3
L	11.4	5.5	7.0	9.1	16.0	7.5	10.3	16.7	10.6	7.0 / 21.2	3.5	-4.8	-5.4	-2.6	-3.5	-4.7	-3.4	1.0	-1.7	-1.3
M	13.0	6.7	7.2	10.7	14.8	8.8	10.7	15.6	12.0	17.6	10.9 / 25.0	-3.4	-6.4	0.0	-3.3	-3.0	-1.9	0.0	-0.3	-0.8
N	12.6	7.1	16.4	13.5	9.0	13.1	14.8	8.5	14.2	9.3	10.7	8.0 / 22.1	-2.9	-0.1	-0.8	0.7	-0.1	-4.9	-6.8	-2.8
P	13.0	0.0	12.7	13.0	8.5	10.8	11.1	9.2	12.7	8.7	7.7	11.2	10.4 / 24.6	-2.5	-2.7	-1.2	-1.6	-4.6	-7.2	-4.4
Q	13.0	6.8	13.7	16.2	9.8	11.3	15.6	9.1	15.5	11.5	14.1	14.0	11.6	8.3 / 22.5	1.2	-0.3	-0.6	-3.2	-4.4	-3.1
R	11.8	9.0	11.4	13.8	8.5	9.9	13.9	9.9	17.2	10.6	10.8	13.3	11.4	15.3	10.0 / 24.1	-1.1	-1.5	-3.9	-2.2	-3.0
S	14.4	4.1	13.7	13.2	9.5	12.7	11.7	8.6	13.0	9.4	11.1	14.8	12.9	13.8	13.0	6.3 / 20.5	2.0	-3.5	-6.7	-3.4
T	13.1	5.1	12.4	13.1	9.3	10.3	11.5	11.5	13.7	10.7	12.2	14.0	12.5	13.5	12.6	16.1	7.0 / 21.2	-1.4	-7.1	-3.2
V	13.6	3.1	8.6	10.4	12.7	8.0	10.0	17.9	10.8	15.1	14.1	9.2	9.5	10.9	10.2	10.6	12.7	6.7 / 20.9	-4.3	-2.6
W	9.3	8.2	6.3	7.6	16.3	8.7	9.3	11.5	9.8	12.4	13.8	7.3	6.9	9.7	11.9	7.4	7.0	9.8	15.5 / 29.7	3.1
Y	10.5	3.9	9.1	10.5	18.4	8.5	14.5	11.9	10.8	12.8	13.3	11.3	9.7	11.0	11.1	10.7	10.9	11.5	17.2	10.6 / 24.7

[a] Structure-based (96 aligned families of 3D structures) amino acid scoring table (upper triangle) and all-positive table (lower triangle). All values have been multiplied by 10.

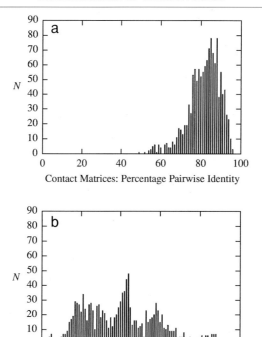

FIG. 3. Conservation of (a) residue–residue contacts (percentage contact map identity) and (b) sequence (percentage sequence identity) in 1261 pairs of homologous protein structures contributing to the 96-family structural alignments (M. A. Rodionov and M. S. Johnson, 1995, unpublished results).

contact matrix for a 3D structure, elements of the contact matrix, $a_{i,j}$, are set equal to 1 if residues i and j in the protein are in contact with one another; otherwise, $a_{i,j}$ is set equal to 0. Two residues are said to be in contact when the Cβ atoms (Cα, in the case of glycine) of these residues are less than the distance $r_{i,j}$:

$$r_{i,j} = r_i + r_j + 3(\sigma_i + \sigma_j) \tag{1}$$

The effective residue radii (Table II), r_i, for residue i and the corresponding standard deviations, σ_i, were obtained by optimizing the correspondence between authentic X-ray structures and models reconstructed using distance geometry and contact maps derived from the structures.[36]

[36] M. A. Rodionov and S. G. Galaktionov, *Mol. Biol.* **26,** 773 (1992); translated from *Molekular-naya Biologiya* (*Kiev*) **26,** 1160 (1992).

TABLE II
CONTACT DISTANCE $C\beta$ ATOMS[a]

Residue type		r_i (Å)	σ_i (Å)
Ala	A	2.45	0.28
Cys	C	2.36	0.32
Asp	D	2.69	0.29
Glu	E	2.78	0.28
Phe	F	2.92	0.33
Gly	G	2.01	0.18
His	H	2.88	0.38
Lys	K	2.88	0.29
Ile	I	2.87	0.31
Leu	L	2.86	0.35
Met	M	2.87	0.36
Asn	N	2.76	0.35
Pro	P	2.76	0.35
Gln	Q	2.80	0.28
Arg	R	2.89	0.28
Ser	S	2.45	0.28
Thr	T	2.67	0.28
Val	V	2.76	0.31
Trp	W	2.96	0.34
Tyr	Y	2.90	0.39

[a] Amino acid side-chain radii (r_i) and standard deviations (σ_i) chosen for detection of contacts. Adapted from Ref. 36. The contact atom is $C\beta$, except $C\alpha$ for glycine.

Given that contact matrices specify those residues that are in close proximity in the protein fold, and that an alignment between two protein structures indicates those positions which are spatially equivalent in each, contact-pair substitutions can be tabulated within a family of structures. In other words, a contact between two residues k and l within one fold (p_1) is compared with the residues x and y in contact at those equivalent positions in another fold (p_2). Furthermore, substitutions can be segregated according to their residue solvent accessibility at the 7% level (relative side-chain solvent accessibility ≤7% has been used to define the solvent-inaccessible core of a protein[31]). The choices are both residues are completely buried, both residues are exposed to solvent, or one of the pair is buried and the other is not; three 400 by 400 matrices of residue-pair substitutions result. Elements of the substitution matrix, $S_{k,l \rightarrow x,y}$ are summed over all pairs of structures (Pr) within each of the 96 families of tertiary structures (F):

$$S_{k,l \to x,y} = \sum_1^F \sum_{p_2=p_1}^{Pr} \sum_{p_1=1}^{Pr} \sum_{k,l} a_{k,l}^{p_2} \tag{2}$$

and lead to a total of 3,299,027 residue–residue contact substitutions (~2500 data points for each set of residue-pair substitutions) from the 2986 individual contact maps from the 96 families.

Discrimination Function

To compare one protein structure with a sequence whose structure is not yet known, there are two essential requirements. First, one 3D structure must be known in order to specify a contact map. (For an aligned set of structures a family-average contact matrix may be constructed.) Second, an alignment between the known 3D structure and the sequence of unknown structure must be made in order to map contacts from the one to the other. This is accomplished using the structure-based templates described above. It is then a matter of calculating the similarity between the pairs of contacts over the whole structure–sequence alignment.

We have chosen to combine two measures of similarity to evaluate contact-pair matches.[33] First, the correlation coefficient is used to compare distributions of substitutions and helps to overcome poor statistics for some tabulated substitutions:

$$q_{k,l \to m,n} = \frac{\sum_{k,l} (S_{k,l \to x,y} - \bar{S}_{k,l})(S_{m,n \to x,y} - \bar{S}_{m,n})}{\sigma_{k,l} \sigma_{m,n}} \tag{3}$$

Values of q will be close to 1 if the contacts for k, l and for m, n have a similar pattern of substitution to the 400 possible amino acid pairings, and q will be 0 or negative if substitution patterns are different. For example, the correlation computed between the distribution for Ser–Leu (to all other pairs of contacts x–y) and the distribution for Pro–Val (to all other pairs of contacts x–y), can be a measure of the potential for substitution of Ser–Leu for Pro–Val or Pro–Val for Ser–Leu.

Second, we can express the contact substitution probabilities, $p_{k,l \to x,y}$, directly, normalized to 1.0,

$$p_{k,l \to x,y} = \frac{S_{k,l \to x,y}}{\sum_{x,y} S_{k,l \to x,y}} \tag{4}$$

and combine them with their estimates obtained from the correlation coefficients, $q_{kl \to x,y}$:

$$w_{k,l \to x,y} = \frac{1}{2}(p_{k,l \to x,y} + q_{k,l \to x,y}) \tag{5}$$

To reduce the influence of the unequal numbers of contacts for different residue pairs, we have divided values of $w_{k,l \to x,y}$ by the number of substituted contacts:

$$w'_{k,l \to x,y} = \frac{w_{k,l \to x,y}}{\displaystyle\sum_{kl} S_{k,l \to x,y}} \tag{6}$$

The sequence–structure match is then evaluated in terms of the contacts with a discrimination function of the form

$$C_S = \left(\sum_{i,j \subset G_{k,l \to x,y}} a_{i,j} \ln w'_{k,l \to x,y} \right) \Big/ n \tag{7}$$

In Eq. (7), $G_{k,l \to x,y}$ is the set of i, j positions in the contact matrix, $a_{i,j}$, of a protein with known structure corresponding to the contacts between residue types k and l in the protein with the known 3D structure and the implied contacts between residues x and y in the protein sequence under comparison; n is the number of contacts contributing to C_S.

In Fig. 4 we show the discrimination of related folds from unrelated

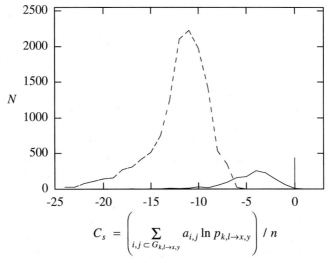

$$C_s = \left(\sum_{i,j \subset G_{k,l \to x,y}} a_{i,j} \ln p_{k,l \to x,y} \right) / n$$

FIG. 4. Frequency (N) of contact evaluation scores C_S obtained for comparisons of structures with sequences (M. A. Rodionov and M. S. Johnson, 1995, unpublished results). For all pairs of structures within each of the 96 families, the structure of one is compared with the sequence of the other and vice versa: the line at $C_S = 0$ corresponds to a structure compared with its own sequence; the solid curve represents a total of 1272 comparisons within the 96 protein families; and the dashed curve corresponds to comparisons where the 3D structure and 10 random permutations of the sequence are evaluated for each pair of proteins.

comparisons. Here, we have only considered the contribution to C_S [Eq. (7)] provided by the contact substitution probabilities as described by Eq. (4). Three curves are shown. The first, which appears as a line (because of the natural log scale) centered about $C_S = 0$, is for comparison of a 3D structure with its own sequence. The second curve, which extends over the middle range of values, is for 1272 authentic comparisons within a related protein family. The third curve extends to low negative values and corresponds to 10 comparisons for each of the authentic structure–sequence alignments, but where the sequence is a random permutation of the original.

Indels in Protein Structure Families

Thus far, we have discussed the use of environment-dependent residue substitutions and residue–residue contact substitutions in the alignment of proteins. We have also examined characteristics of indels and "loop" regions (loops refer to those portions of a protein structure that link elements of regular secondary structure such as α helices and β strands) within these families of aligned protein structures[37] in an effort to assign more adequately gap penalties during sequence or sequence–structure comparisons.

For sequence comparisons, a gap penalty is typically assigned as either a constant value or one that consists of both a constant component (for initiating a gap) and a length-dependent component (to control the gap length). Others have chosen to use more elaborate penalty functions.[21,22] None of these strategies is ideal: each implies that a gap can occur anywhere along the sequence and for roughly the same cost, although their location is controlled to large degree by the interplay of the residue match scores. When sequence similarity is low, it is likely that regions of high sequence similarity will not stand out to prevent gaps from appearing within structurally unreasonable locations. Gaps do not occur with equal frequency anywhere within the protein fold: for example, they are rarely found within elements of regular secondary structure,[37] but rather most often within loop regions that are exposed to solvent at the surface of a protein (Figs. 5 and 6a), and can thus more readily accommodate structural change.

Residues matched to structurally defined gaps in family structure alignments were examined in comparison with those positions that fall outside of gaps,[37] and where the following features were considered: (1) residues matched to gaps, (2) residue secondary structure, (3) residue percentage solvent accessibility, and (4) the local density of packing as described by the number of Cα atoms within a sphere of 14 Å centered about the Cα atom of the residue of interest. It was clear from our analysis[37] that regions

---

[37] M. S. Johnson, A. C. W. May, R. Sowdhamini, and J. P. Overington, submitted.

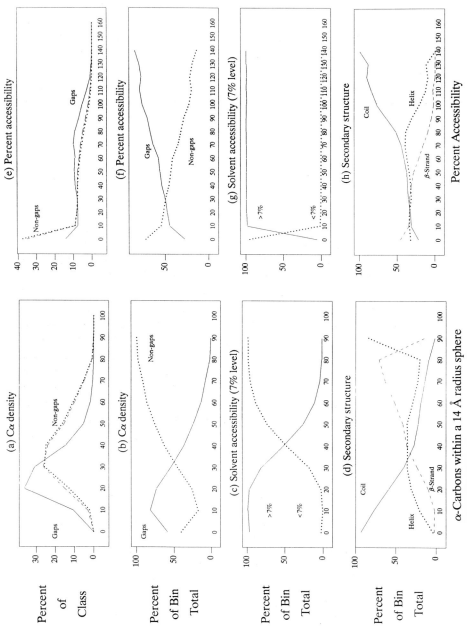

of a protein where gaps are more likely to appear can be identified (Fig. 5). In particular, residues within gaps prefer coil regions of proteins, relative solvent accessibility ≥80%, and a surrounding $C\alpha$ density ≤40.[37] Note in Fig. 6 the correspondence of gaps, both insertions and deletions observed in the alignment of 11 3D structures in the aspartic proteinase family, to the features based on the single unaligned 3D structure.

For structure–sequence comparisons, gap penalties have been assigned according to features of each residue including the three-residue average of the percentage solvent accessibility, the density of main-chain packing, and the residue secondary structure (Fig. 6b). Observed probabilities of amino acids matching a gap under different local environments have also been tabulated: gap penalties inversely proportional to the three-residue average of these values correlate well with positions of observed gaps in families of aligned 3D structures (Fig. 6c).

For sequence comparisons, structural details are not available to help pinpoint positions where gaps are more likely to occur. However, residue propensities can be calculated for different alternative environmental features, E and E', observed in the family structure alignments (Table III): (1) Nongap positions (E) versus gap positions (E'), (2) helix or strand versus coil, (3) solvent accessibility <80% versus ≥80%, (4) $C\alpha$ density >40 versus $C\alpha$ density ≤40, (5) helix or strand and accessibility <80% versus coil and accessibility ≥80%, (6) helix or strand and $C\alpha$ density >40 versus coil and $C\alpha$ density ≤40, (7) solvent accessibility <80% and $C\alpha$ density >40 versus solvent accessibility ≥80% and $C\alpha$ density ≤40, and (8) helix or strand and accessibility <80% and $C\alpha$ density >40 versus coil and accessibility ≥80% and $C\alpha$ density ≤40. The residue solvent accessibility cutoff of 80% was chosen on the basis of the distribution of accessibilities for coil residues in comparison to those for helices and strands (Fig. 5). The environmental propensities (P_i) were calculated from the frequency

FIG. 5. Characteristics of positions in 372 3D structures (derived from 94 homologous families). In (a–d) the surrounding $C\alpha$-atom density of residues is plotted; in (e–h) the relative percentage solvent accessibility is shown. Residues are broken down into categories that include gap versus nongap positions (a, b, e, and f), solvent accessibility at the 7% level (c and g), and (d and h) the secondary structure (helix, β strand, and "coil"). In (a and e) the percentage of class is used in which the sum of the gap and nongap curves each equal 100%. Because of the large difference in the number of gap positions relative to the nongap positions, in (b–d and f–h) we have used a percentage of bin total in which the sum of values of all curves is 100% at every vertical position on the graph. Although shown as discrete values on the x axis, the $C\alpha$ density and residue percent solvent accessibility values have been grouped into bins of 10. This means that a $C\alpha$ density value of 10 actually represents the range between 10 and 20. Used with permission.[37]

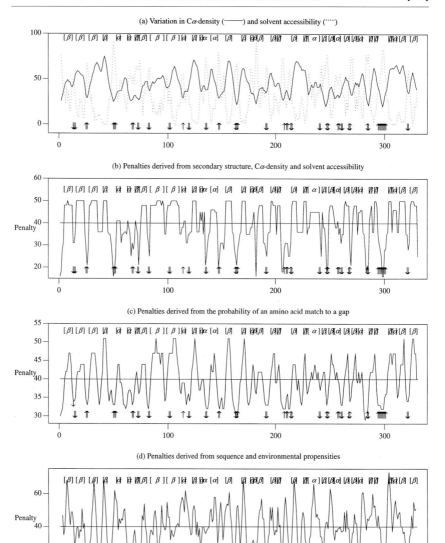

FIG. 6. Variation of different features for the 3D structure and amino acid sequence of the aspartic proteinase mouse renin and application to assignment of gap penalties. (a) Variation in Cα density and percentage residue solvent accessibility (three-residue average) calculated for the known structure. Helices and strands (delimited by brackets and α and β, respectively) and gaps [up arrows (↑) indicate insertions in the renin structure relative to one or more

(N) of residues of type i (of the 21 amino acid types) classified according to these different combinations of environments (Table III):

$$P_i = \frac{N_{i,E}/(N_{i,E} + N_{i,E'})}{\sum\limits_{i=0}^{<21} N_{i,E} \Big/ \Big(\sum\limits_{i=0}^{<21} N_{i,E} + \sum\limits_{i=0}^{<21} N_{i,E'}\Big)} \tag{8}$$

Sequences can then be examined over a seven-residue window and the difference in propensity scores for the two alternative environments summed:

$$\text{pen } \alpha \sum_{k=0}^{<7} \Delta P_k \qquad \text{where } \Delta P = P_E - P_{E'} \tag{9}$$

The resulting value is attributed to the central position of the segment. Thus, for sequence comparisons where no structural details are available a gap penalty (pen) can be assigned to every residue position and where it is proportional to ΔP_k (Fig. 6d).

Computer Programs and Data

To review, the template procedure is used to generate an alignment between structure and sequence. The alignment is then used to map contacts from the known 3D fold to the sequence of unknown fold, which is then evaluated in terms of the suitability of the residues in contact that have apparently been replaced during evolution. This approach can be used to search a sequence against the known folds; alternatively, a known fold can be searched against sequence data banks. Both multiple alignments of sequences or multiple alignments of 3D folds can be used to provide a more general family description within these search procedures. Scores from searches are sorted from high to low based on the normalized alignments score; the contact discrimination function is used as a secondary screen of the sorted comparisons.

---

members in the family structure alignment; down arrows show (↓) deletions in the renin structure relative to one or more family members]. (b) Penalties derived from consideration of residues evaluated as helix, strand, and coil, the Cα density of the residue, and the three-residue average of the solvent accessibility. The default constant gap penalty used by Q/PSLAVE is 40 (indicated by a horizontal line). (c) Penalties derived from the probability of an amino acid matching a gap in a particular environment; values were obtained from 93 aligned families (the aspartic proteinase family was not included). (d) Penalties based on amino acid propensity values from the last two columns of Table III and the renin sequence (information from the known structure was not considered here). Used with permission.[37]

TABLE III
PROPENSITIES FOR GAPS[a]

Amino acid	Position[b]		Sec[c]		Acc[d]		Cα density[e]		Sec/Acc[c,d]		Sec/Cα density[c,e]		Acc/Cα density[d,e]		Sec/Acc/Cα density[c–e]	
	Nongap	Gap	αβ	Coil	<80%	≥80%	>40	≤40	αβ/<80%	Coil/≥80%	αβ/>40	Coil/≤40	<80%/>40	≥80%/≤40	αβ/<80%/>40	Coil/≥80%/≤40
A	1.00	0.97	1.16	0.78	1.00	0.96	1.05	0.94	1.03	0.74	1.18	0.76	1.01	0.91	1.04	0.72
C	1.05	0.52	0.90	1.12	1.02	0.70	1.16	0.82	1.01	0.89	1.03	0.96	1.04	0.64	1.03	0.77
D	0.96	1.34	0.76	1.32	0.95	1.60	0.67	1.36	0.91	1.90	0.55	1.61	0.86	2.18	0.80	2.49
E	0.97	1.25	1.12	0.84	0.99	1.15	0.62	1.43	1.01	0.88	0.80	1.27	0.91	1.78	0.94	1.43
F	1.04	0.64	1.18	0.77	1.03	0.58	1.28	0.69	1.05	0.48	1.28	0.61	1.08	0.36	1.09	0.34
G	0.97	1.26	0.63	1.49	0.91	2.18	0.92	1.08	0.79	3.04	0.66	1.46	0.86	2.15	0.75	2.83
H	1.00	1.02	0.95	1.06	1.02	0.67	0.94	1.07	1.03	0.73	0.92	1.11	1.03	0.73	1.03	0.81
I	1.04	0.57	1.28	0.62	1.04	0.48	1.45	0.50	1.06	0.41	1.45	0.39	1.08	0.28	1.10	0.26
K	0.98	1.19	1.01	0.98	1.00	0.99	0.60	1.45	1.00	0.97	0.70	1.41	0.93	1.56	0.93	1.48
L	1.03	0.69	1.23	0.69	1.05	0.38	1.36	0.60	1.07	0.31	1.37	0.50	1.08	0.31	1.10	0.27
M	1.01	0.90	1.25	0.67	1.03	0.64	1.22	0.75	1.04	0.60	1.35	0.53	1.06	0.45	1.07	0.46
N	0.99	1.13	0.68	1.42	0.96	1.49	0.79	1.23	0.89	2.09	0.59	1.55	0.91	1.79	0.81	2.38
P	0.95	1.45	0.55	1.60	0.97	1.43	0.73	1.30	0.86	2.34	0.44	1.76	0.92	1.71	0.80	2.47
Q	1.00	0.97	1.06	0.92	0.99	1.15	0.80	1.23	0.99	1.07	0.90	1.13	0.94	1.48	0.94	1.44
R	0.99	1.09	1.01	0.98	1.01	0.83	0.90	1.11	1.02	0.82	0.94	1.08	1.01	0.90	1.01	0.90
S	0.96	1.34	0.81	1.25	0.98	1.31	0.82	1.20	0.95	1.50	0.72	1.38	0.93	1.57	0.90	1.70
T	0.99	1.06	0.96	1.06	1.01	0.89	0.87	1.14	1.00	1.01	0.87	1.17	1.00	1.00	0.98	1.13
V	1.04	0.64	1.27	0.64	1.04	0.40	1.40	0.56	1.06	0.37	1.42	0.43	1.08	0.27	1.10	0.27
W	1.04	0.60	1.21	0.72	1.05	0.30	1.39	0.56	1.08	0.26	1.46	0.38	1.09	0.26	1.10	0.26
Y	1.03	0.74	1.10	0.87	1.04	0.50	1.22	0.75	1.05	0.48	1.23	0.69	1.07	0.40	1.08	0.41
J	1.07	0.28	1.08	0.89	1.04	0.42	1.23	0.74	1.07	0.34	1.20	0.72	1.09	0.26	1.09	0.31

[a] Propensities (P_i) for different environmental conditions in families of aligned 3D structures. From Ref. 37.

[b] Nongap and gap positions as defined in the Methods section.

[c] Sec: Secondary structure; helical or strand residues (αβ) and coil residues.

[d] Acc: Average (over 3 residues) residue accessibility to solvent.

[e] Cα density: Residue Cα density number, either >40 Cα atoms within 14 Å distance or ≤40 Cα atoms.

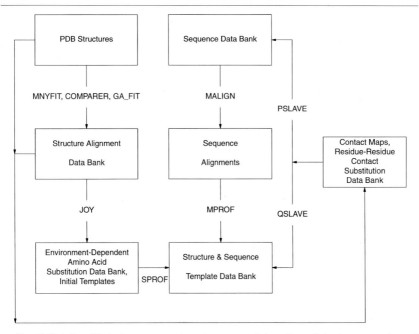

Fig. 7. Relationship between computer programs and data (boxed) in the comparison of proteins and search for common protein folds. Adapted from Ref. 32.

Comparisons between sequences and 3D structures are made with a set of computer programs (Fig. 7). These include those for making sequence alignments (MALIGN) and structure comparisons (MNYFIT, COM-PARER, and GA_FIT). GA_FIT employs a genetic algorithm combined with dynamic programming to produce totally automatic structure compari-sons.[38,39] JOY is used both to analyze the collection of aligned families of protein structures and to generate intermediate templates for all individual structures in the Brookhaven Protein Data Bank (PBD)[40] and all aligned families of 3D structures. These templates specify details of the protein structure at each position along the fold and are used by SPROF to construct a template of scores used to match the fold with sequences. SPROF encodes details necessary to provide position-by-position assignment of gap penal-ties. (Sequence-based templates can also be created using the program MPROF and scoring of the type shown in Table I.) Two very similar

[38] A. C. W. May and M. S. Johnson, *Protein Eng.* **7**, 475 (1994).
[39] A. C. W. May and M. S. Johnson, *Protein Eng.* **8**, 873 (1995).
[40] F. C. Bernstein, T. F. Koetzle, J. B. Williams, E. F. Meyer, Jr., M. D. Brice, J. R. Rodgers, O. Kennard, T. Shimanouchi, and M. Tasumi, *J. Mol. Biol.* **112**, 535 (1977).

programs are used for comparisons (Fig. 7): QSLAVE compares sequences against one or more structural templates, whereas PSLAVE compares a structural template against one or more sequences. These two programs employ dynamic programming to make alignments. They construct a contact map from the coordinates of the known fold and evaluate the alignment based on compatibility of equivalent contacts at aligned positions.

Comparisons

To illustrate the utility of this approach, we consider only a single search structure and compare it against more than 21,000 primary structures. In the first case we consider a search of the 3D structure of the erythrocruorin (a globin) of *Chironomous thummi thummi*; in the second we show that a single bacterial serine proteinase can detect members of the distantly related mammalian group.

Globin Searches

A search[16] of a structural template created for the 3D structure of the erythrocruorin from the midge larva *Chironomous thummi thummi* is depicted in Fig. 8. A total of 537 globins were detected prior to the first nonglobin, which was the chicken B-G antigen (CHKBGAF). The last of 618 globin sequences, an extracellular earthworm globin (GLBI_PHESE), is found at ranked position 901. Between the last of the contiguous high-scoring globin matches at position 537 and the final globin at position 901 there are a total of 283 nonglobins. Note the overlap between the nonglobin and the authentic globin matches in Fig. 8a.

If the alignments are reevaluated in terms of residue–residue contact substitutions using a cutoff value of $C_S = 9.0$, not only does the number of globin matches before the first observed nonglobin increase to 576 (Ref. 33), but the number of contaminating matches before the last globin at position 901 reduces to only 11! Thus, only 42 globins had contact scores less than 9.0, and many of these were within the top matches on the basis of the alignment scores and would have been detected on the basis of the A_S alone. Note the reduction of nonglobin matches when matches have been eliminated according to this criterion (Fig. 8b).

When we consider a cutoff value set to the average of three scores, namely, the alignment score (A_S), the alignment score normalized to the largest attainable score (A_S'), and the contact score (C_S), 604 of the 618 globin matches have a score larger than a cutoff of 7.5 and only 14 globins were eliminated from the qualifying matches because of scores lower than the cutoff (Fig. 8c). Although the first nonglobin was detected at ranked

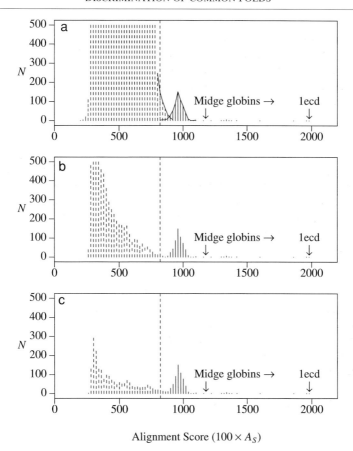

Alignment Score ($100 \times A_S$)

FIG. 8. Search of a single 3D structure against each of the 21,792 sequences (≥55 residues) in a sequence data bank. The erythrocruorin of *Chironomous thummi thummi* (1ecd) was encoded as a structural template with the program SPROF. The search was made with the program PSLAVE adapted to evaluate each alignment in terms of the match of residue-pair substitution with those collected from families of aligned 3D structures; no globins were present among the families contributing to the residue-contact substitution data. The frequency of globin alignment scores, normalized for the length of the alignment (A_S), are depicted with solid vertical lines; the nonglobins are indicated with dashed lines. Frequencies of scores greater than 500 [7566 scores in (a); 1047 in (b); 0 in (c)] have been removed for the sake of clarity. In (a) all scores are depicted (less those with frequencies >500). In (b) only alignments where a minimum contact score (C_S) of 9.0 was obtained are shown; 69% of the nonglobins are eliminated. In (c) a cutoff score of 7.5 was required (average of the combination of A_S, C_S, and A'_S, where the latter is calculated as the raw alignment score relative to the maximum possible score obtained from the largest values at each position along the structural template); 84% of the nonglobin sequences are eliminated by this screen. The full-length dashed vertical line indicates the position of the globin–last globin match at position 901 of the original ranked alignment scores (A_S), namely, to the extracellular earthworm globin (GLBI_PHESE). Adapted from Ref. 33.

Rank		Code	A_s	A_s'	C_s	$(A_s+A_s'+C_s)/3$	Title
1	↓	MXBPALP	44.69	44.69	100.00	63.13	MXBPALP prepro-alpha-lytic protease - Lysobacter enzymogenes
2	↓	N_2ALP1	44.69	44.69	100.00	63.13	ALPHA-LYTIC PROTEASE (E.C. NUMBER NOT ASSIGNED) - (LYSOBACTER ENZYMO
3	↓	MXBPALPA	44.69	44.69	100.00	63.13	MXBPALPA pre-pro alpha-lytic protease - Lysobacter enzymogenes
4	↓	PRLA_LYSEN	44.69	44.69	100.00	63.13	ALPHA-LYTIC PROTEASE PRECURSOR (EC 3.4.21.12) (GENE NAME
5	↓	N_1SGC1	20.01	19.62	33.77	24.47	PROTEINASE A COMPLEX WITH CHYMOSTATIN - (STREPTOMYCES GRISEUS, STRA
6	↓	PRSMBG	19.98	18.84	24.15	20.99	Protease B (EC 3.4.21.-) - Streptomyces griseus
7	↓	N_3SGB1	19.96	18.81	25.80	21.52	PROTEINASE B FROM STREPTOMYCES GRISEUS (SGPB) (E.C. NUMBER NOT ASSIG
8	↓	PRSMAG	19.89	19.51	38.25	25.88	Protease A (EC 3.4.21.-) - Streptomyces griseus
9	↓	PRTA_STRGR	19.85	19.38	30.42	23.21	PROTEASE A PRECURSOR (EC 3.4.21.-) (GENE NAME
10	↓	STMSPRB	18.50	16.14	19.41	18.02	STMSPRB protease B precursor (gtg start codon) - Streptomyces griseus
11	↓	STMSPRA	15.72	12.15	19.11	15.66	STMSPRA protease A precursor (gtg start codon) - Streptomyces griseus
12	↓	CHB1_BOMMO	14.18	13.60	0.58	9.45	CHORION CLASS B PROTEIN L11 PRECURSOR. - SILK MOTH (BOMBYX MORI).
13	↓	CHB4_BOMMO	14.10	13.59	0.60	9.43	CHORION CLASS B PROTEIN B.L1 (410) (FRAGMENT). - SILK MOTH (BOMBYX MORI)
14	↓	CHKIGLAM	13.88	13.75	0.50	9.38	CHKIGLAM Ig lambda-chain V-J region (AA at 2) - Gallus gallus
15	↓	GRP1_PHAVU	13.88	12.32	1.96	9.39	GLYCINE-RICH CELL WALL STRUCTURAL PROTEIN GRP 1.0 PRECURSOR. - KIDNE
16	↑	TRPGTR	13.77	10.98	10.46	11.74	Trypsinogen (EC 3.4.21.4) - Pig
17	↑	TRFF	13.45	10.70	8.57	10.91	Trypsinogen-like (EC 3.4.21.-) proenzyme precursor - Fruit fly
18	↑	KCUF	13.38	10.75	8.17	10.76	Collagenolytic protease (EC 3.4.21.32) - Atlantic sand fiddler crab
19		CHKIGLAN	13.24	12.94	0.34	8.84	CHKIGLAN Ig lambda-chain V-J region (AA at 2) - Gallus gallus
20		KRBO23	13.22	12.28	1.03	8.84	Keratin, 60K type II cytoskeletal, component III - Bovine (fragment)
21	↑	N_1NTP1	13.22	10.31	6.66	10.06	MODIFIED BETA TRYPSIN (MONOISOPROPYLPHOSPHORYL INHIBITED) (E.C.3.4.21.
22	↑	N_1TGB1	13.21	10.30	6.94	10.15	TRYPSINOGEN-CA FROM PEG - BOVINE (BOS TAURUS) PANCREAS
23		BASE77	13.20	12.30	0.55	8.68	CHORION (EGG SHELL) C PROTEIN 10A - SILKMOTH (ANTHERAEA POLYPHEMUS)
24	↑	N_1TGS1	13.16	10.23	7.46	10.28	TRYPSINOGEN COMPLEX WITH PORCINE PANCREATIC SECRETORYTRYPSIN INHI
25	↑	TRBOTR	13.16	10.23	7.54	10.31	Trypsinogen (EC 3.4.21.4) - Bovine
26	↑	EZEC540	13.11	10.20	7.65	10.32	TRYPSINOGEN EC 3.4.21.4 BOVINE
27	↑	TRDGC	13.10	10.59	6.92	10.20	Trypsinogen (EC 3.4.21.4), cationic, precursor - Dog
28	↑	EL1_PIG	13.10	10.34	7.79	10.41	ELASTASE 1 PRECURSOR (EC 3.4.21.36). - PIG (SUS SCROFA).
29	↑	ELPG	13.07	10.34	7.32	10.24	Elastase (EC 3.4.21.11) - Pig
30	↑	N_1EST1	13.07	10.34	7.32	10.24	TOSYL-ELASTASE (E.C.3.4.21.11) - PORCINE (SUS SCROFA) PANCREAS
31	↑	N_1TGN1	13.07	10.17	6.17	9.80	TRYPSINOGEN - BOVINE (BOS TAURUS) PANCREAS
32		B25297	13.07	12.02	0.91	8.67	class B protein m2G12 precursor - Silkworm (fragment)
33	↑	TRY1_HUMAN	13.03	10.59	6.94	10.19	TRYPSINOGEN I PRECURSOR (EC 3.4.21.4) - HUMAN (HOMO SAPIENS).
34	↑	TRDFS	13.01	10.96	8.88	10.95	Trypsinogen (EC 3.4.21.4) - Spiny dogfish
35	↑	TRY2_HUMAN	12.94	10.51	6.89	10.11	TRYPSINOGEN II PRECURSOR (EC 3.4.21.4). - HUMAN (HOMO SAPIENS).
36		A24713	12.90	11.72	1.26	8.63	Sericin
37	↑	N_2TPI1	12.87	10.06	5.85	9.59	TRYPSINOGEN -- PANCREATIC TRYPSIN INHIBITOR -- ILE-VAL COMPLEX (2.4 M MA
38	↑	TRY3_RAT	12.84	9.74	7.29	9.96	TRYPSINOGEN III CATIONIC PRECURSOR (EC 3.4.21.4). - RAT (RATTUS NORVEGICU
39		SODC_XENLA	12.83	12.36	0.55	8.58	SUPEROXIDE DISMUTASE 1 (CU-ZN) (EC 1.15.1.1). - AFRICAN CLAWED FROG (XENO
40	↑	EZEC562	12.83	10.19	4.41	9.14	KALLIKREIN EC 3.4.21.34 PIG - PIG PANCREAS
41		CHHA_BOMMO	12.80	12.69	0.29	8.51	CHORION CLASS HIGH-CYSTEINE HCA PROTEIN 12 PRECURSOR. - SILK MOTH (BO

position 538, there are 35 remaining nonglobins between the globin at ranked position 537 and the earthworm globin at position 901. At the higher cutoff score of 9.0 this number is reduced to 6, the first nonglobin appears at ranked position 595 (TPC_TACTR, horseshoe crab troponin C), but 46 globins score less than the cutoff and 572 globins achieve the cutoff value.[33]

In a search with a template constructed from two aligned 3D structures, human globin α chain and erythrocruorin, all of the globins were detected prior to the first high-scoring nonglobin sequence.[33]

Serine Proteinases

Figure 9 shows the results of a search of the structural template for the α-lytic proteinase from *Lysobacter enzymogenes*.[33] The first 41 top-scoring hits are shown (ranked by values of A_S), where arrows indicate members of the bacterial serine proteinase family (left pointing) and the mammalian serine proteinase family (right pointing); the remaining matches are to proteins unrelated to the serine proteinases, and are all characterized by unusual sequences with compositions exceptionally high (between 17 and 63%) in a few amino acids such as glycine, serine, and cysteine. On the basis of the contact score alone, C_S, the values for the unrelated protein matches range from 0.29 to 1.96; for the mammalian serine proteinase matches, the contact scores range between 4.41 and 10.46. From the combined average score, the largest value seen in Fig. 9 for a nonserine proteinase corresponds to a match with the sequence of the silk moth chorion protein (9.45). All of the authentic serine proteinases shown in Fig. 9 exceed this value except one, namely, the sequence of pig pancreas kallikrein (9.14).

Concluding Remarks

Given that there is a finite number of protein family folds[4] and a rapid (exponential) increase in the number of new structures appearing each year, we should soon have a representative structure for most of the common folds. If a protein of known sequence, but unknown structure, can be matched to one of the common folds, it is possible to model the 3D structure of the protein and often learn more about its function.[5] We have described

---

FIG. 9. Top 41 hits in a search of the α-lytic proteinase structure (2alp) against the 21,792 sequences in a sequence data bank. All 21,792 matches were ranked according to their alignment scores A_S and A_S'; in addition, the contact probability score C_S and the average contact alignment score is shown for each match. In column †, bacterial serine proteinases closely related to the search structure are indicated with arrows ($\leftarrow$) along with those that belong to the mammalian group ($\rightarrow$). Used with permission.[33]

an integrated and knowledge-based approach to the prediction of protein structure and function that can maximize the value of genome sequence data.

Programs and Data Availability

Computer programs and data are available via the World Wide Web (Molecular Modelling and Biocomputing Group) from WWW.btk.utu.fi and via anonymous FTP from FTP@btk.utu.fi. Programs are written in both FORTRAN and C language; some programs involve translations of FORTRAN to C language using the GNU f2c utility. These programs require the f2c compiled libraries and definitions in f2c.h. Programs are developed under the UNIX operating system (LINUX and Silicon Graphics).

Acknowledgments

This research has been supported by the American Cancer Society, Pfizer Central Research (UK), Imperial Cancer Research Fund (UK), Zeneca Agrochemicals (UK), the Royal Society (London), the European Molecular Biology Organization, and the Academy of Finland.

[35] Three-Dimensional Profiles for Measuring Compatibility of Amino Acid Sequence with Three-Dimensional Structure

By James U. Bowie, Kam Zhang, Matthias Wilmanns,
and David Eisenberg

Introduction

The three-dimensional (3D) profile method is a general approach to assessing whether a given sequence is compatible with a particular three-dimensional structure. The method has been used in structure identification, structure verification, *de novo* protein folding, and secondary structure prediction.[1–7] In this chapter we focus on approaches to fold identification and verification of the correctness of structures using 3D profiles.[1–7]

[1] J. U. Bowie, R. Lüthy, and D. Eisenberg, *Science* **253**, 164 (1991).
[2] J. Bowie and D. Eisenberg, *Proc. Natl. Acad. Sci. U.S.A.* **91**, 4436 (1994).
[3] R. Luthy, J. U. Bowie, and D. Eisenberg, *Nature (London)* **356**, 83 (1992).
[4] T.-M. Yi and E. Lander, *J. Mol. Biol.* **232**, 1117 (1993).
[5] K. Zhang and D. Eisenberg, *Protein Sci.* **3**, 687 (1994).

Copyright © 1996 by Academic Press, Inc.
All rights of reproduction in any form reserved.

The first step in the method is to create a 3D profile as outlined in Fig. 1. We start with a three-dimensional structure. The structure could be derived experimentally, by X-ray crystallography or nuclear magnetic resonance (NMR) spectroscopy, or computationally, by modeling studies. The environment of every residue in the three-dimensional structure is defined by a set of parameters, such as area of the side chain that is buried, the secondary structure, or what types of side chains are nearby. These parameters define what types of residues would be favorable at each position. For example, if the residue is buried, hydrophobic residues would be favored and hydrophilic residues disfavored. These preferences are quantified by assigning a score for placing each of the 20 amino acids at each position in the structure. The scores are listed in a sequence-specific scoring table called a 3D profile. A representation of a 3D profile is shown in Fig. 1. Each row represents a position in the three-dimensional structure, and scores for placing each of the amino acids at that position are listed in 20 columns. Using these scores, then, it is possible to find the best scoring alignment of any sequence to positions in the structure using a dynamic programming algorithm.[8–11] The last two columns list penalties for opening and extending a gap in an alignment of a sequence to the profile, and these, too, can be different at each position based on structural parameters. The overall alignment score for an amino acid sequence and a 3D profile is a measure of the compatibility of the sequence with the three-dimensional structure represented by the 3D profile. The format of the 3D profile is identical to the sequence or one-dimensional (1D) profiles developed by Gribskov and Eisenberg.[10,11]

One of the uses of 3D profiles in structure prediction is to determine whether a sequence of unknown structure is compatible with a known three-dimensional structure. As outlined in Fig. 2, every sequence in a database of protein sequences is aligned to a profile to obtain an alignment score. This alignment score is then converted to a Z score, which is the number of standard deviations the alignment score is from the mean alignment score for other proteins of similar length. The highest scoring proteins in the database are the ones that are most compatible with the starting structure, as judged by the scoring criteria used to make the 3D profile,

[6] M. Wilmanns and D. Eisenberg, *Proc. Natl. Acad. Sci. U.S.A.* **90,** 1379 (1993).
[7] M. Wilmanns and D. Eisenberg, *Protein Eng.* in press (1995).
[8] S. B. Needleman and C. D. Wunsch, *J. Mol. Biol.* **48,** 443 (1970).
[9] T. F. Smith and M. S. Waterman, *Adv. Appl. Math.* **2,** 482 (1981).
[10] M. Gribskov, A. D. McLachlan, and D. Eisenberg, *Proc. Natl. Acad. Sci. U.S.A.* **84,** 4355 (1987).
[11] M. Gribskov, R. Lüthy, and D. Eisenberg, this series, Vol. 183, p. 146.

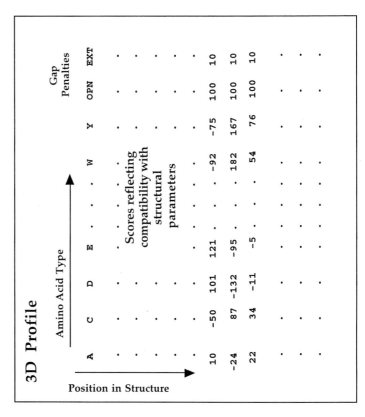

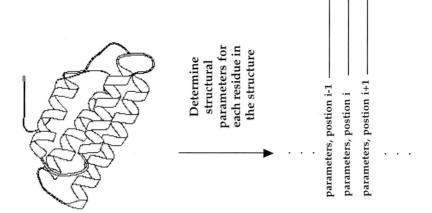

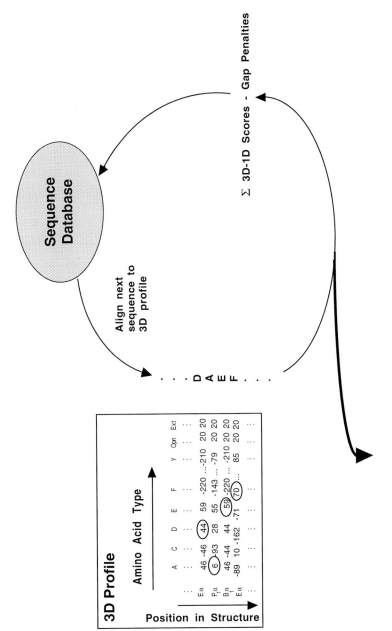

List of Top Scoring Sequences

FIG. 2. Using a 3D profile to find sequences compatible with the starting structure.

and may adopt a fold similar to that structure. Database searches of this type are called compatibility searches.

Another application of 3D profiles is to verify that a proposed model of a protein structure is correct. This is done by testing whether the sequence is compatible with its own structure as outlined in Fig. 3. Consider two possible structural models for a sequence, one correct and one incorrect. Because the structures are different, the environments of the residues in the three-dimensional structure will be different. Consequently, the 3D profiles calculated from the structures will be different. Because the amino acid sequence is more compatible with the environments in the correct structure than the environments of the incorrect structure, the 3D profile score obtained from an alignment of the sequence for the correct structure should be higher than the 3D profile score for the incorrect structure. Locally incorrect regions of a model can also be identified by plotting the average 3D profile score versus regions of sequence. Such a plot is called a 3D profile window plot.[3]

In this chapter we outline three different methods for generating the 3D profile and the application of these methods to fold identification and structure verification.

Three-Dimensional Profiles with Discrete Environment Classes

In the original method, environments were defined by (1) the area of the side chain that is buried, (2) the fraction of the surface of the residue covered by polar atoms, and (3) the secondary structure of the position.[1] On the basis of these three parameters the residue was classified into discrete environment classes. Residues were divided into three types according to the surface area buried: buried, partially buried, and exposed. The buried class was then further subdivided into three subclasses, and the partially buried class was subdivided into two subclasses based on the fraction of the surface covered by polar atoms. These six classes were then further subdivided according to secondary structure (α, β, or other) to give a total of 18 different environment classes. The preference of residue type a at position i, with secondary structure state x, area buried b, and fractional polarity p is defined[12] as the information value $S_{a_i x}(b, p)$:

$$S_{a_i x}(b, p) = \ln[P(a|x, b, p)/P(a)]$$

where $P(a|x, b, p)$ is the conditional probability of finding residue a at secondary structure state x in environment (b, p) and $P(a)$ is the *a priori*

---

[12] R. Fano, "Transmission of Information." Wiley, New York, 1961.

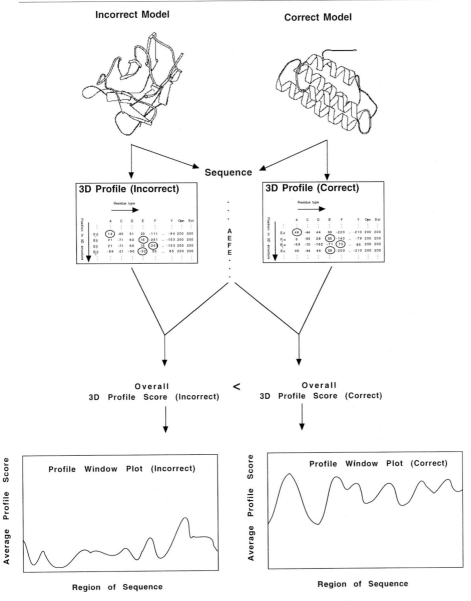

FIG. 3. Using a 3D profile to verify the correctness of a model structure.

probability of residue a derived from amino acid compositions. The measure $S_{a_ix}(b, p)$ is referred to as the 3D–1D score.

Figure 4a shows the discrete scoring surface for a tyrosine residue in a β-sheet secondary structure. As might be expected, tyrosine scores well in a highly buried environment and prefers some polar groups in its environment to satisfy the hydrogen bonding potential of the hydroxyl group.

Continuous Three-Dimensional Profiles

A significant weakness with discrete environment class definitions is that, because of sharp class boundaries, an infinitesimal change in the buried area of a residue can change its environmental class and thus can alter the residue preferences dramatically. To overcome this shortcoming, we introduced a continuous representation of the residue preferences as a function of the independent variables area buried and fraction polar.

The 3D–1D score in a given secondary structure state can be represented as a continuous function of its environments (b, p) using a two-dimensional Fourier series[5]:

$$S_{a_ix}(b, p) = \sum_{k,l} f_{kl} \, e^{-2\pi i(kb + lp)}$$

If residue preferences $S^\circ_{a_ix}(b, p)$ are observed at some sampling points of (b, p) as described above (the discrete 3D–1D scores), a set of Fourier coefficients f_{kl} for function $S_{a_ix}(b, p)$ that best represents the observed values $S^\circ_{a_ix}(b, p)$ can be evaluated by a least squares minimization method.

A smoothing factor, analogous to the Debye–Waller thermal factor (or B factor) in X-ray crystallography, can be used to average over a small area near the calculated area buried and fraction polar. The smoothing can be conveniently performed by multiplying the Fourier coefficients by an exponential term:

$$S_{a_ix}(b, p) = \sum_{k,l} f_{kl} \, e^{-2\pi i(kb + lp)} \, e^{-B(k^2 + l^2)}$$

where B is the smoothing factor.

We divided the area buried and fraction polar each into 32 equal bins for each of the three secondary structure classes of each of the 20 residues. The score value was from interpolating the 3D–1D scoring table of Bowie et al.[1] We chose to use a fourth order Fourier series [$k = l = 4$] to represent the residue preference surface. The Fourier coefficients were evaluated by a least squares minimization method. There are altogether 60 sets of 64-term Fourier coefficients to represent the preferences of the 20 residues in each of the three secondary structure states.

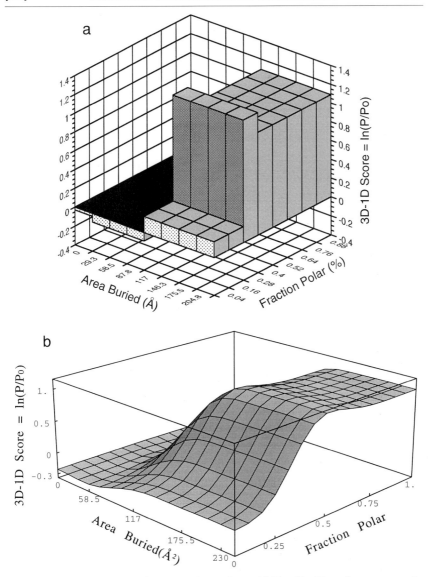

FIG. 4. Discrete versus continuous scoring surfaces. (a) The 3D–1D preference scores for tyrosine in β sheet for six discrete environmental classes. (b) The 3D–1D preference for tyrosine in β sheet as a continuous function of the environmental variables, area buried and fraction polar.

Figure 4b shows the scoring surface for tyrosine in the β-sheet state as a continuous function of area buried and fraction polar, represented by a Fourier series. This can be compared with Fig. 4a, which represents the discrete scoring surface of tyrosine. Both these scoring surfaces share the same general features of favorable and unfavorable regions of area buried and fraction polar. However, the smooth scoring surface in the Fourier series representation avoids abrupt changes of score when the area buried or fraction polar moves across class boundaries.

Given the Fourier coefficients, the 3D–1D scores of the 20 amino acid types for a position can be evaluated from the secondary structure and the values of the environmental variables area buried and fraction polar of that position. A 3D profile can thus be created using this continuous representation of residue preferences in place of the original discrete representation. This new 3D profile, which has exactly the same matrix form as the original profile, is referred to as a continuous 3D profile.

Residue Pair Preference Profiles

The residue pair preference profile (R3P) method generates profiles using preferences of residues to be in the local environment of each residue position of the profiled structure. The use of distance-dependent residue pairs was pioneered by the group of Sippl.[13] Jones et al. combined the pair potentials of Sippl with a solvent potential, similar to the local environment descriptors in the 3D profile method.[14] This group made effective use of the combined potentials for detection and alignment of distantly related proteins by using a multiple layer alignment technique.[15] Godzik et al. developed a method for detection of distantly related sequences based on distance-dependent potentials for residue pairs and triplets.[16] More recently developed inverse folding applications employing pair potentials have been reviewed.[17,18] A significant advantage of the R3P method over other 3D profile types is an improved iterative alignment procedure as described below. In the following we summarize the selection of interacting residue pairs, the types of residue pair preferences, the generation of profiles, and the iterative refinement procedure used by the R3P method. Details of the method are discussed elsewhere.[6,7]

[13] M. Sippl, *J. Mol. Biol.* **213,** 859 (1990).
[14] D. T. Jones, W. R. Taylor, and J. M. Thornton, *Nature (London)* **358,** 86 (1992).
[15] W. Taylor and C. Orengo, *J. Mol. Biol.* **208,** 1 (1989).
[16] A. Godzik, A. Kolinksi, and J. Skolnick, *J. Mol. Biol.* **227,** 227 (1992).
[17] J. Bowie and D. Eisenberg, *Curr. Opin. Struct. Biol.* **3,** 437 (1993).
[18] M. Sippl, *Proteins* **17,** 355 (1993).

Residue Pair Selection

A given protein 3D structure is characterized in terms of interacting residue pairs. At each residue position a local environment is created with a radius of 12 Å centered around its $C\beta$ atom. A cylinder with radius of 1.6 Å is created between the $C\beta$ positions of each residue of the pair, formed by the profiled residue and a residue in the local environment of the profiled residue. The residue pair is defined as an interacting pair if no nonhydrogen atoms of any other residue fall within this cylinder. For proteins with sequences exceeding 100 residues, about eight residue pairs per profiled residue position are found.

Residue Pair Characterization and Scoring

Each interacting residue is characterized by its amino acid type and a structural property. The residue pair preference PP_{a_i,x_i,a_j,x_j}, is given by the logarithmic likelihood ratio

$$PP_{a_i,x_i,a_j,x_j} = \ln[P(a_i, x_i, a_j, x_j)/P(a_i, a_j)]$$

where $P(a_i, x_i, a_j, x_j)$ is the probability of a residue pair at positions i and j of amino acid types a_i, a_j and structural properties x_i, x_j. The three types of structural properties used are (1) dihedral angles (ϕ, ψ), (2) secondary structure, and (3) number of neighboring residues in the local environment, called OOI numbers after Nishikawa and Ooi from where it is derived.[19] Each structural property is grouped into three classes, for example, helix, sheet, and other for secondary structure. Thus, for each structural property used, there are 3×20 types of interacting classes of residues.

The 3D–1D score for a position, S_{a_i}, is then simply a weighted average of pair preference scores for all the neighboring residues:

$$S_{a_i} = \sum_{j=1, j\neq i, i+1}^{n} w_j PP_{a_i,x_i,a_j,x_j} \bigg/ \sum w_j$$

The weights applied to the interacting residues, w_j, are described below.

Three types of profiles can be generated from a three-dimensional structure, one for each structural property (dihedral angles, secondary structure, or OOI number profiles). The three types of R3Ps are combined by averaging the scores for elements of the profiles to give the final R3P profile. Since the format of the R3P is identical to the profile format used in the 1D profile method,[10,11] the discrete profile method,[1] and the continuous profile method,[5] it can also be merged with profiles generated with those methods.

[19] K. Nishikawa and T. Ooi, *J. Biochem.* (*Tokyo*) **100**, 1043 (1986).

Iterative Alignment

One of the main advantages of the R3P method is that it provides a simple method for improving the alignments iteratively. An alignment implies that the residues in the aligned sequence would assume particular positions in the structure. Since these residues may be different, the environments created by the new sequence could be different. With the R3P method, it is a simple matter to replace the residue identities with those of the aligned sequence and recalculate 3D–1D scores based on the new sequence. The new sequence should produce better representations of the environments in the new structure and should therefore yield a more accurate alignment.

Iterative refinement of R3P alignments is carried out by regeneration of R3Ps derived from the previous alignment. The original residue types in the profiled structure are replaced by the corresponding residue types in the aligned sequence. The selection of interacting residue pairs and structural characterization of residue pairs remains unchanged. The residue pair preferences only change through modified residue types of the local environmental residues.

Pair Weighting

Because some residues in any given alignment will not be aligned properly, and this could have an adverse effect on the scoring of neighboring residues, we chose to weight the contribution of environmental residues to the scoring by some measure of the likelihood that the placement of the residue is correct. This is assessed by how compatible each residue is with the environment into which it is placed. For example, if the alignment places a polar residue at a buried position, it is reasonable to expect that an alignment error has occurred. This error will be reflected in a low 3D–1D score for that position. Each pair preference value, PP_{a_i, x_i, a_j, x_j}, is given a weight, w_j, according to

$$w_j = \exp[S_{a_j}(b, p)]$$

where $S_{a_j}(b, p)$ is the discrete 3D–1D score of position j. The exponent of $S_{a_j}(b, p)$ is equivalent to a normalized conditional probability.

The weighting of each residue pair score by the fit of the environmental residue within the current model is an essential component for improving the alignment during refinement. The effect of residue pair score weighting is to reduce the contribution of locally unfavorable alignments that are on average reflected by low fits of aligned residue types in local environments. Correct parts of alignments are enhanced by this weighting scheme.

Continuous Three-Dimensional Profiles and Compatibility Searches

Determining Best B Value to Use

In comparing continuous and discrete 3D profiles for the structure of sperm whale myoglobin (1MBO), we first determined the optimal value of the smoothing factor B. Three-dimensional profiles made with different B values were used to score all protein sequences in a sequence database containing 59,091 nonidentical sequences.

The relative effectiveness of different profiles is shown in Fig. 5. In a compatibility search with a given profile, each sequence receives a profile score for the optimal alignment of that sequence with the profile. These profile scores are then expressed as Z scores. In this plot, a perfect profile would be represented as a line of slope 1 extending to the number of globin sequences in the database (691 in this example), and then turn horizontal, because each additional sequence of lower Z score would be a nonglobin. For less than perfect profiles, some nonglobins will have higher Z scores than some globins, and the trace will fall below that for a perfect profile.

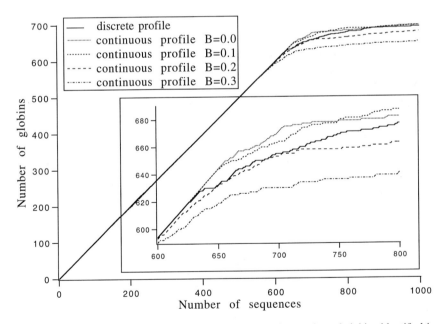

FIG. 5. Effectiveness of different smoothing factors. The number of globins identified is plotted as a function of number of high scoring sequences in a 3D profile compatibility search using the different profiles. The inset shows a magnification of the upper right-hand portion of the graph.

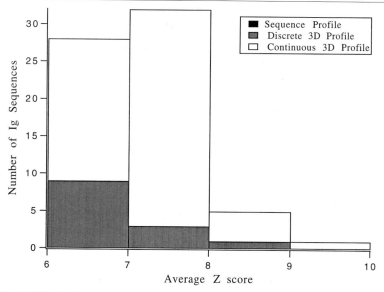

F<small>IG</small>. 6. Distribution of Z scores for the immunoglobulin sequences identified by the sequence profile and the discrete and continuous 3D profiles of 2SOD. The ordinate is the average Z score for the immunoglobulin sequences. The abscissa is the number of immunoglobulin (Ig) sequences at a given Z score.

It appears that values between 0.0 and 0.1 for the smoothing factor B are the most discriminating for the recognition by the profile of its own and very similar sequences, and yet still sensitive in identifying remotely related sequences.

Continuous Profile Compared with Discrete Profile and Sequence Profile for Bovine Superoxide Dismutase

The greater sensitivity of the continuous profile in detecting distant relationships than can be achieved with either the sequence profile[10] or the discrete profile is demonstrated further in the case of bovine Cu,Zn-superoxide dismutase (2SOD).[20] The continuous, discrete, and sequence profiles all identified 51 closely related superoxide dismutase sequences. However, the continuous profile and the discrete profile were able to identify many immunoglobulin sequences that have a similar greek-key fold as that of 2SOD.[21] Figure 6 shows the distribution of Z scores for those

---

[20] J. Tainer, E. Getzoff, and K. Beem, *J. Mol. Biol.* **160**, 181 (1982).

[21] J. Richardson, D. Richardson, K. Thomas, E. Silverton, and D. Davies, *J. Mol. Biol.* **102**, 221 (1976).

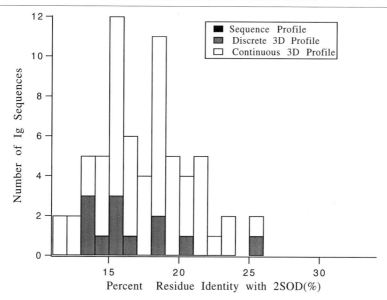

FIG. 7. Histogram of the sequence identities for all immunoglobulin sequences identified by the sequence profile and the discrete and continuous 3D profiles of 2SOD with Z scores above 6.

immunoglobulin sequences identified by the three profiles. The sequence profile failed to detect any immunoglobulin sequences with a Z score above 6. The discrete profile identified only 13 immunoglobulin sequences, as compared with 66 identified by the continuous profile. The sequence identities of these immunoglobulin sequences detected by the continuous profile with the sequence of 2SOD range from 11 to 26%, as shown in Fig. 7. Their average sequence identity is around 16%. This shows that the continuous profile can be more sensitive in detecting distant sequence–structure relationships than either the discrete profile or the sequence profile.

Continuous Three-Dimensional Profiles and Structure Verification

Three-Dimensional Profile Window Plots

The 3D profile window plot[3] is a tool to assess the accuracy of a 3D protein structure, based on the compatibility of a structure for its own sequence. The average profile score for each 21-residue segment is plotted versus sequence number. Segments scoring poorly have environments incompatible with the sequence, and they suggest errors in the structure of these segments. However, the details of a profile window plot prepared from a discrete 3D profile can vary significantly with the orientation of coordinates of the same structure. The reason for the changes in the discrete profiles is the sampling error in estimating the area buried and fraction polar. Different orientations of coordinates yield slightly different values of area buried and fraction polar. Small as they might be, if they are near the border between two classes, only a small difference is enough to change the classification of the residue environment. In contrast, a continuous profile prevents an abrupt change of preference with a small change in orientation, because the residue preference varies smoothly with the area buried and fraction polar.

We tested both the discrete and continuous profiles created from the same structure of myoglobin (1MBO) and randomly rotated the coordinates nine times. A window plot for all the nine coordinates was generated using discrete profiles as shown in Fig. 8a. The discrete profile showed significant variation in almost all regions. In contrast, the window plot created with the continuous profile (Fig. 8b) is largely unaffected by changes in orientation.

Detecting Wrong Folds and Monitoring Progress of Structure Refinement

The continuous 3D profile can be used to detect wrong folds or an incorrect segment in an otherwise correct structure in the same way as the discrete profile.[3] An incorrectly modeled region tends to have a score near or below zero in a profile window plot. Similarly, the 3D profile can also be used to monitor the progress of structure refinement. Generally, a well-refined structure has a higher profile score than an unrefined structure.

The detection of an incorrect segment of the model and monitoring the progress of refinement is illustrated for diphtheria toxin (DT) in Fig. 9. Diphtheria toxin is a member of ADP-ribosylation toxin family. The crystal structure of DT was solved initially at 2.5 Å by Choe *et al.*[22] and refined to 2.0 Å by Bennett *et al.*[23] In a preliminary trial model built into a 3.0 Å electron density map, much of the C-terminal receptor binding domain (R-domain) was at first reversed. However, it was rebuilt as the quality of the electron density map was improved by phase refinement and extension to 2.5 Å. Figure 9 shows the profile window plots of the trial model with reverse-traced R-domain, the 2.5 Å model, and the 2.0 Å refined model. The profile scores in the R-domain (residues 380–518) are near or below zero, suggesting that this region was incorrect. The R-domain in the 2.5 Å model has a significantly higher score, confirming that the trial model was

[22] S. Choe, *et al.*, *Nature* (*London*) **357**, 216 (1992).
[23] M. Bennett, S. Choe, and D. Eisenberg, *Protein Sci.* **3**, 1444 (1994).

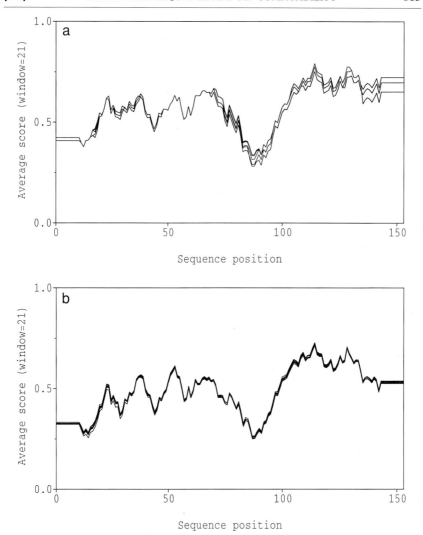

FIG. 8. Sensitivity of different 3D profile window plots to errors in calculating environmental parameters. For both plots, the vertical axis shows the average 3D–1D score for residues in a 21-residue sliding window, the center of which is at the sequence position indicated by the horizontal axis. (a) A 3D profile window plot for 1MBO with discrete 3D profiles created from coordinates rotated randomly nine times. (b) A 3D profile window plot for 1MBO with continuous 3D profiles created from coordinates rotated randomly nine times. The fluctuation of the profile scores is significantly reduced when using the continuous profile.

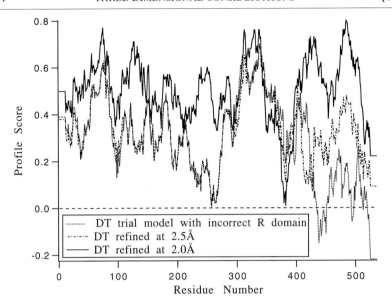

FIG. 9. Continuous 3D profile window plot for three models of diphtheria toxin. The profile window plots for an initial trial model and for 2.5 Å and 2.0 Å models are represented by dotted, dashed, and solid lines, respectively.

incorrect. The scores in profile window plot for the 2.0 Å refined model are generally higher than those for the 2.5 Å model, reflecting the improved structure of the 2.0 Å refined model. The regions with the greatest improvement reflect an initial misregistration of the sequence with the electron density, corrected during rebuilding. The profile score for the 2.0 Å refined model has one noticeable dip near residues 380–387. These residues are in the hinge loop, which changes conformation when DT dimerizes by domain swapping.[24]

Residue Pair Preference Profile Method and Sequence–Structure Alignments

Two aspects are crucial in 3D compatibility searches by inverse folding methods: sensitivity of detection of related sequences and correctness of the alignment. A compatible candidate is identified through a statistical measure of the fit of the alignment, which is normally expressed by a Z score.[10] However, for successful detection there is no *a priori* requirement for an overall correct alignment as long as part of the fold or even only a

[24] M. Bennett, S. Choe, and D. Eisenberg, *Proc. Natl. Acad. Sci. U.S.A.* **91**, 3127 (1994).

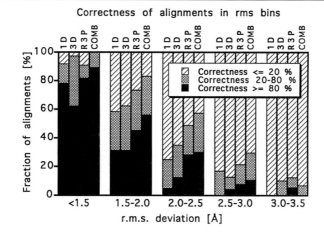

FIG. 10. Correctness of alignments as a function of structural divergence. The analysis uses 120 protein pairs with known 3D structures to produce 240 alignments with each profile method. The correctness of the alignments is evaluated from structural superposition alignments.[7] Each rms bin contains about the same number of alignments. The four bars represent the percentage of mostly correct (≥80% correctness), intermediate (20% < correctness < 80%), and mostly incorrect (≤20% correctness) alignments by different fill patterns for 1D sequence alignments,[10] 3D profile alignments,[5] R3P alignments, and combined alignments.

motif is compatible enough to score with statistical significance. For example, the original 3D profile method was able to detect the structural relationship of actin and heat-shock protein 70 (Hsp70).[1] The score, however, resulted from a short 30-residue segment covering the ATP binding site of both proteins. In those cases additional tools are needed to improve the alignment which can in turn serve as basis for building a model of the identified sequence. In a recent test analysis on distantly related pairs of sequences with known 3D structures, we have shown that the R3P method, in combination with 1D and 3D profiles, can improve the correctness of alignments.[7] Figure 10 shows the alignment quality as a function of structural similarity for different types of profiles. On average, more than 50% of the alignments are correct for structure pairs that are superimposed with a root-mean-square (rms) deviation of ≤2.12 Å if combined R3Ps are used. In R3P or combined profile–sequence alignments of the pair Hsp70–actin, most of the core regions of the two sequences are correctly aligned (not shown).

Residue Pair Preference Profiles and Compatibility Searches

The R3P method has also been used for 3D compatibility searches either of a sequence against a profile database or of profile against a sequence database. We have applied a preliminary version of the R3P method in combination with the 3D profile method for the prediction of several sequences to be folded as β/α barrels by sequence database searches with a number of β/α barrel profiles.[6]

Program Use and Availability

As described above there are a large number of ways to generate a profile. The method one uses depends on the application. For fold identification, there is no absolute favorite. No profile works best in all cases, but best results are generally obtained with continuous profiles or an R3P profile. For structure verification, the clear choice is continuous profiles. For alignment of an identified structure with the sequence, the clear choice is the R3P method. The following programs are available from the authors.

ENVIRON: Calculates area buried and fraction polar for each residue in a structure

3D_PROFILER: Reads the output file from ENVIRON and generates a discrete profile

PROFGEN: Reads the output file from ENVIRON and generates a continuous profile of the structure

MAKER3P: Produces an R3P profile directly from a coordinate file

ALIGNR3P: Produces a sequence–R3P profile alignment by an iterative refinement procedure

PROFILESEARCH: Determines optimal alignment scores for a database of sequences with a 3D profile; the alignment scores are converted to Z scores, and the output is a sorted list of sequences and alignment scores

VERIFY_3D: Determines the overall 3D profile score for a structure and the average score in regions of sequence along the structure

[36] SSAP: Sequential Structure Alignment Program for Protein Structure Comparison

By CHRISTINE A. ORENGO and WILLIAM R. TAYLOR

Introduction

Since the 1970s, protein structure comparison methods have become increasingly sophisticated. Early rigid-body techniques[1] have been used to study different mutant and ligand-bound forms. They are fast and extremely efficient for superposing very similar structures, but as structures diverge these methods cannot always identify equivalent positions because of insertions and deletions (indels).

A major incentive to developing more robust methods has been the need to analyze protein fold families, extracting information that can improve structure prediction and modeling. Because there are nearly 30 times more known sequences than structures, this is an important consideration. During evolution, the sequence of a protein may change, but the overall fold is much more conserved, remaining the same even if 70% of the sequence changes.[2,3] This gives rise to families of related structures and means that a new structure can be modeled on a known one if the proteins have similar sequences.

Analyses of protein families can help in protein structure modeling by setting tolerances on variability at different positions in the fold. Similarly, for structure prediction, information from protein families can improve template or profile-based methods by incorporating residue preferences in specific structural locations in the fold.[4,5]

In some protein families (e.g., dinucleotide binding proteins), very low sequence similarities (<10%) have been found and there are often very extensive indels, usually in the loops, whereas the core of the fold is much more conserved. For these much broader fold families, very sensitive structure comparison methods are needed. In particular these should be able to identify conserved structural regions that may be associated with specific sequence patterns. Such regions might be important for the folding pathway or for stabilizing the fold. To meet this challenge, a wealth of new compari-

---

[1] B. W. Matthews and M. G. Rossmann, this series, Vol. 11, p. 397.
[2] C. Sander and R. Schneider, *Protein Eng.* **1,** 159 (1991).
[3] T. P. Flores, C. A. Orengo, D. M. Moss, and J. M. Thornton, *Protein Sci.* **2,** 1811 (1993).
[4] W. R. Taylor, *J. Mol. Biol.* **188,** 233 (1986).
[5] M. S. Johnson, J. P. Overington, and T. L. Blundell, *J. Mol. Biol.* **231,** 735 (1993).

METHODS IN ENZYMOLOGY, VOL. 266

Copyright © 1996 by Academic Press, Inc.
All rights of reproduction in any form reserved.

son methods have been developed, some able to cope with very distantly related proteins.

There are now over 30 methods for comparing structures. This chapter discusses those flexible enough to align distantly related structures and therefore most suitable for identifying and analyzing protein fold families. To illustrate ways of overcoming the various difficulties encountered, we have focused on our method, sequential structure alignment program (SSAP), and describe the various modifications that have been required to handle more complex similarities. In particular, the need to identify similar motifs between proteins and the development of a multiple comparison method that can identify the consensus structure for a family of related proteins are discussed.

Different Approaches

There are two main approaches to structure alignment, both based on comparing the global protein geometry (reviewed in Ref. 6). Rigid-body techniques superpose structures in a common external frame of reference and measure distances between equivalent positions. Alternatively, the internal geometry of two proteins can be compared, that is, distances or vectors between residues in the same protein. Both types need strategies for coping with insertions and deletions. Some include information about local residue features, such as torsional angles, to enhance accuracy.

Although early rigid-body methods[1] had problems with indels, more recent solutions[7-9] include using dynamic programming to locate equivalent positions for superposition. Initial alignments are often obtained by comparing sequences or torsional angles. The fit can then be optimized by iterating through cycles of superposition followed by alignment based on distances between superposed positions.

The effect of indels on superposition methods can also be minimized by removing the variable loops and comparing secondary structures[10] or similarly chopping the protein into fragments.[11] Although some of these approaches have been able to identify quite remote similarities, none have yet been used for automatically clustering structures into fold families.

Many methods used for identifying structural families have been based on comparing internal global geometry. Different algorithms can be used

[6] C. A. Orengo, *Curr. Biol.* **4,** 429 (1994).
[7] J. Rose and F. Eisenmenger, *J. Mol. Evol.* **32,** 340 (1991).
[8] R. B. Russel and G. J. Barton, *Proteins* **14,** 309 (1993).
[9] S. Subbiah, D. V. Laurents, and M. Levitt, *Curr. Biol.* **3,** 141 (1993).
[10] R. A. Abagayan and V. N. Maiorov, *J. Biomol. Struct. Dynam.* **6,** 1045 (1989).
[11] G. Vriend and C. Sander, *Proteins* **11,** 552 (1991).

(e.g., graph theory,[12] geometric hashing,[13] distance plot comparison[14,15]) along with various ways of coping with indels. In some graph theoretical approaches, graphs are based on distances and angles between secondary structure elements. Use of subgraph isomorphism algorithms[12] allows partial matches between proteins, thereby accommodating indels. Some very interesting similarities have been identified using these techniques, and they are fast allowing new structures to be scanned against all those known to search for matches.

Distance plot-based methods compare all the interresidue distances in one protein, to equivalent distances in another. Distance plots were originally suggested by Phillips[16] in 1970, and are two-dimensional matrices whose axes correspond to all the residue positions. Cells can be shaded according to distances between residues. Nearly identical structures can be compared simply by overlaying their distance plots. Early strategies for indels included chopping out apparently nonequivalent residue positions in order to be able to generate conformant plots that could be overlaid.[17]

In the method of Holm and Sander,[15,18] proteins are chopped into hexapeptide fragments to limit the effect of indels, and the distance plots are compared to find matching fragments. These are then recombined using simulated annealing. This flexible approach has been used to cluster all the known structures into protein families automatically. It can also be used to search for structural matches with different topologies, although this can be very time-consuming because all possibilities are explored. Few such instances have been identified to date.

A method that includes information about other relationships between residue positions (e.g., hydrogen bonding patterns) has been developed by Sali and Blundell.[19] Relationships are compared using simulated annealing, to identify possible equivalences between the two proteins. This information and that of similarities for a range of residue features (e.g., accessibility, torsional angles, and volume) are accumulated in a two-dimensional similarity matrix, except axes are labeled with positions from each protein. In a final step, corresponding residues are determined using single dynamic programming. Most families in the protein structure data bank have been compared and aligned using this approach. Contributions of scores from

[12] H. M. Grindley, P. J. Artymiuk, D. W. Rice, and P. Willett, *J. Mol. Biol.* **229,** 707 (1993).
[13] R. Nussinov and H. J. Woolfson, *Proc. Nat. Acad. Sci.* **88,** 10495 (1989).
[14] W. R. Taylor and C. A. Orengo, *J. Mol. Biol.* **208,** 1 (1989).
[15] L. Holm and C. Sander, *J. Mol. Biol.* **233,** 123 (1993).
[16] D. C. Phillips in "Development of Crystallographic Enzymology," Vol. 31, p. 11 (1970).
[17] E. A. Padlan and D. R. Davies, *Proc. Nat. Acad. Sci.* **72,** 819 (1975).
[18] L. Holm and C. Sander, *Methods Enzymol.* **266,** Chap. 39, 1996 (this volume).
[19] A. Sali and T. L. Blundell, *J. Mol. Biol.* **212,** 403 (1990).

comparing different features and relationships have to be carefully weighted, and this makes the approach less suited to data bank searching and automatic clustering of protein families.

Sequential Structure Alignment Program: A Distance Plot-Based Method for Comparing Protein Structures

Some of the problems encountered in comparing distantly related proteins and ways of overcoming them can be illustrated for the SSAP method of Taylor and Orengo.[14] This compares internal geometry between proteins using the Needleman–Wunsch dynamic programming algorithms developed for sequence alignment. Instead of residue identities or physicochemical properties, three-dimensional geometry is compared to identify equivalent positions.

This is done by describing a structural environment or view for each residue which is the set of vectors from the $C\beta$ atom to $C\beta$ atoms of all other residues in the protein (Fig. 1). The view is defined within a common frame of reference for each residue based on the tetrahedral geometry of the $C\alpha$ atom. Vectors give more information on relative positions in a view than simple distances. Similarly, using $C\beta$ atoms gives more information

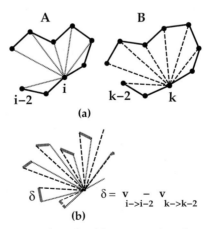

FIG. 1. Schematic representation of residue structural environments or views that are compared in the SSAP method.[14] In (a), A and B represent fragments of protein structure, and the dotted lines are vectors from residues i (in A) and k (in B). Common frames of reference are used to derive these vectors, based on $C\alpha$ geometry. This means that views from i and k can simply be compared by calculating the difference between equivalent vectors.

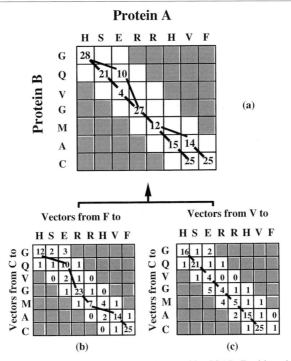

FIG. 2. Double dynamic programming algorithm used by SSAP. Residue views are scored in lower level, vector matrices (b, c). The alignment paths through these matrices, found by dynamic programming, are added to an upper level, summary matrix (a). Once views from all selected residue pairs have been compared, the optimal path through the summary matrix, again identified by dynamic programming, gives the alignment of residues in the two structures. Shaded cells are outside the window of selected residue pairs.

than Cα atoms, particularly for alternating positions along a β strand. Because views are defined in the coordinate frame of the Cα atom, they are rotationally invariant, which makes their comparison insensitive to the displacement of substructures.

If proteins are nearly identical, residue views can be compared by simply subtracting equivalent vectors (Fig. 1). However, as with distance plots, insertions and deletions make it difficult to identify equivalent positions. This problem is solved by using dynamic programming to align residue views (Fig. 2). As with sequence alignment, a two-dimensional matrix is constructed (vector matrix). The axes are the vector sets of the two proteins, and cells are scored by subtracting the associated vectors. For example, the score for comparing vector $v_{i\rightarrow i-2}$ in protein A with vector $v_{k\rightarrow k-2}$ in protein B (see Fig. 1) is

$$Svect_{i \to i-2, k \to k-2} = a/(b + \delta) \tag{1}$$

where

$$\delta = v_{i \to i-2} - v_{k \to k-2} \tag{2}$$

In Eq. (1) δ is the distance between the vectors, and a (=500) and b (=10) are parameters that were optimized using a large set of structure comparisons from the protein data bank. Parameter b softens the contribution of local distances and prevents residue pairs scoring too highly for similar local geometry (e.g., if they are both in helices) regardless of their relative positions in the structures.

The optimal pathway aligning the views is obtained by dynamic programming and added to a summary similarity matrix (Fig. 2). All residue views are compared in this way, and a final dynamic programming evaluation of the summary matrix gives the optimal alignment of the two structures. Use of dynamic programming at two levels, that is, to align all residue views and finally to align the residues, has been described as double dynamic programming.

The advantage of summing the whole path, rather than putting a single value in the residue pair cell, is that information about structurally similar regions tends to be reinforced with the addition of each path. Consequently, cells in the matrix corresponding to structurally equivalent positions have much higher scores than the average (typically 1000-fold) depending on the sizes of the proteins being compared. This makes the method relatively insensitive to the value of the gap penalty and very robust, as equivalent regions can be easily recognized. A nominal gap penalty of 50 is chosen for all types of comparisons. To reduce noise caused by adding paths from nonequivalent residue positions, a cutoff of 10 on the path score prevents low scoring pathways from being accumulated.

An upper limit can be placed on the number of insertions or deletions expected between the proteins being compared. This is equivalent to placing a window on the vector and summary score matrices (Figs. 2 and 3). Only vector pairs or residue pairs within the window are compared and scored, reducing the time required for an alignment. Default window sizes are typically set to be the difference in the number of residues in the proteins being compared, plus 50.

Including Other Information besides Global Geometry

In addition to comparing structural environments, other residue information can be included[20] (e.g., accessibility, torsional angles, volume). Dif-

---

[20] W. R. Taylor and C. A. Orengo, *Protein Eng.* **2,** 505 (1989).

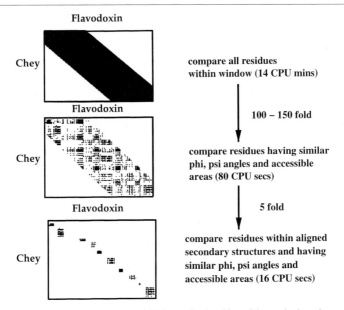

Fig. 3. Increases in speed for the SSAP method achieved by reducing the number of residue pair comparisons performed. Central processing unit (CPU) times are given for running SSAP on an Indigo R4000 computer.

ferences in these properties are scored in the summary similarity matrix. Information about other structural relationships, such as hydrogen bonding patterns or disulfide bonds, can also be included in the residue views. Weights for the contribution of these different features and relationships were optimized on a set of comparisons in the immunoglobulin and dinucleotide-binding families of proteins. Vector views provide most information and are weighted highly, but for some of the more remote comparisons improvements can be obtained by including information about hydrogen bonding and accessibility.

Increasing Speed

Various modifications have been implemented to speed up the method. In the first,[21] views are only compared if the residues appear to be in similar structural locations, that is, they have similar accessibilities and torsional angles. A parameter δtot controls the allowed difference in these properties. The selection criterion expressed by Eq. (3) below has been found to give

---

[21] C. A. Orengo and W. R. Taylor, *J. Theor. Biol.* **147,** 517 (1990).

optimal increases in speed and quality of alignments:

$$\delta\text{bur}_{ij} + A_i + A_j + (\delta\text{phi}_{ij} + \delta\text{psi}_{ij})/2 < \delta\text{tot} \tag{3}$$

where δbur_{ij} is the difference in buried area, δphi_{ij} and δpsi_{ij} are differences in the torsional angles phi (ϕ) and psi (ψ), respectively, and A_i and A_j are the accessibilities of residues i and j. Including information about accessibility in this way favors the selection of buried residues, which are generally in more conserved structural locations. The default value of δtot is 150. For very distantly related structures, this may need to be increased to 200. Using $\delta\text{tot} = 150$ usually results in less than 2% of the residue pairs being selected, which can give up to a 150-fold increase in speed for two medium-sized proteins of about 120 residues (see Fig. 3).

To improve accuracy with this fast version of SSAP, a second pass is performed (Fig. 4). After all the selected residue pairs have been compared and their paths added to the summary matrix, the 20 highest scoring pairs are identified. The summary matrix is reset to zero and the selected pairs recompared to obtain a new alignment of the structures. This second pass significantly reduces noise in the matrix and takes an insignificant time compared to the first pass as so few residue pairs are involved.

Another version of SSAP has been developed for data bank searching[22] (SSAPc) and identification of protein families. To reduce search time, comparisons between unrelated structures are filtered out by first comparing secondary structures. Since there are far fewer of these than residues this can be very quick. Secondary structures are represented as linear segments calculated using principal components analysis. Local coordinate frames are centered on the midpoint and based on the secondary structure vector and the vector to the next successive midpoint along the sequence. Views then consist of sets of vectors from the midpoint to midpoints of other segments. Information about tilt/rotation angles for secondary structure pairs is also included.

As with residue comparison, dynamic programming aligns both secondary structure views and finally the secondary structures themselves. The score returned determines whether the proteins are sufficiently related to compare their residues. If so, the procedure is the same as for SSAP. Residue pairs are selected if buried area and torsional angle differences are within the cutoff. Additionally, selected residue pairs must be from equivalent secondary structures. This reduces the number of residue pairs selected, giving a further increase in speed (Fig. 3). As for SSAP, a second pass is performed, recomparing the 20 highest scoring residues pairs to refine the alignment path.

---

[22] C. A. Orengo, N. P. Brown, and W. R. Taylor, *Proteins: Struct. Funct. Genet.* **14,** 139 (1992).

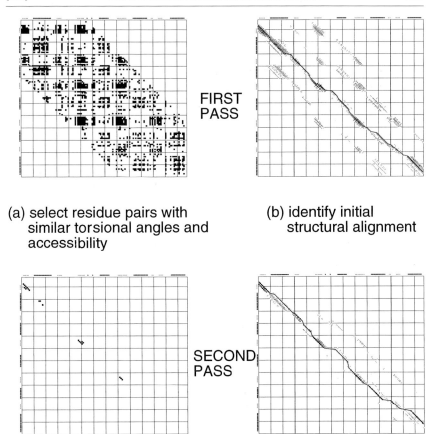

(a) select residue pairs with similar torsional angles and accessibility

(b) identify initial structural alignment

FIRST PASS

(c) select 20 highest scoring residue pairs

(d) refine structural alignment

SECOND PASS

FIG. 4. Two-pass fast version of SSAP, showing in (a, c) the residue pairs selected for each pass and in (b, d) the final alignment path through the summary matrix. Shading in (a, c) corresponds to selected residue pairs. In (b, d), cells are shaded from white to black according to their accumulated structural similarity score. The 20 highest scoring residue pairs from the first pass are recompared in the second pass.

Normalizing SSAP Score for Data Bank Searches

To generate a significant score for data bank searching, the SSAP score is normalized to be independent of protein sizes. Extensive trials[22] involving comparisons with both related and unrelated structures established a robust scoring scheme for identifying related structures. This calculates the logarithm of the average similarity score for equivalent vectors. It is measured over all pairs of equivalent vectors from all equivalent residue pairs. Vectors to adjacent positions (± 5 residues) are excluded to reduce the effect of high scores from local secondary structure similarity:

$$S_{SSAP} = \left(\sum_{i=1,i'=1}^{aln} \sum_{j=1,j'=1}^{aln} Svect_{i \to j, i' \to j'} \right) \Big/ (\text{maxequivs} * (\text{maxequivs} - 11)) \quad (4)$$

where $Svect_{i \to j, i' \to j'}$ is the vector similarity score for comparing vectors from equivalent residues i and i' in proteins A and B to equivalent residues j and j', respectively. In Eq. (4), aln is the number of aligned residue pairs, and maxequivs is the number of residues in the smallest protein and therefore the maximum possible number of equivalent positions between the proteins. Since the maximum score from comparing two identical vectors is 50, the final SSAP score is set to have a maximum value of 100 as follows:

$$S'_{SSAP} = \ln(S_{SSAP}) * 100/\ln(50) \quad (5)$$

Taking logarithms gave a better resolution of scores between structurally related and unrelated protein pairs when different scoring schemes were tested,[22] largely because differences between equivalent vectors increase exponentially as the similarity between proteins decreases.

SSAP scores above 80 are associated with highly similar structures. In many cases matching proteins have similar sequences or functions, suggesting they are homologous and have diverged from a common ancestor. Scores in the range of 70 to 80 indicate a similar fold but with more variation in the loops and larger shifts in secondary structure orientation. Often there is no sequence similarity or common function, and the relationships between the proteins are not clear. Either they have diverged a long way from the same ancestor or converged toward the same fold. As well as the similarity score, SSAP outputs the number of equivalent residue pairs between proteins. To ensure global similarity between folds, at least 60% of residue positions in the two proteins should be equivalent.

With scores between 60 and 70, proteins do not have the same global fold but usually belong to the same protein class and often have structural motifs in common. For example, comparisons of TIM barrels and Rossmann folds in the alternating α/β class of proteins often return scores in this range because of matching $\beta\alpha\beta$ motifs.

Identifying Domains and Local Structural Motifs

Comparison methods also need to be able to identify common structural motifs. For some folds these may be more strongly determined than the rest of the structure, with clearer sequence patterns, and hence easier to predict. Several such motifs have already been observed (e.g., β hairpins, $\beta\alpha\beta$ motifs), and templates expressing their recurrence within particular structures have improved prediction.[23] In the $\alpha+\beta$ class, folds are often asymmetric and complex, using motifs from all other classes which are then packed together in many different ways. Prediction of these structures will probably need an approach based more on recognizing motifs and understanding ways in which they prefer to pack.

For this reason, another version of SSAP was developed[24] (SSAPl) that finds conserved structural motifs between proteins. As for the original SSAP method, a summary matrix between two proteins is scored by comparing views from all residue pairs between the structures. Unlike the global method, where only the optimal vector alignment path is added to the summary matrix, in SSAPl the complete vector matrix is added to accumulate information about locally similar regions. Subsequently local paths within the matrix are extracted using a Smith–Waterman dynamic programming algorithm.

A fragment nucleus is first extracted from the summary matrix by tracing back a path from the highest scoring cell, until a running score (S_{run}) calculated at each position falls below a cutoff:

$$S_{run} = (S_{run} + Sres_{ij})/2 \qquad (6)$$

where $Sres_{ij}$ is the residue similarity score for cell ij in the summary matrix. The 20 highest scoring residue pairs from the fragment path are then recompared and their views aligned. Scores from along these alignment paths are accumulated in a separate fragment matrix (Fig. 5). The best local path through this fragment matrix is then sought by growing a path from the highest scoring cell and again truncating this once the running score falls below a cutoff. A softer cutoff is used to allow fragments to grow slightly. As with the global SSAP, the procedure of regenerating the path by recomparing high scoring pairs improves alignments by reducing noise. Once the optimal fragment path has been identified, it is removed from the summary matrix together with adjacent cells by resetting scores to zero. This prevents subsequent paths crossing previously identified fragment pairs. Cells in secondary structures crossed by the path are also reset (Fig. 5). Alignment scores for each path are normalized to be independent of fragment size and in the range of 0–100.

The SSAPl program can also be used to search for a particular motif within a set of structures. Although slower than SSAP, requiring approxi-

[23] W. R. Taylor and J. M. Thornton, *J. Mol. Biol.* **173**, 487 (1984).
[24] C. A. Orengo and W. R. Taylor, *J. Mol. Biol.* **233**, 488 (1993).

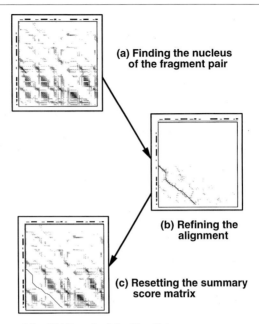

(a) Finding the nucleus of the fragment pair

(b) Refining the alignment

(c) Resetting the summary score matrix

Fig. 5. Overview of the SSAP method for identifying common structural motifs. A path corresponding to the nucleus of the motif is first identified in the summary matrix (a). Recomparison of views for residue pairs along this path allows refinement of the alignment path in a separate fragment matrix (b). Once identified, the path associated with the aligned structural motifs is removed from the summary matrix together with adjacent cells belonging to equivalent secondary structures (c).

mately 5 CPU minutes on an Indigo R4000 computer to identify all greek-key motifs in two immunoglobulin structures, it is feasible to use the method to search for common motifs among a set of proteins with similar architectures, for example, all the mainly α proteins which have predominantly orthogonal helix–helix interactions.

Using SSAP for Data Bank Searches and for Identifying
   Protein Families

Protein fold families can be identified by using fast SSAPc to compare protein structures. In the June 1994 release of the Protein Data Bank[25] (PDB) there are over 3000 well-resolved structures (<3.0 Å) which cluster

[25] F. C. Bernstein, T. F. Koetzle, G. J. D. Williams, E. F. Meyere, M. D. Brice, J. R. Rodgers, O. Kennard, T. Shimanochi, and M. Tasumi, *J. Mol. Biol.* **112,** 535 (1977).

into 430 families on the basis of sequence similarity.[26] Sequence alignment is at least 10 times faster than structure comparison, and if more than 30% of the sequences correspond the folds will be the same.[2,3] Representatives from each family can then be structurally compared using SSAPc and families merged if their representatives match with SSAP scores above 80. This gives 274 homologous families. If the SSAP cutoff is softened to 70, 206 fold families are obtained.[26] Relationships between structures in these broader families is less clear, as functions and sequences can differ substantially. In view of this uncertainty, the families can be described as analogous fold families.

A newly determined protein structure can be scanned against a set of representative structures from each of the 274 homologous fold families to check whether it is a novel fold or can be assigned to a particular family. The distribution of SSAP scores across the set enables structural matches to be expressed as standard deviations from the mean, so that their significance can be assessed (Fig. 6).

Multiple Structure Comparisons across Protein Family to Identify Conserved Positions

In the current structure data bank there are several fold families that are very highly populated with proteins having no sequence or functional similarity.[27] Nearly 30% of known structures adopt one of these folds, and this increases to 50% if only nonhomologous structures are considered [where homologous proteins have >25% sequence identity or high SSAP scores (>80) and related functions]. Because of their ability to support a wide range of sequences and functions, they can be described as superfolds. Outside the superfolds, proteins in other fold families all have related functions. Most of the superfolds are very familiar structures (e.g., TIM barrels and globin folds) and have been known since the early days of crystallography. Relationships between the structures in the families are unclear. They may have diverged a long way from a common ancestor. Alternatively, the folds may be particularly stable arrangements of secondary structures toward which many structures converge. Whatever the cause, the existence of these superfold families has important implications for structure prediction and fold recognition. If a protein has no sequence or functional similarity to any known structure, then there is a 50% probability that it will either belong to one of the superfold families or have a unique fold.

[26] C. A. Orengo, T. P. Flores, W. R. Taylor, and J. M. Thornton, *Protein Eng.* **6,** 485 (1993).
[27] C. A. Orengo, D. T. Jones, and J. M. Thornton, *Nature* (*London*) **370,** 631 (1994).

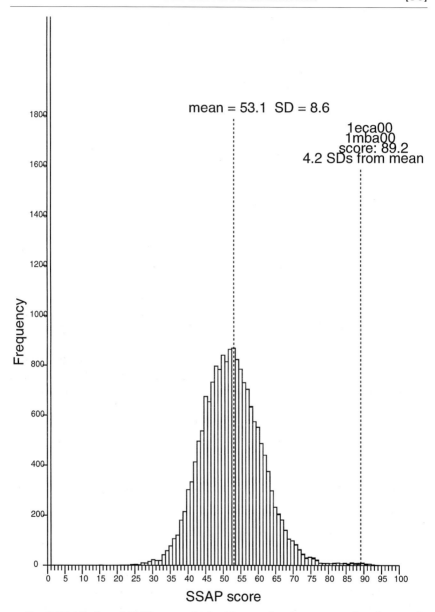

mean = 53.1 SD = 8.6

1eca00
1mba00
score: 89.2
4.2 SDs from mean

Frequency

SSAP score

FIG. 6. Distribution of SSAP scores obtained by scanning the structure of erythrocruorin (1eca) against a set of 274 nonhomologous structures. A match between erythrocruorin and myoglobin (1mba) returns a score 4.2 standard deviations above the mean and is considered significant.

The most recent version of SSAP (SSAPm) was developed[28] with the aim of analyzing fold families and is particularly suited to broad structural families such as those of the superfolds. SSAPm multiply aligns a fold family to find conserved regions which can be used as structural fingerprints in prediction and recognition methods. All pairs of proteins in the family are compared using SSAP. The alignment of the highest scoring pair is then used to seed the multiple alignment, and a consensus structure is calculated consisting of average views at each residue position. Both average vectors and information about their variability are calculated. All remaining structures in the set are then compared to the consensus and the highest scoring structure added next. This cycle of addition, construction of a new consensus, and pairwise comparisons against the consensus is repeated until all structures have been merged into the alignment. Information on variability is used to set weights for scoring vector comparisons. Similarly, a structure conservation measure, calculated at each position, weights vector comparison scores for that position and enhances the alignment of conserved core regions in the fold.

The program CORA takes the multiple alignment given by SSAPm and generates a structural template for the family. Only positions with individual SSAP scores above a cutoff are written to the template, and average views to other selected positions are described together with vector variability information. The cutoff can be adjusted depending on the similarity of structures compared. Structural templates can be used both for fold recognition by threading type algorithms and in fold identification. Considerable improvements in resolution between related and unrelated structures can be obtained by searching against the structure database with a fold template (Figs. 7 and 8). The program CORALIGN is a modified version of SSAP that aligns a protein structure against a template generated by CORA. Again, individual SSAP conservation scores and information on vector variability are used to set weights for matching key conserved positions in the fold.

Conclusion

In summary, there are now several robust methods for comparing distantly related protein structures, some[8,19,28] performing multiple comparisons with protein fold families to identify and characterize conserved structural features. Already, the data generated by these methods offer considerable hope for improvements in fold recognition.[4,5]

[28] W. R. Taylor, T. P. Flores, and C. A. Orengo, *Protein Sci.* **3,** 1858 (1994).

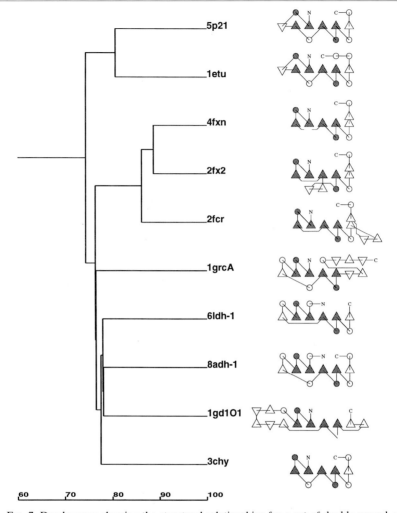

FIG. 7. Dendrogram showing the structural relationships for a set of doubly wound α/β proteins. The x axis gives the SSAP score. On the right-hand side, schematic topology representations, drawn by the program TOPS,[33] are shown for each structure. Helices are represented by circles and β strands by triangles. The common core of the fold, identified by SSAPm, is shaded and consists of a four-stranded β sheet with an α helix packed on each side.

This is timely as there are still far more protein sequences known (~100,000) than structures determined (~4000). By the year 2005, we may expect thousands more sequences from the various genome mapping projects. The ultimate goal is to understand the functions of these proteins, and knowledge of the protein structure is an essential step in this process.

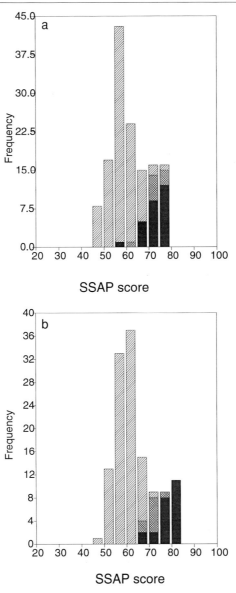

SSAP score

SSAP score

Fig. 8. Improvements in fold identification obtained by scanning through the dataset of nonhomologous structures using a structural template based on the common core of the α/β doubly wound folds. (a) Distribution of SSAP scores obtained by scanning against the data set with a representative doubly wound α/β structure (Chey). (b) Distribution of SSAP scores using the core template. Solid bars are correct hits (other alternating β/α type proteins with a similar fold), whereas crosshatched bars represent hits on TIM-barrel proteins (alternating β/α type proteins with a different fold). Hatched bars are unrelated folds.

We can expect that analysis of protein structural families and identification of consensus structural templates will improve both template- and threading-based prediction algorithms for identifying the fold of a protein, particularly for the superfolds. For more complex and less frequently observed folds, characterization of common structural motifs and any associated sequence patterns would be expected to improve prediction.

Running SSAP and Generating Superposition of Structures

All versions of SSAP require an input WOLF file. This can be created by running the program WOLF, a modified version of DSSP[29] that generates information about hydrogen bonding patterns and secondary structure assignments. The input file for WOLF is the Protein Data Bank[25] file for the structure. WOLF files also contain residue accessibilities and torsional angles, together with frames of reference for each residue, centered on the $C\beta$ position.

Additional SEC files are needed for running SSAPc. These contain information about the line vectors representing secondary structures, together with pairwise distances and angles and overlap information for each secondary structure pair. They are generated by running the program SECLINE.

Besides returning the normalized global score, SSAP outputs an alignment of the two structures together with individual scores for each residue position. These are in the range of 0 to 100 and measure similarity in views for the residue pair. Scores are generally higher for secondary structure regions and around active sites.

SSAP also outputs two superposition (SUP) files, listing equivalent positions and scores. These are input for the program SUPRMS,[30] which performs a weighted superposition using the SSAP residue scores as weights. SUPRMS is based on the McLachlan least squares method[31] and returns the root-mean-square (rms) deviation between the structures together with the number of residue pairs superposed for a 3.0 Å cutoff. It also outputs a QUANTA graphics file for coloring the superposed structures according to similarity in views at each position.

For multiple comparisons across a family of structures, the program SSAPm generates a MULTISUP file listing equivalent positions and information about conservation of residue views. This can be input to the pro-

[29] W. Kabsch and C. Sander, *Biopolymers* **22,** 2577 (1983).
[30] F. Rippmann and W. R. Taylor, *J. Mol. Graph.* **9,** 169 (1991).
[31] A. D. McLachlan, *Acta Crystallogr.* **A28,** 656 (1979).

gram MULTISUP,[32] which performs weighted multiple superposition of the structures.

Availability

The SSAP package can be obtained by sending a request to orengo@ bsm.bioc.ucl.ac.uk. Programs are distributed as binary files compiled for execution on a Silicon Graphics machine running UNIX.

[32] T. P. Flores, personal communication.
[33] T. P. Flores, D. S. Moss, and J. M. Thornton, *Protein Eng.* **7,** 31 (1993).

[37] Understanding Protein Structure: Using Scop for Fold Interpretation

By Steven E. Brenner, Cyrus Chothia, Tim J. P. Hubbard, and Alexey G. Murzin

Introduction

The structure of a protein can elucidate its function, in both general and specific terms, and its evolutionary history. Extracting this information, however, requires a knowledge of the structure and its relationships with other proteins. These two aspects are not independent, for an understanding of the structure of a single protein requires a general knowledge of the folds that proteins adopt, while an understanding of relationships requires detailed information about the structures of many proteins.

Fortunately, this complex problem with its intertwined requirements is not insurmountable, for two reasons. First, protein structures can be fundamentally understood in ways that most of their sequences cannot. The comprehensibility of protein structures derives from the relatively few secondary structure elements in a given domain and the fact that the arrangement of these elements is greatly restricted by physics and probably by evolution. Second, resources are now available to aid recognition of the relationships between protein structures. The structural classification of proteins (scop) database hierarchically organizes proteins according to their structures and evolutionary origin.[1] As such, it forms a resource that allows researchers to learn about the nature of protein folds, to focus their investi-

[1] A. G. Murzin, S. E. Brenner, T. Hubbard, and C. Chothia, *J. Mol. Biol.* **247,** 536 (1995).

Copyright © 1996 by Academic Press, Inc.
All rights of reproduction in any form reserved.

gation, and to rely on expert-defined relationships when seeking new ones. In addition, automated methods can compare and recognize structures that are similar by some criteria and therefore provide a method for associating new experiments with the scop hierarchy.

Scop Hierarchy

An explanation of the classification of folds in scop provides a tutorial in the understanding of protein structures, and vice versa. Thus, when we discuss the way proteins are organized in scop, we intend to indicate simultaneously how one can examine any protein structure. Because the whole of scop is an extended collection of many thousand classifications, relatively few examples are provided here. Likewise, scop contains hundreds of references, and therefore the descriptions of the classification will note only highlights.

To illustrate the scop hierarchy, we have chosen proteins that bind NAD or NADP (Table I), in particular those whose NAD(P)-binding domains share the so-called Rossmann-fold (Table II). The discovery of this shared fold in 1975[2] raised the question that is fundamental to our understanding of protein structure and evolution: What are the origins of structural similarity in proteins whose sequences show no significant sequence similarity? The many structures for the NAD(P)-binding proteins which have been determined since that time provide us with the clear examples of divergent and convergent evolution highlighted here.

The scop database is organized on a number of hierarchical levels, with the principal ones being family, superfamily, fold, and class. Within the hierarchy, the unit of categorization is the protein domain, rather than whole proteins, since protein domains are typically the units of protein evolution, structure, and function. Thus, different regions of a single protein may appear in multiple places in the scop hierarchy under different folds or, in the case of repeated domains, several times under the same fold.

In scop, families contain protein domains that share a clear common evolutionary origin, as evidenced by sequence identity or extremely similar function and structure. Superfamilies consist of families whose proteins share very common structure and function, and therefore there are compelling reasons to believe that the different families (with low interfamily sequence identities) are evolutionary related. Folds consist of one more superfamilies that share a common core structure (i.e., same secondary structure elements in the same arrangement with the same topological

---

[2] M. G. Rossmann, A. Liljas, C.-I. Brändén, and L. J. Banaszak, *in* "The Enzymes" (P. D. Boyer, eds.), 3rd ed., Vol. 11, p. 61. Academic Press, New York, 1975.

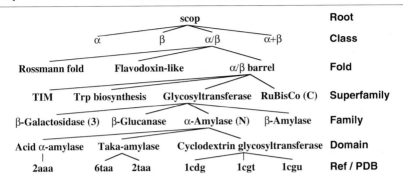

FIG. 1. Region of scop hierarchy. All the major levels, including class, fold, superfamily, and family, are shown. Also shown are individual proteins and, at the lowest level, either the Protein Data Bank (PDB) coordinate identifier or a literature reference.

connections). Frequently, proteins clustered together at this level will have considerable elaboration on the shared fold topology. Finally, folds are grouped into one of four classes depending on the type and organization of the secondary structure elements: all-α, all-β, α/β, and $\alpha+\beta$. In addition, there are several other classes for proteins that are very atypical and therefore difficult to classify. Figure 1 depicts a region of the scop hierarchy, with examples of items at all levels.

The following discussion describes how to classify a protein structure at each of the different levels given above, and to make use of that information.

Class

Before a protein structure can be properly classified, it needs to be divided into domains. The basic idea of a domain is simple: it is a region of the protein that has its own hydrophobic core and has relatively little interaction with the rest of the protein, so that it is essentially structurally independent. In practice, however, identification of domains is not a trivial task and can frequently be done correctly only by using evolutionary information to see, for example, how domains have been "shuffled" in different proteins. Typically domains are colinear in sequence, which aids their identification, but occasionally one domain will have another "inserted" into it, or two homologous domains will intertwine by swapping some topologically equivalent parts of their chains. Because of the problem of identifying domains on the basis of a single protein structure, it is usually best to iteratively refine domain definitions: first make a tentative set of assignments and, as understanding of the structure grows, refine as necessary. An application of this approach to the domain definition in NAD(P)-binding proteins is discussed in following sections.

Fortunately, the first real step in interpreting a domain structure, namely, placing it in the appropriate class, is usually a straightforward task. It should be readily apparent whether a domain consists exclusively of α helices, β sheets, or some mixture thereof. One caveat is that only the core of the domain should be considered: it is possible for an all-β protein to have very small adornments of α or 3_{10} helix, whereas so-called all-α structures may actually have several regions of 3_{10} helix, and in rare cases, small β sheet outside the α-helical core.

Domains with a mixture of helix and sheet structures are divided into two classes, α/β (alpha and beta), and $\alpha+\beta$ (alpha plus beta). The α/β domains consist principally of a single β sheet, with α helices joining the C terminus of one strand to the N terminus of the next. Commonly these proteins are divided into two subclasses: in one subclass, the β sheet is wrapped to form a barrel surrounded by α helices; in the other, the central sheet is more planar and is flanked on either side by helices. Domains that have the α and β units largely separated in sequence fall into the $\alpha+\beta$ class. Because the strands in these structure do not have the intervening helices, they are typically joined by hairpins, leading to antiparallel sheets much as are found in all-β class folds. However, $\alpha+\beta$ structures may have one, and often a small cluster, of helices packing tightly and integrally against the sheet.

When the four classes were originally defined on the basis of a handful of structures,[3] the distinction between the α/β and $\alpha+\beta$ classes was very clear. Since then, the picture has become more cloudy, and although most mixed-structure domains can be clearly placed in one class or the other, some domains defy easy classification. In the event that a protein with mixed secondary structure does not clearly fall into one of these two classes, it is best to hold this decision in abeyance and proceed to the more important task of identifying other structures with the same fold.

In addition to the four classes of globular protein structures, scop contains a few other classes. Most important is the multidomain class. Proteins here have multiple domains that would ordinarily be placed in different classes. However, the different domains of these proteins have never been seen independently of each other, so accurate determination of their boundaries is not possible and perhaps not meaningful or useful. Another important class contains the many small proteins having structures stabilized by disulfide bridges or by metal ligands rather than by hydrophobic cores. Membrane proteins frequently have unique structures because of their unusual environment, and therefore they also have their own class. The scop

---

[3] M. Levitt and C. Chothia, *Nature* (*London*) **261,** 552 (1976).

database also contains classes for the short peptides, theoretical models, and nonproteins (such as nucleic acids) in the Protein Data Bank.

Fold

Identification of the fold of a protein is one of the most difficult stages of classification, so much so that papers about new structures often fail to report structural similarity with other proteins. The problem is in part due to the fact that there are more than just a handful of different folds used in nature; about 50 are currently known for each of the four classes of globular proteins. The best way to characterize a fold is to look first at the major architectural features, and then identify more subtle characteristics, as described below for NAD(P)-binding proteins.

In addition, there is a shortcut that will aid identification of the fold of a large number of proteins. This is because there are about a dozen folds, such as the β/α barrel, that occur very frequently (as identified by the number of superfamilies they contain), so first comparing a structure of interest with each of these will often be expedient. For example, at least three such folds—the β/α barrel (currently contains 11 superfamilies), the SH3-like β barrel (6 superfamilies), and the ferredoxin-like $\alpha+\beta$ fold (16 superfamilies)—contain NAD(P)-binding proteins (Table I).

So far no all-α protein or domain that binds NAD or NADP has been found, but in the structure of catalase, currently classified as multidomain, the NADP molecule binds between an all-α and a β-barrel domain. The other known NAD(P)-binding structures are distributed between the three other major classes: all-β, $\alpha+\beta$, and α/β. Apart from the ADP-ribosylation toxins, which utilize NAD as substrate, these proteins are oxidoreductases, which use NAD or NADP as cofactor. The different folds and classes into which these enzymes fall clearly indicate multiple origins of their similar function. Nevertheless, their catalytic sites can be very similar. A particularly striking example of functional convergence can be seen in the active-site structures of NAD(P) oxidoreductases of the β/α-barrel fold and a family of the Rossmann-fold domains, which have their substrate-binding cavities and the catalytically essential tyrosine and lysine residues similarly located relative to the cofactor. There is also convergence of a different kind. The NADP-binding domain of ferredoxin reductase and related enzymes is topologically similar to the Rossmann-fold domain, and this similarity extends to the locations of the coenzyme-binding sites. However, because the protein structures do not superimpose well and the binding modes are very different, these proteins cannot be classified as having the same fold or superfamily (see below). Topological similarities of this kind are on intermediate level between class and fold, and, in the current version of scop, they are silently indicated by listing folds with similar topologies together on the class page. This approach is also used to segregate different architectural motifs, like two-sheet sandwiches and single-sheet barrels in the all-β class. Future versions of scop will include the necessary additional levels of classification to make such distinctions explicit.

Superfamilies

Protein structures classified in the same superfamily are probably related evolutionarily, and therefore they must share a common fold and usually perform similar functions. If the functional relationship is sufficiently strong, for example, the conserved interaction with substrate or cofactor molecules, the shared fold can be relatively small, provided it includes the active site. This is in contrast with classification on the fold level, which ordinarily requires greater structural similarity.

We have already mentioned that NAD(P)-binding domains are classified in scop in several distinct superfamilies. The largest of these superfamilies includes all the original members of the Rossman-fold, hence its full name, the NAD(P)-binding Rossmann-fold domain (Tables I and II). All members of this superfamily bind the cofactor molecule in the same way; that is, they will be positioned identically when the whole protein structures are superimposed. This superfamily also includes a domain of succinyl-CoA synthetase that binds a different cofactor, coenzyme A (CoA), but shows good structural similarity and recognizes the common part of CoA

TABLE I
NAD(P)-BINDING PROTEINS IN SCOP DATABASE

Class	Fold[a]	Superfamily
All-β	SH3-like	R67 dihydrofolate reductase
α/β	β/α barrel	NAD(P) oxidoreductases
	—	FAD (also NAD)-binding motif
	—	NAD(P)-binding Rossmann-fold domain[b]
	—	Ferredoxin reductase-like, C-terminal domain
	—	Dihydrofolate reductases
	—	Isocitrate and isopropylmalate dehydrogenase
$\alpha+\beta$	Ferredoxin-like	HMG-CoA reductase, N-terminal domain
	—	ADP-ribosylation toxins
Multidomain	—	Heme-linked catalases

[a] If there is only one superfamily in the fold, a — is shown.
[b] The families of this superfamily are shown in Table II.

TABLE II
SUPERFAMILY OF NAD(P)-BINDING ROSSMANN-FOLD DOMAINS[a]

Family and protein	Specific features

Tyrosine-dependent oxidoreductases
  Short-chain dehydrogenases
  Dihydropteridin reductase
  UDP-galactose epimerase
  Enoyl-ACP reductase
	Extensive structural similarity; coenzyme-binding and catalytic site are in one domain; rare left-handed $\beta\alpha\beta$ unit in extension of superfamily fold

Lactate and malate dehydrogenases
  Lactate dehydrogenase
  Malate dehydrogenase
	Sequence similarity, extensive structural similarity; C-terminal catalytic domain has an unusual $\alpha+\beta$ fold

Alcohol dehydrogenase
  Alcohol dehydrogenase
  Glucose dehydrogenase
  Quinone reductase
	Extensive structural similarity; N-terminal catalytic domain has a GroES-like all-β fold

Formate dehydrogenase
  Formate dehydrogenase
  D-Glycerate dehydrogenase
  Phosphoglycerate dehydrogenase
	Extensive structural similarity; catalytic domain is formed by N- and C-terminal regions and has common flavodoxin-like α/β fold

Glyceraldehyde-3-phosphate dehydrogenase
  Glyceraldehyde-3-phosphate dehydrogenase
  Glucose-6-phosphate dehydrogenase
  Dihydrodipicolinate reductase
	Common $\alpha+\beta$ fold in the catalytic domain, inserted in coenzyme domain in same topological location

6-Phosphogluconate and acyl-CoA dehydrogenases
  6-Phosphogluconate dehydrogenase
  Acyl-CoA dehydrogenase
	Superfamily fold is similarly extended; common all-α fold in the catalytic C-terminal domain

[a] The superfamily of NAD(P)-binding Rossmann-fold domains consists of a number of families. In addition to the fold and the cofactor-binding mode common to all members of the superfamily, there are some family-specific similarities in either sequence, structure, or domain organization.

and NADP molecules in very similar way. The scop classification differs from a traditional classification of the nucleotide-binding Rossmann-fold which contains all proteins that show topological (but not close structural) similarity and bind nucleotides, often in a very different way.

The NAD(P)-binding Rossmann-fold domain is well defined by its shared fold and the coenzyme-binding site. It is usually conjoined with another domain that contains the catalytic site of the whole enzyme. The catalytic domain may precede or follow the coenzyme domain, or it may interrupt or be interrupted by this domain. The protein folds of catalytic domains in different enzymes may be very different and fall in all four major classes and in a number of different folds (Table II).

Families

Most scop families cluster together homologous proteins with high sequence similarity. The structures of these proteins are also very similar (e.g., lactate and malate dehydrogenases). However, in extraordinary cases, extensive structural similarity and strong functional relationships can define families in the case of low sequence similarity, as in the case of the tyrosine-dependent oxidoreductases.[4] This family seems to be the largest family of NAD(P)-dependent enzymes, as evidenced by sequence similarity among its members. It could be further divided into a number of subfamilies, according to the extent of their sequence similarity.

A small number of scop families currently embrace the relationships which are above the standard family definition but below the superfamily level. It can be suggested that proteins that have similar domain organization and share a common fold in the catalytic domain, like dihydrodipicolinate reductase and the glyceraldehyde-3-phosphate and glycose-6-phosphate dehydrogenases, are likely to be even closer related than those sharing the common fold in their coenzyme domain only. Such similarities currently lead to coenzyme domains being assigned to the same family and the catalytic domains to the same family or superfamily, depending on the extent of structural similarity of the catalytic domains.

With future releases of the scop database, additional levels of classification around the family level will be introduced so that each will have a unique, well-defined meaning.

Role of Automated Systems

Computational approaches are now beginning to play an important role in the understanding of protein structures and can be fruitfully used in conjunction with the scop classification. At present there are two particularly valuable types of programs, both described in detail elsewhere in this volume.

Sequence comparison is a simple and reliable way of learning about the structural and evolutionary relationships of a protein. Two proteins with 40% sequence identity with each other will have very similar structures, and if a sequence has 30% identity to a protein of known structure, then an outline of its configuration can also be deduced. If there is significant similarity between a sequence and a protein in scop, then that sequence can be put into the appropriate family, which then defines its superfamily, fold, and class.

---

[4] L. Holm, C. Sander, and A. Murzin, *Nat. Struct. Biol.* **1,** 146 (1994).

The major limitation of sequence comparison is that it fails to identify many of the structural relationships in scop either because the sequence relationship has become too weak (for evolutionarily related proteins) or never existed (for evolutionarily unrelated proteins with similar folds). Structure–structure comparison programs use various methods to recognize similar arrangements of atomic coordinates and thus identify domains of similar structure. Although these methods lack complete accuracy, they can be used to suggest a shared fold between a protein of interest and others in scop. Manual inspection must then be used to verify the choice of fold and to select an appropriate superfamily. The selection of superfamily is the most challenging and most scientifically rewarding step of protein classification, for it ascribes a biological interpretation to chemical and physical data. For this reason, the assignment of all proteins of known structure to evolutionarily related superfamilies is perhaps the single most powerful and important feature of the scop database.

Resources

The scop database can be accessed on the World Wide Web, at the URL http://scop.mrc-lmb.cam.ac.uk/scop/. For improved access, mirrors of scop are available worldwide, and their addresses can be found from the above location.

[38] Detecting Structural Similarities: A User's Guide

By Mark Basil Swindells

Introduction

If I were to tell you that I had just discovered a new similarity between two structures, which consisted of a single equivalenced α helix, you would probably conclude that my future contributions to the field were likely to be limited. However, if I were to tell you that I had detected an unanticipated similarity in which 700 residues could be superimposed and that intriguing functional similarities were retained, you would, I would hope, be impressed. In reality, most similarities revealed by a typical database search fall between these two extremes and as a result it is frequently difficult to know whether they merely reflect the features of proteins per se, in which helices and strands are packed in a compact manner, or whether they have some additional functional or even evolutionary meaning.

METHODS IN ENZYMOLOGY, VOL. 266
Copyright © 1996 by Academic Press, Inc.
All rights of reproduction in any form reserved.

Before looking at structural comparisons, let us first consider the humble sequence alignment. In its simplest form, two sequences are aligned and the percentage sequence identity calculated. This is then (or should be) compared with the results derived from aligning randomized sequences of the same composition. If our real alignment score is higher than the randomized score, we may infer an evolutionary and hence functional relationship. This procedure provides an elegant way of provisionally dividing sequences into families, even though we may later find that some of the families are also related while being undetectable from sequence alone.

However, equivalent baselines for assessing structural similarities do not yet exist. Although we frequently make use of root-mean-square deviation (RMSd) as a quality indicator, we cannot compare the observed RMSd with a randomized score, as we do in sequence alignments. As a result we search for RMSds that are comparable with those previously observed in proteins known to be similar.[1] Further complications arise because the RMSd varies, both with the number of atoms superimposed (usually the Cα atom of each residue) as well as the secondary structures that they occupy. To appreciate the latter, consider the superposition of two proteins in which a single, 40-residue helix is superimposed. Now consider a similar superposition in which 40 residues are distributed over a four-helix bundle. It is much more likely that the first case will yield a lower RMSd than the second. In addition, we have little idea of how the RMSd can vary for superpositions of different sizes. Thus, if we find a structural similarity, we can only infer an evolutionary/functional relationship if there is additional supporting evidence.

Searching the Protein Data Bank for Global Structure Similarities

To see how all of these factors come into play, consider the following example. The replication terminator protein (RTP) structure was solved by X-ray crystallography[2] and the protein was found to exist as a homodimer, with dimerization facilitated by the interaction of a C-terminal helix from each unit (Fig. 1). Each monomer has 116 residues and includes a helix–turn–helix motif similar to those found in many DNA-binding proteins. As anticipated, DNA interactions are made using this element. No report was made of a global structural similarity between RTP and any other protein. So, is it a unique topology, or does it have an underlying structural similarity that went unnoticed? With a good memory of protein

[1] C. Chothia and A. M. Lesk, *EMBO J.* **5,** 823 (1986).
[2] D. E. Bussiere, D. Bastia, and S. W. White, *Cell* (*Cambridge, Mass.*) **80,** 651 (1995).

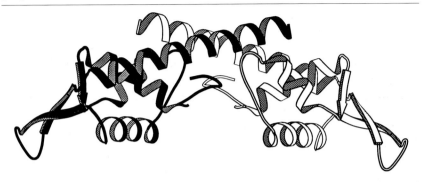

Fig. 1. Ribbon diagram of the RTP dimer. This and subsequent diagrams were produced using Molscript [P. J. Kraulis, *J. Appl. Crystallogr.* **24,** 946 (1991)].

structures one might be able to work out the answer just by looking at Fig. 1, but for most this is impractical.

With one of the many structural comparison methods now available (we use the latest version of the sarf algorithm[3] here), we can search a nonhomologous set of protein structures and see what similarities appear. Figure 2 shows the results of comparing the RTP structure with 450 protein chains from the Protein Data Bank (PDB),[4] whose sequence similarities are all below 35%. For each comparison we basically select the alignment that gives the highest number of superimposed $C\alpha$ atoms, while ensuring that the RMSd does not exceed certain thresholds. In addition, secondary structural elements must maintain the same connectivity as RTP for a comparison to be valid. From Fig. 2, it is apparent that most of the structural similarities consist of a combination of less than 45 superimposed $C\alpha$ atoms and an RMSd worse than 1.8 Å. Presumably, points scattered outside these limits are the most likely to reveal biologically interesting results, and so in the next section we show what kinds of similarities are being picked up. The arc on the graph in Fig. 2 gives an idea of which similarities may be worth looking at, but it should be emphasized that this does not represent any formal relationship between the number of residues superimposed and the RMSd.

Starting with the lowest RMSd similarities, pyruvoyl-dependent histidine decarboxylase (PDB code 1pya) has only a single equivalenced α helix and is too small to have any biological meaning. T4 endonuclease V initially seems more promising, as the superposition is larger (40 residues superimposed with a RMSd of 1.6 Å) and both proteins are known to bind to

[3] N. N. Alexandrov, K. Takahashi, and N. Go, *J. Mol. Biol.* **225,** 5 (1992).
[4] F. C. Bernstein, *et al., J. Mol. Biol.* **122,** 535 (1977).

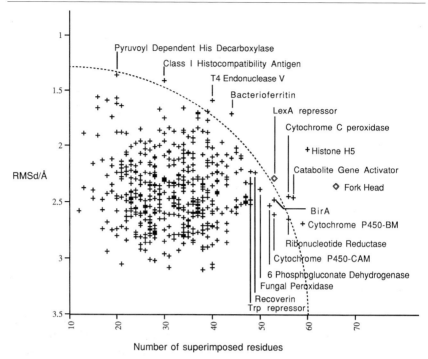

Fig. 2. Comparison of RTP with 450 protein chains from the PDB. All protein structures were determined by X-ray crystallography and had resolutions of 3.0 Å or better. The fork head protein (not available from PDB) and the LexA repressor (determined by NMR spectroscopy) were subsequently added as they are known to be similar to histone H5.

DNA. However, on viewing this similarity (Fig. 3), we find that only two helices from RTP have been superimposed, one being the long dimerization helix. This case emphasizes the problem of long elements of regular secondary structure contributing a large number of residues to a superposition. A similar situation is observed in the superposition with bacterioferritin (1bcf; 44 residues over 1.7 Å).

The next potentially interesting similarity is with histone H5 (1hst), where 60 superimposed residues yield an RMSd of 2 Å. This similarity, which is shown in Fig. 4a, represents an extremely good match, and with the exception of the dimerization helix, all secondary structural elements are equivalenced. Moreover, histone H5 has a clear DNA-binding function and is already known to belong to a structural superfamily consisting of fork head[5] (not in PDB), LexA repressor (1leb), catabolite gene activator

[5] K. L. Clark, E. D. Halay, and E. Lai, *Nature* (*London*) **364**, 412 (1993).

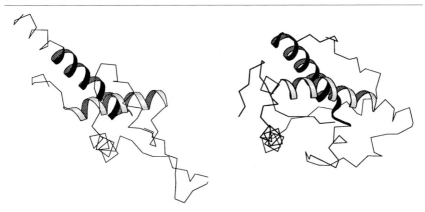

Fig. 3. RTP (left) and T4 endonuclease V (right) with the superimposed helices shown as ribbons.

protein (1cgp), and the biotin repressor (1bia). If we look at a structurally derived sequence alignment of these proteins, not only can we appreciate the full extent of the similarity, but it also seems that there is a region of surprisingly high sequence conservation between RTP and histone H5, involving the DNA-binding helix and subsequent β strand (Fig. 4b).

Of the remaining hits, cytochrome P450 structures seem to crop up rather frequently together with the peroxidase structures. In fact, cyto-chrome P450-BM (2bmh) has almost as many residues superimposed as histone H5, but the RMSd is much worse (59 residues over 2.7 Å). Figure 5 shows the similarity. The topology of the entire winged helix motif is retained (dimerization helix excluded), while the higher RMSd reflects significant differences in the orientation of the structural elements. How-ever, from the higher RMSd and the absence of any apparent functional similarity, we must conclude that this similarity arises from the preferred packing arrangements of secondary structural elements in proteins.

So, our search has shown that RTP is surprisingly similar to the winged helix-type proteins and that the DNA-binding region has surprisingly high sequence similarity with histone H5. It seems clear from these data that there is also a functional similarity, and it is possible that there could be an evolutionary relationship. In contrast, the other hits are associated either with long pieces of regular secondary structure being equivalenced or with the detection of general similarities in helix and strand packing arrange-ments.

There are many other examples of structural similarities in newly solved protein structures, and many of these initially escape detection. For exam-ple, it was claimed that the first C_2 domain of synaptotagmin, a protein

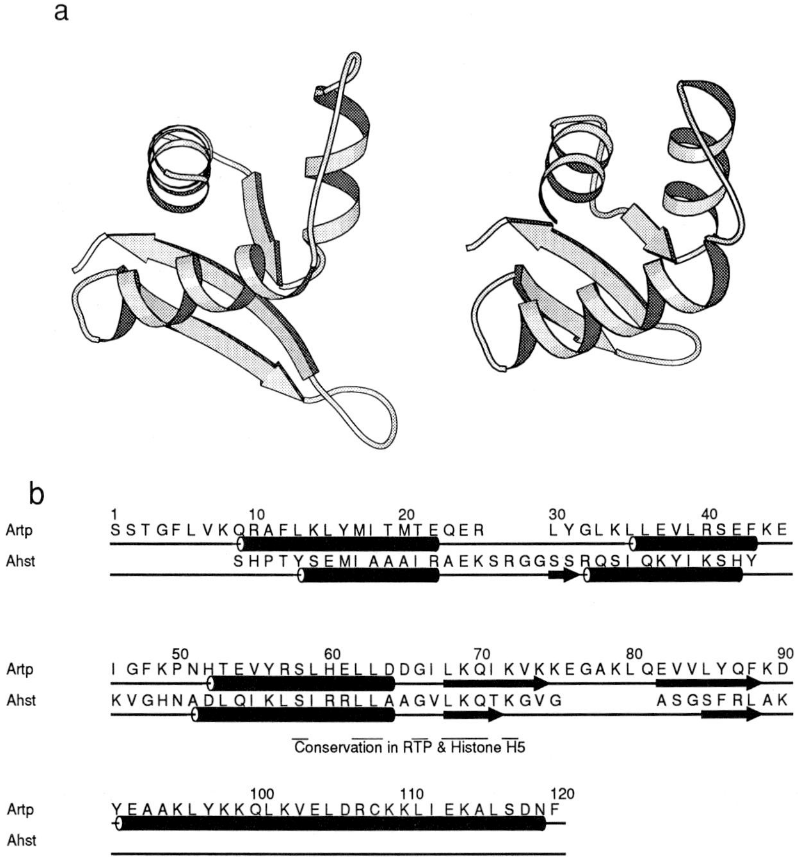

a

b

Conservation in RTP & Histone H5

Fig. 4. (a) Ribbon diagrams of RTP (left, dimerization helix not shown) and histone H5 (right) showing the structural equivalence of these proteins. (b) Structurally derived sequence alignment showing pattern of sequence conservation around the DNA-binding helix.

which acts as a calcium sensor, had a unique topology,[6] but in fact it is reminiscent of the superfamily typified by tenascin (1ten). In another case dihydroxybiphenyl dioxygenase[7] has clear topological similarities with the bleomycin resistance protein,[8] even though it was purported to be a novel fold. In all these cases the similarities can be identified by available structure

[6] R. B. Sutton, B. Davletov, A. M. Berhuis, T. C. Sudhof, and S. R. Sprang, *Cell* (*Cambridge, Mass.*) **80,** 929 (1959).

[7] K. Sugiyama, *et al., Proc. Jpn. Acad.* **71,** 32 (1995).

[8] P. Dumas, M. Bergdoll, C. Cagnon, and J.-M. Masson, *EMBO J.* **13,** 2483 (1994).

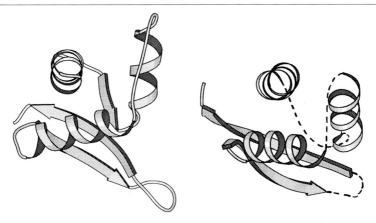

FIG. 5. RTP and superimposable regions from cytochrome P450-BM.

comparison methods, provided that the structures are publicly available in the PDB.

Identifying Common Loop Motifs

Common Loops in Different Topologies

Structural similarities are dominated by three-dimensional arrangements of helices and strands. This reflects both their dominance in protein structures as well as the ease with which regularly repeating elements can be superimposed. However, loop regions also play a vital functional role, and in certain cases the same loop conformation recurs in otherwise unrelated proteins. This is particularly true for the phosphate binding loops of many α/β proteins. In this section we describe how to make loop searches,[9] show under what conditions they are feasible, and give some examples of the results.

The nucleoside phosphate-binding loop of adenylate kinase is frequently referred to as the P-loop and has a preference for the sequence motif GxxxxGKT.[10] Many other proteins contain this sort of loop, two of which are the ras P21 oncogene product and recA. These examples are important because in each case the P-loop connects strands with different connectivi-

---

[9] M. B. Swindells, *Protein Sci.* **2,** 2146 (1993).
[10] J. E. Walker, M. Saraste, M. Runswick, and N. J. Gay, *EMBO J.* **1,** 945 (1982).

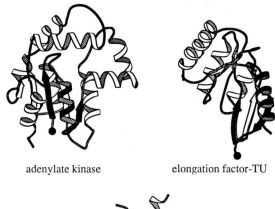

adenylate kinase   elongation factor-TU

rec A

FIG. 6. Three P-loop-containing proteins in which each loop occurs between strands with different connectivities. These regions are emphasized in black. Adenylate kinase has a $+2x$ connectivity, whereas in EF-Tu the connectivity is $+2$ and in recA it is $+3x$.

ties. This can easily be appreciated by looking at Fig. 6. Using the Richardson nomenclature for strand connectivity,[11] the adenylate kinase P-loop[12] is located in a $+2x$ connection, whereas in ras p21[13] and elongation factor-Tu (EF-Tu)[14] the connectivity is $+2$ and in recA[15] it is $+3x$. Because these connectivities differ, we would not be able to find all P-loop-containing proteins by searching for global structural similarities.

Nevertheless, there is a way around this problem, which involves using the loop itself as the probe. This may seem surprising, because in the previous example with RTP, it became harder to conduct meaningful searches as the probe structure became smaller. However because the

[11] J. S. Richardson, *Adv. Protein Chem.* **34,** 167 (1981).

[12] D. Dreusicke, P. A. Karplus, and G. E. Schulz, *J. Mol. Biol.* **199,** 359 (1988).

[13] E. Pai, *et al., Nature (London)* **341,** 209 (1989).

[14] M. Kjeldgaard and J. Nyborg, *J. Mol. Biol.* **223,** 721 (1992).

[15] R. Story and T. A. Steitz, *Nature (London)* **355,** 374 (1992).

P-loop contains residues which adopt glycine-only ϕ,ψ conformations, the loop structure is noticeably different from other structures of the same length in the database. Thus, by searching the database for other loops of the same length (no insertions or deletions permitted) that superimpose extremely well with the probe, we can successfully locate all relevant P-loops in the database. Figure 7 shows the result of conducting this type of automated loop search using a 12-residue probe on a nonredundant set of structures from the Protein Data Bank. All of the functionally related loops are restricted to those similarities with RMS deviations of less than 1 Å, and there is good discrimination between these and all of the other unrelated comparisons. This shows how automated loop searches can complement global superposition strategies; not only are they fast, but they can identify common functions in proteins with different topologies.

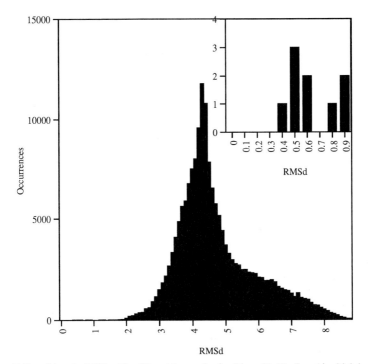

FIG. 7. Searching the PDB with a 12-residue probe (residues 64–75 of recA) which includes the P-loop motif. The distribution shows that while most fragments align with an RMSd greater than 2.0 Å, those with a small RMSd (inset) are all P-type phosphate-binding loops: 0.4 Å, uridylate kinase (1ukz); 0.5 Å, adenylate kinase (1ak3, 2 loops), GIα1 (1gia); 0.6 Å, ras p21 (5p21), elongation factor-Tu (left); 0.8 Å, guanylate kinase (1gky); 0.9 Å, nitrogenase iron protein (1nip, 2 loops).

Using Loop Motifs to Distinguish Common Topologies

So far in this review I have said that if one finds two proteins with similar topologies, it is likely that they will have a related function. In addition, I have said that loop searches can identify similarities in proteins whose topologies differ. Now I would like to discuss the opposite case, where proteins have the same topology but their functions differ. Orengo *et al.* elegantly showed that the distribution of topologies observed in the PDB is distinctly skewed,[16] with structures like TIM barrels and Rossmann-type folds occurring far more often than one would expect by chance. This suggests that certain topologies are easier to form than others, and as a result they called these recurrent topologies superfolds. In this case, loop similarities can once more play an important role, by providing an extra structural filter for the comparisons. A simple example of this situation occurs in the α subunit of integrin CR3,[17] where the overall topology bears an uncanny resemblance to the dinucleotide binding Rossmann fold (i.e., both have $+1x$, $+1x$, $-3x$, $-1x$, $-1x$ topologies). However, although they are globally similar, we can easily distinguish them by their loop motifs as only the Rossmann fold has the nucleoside phosphate-binding loop conformation typified by a GxGxxG sequence.

Identifying New Loop Motifs

Not only do loop searches enable all examples of known motifs to be rapidly identified, but it also facilitates the detection of new motifs. In a previous publication we showed that the phosphate-binding loop from flavodoxin is essentially the same as the phosphate-binding loop from tryptophan synthase, even though globally these protein structures are quite different.[9] In addition, a similarity search using the phosphate-binding loop from orotate phosphoribosyltransferase (1sto) identifies the equivalent phosphate-binding loop from the C-terminal domain of glutamine amido-transferase (1gph), confirming that this too is a new prototype loop conformation. (Although hypoxanthine–guanine phosphoribosyltransferase also contains this loop structure, its coordinates have not been deposited.)

Conclusion

In this chapter we have described how structural similarities can be identified and their importance assessed. Assessment remains the most time-consuming task, and this is where familiarity with protein structures

[16] C. A. Orengo, D. T. Jones, and J. M. Thornton, *Nature* (*London*) **372**, 631 (1994).
[17] J.-O. Lee, P. Rieu, M. A. Arnaout, and R. Liddington, *Cell* (*Cambridge, Mass.*) **80**, 631 (1995).

per se remains a distinct advantage. Many structures are still not deposited in the PDB, and for these cases the only solution is to have good memory (although the SCOP database is now extremely useful for these cases[18]).

Nevertheless, it is now more likely for a newly determined structure to adopt a familiar fold rather than a novel one.[16] Researchers who fail to use these procedures not only run the risk of missing structural similarities with their own structure, but also miss the concomitant functional and evolutionary relationships as well. There are now many programs available that have been shown to be effective and that provide the novice with an ideal introduction to the field. In addition there are a number of groups specializing in this work, who are always willing to collaborate, as well as a growing number of information sources available on the World Wide Web. With these facilities in place, we are now in an ideal position to exploit the mass of structural data anticipated in the near future.

Availability

The SCOP database of manually detected structural similarities is described in Chapter 37. The automatically classified CATH database is at http://www.biochem.ucl.ac.uk/bsm. In addition, database searches can be made using sarf2 at http://www.ncifcrf.gov/~nicka/info.html. I have additional data available at http://www.biochem.ucl.ac.uk/~swintech.

[18] A. G. Murzin, S. E. Brenner, T. Hubbard, and T. Chothia, *J. Mol. Biol.* **247,** 536 (1995).

[39] Alignment of Three-Dimensional Protein Structures: Network Server for Database Searching

By LIISA HOLM and CHRIS SANDER

Introduction

Access to computational services and biological databases over the Internet, in particular through the World Wide Web, is an increasingly important research tool for the biochemist. A major use of molecular biology databases involves searching for evolutionary links that allow transfer of functional information about one protein family to another. An increasing number of distant evolutionary relationships that are not evident by sequence comparison are being revealed by similarity of three-dimen-

Copyright © 1996 by Academic Press, Inc.
All rights of reproduction in any form reserved.

sional (3D) protein structures, both because of a rapid increase in the number of known structures and because of improved methods of detection. For example, structure comparison has revealed surprising biochemical similarities between urease and adenosine deaminase[1] and between glycogen phosphorylase and a DNA glucosyltransferase from phage T4[2] that were not detected by sequence comparison. These are just two of a long list of examples[3] to illustrate the evolutionary principle of adapting structural motifs that support particular active-site constellations to different functional roles in diverse cell types and organisms.

A large number of automated methods for protein structure comparison that use different representations of structure, different definitions of similarity, and various optimization algorithms have been developed (reviewed in Ref. 3). This chapter describes the Dali method,[4] which is a general approach for aligning a pair of proteins represented by two-dimensional matrices. The implementation of prefilters to speed up database searches has enabled us to provide Internet access using either World Wide Web software addressing http://www.embl-heidelberg.de/dali/ or electronic mail to dali@embl-heidelberg.de.

Formulation of Problem

The utility of distance matrices, also called distance plots or distance maps, in describing and comparing protein conformations has been recognized for a long time. A distance matrix is a two-dimensional (2D) representation of 3D structure. The matrix is independent of the coordinate frame and contains more than enough information to reconstruct the 3D structure, except for overall chirality, by distance geometry methods. The most commonly used variant is that containing all pairwise distances between residue centers (i.e., $C\alpha$ atoms).

Distance matrices are useful in structure comparison because similar 3D structures have similar interresidue distances. Imagine a (transparent) distance map of one protein placed on top of that of another protein and then moved vertically and horizontally. Depending on the relative displacement of the matrices, matching substructures appear as patches (submatrices) in which the difference of distances is small. Matching patches centered on the main diagonals correspond to locally similar backbone

[1] E. Jabri, M. B. Carr, R. P. Hausinger, and P. A. Karplus, *Science* **268**, 998 (1995).
[2] L. Holm and C. Sander, *EMBO J.* **14**, 1287 (1995).
[3] L. Holm and C. Sander, *Proteins* **19**, 165 (1994).
[4] L. Holm and C. Sander, *J. Mol. Biol.* **233**, 123 (1993).

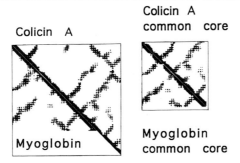

FIG. 1. Comparison of distance matrices. Structures similar in three dimensions necessarily have a similar set of intramolecular distances. The common folding pattern of colicin A and myoglobin is highlighted in the collapsed matrix (right) that brings the two sequences into register through deleting incompatible rows and columns from the full distance matrices. Formally, the Dali score [see Eq. (1)] is a weighted sum over similarities of distances between residue centers in a common core. The distance matrices depicted here have a black dot for $C\alpha$–$C\alpha$ distances shorter than 12 Å. Helices show up as thick bands along the diagonal and helix pairs as black bands parallel or orthogonal to the diagonal.

conformations (i.e., secondary structures). Matches of short distances found off the main diagonals reveal similar tertiary structure contacts. The presence of a common structural motif made up of several disjoint regions of the backbone becomes visible at one glance in a pair of "collapsed" submatrices that are obtained by deleting residues with no structural equivalent in the other structure (Fig. 1). Allowing permutations in the order of rows and columns leads to detection of spatial similarities in protein structures when topological connectivities differ. An advantage of the 2D representation used in the Dali method over rigid-body 3D superimposition is that local conservation of structure is not masked by shifts in the relative positions of structural elements, for example, as a result of hinge motion of domains.

A quantitative solution to the geometrically complicated problem of comparing protein shapes requires a precise definition of similarity of protein structures. In the Dali method, the structural similarity S (Dali score) is defined as the following weighted sum:

$$S = \sum_i \sum_j \left[\left(\partial - \frac{|d_{ij}^A - d_{ij}^B|}{d_{ij}^*} \right) e^{-(d_{ij}^*/\mu)^2} \right] \tag{1}$$

where the summation is over all residues i, j of the common core and d_{ij}^* denotes the arithmetic average of the $C\alpha$–$C\alpha$ distances d_{ij}^A and d_{ij}^B in

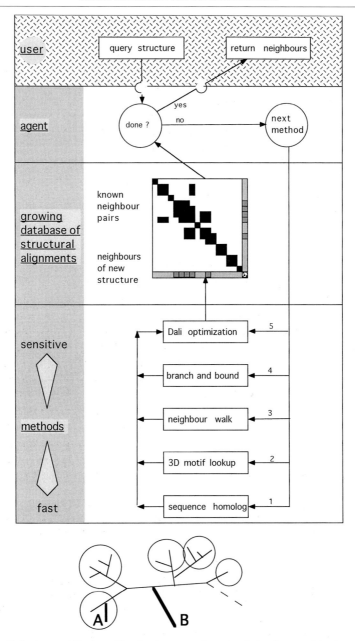

FIG. 2. System architecture. The database search system is implemented in three layers. At the top of the system is an "intelligent agent," actually a Perl script, that receives a query structure from the user and returns the list of structural neighbors of the query structure.

proteins A and B. In Eq. (1) ∂ is the threshold of similarity and is set to a relative deviation of 0.20 (20%). This means that, for example, adjacent strands in a β sheet (typical distance 4–5 Å) should match to within 1 Å, while 2–3 Å displacements are well tolerated for strand–helix or helix–helix contacts (typical distances 8–15 Å). The exponential factor downweights contributions from pairs in the long-distance range. We chose $\mu = 20$ Å calibrated on the size of a typical domain.

The set of equivalences between residues in A and B that maximizes S defines the common core of proteins A and B. Optimization of several nonoverlapping alignments in parallel leads to automatic detection of, for example, internal repeats. As similarity is quantified at the residue level, the resulting structural alignments can be directly linked to sequence and evolutionary comparisons.

Database Searches

The optimization of assignments for equivalent residue pairs looks simple graphically (Fig. 1) but is computationally hard because of the complicated combinatorics. In the original Dali method, matches are built up by combining small submatrices with similar distance patterns and using a Monte Carlo algorithm for optimization. The approach was shown to be robust and to yield accurate alignments.[4] In database searching, one is generally only interested in a few top hits so that sensitive pairwise comparison against the bulk of the database is unnecessarily costly. For the network server version of Dali, we have implemented efficient screening steps that work with approximations at the level of secondary structure elements[5] and as a result increase the speed of comparison from 5–10 min per protein pair to, in favorable cases, 5–10 min for scanning one structure against all structures in the protein database. The increase in speed is approximately 500-fold. The complete database search system is outlined in Fig. 2.

[5] L. Holm and C. Sander, in "Proceedings of the Third International Conference on Intelligent Systems for Molecular Biology" (C. Rawlings, D. Clark, R. Altman, L. Hunter, T. Lengauer, and S. Wodak, eds.), pp. 179–187. AAAI Press, Menlo Park, California, 1995.

The task can be performed efficiently using stored knowledge of the structural neighborhoods of all proteins in the PDB (FSSP database) and a hierarchy of different methods (mostly FORTRAN programs). The goal is to place a query structure in the proper neighborhood of fold space, which is illustrated in the form of a tree at the bottom. Fast filters efficiently detect "trivial" similarities (branch labeled A). Only if the query structure cannot be mapped to a known neighborhood (branch labeled B) is it necessary to test its similarity to each one of the known structures in order to position properly the new structure, a more time-consuming task.

Finding all structural neighbors in a database of thousands of proteins is simplified by defining a hierarchy of neighborhoods in protein fold space. The tightest clusters of folds are formed by sequence homologs (sequence identity above 25%[6]), and each such family is represented by a single member. Remote functional homologs (proteins with sequence identity below 25%) typically are more conserved in structure than pairs of proteins that are unrelated in function and only have similar folding topology. An example is the case of the functionally related myo-, hemo-, and leghemoglobins, which form one group among three about equally distant structural classes, where the other two are phycocyanins and colicin A.[7]

The database search system exploits these observations on clusters in fold space by first trying so-called cheap and quick filters for identifying trivial hits. For example, a new globin structure or another mutant of T4 lysozyme would map to an already characterized neighborhood of fold space. Sophisticated and sensitive methods are reserved for the potentially unique structures that require charting new regions of fold space. Because the screening methods use different approximations of the proper Dali score, it is important to put all pairs, whether they come via the fast or slow route, on the same footing for consistency of the final result. This is done by passing all prealignments produced by the different methods through a Monte Carlo algorithm that optimizes complete alignments with respect to the Dali score.

Biological Meaning of Structural Similarity

We have empirically determined the background strength of similarity as a function of chain length. The statistical significance of a database hit relative to the background is reported as a Z score (score minus mean divided by standard deviation). In particular, the rescaling so obtained provides a general and quantitative definition of structural neighborhoods. For example, using Z scores raises a database match to the SH3-like domain of biotin repressor/biotin holoenzyme synthetase from rank 58 in Dali score to rank 3 in Z score, compensating for the effects of very different domain sizes (Table I). In reporting results from structure database searches, we list pairs of proteins or domains[8] for which the Z score is above 2.

It is well established that protein folds are better conserved in the course

[6] C. Sander and R. Schneider, *Proteins* **9**, 56 (1991).
[7] L. Holm and C. Sander, *FEBS Lett.* **315**, 301 (1993).
[8] L. Holm and C. Sander, *Proteins* **19**, 256 (1994).

TABLE I
IDENTIFICATION OF SIMILAR DOMAIN FOLDS

Domain description	Size of match/ size of domain, residues	Best database hit[a]	Dali score (rank[b])	Z score (rank[b])
N-terminal, DNA-binding	59/64	LexA repressor	258 (14)	6.7 (1)
Middle, catalytic	129/181	Seryl-tRNA synthetase	523 (1)	3.6 (6)
C-terminal, SH3-like fold	46/47	Photosystem I accessory protein (psaE)	152 (58)	4.2 (3)

[a] Structurally most similar protein in a sequence-representative set of 557 3D structures.
[b] Ranks in structure comparison against sequence-representative set of 557 3D structures are given in parentheses.

of evolution than amino acid sequences. However, the smaller and simpler a folding motif is, the more frequent its recurrence between protein families without any apparent biological connection. This raises the question under which circumstances it is justified to base inferences of, say, biochemical mechanism on structural resemblance. Indicators of common descent include conserved active-site residues, a conserved structural framework

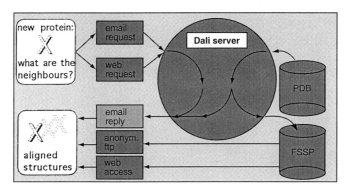

FIG. 3. Server traffic chart. The Dali server compares a query structure against the database of known structures (Protein Data Bank, PDB) and returns a list of structurally aligned neighbors in fold space. Newly solved proteins can be submitted for comparison by E-mail (dali@embl-heidelberg.de) or interactively from the WWW (home page URL http:// www.embl-heidelberg.de/dali/dali.html). The results for exhaustive structure comparisons for proteins already deposited in the PDB are stored in the FSSP database of structural neighbors.[8] To get at this information over the WWW, point to the FSSP home page at URL http:// www.embl-heidelberg.de/dali/fssp/. The neighbor lists in the FSSP database are updated whenever new structures are released by the PDB.

(a)

```
STRID2   Z      RMSD   LALI   LSEQ2   %IDE   PROTEIN
1kan-A   38.8   0.0    253    253     100    KANAMYCIN NUCLEOTIDYLTRANSFERASE
1kan-B   34.7   0.4    253    253     100    KANAMYCIN NUCLEOTIDYLTRANSFERASE
1bpb     5.8    4.0    106    242     9      DNA POLYMERASE BETA (BETA POLYME
1rpo     2.1    4.1    53     61      6      ROP (COLE1 REPRESSOR OF PRIMER)
```

(b)

```
1kan-A  MNGPIIMTREERMKIVHEIKERILDKYGDDVKAIGVYGSLGRQTDGPYSDIEMMCVMSTEEAEFSHEW
                    hhhhhhhhhhhhhhhhhhs sseeeeeeegggttt    tt  eeeeeee stt ee
1kan-B  MNGPIIMTREERMKIVHEIKERILDKYGDDVKAIGVYGSLGRQTDGPYSDIEMMCVMSTEEAEFSHEW
                    hhhhhhhhhhhhhhhhhhhtttteeeeeebgggtss    ss    eeeees ss eee  e
1bpb    FEDFKRIPREEMLQMQDIVLNEVKKL.DPEY.IATVCGSFRRGAES..GDMDVLLTHPNFT..TKFMG
                    ttggs eehhhhhhhhhhhhhhhhhh  tt  eeee hhhhtt se  s eeeeee tt    seeee
1rpo    ...............................................................
```

(c)

```
swiss           FSSP
kanu_staau      1kan-A  MNGPIIMTREERMKIVHEIKERILDKYGDDVKAIGVYGSLGRQTDGPYSDIEMMCVMSTEEAEF
kanu_bacsp      1kan-A  MNGPIIMTREERMKIVHEIKERILDKYGDDVKAIGVYGSLGRQTDGPYSDIEMMCVMSTEEAEF
dpob_rat        1bpb    FEDFKRIPREEMLQMQDIVLNEVKKL~DPEY~IATVCGSFRRGAES~~GDMDVLLTHPNFT~~T
dpob_human      1bpb    FGDFKRIPREEMLQMQDIVLNEVKKV~DSEY~IATVCGSFRRGAWS~~GDMDVLLTHPSFT~~T
```

(d)

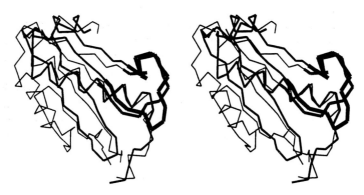

FIG. 4. Structures aligned with structures and structures aligned with sequences. (a) Ranked list of structural neighbors as a result from the database search using kanamycin nucleotidyltransferase (1kan-A) as the query structure. Hits in the database are identified by PDB code and chain (STRID2). The positional root-mean-square deviation of Cα atoms corresponding to the optimal Dali alignment (RMSD), the number of aligned residues (LALI), the length of the matched protein (LSEQ2), sequence identity among equivalenced residues (%IDE), and the name of the matched protein (PROTEIN) are given for each pair. The strongest structural similarity by Z score (column labeled Z) reveals a distant evolutionary connection between DNA polymerase beta (1bpb) and kanamycin nucleotidyltransferase.[10] (b) Structural alignment view loaded on the fly by the FSSP web server. The amino acid sequence and secondary structure (h, helix; e, strand; t, turn) are shown for each chain. Walking in fold space is possible by clicking underlined links. Only the structurally equivalent blocks are

around the active site, and similar biochemical function. The structure signal is usually captured by the Z-score ranking. The sequence signal often stands out when the structural alignment is expanded by sequence homologs from the HSSP database of sequence neighbors of structurally known proteins.[6]

Availability

The pool of known protein 3D structures is growing by hundreds of new ones each year, and systematic comparisons are needed that can keep track of all the interesting similarities. The Dali server can be used by X-ray crystallographers and nuclear magnetic resonance (NMR) spectroscopists at the last stage of structure determination to detect possible structural similarities with structures currently in the Protein Data Bank (PDB). Coordinates of new structures can be sent either by E-mail or interactively via the World Wide Web (WWW). The list of structural alignments of the query structure against all significantly similar neighbors in protein fold space is returned. The Dali search engine is also used to maintain the FSSP database[9] of precalculated structural neighbors for all structures released to the public through the Protein Data Bank (Fig. 3).

In time, the work of structural biologists will result in a complete survey of the role of protein structures in the evolution of biochemical complexity. The FSSP database of multiple structure comparisons provides a continuously updated structural classification.[9] The rich information contained in multiple structural alignments is best appreciated in graphical form. Public software (e.g., Rasmol) can be downloaded to look at superimposed 3D structure pairs and generate interactive pictures on a personal computer, and a web browser can be used to follow links from structure alignment

[9] L. Holm and C. Sander, *Nucleic Acids Res.* **22**, 3600 (1994).

---

shown (dots are gaps and trailing ends). Repressor of primer (ROP) (1rpo) matches to a helical domain at the C terminus and not to the N-terminal catalytic domain shown here. (c) Combining the power of structure comparison with that of multiple sequence alignment. DNA polymerase beta has been selected from the list of neighbors of kanamycin nucleotidyltransferase (1kan-A), and the structural alignment of these proteins is viewed in combination with sequence alignments from the HSSP database[6] (the parent structure is given in the FSSP column). The sequence identifiers (column labeled swiss) are linked to the SWISS-PROT database via the Sequence Retrieval System (SRS).[11] Only structurally equivalent segments are shown; excluded segments are marked by ~. (d) Structural superimposition of the Cα traces of kanamycin nucleotidyltransferase (thick lines) and DNA polymerase beta (thin lines).[10]

to sequence families, annotations of function, or literature references, facilitating a closer look at protein evolution (Fig. 4[10,11]).

Acknowledgments

We thank Antoine de Daruvar for writing the scripts managing the server traffic and Protein Data Bank updates.

[10] L. Holm and C. Sander, *Trends Biochem. Sci.* **20,** 345 (1995).
[11] T. Etzold and P. Argos, *Comput. Appli. Biosci.* **9,** 49 (1993).

[40] Converting Sequence Block Alignments into Structural Insights

By OLIVIER POCH and MARC DELARUE

Introduction

In the past decade, the number of available protein sequences has grown exponentially. In the same time, sequence analysis has become a major tool to gain insight on protein function in cellular processes from the sequence information. Many aspects of this sequence–structure–function relationship problem are not fully understood and are far from being solved by computational biology, but it is now possible to try to address problems as diverse as the localization of a protein in the cell (nucleus, organelles, membrane), the rate of protein degradation, putative posttranslational modifications, as well as some functional aspects such as binding properties (to other proteins, nucleic acids, cofactors, ions, etc.). Obviously, this list is far from being exhaustive, but it clearly highlights the fact that biologists can now expect major information about biological processes from even a single sequence.

One of the first issues to be addressed when undertaking structural and functional studies of a particular protein is to find out how many related sequences can be identified in a protein sequence database search. Even if structural data are not available for any member of a protein family, a multiple alignment will often offer much information that has to be analyzed carefully for future biological studies. The chance of hitting at least a homologous protein is constantly increasing as genome-sequencing projects proceed toward completion in different organisms (*Saccharomyces cerevisiae, Arabidopsis thaliana, Homo sapiens, Caenorhabditis elegans*). In addi-

Copyright © 1996 by Academic Press, Inc.
All rights of reproduction in any form reserved.

tion to sequences of the same protein in different species, sequence homology between proteins otherwise apparently unrelated is very frequent. This opens new fields in the comprehension of a protein in the context of the different metabolic pathways.

With the growing number of three-dimensional protein structures available and comparisons of these structures, it has been possible to deduce some of the principles that govern the establishment of secondary, supersecondary, or tertiary structures from a given amino acid sequence. However, the prediction of protein structure from its sequence alone is still a largely unsolved problem. Conversely, it can be stated that sequence analysis has become an integral part of structural biology in that a discussion of the functional implications of a new protein structure solved by X-ray crystallography has to take into account all the sequence data available and try to rationalize them.

As more and more protein structures are solved and deposited in the Protein Data Bank (PDB),[1] clear structural homology is often revealed between proteins with no sequence similarity[2]; in other words, proteins whose sequences cannot be aligned in a reliable way using any of the available sequence comparison programs often appear to have the same folding topology. This is often observed for proteins using the same catalytic mechanism and one common substrate (e.g., Class II aminoacyl-tRNA synthetases and BirA[3] or the different thioredoxins and D-alanine-D-alanine ligase).[4] This emphasizes the notion of the plasticity in sequence space for any given three-dimensional fold. For a large family of widely divergent proteins, one should be prepared to give up an alignment over the entire length of the proteins and to expect only limited sequence similarity, localized in a few blocks and representing mainly the catalytic site, with the only invariant remaining positions being the catalytic ones. This level of homology is often referred to as the twilight zone.[5]

The notion of plasticity of a given folding type to many different sequences has led to a new strategy to predict protein structure, the so-called inverse protein folding (reviewed in Ref. 6). This method tries to determine the compatibility of the sequence under study to known three-dimensional structures. This relies on the assumption that the number of different folds

[1] F. C. Berstein, T. F. Koetzle, G. J. B. Wiliams, E. F. Meyer, M. D. Brice, J. R. Rodgers, O. Kennard, T. Shimanouchi, and M. Tasumi, *J. Mol. Biol.* **162,** 535 (1977).
[2] R. B. Russell and G. J. Barton, *J. Mol. Biol.* **244,** 332 (1994).
[3] P. J. Artymiuk, D. W. Rice, A. R. Poirette, and P. Wilett, *Nat. Struct. Biol.* **1,** 758 (1994).
[4] C. Fan, P. C. Moews, Y. Shi, C. T. Walsh, and J. R. Knox, *Proc. Natl. Acad. Sci. U.S.A.* **92,** 1172 (1995).
[5] R. F. Doolittle, *Sci. Am.* **253,** 88 (1985).
[6] P. Argos, *Curr. Opin. Biotechnol.* **5,** 361 (1994).

in nature is limited (it has been estimated at about 1000 by Chothia[7]) and also assumes that the correct folding type is already present in the protein structure database,which is scanned using empirically defined energy criteria (see [35] in this volume).

Another way of predicting protein structure is "homology model building"[8] but we will not be concerned with this aspect in this chapter. Homology model building is restricted to a situation in which there is strong sequence similarity with a sequence of a protein known at the structural level, a situation not commonly encountered.

Here, we discuss the methods used in our laboratory to extract maximal structural information from a set of distantly related protein sequences that ultimately share only isolated weakly conserved regions (blocks). Fully automated computer programs are not available to achieve this goal, implying that human intervention, empirical choices, and approximations are frequently used in the process of sequence alignments and structure predictions. However, we will try to describe precisely the "nonobjective" rules applied to ensure reliable results and will discuss which of the recent improvements in the field of computational biology are covering some of them. Gaining structural insight from sequences alone necessitates a patient analysis based on a progressive scanning of sequence databases, sequence alignments, and secondary structure predictions combining all the available computational tools with a constant reference to all available experimental data. Protein evolutionary information implied by sequence alignments has proved useful in improving the accuracy of secondary and tertiary structure predictions, and these methods are under constant development, use, and evaluation.[9]

Sequence Alignments and Block Delineation

Accurate overall sequence alignment of related sequences is a major prerequisite to ensure accurate block(s) delineation, subsequent additional searches, and correct structure predictions. Here, we survey some of the aspects of the overall sequence alignment technique considered important, as most of them are widely discussed elsewhere in this volume and in Volume 183 of this series. Sequence alignment can be partitioned in two distinct steps: (1) the collection of related sequences and (2) their proper alignment.

[7] C. Chothia, *Nature* (*London*) **357**, 543 (1992).
[8] W. Taylor, *Nature* (*London*) **356**, 478 (1992).
[9] S. A. Benner, *J. Mol. Recognit.* **8**, 9 (1995).

Collection of Related Sequences

Among the various programs available for the detection of a family of related proteins within the database, the two rapid pairwise comparison programs BLAST[10] and FASTA[11] are the most commonly used, and they look for local zones of sequence similarity. They are based on two different algorithms: BLAST is faster but does not take into account gap penalties, whereas Fasta looks for dense regions of similarity along the diagonals.

However, it should be stressed that, in the process of sequence alignments and block delineation, sequence database searching via such programs is just a tool to obtain a list of the protein families. Even though the comparison programs are highly efficient and sequence databases mostly exhaustive, obtaining sequences by personal communication prior to publication should always be considered; in addition, access to biochemical data on the identified sequences through literature scanning should always be an integral part of sequence data collection. Partial protein sequences must be taken into account as they are frequently revealed, via the tBLASTN and TFASTA programs, in the growing sequence database of expressed sequence tags[12] (dbEST) or on the borders of reported open reading frames (ORFs) present in the DNA database. The main idea is that the more sequences, the higher the signal-to-noise level when sequences are compared. Whatever the total final number of related sequences, a single amino acid sequence can always constitute a major clue, as has been clearly highlighted for DNA-dependent polymerases, in which the inclusion of one single sequence, that from phage SPO2, considerably narrows the delineation of important and conserved regions in this family.[13]

Sequence Alignment

Typically, two major techniques can be used in the process of aligning a set of related sequences: simultaneous alignment of several sequences using one of the numerous and potent multiple alignment programs [e.g., CLUSTALW[14] or the program PILEUP from the constantly updated Genetics Computer Group package[15]] or pairwise alignments[16,17] of each mem-

[10] S. F. Altschul, W. Gish, W. Miller, E. W. Myers, and D. J. Lipman, *J. Mol. Biol.* **215,** 403 (1990).

[11] W. R. Pearson, this series, Vol. 183, p. 63.

[12] M. S. Boguski, T. M. J. Lowe, and C. M. Tolstoshev, *Nat. Genet.* **4,** 332 (1993).

[13] M. Delarue, O. Poch, N. Tordo, D. Moras, and P. Argos, *Protein Eng.* **3,** 461 (1990).

[14] J. D. Thompson, D. G. Higgins, and T. J. Gibson, *Nucleic Acids Res.* **22,** 4673 (1994).

[15] J. Devereux, P. Haeberli, and O. Smithies, *Nucleic Acids Res.* **12,** 387 (1984).

[16] T. F. Smith and M. S. Waterman, *Adv. Appl. Math.* **2,** 482 (1981).

[17] S. B. Needleman and C. D. Wunsch, *J. Mol. Biol.* **48,** 443 (1970).

ber of the set with each other, followed by analysis and delineation of the corresponding conserved regions (DOTPLOT representations can be used for a quick evaluation of the results[18]). These two approaches are better viewed as complementary tools that allow distinct problems to be addressed, such as the alignment of related proteins via multiple alignment and the comparison of more divergent ones by pairwise sequence comparisons. In our experience, we strongly favor pairwise sequence comparisons and view the results of direct multiple alignment programs with a critical eye.

Indeed, regardless of the accuracy of the alignment obtained by the multiple alignment programs, which have been greatly improved by the development of various algorithms,[14,19] simultaneous alignment of sequences may unintentionally conceal fruitful information. This is best exemplified by comparing the identity profiles obtained after multiple alignment of the extensively conserved sequences of (i) vertebrate poly(ADP)-ribose (PARP) sequences or (ii) PARP sequences of vertebrates plus one insect (invertebrate) sequence (Fig. 1).[20] As clearly highlighted by these two conservation profiles, all PARP enzymes share an organization in concatenated functional elements and a perfect coincidence of the lowest conserved regions (indicated by arrows) with the major interdomains previously determined by limited proteolysis experiments. However, in addition to promoting a dramatic decrease in the overall sequence conservation (from 62 to 32%), inclusion of the *Drosophila* sequence results in the erasing of the peaks corresponding to domain C (unknown function) and to domain D (automodification and putative dimerization functions). This information, lost in the multiple alignment containing all sequences, may be indicative of the existence of a distinct function(s) in domains C and/or D specific to the invertebrate enzyme.

In addition, as noted by Thompson *et al.*,[21] the alignment strategy developed in most of the multiple alignment programs uses progressive alignments following the branching order of a phylogenetic tree based on a series of pairwise alignments, and this cannot guarantee that the resulting global alignment is optimal. Indeed, misaligned regions are frequently observed, and they generally proceed from errors made early on in the alignment process that cannot be corrected later by the subsequent alignment of additional sequences. Thus, multiple alignment is better viewed as a good starting point to locate the main peaks of conservation between related

[18] M. Vingron and P. Argos, *J. Mol. Biol.* **218,** 33 (1991).

[19] O. Gotoh, *Comput. Appl. Biosci.* **9,** 361 (1993).

[20] G. de Murcia, V. Schreiber, M. Molinete, B. Saulier, O. Poch, M. Masson, C. Niedergang, and J. Menissier de Murcia, *Mol. Cell. Biochem.* **138,** 15 (1994).

[21] J. D. Thompson, D. G. Higgins, and T. J. Gibson, *Comput. Appl. Biosci.* **10,** 19 (1994).

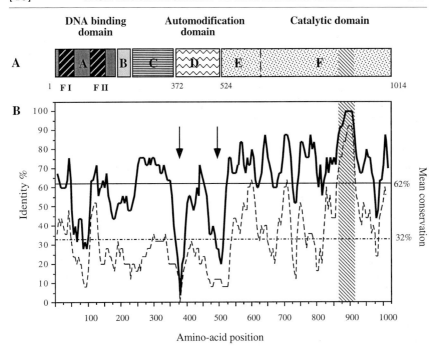

FIG. 1. (A) Schematic representation of the three functional domains of human poly(ADP)-ribose polymerase (PARP), obtained after mild trypsin and papain digestion. (B) Identity profiles obtained after alignment of the human, murine, bovine, chicken, *Xenopus*, and *Drosophila* PARP sequences: the solid line corresponds to the profile calculated with the vertebrate PARP sequences, and the dashed line corresponds to that of the vertebrates plus the invertebrate sequences. Arrows indicate regions of lowest conservation. Adapted from Ref. 20.

sequences, while further refinements are performed by pairwise sequence comparisons using either global alignment algorithms or dot-matrix comparisons.

The other main advantage of pairwise sequence comparisons is the possibility to change the comparison matrices used according to the amino acid sequences compared. Indeed, it is now well established that the best matrices are based on the actual substitution rates observed from matched sets of proteins, rather than procedures based on chemical similarity and physical characteristics of amino acids.[22] The hope that one single matrix may account for all sequence similarities regardless of the evolutionary distances between the compared protein sequences appears somewhat presumptuous. The usefulness of different matrices for aligning very closely

[22] M. S. Johnson and J. P. Overington, *J. Mol. Biol.* **233,** 716 (1993).

or distantly related sequences has been discussed,[23] and algorithms used in multiple alignment programs have dealt with this problem by changing the weight matrices, as the alignment proceeds, depending on the estimated divergence of the sequences. This strategy leads to a substantial enhancement in the accuracy of sequence alignments, and we agree with Thompson *et al.*[21] in the use of distinct matrices of the BLOSUM[24] and PAM[25] series in different cases and even, for closely related sequences, the use of the unitary matrix.

The strategy discussed previously has proved to be fruitful in delineating regions of strong conservation (blocks) in different sets of related proteins. Here, we discuss what additional information can be taken into account in the process of block delineation, using the example of class II aminoacyl-tRNA synthetases (aaRSs).[26] Clearly, in this family, a global multiple alignment of all aaRSs sequences is impossible without further information. Therefore, one must proceed gradually, first aligning closely related sequences; one such set of sequences included ThrRS, SerRS, and ProRS. Two major points could be deduced from the three-way alignment of these sequences (Fig. 2).

First, there are mainly three regions of high homology. Furthermore, the alignment confirmed that the number of amino acids separating the conserved regions II and III were widely different (from 83 to 268 residues) even between proteins otherwise highly related; this information was already apparent from the alignment of AspRS, AsnRS, and LysRS, which also form a clear, different subgroup (see Ref. 26). This allows us to analyze the subsequent pairwise sequence comparisons performed between aaRSs sequences with a less stringent prerequisite on the distances existing between motifs II and III, while distances between motifs I and II appeared less variable. At the practical level, this means that the comparisons performed with a very low gap extension penalty (GEP, which gives the cost of every additional residue introduced in a gap) may be more accurate in aligning motif II relative to motif III, whereas alignment of motif I relative to motif II may be more accurate by using a higher GEP. There is a growing literature addressing the major problem of the gap opening penalty (GOP, which gives the cost of opening a new gap) and the GEP.[27] Cleary, as already discussed for substitution matrices, GOP or GEP must be adjusted

[23] S. A. Brenner, M. A. Cohen, and G. H. Gonnet, *Protein Eng.* **7**, 1323 (1994).

[24] S. Henikoff and J. G. Henikoff, *Proc. Natl. Acad. Sci. U.S.A.* **89**, 10915 (1992).

[25] M. O. Dayhoff, R. M. Schwartz, and B. C. Orcutt, *in* "Atlas of Protein Sequence and Structure" (M. O. Dayhoff, ed.), p. 345. National Biomedical Research Foundation, Washington, D.C., 1978.

[26] G. Eriani, M. Delarue, O. Poch, J. Gangloff, and D. Moras, *Nature (London)* **347**, 203 (1990).

[27] M. Vingron and M. S. Waterman, *J. Mol. Biol.* **235**, 1 (1994).

```
T  78  LLGHAIKQLWPHTKMAIGPVIDNGFYYDVDLDRTLTQEDVEALEKRMHELAEKNYDVIKK
S   1  MLDPNL..LRNEPDAVAEKLARRGFKLDVD.......KLGALEERRKVLQVKTENLQAE

T 138  KVSWHEARETFANRGES.....YKVSILDENIAHDDKPGLYFHEEYVDMC.RGPHVPNMR
S  51  RNSRSKSIGQAKARGEDIEPLRLEVNKLGEELDAAKAELDALQAEIRDIALTIPNLPADE

                                             D    φλ
P   1                                MRTSQYLLSTLKETPADAEVISHQL
T 192  FCHRFKLMKTAGAYWRGDSNNKMLQRIYGTAWADKKALNAYLQRLEEAAKRDHRKIGKQL
S 111  VPVGKDENDNVEVSRWGTP.....................REFDFEVRDHVTLGEMH

         φ   φ     λG  φ    G        -  φφ      -   φ  E   Pφφ        φ
P  26  .IVRAGMIRKLASGLYTWLPTGVR...VLKKVENIVREEMNNAGAIEVSMPVVQPADLWQ
T 252  ...DLYHMQEEAPGMVFWHNDGWT...IFRELEVFVRSKLKEYQYQEVKGPFMMDRVLWE
S 147  SGLDFAAAVKL.TGSRFVVMKGQIARMHRALSQFMLDLHTEQHGYSENYVPYLVNQDTLY

        λG  φ    φ    φφ              φ  φ  P       -φφ    φ     -LPφ  φ       λ
P  82  ESGRWEQYGPELLRFV....DRGERPFVLGPTHEEVITDLIRNELSSYKQLPLNFYQIQT
T 306  KTGHWDNYKDAMFTTS.....SENREYCIKPMNCPGHVQIFNQGLKSYRDLPLRMAEFGS
S 206  GTGQLPKFAGDLFHTRPLEEEADTSNYALIPTAEVPLTNLVRGEIIDEDDLPIKMTAHTP

          RE  λ       Gφφ R + F   -φ   φ      --         φ    φ   φφ   φGφ
P 138  KFRDEVRPRF....GVMRSREFLMKDAYSFHTSQESLQETYDAMYAAYSKIFSRMGLDFR
T 361  CHRNEPSGSLH...GLMRVRGFTQDDAHIFCTEQIRDEVNGCIRLVY.DMYSTFGFEKI
S 266  CFRSEAGSYGRDTRGLIRMHQFDKVEMVQIVRPEDSMA.ALEEMTGHAEKVLQLLGLPYR

         φ  φ        T
P 193  AVQA...DTGSIGGSASHEFQVCAQSGEDDVVFSDTSDYAANIELAELIAPKEPRAATQE
T 417  VVKLVKLSTRPEKRIGSDEMWDRAEADLAVALEENNIPFEYQLGEGAFYGPK........
S 325  KIIL...CTGDMGFGA.............................................

                           Y             T
P 362  - 110 aa - GANIDGKHYFGINWDRDVATPEVADIRNVVAGDPSPDGQGRLLYKRGIE
T 466  .............IEFTLYDCLDRAWQCGT...............................
S 338  ...............CKTYDLEVWIPAQNT....................YREISS

          φ -φ            S    λ-               φ   Rφφ  φφ   E-           φP
P 411  VGHIFQLGTKYSEALKASVQGEDG..RNQILTMGCYGIGVTRVVAAAIEQNYDERG.IVWP
T 483  ....VQLDFSLPSRLSASYVGEDNERKVPVMIHRAILGSMERFIGILTEEFAG.....FFP
S 359  CSNVWDFQARRMQARCRS.KSDKK..TRLVHTLNGSGLAVGRTLVAVMENYQQADGRIEVP

          φφ  Pφ     φ   φ
P 469  DAIAPFQVAILPMNMHKSFRVQELAEKLYSELRAQGIEVLLDDRKERPGVMFADMELIG
T 535  TWLAPVQVVIMNITDSQS....EYVNELTQKLSNAGIRVKADLRNEKIGFKIREHTLRR
S 417  EVLRPYMNGLEYIG      430

P 528  IPHTIVLGDRNLNDNDDIEYKYRRNGEKQLIKTGDIVEYLVKQIKG          572
T 590  VPYMLVCGDKEVESGKVAVRTRRGKDLGSMDVNEVIEKLQQEIRSRSLKQLEE      642
```

FIG. 2. Total alignment of the *Escherichia coli* ProRS (P), SerRS (S), and ThrRS (T). Conserved residues are indicated at the top line by λ for small amino acid (P, G, S, T), − for negatively charged amino acids (D, E, Q, N), + for positively charged amino acids (H, R, K), and φ for hydrophobic residues (F, Y, W, I, L, V, M, A). Motifs I, II, and III are in boldface type and underlined. Reprinted from Ref. 26.

with respect to the set of compared proteins whatever the procedure used, namely, pairwise comparison or multiple alignment.

Second, careful analysis of motif II for SerRS, ThrRS, and ProRS (Fig. 2) indicated that it is composed of two distinct subregions, separated by a gap, each preserving invariant residues R and E in the amino-terminal part of the motif and G, R, and F in the carboxy–terminal part. The subsequent sequence comparisons performed revealed that, although the chemical nature of the residues surrounding these invariant residues was preserved, the arginine residue was the only one strictly conserved; conservative substitutions were observed for the other residues. However, the finding that two distinct subregions can exist in motif II constituted a main factor in delineating and defining this motif.

We have just emphasized how indels (insertion/deletions) can be important insights in the process of sequence alignment. Nevertheless, it is obvious that the strictly or conservatively maintained residues observed within each motif are also of primary importance. Considering the alignment of the class II aaRSs, the alignment of AspRS, AsnRS, and LysRS was then analyzed keeping in mind the previously discussed rules concerning the indels and with particular attention to highly conserved regions containing conserved residues similar to those identified in the other subfamily (ProRS, SerRS, and ThRS). This in fact should be stressed as a general rule: in widely divergent enzymes, aligning blocks of sequences of different subfamilies to build profiles should only be considered if they are highly conserved in the respective subfamilies. This indeed allowed us to propose the three motifs described in Ref. 26, which were used to define profiles (fuzzy probes describing the information present in a group of aligned sequences) to screen the entire protein database and to test them for consistency, in a quantitative way. We now discuss the uses of profiles: the profile method developed by Gribskov et al.[28] constitutes, in our opinion, one of the most powerful tools to detect and validate distant relationships between amino acid sequences.

Using Profile Tool

Profiles are derived from multiple sequence alignments and can be viewed as position-specific substitution matrices or fuzzy probes that take into account, for each position of a multiple alignment, the frequency of occurrence of a given residue and the substitutions observed at that position, weighted according to evolutionary substitution matrices such as the PAM or BLOSUM matrices. The principles and general uses of the profile sequence method have been widely discussed and the method has been greatly improved.[21,29] One of the major insights provided by the profile method is the possibility of testing the accuracy of an alignment by a quantitative "validation" of the profile. In short, validation of a profile is obtained when

[28] M. Gribskov, A. D. McLachlan, and D. Eisenberg, *Proc. Natl. Acad. Sci. U.S.A.* **84**, 4355 (1987).

[29] R. Lüthy, I. Xenarios, and P. Bucher, *Protein Sci.* **3**, 139 (1994).

all the protein sequences used to construct the profile are detected at score levels higher than 2.2σ, during profile searches through the entire protein database. In a multiple alignment involving concatenated motifs, validation procedures can be performed for each motif individually or for all the concatenated motifs.

In this sense, we emphasize that the two signature sequences of class I aaRSs (HIGH and KMSKS) are not motifs, in the sense that they cannot be validated by the procedure described by Gribskov *et al.*[28] Validation allows to verify the accuracy of the proposed alignment and also gives information about the specificity of each motif with respect to the sequences present in the database. Such a procedure was used in the class II aaRS sequence alignment, allowing clear discrimination of class II aaRS sequences from the class I aaRS sequences and from the rest of the database sequences. It should be stressed that a major problem can be encountered with profiles derived from concatenated motifs. Indeed, a GOP and GEP are defined in the profile for each position including the widely variable regions separating consecutive motifs. The profile searches we performed have proved to be very sensitive to the GOP and GEP assigned to these variable regions, and values close to 0 generally give the best results. The importance of GOP and GEP in profile alignments has been discussed, highlighting the usefulness of locally reduced gap penalties for improved accuracy of the profile method.[21,29]

Once validated, the profile itself is also very powerful in detecting distantly related sequences that cannot be detected at the sequence level by the currently available amino acid pairing programs. This is generally performed by an iterative process involving constant improvement of the profiles: newly discovered sequences are added to the initial profile as they appear in subsequent profile database scans. This approach was used in the alignment of the RNA-dependent polymerase sequences,[30] and we now briefly discuss some of the factors which must be taken into account to perform such iterative screens.

One of the major points for success in profile searches is the original selection of the sequences to be included in the initial profile. Indeed, the general rule implying that the greater the number of sequences, the more information and the better the signal-to-noise ratio is somewhat erroneous in this case because several highly related sequences can quickly bias the construction of the profile by occulting the information on sequence variability produced by a single highly divergent sequence. Distinct algorithms have attempted to solve this problem by weighting each sequence according to their evolutionary distances. Alternatively, the weights can be chosen

[30] O. Poch, I. Sauvaget, M. Delarue, and N. Tordo, *EMBO J.* **8**, 3867 (1989).

to be proportional to the distance to the consensus sequence, taken to be the profile one (M. Delarue, 1991, unpublished results). Obviously, this must be done in an iterative and self-consistent way, as modifying the weights will modify the resulting profile and consensus sequence; this procedure converges in a few cycles. Published protocols automatically assign higher weights for sequences distantly related and lower weights for closely related ones.

In the same spirit, we manually selected, as a starting point, a set of disparate reverse transcriptase sequences (Fig. 3) that were chosen to reflect the variability occurring in the conserved sequence motifs. This was mainly achieved by using the information provided by a phylogenetic tree based on the alignment of the reverse transcriptase sequences and by excluding the highly related sequences from the initial set. In addition, careful analysis of the sequence alignment revealed some positions (shown by arrows in Fig. 3) of particular interest, with differential conservations linked to the group of aligned sequences. In these two positions, the sequences belonging to the linelike group exhibit highly conserved residues (D and G). The two preserved residues are chemically widely divergent from the residues highly preserved in the other reverse transcriptase sequences, where an aromatic amino acid (instead of the aspartate residue) or a residue with an important molecular bulk (instead of the glycine residue) are preserved. The strong conservation of the residues present at these positions was taken as an indication of the importance of these positions for catalytic activity, whereas the widely different chemical nature at this position in the linelike group and the rest of the reverse transcriptase sequences was interpreted to reflect a possible change of the structure of the active site between these two enzyme families. Therefore, in an effort to try to reflect such information in the initial profile, the linelike sequences were overrepresented in the initial set of sequences defining the profile, even though they are highly evolutionary related.

Subsequent profiles derived and used mainly in the same way allowed us to delineate four motifs that were preserved in all the RNA-dependent polymerases.[30] The subsequent sequence alignments and analysis that resulted in discovery of conservation of two of the four motifs in all the monomeric DNA-dependent polymerases have been widely described[13] and were based on the same sequence alignment procedures discussed above.

It clearly stands out that accurate sequence alignment allows delineation of the regions and residues that are important for the function of a protein family and can be used as powerful guides in performing site-directed mutagenesis in order to better define the distinct roles of the conserved residues. Also, biochemical data such as photochemical and/or chemical cross-linking studies should be taken into account, when available.

In our opinion, determination of the structural information that can be deduced from any sequence alignment must now be considered as an integral part of the process of sequence analysis. With the growth of the protein structure data bank, appreciation of structural information from multiple sequence alignments may eventually lead to the assignment of a family of related proteins to a given and known fold.

Structural Insights That Can Be Gained from Multiple Alignments

Insertion/Deletion

Obviously, multiple alignments give clear hints on loop or turn regions, as it is well known that they are frequently linked to variable regions containing a high proportion of small polar residues (P, A, G, S, T) and/ or to regions exhibiting insertion/deletion (indels). Overall lengths of the indels are also informative since preservation of indels of limited extent may be indicative of the existence of two structural regions that are probably linked by a loop or turn region but must be preserved in close spatial conformation.

The relative distance between the motifs, or blocks of aligned sequences, should be studied carefully. Indeed, in the case of RNA-dependent DNA polymerases, the minimal distance between motifs A and C was too short to accommodate a tertiary structure like the one of the Klenow fragment, even when allowing for many deletions of secondary structure elements and loops that did not appear to be essential for function.[13] Therefore, we reasoned that motif B of RNA-dependent polymerases was probably structurally different from motif B of DNA-dependent polymerases (which should really be called motif B'). Consequently, this region was a good candidate for bearing the template specificity; this turned out to be case, when the structure of reverse transcriptase was revealed (for reviews, see Refs. 31 and 32).

Invariant Residues

The strictly conserved residues of different blocks can be predicted with almost certainty to be responsible for the catalytic functions and should be close in space; this is a strong prediction to test various structural models. In the case of DNA polymerases, we used this kind of reasoning to align motifs A and C of RNA-dependent polymerases onto motifs A and C of

---

[31] C. M. Joyce and T. A. Steitz, *Annu. Rev. Biochem.* **63,** 777 (1994).
[32] E. Arnold, J. Ding, S. H. Hughes, and Z. Hostomsky, *Curr. Opin. Struct. Biol.* **5,** 27 (1995).

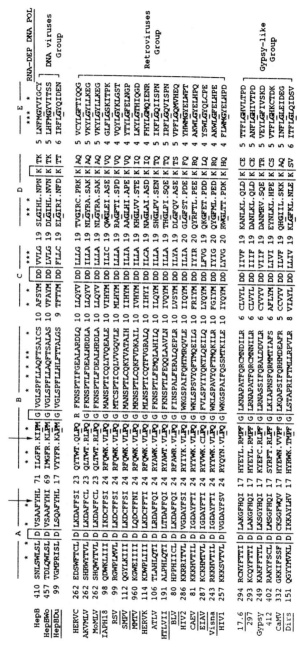

Fig. 3. Alignments of the five conserved motifs (from A to E) within RNA-dependent polymerases. Names underlined are those of the sequences that were used to construct initial profiles based on reverse transcriptase sequences. Some of the RNA-dependent polymerase sequences from the plus-strand RNA viruses detected by the initial profile scan of the database are indicated at bottom. The nearly invariant residues are boxed, and conserved residues are indicated by stars at the top line. Arrows indicate positions of particular interest discussed in the text. Adapted from Ref. 30.

```
                                                                                                                          ┐
TY912  911  NNYYITQL │D│ ISSAYLYA 24 KSLYE.LKQS │G│ ANWYETIKSYLIQQCGM 16 ICLFV │DD│ MVLF  5 SNKRII.EKL │K│ MQ 16 YDILGLEIKYQ ├ Ty-like
1731   619  QLYLLHHM │D│ VCTAYLNS 27 KAIYG.IKQS │G│ REWNSKLDGVIKDLGFA 19 ILVVV │DD│ LILA 13 ISESFE.CTD │K│ GP  1 HLFGMEVQRD  │   Group
Copla  994  YNLKVHQM │D│ VKTAFLWG 25 KAIYG.LKQA  A  RCWFEVFEQALKECEFV 21 VLLVV │DD│ VVIA  7 NNFKRY.LME │K│ FR  7 KHFIGLRIEMQ ┘

MauP   189  DSQNIYEF │D│ LKNFFPSV 81 DIATN.GVPQ │G│ ASTSCGLATYNVKELFK  4 FTIYA │DD│ GILC 12 EAGVVQ.EPA │K│ SG 11 VKFLGLEFIPA  ┐
RTChla 177  QQAVLVTF │D│ LQAAYNSV 33 NAGIN.GLAQ │G│ YAVSPTLFAWTVDQLVG  4 FTIYA │DN│ FAGV 10 VKEAQT.LLQ │K│ SG 20 LNWLGHKVLFP  │
Ingi   272  YRTGAVFV │D│ YEKAFDTV 42 RTFER.GVPQ │G│ TVPGSIMFIIVMNSLSQ  9 HGFFA │DD│ LTLL 25 EYFMSV.NVA │K│ TK 26 PKLLGVTFQCL  │
Ffac   575  EYCTAVFL │D│ VSQAFKV  41 HTIEA.GVPQ │G│ SVLGPTLYLIYTADIPT  5 VSTFA │DD│ TAIL 25 DWRIKV.NEQ │K│ CK 26 VTYLGVHLDRR  │
CIN4    74  QAALFVKL │D│ ISKAFDSL 42 IKHMR.GVRQ │G│ DPLSPFLFILAMDPLQR 22 CSLYA │DD│ AGVF 20 CSGLXI.NFE │K│ TE 26 GKYLGLPL     │ Line-like
Ifac   435  MHTSLVTL │D│ FSRAFDRV 42 LPLFN.GTPQ │G│ SPISVILFLIAFNKLSN  8 FNAYA │DD│ FFLI 25 YSGASL.SLS │K│ CQ 26 LKILGITLNNK  │   Group
IntSp  362  GCTWWIEG │D│ IKACFDSI 36 KYDIV.GTPQ │G│ SIVSPILAWYLHQLDE  66 YVRYA │DD│ WIVA 20 SIGLTV.SPT │K│ TK  8 ILFLGVNISHS  │
Int31  377  GSNWFIEV │D│ LKKCFDTI 37 HKPML.GLPQ │G│ SLISPILCWYVMTLVDN 61 FVRYA │DD│ ILIG 20 SLGLTM.NEE │K│ TL  8 ARFLGVNISIT  │
Int32  345  YCNWFIKV │D│ LNKCFDTI 37 HNTTL.GIPQ │G│ SVVSPILCWYFLDKLDK 64 FVRYA │DD│ IIIG 21 NLGMSI.NMD │K│ SV  7 VSFLGVDKVKT  │
LiMd   618  KNHMIISL │D│ AEKAFDKI 41 AIPLKSGTRQ │G│ CPLSPYLFNYVLEVLAR 19 ISLLA │DD│ MIVV 20 VVGYKI.NSN │K│ SM 26 IKYLGVTLTKE  │
LiHu   591  TNHMIISI │D│ AEKAFDKI 41 APPLKTGTRQ │G│ CPLSPLLFNYVLEVLAR 19 LSLFA │DD│ MIVV 20 VSGYKI.NVQ │K│ SQ 26 IKYLGLQLTRD  │
LiSi   591  KDHMILSI │D│ AEKAFNI  41 SFPLRSGTRQ │G│ CPLSPLLFNYMEVLAI  19 LSLFA │DD│ MIVV 20 VSGYKI.NTH │K│ SV 26 MKYLGVYLTKD  ┘

                                                                                                                          ┐
MS2V   249  VDGSLATI │D│ LSSASDSI 30 TIRWELFSTM │G│ NGFT.FELESMIFWAIV 12 IGIYG │DD│ IICP 12 TYGFKP.NLR │K│ TF  5 RESCGAHFYRG  │ RNA-DEP RNA POL
GaV    252  IDGSLATI │D│ LSSASDSI 30 LHKWGLFSTM │G│ NGFT.FELESMIFWALS 12 LGIYG │DD│ IIVP 12 AVNFLP.NEE │K│ TF  5 RESCGAHFFKD  │
QBetaV 266  VTNNLATV │D│ LSAASDSI 31 VVTYEKISSM │G│ NGYT.FELESLIFASLA 13 VTVVG │DD│ IIILP 12 TVGFTT.NTK │K│ TF 5 RESCGKHYYSG  │
PolV  1973  MEEKLFAF │D│ YTG.YDAS 38 TYCVKGGMPS │G│ CSGT.SIFNSMINNLII 17 VIAS  │DD│ VIAS 15 DYGLTMTPAD │K│ SA  9 VTFLKRFFRAD  │
CoxV  1948  LDGHLIAF │D│ YSG.YDAS 39 HYFVRGGMPS │G│ CSGT.SIFNSMINNIII 17 VIAS  │DD│ VIAS 15 GYGLIMTPAD │K│ GE  9 VTFLKRYFRAD  │ Plus-strand
HRV14 1944  MDGHLMAF │D│ YSN.FDAS 37 IVVVEGGMPS │G│ CSGT.SIFNSMINNIII 17 ILAYG │DD│ LIVS 15 NYGLTITPPD │K│ SE  9 LTFLKRYFKPD  │ RNA viruses
HRV2  1916  DDKCIMAF │D│ YTN.YDGS 36 YYEVEGGVPS │G│ CSGT.SIFNTMINNIII 17 IIAYG │DD│ VIFS 15 KYGLTITPAD │K│ SN  9 VTFLKRGFKQD  │
EMCV  2057  GFERVVDV │D│ YSN.FDST 41 RFLITGGLPS │G│ CAAT.SMLNTIMNNII  17 VLSYG │DD│ LLVA 15 KTGYKITPAN  T TS 10 VVFLKRFKKE   │
FMDV  2095  QYRNVWDV │D│ YSA.FDAN 41 RITVEGGMPS │G│ CSAT.SIVNTIMNNIYV 17 MISYG │DD│ IVVA 15 SLGQTITPAD │K│ SD 11 VTFLKRHFHMD  ┘
HAV   1974  FGDVGLDL │D│ FSA.FDAS 44 CYHVCGSMPS │G│ SPCT.ALLKSIINNINL 20 LLCYG │DD│ VLIV 22 KLGWTATSAD │K│ NV  8 LTFLKRSFNLV  ┐
CPMV  1427  CDAEVIET │D│ YSS.FDGL 44 VWRVECGIPS │G│ FPMT.VIVNSIFNEILI 26 LVTYG │DD│ NLIS 20 GGVTITDGKD │K│ TS 10 CDFLKRTFVQR  │
BBV    579  SGWVYCDA │D│ FSN.LDGR 45 RYEPGVGVKS │G│ SSTT.TPHNTQYNGCVE 20 GPKCG │DD│ GLSR  8 RAAKCFGLEL │K│ VE  7 LCFLSRVVFDP  │ Polio-like
TEV   2519  SGWVYCDA │D│ GSQ.FDSS 45 IIKKHKGNNS │G│ QPST.VVDNTLMVIIAM 15 YYVNG │DD│ LLIA 21 KYEFDCTTRD │K│ TQ  0 LWFMSHRALER  │   Group
TVMV  2464  DGWYCDA  │D│ GSQ.FDSS 45 IVKKFKGNNS │G│ QPST.VVDNTLMVVLAM 19 FFANG │DD│ LLIA 21 NYDFSSRTRD │K│ KE  0 LWFMSHRALSK  │
TMEV   366  GFNYVYDV │D│ YSN.FDAS 41 RV.YSWGPAS │G│ CAAT.SMLNTIMNNVII 17 VLSYG │DD│ LLIG 15 PFGYKITPAN │K│ TT 10 VTFLKRRFVRF  ┘
```

Fig. 3. (*continued*)

DNA-dependent polymerases. Also, mean secondary structure predictions of motif C of RNA-dependent polymerases were clearly indicating a β–turn–β supersecondary structure, with negatively charged residues in the turn, as observed in the Klenow structure.[13] The alignment of the strongly conserved regions of DNA-dependent DNA polymerases of the different subfamilies (polI and polα) also made use of this kind of information.

Averaged Secondary Structure Assignment

Averaged secondary structure predictions (ASSP) calculated from a set of aligned sequences are routinely used, and this can be considered as an initial step in the structural analysis of a multiple alignment. Various methods for performing ASSP have been published (see Refs. 33–35 and references therein), and they frequently emphasize the problem of assessing the accuracy of the predictions, for example, evaluation of the percentage of individual amino acids well predicted as well as the percentage of the secondary structures well predicted. The use of an interactive/intuitive process incorporating predictions, expertise on sequence alignments, biochemical data, and refinement of the secondary prediction[9] instead of automatic procedures based on the multiple alignment alone is still a matter of debate. Concerning the former point, there is now growing evidence suggesting that prediction of the approximate location of secondary structures is generally sufficient.[34] Concerning the latter point, the experiments always prompted us to favor an interactive procedure which, as discussed for sequence alignment, is supposed to preserve the maximum sequence information.

At the practical level, ASSPs are mainly performed according to three steps. (i) Use of various methods of secondary structure predictions and combination of their predictions are recommended. Indeed, the different methods are not equally successful in detecting the different types of secondary structures. For example, it has been shown that the Chou and Fasman method predicts helices better than β strands, whereas the GOR method is better in predicting β strands.[33] The combination of distinct methods in conjunction with knowledge of their respective error propensities helps in reducing errors. (ii) Alignment of the predicted secondary structures according to a multiple alignment is advised. As already emphasized, the construction of a multiple alignment using programs often remains imperfect, and, generally a manual adjustment is necessary to ameliorate it. Such an adjustment is frequently fruitfully guided by secondary structure

---

[33] Z. Y. Zhu, *Protein Eng.* **8**, 103 (1995).
[34] B. Rost, C. Sander, and R. Schneider, *J. Mol. Biol.* **235**, 13 (1994).
[35] C. Geourjon and G. Deléage, *Protein Eng.* **7**, 157 (1994).

predictions of the different sequences in the alignment. (iii) For each amino acid position, the percentage of sequences predicted with a given secondary structure (α, β, turn, or coil) should be calculated by distinct prediction methods.

Correlated Mutations

Various methods have been published to try to retrieve from a multiple alignment indications about possible spatial proximity between different positions along the sequence, taking advantage of correlations of mutations at these two positions.[36,37] If the set of aligned sequences is sufficiently large, it should indeed be possible to detect such pairs with simple statistics. The rationale behind this method is that there must be a conserved physical property, when mutations occur, to preserve the stability of the protein. One such physical property for buried residues is the volume, because it is well known that the interior of protein structures is extremely well packed.[38] However, this method has met with only limited success; when assessed on known cases, it has been shown that positions appearing to be highly correlated on a statistical level are in spatial contact in only 50% of the cases. There is no way at the present time to determine if a particular pair of correlated positions is of any significance.

Construction of Binary Profiles: Classification of Positions as Exposed or Buried

The Benner group[9] has also used multiple alignments to predict the buried or accessible character of each position, by simply observing the preferred amino acid types occurring most frequently. Bowie and colleagues[39] have tried to derive a binary map for each position along the sequence, describing the protein as a string of 0 or 1 if the residue is exposed or buried, respectively; they used experimental data (site-directed mutagenesis to probe sequence tolerance at various positions) for this purpose. They then used this "profile" to screen a database of similar binary profiles of protein structures, which were derived from atomic properties of the corresponding models, with encouraging results. This type of methodology and its subsequent refinements are described in more detail elsewhere in this volume (see [35], this volume).

[36] W. R. Taylor and K. Hatrick, *Protein Eng.* **7,** 341 (1994).

[37] I. N. Shindyalov, N. A. Kolchanov, and C. Sander, *Protein Eng.* **7,** 349 (1994).

[38] F. M. Richards, *Annu. Rev. Biophys. Bioeng.* **6,** 151 (1977).

[39] J. U. Bowie, N. D. Clarke, C. O. Pabo, and R. T. Sauer, *Proteins: Struct. Funct. Genet.* **7,** 257 (1990).

However, a caveat must be issued concerning this last method. A detailed study of structurally aligned and closely related protein structures has been published,[2] and it turns out that aligned positions do not conserve such simple properties as solvent accessibility as well as previously thought.

Delineation of Domains Outside Catalytic Domain

Once the catalytic domain has been found, it is often interesting to focus on the remaining ones, as in the case of aminoacyl-tRNA synthetases.[40] Indeed, this family of proteins is clearly made of several domains, one of which is devoted to the task of anticodon recognition. For class II enzymes, it is then straightforward to observe that AspRS, AsnRS, and LysRS all share an additional domain at the N-terminal part of the molecule, which is indeed responsible for anticodon recognition, whereas HisRS, ProRS, and ThrRS have no extension at the N terminus and, on the contrary, share an additional domain in the C terminal region. This domain can be safely predicted to be the anticodon-binding domain; indeed, it is absent in SerRS, because this protein, although a member of the same subfamily, is known not to recognize any anticodon (serine is coded by six different codons in the genetic code).

Vertical versus Horizontal Sequence Conservation Profiles

For aminoacyl-tRNA synthetases, another useful piece of structural information that can be safely deduced from multiple alignments is the location of sequences responsible for the recognition of specific substrates, namely, the amino acid itself. In this case, it is necessary to plot a conservation profile representation of a multiple alignment of all known sequences of the same enzyme from different species: if there are more than five or six sequences from different kingdoms, several blocks of highly conserved regions delineate very accurately the regions responsible for amino acid recognition, once the three motifs common to all aaRSs have been excluded (see Fig. 4). The same type of analysis can be performed for either class I or class II aaRSs.

Conclusion

The extraction of structural insight from aligned sequences is an integrative process that attempts to make use of all the available information. Clearly, one of the most critical steps in the process is to obtain the best possible multiple alignments. We have described empirical rules that have

[40] M. Delarue and D. Moras, *BioEssays* **15,** 675 (1993).

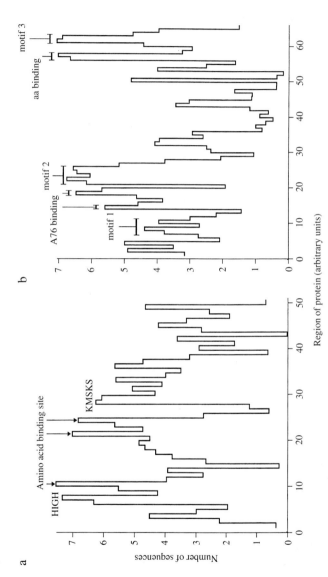

FIG. 4. (a) Conservation profile obtained with the aligned sequences of the TyrRS from *Bacillus subtilis*, *Bacillus stearothermophilus*, yeast mitochondria, *E. coli*, and *Neurospora crassa*. (b) Same construction for AspRS from *E. coli*, *Thermus thermophilus*, yeast, yeast mitochondria, and *Homo sapiens*. The locations of the signature sequences and of the motifs define the positioning of other regions of interest along the sequence. Reprinted from Ref. 40.

been successfully used in different cases and that proceed from the same spirit as the one at work in Steve Benner's group. Clearly, more work is needed to make these methods more general. Indeed, the translation of evolutionary information into structural insights is a major goal for future research in sequence analysis studies.

Acknowledgments

We are deeply indebted to David Sourdive for invaluable help in the preparation of the figures. We thank Eric Westhof for critical reading of the manuscript. We also thank Michel. Zerbib for computer assistance.

Author Index

Numbers in parentheses are footnote reference numbers and indicate that an author's work is referred to although the name is not cited in the text.

Subject Index

A

B

L

Lactate dehydrogenase
iterative template refinement, 333, 335, 337
phylogenetic tree construction, 381–382
LALIGN, repeats, identification in proteins, 254–256
Linguistic analysis, nucleotide sequences
advantages, 281–282
algorithms, *see* CODONTREE; WORDUP
evolution, 281–282
principle, 282–284
Local similarity score
advantages over global scoring, 240
statistics
algorithms, 231
empirical statistics from database searches, 477–478, 480
importance of evaluation, 230
local optimality, 467–468
parameter dependence on search space size, 463–464
parameter estimation, 464–467
parameters for affine gap costs with amino acid substitution matrices, 473–477
probability of normalized score calculation, 231
sum statistics, 468–471, 473, 478
ungapped subalignment scores, 461–463
Low-complexity region, handling in sequence comparison, 146

M

MacClade, parsimony method of phylogenetic tree construction, 422
MEDLINE
browsing in *Entrez*, 150
keyword search, 149
record format, 149
ME_TREE, minimum evolution method of phylogenetic tree construction, 442–443
MOLPHY
availability, 426, 448

distance matrix method of phylogenetic tree construction, 425–426
neighbor-joining method of phylogenetic tree construction, 437–439, 441
MoST, *see* Motif Search Tool
Motif
databases, 167
definition, 163
genomic protein sequence analysis, 303–304
loop motif
common loop identification in different topologies, 649–651
common topology distinction, 652
new loop detection, 652
profile analysis, 672–673
searching
checklist, 171–173
methods, 163–164, 166–167
SSAP searching, 626–628
MOTIF, *see* Blocks Database
Motif Search Tool
genomic protein sequence analysis, 305
motif searching, 138, 140, 170
MOTOMAT, *see* Blocks Database
Multiple sequence alignment
combined DNA and protein alignment algorithm, *see* GenAl
dynamic programming algorithms, 345–346, 352–354, 364–366, 576
history, 383
quality assessment, 366–367
secondary structure prediction, *see* Secondary structure, protein
sequence weighting, 392–394
strategies, 666–668
tertiary structure prediction, *see* Tertiary structure, protein

N

Neural network, *see also* Gene classification
artificial neural system
GRAIL algorithms, 260–262, 273
protein classification, 73
secondary structure prediction from multiple sequence alignment, 525–537
sequence data analysis, 72–73

ISBN 0-12-182167-6

90038

THE

T LIBRARY
University of California, San Francisco

stamped